Notes on

Building Code Requirements for Reinforced Concrete

with Design Applications

Edited by Gerald B. Neville

An organization of cement manufacturers to improve and extend the uses of portland cement and concrete through scientific research, engineering field work, and market development.

5420 Old Orchard Road, Skokie, Illinois 60077-4321

Fourth edition, revised 1984

Printed in U.S.A.

Library of Congress catalog card number 80-83547

ISBN 0-89312-038-3

About the building on the cover

Cover art: line representation derived from posterized photograph
Water Tower Place, Chicago, Illinois
Architects—Loebl Schlossman, Bennett and Dart, Chicago, Illinois
Engineers—C. F. Murphy Associates, Chicago, Illinois

Foreword

The first edition of this reference manual was prepared to aid users in applying the provisions of the 1971 edition of "Building Code Requirements for Reinforced Concrete (ACI 318-71)." The second and third editions updated the material in conformity with the provisions of the 1977 Code edition and the 1980 Code Supplemental Revisions, respectively. Through three editions, much of the first edition chapter material has been revised to better emphasize the subject matter, and new chapters added to assist the engineer and designer in proper application of the ACI 318 Design Standard for Reinforced Concrete.

This fourth edition addresses the 1983 edition of "Building Code Requirements for Reinforced Concrete (ACI 318-83)." New features of the 4th edition include: an updated Part 1 addressing the revisions in concrete mixture proportioning for the '83 Code; expanded treatment on details, development, and splices of reinforcement for Parts 2 and 3 ...including new hooked-bar anchorage provisions; a completely new Part 7 on deflections...with many new sample calculations for deflection control of both nonprestressed and prestressed members; new code provisions for slender column design of unbraced frames addressed in Part 12...with illustrative examples; Part 15 addressing shear-friction design expanded to include what's new for '83 in shear-friction; Part 16 is completely rewritten to address the new design rules for brackets and corbels; new code simplifications for design of two-way slabs by Direct Design Method addressed in Part 20; Part 23 includes design for force transfer at precast column-footing connections...new for '83 Code; Parts 25 and 26 extensively revised to reflect what's new for prestressed concrete in the '83 Code; a new Part 27 is added to the fourth edition on design of post-tensioned slabs; and a completely new and expanded Part 30 on design and detailing of concrete structures for earthquake resistance...addressing the new Appendix A for the '83 Code.

Some of the existing text and design examples have also been revised to reflect, where possible, many of the comments received from users of the "Notes" who suggested improvements in wording, identified some errors, and recommended items for inclusion or deletion.

The primary purpose for publishing this manual is to assist the engineer and designer in the proper application of the ACI 318-83 design standard. The emphasis is placed on "how-to-use" the Code. For complete background information on the development of the Code provisions, the reader is referred to the "Commentary on Building Code Requirements for Reinforced Concrete (ACI 318-83)."

This manual can also be a valuable aid to educators, architects, contractors, materials and products manufacturers, building code authorities, inspectors, and others involved in the design, construction, and regulation of concrete buildings.

A total of 30 separate topics are covered under 30 chapters. The 30 chapters are concerned directly with specific design provisions of the ACI Code. More than 80 design examples are included to illustrate how to apply the code provisions to design practice. Moreover, a code reference index readily correlates each discussion and design application with a particular code section.

About the authors:

This manual was prepared under the direction of Gerald B. Neville, Manager, Structural Codes, Codes and Standards Department. In addition to providing technical guidance to all the individual chapter authors, he served as editor and coordinator for the final manuscript. The work of the authors in all stages of producing this manual is gratefully acknowledged.

Special thanks go to: Dr. David P. Gustafson of the Concrete Reinforcing Steel Institute for his extensive rework of Parts 2 and 3, details of reinforcement and development and splices of reinforcement. Also for Dr. Gustafson's review of all reinforcing details throughout the text for conformance with construction practice;

Professor Dan E. Branson of the University of Iowa for his outstanding new Part 7 on control of deflections;

Mr. Chris Pickett, Jr., consulting engineer in Salt Lake City, for his major update of Part 15 and 16 addressing the new design provisions for Shear-Friction and Brackets and Corbels.

Professor Alan H. Mattock of the University of Washington for his technical guidance on the design of beam ledges in Part 16.

Mr. Clifford L. Freyermuth of the Post-Tensioning Institute for his Part 27 addressing prestressed slab systems, new for the 4th edition;

Dr. Milo S. Ketcham of KKBNA, Inc., consulting engineer in Denver, for his complete update of Part 28 on shells and folded plates; and

Dr. Arnaldo T. Derecho of Wiss, Janney & Elstner, Consulting Engineers, Northbrook, Illinois, and Dr. S. K. Ghosh of the University of Illinois at Chicago for their outstanding work on new Part 30 addressing the new code provisions for seismic design.

These experts from outside the fold of PCA were extremely kind in contributing their specialized knowledge and technical skills to this 4th edition.

The following PCA engineers were especially helpful in providing material for individual chapters in this 4th edition (listed alphabetically):

Messers: Arnold H. Bock (Slenderness Effects); Randall C. Cronin (Torsion); Ken Hanson (Shells); Dr. Alexey Mindich (Prestressed Concrete-Flexure); Elwin M. Pell (Footings); Gary D. Pfuehler (Two-Way Slabs); and Dr. Renata W. Zwiers (Shear in Slabs). Their work is deeply appreciated.

Additionally, we are thankful to the authors of chapters of the 1st, 2nd, and 3rd editions of the "Notes," whose initial work is carried over to this edition, even though their names are no longer identified with the chapter material; to PCA's Communication Materials Department for its assistance in editing and proofreading the manuscript; and most particularly, to PCA's Word Processing Department for its massive support in typing the manuscripts, making possible the timely production of this fourth edition.

James P. Barris
Director, Codes and Standards

Contents

Part 10 - Design for Flexure and Axial Load

Part 11 - Biaxial Loading

Part 12 - Slenderness Effects

Part 13 - Shear

Part 20 - Two-Way Slabs - Direct Design Method

Part 21 - Two-Way Slabs - Equivalent Frame Method

Part 22 - Walls

Part 23 - Footings

Part 24 - Precast Concrete

Part 25 - Prestressed Concrete - Flexure

Code Reference Index

1

Materials, Concrete Quality

MATERIALS

3.1 Tests of Materials

Testing agencies will need to be aware of a wording change to Section 3.1.3. For the '83 Code, the two year period test records must be kept available has been clarified to two years "after completion of the project." The '77 Code wording...two years "thereafter" was ambiguous; if records were destroyed two years after taken, a major project still might not be completed. The Commentary defines completion of the project as the date at which the owner accepts the project or when the certificate of occupancy is issued, whichever date is later. Also note that a similar change has been made to Section 1.3.4 on preservation of records of inspection.

3.2 Cement

The cement used in the work must correspond to the type upon which the selection of concrete proportions for strength and other properties has been based. This may simply mean the same type of cement or it may mean cement from the identical source. The latter would be the case if the standard deviation of strength tests used in establishing the required target strength was based on one particular type of cement from one particular source. In the case of a plant that has determined the standard deviation from tests involving cement from several sources, the former would apply.

3.3 Aggregate

The maximum size of aggregate is limited to one-third the depth of the slab as recommended by ACI Committee 301.[1.1] Note that the limitations on maximum size of the aggregate may be waived if, in the judgment of the engineer, the workability and methods of consolidation of the concrete are such that the concrete can be placed without honeycomb or void. In this instance, the engineer in charge of inspection must decide whether or not the limitations on maximum size of aggregate may be waived.

3.4 Water

Precautions concerning the chloride ion content of water (including that portion of the mixing water contributed as free moisture on the aggregates) to be used in prestressed concrete or in concrete with aluminum embedments are noted. With the '83 Code edition, new limits on chloride ion content contributed from the concrete ingredients including water, aggregates, cement, and admixtures are given in Chapter 4, Table 4.5.4. For an in-depth discussion, see Commentary Section 4.5.4.

3.5 Reinforcement

3.5.3 Deformed Reinforcement

Only deformed reinforcement as defined in Chapter 2 may be used for nonprestressed reinforcement, except that plain bars and smooth wire may be used for spiral reinforcement. Especially note that welded smooth wire fabric is included under the Code definition of deformed reinforcement. Smooth wire fabric bonds to concrete by positive mechanical anchorage at each wire intersection. Deformed wire fabric utilizes wire deformations plus welded intersections for bond and anchorage. This difference in bond and anchorage for the smooth vs. deformed fabric is reflected in the development and lap splice provisions of Chapter 12.

Reinforcing bars rolled to ASTM A615 specifications (billet steel) are the most commonly specified for construction. Rail and axle steels (ASTM A616 and A617) are not generally available, except in a few areas of the country. ASTM A706 covers low-alloy steel deformed bars (Grade 60 only) intended for special applications where welding or bending, or both, are of importance. Note: Be sure to check availability before ordering bars to the A706 specification.

Currently no ASTM specification for deformed bars with a yield strength exceeding 60,000 psi is available. However, this should not be interpreted to preclude the use of higher strength bars that meet the requirements of Sections 3.5.3.3. For welded wire fabric, yield strength above 60,000 psi is available; however, the Code assigns a yield strength value of 60,000 psi, but makes provisions for the use of high yield strengths provided the stress used in design corresponds to a strain of 0.35 percent.

With the '83 Code, Section 3.5.3.2 is deleted. The two exceptions to ASTM A615, A616, and A617, requiring yield strength to be determined by tests on full-sized bars and specifying tighter bend test requirements for bars, are now covered by appropriate reference to the ASTM specifications in Section 3.5.3.1.

For A615 (billet steel) the two exceptions are now contained in the A615 specification as supplementary requirements (S1). Note: the (S1) requirements are optional and apply only when specified by the purchaser. Thus, A615 reinforcing bars used with the ACI Code must be specified as conforming to ASTM Specification A615 plus Supplementary Requirements (S1)...ASTM A615-82(S1). Bars supplied with these requirements will be marked with a letter S instead of the letter N which is normal marking for A615 bars. For Code conformance, the inspector need only check for the letter S marking on bars supplied at the job site.

For A616 (rail steel), the necessary exception requiring tighter bend tests is stated directly in Section 3.5.3.1(b); also, bars meeting the tighter bend tests must be rolled with a letter R to designate rail steel meeting the tighter bend tests of A617. The exception requiring yield strength based on tests of full-sized bars is already part of A616.

For A617 (axle steel), the two exceptions are included directly in the main body of the specification.

Note: Guidelines on welding to existing reinforcing bars is added to Commentary Section 3.5.2.

3.5.3.8 Coated Reinforcement

Appropriate reference to the ASTM specifications for coating reinforcement, A767 (galvanized) and A775 (epoxy-coated), is added to the '83 Code to reflect increased usage of coated reinforcement for corrosion protection. Commentary Section 3.5.3.8 gives designers guidance in specifying galvanized or epoxy-coated reinforcement.

3.5.5 Prestressing Tendons

For the '83 Code, appropriate ASTM specifications for low-relaxation wire and strand is added. Since low-relaxation tendons are addressed in supplements to A416 and A421, which apply only when low-relaxation material is specified, the appropriate ASTM reference for the low-relaxation tendons are listed as a separate entity.

CONCRETE QUALITY

Chapter 4 addressing concrete quality is a complete revision for the '83 code, with significant changes in the Code provisions for selecting concrete proportions and for special exposures as follows:

Requirements for selection and documentation of concrete proportions are expanded to further emphasize field test data as the basis for selecting concrete proportions. Revisions include: new procedures for selecting mixture proportions on the basis of less than 30 test records (Section 4.3.1.2); new expressions for determining a required strength using a standard deviation (Section 4.3.2.1); when test data are not available, new criteria for determining a required strength based on the specified design strength

(Section 4.3.2.2); and expanded requirements for documentation of required strength (Section 4.3.3).

Requirements for special exposures are expanded to include; for resistance to freezing and thawing, new limits on air content based on degree of exposure (Table 4.5.1); for concrete intended to be watertight or subject to deicer salts, new limits on water-cement ratio or minimum strength (Table 4.5.2); for protection of concrete in sulfate exposures, new limits on sulfate and cement type (Table 4.5.3); and for corrosion protection, new limits on chloride ion content (Table 4.5.4).

This is the first major revision to Code Chapter 4 since probabilistic concepts for proportioning concrete mixtures were first introduced in the 1971 Code edition. With new Chapter 4, greater emphasis is placed on the special exposure conditions for improved concrete durability. Sufficient evidence of badly deteriorated concrete in many areas due to severe exposures such as exposure to deicing salts, sulphates, freezing and thawing, and chloride exposure warranted a more positive Code response. The new special exposure requirements of Section 4.5 direct special attention to the need to consider concrete durability in addition to strength. For an in-depth discussion on the new requirements for concrete quality for '83, especially the new special exposure requirements, the reader is referred to the Commentary to new Chapter 4.

The remainder of Part 1 will address selecting concrete mixture proportions for strength based on probabilistic concepts.

4.1 General

It is emphasized in Section 4.1.1 that the average strength of concrete produced must always exceed the specified value of f'_c that was used in the structural design phase. This is based on probabilistic concepts, and is intended to ensure that adequate strength will be developed in the structure.

Note that Section 4.1.3 permits f'_c to be based on tests other than the customary 28 days. If other than 28 days, the test age for f'_c must be indicated on the design drawings or specifications...new for '83.

4.2 Selection of Concrete Proportions

Recommendations for selecting proportions for concrete are given in detail in "Standard Practice for Selecting Proportions for Normal, Heavy Weight, and Mass Concrete" (ACI 211.1)[1.3] and "Standard Practice for Selecting Proportions for Structural Lightweight Concrete" (ACI 211.2).[1.4]

The selected water cement ratio must be low enough, or the compressive strength high enough (for lightweight concrete) to satisfy both the strength criteria (Sections 4.3 or 4.4) and the special exposure requirements (Section 4.5). The Code does not include provisions for especially severe exposures, such as to acids or high temperatures, nor is it concerned with aesthetic considerations such as surface finishes. Items like these, which are beyond the scope of the Code, must be covered specifically in the project specifications. Concrete ingredients and proportions must be selected to meet the minimum requirements stated in the Code and the additional requirements of the contract documents.

The Code emphasizes the use of field experience or laboratory trial batches (Section 4.3) as the preferred method for selecting concrete mixture proportions. When no prior experience or trial batch data are available, estimation of the water cement ratio as prescribed in Section 4.4 is permitted but only when special permission is given.

4.3 Proportioning on the Basis of Field Experience and/or Trial Mixtures

For establishing concrete proportions, emphasis is placed in the use of laboratory trial batches or field experience as the basis for selecting the required water cement ratio. The Code emphasizes a statistical approach to establish the target strength f'_{cr} required to assure attainment of the f'_c used in the structural design. If an applicable standard deviation, s, for strength tests of the concrete is known, this establishes the target strength level from which the concrete must be proportioned. Otherwise, the proportions must be selected to produce an excess of target strength sufficient to allow for a high degree of variability in the strength tests. For background information on statistics as relates to concrete, see "Recommended Practice for Evaluation of Compression Test Results of Concrete"[1.5] and "Statistical Product Control".[1.6]

Where the concrete production facility has a record based on at least 30 consecutive strength tests representing similar materials and conditions to those expected (or a record based on 15 to 29 consecutive tests with the calculated standard deviation modified by the factor of Table 4.3.1.2), the strength used as the basis for selecting concrete proportions must be the larger of:

$$f'_{cr} = f'_c + 1.34s$$

or

$$f'_{cr} = f'_c + 2.33s - 500$$

If the standard deviation is unknown, the required average strength f'_{cr} used as a basis for selecting concrete proportions must be determined from Table 4.3.2.2:

For f'_c less than 3000 psi	$f'_{cr} = f'_c + 1000$
3000 to 5000	$f'_{cr} = f'_c + 1200$
greater than 5000	$f'_{cr} = f'_c + 1400$

Formulas for calculating the required target strengths are based on the following criteria:

(1) A probability of 1 in 100 that an average of 3 consecutive strength tests will be below the specified strength, f'_c;

$$f'_{cr} = f'_c + 1.34s$$

(2) A probability of 1 in 100 that an individual strength test will be more than 500 psi below the specified strength, f'_c;

$$f'_{cr} = f'_c + 2.33s - 500$$

Criterion (1) will produce a higher required target strength than will criterion (2) for low to moderate standard deviations, up to about 500 psi. For higher standard deviations, however, criterion (2) governs (i.e., limiting the expected frequency of test more than 500 psi below the specified f'_c to 1 in 100).

The indicated average strength levels are intended to reduce the probability of concrete strength being questioned on the following usual bases: (1) strength averaging below specified f'_c for an appreciable period (three consecutive tests); or (2) an individual test being disturbingly low (more than 500 psi below specified f'_c).

Concrete for background tests to determine standard deviation is considered to have been 'similar' to that required if it was made with the same general types of ingredients under no more restrictive conditions of control over material quality and production methods than will exist on the proposed work, and if its specified strength did not deviate more than 1000 psi from the f'_c required. A change in the type of concrete or a major increase in the strength level may increase the standard deviation. Such a situation might occur with a change in type of aggregate--i.e., from natural aggregate to lightweight aggregate or vice versa--or a change from non-air-entrained concrete to air-entrained concrete. Also, there may be an increase in standard deviation when the average strength level is raised by a significant amount, although the increment of increase in standard deviation should be somewhat less than directly proportional to the strength increase. When there is reasonable doubt, any estimated standard deviation used to calculate the required average strength should always be on the conservative (high) side.

Statistical methods provide valuable tools for assessing results of strength tests. It is very important that concrete technicians understand the basic language of statistics and be capable of effectively using the tools to evaluate test results.

Fig. 1-1 illustrates several fundamental statistical concepts. Data points represent six strength test results from consecutive tests on a given class of concrete. The horizontal line represents the average of tests which is designated $\bar{X}$. The average is computed by adding all test values and dividing by the number of values summed; i.e., in Fig. 1-1:

$$\bar{X} = \frac{4000 + 2500 + 3000 + 4000 + 5000 + 2500}{6}$$

$$\bar{X} = 3500 \text{ psi}$$

The average $\bar{X}$ gives an indication of the overall strength level of the concrete tested.

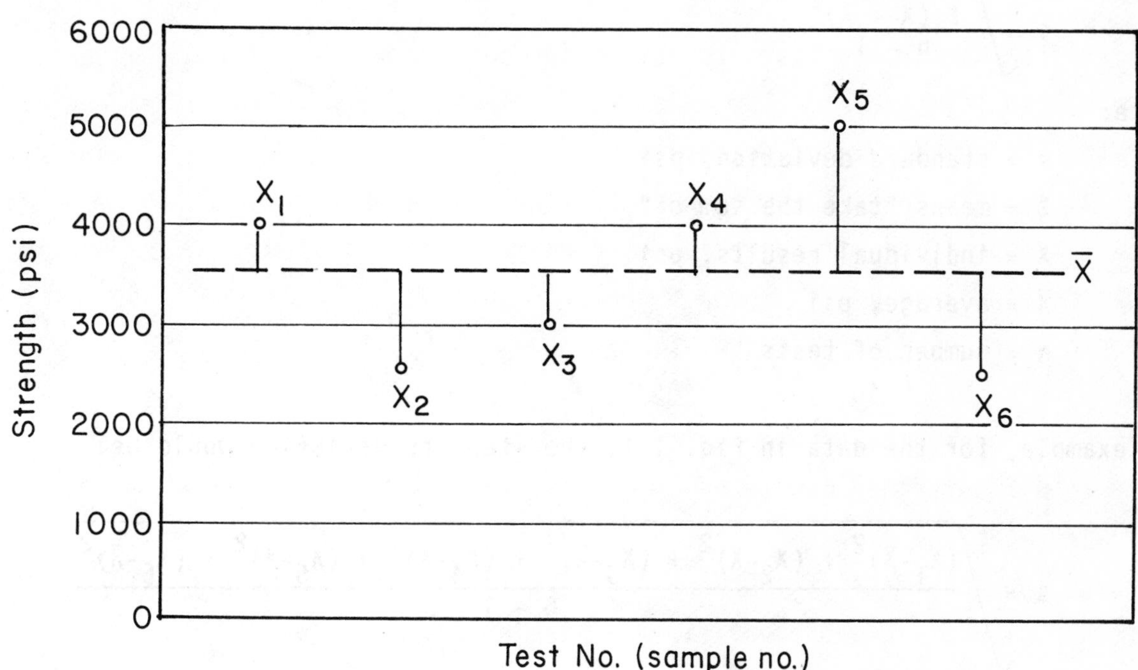

Fig. 1-1 - Illustration of Statistical Terms

It would also be informative to have a single number which would represent the variability of the data about the average. The up and down deviations <u>from the average</u> are given as vertical lines in Fig. 1-1. If one were to accumulate the total length of the vertical lines without regard to whether they are up or down and divide that total length by the number of tests, the result would be the average length, or the average distance from the average strength. (500 + 1000 + 500 + 500 + 1500 + 1000)/6 = 833 psi. This is one measure of variability. If concrete test results were quite variable, the vertical lines would be long. On the other hand, if the test results were close, the lines would be short.

In order to emphasize the impact of a few very high or very low test values, statisticians recommend the use of the square of the vertical line lengths. The square root of the sum of the squared lengths divided by one less than number of tests (some texts use the number of tests) is known as the standard deviation. This measure of variability is commonly designated by the letter s. Mathematically, s is expressed as:

$$s = \sqrt{\frac{\Sigma (X - \bar{X})^2}{n - 1}}$$

where:

s = standard deviation, psi

Σ = means "take the sum of"

X = individual results, psi

$\bar{X}$ = average, psi

n = number of tests

For example, for the data in Fig. 1-1, the standard deviation would be:

$$s = \sqrt{\frac{(X_1-\bar{X})^2 + (X_2-\bar{X})^2 + (X_3-\bar{X})^2 + (X_4-\bar{X})^2 + (X_5-\bar{X})^2 + (X_6-\bar{X})^2}{6 - 1}}$$

or:

Deviation $(X - \bar{X})$ (length of vertical lines)	$(X - \bar{X})^2$ (length squared)
4000 - 3500 = + 500	+ 250,000
2500 - 3500 = - 1000	+ 1,000,000
3000 - 3500 = - 500	+ 250,000
4000 - 3500 = + 500	+ 250,000
5000 - 3500 = + 1500	+ 2,250,000
2500 - 3500 = - 1000	+ 1,000,000
Total	5,000,000

$$s = \sqrt{\frac{5,000,000}{5}}$$

= 1000 psi (a very large value)

Obviously, it would be time consuming to actually calculate s in the manner described above. There are many short-cut methods which can be used. Examples 1.1 and 1.2 illustrate two methods which are relatively easy to use.

The coefficient of variation V is simply the standard deviation expressed as a percentage of the average strength. The mathematical formula would be:

$$V = \frac{s}{\bar{X}} \times 100\%$$

Then, for Fig. 1-1:

$$V = \frac{1000}{3500} \times 100$$

$$= 29\%$$

Standard deviation may be computed either from a single group of successive tests of a given class of concrete or from two groups of such tests. In the latter case, a statistical average value of standard deviation is to be used, calculated by usual statistical methods as follows:

$$s_3 = \sqrt{\frac{(n_1 - 1)(s_1)^2 + (n_2 - 1)(s_2)^2}{n_{total} - 2}}$$

where:

n_1 = number of samples in group 1

n_2 = number of samples in group 2

s_1 or s_2 is calculated as follows:

$$s = \sqrt{\frac{(X_1 - \bar{X})^2 + (X_2 - \bar{X})^2 + \ldots + (X_n - \bar{X})^2}{n - 1}}$$

For ease of computation, when a calculating machine is available:

$$s = \sqrt{\frac{X_1^2 + X_2^2 + X_3^2 + \ldots + X_n^2 - n\bar{X}^2}{n - 1}}$$

or:

$$s = \sqrt{\frac{(X_1^2 + X_2^2 + X_3^2 + \ldots + X_n^2) - \frac{(X_1 + X_2 + X_3 + . + \bar{X}_n)^2}{n}}{n - 1}}$$

Where X_1, X_2, X_3 . . . X_n are the strength results of individual specimens and n is the total number of specimens tested.

ACI REQUIREMENTS

ACI 318 and ACI 301[1.1] contain detailed requirements for judging the adequacy (or acceptability) of test results. They also contain specific requirements and procedures for the mix design proposed for use in the project. These mix approval procedures are considered necessary to insure that the concrete furnished will actually meet the strength requirements.

MIX APPROVAL PROCEDURES

The steps in the procedure can be outlined as follows:

1. Determine the expected standard deviation from past experience.

(a) This is done by submitting a record of (30) consecutive tests made on a similar mix.

(b) If it is difficult to find a job with (30) tests, the standard deviation can be computed from each of two jobs if the total number exceeds (30) tests. The standard deviations are computed separately and then averaged by the statistical average method already described.

2. Use the standard deviation to select the appropriate overdesign from the larger of:

$$f'_{cr} = f'_c + 1.34s$$

or

$$f'_{cr} = f'_c + 2.33s - 500$$

(a) For example, if the standard deviation is 450 psi, then the overdesign should be the larger of 1.34(450) = 603 psi or 2.33(450) - 500 = 549 psi. Thus, for a 3000 psi specified strength, the average strength used as a basis for selecting concrete proportions would be 3600 psi.

(b) Note that if no acceptable test record is available, the average strength should be 1200 psi greater than f'_c (i.e., 4200 psi average for specified 3000 psi concrete). See Table 4.3.2.2.

3. Furnish data to document the fact that the mix proposed for use will give the average strength needed. This can be done by either of two methods.

(a) A record of (30) tests of field concrete. This would generally be the same test record that was used to document the standard deviation, but it could be a different set of (30) results;

(b) or a series of laboratory trial batches.

Section 4.3.2.2(c) permits tolerances on slump and air content when proportioning by laboratory trial batches. The tolerance limits are stated at maximum permitted values because most specifications, regardless of form, will permit establishing a maximum value for slump or air content. The wording also makes it clear that tolerances on slump and air content are to be applied only to laboratory trial batches and not a field record of tests.

4.4 Proportioning by Water-Cement Ratio

The use of Table 4.4 includes Type V cement and a number of the blended hydraulic cements in ASTM C595. The Table is not applicable for IS or IP cements carrying a (MA) suffix, or for Type II cement for which the optional moderate heat of hydration requirements has been involved. Because of the conservative water-cement ratio limits, Table 4.4 is considered usable for all of the cement types listed, despite the fact that minimum specified 3, 7, and 28 day strengths of these cements vary from one cement type to another. Typically, cement strengths exceed the ASTM minimum requirements by significant but different amounts. The following recommendations are given as guidelines to aid in the use of blended hydraulic cements:

(1) The cement used in the work should correspond to that on which the selection of concrete proportions was based.

(2) When Types V and P cements are used, proper recognition should be given to the effects of slower strength gain and lower heat of hydration on concrete proportioning and construction practices.

(3) In concrete made with blended hydraulic cements, if fly ash or other pozzolans are used as admixtures resulting in dilution of the cement component, proper recognition should be given to changes in the properties of concrete, such as strength, durability, and protection against corrosion of reinforcement.

4.7 Evaluation and Acceptance of Concrete

Once the mix is approved, the tests made on job concretes are required to meet the following two criteria to be considered acceptable:

(1) No single test (where a test is the average of two cylinders from a batch) shall be more than 500 psi below the design strength; i.e., 2500 psi for specified 3000 psi concrete.

(2) And the average of any three consecutive tests must equal or exceed the design strength, f'_c.

Selected References

1.1 ACI Committee 301, "Specifications for Structural Concrete for Buildings (ACI 301-72) (Revised 1981)," American Concrete Institute, Detroit, 1981, 36 pp.

1.2 ACI Committee 222, "Corrosion of Metals in Concrete", (SP-49), American Concrete Institute, Detroit, 1975, 142 pp.

1.3 ACI Committee 211, "Standard Practice for Selecting Proportions for Normal, Heavyweight, and Mass Concrete (ACI 211.1-81)", American Concrete Institute, Detroit, 1981, 32 pp.

1.4 ACI Committee 211, "Standard Practice for Selecting Proportions for Structural Lightweight Concrete (ACI 211.2-81)", American Concrete Institute, Detroit, 1981, 18 pp.

1.5 ACI Committee 214, "Recommended Practice for Evaluation of Compression Test Results of Concrete (ACI 214-77)", American Concrete Institute, Detroit, 1977, 14 pp.

1.6 "Statistical Product Control", Portland Cement Association, Skokie, IS172T, 1970, 15 pp.

EXAMPLE 1.1 - Simplified Method to Calculate Standard Deviations

The average of strength results of 46 pairs of cylinders sampled from a particular class of concrete delivered to a project are as follows:

AVERAGES OF 46 PAIRS OF CYLINDERS

3395	2975	4220	3395	3045
3555	3200	3820	2965	2665
3545	3120	3995	3655	3485
3110	3055	3675	3815	4035
3260	3500	3220	4480	3500
3575	3840	3455	3650	3025
3975	3055	2980	3385	3435
3775	2815	3195	3595	3600
3670	3410	3260	3250	3515
3835				

Table 1-1 shows a simplified method to calculate the mean or average, $\bar{X}$, and standard deviation, s, for the above set of test results.

TABLE 1-1 - SIMPLIFIED STATISTICAL COMPUTATIONS

1	2	3	4	5	6	7	8
Cell Boundaries	Tally Count	Mid-cell Values	Column 3 -3500	Column 4 ÷200	Frequency	Column 5 x Column 6	Column 5 x Column 7
2600-2799	1	2700	-800	-4	1	-4	16
2800-2999	4	2900	-600	-3	4	-12	36
3000-3199	7	3100	-400	-2	7	-14	28
3200-3399	8	3300	-200	-1	8	-8	8
3400-3599	11	3500	0	0	11	0	0
3600-3799	6	3700	200	1	6	6	6
3800-3999	6	3900	400	2	6	12	24
4000-4199	1	4100	600	3	1	3	9
4200-4399	1	4300	800	4	1	4	16
4400-4599	1	4500	1000	5	1	5	25
Totals					46	-8	168
Squares of Totals					2116	64	

EXAMPLE 1.1 - Continued

Step 9 - $\frac{-8}{46} \times 200 = -35.$

Step 10 - Mean, $\bar{X} = -35 + 3500 = 3465$ psi, rounded out to 3470 psi.

Step 11 - 46 x 168 = 7728.

Step 12 - 8 x 8 = 64.

Step 13 - 7728 - 64 = 7664.

Step 14 - 7664 ÷ 2116 = 3.62.

Step 15 - Standard deviation, $s = 200 \times \sqrt{3.62} = 380$ psi, or directly from Table 1-2, s = 380 psi.

Steps in developing Table 1-1 are as follows:*

Step 1 - Group the results into cells (groups) of 200 psi intervals and tally them in increasing order (Columns 1, 2).

Step 2 - Tabulate the mid-cell values (Column 3) rounded to the nearest 100 psi.

Step 3 - Examine Column 3 to determine the mid-cell value with the highest frequency (highest count). Call this value the Central Value. In our example, this value is 3,500 psi. Subtract it from each of the mid-cell values and tabulate the results in Column 4.

Step 4 - Divide all values in Column 4 by 200 and tabulate the results in Column 5.

Step 5 - Tabulate in Column 6 the frequencies tallied in Column 2.

*To simplify this discussion, equations and derivations of formulas have been omitted. Interested readers may find them in ACI 214,[1.5] or in any textbook on statistics.

EXAMPLE 1.1 - Continued

Step 6 - Multiply each of the values in Column 5 by the corresponding frequencies in Column 6 and tabulate the results in Column 7 (some values will be positive and some will be negative).

Step 7 - Multiply each of the values in Column 5 by the corresponding values in Column 7 and tabulate the results in Column 8 (all values will be positive).

Step 8 - Add Columns 6, 7, and 8. (The total of Column 6 is the total number of tests.)

Step 9 - Divide the total of Column 7 by the total of Column 6 (if the sum of Column 7 was negative, then this result will also be negative). Do not calculate beyond two figures after the decimal point. Multiply this result by 200.

Step 10 - To obtain the mean, $\bar{X}$, add (or subtract) the result of Step 9 to (from) the Central Value. Round out the result to the nearest 10 psi.

Step 11 - Multiply the total of Column 6 by the total of Column 8.

Step 12 - Square the total of Column 7 (the result will be always positive).

Step 13 - Subtract the result of Step 12 from that of Step 11.

Step 14 - Divide the result of Step 13 by the square of the total of Column 6. Do not calculate beyond two figures after the decimal point.

Step 15 - Calculate the square root of the result of Step 14 rounded to two figures after the decimal point, and multiply it by 200. This is the standard deviation, s. Table 1-2 may be

EXAMPLE 1.1 - Continued

used as a guide to determine s directly from the results of Step 14 (intermediate values will have to be estimated).

TABLE 1-2 - STANDARD DEVIATION, s*

Result of Step 14	s	Result of Step 14	s	Result of Step 14	s	Result of Step 14	s
1.00	200	4.00	400	9.00	600	16.00	800
1.21	220	4.41	420	9.61	620	16.81	820
1.44	240	4.84	440	10.24	640	17.64	840
1.69	260	5.29	460	10.89	660	18.49	860
1.96	280	5.76	480	11.56	680	19.36	880
2.25	300	6.25	500	12.25	700	20.25	900
2.56	320	6.76	520	12.96	720	21.16	920
2.89	340	7.29	540	13.69	740	22.09	940
3.24	360	7.84	560	14.44	760	23.04	960
3.61	380	8.41	580	15.21	780	24.01	980

*Only applicable for cell intervals of 200 psi.

Other methods of computation (see Example 1.2) may yield more accurate results. The alternate methods involve more refinements, require more effort and may necessitate the use of a desk calculator. However, the accuracy of the values obtained by the above described method is adequate for general product control purposes in the concrete industry.

EXAMPLE 1.2 - Control Charts to Compute Strength Test Parameters*

Control charts are a valuable quality control tool to help continually evaluate strength test results. Figure 1-2 shows a control chart and computation form to evaluate $\bar{X}$, s, and V for a specified 4000 psi concrete with 20 strength test results.

Steps in developing Form I are as follows:

1. The values along the strength scale should be set up so that the expected average 28-day strength values will fall in the upper part of the chart and the expected 7-day or accelerated strengths, if obtained, will be plotted in the lower part. Select the proper first digits for the strength column and enter them in the blank (2000, 3000, etc.). Mark any control limits, such as minimum 28-day strength, as a horizontal line in the row corresponding to the limit value.

2. When a test result is received, enter the test date or test identification on the vertical line at the top of the chart. In the column below this line, place an X in the square opposite the strength range which contains the test data. Connect this X to the preceding one to form a graph.

3. If a test value plots outside the control limit line, appropriate action should be taken.

4. Strength test parameters can be calculated at any time using Form II.

*Courtesy National Ready Mixed Concrete Association.

EXAMPLE 1.2 - Continued

Steps in developing Form II are as follows:

1. Place Form II over Form I (Cylinder Strength Control Chart) with the arrows matching the top and bottom of the chart. The left edge of Form II should be slightly to the right of the last column containing strength data.

2. Insert the proper first digit of the mid-cell value (1) on Form II so that it corresponds to the strength range shown to the left on Form I.

3. Count the X's in each row and record the number in Column (2) of Form II. The total for Column (2) is "n" for the calculations.

4. In Column (3), enter the cumulative totals of frequencies from Column (2). For the highest cell, the number in Column (3) will always be the same as the value in Column (2) (representing the number of test results in the highest cell range). But the next value in Column (3) will be the sum from Column (2) for the two highest cells, the third value the sum for the highest 3 cells, and so on. The total of Column (3) is the term "a" in the calculations. As a check, the last value recorded in Column (3) for an individual (the lowest) cell should equal the total "n" of Column (2).

5. In Column (4), enter the <u>cumulative</u> totals from Column (3). As before, the number in Column (4) for the highest cell will always be the same as the value in that cell under Column (3). The next value in Column (4) will be the sum from Column (3) for the two highest cells, the third value the sum for the highest 3 cells, and so on. The total of Column (4) is the term "b" in the calculations. As a check,

EXAMPLE 1.2 - Continued

the last value recorded in Column (4) for an individual (the lowest) cell should equal the total "a" from Column (3).

6. The step-by-step calculation of average strength, standard deviation, and coefficient of variation is self-explanatory, involving step-by-step insertion of numbers previously developed and performing the indicated arithmetic.

FIG. 1-2 FIELD TEST RESULTS ON 4000 PSI CONCRETE

Form I. Cylinder Strength Control Chart

Concrete Classification *6 SX, NON AE*

Date or Cylinder Identification: *14, 15, 16, 18, 19, 22, 21, 25, 26, 27, 29, 31, 32, 33, 37, 38, 41, 43, 45, 46*

$f'_c = 4000$

Form II. Calculation of Strength Test Parameters

Concrete Classification *6 SX NON AE*
Test Age *28* Days

Strength range	Mid-cell Value (1)	No. of Tests in Row f (2)	Cumulative f 1st (3)	Cumulative f 2nd (4)
900-999	950			
800-899	850			
700-799	750			
600-699	650			
500-599	550			
400-499	450			
300-399	350			
200-299	250			
100-199	150			
5000-099	5 050			
900-999	950			
800-899	850			
700-799	750	1	1	1
600-699	650	1	2	3
500-599	550	1	3	6
400-499	450	1	4	10
300-399	350	0	4	14
200-299	250	2	6	20
100-199	150	6	12	32
4000-099	4 050	4	16	48
900-999	950	1	17	65
800-899	850	1	18	83
700-799	750	0	18	101
600-699	650	1	19	120
500-599	550	1	20	140
400-499	450			
300-399	350			
200-299	250			
100-199	150			
3000-099	3 050			
900-999	950			
800-899	850			
700-799	750			
600-699	650			
500-599	550			
400-499	450			
300-399	350			
200-299	250			
100-199	150			
2000-099	2 050			
900-999	950			
800-899	850			
700-799	750			
600-699	650			
500-599	550			
400-499	450			
300-399	350			
200-299	250			
100-199	150			
1000-099	1 050			
	Column Totals	20 (n)	140 (a)	643 (b)

	Calculations
F_1	$a/_n$ = 140/20 = 7.00
F_1^2	49.0
$F_1+F_1^2$	56.0
F_1-1	6.0
F_2	$b/_n$ = 643/20 = 32.15
$2F_2$	64.30
B	Lowest midcell value for which an f greater than zero is recorded 3550 PSI
$\overline{X}$	Average Strength, $\overline{x}$ $\overline{x} = B + 100 (F_1 - 1)$ = 3550 + 100 (6.00) = 3550 + 600 $\overline{x}$ = 4150 PSI
S	Standard Deviation, s $s = 100\sqrt{(2F_2) - (F_1 + F_1^2)}$ = 100 $\sqrt{64.30 - 56.00}$ = 100 $\sqrt{8.30}$ s = 288 PSI
V	Coefficient of Variation, V $V = 100\, s/_{\overline{x}}$ = 100 (288)/4150 = 2880/4150 V = 6.94 %

EXAMPLE 1.2 - Continued

Form I. Cylinder Strength Control Chart

Concrete Classification ____________

Date or Cylinder Identification

900-999
800-899
700-799
600-699
500-599
400-499
300-399
200-299
100-199
000-099

900-999
800-899
700-799
600-699
500-599
400-499
300-399
200-299
100-199
000-099

900-999
800-899
700-799
600-699
500-599
400-499
300-399
200-299
100-199
000-099

900-999
800-899
700-799
600-699
500-599
400-499
300-399
200-299
100-199
000-099

900-999
800-899
700-799
600-699
500-599
400-499
300-399
200-299
100-199
000-099

EXAMPLE 1.2 - Continued

Form II. Calculation of Strength Test Parameters

Concrete Classification ____________
Test Age _________ Days

Mid-cell Value (1)	No. of Tests in Row f (2)	Cumulative f 1st (3)	Cumulative f 2nd (4)
950			
850			
750			
650			
550			
450			
350			
250			
150			
_ 050			
950			
850			
750			
650			
550			
450			
350			
250			
150			
_ 050			
950			
850			
750			
650			
550			
450			
350			
250			
150			
_ 050			
950			
850			
750			
650			
550			
450			
350			
250			
150			
_ 050			
950			
850			
750			
650			
550			
450			
350			
250			
150			
_ 050			
Column Totals	(n)	(a)	(b)

	Calculations
F_1	$a/_n$ = -----/_____ = -----.--
F_1^2	------.---
$F_1 + F_1^2$	------.---
$F_1 - 1$	------.---
F_2	$b/_n$ = -----/_____ = -----.--
$2F_2$	------.---
B	Lowest midcell value for which an f greater than zero is recorded ------ PSI
$\overline{X}$	Average Strength, $\overline{x}$ $\overline{x} = B + 100\,(F_1 - 1)$ = ------ + 100 (----.---) = ------ + ------ $\overline{x}$ = ------ PSI
s	Standard Deviation, s $s = 100\sqrt{(2F_2) - (F_1 + F_1^2)}$ = 100 √(------.--- - ------.---) = 100 √(------.---) s = ------ PSI
V	Coefficient of Variation, V $V = 100\, s/_{\overline{x}}$ = 100 (-----)/_____ = -------/_____ V = ----.--- %

2

Details of Reinforcement

General Considerations

Good structural details are vital to satisfactory performance of reinforced concrete structures. Standard practice for reinforcement details has developed gradually. The Building Code Committee (ACI 318) continually collects reports of research and practice with reinforcing materials, suggests new research needed, receives reports on new research, and translates the results into specific code provisions for details of reinforcement. The ACI Detailing Manual, Reference 2.1, provides recommended methods and standards for preparing design drawings, typical details, and drawings for fabrication and placing of reinforcing steel in reinforced concrete structures. Separate sections of the manual define responsibilities of both engineer and reinforcing bar detailer.

7.1 Standard Hooks

Requirements for standard hooks and minimum finished inside bend diameters for reinforcing bars are illustrated in Tables 2-1 and 2-2. These requirements are based on multiples of nominal bar diameters. The standard hook details for stirrups and ties are slightly revised for the '83 Code edition. Only bar sizes #8 and smaller are addressed, with the larger bar sizes--#6, #7, and #8--required to have an increased length of extension ($12d_b$) if 90-deg bends are used.

TABLE 2-1. Standard Hooks for Primary Reinforcement*

Bar size	Min. finished bend dia.[a]
#3 through #8	6 bar dia.
#9, #10, #11	8 bar dia.
#14 and #18	10 bar dia.

[a]Measured on inside of bar.

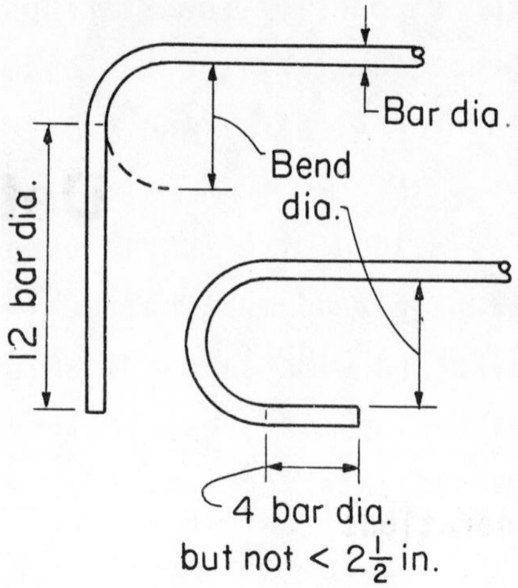

TABLE 2-2. Standard Hooks for Stirrups and Tie Reinforcement*

Bar size	Min. finished bend dia.[b]
#3 through #5	4 bar dia.
#6 through #8	6 bar dia.

[b]Measured on inside of bar.

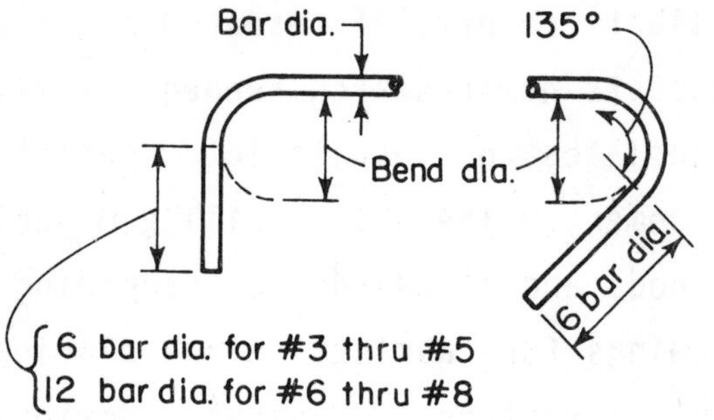

7.2 Minimum Bend Diameters

Minimum bend diameter for a reinforcing bar is defined as "the diameter of bend measured on the inside of the bar." Minimum bend diameters are dependent on bar size and multiples of nominal bar diameters; for #3 to #8, the minimum bend diameter is 6 bar diameters; for #9 to #11, the minimum bend diameter is 8 bar diameters; and for #14 and #18, the minimum bend diameter is 10 bar diameters. Exceptions to these provisions are:

*Table 1 in Part C of Reference 2.1 provides actual bar dimensions for end hooks, and stirrup and tie hooks.

(1) For stirrups and ties, the minimum bend diameter is 4 bar diameters (1-1/2 in. for #3 bars, 2 in. for #4 bars, and 2-1/2 in. for #5 bars). For stirrups and ties #6 through #8, the minimum bend diameter is 6 bar diameters (4-1/2 in. for #6 bars, 5-1/4 in. for #7 bars, and 6 in. for #8 bars).

(2) For welded wire fabric used for stirrups and ties, inside diameter of bend must not be less than four wire diameters for deformed wire larger than D6 and two wire diameters for all other wire. Special restrictions apply within 4 wire diameters of a welded intersection.

7.3 Bending

All reinforcement must be bent cold unless otherwise permitted by the engineer. For unusual bends, special fabrication including heating may be required and the engineer must give approval to the techniques used.

Reinforcing bars partially embedded in concrete must not be field bent without authorization of the engineer. Code Commentary Section 7.3.2 provides guidelines for field bending and heat, if necessary, for bars partially embedded in concrete. The exception that permitted Grade 40 bars with 180-deg bends to have a tighter minimum bend diameter ($5d_b$) is deleted from the '83 Code to reflect current industry practice for bar bending details.

7.5 Placing Reinforcement

Supports for reinforcement are required to adequately support and secure the reinforcement against displacement, but the supports are not required to be of any specific material or type. Welding of crossing bars (tack welding) for assembly of reinforcement is prohibited except as specifically authorized by the engineer.

Tolerances for placing reinforcement are given for minimum concrete cover and for the effective depth, d. Both dimensions are components of the total depth so the tolerances on these dimensions are directly related. The amount of tolerance allowed is dependent on the size of the member expressed as a function of the effective depth d, and minimum concrete cover. These tolerances are illustrated in Table 2-3. Exceptions to these provisions are:

(1) The tolerance for the clear distance to formed soffits must be minus 1/4 in.

(2) Tolerance for cover must not be reduced more than one-third of the specified minimum concrete cover.

TABLE 2-3. Critical Dimensional Tolerances for Locating Reinforcement

Effective Depth d	Tolerance on d	Tolerance on Min. Cover
d = 8 in. or less	± 3/8 in.	- 3/8 in.
d = over 8 in.	± 1/2 in.	- 1/2 in.

Effective depth, d, (±)

Clear cover(±) as specified in Sect. 7.7

For ends of bars and longitudinal location of bends, the tolerance is ± 2 in. except at discontinuous ends of members where the tolerance is ± 1/2 in. These tolerances are illustrated in Fig. 2-1.

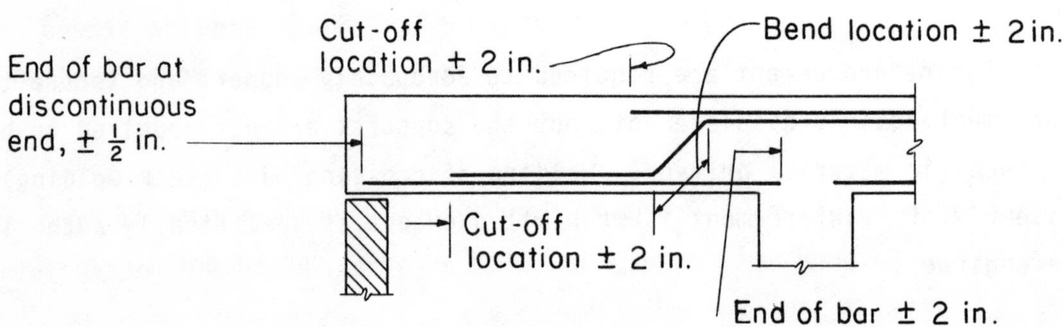

Fig. 2-1 Tolerances for Bar Bend and Cutoff Locations

7.6 Spacing Limits for Reinforcement

Spacing requirements (clear distance between bars) must be as follows: For members with parallel bars in a layer, not less than one nominal bar diameter nor less than 1 in.; and for reinforcement in two or more layers, bars must be directly above one another with at least 1 in. clear vertically. For spirally reinforced and tied reinforced compression members, the clear distance between longitudinal bars must not be less than 1-1/2 nominal bar diameters, nor less than 1-1/2 in. Note that these spacing requirements also apply to clear distance between spliced bars. Also note that Section 3.3.3, which covers spacing requirements based on aggregate size, may be applicable (see Table 2-4, Note 1). In walls and slabs other than concrete joists, primary flexural reinforcement must not be spaced greater than 3 times the wall or slab thickness nor 18 in.

7.6.6 Bundled Bars

Bundling of bars (parallel reinforcing bars in contact, assumed to act as a unit) is permitted, but only if such bundles are enclosed by lateral ties or stirrups. Some limitations are placed on the use of bundled bars in flexural members as follows:

(1) # 14 and #18 bars cannot be bundled in beams and girders.

(2) If individual bars in a bundle are cut off within the span, such cutoff points must be staggered at least 40 bar diameters.

(3) Two bundled bars maximum in any one plane is implied (three or more adjacent bars in one plane are not considered as bundled bars).

(4) For spacing and minimum clear cover, a unit of bundled bars must be treated as a single bar with an area equivalent to the total area of all bars in the bundle.

(5) A maximum of four bars may be bundled (see Fig. 2-2).

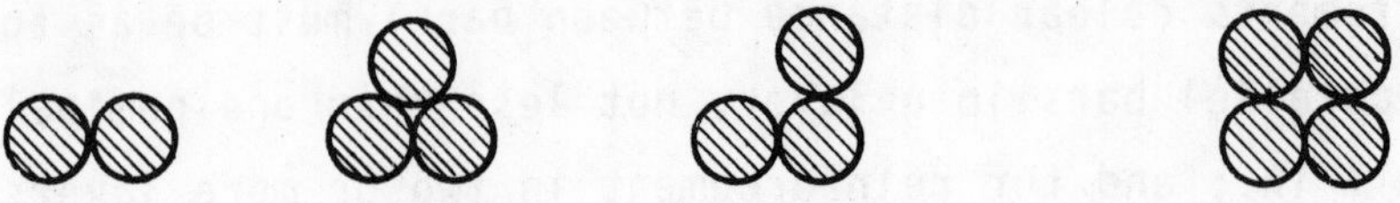

Fig. 2-2. Possible Reinforcing Bar Bundling Arrangements

7.6.7 Prestressing Tendons and Ducts

Clear distance between pretensioning tendons at ends of members is handled separately and is limited to 4 nominal diameters of individual wires, or 3 nominal strand diameters. Closer vertical spacing or bundling of tendons is permitted in the middle portion of the span if special care in design and fabrication is employed. Post-tensioning ducts may be bundled if concrete can be placed and if it is assured that there will be no steel breakout when tensioned. Spacing requirements are illustrated in Table 2-4.

TABLE 2-4. Clear Distances Between Bars, Bundles, or Tendons

Second layer
1" clear
1" clear
Clear distance

Type member	Clear distance
Reinforcement	
Flexural members	1 bar dia. but not < 1 in.
Compression members, tied or spirally reinforced	1.5 bar dia. but not < 1-1/2 in.
Pretensioning tendons	
Wires	4 wire dia.
Strands	3 strand dia.

Notes: (1) Clear distance must also be greater than 4/3 of the maximum size aggregate used (Section 3.3.3). (2) For bundled bars, diameter of a single bar of equivalent total area must be used. (3) Closer vertical spacing and bundling of pretensioning may be allowed (Section 7.6.7.1).

7.7 Concrete Protection for Reinforcement

Concrete cover or protection requirements are specified for members cast against earth, in contact with earth or weather, and for interior members not exposed to weather. Slightly reduced cover or protection is permitted under these same conditions for precast concrete manufactured under plant control conditions, and other values are given for prestressed concrete. The term "manufactured under plant controlled conditions" does not specifically imply that precast members must be manufactured in a plant. Structural elements precast at the job site will also qualify for the lesser cover if the control of form dimensions, placing of reinforcement, quality control of concrete, and curing procedure are equal to that normally expected in a plant operation. Larger diameter bars and bundled bars require slightly greater cover. Corrosive environments or fire protection may also warrant special consideration. The designer should take special note of the Commentary recommendations (Section 7.7.5) for increased cover when concrete will be exposed to external sources of chlorides in service, such as deicing salts and sea water. These recommendations are new for the '83 Code Edition.

7.8 Special Reinforcing Details for Columns

Section 7.8 covers the special detailing requirements for offset bent longitudinal bars and steel cores of composite columns.

When column offsets are necessary, longitudinal bars may be bent, subject to the following limitations:

(1) Slope of the inclined portion of the bars must not exceed 1 in 6 (see Fig. 2-3).

(2) On either side of the offset portion, bars must remain parallel to the axis of the column.

(3) Horizontal support at the bends must be provided by lateral ties, spirals, or part of the floor construction at points not farther than 6 in. from the bend point (see Fig. 2-3). The horizontal support provided must be designed to resist 1-1/2 times the horizontal component of the computed force in the inclined portion of the bars.

(4) Offset bars must be bent before placement in the forms.

(5) When column faces are offset 3 in. or more, longitudinal column bars must not be bent, and must be lap spliced by separate dowels. (See Fig. 2-3). In some cases, a column might be offset 3 in. or more on one face only, which could possibly result in some offset bent longitudinal column bars and some separate dowels being used in the same column.

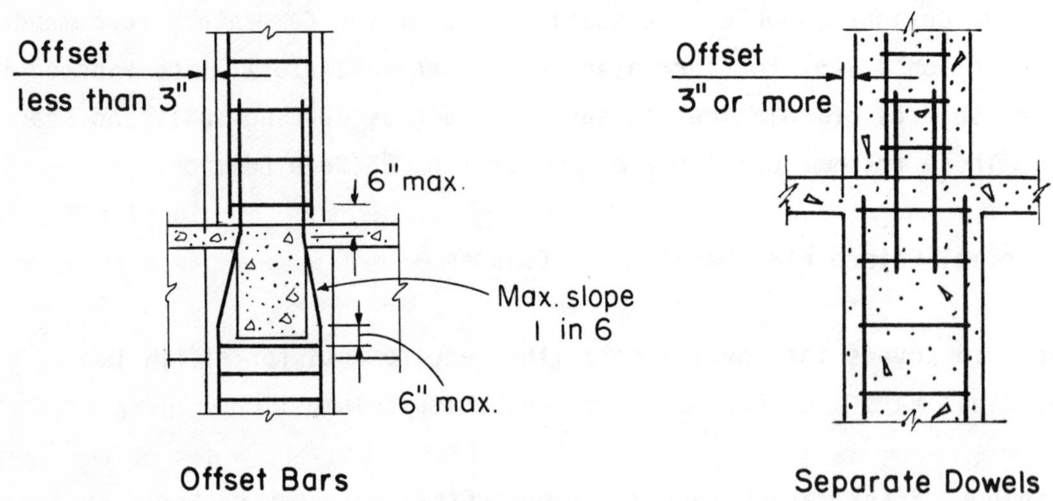

Fig. 2-3. Special Column Details

Steel cores in composite columns can be detailed to allow transfer of up to 50 percent of the compressive load in the core by direct bearing. The remainder of the load must be transferred by welds, dowels, splice plates, etc. The result of this should be to insure a minimum tensile capacity similar to that of a more common reinforced concrete column.

7.9 Connections

Enclosures must be provided for splices of continuing reinforcement, and for end anchorage of reinforcement terminating at beam and column connections. This enclosure may consist of external concrete or internal closed ties, spirals, or stirrups.

7.10 Lateral Reinforcement for Compression Members

Circular spirals must be held firmly in place and true to line by vertical spacers. For spiral reinforcement of less than 5/8 in. diameter, 2 spacers must be used for spirals less than 20 in. diameter, 3 spacers for spirals from 20 in. to 30 in. diameter, and 4 spacers for spirals over 30 in. diameter. If the spiral reinforcement is 5/8 in. diameter or greater, the spacer requirements become 3 spacers for spirals 24 in. or less in diameter, and 4 spacers for spirals greater than 24 in. diameter.

Minimum diameter of spiral reinforcement in cast-in-place construction is 3/8 in. and the clear spacing must be between the limits of 1 in. to 3 in. This requirement does not preclude use of a smaller minimum spiral diameter for precast units. Splices in spirals must be welds or tension lap splices of at least 48 spiral nominal bar or wire diameters but not less than 12 in.

Spiral reinforcement must extend from the top of footing or slab in any story to the level of the lowest horizontal reinforcement in slabs, drop panels, or beams above. If beams or brackets do not frame into all sides of the column, ties must extend above the top of the spiral to the bottom of the slab or drop panel (see Fig. 2-4).

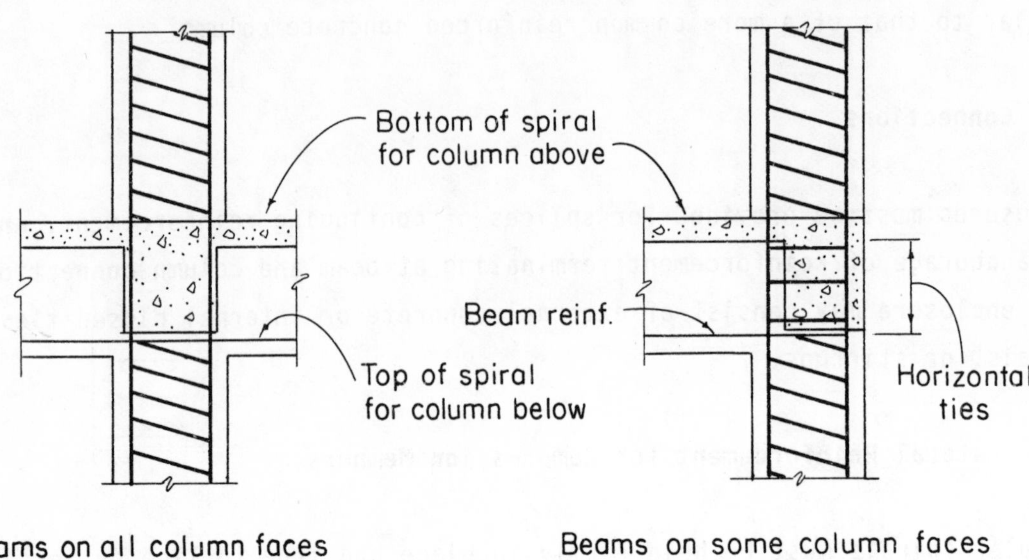

Fig. 2-4. Termination of Spirals

In tied reinforced columns, ties must be located no more than half a tie spacing above floor or footing and no more than half a tie spacing below the lowest horizontal reinforcement in the slab or drop panel above. If beams or brackets frame from four directions into a column, ties may be terminated no more than 3 in. below the lowest horizontal reinforcement in beams or brackets (see Fig. 2-5). Minimum size of lateral ties in tied reinforced columns is related to the size of the vertical bars. Minimum tie sizes are #3 ties for vertical bars #10 and smaller, and #4 ties for #11 vertical bars and larger and for bundled bars. The following conditions also apply: spacing must not exceed 16 longitudinal bar diameters, 48 tiebar diameters, or the least dimension of the column; every corner bar and alternate bar must have lateral support provided by a tie with at least a 45 degree bend; no unsupported bar shall be farther than 6 in. from a supported bar (see Fig. 2-6). Note that the 6-in. clear limitation is measured along the tie.

Welded wire reinforcement of equivalent area may be used for ties. When main reinforcement is arranged in a circular pattern, it is permissible to use only one circular tie per specified spacing. This provision allows the use of circular ties at a spacing greater than that specified for spirally reinforced columns.

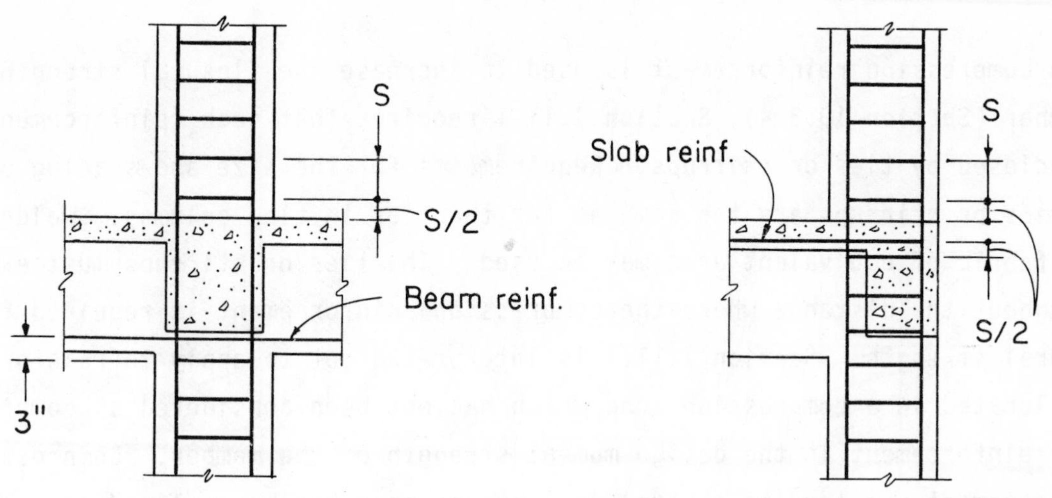

Fig. 2-5. Termination of Column Ties

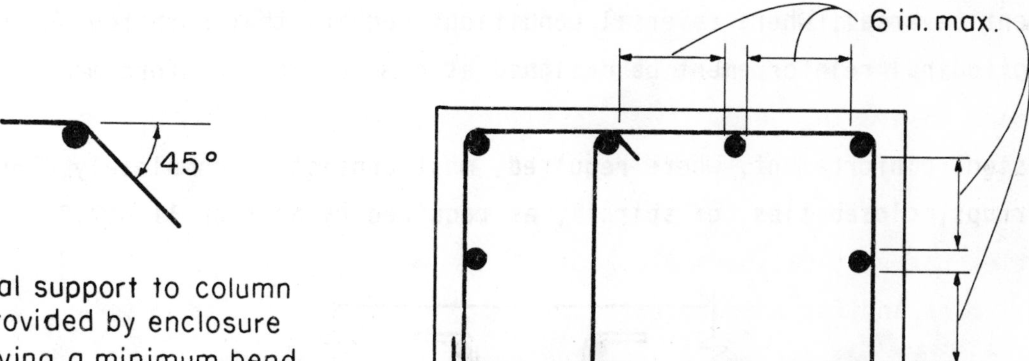

Fig. 2-6. Lateral Support of Column Bars by Ties

7.11 Lateral Reinforcement for Flexural Members

Where compression reinforcement is used to increase the flexural strength of a member (Section 10.3.4), Section 7.11.1 requires that such reinforcement be enclosed by ties or stirrups. Requirements for the size and spacing of the ties or stirrups are the same as for the ties in tied columns. Welded wire fabric of equivalent area may be used. The ties or stirrups must extend throughout the distance where the compression reinforcement is required for flexural strength. Section 7.11.1 is interpreted not to apply to reinforcement located in a compression zone which has not been considered as compression reinforcement in the design moment strength of the member. Compression reinforcement provided to control deflections need not be enclosed as required by Section 7.11.1.

Enclosure required by Section 7.11.1 is illustrated by the U-shaped stirrup in Fig. 2-7; the continuous bottom portion of the stirrup satisfies the enclosure intent of Section 7.11.1 for the two bottom bars shown. A completely closed stirrup is ordinarily not necessary, except in cases of high moment reversal, where reversal conditions require that both top and bottom longitudinal reinforcement be designed as compression reinforcement.

Torsion reinforcement, where required, must consist of completely closed stirrups, closed ties, or spirals, as required by Section 11.6.7.3.

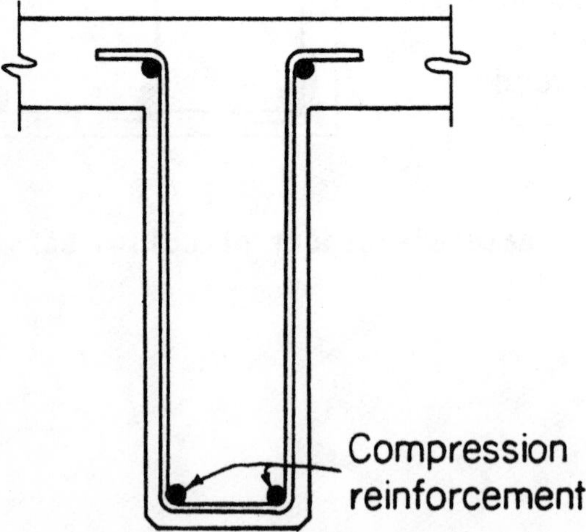

Fig. 2-7 Enclosure for Compression Reinforcement

7.11.3 Closed Stirrups

According to Section 7.11.3, a closed stirrup is formed either in one piece with overlapping 90 deg. or 135 deg. end hooks, or in one or two pieces with a Class C lap splice, as illustrated in Fig. 2-8. The one-piece closed stirrup with overlapping end hooks is not practical for placement. Neither of the closed stirrups shown in Fig. 2-8 is considered effective for members subject to high torsion. Tests have shown that, with high torsion, loss of the concrete covering and subsequent loss of anchorage results if the 90 deg. hook and lap splice details are used where confinement by external concrete is limited. See Fig. 2-9. The ACI Detailing Manual, Reference 2.1, recommends the details illustrated in Fig. 2-10 for closed stirrups used as torsional reinforcement.

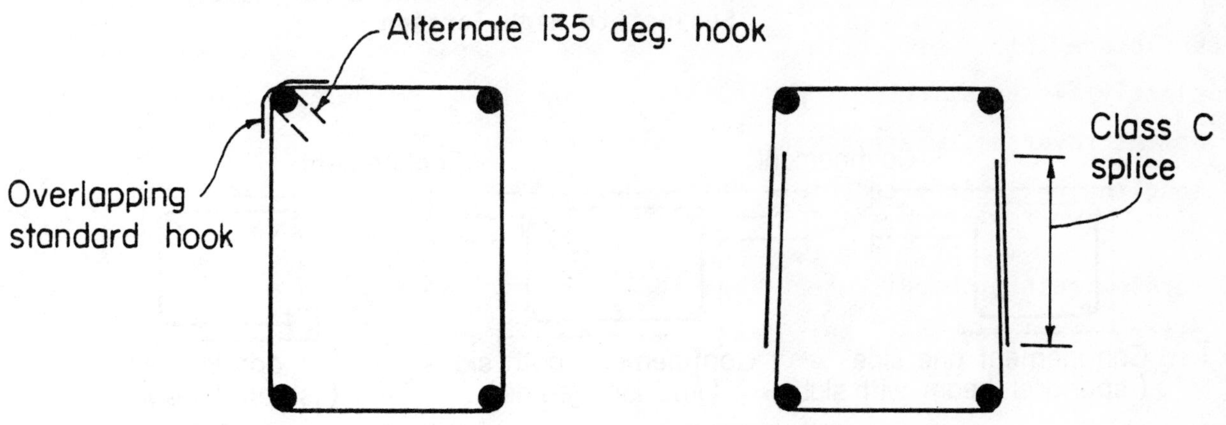

Fig. 2-8 Code Definition of Closed Tie or Stirrup

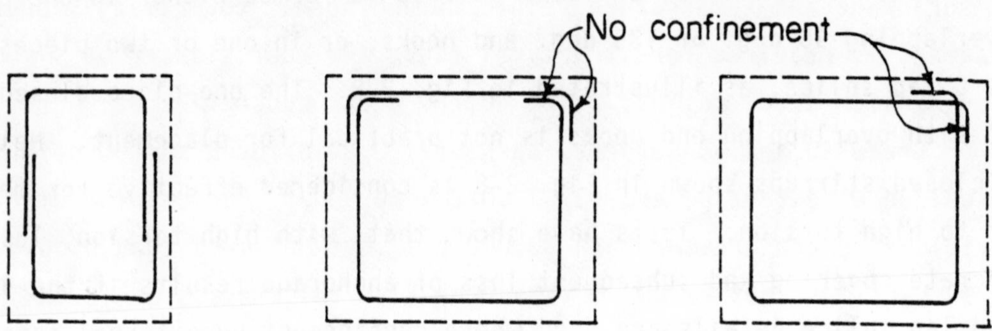

These details are not considered effective for members subject to high torsion. Note lack of confinement when compared to similar members with confinement shown in Fig. 10.

Fig. 2-9 Closed Stirrup Details Not Recommended for Members Subject to High Torsion

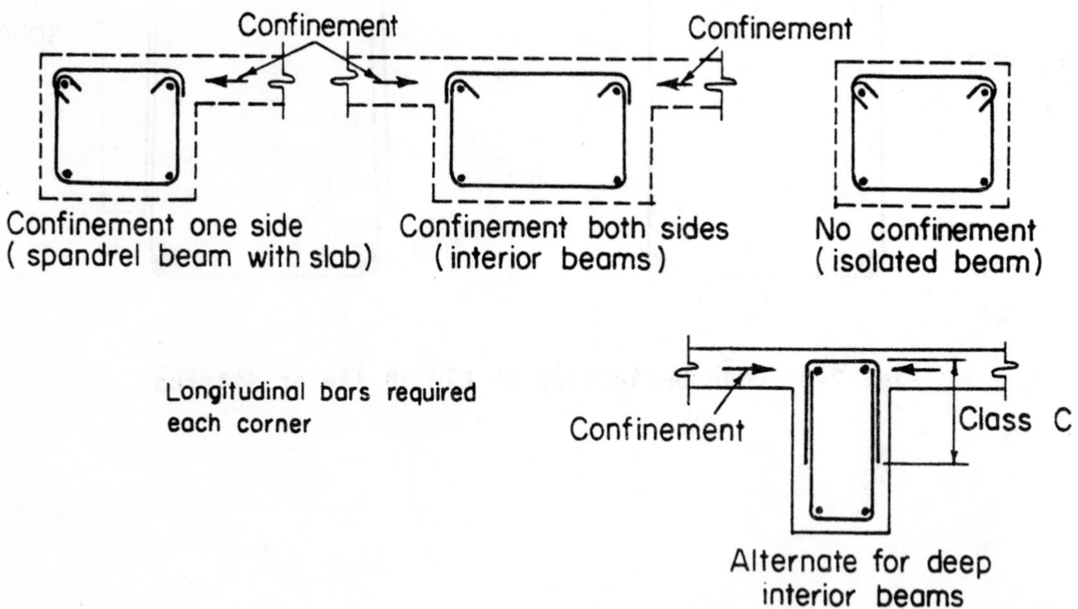

Fig. 2-10 Recommended Two-Piece Closed Stirrup Details[2.1]

7.12 Shrinkage and Temperature Reinforcement

Minimum shrinkage and temperature reinforcement normal to primary flexural reinforcement is required for structural floor and roof slabs (not slabs on ground). Minimum steel ratios, based on the gross concrete area, are: (1) 0.0020 for Grade 40 and 50 deformed bars; (2) 0.0018 for Grade 60 deformed bars or welded wire fabric; and (3) 0.0018 x 60,000/f_y for reinforcement with a yield strength greater than 60,000 psi, but not less than 0.0014. Spacing of bars must not exceed 5 times the slab thicknesses or 18 in. Splices and end anchorages of shrinkage and temperature reinforcement must be designed for the full specified yield strength.

Bonded or unbonded prestressing tendons may be used for shrinkage and temperature reinforcement in structural slabs (Section 7.12.3). The tendons must provide a minimum average compressive stress of 100 psi on the gross concrete area, based on effective prestress after losses. Spacing of tendons cannot exceed 6 ft. When the spacing is greater than 54 in., additional bonded reinforcement must be provided.

Selected Reference

2.1 ACI Detailing Manual - 1980 (SP-66), ACI Committee 315, American Concrete Institute, Detroit, 1980.

3

Development and Splices of Reinforcement

Update for '83 Code

For ACI 318-83, the hooked-bar anchorage provisions of Section 12.5 have been extensively revised to reflect new research. The new anchorage provisions are a major departure from ACI 318-77 in that they uncouple hooked-bar anchorages from straight bar development and give total hooked-bar embedment length directly. The new provisions not only greatly simplify calculations for hook anchorage lengths but also result in a required embedment length considerably less, especially for the larger bar sizes, than that required by the 1977 code provisions. See discussion for Code Section 12.5.

Development Length Concept

The development length concept for anchorage of reinforcement, such as for deformed bars and deformed wires, is based on the attainable average bond stress over the length of embedment of the reinforcement. In application, the development length concept requires the specified minimum lengths or extensions of reinforcement beyond all points of peak stress in the reinforcement. Such peak stresses generally occur in flexural members at the points specified in Section 12.10.2.

The strength reduction factor φ is not used in Chapter 12, because the specified development lengths already include an allowance for understrength.

The required development and lap splice lengths are the same for either the strength design method or the alternate design method of Appendix B (Section B.4).

12.1 Development of Reinforcement - General

Development length or anchorage of reinforcement is required on both sides of a point of peak stress at each section of a reinforced concrete member. In continuous members for example, reinforcement typically continues for a considerable distance on one side of a critical stress point so that detailed calculations are usually required only for the side where the reinforcement is terminated.

12.2 Development of Deformed Bars and Deformed Wire in Tension

Requirements for basic tension development length ℓ_{db} of deformed bars and deformed wire are given in Section 12.2.2:

For bar sizes #3 - #11 . . . . $\ell_{db} = 0.04\,A_b f_y/\sqrt{f'_c}$
but not less than $0.0004\,d_b f_y$

For #14 bars . . . . $\ell_{db} = 0.085\,f_y/\sqrt{f'_c}$

For #18 bars . . . . $\ell_{db} = 0.11\,f_y/\sqrt{f'_c}$

For deformed wire . . . . $\ell_{db} = 0.03 d_b\,f_y/\sqrt{f'_c}$

Section 12.2.5 further requires that the development length ℓ_d (including applicable modification factors) must not be less than 12 in., except in computing the lengths of tension lap splices or development of web reinforcement.

Basic development lengths ℓ_{db} for Grade 60 bars in tension are tabulated in Table 3-1. The tabulated values are for the bars embedded in normal weight concrete with compressive strengths of 3,000, 4,000 and 5,000 psi.

TABLE 3-1 - "Basic" Tension Development Length ℓ_{db} (inches) for Grade 60 Bars* in Normal Weight Concrete

Bar Size	$f'_c = 3000$	$f'_c = 4000$	$f'_c = 5000$
# 3	9**	9**	9**
4	12	12	12
5	15	15	15
6	19	18	18
7	26	23	21
8	35	30	27
9	44	38	34
10	56	48	43
11	68	59	53
14	93	81	72
18	121	104	93

*For Grade 40 bars, basic development lengths are two-thirds of the values tabulated, but not less than 12 in.

**Development length ℓ_d must not be less than 12 in. (including applicable modification factors).

After establishing the basic tension development lengths, consideration must be given to the applicable modification factor or factors in Sections 12.2.3 and 12.2.4. The modification factors are multipliers for the basic development lengths (ℓ_{db} x applicable modification factors) to account for various conditions as follows:

Section 12.2.3.1 reflects the condition that top reinforcement may have reduced anchorage bond due to settlement of the fresh concrete below the reinforcement. The basic ℓ_{db} must be multiplied by a factor of 1.4 to account for the so-called "top bar effect." Top reinforcement is defined as horizontal reinforcement where more than 12 in. of fresh concrete is cast in the member below the reinforcement.

The modification factors in Section 12.2.3.3 are based on the generally lower splitting strength of lightweight aggregate concretes. Note that the 1.33 or 1.18 factors may be reduced, but not less than 1.0, when the splitting tensile strength f_{ct} is specified at an adequate level. The value of f_{ct} is a function of the concrete mix design and lightweight aggregate characteristics.

Section 12.2.4.1 permits a 20 percent reduction in development length under conditions which reduce the splitting tendency of the concrete caused by the lug action of the bar or wire deformations. If bars are spaced laterally at least 6 in. on center with at least 3 in. clear from the face of the member to the edge bar, the basic ℓ_{db} may be multiplied by a factor of 0.8. The required lateral spacing and end clearance dimension are illustrated in Fig. 3.1.

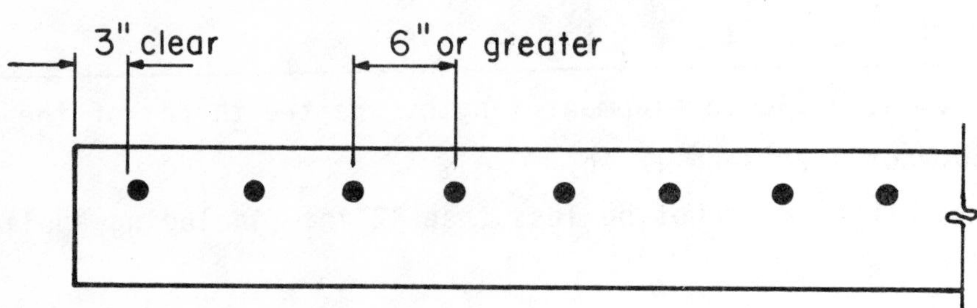

Fig. 3-1 - Spacing and Edge Cover for (0.8) Reduction Factor

The factors of Sections 12.2.3 and 12.2.4 are multiplied together when more than one is applicable. An example of development length computations will illustrate proper application of the various modification factors.

Compute the required tension development length for the #8 bars in the "sand-lightweight" slab shown in Fig. 3-2. Use f'_c = 4000 psi, f_y = 60,000 psi:

basic development length:

$$\ell_{db} = 0.04A_b f_y/\sqrt{f'_c} = 0.04 \times 0.79 \times 60{,}000/\sqrt{4000} = 30 \text{ in.}$$

but not less than

$$0.0004\, d_b f_y = 0.0004 \times 1.0 \times 60{,}000 = 24 \text{ in.}$$

or 12 in. min.

factor for top bars = 1.4
(more than 12 in. of fresh concrete will be cast below the bars)
factor for "sand-lightweight" = 1.18
factor for lateral spacing = 0.8
(spacing for short bars may be considered the same as for long bars for computing required development length since short bars are developed in length BC, while long bars are already developed in length AB.)

$$\text{Required } \ell_d = \ell_{db} \times \text{applicable modification factors}$$
$$= 30 \times 1.4 \times 1.18 \times 0.8 = 40 \text{ in.}$$

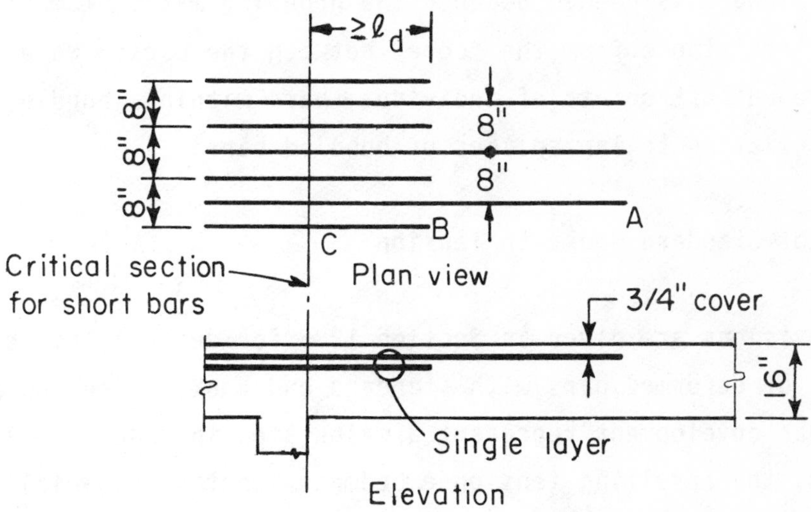

Fig. 3-2 - Example

Two additional modification factors are given in Sections 12.2.4.2 and 12.2.4.3 for reducing basic tension development lengths: (1) when excess reinforcement is provided, and (2) when confining reinforcement in the form of spirals is provided around the bars. It should be noted that the reduction factor for excess bar area (A_s required/A_s provided) does not apply when the full f_y development is required, as for tension lap splices in Section 12.15.1.

12.3 Development of Deformed Bars in Compression

Shorter development lengths are required for bars in compression since the weakening effect of flexural tension cracks in the concrete is not present. The basic compression development length is $\ell_{db} = 0.02 f_y d_b / \sqrt{f'_c}$, but not less than $0.0003 f_y d_b$ or 8 in. The basic development length can be reduced where excess bar area is provided and where a "confining" spiral is provided around the bars (Section 12.3.3).

12.4 Development of Bundled Bars

Increased development length for individual bars, in tension or compression, is required when 3 or 4 bars are bundled together. The increase in development length is 20 percent for a 3-bar bundle, and 33 percent for a 4-bar bundle. The extra length is needed because the grouping makes it more difficult to mobilize resistance from the "core" between the bars. Note Section 7.6.6.4 relating to cut-off points of individual bars within a bundle, and Section 12.14.2.2 relating to lap splices of bundled bars.

12.5 Development of Standard Hooks in Tension

Completely new provisions are given in Section 12.5 for determining the anchorage capacity of deformed bars with standard end hooks. The new provisions for hooked bar development represent a major step in Code simplification. In addition, the resulting tension embedment lengths, especially for the larger bar sizes, are significantly shorter than those calculated under provisions of the 1977 Code. End hooks can only be considered effective in developing bars in tension, and not in compression; see Sections 12.1 and 12.5.5. Only standard end hooks (Section 7.1) are considered in this section. The anchorage capacity of end hooks with larger bend diameters cannot be determined by the provisions in Section 12.5.

Application of the new hook development provisions is essentially parallel to calculating development lengths of straight bars. The first step is to calculate the basic development length of the hooked bar, ℓ_{hb}. For Grade 60 bars, $\ell_{hb} = 1200d_b/\sqrt{f'_c}$. The basic development length is then multiplied by the applicable modification factor or factors to determine the development length of the hook, $\ell_{dh} = \ell_{hb}$ x applicable modification factors. The development length ℓ_{dh} is measured from the critical section to the outside end of the standard hook, i.e., the straight embedment length between the critical section and the start of the hook, plus the radius of bend of the hook, plus one bar diameter. Fig. 3-3 shows ℓ_{dh} and the standard hooks (Section 7.1) for all the bar sizes.

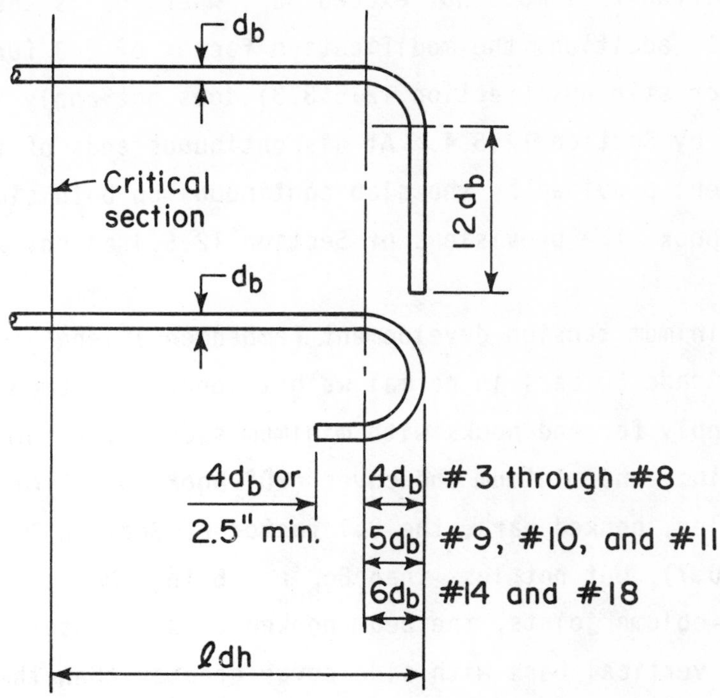

Fig. 3-3 - Development ℓ_{dh} of Standard Hooks

The ℓ_{hb} modification factors listed in Section 12.5.3 account for:

- Bar yield strength
- Favorable confinement conditions provided by increased cover, or transverse ties or stirrups to resist splitting of the concrete.
- More reinforcement provided than required by analysis.
- Lightweight aggregate concrete.

After multiplying the basic development length ℓ_{hb} by the applicable modification factor or factors, the resulting development length ℓ_{dh} must not be less than 8 d_b nor 6 in.

Section 12.5.4 is a special provision for hooked bars terminating at discontinuous ends of members, such as at the ends of simply-supported beams, at free ends of cantilevers, and at ends of members framing into a joint where the member does not extend beyond the joint. If the full strength of a hooked bar must be developed, and both side cover and top (or bottom) cover over the hook is less than 2 1/2 in., Section 12.5.4 requires the hook to be enclosed within ties or stirrup-ties over the development length, ℓ_{dh}. Spacing of the ties or stirrup-ties must not exceed $3d_b$, where d_b is the diameter of the hooked bar. In addition, the modification factor of 0.8 for confinement provided by ties or stirrups (Section 12.5.3.3) does not apply to the special condition covered by Section 12.5.4. At discontinuous ends of slabs with concrete confinement provided by the slab continuous on both sides normal to the plane of the hook, the provisions of Section 12.5.4 do not apply.

Table 3-2 gives minimum tension development (embedment) lengths ℓ_{dh} for standard end hooks on Grade 60 bars in normal weight concrete. Embedment lengths in Table 3-2(a) apply for end hooks with minimum side cover normal to plane of hook of 2 1/2 in., and minimum end cover (90° hooks only) of 2 in. Thus, for #11, and smaller, hooked bars, the 0.7 factor of Section 12.5.3.2 applies; $\ell_{dh} = 1200d_b/\sqrt{f'_c}(0.7)$, but not less than $8d_b$ nor 6 in. Note: for hooked bar anchorage in beam-column joints, the beam hooked bars are usually placed inside the column vertical bars with side cover greater than the 2 1/2 in. minimum required for application of the 0.7 factor. Also, for 90° end hooks, with hook extension located inside the column ties, the 2 in. minimum will usually be satisfied to permit the 0.7 reduction factor. Table 3-2(b) includes the additional 0.8 reduction factor for confining ties or stirrup-ties (Section 12.5.3.3); thus for #11, and smaller, hooked bars, $\ell_{dh} = 1200d_b \sqrt{f'_c}(0.7)(0.8)$, but not less than $8d_b$ nor 6 in.

TABLE 3-2 - Minimum Embedment Length ℓ_{dh} (inches) for Standard End Hooks on Grade 60 Bars in Normal Weight Concrete

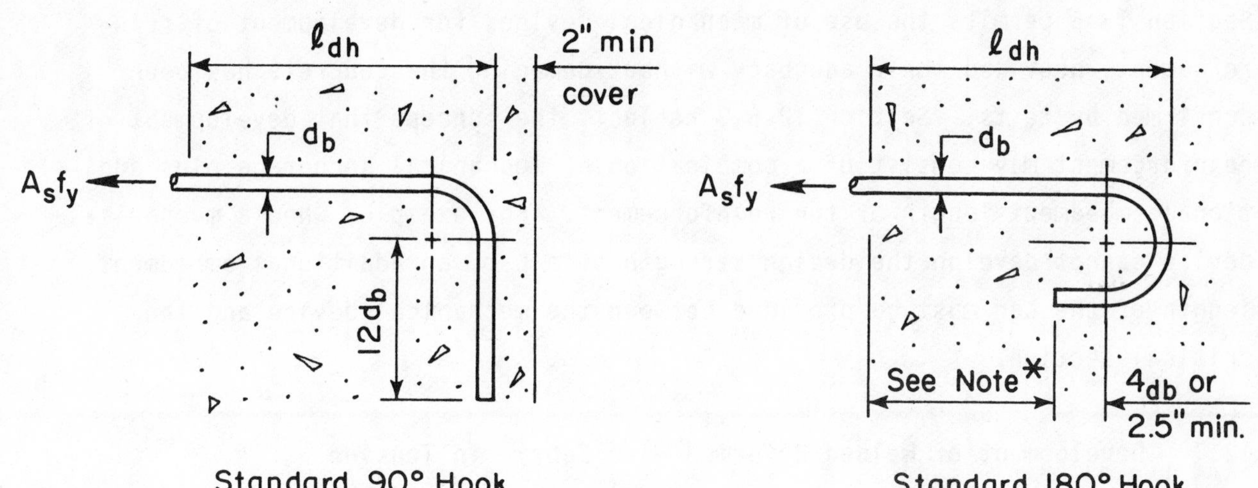

Standard 90° Hook

Standard 180° Hook

Bar Size	(a) General Use • Side Cover ≥ 2-1/2 in. • End Cover (90° hooks) ≥ 2 in.			(b) Special Confinement • Side Cover ≥ 2-1/2 in. • End Cover (90° hooks) ≥ 2 in. • Ties or Stirrups spaced ≥ $3d_b$		
	f'_c = 3000	f'_c = 4000	f'_c = 5000	f'_c = 3000	f'_c = 4000	f'_c = 5000
# 3	6	6	6	6	6	6
4	8	7	6*	7	6*	6*
5	10	9	8	8	7	6*
6	12	10	9	10	8	8
7	14	12	11	11	10	9
8	16	14	12	13	11	10
9	18	15	14	14	12	11*
10	20	17	16	16	14	13*
11	22	19	17	18	15	14*
14	38	33	29	--	--	--
18	50	43	39	--	--	--

*For 180° hooks normal to exposed surfaces, minimum embedment to provide 2 in. (recommended) cover to tail of hook:

Bar size	#3	4	5	6	7	8	9	10	11	14	18
Embedment length (in.)	6	7	7	8	9	10	12	13	15	19	25

12.6 Mechanical Anchorage

Section 12.6 permits the use of mechanical devices for development of reinforcement, provided their adequacy without damaging the concrete has been confirmed by tests. Section 12.6.3 reflects the concept that development of reinforcement may consist of a combination of mechanical anchorage plus additional embedment length of the reinforcement. For example, when a mechanical device cannot develop the design strength of a bar, an additional embedment length of the bar must be provided between the mechanical device and the critical section.

12.7 Development of Welded Deformed Wire Fabric in Tension

For welded deformed wire fabric, development length is measured from the point of critical section to the end of the wire. The provisions in Section 12.7.2 for calculating the basic development length are based on the condition that at least one cross wire is located within the development length, and the cross wire is not less than 2 in. distant from the critical section. The basic $\ell_{db} = 0.03d_b(f_y - 20{,}000)/\sqrt{f'_c}$, but not less than $0.20\ A_w f_y / s_w \sqrt{f'_c}$.

The required development length is established by multiplying the basic development length ℓ_{db} by the applicable modification factor or factors of Sections 12.2.3 and 12.2.4. The resulting development length ℓ_d cannot be less than 8 in., except in computation of lap splice lengths (Section 12.18) and development of web reinforcement (Section 12.13). Fig. 3-4 shows the development length requirements for the condition when none of the modification factors in Sections 12.2.3 and 12.2.4 apply.

If no cross wires are located within the development length, the basic development length for the fabric must be based on the provisions for deformed wire in Section 12.2.

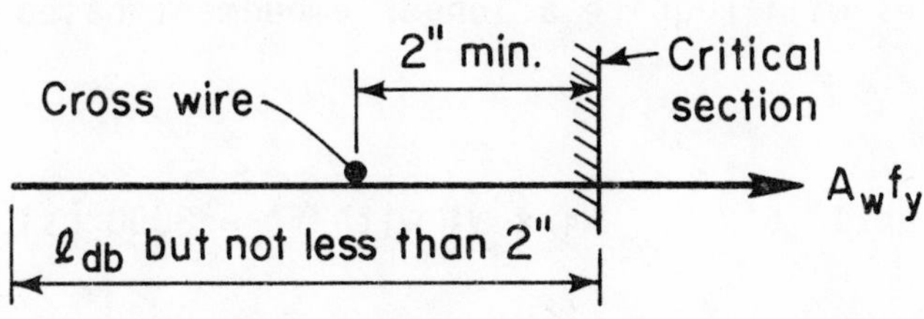

Fig. 3-4 - Development of Welded Deformed Wire Fabric

12.8 Development of Welded Smooth Wire Fabric in Tension

For welded smooth wire fabric, the development length is measured from the point of critical section to the outermost cross wire. Full development of smooth fabric ($A_w f_y$) is achieved by embedment of at least two cross wires beyond the critical section, with the closer cross wire located not less than 2 in. from the critical section. Section 12.8 further requires that the length of embedment from critical section to outermost cross wire must not be less than $\ell_{db} = 0.27\ A_w f_y / s_w \sqrt{f'_c}$, nor 6 in. For lightweight aggregate concrete, the basic development length ℓ_{db} must be modified by the factor in Section 12.2.3. If more reinforcement is provided than that required by analysis, the basic development length ℓ_{db} may be reduced by the ratio of (A_s required)/(A_s provided). The 6 in. minimum development length does not apply to computation of lap splice lengths (Section 12.19). Fig. 3-5 shows the development length requirements for smooth fabric embedded in normal weight concrete.

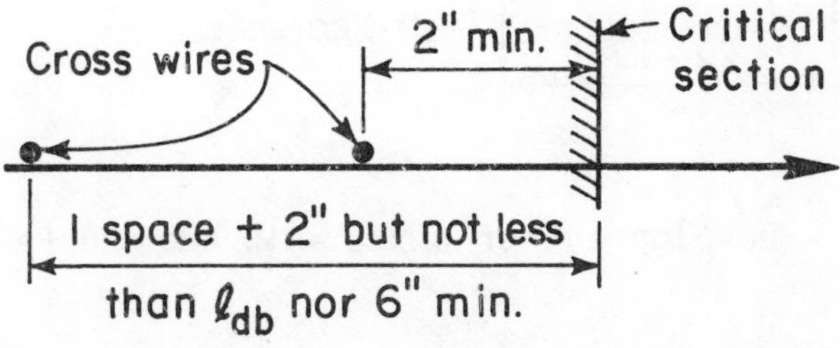

Fig. 3-5 - Development of Welded Smooth Wire Fabric

For fabrics made with smaller wires, the embedment of two cross wires, 2 in. or more beyond the point of critical section, is usually adequate to develop the full yield strength of the anchored wire. Fabrics made with larger (closely spaced) wires will require a longer embedment based on the ℓ_{db} development length.

For example, check fabric 6 x 6 - W4 x W4 with f'_c = 3000 psi.

$$\text{min. } \ell_{db} = 0.27 \times \frac{A_w}{s_w} \times \frac{f_y}{\sqrt{f'_c}} = 0.27 \times \frac{0.04}{6} \times \frac{60,000}{\sqrt{3,000}} = 1.97 \text{ in. or 6 in. min.}$$

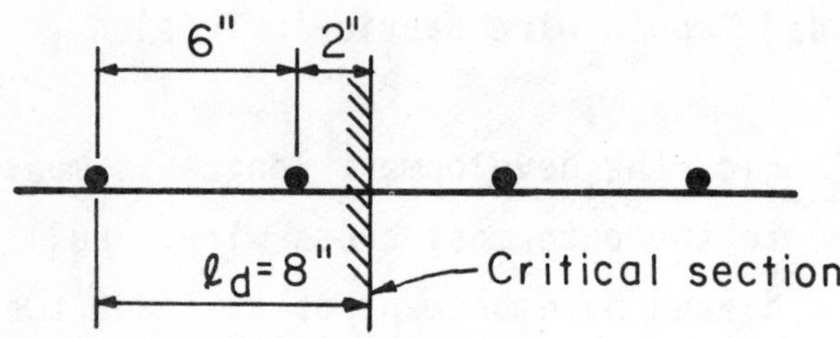

Two cross wire embedment plus 2 in. is satisfactory.

Fig. 3-6 - Development of 6 x 6 - W4 x W4 Fabric

Check fabric 6 x 6 - W20 x W20:

$$\text{Min. } \ell_{db} = 0.27 \times \frac{0.20}{6} \times \frac{60,000}{\sqrt{3,000}} = 9.9 \text{ in.} > 6 \text{ in. min.}$$

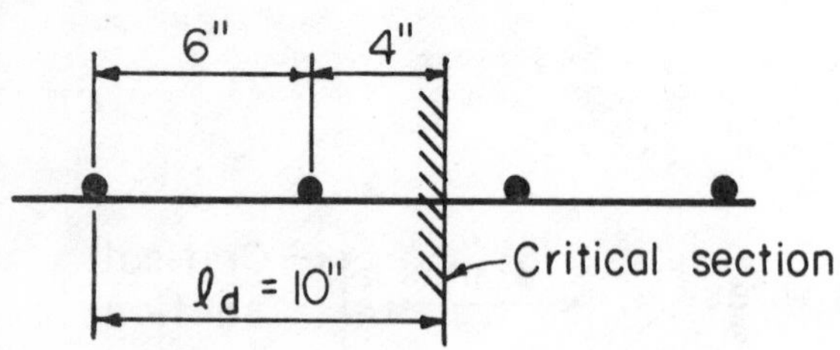

Fig. 3-7 - Development of 6 x 6 - W20 x W20 Fabric

As shown in Fig. 3-7, an additional 2 in. beyond the two cross wires + 2 in. embedment is required to fully develop this W20 fabric. If the longitudinal spacing is reduced to 4 in. (4 x 6 - W20 x W20), a minimum ℓ_d of 15 in. is required for full development -- e.g. 3 cross wires + 3 in. embedment.

References 3.1 and 3.2 provide design aid data in the use of welded wire fabric, including development length tables for both deformed and smooth welded wire fabric.

12.9 Development of Prestressing Strand

The development provisions for prestressing strand are modified slightly for the '83 Code; double development length for "debonded" strands (Section 12.9.3) is required only when the member is designed for calculated tension in the concrete under service load conditions.

12.10 Development of Flexural Reinforcement - General

Section 12.10 gives the basic requirements of providing development length for reinforcement from the points of maximum or critical stress. Requirements are essentially the same as in the 1977 Code.

Fig. 3-8 illustrates typical critical sections for development of flexural reinforcement. The point of maximum positive factored moment ($+M_u$) is a critical section for the positive moment reinforcement, from which adequate anchorage ℓ_d must be provided. The critical section for positive moment bars "a" is at the theoretical cut-off point; bars "a" must have adequate anchorage ℓ_d from this location. Note also that the terminated bars must be extended beyond the theoretical cut-off point in accordance with the provisions of Section 12.10.3. These requirements are to guard against possible shifting of the moment diagram due to load variations, settlement of supports, and other unforeseen changes in the moment conditions.

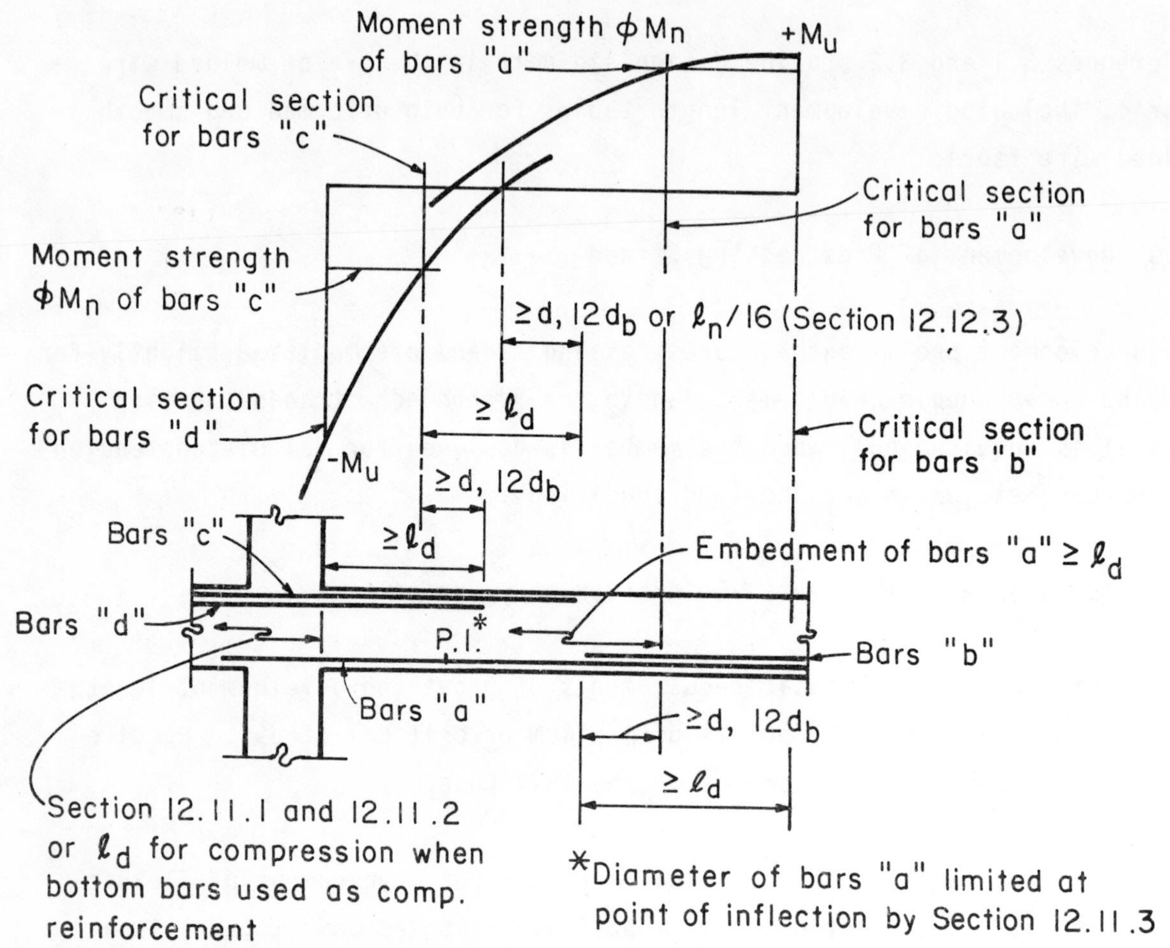

Fig. 3-8 - Development of Flexural Reinforcement in a Typical Continuous Beam

The anchorage and development requirements for the negative moment reinforcement are also illustrated in Fig. 3-8.

Sections 12.10.1 and 12.10.5 acknowledge the well-established preference for anchoring tension reinforcement in a compression zone. Research has confirmed the need for restrictions on terminating bars in a tension zone. When flexural bars are cut off in a tension zone, flexural cracks tend to open early. If the shear stress in the area of bar cut-off and tensile stress in the remaining bars at the cut-off location are near the permissible limits, diagonal tension cracking tends to develop from the flexural cracks. One of the three alternates of Section 12.10.5 must be satisfied to reduce the possible occurrence of diagonal tension cracking near bar cut-offs in a tension zone. Section 12.10.5.2 requires excess stirrup area over that required for shear and torsion. Requirements of Section 12.10.5 are not intended to apply to tension splices.

Section 12.10.6 is for end anchorage of tension bars in special flexural members such as brackets, members of variable depth, and others where f_s does not decrease linearly in proportion to a decreasing moment. In Fig. 3-9, the ℓ_d from the support is probably less critical than the required development length for a slightly smaller f_s existing near the load point. In such a case, safety depends largely on the outer end anchorage provided. A welded cross bar of equal diameter should provide an effective end anchorage. A standard end hook in the vertical plane may not be effective because an essentially plain concrete corner might exist near the load and could cause localized failure. Where brackets are wide and loads are not applied too close to the corners, U-shaped bars in a horizontal plane provide effective end hooks.

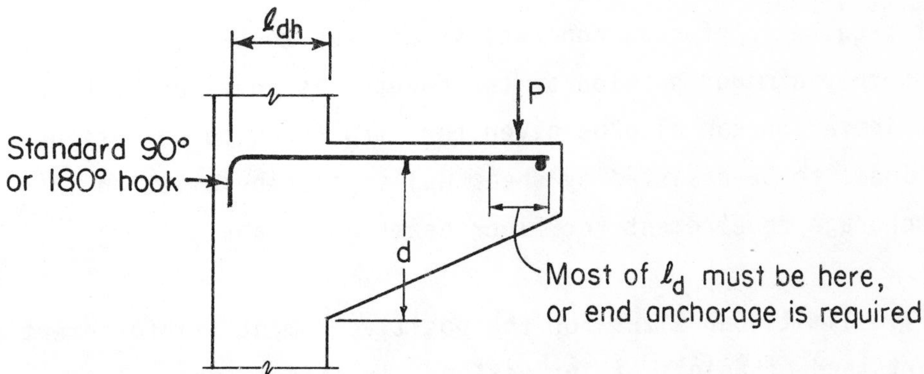

Fig. 3-9 - Special Member Largely Dependent on End Anchorage

12.11 Development of Positive Moment Reinforcement

To further guard against possible shifting of moments due to various causes, Section 12.11.1 requires specific amounts of positive moment reinforcement to be extended along the same face of the member, and for beams, to be embedded into the support at least 6 in. The specified amounts are one-third for simple members and one-fourth for continuous members. In Fig. 3-8 for example, the area of bars "a" would have to be at least one-fourth of the area of reinforcement required at the point of maximum $+M_u$.

Section 12.11.2 is intended to assure ductility in the structure under severe overload, as might be experienced in a severe earthquake. In a lateral load resisting system, full anchorage of the reinforcement extended into the support provides for possible stress reversal under such overload. Anchorage must be provided to develop the full yield strength in tension at the face of the support. The provision will require such members to have bottom bars lapped at interior supports or hooked at exterior supports. The full anchorage requirement does not apply to any excess reinforcement provided at the support.

It is the opinion of many structural engineers that this full anchorage requirement is necessary for designs in higher seismic zones only. Building codes do not require reinforced concrete structures to be designed for blast overloads and only a minor portion of the country is in zones of high seismicity. Consideration can also be given for structures to be designed with all lateral loads to be resisted by shear walls, thereby omitting this reinforcement anchorage requirement for floor beams and slabs.

Section 12.11.3 limits bar sizes for the positive moment reinforcement at simple supports and at points of inflection. In effect, this places a design restraint on flexural bond stress in areas of small moment and large shear. Such a condition could exist in a heavily loaded beam of short span, thus requiring large size bars to be developed within a short distance. Bars should be limited to a diameter such that the development length ℓ_d computed for f_y according to Section 12.2 does not exceed $(M_n/V_u) + \ell_a$. A new Code provision waives the limit on bar size at simple supports if the bars have standard end hooks or mechanical anchorages terminating beyond the centerline of the support. Mechanical anchorages must be equivalent to standard hooks.

The length M_n/V_u corresponds to the development length of the maximum size bar permitted by the previously used flexural bond equation $\Sigma_o = V/ujd$. The length M_n/V_u may be increased 30% when the ends of the bars are confined by a compressive reaction, such as provided by a column below, but not when a beam frames into a girder.

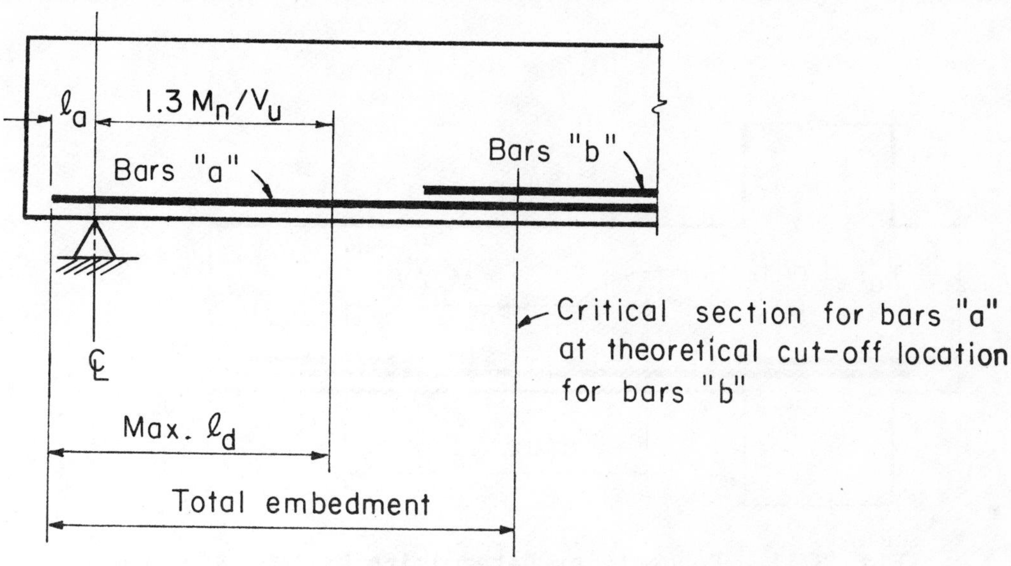

Fig. 3-10 - Development Length Requirements at Simple Support (straight bars)

For the simply-supported beam shown in Fig. 3-10, the maximum permissible ℓ_d for bars "a" is $1.3\ M_n/V_u + \ell_a$. This has the effect of limiting the size of bar to satisfy flexural bond. Even though the total embedment length from the critical section for bars "a" is greater than $1.3\ M_n/V_u + \ell_a$, the size of bars "a" must be limited so that $\ell_d \leq 1.3\ M_n/V_u + \ell_a$. Note that M_n is the nominal moment strength of the cross section (without the φ factor). As noted previously larger bar sizes can be accommodated by providing a standard hook or mechanical anchorage at the end of the bar within the support.

At a point of inflection (see Fig. 3-11), the positive moment reinforcement must have a development length ℓ_d, as computed by Section 12.2, not to exceed the value of $(M_n/V_u) + \ell_a$, with ℓ_a not greater than d or $12d_b$. For example, a #11 bar requires a basic tension development length of $0.04 A_b f_y/\sqrt{f'_c}$ = 59 in. for 60,000 psi steel and 4,000 psi normal weight concrete. For a short span beam, the #11 bar may be too large to satisfy the flexural bond requirement.

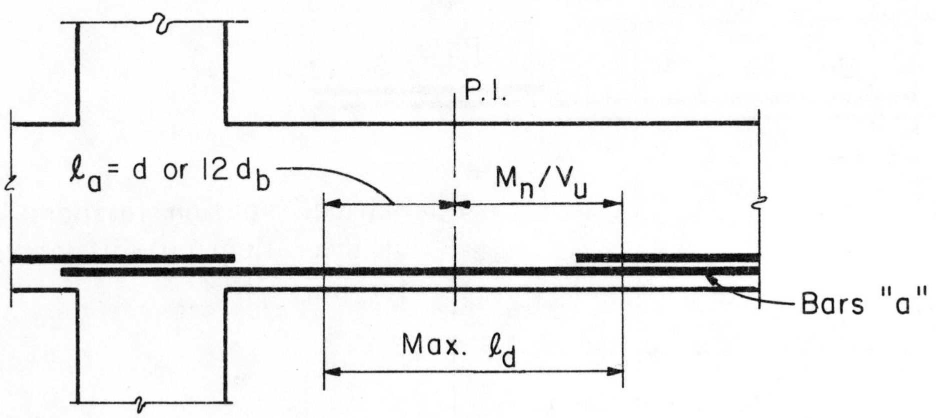

Fig. 3-11 - Concept for Determining Maximum Size of Bars "a" at Point of Inflection

12.12 Development of Negative Moment Reinforcement

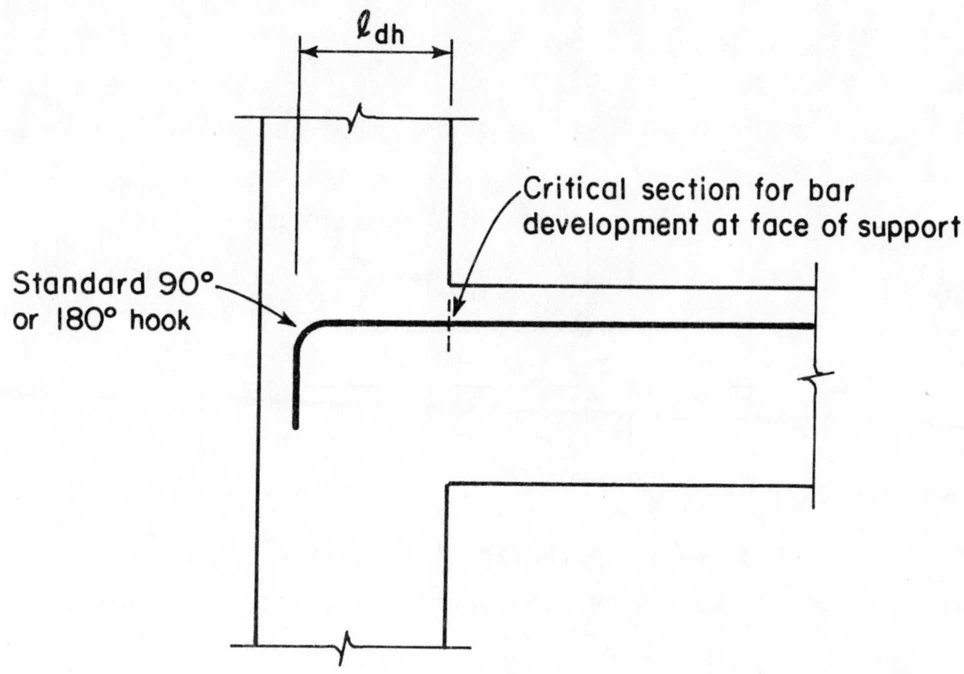

Fig. 3-12 - Anchorage into Exterior Support with Standard Hook

The requirements in Section 12.12.3 guard against possible shifting of the moment diagram at points of inflection. At least one-third of the negative moment reinforcement provided at a support must be extended a specified embedment length beyond a point of inflection. The embedment length must be the effective depth of the member d, $12d_b$, of 1/16 the clear span, whichever is greater -- as shown in Figs. 3-8 and 3-13. The area of bars "c" in Fig. 3-8 must be at least one-third the area of reinforcement provided for $-M_u$ at the face of the support.

Standard end hooks are an effective means of developing top bars in tension at exterior supports as shown in Fig. 3-12. The Code requirements for development of standard hooks were discussed in Section 12.5.

Anchorage of top reinforcement in tension beyond interior supports of continuous members usually becomes part of the adjacent span's top reinforcement -- as shown in Fig. 3-13.

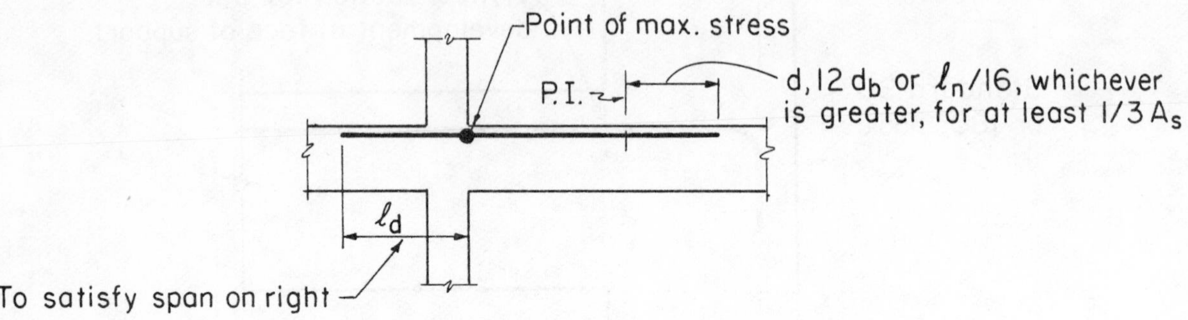

Fig. 3-13 - Anchorage into Adjacent Beam.

(Usually such anchorage becomes part of adjacent beam reinforcement)

12.13 Development of Web Reinforcement

It is important for stirrups to be anchored as close to the compression face of the member as possible because flexural tension cracks penetrate deeply as the strength of the member is approached. Fig. 3-14 illustrates the anchorage requirements for U-stirrups fabricated from deformed bars and deformed wire. Provisions covering the use of welded smooth wire fabric as simple U-stirrups are shown in Fig. 3-15.

For the '83 Code, a new stirrup anchorage detail is added (Section 12.13.2.5) for single leg stirrups formed with welded smooth or deformed wire fabric. Use of welded wire fabric for shear reinforcement has become commonplace in the precast, prestressed concrete industry. The anchorage details are shown in Fig. 3-16. Anchorage of the single leg is provided primarily by the longitudinal cross wires.

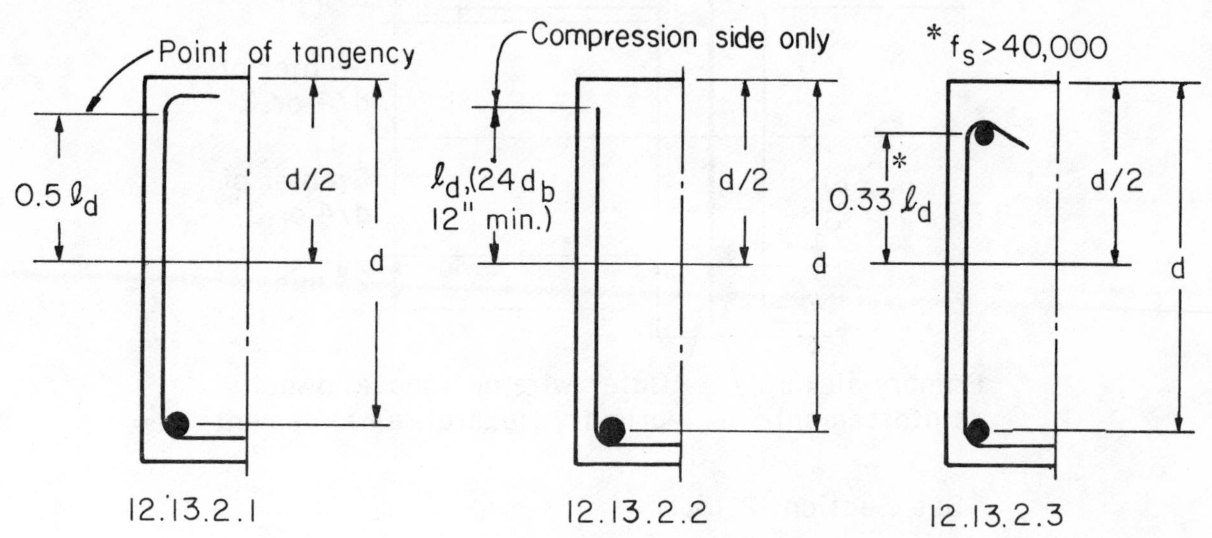

Fig. 3-14 - Anchorage Details for U-Stirrups (deformed bars and deformed wire)

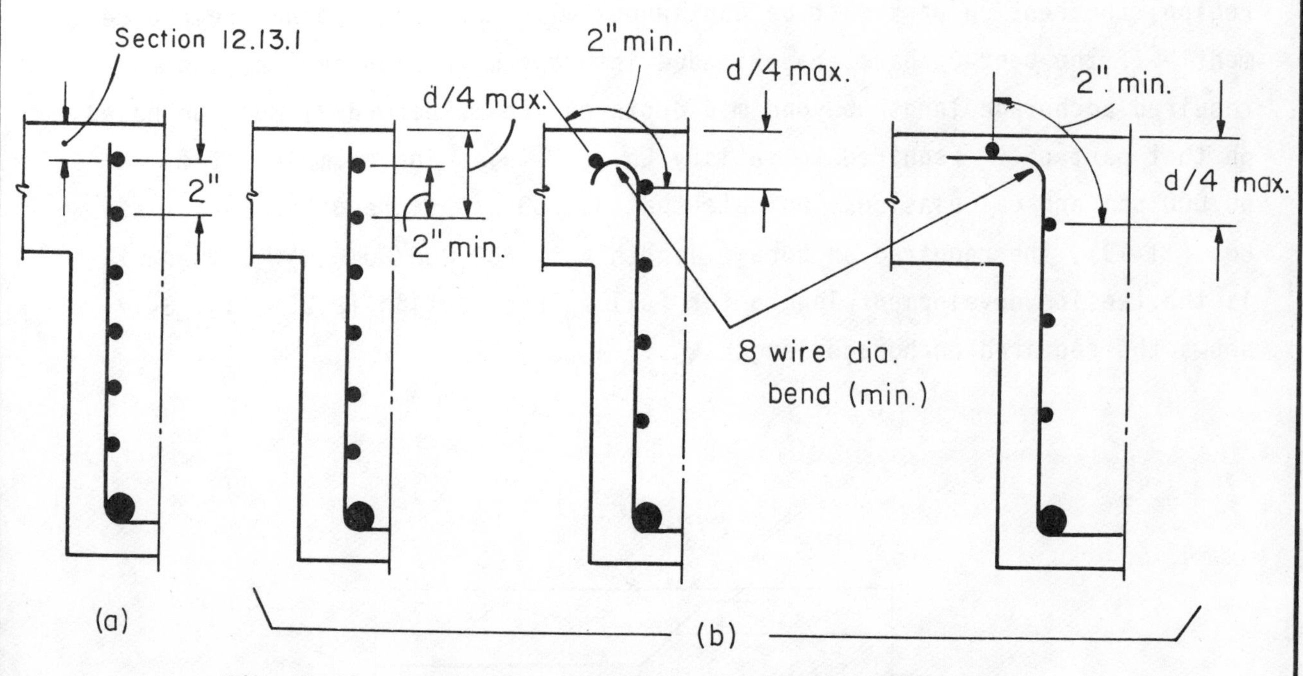

Fig. 3-15 - Anchorage Details for Welded Smooth Wire Fabric U-Stirrups [Section 12.13.2.4]

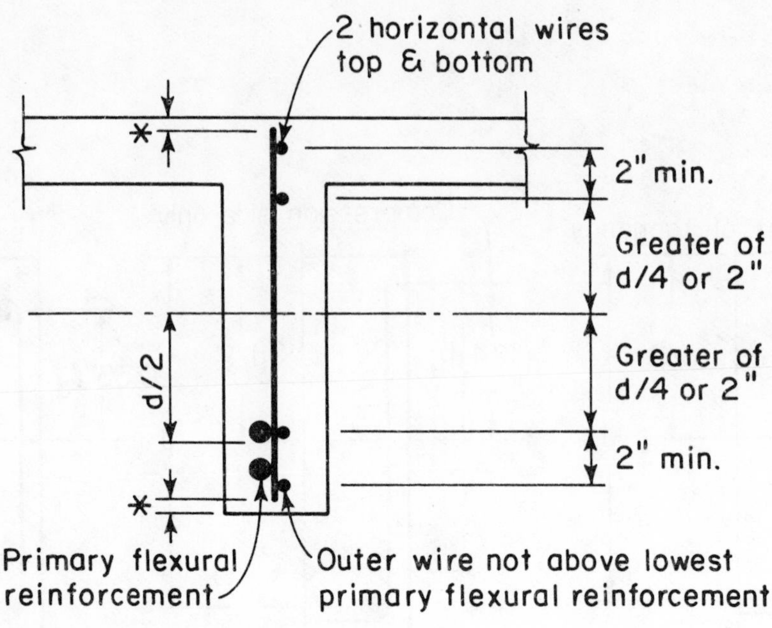

*See Section 12.13.1

Fig. 3-16 - Anchorage Details for Welded Wire Fabric Single Leg Stirrups

Section 12.13.4 gives anchorage requirements for longitudinal (flexural) bars bent up to resist shear. If the bent-up bars are extended into a tension region, the bent-up bars must be continuous with the longitudinal reinforcement. If the bent-up bars are extended into a compression region, the required anchorage length beyond mid-depth of the member (d/2) must be based on that part of f_y required to satisfy Eq. (11-19). For example, if f_y = 60,000 psi and calculations indicate that 30,000 psi is required to satisfy Eq. (11-19), the required anchorage length ℓ_d' = (30,000/60,000)ℓ_d, where ℓ_d is the tension development length for full f_y per Section 12.2. Fig. 3-17 shows the required anchorage length ℓ_d'.

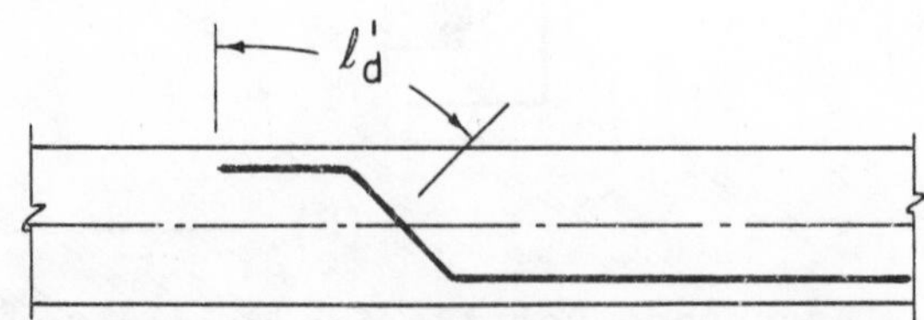

Fig. 3-17 - Anchorage for Bent-Up Bars

Section 12.13.5 gives requirements for lap splicing double U-stirrups or ties (without hooks) to form a closed stirrup. Legs are considered properly spliced when the laps are 1.7 ℓ_d as shown in Fig. 3-18.

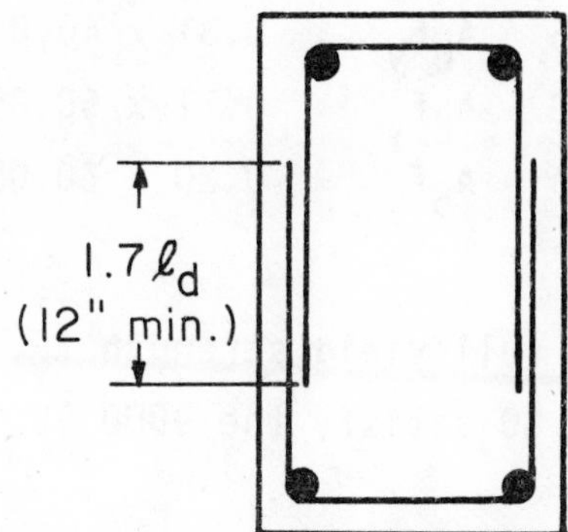

Fig. 3-18 - Overlapping U-Stirrups to Form Closed Unit

Alternatively, for members at least 18 in. deep, the double U-stirrup may be used if each U portion extends the full available depth of the member and the force in each leg does not exceed 9,000 pounds; $A_b f_y \leq 9{,}000$ lbs. (See Fig. 3-19).

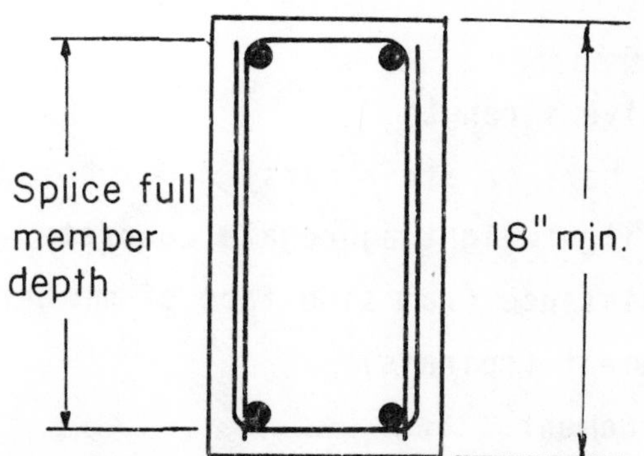

Fig. 3-19 - Lap Splice Alternative

For Grade 40	#3 - $A_b f_y$	=	0.11 x 40,000	=	4,400 lb
	#4 - $A_b f_y$	=	0.20 x 40,000	=	8,000 lb
	#5 - $A_b f_y$	=	0.31 x 40,000	=	12,400 lb
For Grade 60	#3 - $A_b f_y$	=	0.11 x 60,000	=	6,600 lb
	#4 - $A_b f_y$	=	0.20 x 60,000	=	12,000 lb

If stirrups are designed for the full yield strength f_y, #3 and #4 stirrups of Grade 40 and only #3 of Grade 60 satisfy the 9000 lb limitation.

Use of the stirrup detail of Section 12.13.5 is discussed in Part 2, page 2-13.

12.14 Splices of Reinforcement - General

The splice provisions require the engineer to show clear and complete splice details in the Contract Documents. The structural drawings, notes and specifications should clearly show or describe all splice locations, types permitted or required, and for lap splices, length of lap required. The engineer cannot simply state that all splices shall be in accordance with the ACI-318 Code. This is because many factors affect splices of reinforcement, such as the following for tension lap splices of deformed bars:

- bar size
- bar yield strength
- concrete compressive strength
- bar position (top bars or other bars)
- normal weight or lightweight aggregate concrete
- bar spacing and distance from side face of member
- lateral reinforcement (spirals)
- stress in reinforcement
- number of bars spliced at one location
- a 12 in. minimum lap length

It is virtually impossible for a reinforcing bar detailer to know what splices are required at a given location in a structure, unless the engineer explicitly illustrates or defines the splice requirements. Section 12.14.1 states: "Splices of reinforcement shall be made only as required or permitted on the design drawings, or in specifications, or as authorized by the engineer."

Two industry publications are suggested as design reference material for proper splicing of reinforcement. Reference 3.1 provides design aid data in the use of welded wire fabric, including development length and splice length tables for both deformed and smooth wire fabric. Reference 3.2 provides accepted practices in splicing reinforcement. Use of lap splices, welded splices and splice devices are described, including simplified design data for lap splice lengths.

12.14.2 Lap Splices

Lap splices are not permitted for bars larger than #11, either in tension or compression, except:

- #14 and #18 bars in compression only may be lap spliced to #11 and smaller bars (Section 12.16.2), and
- #14 and #18 bars in compression only may be lap spliced to smaller size footing dowels (Section 15.8.2.4).

Section 12.14.2.2 gives the provisions for lap splicing of bundled bars (tension or compression). The lap lengths required for individual bars within a bundle must be increased by 20 percent and 33 percent for 3- and 4-bar bundles, respectively. As noted in Commentary Section 12.14.2.2, the comparable increase for development length of bundled bars (Section 12.4) should not be included (twice) when computing ℓ_d for lap lengths of bundled bars. Overlapping of individual bar splices within a bundle is not permitted.

Bars in lap splices may be spaced or in contact. To prevent a possible unreinforced section in a spaced lap splice, Section 12.14.2.3 limits the maximum distance between bars in a splice to one-fifth the lap length, or

6 in. Contact lap splices are preferred for the practical reason that when the bars are wired together, they are more easily secured against displacement during concrete placement.

12.14.3 Welded Splices and Mechanical Connections

Section 12.14.3 permits the use of welded splices or other mechanical connections. In a full welded splice, the bars must be butted and the splice must develop in tension at least 125 percent of the specified yield strength of the bar. Likewise, a full mechanical connection must develop, in tension or compression, at least 125 percent of the specified yield strength of the bar. The Code permits the use of welded splices or mechanical connections having less than 125 percent of the specified yield strength of the bar in regions of low computed stresses as specified in Section 12.15.4.

Section 12.14.3.2 requires all welding of reinforcement to conform to "Structural Welding Code-Reinforcing Steel" (AWS D1.4). Section 3.5.2 requires that the reinforcement to be welded must be indicated on the drawings, and the welding procedure to be used must be specified. To carry out these Code requirements properly, the engineer should be familiar with provisions in AWS D1.4 and the ASTM specifications for reinforcing bars.

Since the standard rebar specifications ASTM A 615, A 616 and A 617 specifically state that "weldability of the steel is not part of this specification," that are no limits on the chemical elements that affect weldability of the steels. A key item in AWS D1.4 is carbon equivalent (C.E.). The minimum preheat and interpass temperatures specified in AWS D1.4 are based upon C.E. and bar size. Thus, as indicated in Section 3.5.2 and in Commentary Section 3.5.2, when welding is required, the ASTM A 615, A 616 and A 617 rebar specifications must be supplemented to require a report of the chemical composition to assure that the welding procedure specified is compatible with the chemistry of the bars.

ASTM A 706 reinforcing bars are intended for welding. The A 706 specification contains restrictions on chemical composition, including carbon, and

C.E. is limited to 0.55 percent. The chemical composition and C.E. must be reported. By limiting C.E. to 0.55 percent, little or no preheat is required by AWS D1.4. Thus, the engineer does not need to supplement the A 706 specification when the bars are to be welded. However, before specifying ASTM A 706 reinforcing bars, local availability should be investigated.

Reference 3.2 contains an excellent discussion of welded splices. Included in the discussion are requirements for other important items such as field inspection, supervision, and quality control.

Note that careful review of AWS D1.4 reveals that the document essentially covers the welding of reinforcing bars only. For welding of wire to wire, and of wire or welded wire fabric to reinforcing bars or structural steels, such welding should conform to applicable provisions of AWS D1.4 and to supplementary requirements specified by the engineer. Also, the engineer should be aware that there is a potential loss of yield strength of low carbon cold-drawn wire if wire is welded by a process other than controlled resistance welding used in the manufacture of welded wire fabric.

In the discussion of Section 7.5 in Part 2, it was noted that welding of crossing bars (tack welding) is not permitted for assembly of reinforcement unless authorized by the engineer. An example of tack welding would be a column cage where the ties are secured to the longitudinal bars by small arc welds. Such welding can cause a metallurgical notch effect in the longitudinal bars, which may affect the strength of the bars. Tack welding seems to be particularly detrimental to ductility (impact resistance) and fatigue resistance. Reference 3.2 recommends, "Never permit field welding of crossing bars ('tack' welding, 'spot' welding, etc.). Tie wire will do the job without harm to the bars."

12.15 Splices of Deformed Bars and Deformed Wire in Tension

Tension lap splices of deformed bars and deformed wire are designated as Classes A, B, and C with the length of lap being a multiple of the tension development length ℓ_d. The development length ℓ_d (Section 12.2) used in the calculation of lap length must be that for the full f_y because the splice classifications (Table 12.15) already reflect any excess reinforcement at the splice location (factor of Section 12.2.4.2 for excess A_s must not be used). The increasing ℓ_d factors of Section 12.2.3 for top reinforcement, for f_y greater than 60,000 psi and for lightweight concrete must be applied where appropriate. The reduction factors for ℓ_d given in Section 12.2.4 for wide spacing and use of spirals may be applied where the specified conditions are satisfied. Stirrups or ties, unless closely spaced, are only partially as effective as spirals and are not covered in the Code. The minimum length of lap is 12 in.

Splices in tension tie members are required to be made with a full welded splice or full mechanical connection; see Section 12.15.5. With the 1983 Code, the 1.7 ℓ_d stagger between adjacent bar splices (welded splices or mechanical connections) for the tension tie members is reduced to 30 in. The lesser stagger is considered adequate for the full welded or full mechanical splice. The original 1.7 ℓ_d stagger was primarily intended for lap splices in the tension tie, which was deleted in the 1977 Code edition.

In accordance with Table 12.15, the design engineer must specify the class of tension lap slice to be used. The class of splice depends on the magnitude of tensile stress in the reinforcement and the percentage of total steel area to be lap spliced within any given splice length. If the area of tensile reinforcement provided at the splice location is more than half that required for strength (low tensile stress in the reinforcement) and less than 3/4 of the total steel area is lap spliced within the required splice length, a Class A splice may be used. If more than 3/4 of the total area is to be spliced within the lap length, a Class B splice is required.

If the area of reinforcement provided at the splice location is less than twice that required for strength (high tensile stress in the reinforcement), either a Class B or Class C splice will be required. If less than half the total steel area is lap spliced within the required lap length, a Class B splice may be used. If more than half the total area is lap spliced, then a Class C splice must be used, with a lap length equal to 1.7 ℓ_d.

Welded splices or mechanical connections conforming to Section 12.14.3.3 may be used in lieu of lap splices. Section 12.15.4 allows a reduction in the requirements of Section 12.14.3.3 if certain conditions are met.

12.16 Splices of Deformed Bars in Compression

Since bond behavior of reinforcing bars in compression is not complicated by the potential problem of transverse tension cracking in the concrete, compression lap splices do not require such strict provisions as those specified for tension lap splices. Tests have shown that the strength of compression lap splices depends primarily on end bearing of the bars on the concrete, without a proportional increase in strength even when the lap length is doubled. Thus, the Code requires significantly longer lap lengths for bars with a yield strength greater than 60,000 psi.

For compression lap splices, Section 12.16.1 requires the minimum lap length to be the compression development length (Section 12.3), but not less than $0.0005\ f_y d_b$, nor 12 in. Also, a minimum lap length of $(0.0009\ f_y - 24)\ d_b$ is specified for reinforcing bars with a yield strength greater than 60,000 psi. Lap splice lengths must be increased by one-third for concrete with a compressive strength less than 3,000 psi.

Lap splice lengths may be reduced when the splice is enclosed throughout its length by ties, meeting the requirements of Section 12.16.3, or by a spiral (Section 12.16.4). Spirals must meet the requirements of Sections 7.10.4 and 10.9.3. The 12 in. minimum lap length also applies to these permitted reductions. When ties are used to reduce the lap splice length, the ties must have a minimum effective area of 0.0015 hs. The tie legs in both directions must provide the minimum effective area to permit the 0.83 reduction factor. See Fig. 3-20.

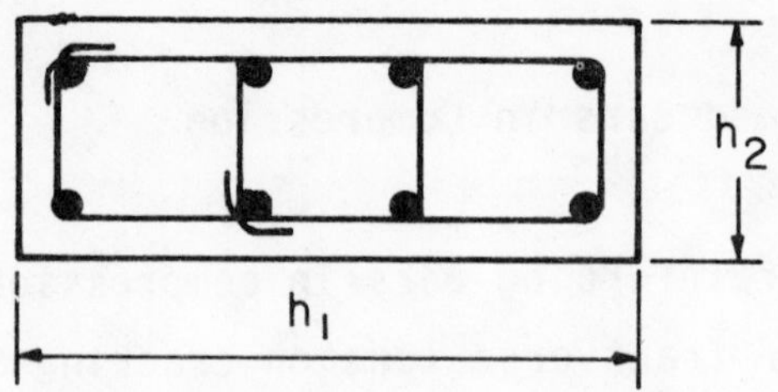

(perpendicular to h_1 direction) $0.0015\ h_1 s \geq 4$ tie bar areas
(perpendicular to h_2 direction) $0.0015\ h_2 s \geq 2$ tie bar areas

Fig. 3-20 - Application of Section 12.16.3

Since the Code provisions for compression lap splices involve only a few variables and other considerations, it is possible to establish a few basic relationships for the required compression lap lengths. For splices not enclosed within either ties or spirals, the required lap length for Grade 60 bars in concrete with a compressive strength $\geq$ 3,000 psi is simply 30-bar diameters (30 d_b), but not less than 12 in. If the splices are enclosed within ties or spirals, conforming to Section 12.16.3 or 12.16.4, the lap lengths can be reduced to $25d_b$ and 22.5 d_b, respectively, but not less than 12 in.

As noted in the discussion of Section 12.14.2, #14 and #18 bars may be lap spliced, in compression only, to #11 and smaller bars or to smaller size footing dowels. Section 12.16.2 requires that when bars of a different size are lap spliced in compression, the length of lap must be the compression development length of the larger bar, or the compression lap splice length of the smaller bar, whichever is the longer length.

Section 12.16.5 specifies the requirements for end-bearing compression splices. End-bearing splices are only permitted in members containing closed ties, closed stirrups or spirals. With the '83 Code, a new Commentary Section 12.16.5.1 is added to caution the engineer in the use of end-bearing splices for bars inclined from the vertical. End-bearing splices for compression bars have been used almost exclusively in columns and the intent is to limit use to essentially vertical bars because of the field difficulty of getting adequate end bearing on horizontal bars or bars significantly inclined from the vertical. Welded splices or mechanical connections are also permitted for compression splices and must meet the requirements of Section 12.14.3.3 or 12.14.3.4.

12.17 Special Splice Requirements for Columns

This section gives the requirements for the tensile strength of spliced longitudinal bars in columns. The provisions are based on the magnitude of calculated factored load stress. A minimum tensile strength of spliced bars is required even when there is no calculated tension.

Section 12.17.1 covers the condition when the calculated factored load stress in the longitudinal bars varies from f_y in compression to $1/2f_y$ in tension. For this condition, each face of the column must have a tensile strength of at least twice the calculated tension in that face. Tensile strength of the splices alone, or the tensile strength of splices in combination with continuing unspliced bars, must provide the required total tensile strength. Lap splices, butt-welded splices, mechanical connections or end-bearing splices may be used. The tensile strength provided cannot be less than the minimum strength required by Section 12.17.3.

The provisions in Section 12.17.2 cover higher tensile stress conditions. When the calculated factored load stress is greater than $1/2f_y$ in tension, lap splices must be designed to develop the yield strength of the bars, or full welded or full mechanical connections must be used.

Section 12.17.3 requires a minimum tensile strength of spliced longitudinal bars in each face of a column. Each face of a column must have a tensile strength of at least $A_sf_y/4$, where A_s is the total area of bars in each face.

12.18 Splices of Welded Deformed Wire Fabric in Tension

For tension lap splices of deformed wire fabric, the Code requires a minimum lap length of $1.7\ell_d$, but not less than 8 in. Lap length is measured between the ends of each fabric sheet. The development length ℓ_d is the value calculated by the provisions in Section 12.7 -- it is not the minimum required development length. The Code also requires that the overlap measured between the outermost cross wires be at least 2 in. Fig. 3-21 shows the lap length requirements.

If there are no cross wires within the lap splice length, the provisions in Section 12.15 for deformed wire must be used to determine the length of lap.

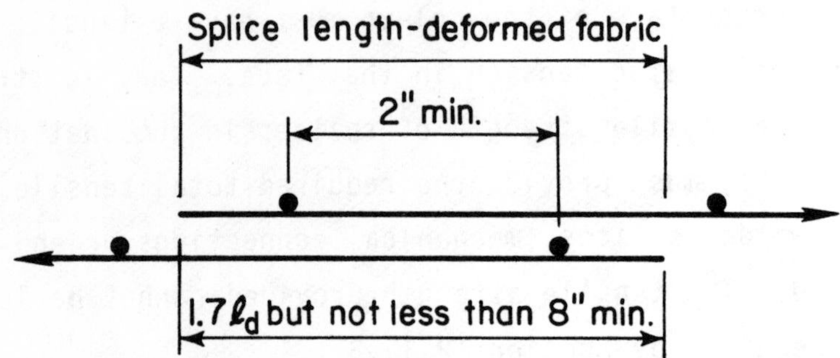

Fig. 3-21 - Lap Splice Length for Deformed Wire Fabric

12.19 Splices of Welded Smooth Wire Fabric in Tension

The minimum length of lap for tension lap splices of smooth wire fabric is dependent upon the ratio of the area of reinforcement provided to that required by analysis. Lap length is measured between the outermost cross wires of each fabric sheet. The required lap lengths are shown in Fig. 3-22.

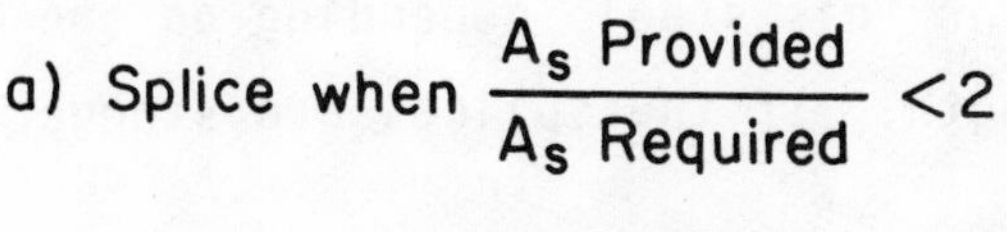

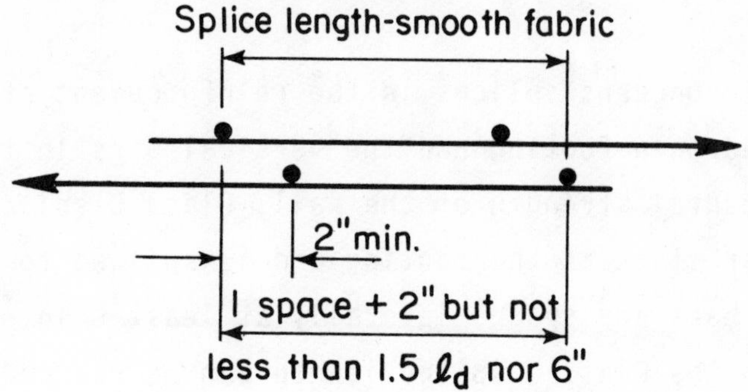

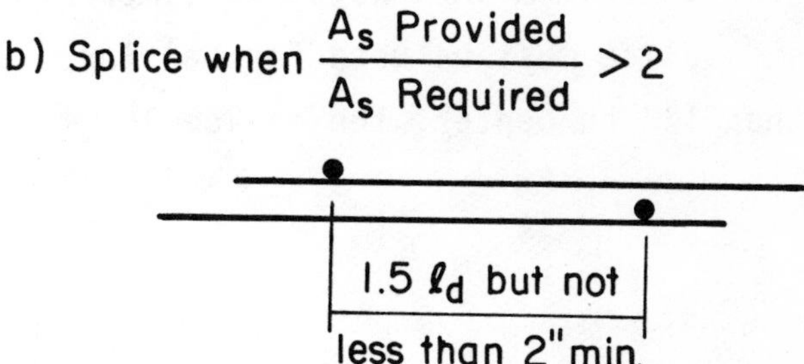

Fig. 3-22 - Lap Splice Length for Smooth Wire Fabric (Splice length is the largest of the values shown in the figures.)

SUMMARY

A few additional comments will be helpful to clarify the use of the splice provisions. The first concerns temperature and shrinkage reinforcement at the exposed surfaces of wall or slabs. One must assume all temperature and shrinkage reinforcement to be stressed to the full specified yield strength f_y. The purpose of this reinforcement is to prevent excess cracking. At some point in the member, it is likely that cracking will occur, thus fully stressing the temperature and shrinkage reinforcement. Therefore, all splices in temperature and shrinkage reinforcement must be assumed to be those required for full strength. An engineer must provide either a Class B or Class C tension splice for this steel, depending on the amount spliced within the required lap length. If the splices are staggered, Class B splices may be used.

A second comment concerns splices in the reinforcement of a wall. If the wall is supported on a footing and the vertical bars in the wall are those required for flexural strength of the wall, Class C splices are required if each vertical bar stops at the footing and is spliced to a footing dowel. However, if the bars are spaced laterally at least 6 in. on center (a common situation), then the Class C splice length can be reduced by 20% in accordance with Section 12.2.4.1. The requirement that the bars must also be "at least 3 inches from the side face of the member" refers to the end of the wall and not to the cover on the reinforcement. Horizontal reinforcement in a wall normally requires minimum shrinkage and temperature reinforcement, and a Class B or Class C splice must be used for bar laps. If the horizontal bars are spaced more than 12" on center, the splice length may be reduced by 20%.

Selected References

3.1 "Welded Wire Fabric Manual of Standard Practice," 3rd Edition, Wire Reinforcement Institute, McLean, Virginia, 1979, 32 pp.

3.2 "Reinforcement Anchorages and Splices," 2nd Edition, Concrete Reinforcing Steel Institute, Schaumburg, Illinois, 1984, 40 pp.

EXAMPLE 3.1 - Development of Reinforcement

Determine lengths of top and bottom bars for the exterior span of the continuous beam shown below. Concrete is normal weight and bars are Grade 60. Total uniformly distributed factored load on beam is w_u = 6.0 klf (including weight of beam).

f'_c = 4,000 psi
f_y = 60,000 psi
b = 16 in.
h = 22 in.
Concrete cover = 1-1/2 in.

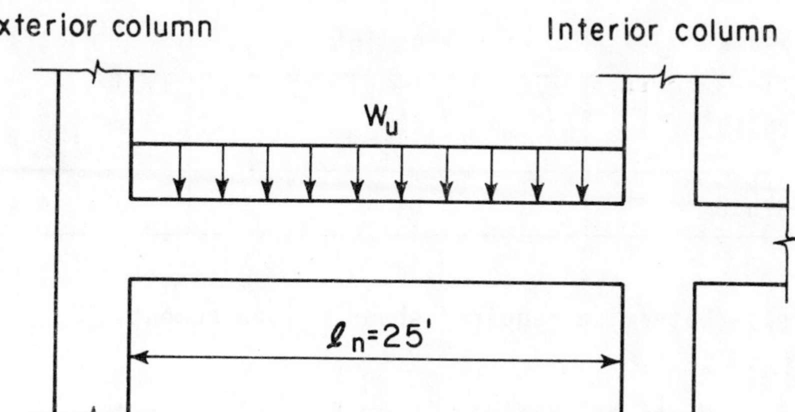

Beam Elevation

Calculations and Discussion	Code Reference
1. Preliminary design for moment and shear reinforcement	
(a) Use approximate analysis for moment and shear values	8.3.3

Location	Factored moments & shears
interior face of exterior support	$-M_u = w_u\ell_n^2/16 = 6 \times 25^2/16 = 234.4'^k$
End span positive	$+M_u = w_u\ell_n^2/14 = 6 \times 25^2/14 = 267.9'^k$
Exterior face of first interior support	$-M_u = w_u\ell_n^2/10 = 6 \times 25^2/10 = 375.0'^k$
Exterior face of first interior support	$V_u = 1.15w_u\ell_n/2 = 1.15 \times 6 \times 25/2 = 86.3^k$

EXAMPLE 3.1 - Development of Reinforcement

(b) Determine required flexural reinforcement using procedures of Part 9. With 1.5 in. cover, #4 bar stirrups, and #9 or #10 flexural bars; $d \simeq 19.4$ in.

M_u	A_s required	Bars	A_s provided
$-234.4^{'k}$	2.93 in^2	4 # 8	3.16 in^2
+267.9	3.40	2 # 8 2 # 9	3.58
-375.0	5.01	4 # 10	5.08

(c) Determine required shear reinforcement

V_u at "d" distance from face of support:

$$V_u = 86.3 - 6(19.4/12) = 76.6^k$$

$$\varphi V_c = \varphi 2\sqrt{f'_c}\, b_w d = 0.85 \times 2\sqrt{4000} \times 16 \times 19.4/1000 = 33.4^k$$

with $s_{max} = d/2 = 19.4/2 = 9.7$ in., try #4 U-stirrups @ 9 in. spacing.

$$\varphi V_s = \varphi A_v f_y d/s = 0.85 \times 0.40 \times 60 \times 19.4/9 = 44.0^k$$

$$\varphi V_n = \varphi V_c + \varphi V_s = 33.4 + 44.0 = 77.4^k > 76.6 \qquad \text{OK}$$

<u>Use #4 U-stirrups @ 9"</u>(both ends of span)

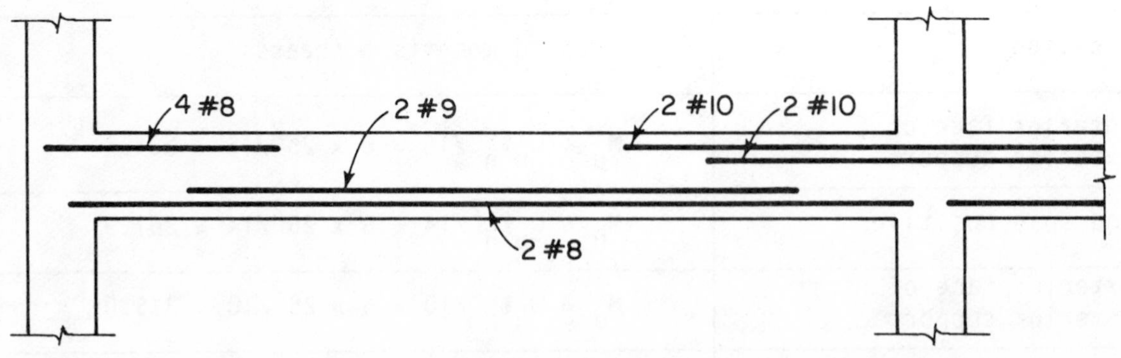

EXAMPLE 3.1 - Continued

Calculations and Discussion	Code Reference

2. Bar lengths for bottom reinforcement

(a) Basic tension development lengths from Table 3-1: 12.2.2

For #8 bars, ℓ_{db} = 30 in.

For #9 bars, ℓ_{db} = 38 in.

(b) Required number of bars to be extended into supports. 12.11.1

One-fourth of ($+A_s$) must be extended at least 6 in. into the supports. One #9 bar could be full span length with the other #9 and the 2 #8 bars cut off within the span. With a longitudinal bar required at each corner of the stirrups (Section 12.13.3), at least 2 bars should be extended full length.

Consider extending the 2 #8 bars full span length (plus 6 in. into the supports) and cut off the 2 #9 bars within the span.

(c) If the beam were part of a primary lateral load resisting system, the 2 #8 bars extending into the supports would have to be anchored to develop the bar yield strength at the face of supports. At the exterior column, anchorage can be provided by a standard end hook. Minimum width of support (overall column depth) required for anchorage of the #8 bar with a standard hook, from Table 3-2: 12.11.2

- 16 in. for 90° hook with 2 in. end cover on hook 12.5.3.2
- 13 in. for 90° hook with 2 in. end cover and hook enclosed within ties or stirrup-ties spaced not greater than $3d_b$. 12.5.3.3

EXAMPLE 3.1 - Continued

Calculations and Discussion	Code Reference

At the interior column, the 2 #8 bars could be extended ℓ_d = 30 in. beyond the face of support into the adjacent span to satisfy the special anchorage requirement of Section 12.11.2.

(d) Determine cut-off locations for the 2 #9 bars and check other development requirements.

Shear and moment diagrams for loading condition causing maximum factored positive moment are shown below.

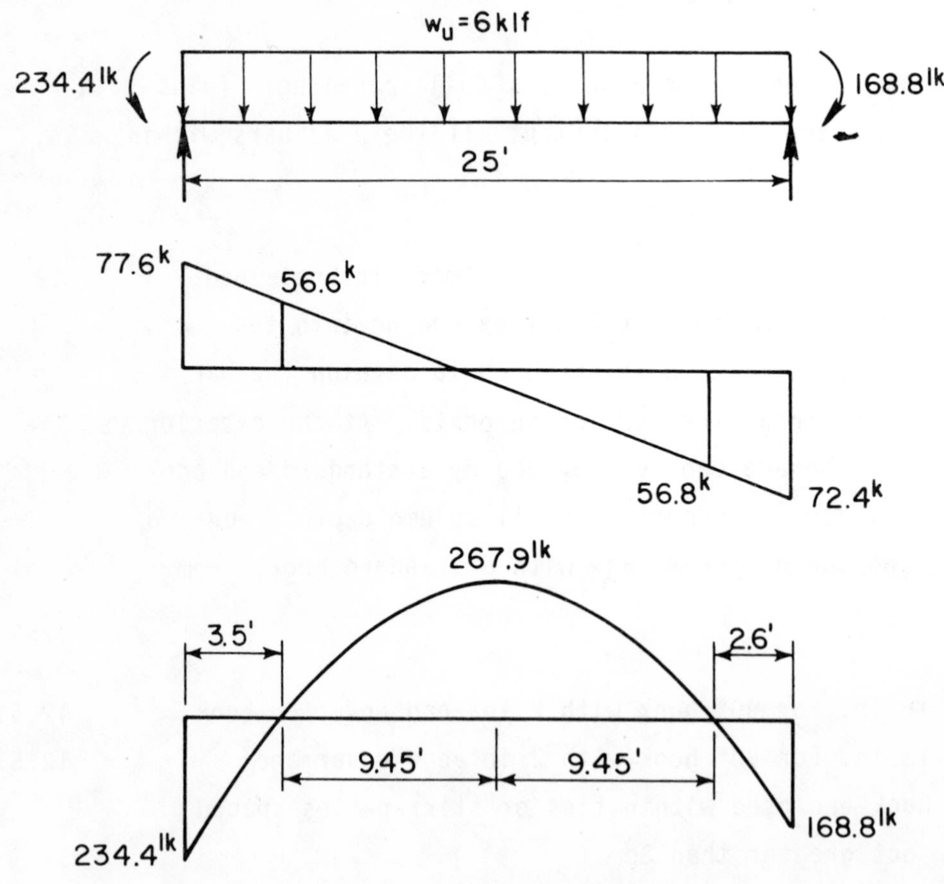

EXAMPLE 3.1 - Continued

Calculations and Discussion	Code Reference

The positive moment portion of the M_u diagram is shown below at a larger scale, incuding the design moment strengths φM_n for the total positive A_s (2 #8 and 2 #9) and for (2 #8 bars) separately; and the necessary dimensions. For 2 #8 and 2 #9, $\varphi M_n = 280.7^{'k}$. For 2 #8, $\varphi M_n = 131.8^{'k}$.

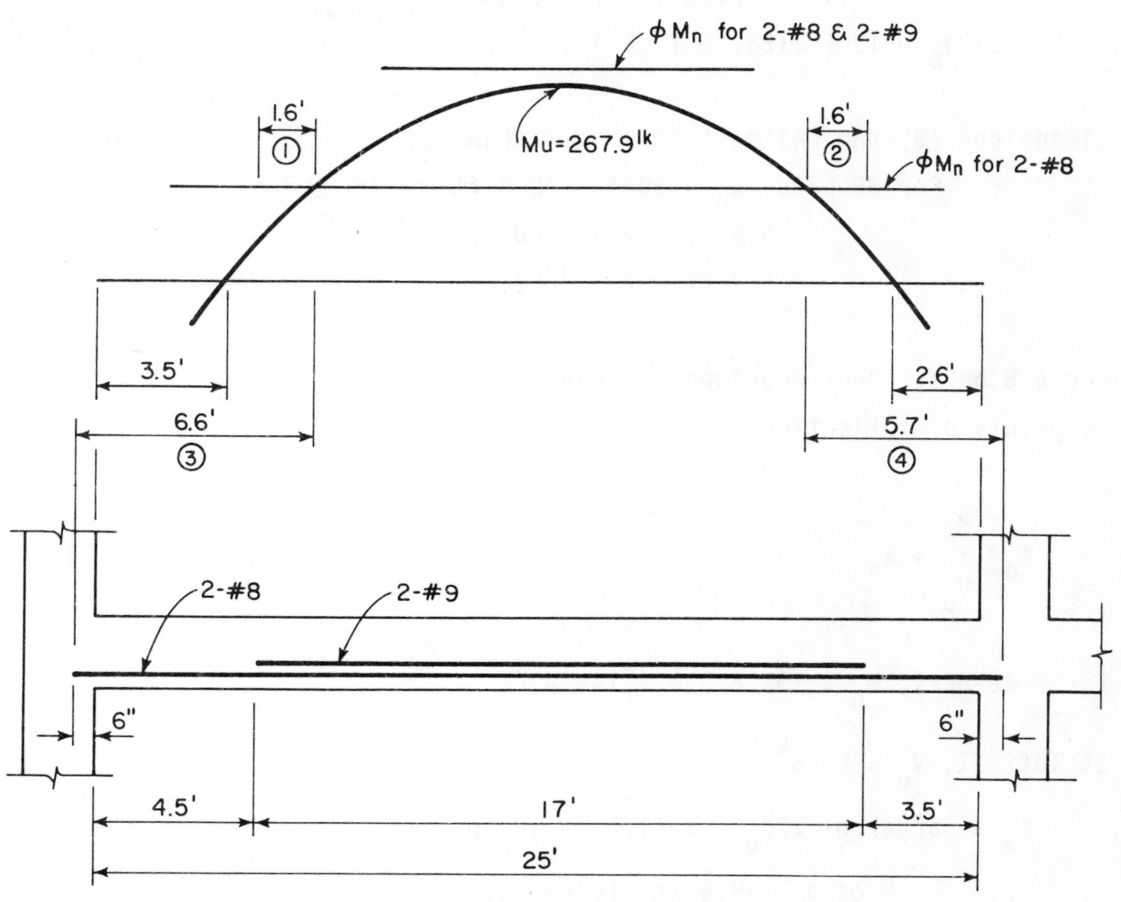

EXAMPLE 3.1 - Continued

Calculations and Discussion	Code Reference

As shown, the 2 #8 bars extend full span length plus 6 in. into the supports. The 2 #9 bars are cut off tentatively at 4.5 ft and 3.5 ft from the exterior and interior supports respectively. These tentative cut-off locations are determined as follows:

dimensions (1) and (2) must be the larger of d or $12d_b$: 12.10.3

d = 19.4 in. (1.6 ft) (governs)

$12d_b$ = 12 (1.128) = 13.5 in.

dimensions (3) and (4) must be larger than ℓ_d: 12.10.4

For #8 bars, ℓ_d = 30 in. (2.5 ft)

6.6 ft > 2.5 OK

5.7 ft > 2.5 OK

For # 8 bars, check development requirements at points of inflection: 12.11.3

$$\ell_d \leq \frac{M_n}{V_u} + \ell_a$$ Eq. (12-1)

For 2 #8 bars, M_n = 131.8/0.9 = 146.4$^{'k}$

at left PI, V_u = 56.6^k

ℓ_a = larger of $12d_b$ = 12(1.0) = 12 in.

or d = 19.4 in. (governs)

$$\ell_d \leq \frac{146.4 \times 12}{56.6} + 19.4 = 50.4 \text{ in.}$$

For #8 bars, ℓ_d = 30 in. < 50.4 OK

at right PI, V_u = 56.8^k; by inspection, the #8 bars are OK.

EXAMPLE 3.1 - Continued

Calculations and Discussion	Code Reference

With both tentative cut-off points located in a zone of flexural tension, one of the three conditions of Section 12.10.5 must be satisfied.

At left cut-off point (4.5 ft from support):

$V_u = 77.6 - 4.5 \times 6 = 50.6^k$

$\varphi V_n = 77.4$ (#4 U-stirrups @ 9")

$2/3(77.4) = 51.6^k > 50.6$ OK — 12.10.5.1

For illustrative purposes, determine if the condition of Section 12.10.5.3 is also satisfied:

$M_u = 54.1^{'k}$ at 4.5 ft from support

A_s required = 0.63 $in.^2$

For 2 #8 bars, A_s provided = 2.00 $in.^2$

$2.00 > 2(0.63) = 1.26\ in.^2$ OK — 12.10.5.3

$3/4(77.4) = 58.1^k > 50.6$ OK — 12.10.5.3

Therefore, Section 12.10.5.3 is also satisfied at cut-off location.

At right cut-off point (3.5 ft from support):

$V_u = 72.4 - 3.5 \times 6 = 51.4^k$

$2/3(\varphi V_n) = 51.6^k > 51.4$ OK — 12.10.5.1

EXAMPLE 3.1 - Continued

Calculations and Discussion	Code Reference

Summary: The tentative cut-off locations for the bottom reinforcement meet all code development requirements. The 2 #9 bars x 17'-0 would have to be placed unsymmetrically within the span. To assure proper placing of the #9 bars, it would be prudent to specify a 19'-0 length for symmetrical bar placement within the span, i.e., 3.5 ft from each support. The ends of the cut-off bars would then be at or close to the points of inflection.....thus, eliminating the need to satisfy the conditions of Section 12.10.5 when bars are terminated in a tension zone. The recommended bar arrangement is shown at the end of the example.

3. Bar lengths for top reinforcement

(a) Tension development lengths:

For #8 bars, ℓ_{db}= 30 in. (Table 3-1) — 12.2.2

including top bar effect, ℓ_d = 30 x 1.4 = 42 in. — 12.2.3.1

For #10 bars, ℓ_{db}= 48 in. (Table 3-1)

including top bar effect, ℓ_d = 48 x 1.4 = 67 in.

EXAMPLE 3.1 - Continued

Calculations and Discussion	Code Reference

(b) Shear and moment diagrams for loading condition causing maximum factored negative moments are shown below.

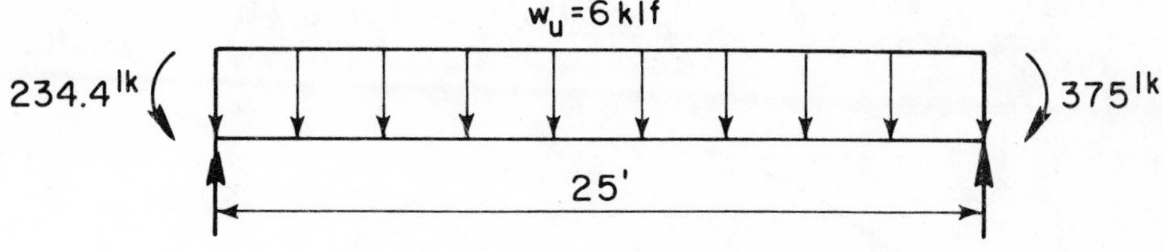

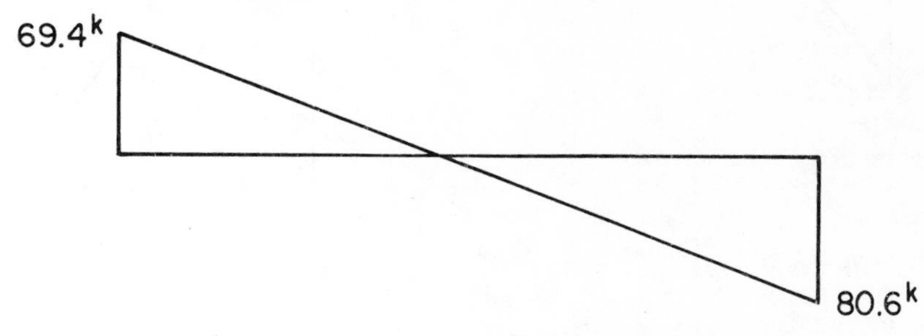

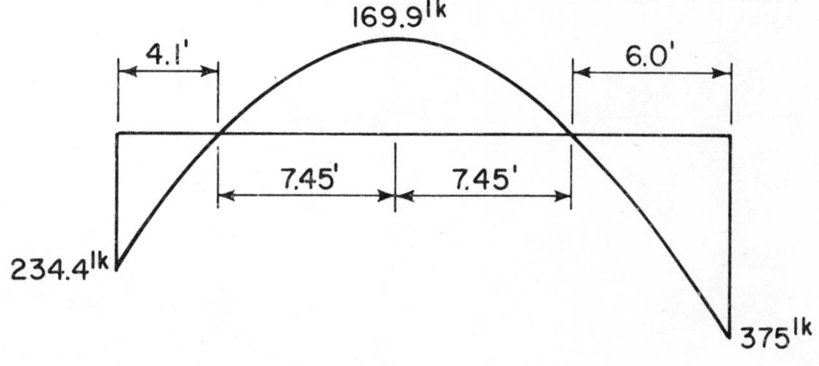

EXAMPLE 3.1 - Continued

Calculations and Discussion	Code Reference

The negative moment portions of the M_u diagram are shown below at a larger scale, including the design moment strengths φM_n for the total negative A_s at each support (4 #8 and 4 #10) and for 2 #10 bars at the interior support; and the necessary dimensions. For 4 #8, $\varphi M_n = 251.1^{'k}$. For 4 #10, $\varphi M_n = 379.5^{'k}$. For 2 #10, $\varphi M_n = 194.3^{'k}$.

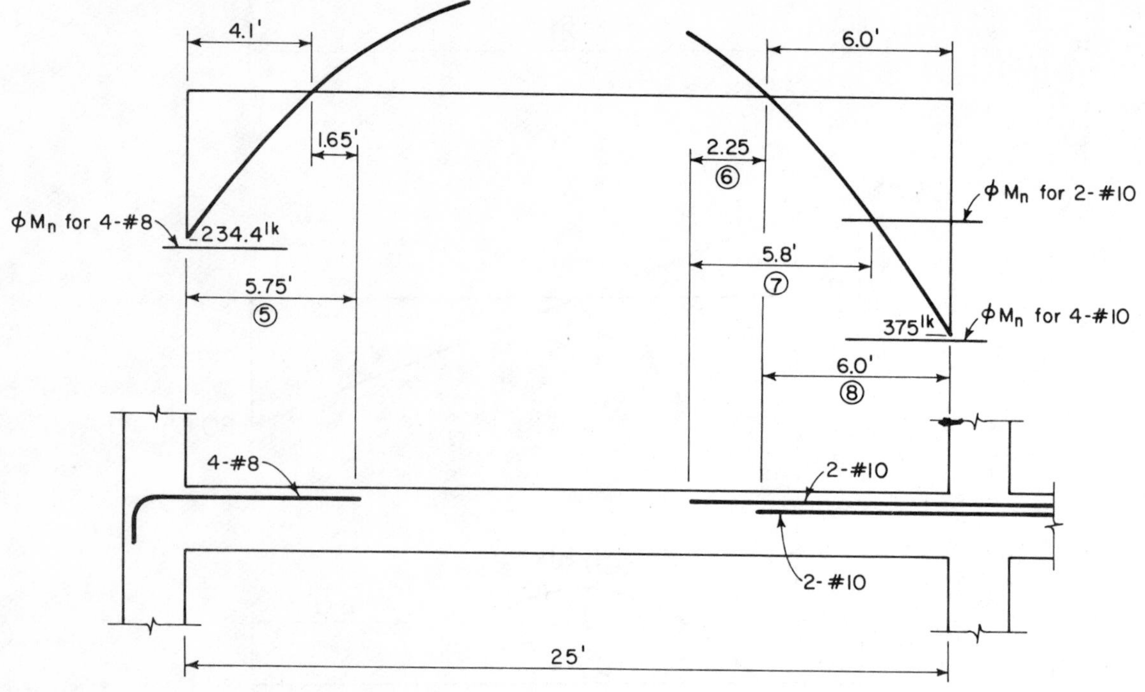

EXAMPLE 3.1 - Continued

Calculations and Discussion	Code Reference

4. Development requirements for 4 #8 bars

(a) Required number of bars to be extended. — 12.12.3

One-third of $(-A_s)$ provided at supports must be extended beyond the point of inflection a distance equal to the greater of d, $12d_b$, or $\ell_n/16$.

d = 19.4 in. or 1.6 ft (governs)

$12d_b$ = 12(1.0) = 12.0 in.

$\ell_n/16$ = 25 x 12/16 = 18.75 in.

Since the inflection point is located only 4.1 ft from the support, total length of the #8 bars will be relatively short even with the required 1.6 ft extension beyond the point of inflection. Check required development length ℓ_d for a cut-off location at 5'-9" from face of support.

dimension (5) must be at least equal to ℓ_d — 12.1

For #8 top bars, ℓ_d = 42 in. < 69 in.   OK

(b) Anchorage into exterior column.

The #8 bars can be anchored into the column with a standard end hook. From Table 3-2, a 90° hook with side cover ≥ 2-1/2 in. and end cover ≥ 2 in. requires a total embedment ℓ_{dh} of 14 in. for a #8 hooked bar. Overall depth of column required would be 16 in. The required ℓ_{dh} for the hook could be reduced by 1 in. (a refinement) if excess reinforcement is considered.

EXAMPLE 3.1 - Continued

Calculations and Discussion	Code Reference
$\dfrac{(A_s \text{ required})}{(A_s \text{ provided})} = \dfrac{2.93}{3.16} = 0.93$	12.5.3.4
$\ell_{dh} = 14 \times 0.93 = 13$ in.	
5. Development requirements for 4 #10 bars	
(a) Required extension for one-third of $(-A_s)$	12.12.3
d = 19.4 in. (1.6 ft) (governs) $12d_b = 12(1.27) = 15.24$ in. $\ell_n/16 = 18.75$ in.	
For illustrative purposes, consider two top bar lengths. Extend 2 #10 bars 8'-3" into the span (2.25 ft beyond inflection point). Extend 2 #10 bars 6'-0" into the span (cut-off at point of inflection)	
Check dimensions:	
dimension (6) = 2.25 ft > 1.6 OK	12.12.3
dimension (7) = 5.80 ft > ℓ_d = 67 in. (5.6 ft) OK	
dimension (8) = 6.0 ft > ℓ_d = 67 in (5.6 ft) OK	

EXAMPLE 3.1 - Continued

Calculations and Discussion	Code Reference

Summary: Selected bar lengths for the top and bottom reinforcement are shown below.

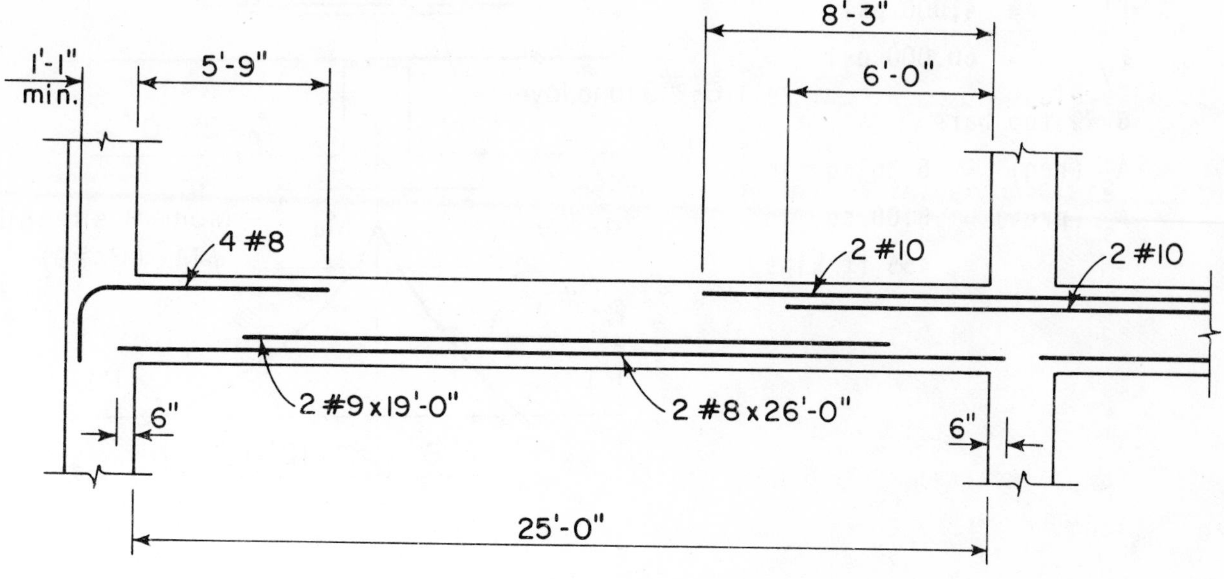

EXAMPLE 3.2 - Development of Reinforcement

Determine for the partial beam shown below the lengths of top bars over the interior support. Both spans are continuous.

$h = 24"$

$d = 21.5"$

$b = 18"$

$f'_c = 4{,}000$ psi

$f_y = 60{,}000$ psi

6-#9 top bars

A_s (req) = 5.36 sq. in.

A_s (prov) = 6.00 sq. in.

$-M_u$ = 455 ft kips

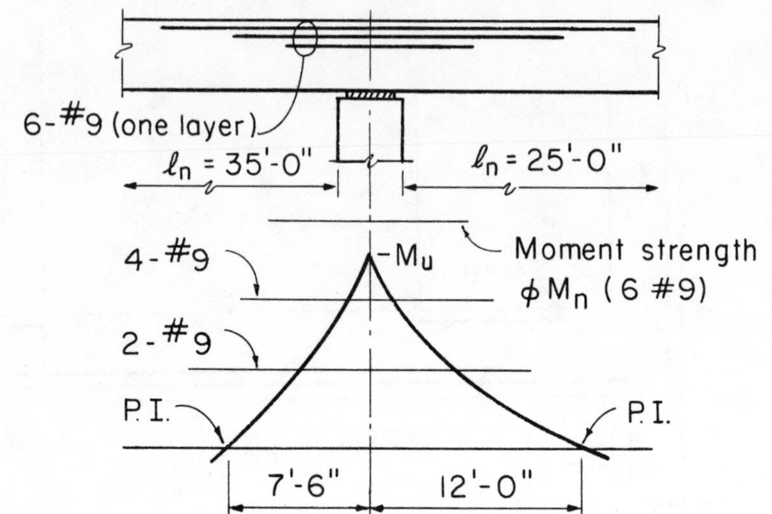

Calculations and Discussion	Code Reference
1. Basic tension development length for #9 bars:	12.2.2

$$\ell_{db} = \frac{0.04A_b f_y}{\sqrt{f'_c}} = \frac{0.04 \times 1.0 \times 60{,}000}{\sqrt{4{,}000}} = 38"$$

but not less than

$0.0004 d_b f_y = (0.0004)(1.128)(60{,}000) = 27"$

or from Table 3-1: $\ell_{db} = 38"$

EXAMPLE 3.2 - Continued

Calculations and Discussion	Code Reference
For top reinforcement ℓ_d = 38" x 1.4 = 53"	12.2.3.1
For excess reinforcement at $(-M_u)$:	12.3.3.1
ℓ_d = 53(5.36/6.00) = 47"	
2. Required number of bars to be extended beyond the point of inflection:	
1/3 A_s = 2 bars extended beyond the point of infection the larger of the following distances:	
d = 21.5 in. (governs for 25' span)	
$12d_b$ = 13.5 in.	
$\ell_n/16$ = 25 x 12/16 = 19 in.	
$\ell_n/16$ = 35 x 12/16 = 26 in. (governs for 35 ft span)	
3. Bar extension beyond the point where it is no longer required to resist flexure: d or $12d_b$, whichever is larger.	12.10.3
d = 21.5 in. (governs)	
$12d_b$ = 13.5 in.	

Calculations and Discussion	Code Reference

Summary:

Anchorage requirements for locating termination points are shown below. The provisions of Section 12.10.5 must also be satisfied to terminate bars in a tension zone. This is not considered here.

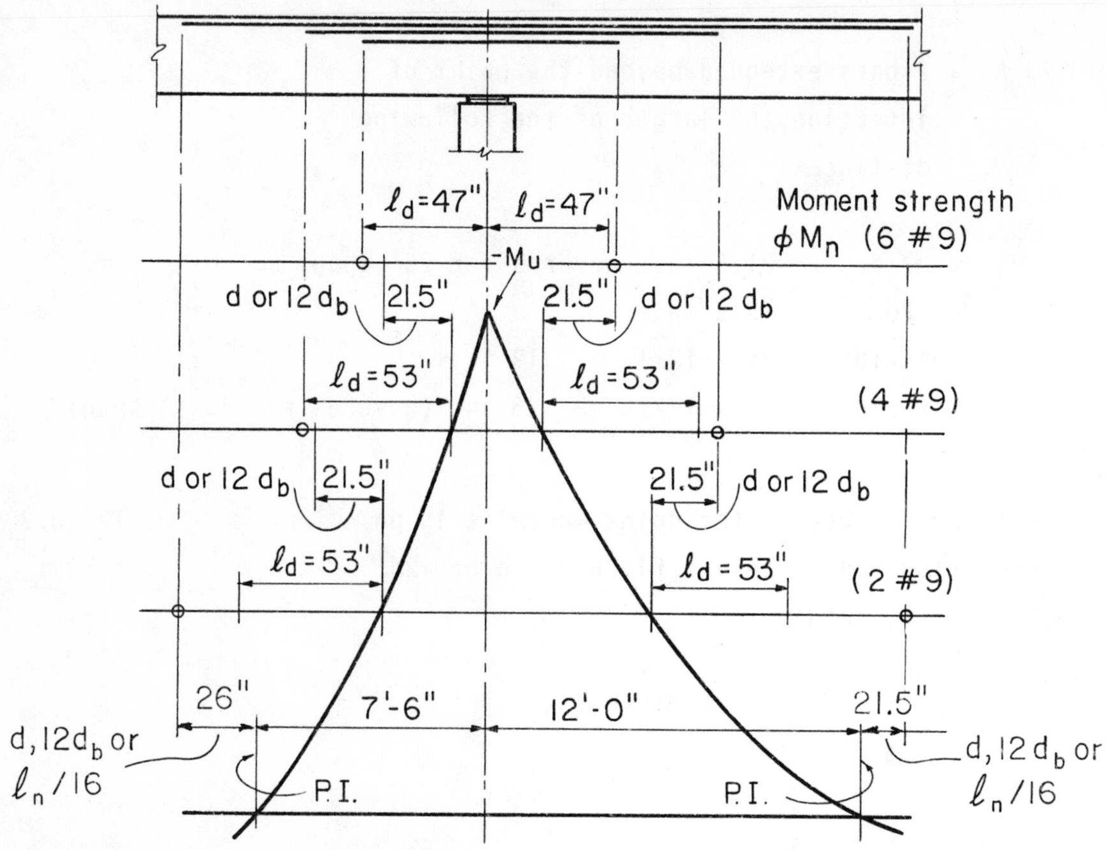

Bar lengths and their extensions into each span must be indicated on the design drawings. With the unsymmetrical bar extensions required to satisfy the development requirements for the different moment variation in the two adjacent spans, there is the possibility of misplacing the bars in the field. To minimize chances of field errors, it would be prudent to specify a symmetrical arrangement, i.e., with the same longer bar extensions into each span.

EXAMPLE 3.3 - Lap Splices in Tension

Design the tension lap splices for the grade beam shown below.

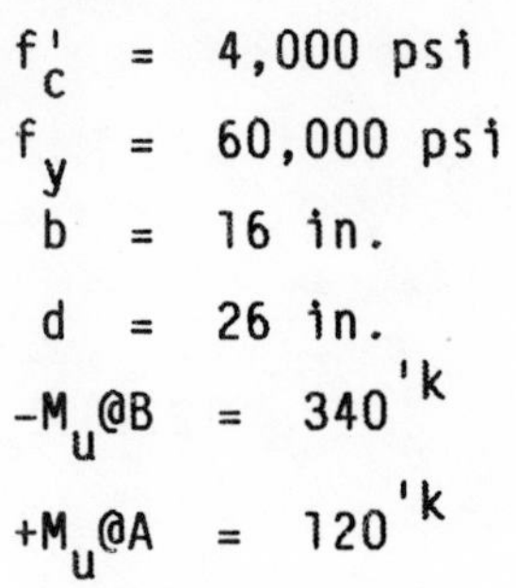

f'_c = 4,000 psi

f_y = 60,000 psi

b = 16 in.

d = 26 in.

$-M_u$@B = 340'k

$+M_u$@A = 120'k

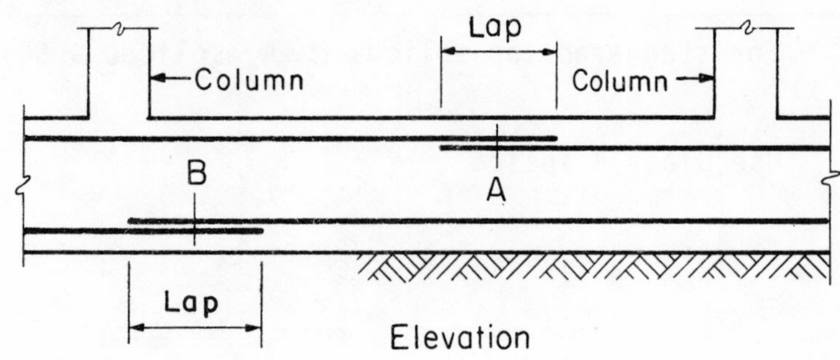

4-#9 bars top and bottom (continuous)

Preferably, splices should be located away from zones of high tension. For a typical grade beam, top bars should be spliced under the columns, and bottom bars about midway between columns. Even though, in this example, the splice at A is not a preferred location, the moment at A is relatively small. Assume for this example that the splices must be located as shown.

Calculations and Discussion	Code Reference
1. Determine required lap splice for bottom bars at B. Bars are spaced closer than 6".	12.14.2.3
A_s required ($-M_u$ @ B = 340'k) = 3.11 sq in.	
As provided (4-#9) = 4.0 sq in.	
$\frac{A_s \text{ provided}}{A_s \text{ required}} = \frac{4.0}{3.11} = 1.29 < 2$	12.15.2
For all bars spliced at same location:	
Use Class C Splice	
$1.7\, \ell_d = 1.7\,(0.04\, A_b f_y/\sqrt{f'_c})$	
$= 1.7\,(0.04 \times 1.00 \times 60{,}000/\sqrt{4{,}000}) = 65$ in.	

EXAMPLE 3.3 - Continued

Calculations and Discussion	Code Reference

For staggered lap splices (% A_s spliced = 50)

Use Class B splice

$1.3\ \ell_d$ = 49" lap — 12.15.2

2. Determine required lap splice for top bars at A.

A_s required (+M_u @ A = $120^{'k}$) = 1.05 sq. in.

$$\frac{A_s \text{ provided}}{A_s \text{ required}} = \frac{4.0}{1.05} = 3.81 > 2$$

Type of lap splice required for (A_s provided/A_s required) > 2 and

not staggered, use Class B splice = $1.3\ \ell_d$

= 69" lap* — 12.15.2

for staggered, use Class A splice = $1.0\ell_d$

= 53" lap*

*For top reinforcement, the basic development length ℓ_{db} is increased by a factor of 1.4 in accordance with Section 12.2.3.1.

EXAMPLE 3.4 - Lap Splices in Compression

The following two examples illustrate typical calculations for compression lap splices.

Calculations and Discussion	Code Reference
1. Design the compression lap splice for the tied column shown below.	
$b = 16"$	
$h = 16"$	
$f'_c = 4,000$ psi	
$f_y = 60,000$ psi	
8-#9 bars	
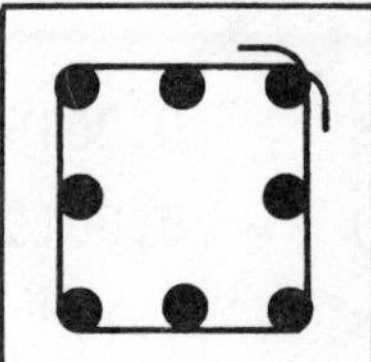	
a) Determine lap splice length:	12.16.1
$\ell_d = 0.02\, d_b f_y/\sqrt{f'_c}$	12.3.2
$= 0.02 \times 1.128 \times 60,000/\sqrt{4,000} = 21"$	
but not less than:	
$= 0.0005\, d_b f_y$ nor 12 in. (min.)	12.16.1
$= 0.0005 \times 1.128 \times 60,000 = 34"$ (governs)	
If $f_y = 75,000$ psi*	
$\ell_d = 0.02 \times 1.128 \times 75,000/\sqrt{4,000} = 27$ in.	12.3.2
but not less than:	
$= (0.0009\, f_y - 24)\, d_b$ nor 12 in. (min.)	12.16.1
$= (0.0009 \times 75,000 - 24)\, 1.128 = 49"$ (governs)	
If $f'_c < 3,000$ psi, calculate ℓ_d as above and increase by 1/3.	

*See Code Commentary Section 3.5.3.2 for discussion of f_y exceeding 60,000 psi.

EXAMPLE 3.4 - Continued

Calculations and Discussion	Code Reference
b) Determine column tie requirements to allow a 0.83 reduced lap length:	12.16.3
Required column ties, #3 @ 16" o.c.	7.10.5
Required spacing of #3 ties for reduced lap length:	
effective area of ties $\geq$ 0.0015hs	
(2 x 0.11) = 0.0015 x 16s	
s = 9.17"	
Spacing of the #3 ties must be reduced to 9" o.c. throughout the lap splice length to allow a lap length of 0.83 ℓ_d = 28".	
2. Determine compression lap splice for spiral column shown.	
f'_c = 4000 psi	
f_y = 60,000 psi	
8-#9 bars	
#3 Spiral	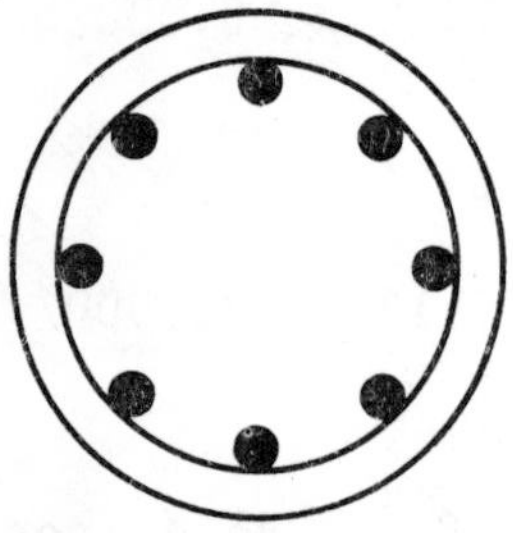
Allowable lap = 0.75 (ℓ_d)	12.16.4
for $f_y \leq$ 60,000 psi, lap = 0.75 x 34" = 26"	
for f_y > 60,000 psi (say 75,000), lap = 0.75 x 49.0 = 37"	
End bearing, welded, or mechanical connections may also be used.	12.16.5

EXAMPLE 3.5 - Lap Splices in Columns

Design the lap splices for the tied column shown below

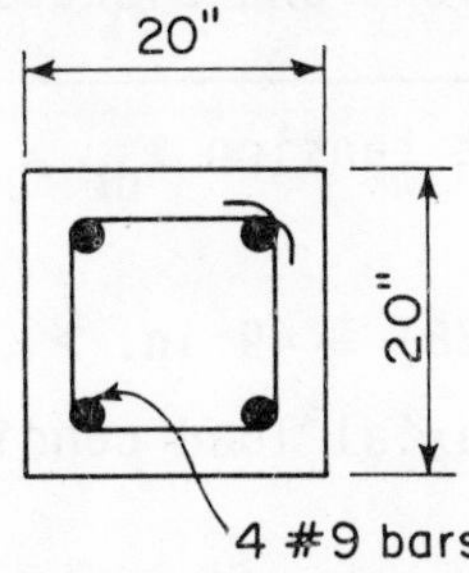

f'_c = 4,000 psi (normal weight)

f_y = 60,000 psi

b = h = 20 in.

4 #9 bars.

Lap splices to be designed for two loading conditions:

(1) Axial load.....calculated factored load stress equal to f_y in compression.

(2) Combined bending and axial load....calculated factored load stress in 2 #9 bars (either direction) equal to 24,000 psi tension.

Calculations and Discussion	Code Reference
(1) For the axial load condition, required lap length in compression is $30d_b$ or 34 in.	12.16.1
(2) For the combined bending and axial load condition, splices must provide twice the calculated tension in each column face, but not less than $A_s f_y/4$.	12.17.1
Tensile force in each face = 2.0 x 24 = 48^k	
A_s required = 48/60 = 0.8 in.^{2}	
$\dfrac{A_s \text{ provided}}{A_s \text{ required}} = \dfrac{2.0}{0.80} = 2.5 > 2$	
If all bars are spliced at the same location, a Class B tension lap splice is required. Lap length = $1.3\ell_d$	12.15.2

EXAMPLE 3.5 - Continued

Calculations and Discussion	Code Reference
For #9 bars, basic tension ℓ_{db} = 38 in. (Table 3-1)	
Lap length = 1.3(38) = 49 in. > 34 in. required for compression (axial load condition)	
The 49 in. lap length required for the Class B tension lap splice exceeds the minimum lap length required for 25% tension (49/4 = 12 in.)	12.17.1
Summary: If all bars are lap spliced at the same location, a lap length of 49 in. is required.	

4

Design Methods and Strength Requirements

8.1 Design Methods

Two philosophies of design for reinforced concrete have long been prevalent. "Working Stress Design" was the principal method used from the early 1900's until the early 1960's. Since publication of the 1963 edition of the ACI Code, there has been a rapid transition to "Ultimate Strength Design," largely because of its more rational approach. Ultimate strength design (referred to in the Code as the "Strength Design Method") is conceptually more realistic in its approach to structural safety.

The 1956 ACI Code (ACI 318-56) was the first code edition which officially recognized and permitted the ultimate strength method of design. Recommendations for the design of reinforced concrete structures by ultimate strength theories were included in an appendix.

The 1963 ACI Code (ACI 318-63) treated the working stress and the ultimate strength methods on an equal basis. However, the major portion of the working stress method was modified to reflect ultimate strength behavior. The working stress provisions of the 1963 Code, relating to bond, shear and diagonal tension, and combined axial compression and bending, had their basis in ultimate strength.

The 1971 ACI Code (ACI 318-71) was based entirely on the strength approach for proportioning reinforced concrete members, except for a small section devoted to what was called an "Alternate Design Method." Even in that section, the service load capacities (except for flexure) were given as various percentages of the ultimate strength capacities of other parts of the Code. This transition to ultimate strength theories for reinforced concrete design was essentially complete in the 1971 ACI Code, with ultimate strength design definitely established as being preferred.

In the 1977 ACI Code (ACI 318-77) the "Alternate Design Method" (ADM) was relegated to Appendix B, just as the strength design method was introduced by way of an appendix in ACI 318-56. The appendix location served to separate and clarify the two methods of design, with the main body of the Code devoted exclusively to the strength design method.

For those who still prefer ADM, the reworded and expanded appendix format for ADM that was adopted in ACI 318-77 makes the provisions more usable; Section 8.10 of ACI 318-71 was found less than satisfactory by many designers preferring ADM, who went on using the 1956 or the 1963 edition of the Code. Hopefully, the removal of ADM from the main body of the Code will encourage the more reluctant to finally accept strength design.

Since an appendix location is sometimes not considered to be an official part of a legal document (unless specifically adopted), specific reference is made in the main body of the code (Section 8.1.2) to make Appendix B a legal part of the Code.

Regardless of whether the strength design method of the Code or the alternate design method of Appendix B is used in proportioning for strength, the general serviceability requirements of the Code, such as the provisions for deflection control and crack control, must always be satisfied.

8.1.1 Strength Design Method

The strength design method requires that the computed nominal strengths reduced by specified strength reduction factors, i.e., design strengths, equal or exceed the service load effects (internal forces and moments) increased by specified load factors, i.e. required strengths.

Since the distinction between "design strength" and "required strength" is crucial to an understanding of the strength design method, the definitions and notations used with the strength design method are summarized as follows:

Definitions

Service load - load specified by general building code (without load factors)

Factored load - load multiplied by appropriate load factors, used to proportion members by the strength design method

Required strength - strength of a member or cross section required to resist factored loads or related internal moments and forces in such combinations as are stipulated

Nominal strength - strength of a member or cross section calculated in accordance with provisions and assumptions of the strength design method before application of any strength reduction factors

Design strength - nominal strength multiplied by a strength reduction factor

Notations

Required Strength:

M_u = factored moment at section
P_u = factored axial load at given eccentricity
V_u = factored shear force at section
T_u = factored torsional moment

Nominal Strength:

M_n	=	nominal moment strength at section
M_b	=	nominal moment strength at balanced strain conditions
P_n	=	nominal axial load strength at given eccentricity
P_o	=	nominal axial load strength at zero eccentricity
P_b	=	nominal axial load strength at balanced strain conditions
V_n	=	nominal shear strength
V_c	=	nominal shear strength provided by concrete
V_s	=	nominal shear strength provided by shear reinforcement
T_n	=	nominal torsional moment strength
T_c	=	nominal torsional moment strength provided by concrete
T_s	=	nominal torsional moment strength provided by torsion reinforcement

Design Strength:

φM_n	=	design moment strength at section
φP_n	=	design axial load strength at given eccentricity
φV_n	=	design shear strength = $\varphi (V_c + V_s)$
φT_n	=	design torsional moment strength = $\varphi (T_c + T_s)$

The following discussion is essentially reproduced from the Commentary on Chapter 2 of ACI 318-83:

A number of definitions for loads are given as the Code contains requirements that must be met at various load levels. The terms "dead load" and "live load" refer to the unfactored loads (service loads) specified or defined by a local building code. Service loads (loads without load factors) are to be used where specified in the Code to proportion or investigate members for adequate serviceability. Loads used to proportion a member for adequate strength are defined as "factored loads." Factored loads are service loads multiplied by the appropriate load factors specified for required strength. The term "design loads" is not used in ACI 318 to avoid confusion with the design load terminology used in general building codes to denote service loads, or posted loads in buildings. The factored load terminology used in ACI 318 clearly indicates whether load factors are applied to a particular load, moment, or shear value as used in the Code provisions.

The required axial load, moment, and shear strengths used to proportion members are referred to either as factored axial loads, factored moments, and factored shears, or required axial loads, moments, and shears. The factored load effects are calculated from the applied factored loads and forces in such load combinations as are stipulated in the Code (Section 9.2).

The subscript "u" is used to denote the required strengths: required axial load strength (P_u), required moment strength (M_u), required shear strength (V_u), and required torsional moment strength (T_u) calculated from the applied factored loads and forces.

Strength of a member or cross section calculated using standard assumptions and strength equations, and nominal (specified) values of material strengths and dimensions, is referred to as "nominal strength." The subscript "n" is used to denote the nominal strengths: nominal axial load strength (P_n), nominal moment strength (M_n), nominal shear strength (V_n), and nominal torsional moment strength (T_n).

"Design strength" or usable strength of a member or cross section is the nominal strength reduced by the strength reduction factor φ, and may be denoted as φP_n, φM_n, φV_n, φT_n.

9.1 Strength Requirements

9.1.1 - The basic criterion for strength design may be expressed as follows:

Required Strength $\leq$ Design Strength

or, [Load Factor] [Service Load Effects] $\leq$
[Strength Reduction Factor] [Nominal Strength]

All members and all sections of members must be proportioned to meet the above criterion under the most critical load combination and under all possible states of stress (flexure, axial load, shear, etc.):

$$P_u \leq \varphi P_n$$
$$M_u \leq \varphi M_n$$
$$V_u \leq \varphi V_n$$
$$T_u \leq \varphi T_n$$

The above criterion provides for the margin of structural safety in two ways:

(1) The required strength is computed in terms of factored loads or the related internal moments and forces. Factored loads are defined in Section 2.1 as service loads multiplied by the appropriate load factors. The loads to be used are described in Section 8.2. Thus, for example, the required flexural strength for dead and live loads is:

$$M_u = 1.4M_d + 1.7\ M_\ell$$

where M_d and M_ℓ are the moments due to service dead and live loads, respectively.

(2) The design strength is computed by multiplying the nominal strength with the appropriate strength reduction factor. The nominal strength is computed by the Code procedures assuming that the member or the section will have the exact dimensions and material properties assumed in the computations. Thus, for example, the design moment strength for a singly reinforced cross section is:

$$\varphi M_n = \varphi[A_s f_y(d-a/2)]$$

For the example section without compression reinforcement and subjected to flexure, the basic criterion for strength design reduces to:

$$1.4M_d + 1.7M_\ell \leq \varphi[A_s f_y(d - a/2)]$$

Similarly, for shear strength of a beam, the basic criterion for strength design can be stated as:

$$V_u \leq \varphi V_n$$
$$\leq \varphi(V_c + V_s)$$
$$1.4\ V_d + 1.7\ V_\ell \leq \varphi \left[2\sqrt{f'_c}\ b_w d + \frac{A_v f_y d}{s} \right]$$

The following reasons for requiring load and strength reduction factors in structural design are given by MacGregor in Reference 4.1:

1. The strength of materials or elements may be less than expected. The following factors contribute:

(A) Material strengths may differ from those assumed in design because of:

- Variability in material strengths--the compression strength of concrete as well as the yield strength and ultimate tensile strength of reinforcement are variable.
- Effect of speed of testing--the strengths of both concrete and steel are affected by the rate of loading.
- In situ strength vs. specimen strength--the strength of concrete in a structure is somewhat different from the strength of the same concrete in a control specimen.
- Effect of variability of shrinkage stresses or residual stresses--the variability of the residual stresses due to shrinkage may affect the cracking load of a member, and is significant where cracking is the critical limit state. Similarly, the transfer of compression loading from concrete to steel due to creep and shrinkage in columns may lead to premature yielding of the compression steel, possibly resulting in instability failures of slender columns with small amounts of reinforcement.

(B) Members may vary from those assumed, due to fabrication errors. The following are significant:

- Rolling tolerances in reinforcing bars.

- Geometric errors in cross section and errors in placement of reinforcement.

(C) Simplified assumptions and equations, such as use of the rectangular stress block and the maximum usable strain of concrete equal to 0.003, introduce both systematic and random errors.

(D) The use of discrete bar sizes leads to variations in the actual capacity of members.

2. Overloads may occur.

(A) Magnitudes of loads may vary from those assumed. Dead loads may vary because of:
- Variations in member sizes.
- Variations in material density.
- Structural and nonstructural alterations.

Live load varies considerably from time to time and from building to building.

(B) Uncertainties exist in the calculation of load effects--the assumptions of stiffnesses, span lengths, etc., and the inaccuracies involved in modeling three-dimensional structures for structural analysis lead to differences between the stresses which actually occur in a building and those estimated in the designer's analysis.

3. Consequences of failure may be severe. A number of factors ought to be considered:

(A) The type of failure, warning of failure, and existence of alternative load paths.
(B) Potential loss of life.
(C) Costs to society in lost time, lost revenue, or indirect loss of life or property due to failure.
(D) The importance of the structural element in the structure.
(E) Cost of replacing the structure.

By way of background to the numerical values of load factors and strength reduction factors specified in the Code, it may be worthwhile reproducing the following paragraph from Reference 4.1:

> "The ACI...design requirements...are based on an underlying assumption that if the probability of understrength members is roughly 1 in 100 and the probability of overload is roughly 1 in 1000, the probability of overload on an understrength structure is about 1 in 100,000. Load factors were derived to achieve this probability of overload. Based on values of concrete and steel strength corresponding to probability of 1 in 100 of understrength, the strengths of a number of typical sections were computed. The ratio of the strength based on these values to the strength based on nominal strengths of a number of typical sections were arbitrarily adjusted to allow for the consequences of failure and the mode of failure of a particular type of member, and for a number of other sources of variation in strength."

An Appendix to Reference 4.1 traces the history of development of the current ACI load and strength reduction factors.

9.1.2 - The provision of adequate strength does not necessarily ensure acceptable behavior at service load levels. Therefore, the Code includes additional requirements designed to provide satisfactory service load performance.

There is not always a clear separation between the provisions for strength and those for serviceability. For actions other than flexure, the detailing provisions in conjunction with the strength requirements are meant to ensure adequate performance at service loads.

For flexural action, there are special serviceability requirements concerning deflection, distribution of reinforcement, and permissible stresses in prestressed concrete. A consideration of service load deflections is particularly important in view of the extended use of high-strength materials and sophisticated methods of design which result in increasingly slender reinforced concrete members.

9.2 Required Strength

As previously stated, the required strength U is expressed in terms of factored loads, or their related internal moments and forces. Factored loads are the loads specified in the general building code, multiplied by appropriate load factors.

While considering gravity loads (dead and live), a designer using the Code moment coefficients (same coefficients for dead and live loads--Section 8.3.3) has three choices: (1) multiplying the loads by the appropriate load factors, adding them into the total factored load, and then computing the forces and moments due to the total load, (2) computing the effects of factored dead and live loads separately, and then superimposing the effects, or (3) computing the effects of unfactored dead and live loads separately, multiplying the effects by the appropriate load factors, and then superimposing them. Under the principle of superposition, all three procedures yield the same answer. To a designer doing a more exact analysis using different coefficients for dead and live loads (pattern loading for live loads), choice (1) does not exist. While considering gravity as well as lateral loads, load effects (due to factored or unfactored loads), of course, have to be computed separately before any superposition can be made.

The Code prescribes load factors for specific combinations of loads. A list of these combinations is given in Table 4-1. The numerical value of the load factor assigned to each type of load is influenced by the degree of accuracy with which the load can usually be assessed, and by the variation which may be expected in the load during the lifetime of a structure. Hence, dead loads, because they can usually be more accurately determined and are less variable, are assigned a lower load factor than live loads. For weight and pressure of liquids with well-defined densities and controllable maximum heights, a reduced load factor of 1.4 is permitted (Section 9.2.5) recognizing the lesser probability of overloading with such liquid loads. A higher load factor of 1.7 is required (Section 9.2.4) where there is considerable uncertainty of pressures such as earth and groundwater pressures and ponding of water.

Table 4-1 Required Strength for Combinations of Loadings

9.2.1 - Dead & Live Load

$$U = 1.4D + 1.7L \tag{9-1}$$

9.2.2 - Dead, Live & Wind Load

$$U = 1.4D + 1.7L$$

or

$$U = 0.75\,(1.4D + 1.7L + 1.7W) \tag{9-2}$$
$$= 1.05D + 1.275L + 1.275W$$

or

$$U = 0.9D + 1.3W \tag{9-3}$$

9.2.3 - Dead, Live & Earthquake Load

$$U = 1.4D + 1.7L$$

or

$$U = 0.75\,(1.4D + 1.7L + 1.87E)$$
$$= 1.05D + 1.275L + 1.402E$$

or

$$U = 0.9D + 1.43E$$

9.2.4 - Dead & Live Load Plus Earth and Groundwater Pressure*

$$U = 1.4D + 1.7L$$

or

$$U = 1.4D + 1.7L + 1.7H \tag{9-4}$$

or (D reducing H)

$$U = 0.9D + 1.7L + 1.7H$$

or (L reducing H)

$$U = 1.4D + 1.7H$$

or (D & L reducing H)

$$U = 0.9D + 1.7H$$

*Weight and pressure of soil and water in soil.

Table 4-1 CONTINUED

9.2.5 - Dead & Live Load Plus Liquid Pressure*

$$U = 1.4D + 1.7L$$

or

$$U = 1.4D + 1.7L + 1.4F$$

or (D reducing F)

$$U = 0.9D + 1.7L + 1.4F$$

or (L reducing F)

$$U = 1.4D + 1.4F$$

or (D & L reducing F)

$$U = 0.9D + 1.4F$$

*Weight and pressures of liquids with well-defined densities and controllable maximum heights.

9.2.6 - Impact

In all equations substitute (L + Impact) for (L) when impact must be considered.

9.2.7 - Dead & Live Load Plus Differential Settlement, Creep, Shrinkage or Temperature Change

$$U = 1.4D + 1.7L$$

or

$$U = 0.75\,(1.4D + 1.4T + 1.7L) \qquad (9\text{-}5)$$
$$= 1.05D + 1.05T + 1.275L$$

or

$$U = 1.4D + 1.4T \qquad (9\text{-}6)$$

With the 1983 Code the definition for H was clarified to state specifically that groundwater pressure is to be considered part of earth pressure with a 1.7 load factor.

While most usual combinations of loading are included, it should not be assumed that all cases are covered. In assigning factors to combinations of loading, some consideration is given to the likelihood of simultaneous occurrence.

In determining the required strength for combinations of loading, due regard must be given to the proper sign (positive or negative), since one type of loading may produce effects which are opposite in sense to those produced by another type. Eq. (9-3) for wind and earthquake loads is specifically included for the case where E or W produces effects opposite in sense to those caused by D and L. Typical cases are uplift on the windward columns and the reversal of moments in beams due to W or E.

Consideration must be given to various combinations of loading in determining the most critical design combination. This is of particular importance when strength is dependent on more than one load effect, such as strength under combined moment and axial load, or the shear strength of members carrying axial load.

9.3 Design Strength

9.3.1 - The design strength provided by a member, its connections to other members, and its cross sections, in terms of flexure, axial load, shear, and torsion, is equal to the nominal strength calculated in accordance with the provisions and assumptions stipulated in the Code, multiplied by a strength reduction factor φ which is less than unity. The wording of Section 9.3.1 is revised with the 1983 Code to alert the designer in specific code language to the importance of connection design. The rules for computing the nominal strength are based generally on conservatively chosen limiting states of stress, strain, cracking or crushing, and conform to research data for each type of structural action. An understanding of all aspects of the strengths computed for various actions can only be obtained by reviewing the background to the Code provisions.

The purpose of the strength reduction factor φ is (1) to define a design strength level that is somewhat lower than would be expected if all dimensions and material properties were those used in computations, (2) to reflect the degree of ductility, toughness, and reliability of the member under the load effects being considered, and (3) to reflect the importance of the member. For example, a lower φ is used for columns than for beams because columns generally have less ductility, are more sensitive to variations in concrete strength, and carry larger loaded areas than beams. Futhermore, spiral columns are assigned a higher φ than tied columns because the former have greater ductility.

9.3.2 - The φ factors prescribed by the Code for different types of action are listed in Table 4-2.

For members subject to flexure and axial load, the design strengths are determined by multiplying both P_n and M_n by the appropriate single value of φ. For members subject to flexure with axial tension, the value of φ given in Section 9.3.2.2(a) is used. For members subject to flexure with axial compression, the value of φ given in Section 9.3.2.2(b) is used for both P_n and M_n.

The wording of Section 9.3.2.2 was revised with the 1983 Code to clarify that, for members subject to flexure and axial load, both load and moment must be multiplied by the appropriate single value of φ.

Table 4-2 Strength Reduction Factor

ACTION	φ
Flexure, without axial load	0.90
Axial tension, and axial tension with flexure	0.90
Axial compression, and axial compression with flexure: Members with spiral reinforcement conforming to Section 10.9.3 Other reinforced members	 0.75* 0.70*
Shear and torsion	0.85
Bearing on concrete	0.70**

*May be increased linearly to 0.90 as φP_n decreases from $0.10 f'_c A_g$ or φP_b, whichever is smaller, to zero. See Fig. 4-1.

** Does not apply to post-tensioning anchorage bearing plates. See Section 18.13.

For members subject to relatively small axial loads (and flexure) it is reasonable to permit an increase in the φ factor from that required for compression members, so that when the axial load reduces to zero and the member is subjected to pure flexure, the strength reduction factor equals 0.90. This is also justified in view of the fact that failure under flexure and small axial loads is initiated by yielding of the tension reinforcement and takes place in an increasingly more ductile manner as the axial load decreases. At the same time, the variability of the strength also decreases. Thus, a varying φ factor is permitted in the Code for members subjected to bending and small axial loads. The φ may be increased from that for compression members to the 0.90 for flexure, as the axial load decreases from a specified value to zero.

The value of the axial load strength φP_n below which an increase in φ can be made is 0.10 $f'_c A_g$ or φP_b, whichever is less. For sections with symmetrical reinforcement and with $f_y \leq 60{,}000$ psi, in which the distance γh (distance between A_s and A'_s) is not less than 0.7h, φP_b will always be greater than 0.10 $f'_c A_g$. Thus, for such sections, the computation of P_b is not required. The procedure outlined in Fig. 4-1 may be used in applying the provisions of Section 9.3.2.2(b).

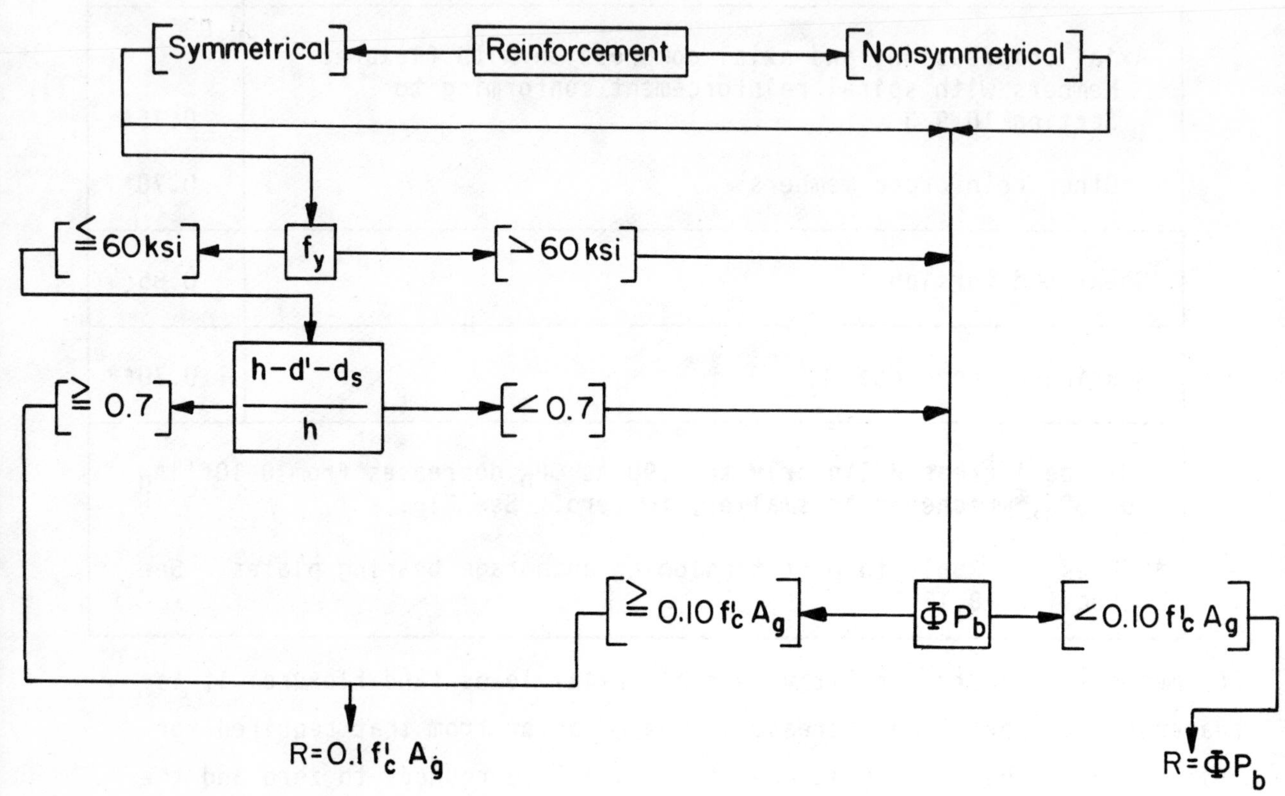

$$\varphi \text{ (Spiral)} = 0.9 - 0.15\ \varphi P_n/R \geq 0.75$$

$$\varphi \text{ (Others)} = 0.9 - 0.20\ \varphi P_n/R \geq 0.70$$

Fig. 4-1 Varying φ Factor--Section 9.3.2.2(b)

9.3.3 - Development lengths for reinforcement, as specified in Chapter 12, do not require a strength reduction modification. Likewise, φ factors are not required for splice lengths, since these are expressed in multiples of development lengths.

9.4 Design Strengths for Reinforcement

An upper limit of 80,000 psi is placed on the yield strength of reinforcing steels other than prestressing tendons. A steel strength above 80,000 psi is not recommended because the yield strain of 80,000-psi steel is about equal to the maximum usable strain of concrete in compression.

No ASTM specification for deformed bars with a yield strength f_y exceeding 60,000 psi is currently available. The reader is referred to the Code Commentary Section 3.5.3.2 for discussion on use and availability of reinforcing bars with f_y exceeding 60,000 psi.

In accordance with Code Section 3.5.3.2, use of reinforcing bars with a specified yield strength f_y exceeding 60,000 psi requires that f_y be measured at a strain of 0.35 percent. This special code requirement also applies to use of welded wire fabric for wire with a specified yield strength greater than 60,000 psi. Higher-yield-strength wire is available and a value of f_y greater than 60,000 psi can be used in design provided compliance with the 0.35 percent strain requirement is certified.

There are limitations on the yield strength of reinforcement in other sections of the Code:

(1) Sections 11.5.2, 11.6.7.4, and 11.7.6: The maximum f_y that may be used in design for shear and torsion reinforcement is 60,000 psi.

(2) Sections 19.3.2 and A.2.5.1: The maximum specified f_y is 60,000 psi in shells, folded plates and structures governed by the special seismic provisions of Appendix A.

(3) Appendix B: The useful f_y is controlled by permissible stresses in the Alternate Design Method.

In addition, the deflection provisions of Section 9.5 and the limitations on distribution of flexural reinforcement of Section 10.6 will become increasingly critical as f_y increases.

8.1.2 Alternate Design Method

An alternate method of design employing load factors and strength reduction factors equal to unity (i.e., service load effects and allowable service load stresses) is permitted for nonprestressed members. The method is outlined in Appendix B.

The Alternate Design Method requires that a structural member (in flexure) be so proportioned that the stresses resulting from the action of service loads (without load factors) and computed by the straight line theory for flexure do not exceed permissible service load stresses. The permissible stresses are limited to values well within the elastic range of the materials, so that the linear relationship between stress and strain is applicable.

The method is similar to the working stress design method of previous ACI Codes. For members subjected to flexure without axial load, the method is identical to that given in the 1963 Code.

Differences in procedure occur in all other cases, including the design of columns, and the design for shear, anchorage length, and splices. In view of the simplifications permitted, the Alternate Design Method will generally result in designs that are more conservative than those based on the Strength Design Method.

Although prestressed members may not be designed for strength under the provisions of Appendix B, Chapter 18 permits linear stress-strain assumptions in the computation of service load stresses and of transfer stresses for use in serviceability control.

It should be noted that all relevant provisions of the Code, except those permitting moment redistribution, apply also to members designed by the

Alternate Design Method. These include control of deflections and distribution of flexural reinforcement, as well as the provisions related to slenderness effects in compression members.

Selected Reference

4.1 MacGregor, J.G., "Safety and Limit States Design for Reinforced Concrete," Canadian Journal of Civil Engineering, Vol. 3, No. 4, December 1976, pages 484-513.

5

General Principles for Strength Design

Introduction

Historically, ultimate strength was the earliest method used in design since the ultimate load could be measured by test without a knowledge of the magnitude or distribution of internal stresses. Since the early 1900's, experimental and analytical investigations have been conducted to develop ultimate strength design theories that would predict the ultimate load measured by test. Some of the early theories that resulted from the experimental and analytical investigations are reviewed in Fig. 5-1.

Structural concrete and reinforcing steel both behave inelastically as ultimate strength is approached. In theories regarding the ultimate strength of reinforced concrete, the inelastic behavior of both materials must be considered and must be expressed in mathematical terms. For reinforcing steel, with a distinct yield point, the elastic behavior may be expressed by a **trapezoidal stress-strain relationship** (see Fig. 5-4). For concrete, the inelastic stress distribution is more difficult to measure experimentally and to express in mathematical terms.

Studies of inelastic concrete stress distribution have resulted in numerous proposed stress distributions as outlined in Fig. 5-1. The development of our present ultimate strength design procedures has its basis in these early experimental and analytical studies. Ultimate strength of reinforced concrete in American design specifications is based primarily on the 1912 and 1932 theories.

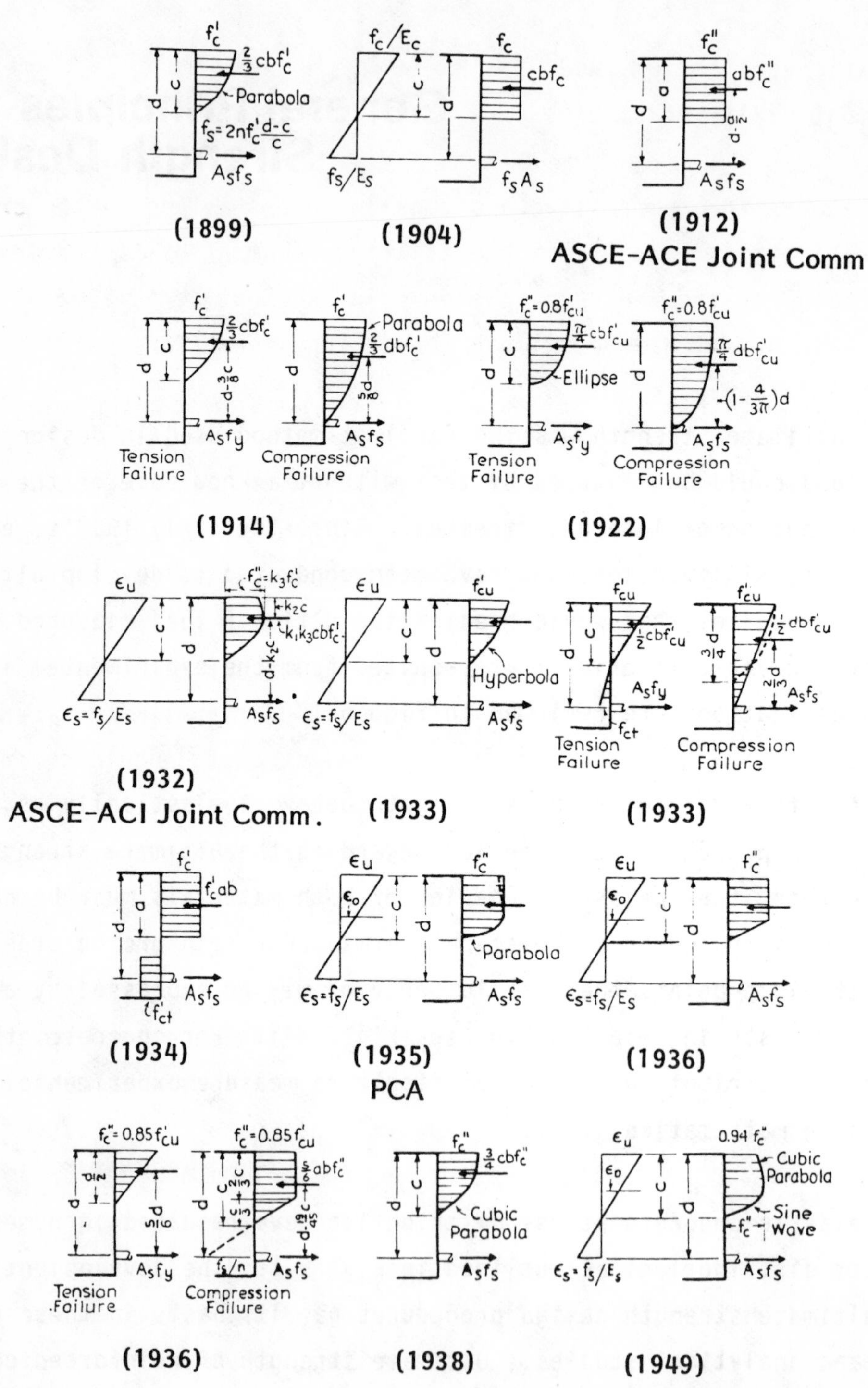

Fig. 5-1 Development of Ultimate Strength Flexure Theories

10.2 Design Assumptions

10.2.1 - The strength of a member or cross section computed by the Strength Design Method requires that two basic conditions be satisfied:

(1) static equilibrium, and
(2) compatibility of strains.

Equilibrium between the compressive and tensile forces acting on the cross section at "ultimate" strength must be satisfied. Compatibility between the stress and strain for the concrete and the reinforcement at "ultimate" conditions must also be established within the design assumptions permitted by the Code (Section 10.2).

The term "ultimate" is used frequently in reference to the Strength Design Method; however, it should be realized that the "nominal" strength computed under the provisions of the Code may not necessarily be the actual ultimate value. Within the design assumptions permitted, certain properties of the materials are neglected and other conservative limits are established for practical design. These contribute to a possible lower "ultimate strength" than that obtained by test. The computed nominal strength should be considered a Code-defined strength only. Accordingly, the term "ultimate" is not used when defining the computed strength of a member. (The term "nominal" strength is used instead).

Furthermore, in discussing the strength method of design for reinforced concrete structures, attention must be called to the difference between loads on the structure as a whole and loads on the cross sections of individual members. Elastic methods of structural analysis are used first to compute service loads on the individual members due to the action of service loads on the entire structure. Only then are the load factors applied to the service loads acting on the individual cross sections. Inelastic (or limit) methods of structural analysis, in which design loads on the individual members are determined directly from the ultimate loads acting on the whole structure, are not considered. Section 8.4 does, however, permit a limited redistribution of negative moments in continuous members. The provisions of Section

8.4 recognize the inelastic behavior of concrete structures and constitute a move toward "limit design." This subject is presented in Part 8.

The computed "nominal strength" of a member must satisfy the design assumptions given in Section 10.2.

Assumption (10.2.2)
"Strain in reinforcement and concrete shall be assumed directly proportional to the distance from the neutral axis."

10.2.2 - In other words, plane sections normal to the axis of bending remain plane after bending. Many tests have confirmed that the distribution of strain is essentially linear across a reinforced concrete cross section, even near ultimate strength. For reinforcement, this assumption has been verified by numerous tests to failure of eccentrically loaded compression members and members subjected to bending only. However, for concrete, the ultimate strain, ϵ_{cu}, at the extreme fiber in compression may vary considerably. Generally, as the compressive strength of the concrete increases, the ultimate strain decreases. This relationship is further discussed under Assumption (10.2.3).

The assumed strain condition at ultimate strength is illustrated in Fig. 5-2. Both the strain in the reinforcement and in the concrete are directly proportional to the distance from the neutral axis. Actually this assumption applies for the full range of loading--zero to ultimate. As shown in Fig. 5-2, this assumption is of primary importance in design for determining the strain (and corresponding stress) in the reinforcement.

Assumption (10.2.3)
"Maximum usable strain at extreme concrete compression fiber shall be assumed equal to $\epsilon_{cu} = 0.003$."

10.2.3 - The maximum concrete compressive strain at crushing of the concrete varies from 0.003 to as high as 0.008; however, the maximum strain for practical cases is 0.003 to 0.004. (Note stress-strain curves shown in Fig. 5-5).

The ultimate strain has been measured in many tests of both plain and reinforced concrete members. The test results of a series of reinforced concrete beam and column members are shown in Fig. 5-3.

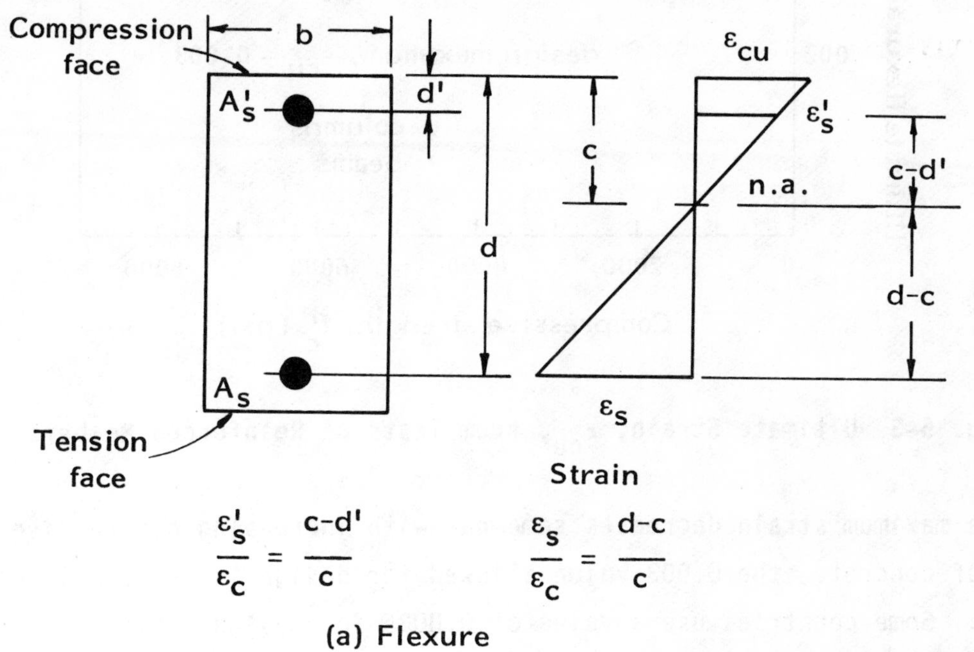

$$\frac{\varepsilon'_s}{\varepsilon_c} = \frac{c-d'}{c} \qquad \frac{\varepsilon_s}{\varepsilon_c} = \frac{d-c}{c}$$

(a) Flexure

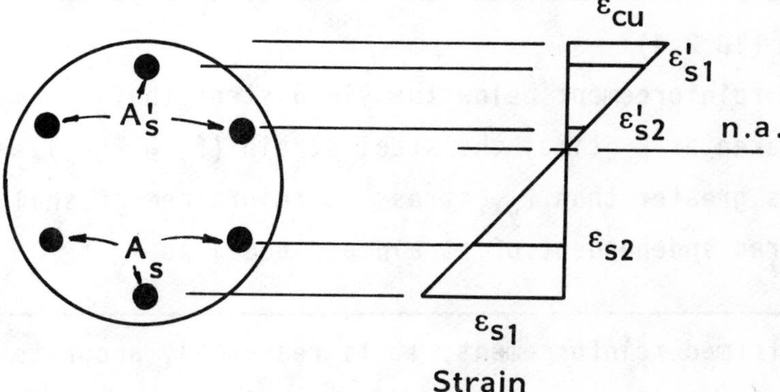

(b) Flexure and axial load

Fig. 5-2 Assumed Strain Variation

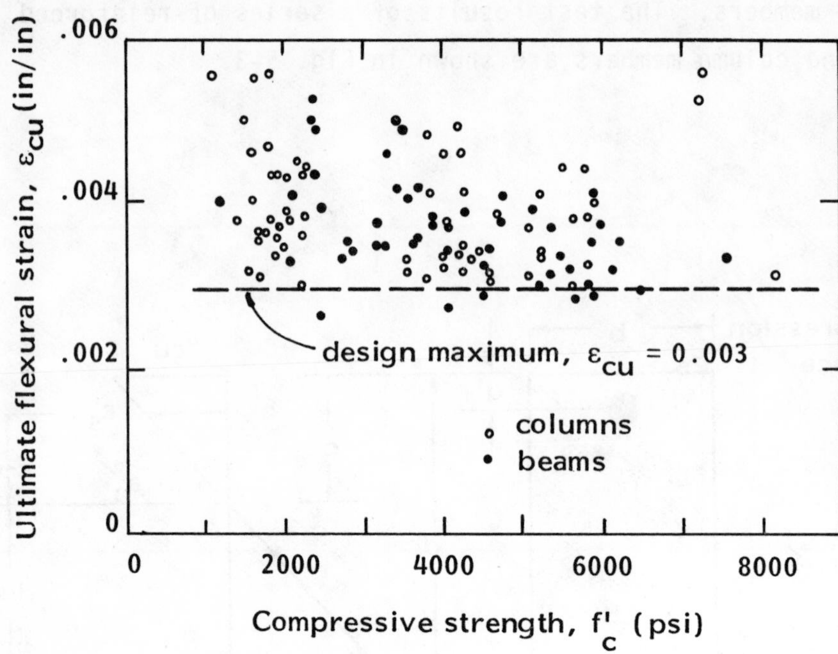

Fig. 5-3 Ultimate Strain, ε_{cu}, from Tests of Reinforced Members

Though the maximum strain decreases somewhat with increasing compressive strength of concrete, the 0.003 value allowed for design is reasonably conservative. Some countries use a value of 0.0035 for design, which makes little difference in the computed strength of a member.

> **Assumption (10.2.4)**
> "Stress in reinforcement below the yield strength f_y shall be taken as E_s times the steel strain ($f_s = E_s \varepsilon_s$). For strains greater than f_y, stress in reinforcement shall be considered independent of strain and equal to f_y."

10.2.4 - For deformed reinforcement, it is reasonably accurate to assume that the stress in reinforcement is proportional to strain below the yield strength, f_y. For practical design, the increase in strength due to the effect of strain hardening of the reinforcement is neglected for strength computations. The actual vs. the design (trapezoidal) stress-strain relationship is shown in Fig. 5-4.

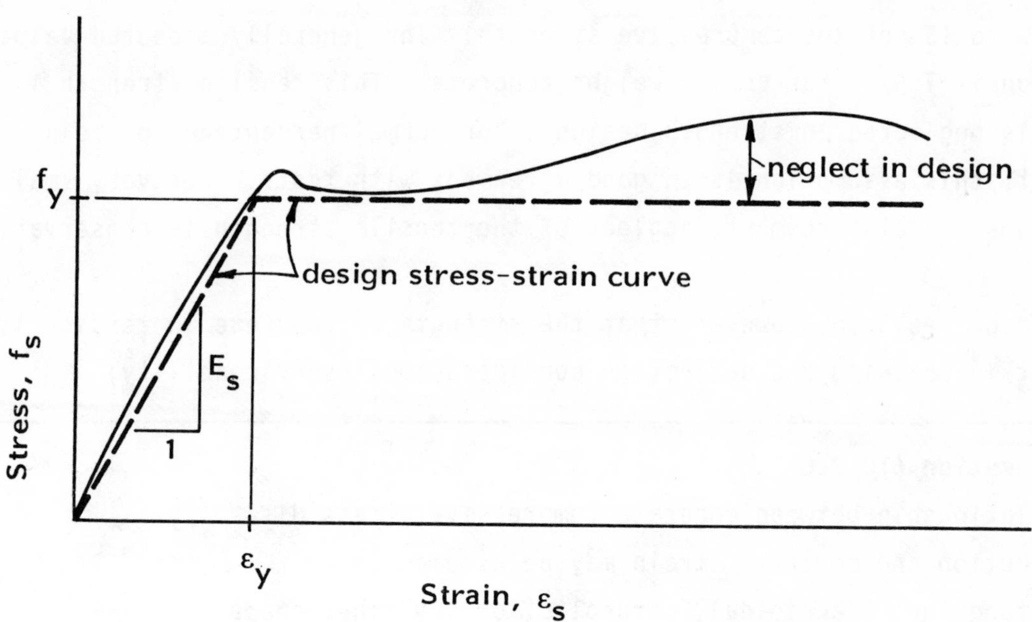

Fig. 5-4 Stress-Strain Relationship for Reinforcement

In strength computations, the force developed in tensile or compressive reinforcement is computed as,

when $\varepsilon_s \leq \varepsilon_y$ (yield strain)

$$A_s f_s = A_s E_s \varepsilon_s$$

when $\varepsilon_s \geq \varepsilon_y$

$$A_s f_s = A_s f_y$$

where ε_s is the value from the strain diagram at the location of the reinforcement (see Fig. 5-2). For design, the modulus of elasticity of steel reinforcement, E_s, may be taken as 29,000,000 psi. (Section 8.5.2.)

Assumption (10.2.5)
"Tensile strength of concrete shall be neglected in flexure calculations of reinforced concrete."

10.2.5 - The tensile strength of concrete in flexure, known as modulus of rupture, is a more variable property than the compressive strength and is about 10% to 15% of the compressive strength. The generally accepted value for design is $7.5\sqrt{f'_c}$ for normal-weight concrete. This tensile strength in flexure is neglected in strength design. For normal percentages of reinforcement, this assumption is in good agreement with tests. For very small percentages of reinforcement, neglect of the tensile strength is conservative.

It should be realized, however, that the strength of concrete in tension is important in cracking and deflection considerations (serviceability).

Assumption (10.2.6)

"Relationship between concrete compressive stress distribution and concrete strain may be assumed to be rectangular, trapezoidal, parabolic, or any other shape that results in prediction of strength in substantial agreement with results of comprehensive tests."

10.2.6 - This assumption recognizes the inelastic stress distribution of concrete at high stress. As maximum stress is approached, the stress-strain relationship for concrete is not a straight line but some form of a curve (stress is not proportional to strain). The general stress-strain behavior is shown in Fig. 5-5. The shape of the curves is primarily a function of concrete strength and consists of a rising curve from zero to a maximum at a compressive strain between 0.0015 and 0.002 followed by a descending curve to an ultimate strain (crushing of the concrete) from 0.003 to as high as 0.008. As discussed under Assumption (10.2.3), the Code sets the maximum usable strain at 0.003 for design. The curves show that the stress-strain behavior for concrete is nonlinear at stress levels above about 0.5 f'_c.

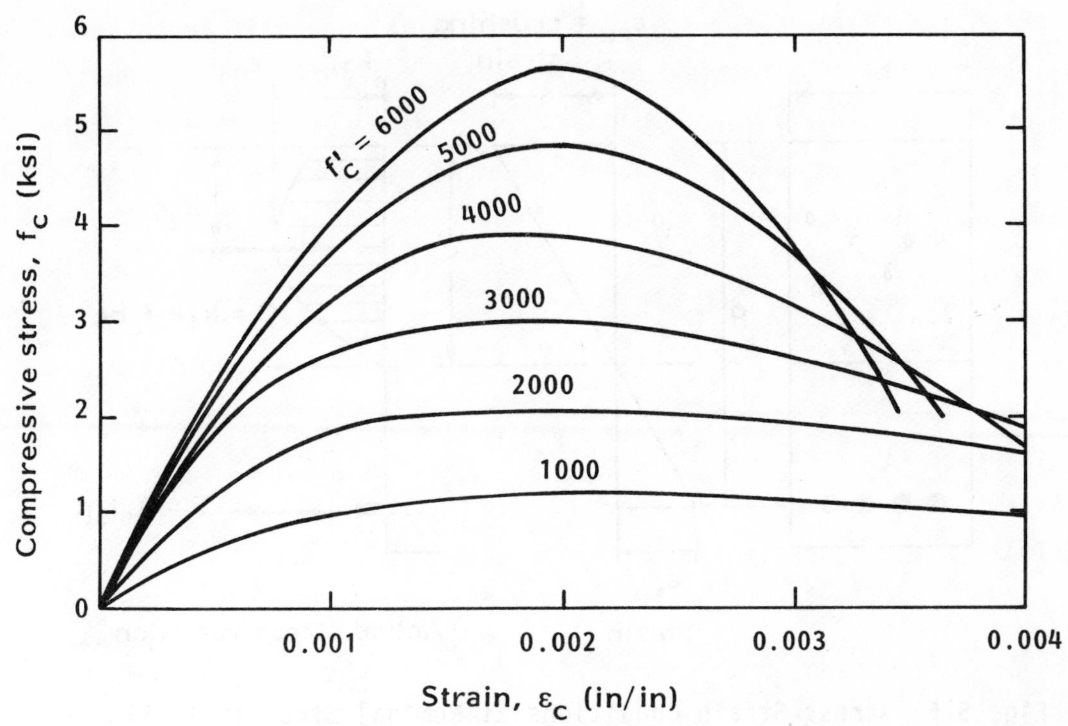

Fig. 5-5 Typical Stress-Strain Curves for Concrete

The actual distribution of concrete compressive stress in a practical case is complex and usually not known. However, research has shown that the important properties of the concrete stress distribution can be approximated closely using any one of several different assumptions as to the form of stress distribution. Many stress distributions have been proposed (see Fig. 5-1). The three most common are the parabola, trapezoid, and rectangle. All yield reasonable results, but a parabolic stress distribution conforms more closely to the actual variation of stress and strain as shown in Fig. 5-5. At the theoretical strength of a member in flexure, the general form of the compressive stress distribution should have a similar stress variation, as shown in Fig. 5-6. The maximum stress is indicated by $k_3f'_c$, the average stress is indicated by $k_1k_3f'_c$, and the centroid of the approximate parabolic distribution by k_2c, where c is the neutral axis location.

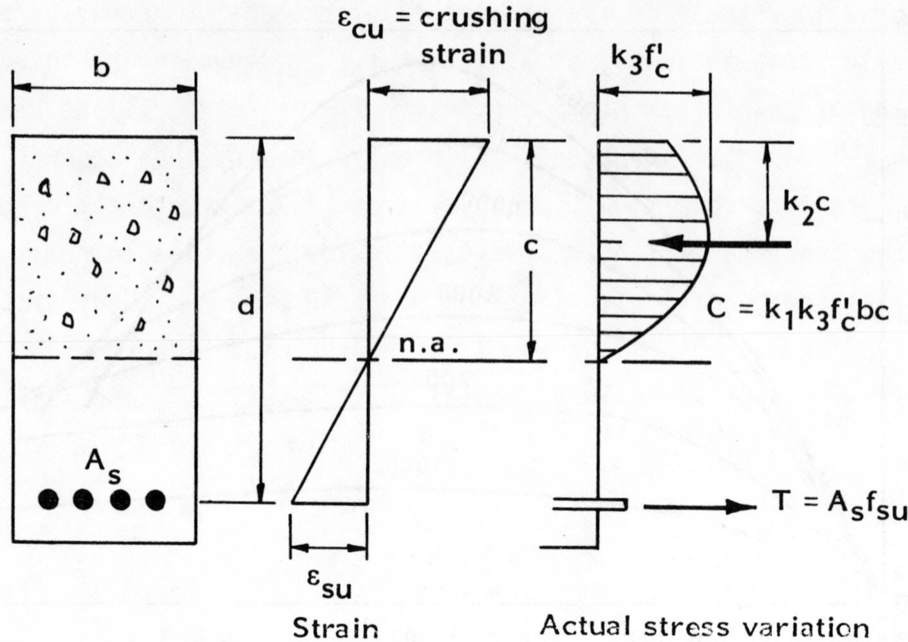

Fig. 5-6 Stress-Strain Conditions at Nominal Strength in Flexure

For the stress conditions at ultimate, the nominal moment strength, M_n, may be computed by equilibrium of forces and moments.

For force equilibrium:

$$C = T$$

$$k_1 k_3 f'_c bc = A_s f_{su}$$

$$c = \frac{A_s f_{su}}{k_1 k_3 f'_c b}$$

For moment equilibrium:

$$M_n = (C \text{ or } T)(d - k_2 c)$$

$$M_n = A_s f_{su} \left(d - \frac{k_2}{k_1 k_3} \frac{A_s f_{su}}{f'_c b}\right) \qquad (1)$$

The maximum strength is assumed to be reached when the strain in the extreme compression fiber is equal to the crushing strain of the concrete, ϵ_{cu}. When crushing occurs, the strain in the tension reinforcement, ϵ_{su}, may be either larger or smaller than the yield strain, $\epsilon_y = f_y/E_s$, depending on the relative proportion of reinforcement to concrete. If the reinforcement amount is low enough, yielding of the steel will occur prior to crushing of the concrete (ductile failure condition). With a larger quantity of reinforcement, crushing of the concrete will occur first, allowing the steel to remain elastic (brittle failure condition). The Code has provisions which are intended to ensure a ductile mode of failure by limiting the amount of tension reinforcement. For the ductile failure condition, f_{su} equals f_y and Equation (1) becomes:

$$M_n = A_s f_y \left(d - \frac{k_2}{k_1 k_3} \frac{A_s f_y}{f'_c b} \right) \qquad (2)$$

If the quantity $k_2/(k_1k_3)$ is known, the moment strength can be computed directly from Equation (2). It is not necessary to know values for k_1, k_2, and k_3 individually. Values for the combined term, as well as the individual k values, have been established from tests and are shown in Fig. 5-7. As shown in the figure, $k_2/(k_1k_3)$ varies from about 0.55 to 0.63.

The Portland Cement Association has adopted the parabolic stress-strain relationship shown in Fig. 5-8 for much of its experimental and analytical research work. All PCA published strength design aids and computer programs are based on the parabolic stress variation shown. Such "more exact" stress distributions have their greatest application with electronic computers and are not recommended for longhand calculations.

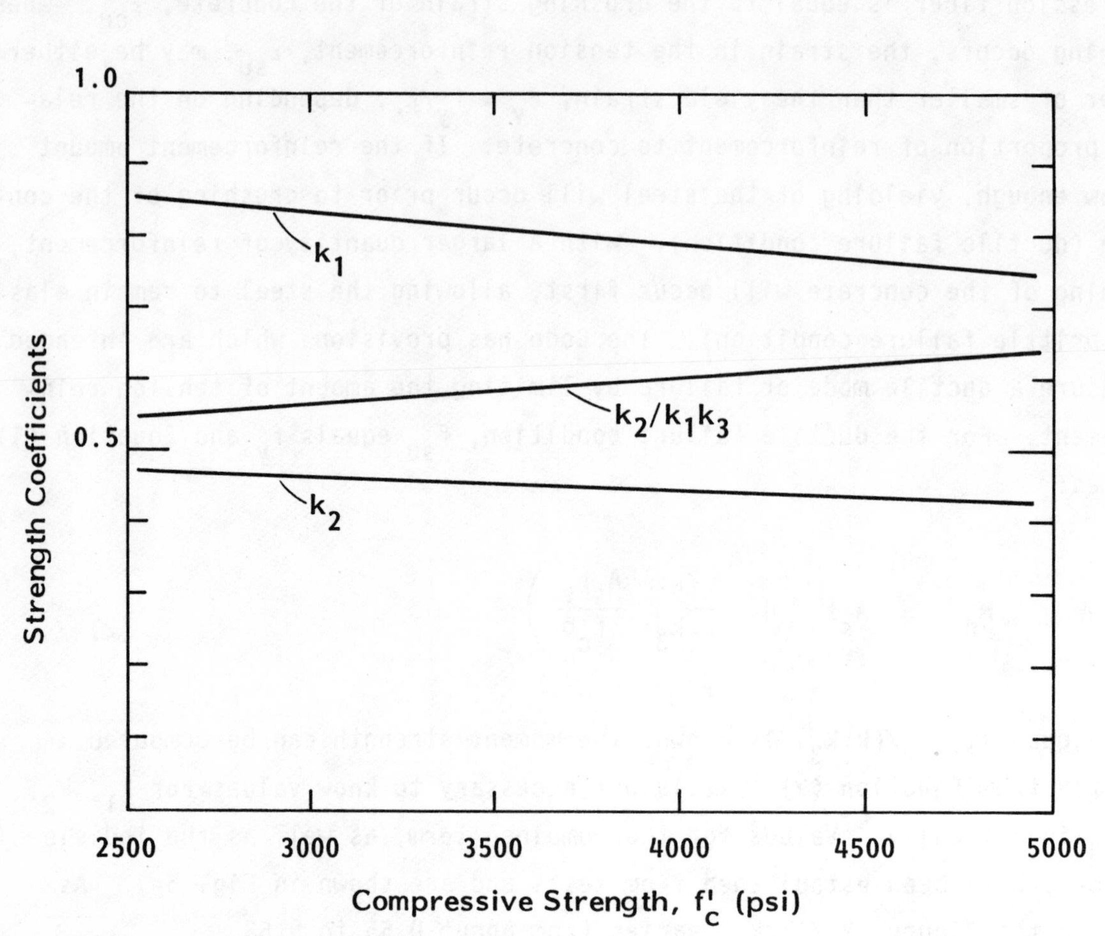

Fig. 5-7 "Stress-Block" Parameters

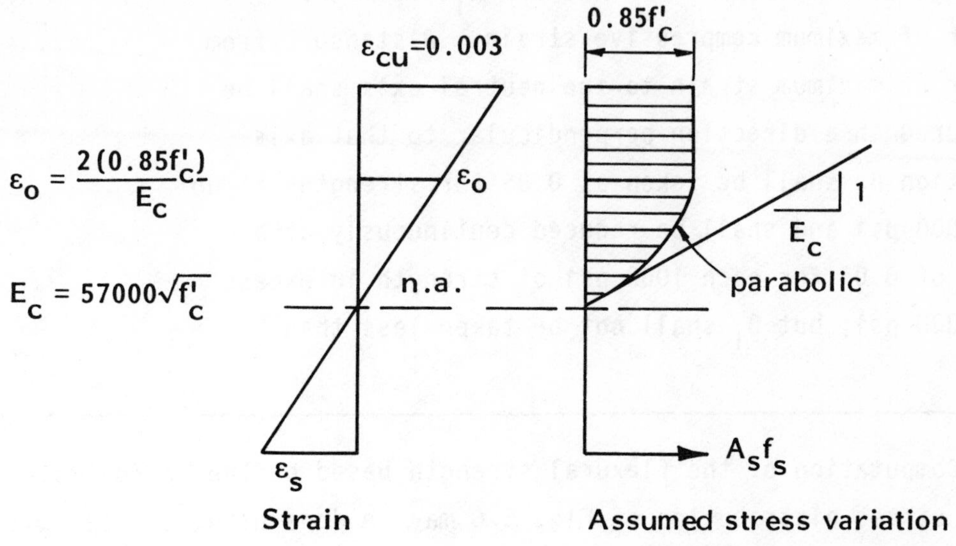

Concrete

$0 < \varepsilon_c < \varepsilon_o \quad f_c = 0.85f'_c\left[2\left(\frac{\varepsilon_c}{\varepsilon_o}\right) - \left(\frac{\varepsilon_c}{\varepsilon_o}\right)^2\right]$

$\varepsilon_c > \varepsilon_o \quad f_c = 0.85f'_c$

Steel

$\varepsilon_s \leq \varepsilon_y \quad f_s = \varepsilon_s E_s$

$\varepsilon_s > \varepsilon_y \quad f_s = f_y$

$E_s = 29{,}000{,}000$

Fig. 5-8 PCA Stress-Strain Relationship

Assumption (10.2.7)

"Requirements of Assumption (10.2.6) may be considered satisfied by an equivalent rectangular concrete stress distribution defined as follows: A concrete stress of 0.85 f'_c shall be assumed uniformly distributed over an equivalent compression zone bounded by edges of the cross section and a straight line located parallel to the neutral axis at a distance $a = \beta_1 c$ from the fiber of maximum compressive strain. Distance c from fiber of maximum strain to the neutral axis shall be measured in a direction perpendicular to that axis. Fraction β_1 shall be taken as 0.85 for strengths f'_c up to 4000 psi and shall be reduced continuously at a rate of 0.05 for each 1000 psi of strength in excess of 4000 psi, but β_1 shall not be taken less than 0.65."

10.2.7 - Computation of the flexural strength based on the approximate parabolic stress distribution of Fig. 5-6 may be done using Eq. (2) with given values of $k_2/(k_1k_3)$. However, for practical design purposes, a method based on simple static equilibrium is desirable. The Code allows the use of a rectangular compressive "stress block" to replace the more exact stress distribution of Fig. 5-6 (or Fig. 5-8). In this equivalent rectangular stress block, as shown in Fig. 5-9, an average stress of 0.85 f'_c is used with a rectangle of depth $a = \beta_1 c$, determined so that $a/2 = k_2 c$. A β_1 of 0.85 for concrete with $f'_c < 4000$ psi and 0.05 less for each 1000 psi of f'_c in excess of 4000 was determined experimentally to agree with test data. For high-strength concretes, above 8000 psi, a lower limit of 0.65 is placed on the β_1 factor. Variation in β_1 vs. concrete strength f'_c is shown in Fig. 5-10. Effect of the β_1 limit of 0.65 for high-strength concretes is illustrated in Example 5.1.

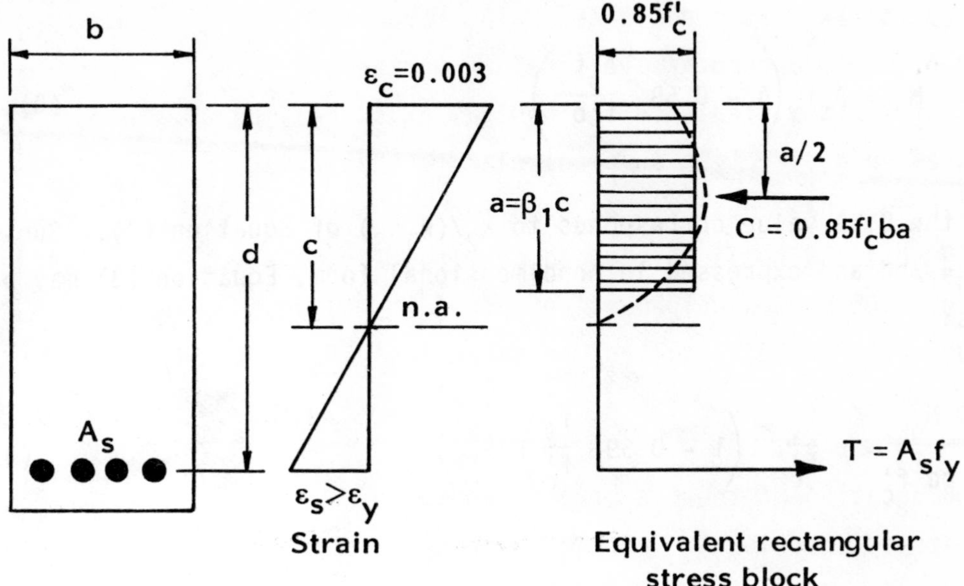

Fig. 5-9 Equivalent Rectangular Concrete Stress Distribution

Using the equivalent rectangular stress distribution, and assuming that the reinforcement yields prior to crushing of the concrete ($\epsilon_s > \epsilon_y$), the nominal moment strength M_n may be computed by equilibrium of forces and moments.

For force equilibrium:

$$C = T$$

$$0.85\ f'_c ba = A_s f_y$$

$$a = \frac{A_s f_y}{0.85\ f'_c b}$$

For moment equilibrium:

$$M_n = (C \text{ or } T)(d - a/2)$$

$$M_n = A_s f_y (d - a/2)$$

substituting "a" from force equilibrium,

$$M_n = A_s f_y \left(d - 0.59 \ \frac{A_s f_y}{f'_c b} \right) \qquad (3)$$

Note that the 0.59 value corresponds to $k_2/(k_1 k_3)$ of Equation (2). Substituting $A_s = \rho bd$ and expressed in nondimensional form, Equation (3) may be written as:

$$\frac{M_n}{bd^2 f'_c} = \rho \ \frac{f_y}{f'_c} \ \left(1 - 0.59 \rho \ \frac{f_y}{f'_c} \right) \qquad (4)$$

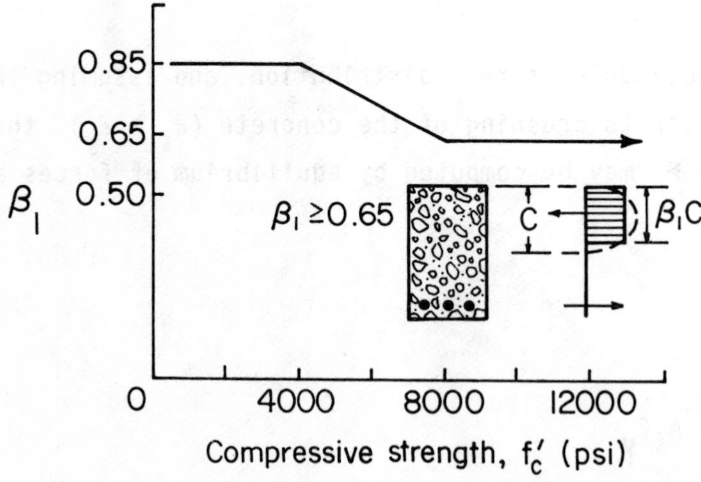

Fig. 5-10 Strength Factor β_1

As shown in Fig. 5-11, Equation (4) is "in substantial agreement with the results of comprehensive tests." It must, however, be realized that the rectangular stress distribution does not represent the actual stress distribution in the compression zone at ultimate, but does provide essentially the same results as those obtained in tests.

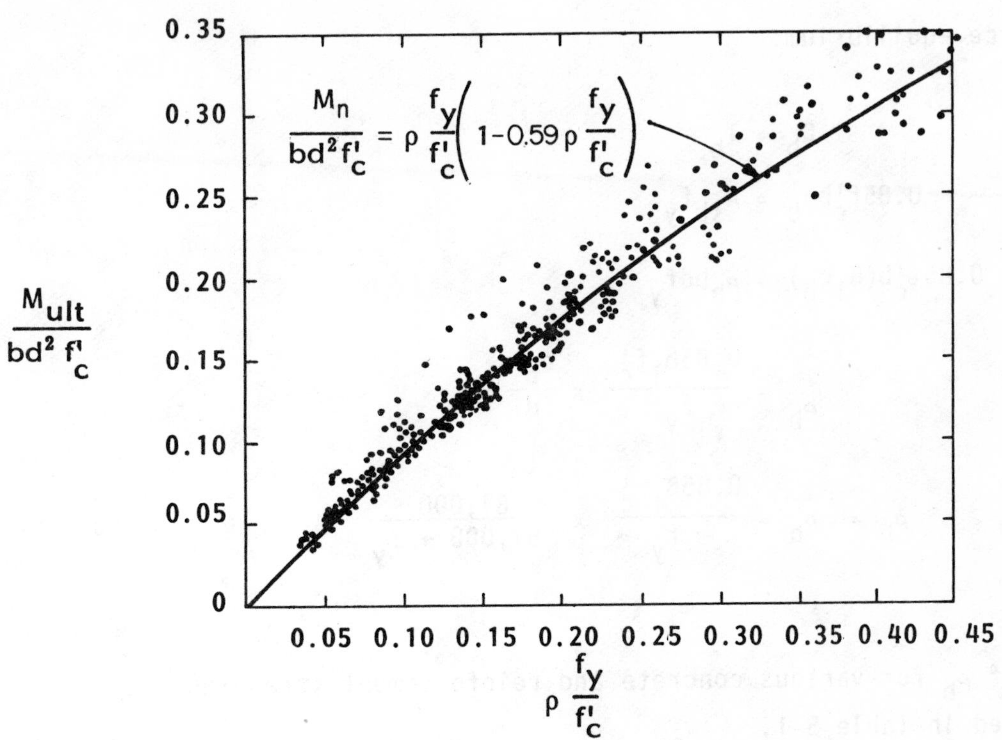

Fig. 5-11 Tests of 364 Beams Controlled by Tension ($\varepsilon_s > \varepsilon_y$)

10.3 General Principles and Requirements

10.3.2 - A <u>balanced strain condition</u> exists at a cross section when the maximum strain at the extreme compression fiber just reaches $\varepsilon_{cu} = 0.003$ simultaneously with the first yield strain of $\varepsilon_s = \varepsilon_y = f_y/E_s$ in the tension reinforcement. This "balanced" strain condition is shown in Fig. 5-12.

The required reinforcement ratio, ρ_b, to produce a balanced condition for a rectangular section with tension reinforcement only may be obtained by applying the equilibrium and compatibility condition. Referring to Fig. 5-12,

For the linear strain condition:

$$\frac{c_b}{d} = \frac{\varepsilon_c}{\varepsilon_c + \varepsilon_y}$$

$$= \frac{0.003}{0.003 + f_y/29{,}000{,}000} = \frac{87{,}000}{87{,}000 + f_y}$$

For force equilibrium:

$$C_b = T_b$$

$$0.85f'_c b a_b = A_{sb} f_y$$

$$0.85f'_c b(\beta_1 c_b) = \rho_b b d f_y$$

$$\rho_b = \frac{0.85\beta_1 f'_c}{f_y} \times \frac{c_b}{d}$$

$$\rho_b = \overline{\rho_b} = \frac{0.85\beta_1 f'_c}{f_y} \times \frac{87{,}000}{87{,}000 + f_y} \qquad \text{Eq. (8-1)}$$

Value of ρ_b for various concrete and reinforcement strengths are tabulated in Table 5-1.

Table 5-1. Balanced Ratio of Reinforcement ρ_b for Rectangular Sections with Tension Reinforcement Only

f_y	$f'_c = 3000$ $\beta_1 = 0.85$	$f'_c = 4000$ $\beta_1 = 0.85$	$f'_c = 5000$ $\beta_1 = 0.80$	$f'_c = 6000$ $\beta_1 = 0.75$	$f'_c = 8000$ $\beta_1 = 0.65$
40,000	0.0371	0.0495	0.0582	0.0655	0.0757
60,000	0.0214	0.0285	0.0335	0.0377	0.0436

The balanced reinforcement ratio ρ_b for flanged sections and rectangular sections with compression reinforcement may be obtained by applying the equilibrium and compatibility conditions in a similar manner,

For flanged section with tension reinforcement only:

$$\rho_b = \frac{b_w}{b} (\overline{\rho_b} + \rho_f)$$

where

$$\rho_f = \frac{A_{sf}}{b_w d} \quad \text{and} \quad A_{sf} = 0.85 \frac{f'_c}{f_y} (b - b_w)\, h_f$$

For rectangular section with compression reinforcement:

$$\rho_b = \overline{\rho_b} + \rho' \frac{f'_{sb}}{f_y}$$

where

f'_{sb} = stress in compression reinforcement at balanced strain condition

$= 87{,}000 - \frac{d'}{d} (87{,}000 + f_y) \leq f_y$

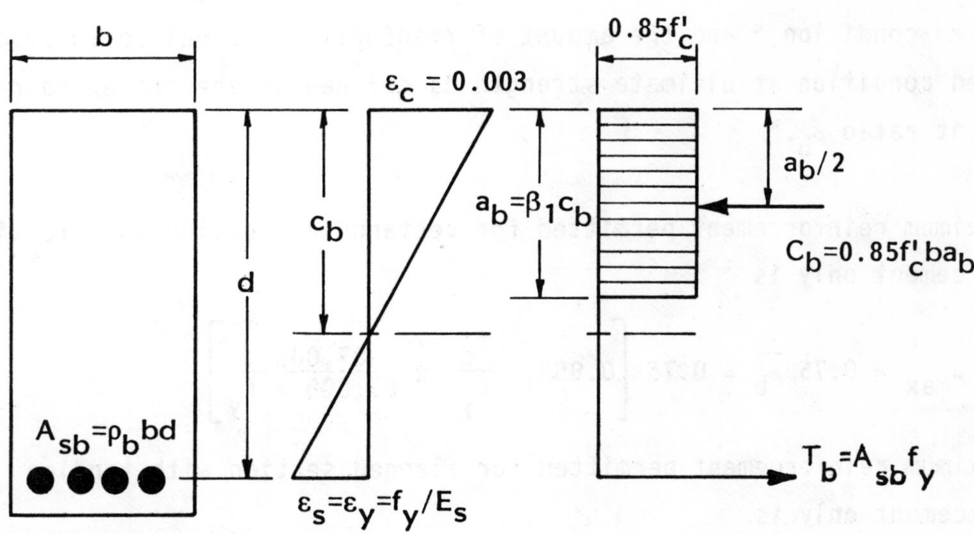

Fig. 5-12 Balanced Strain Condition in Flexure

10.3.3 - The flexural strength of a member is ultimately reached when the strain in the extreme compression fiber reaches the ultimate (crushing) strain of the concrete, ϵ_{cu}. At ultimate ϵ_{cu}, the strain in the tension reinforcement could just reach the strain at first yield ($\epsilon_s = \epsilon_y = f_y/E_s$), be less than the yield strain (elastic), or exceed the yield strain (inelastic). Which steel strain condition exists at ultimate concrete strain depends on the relative proportion of reinforcement to concrete. If the steel amount is low enough, the strain in the tension steel will greatly exceed the yield strain ($\epsilon_s \gg \epsilon_y$) when the concrete strain reaches ϵ_{cu} with large deflection and ample warning of impending failure (ductile failure condition). With a larger quantity of steel, the strain in the tension steel may not reach the yield strain ($\epsilon_s < \epsilon_y$) when the concrete strain reaches ϵ_{cu} with small deflection and little warning of impending failure (brittle failure condition). For design it is considered more conservative to restrict the ultimate strength condition so that a ductile failure mode can be expected.

The Code has provisions which are intended to ensure a ductile mode of failure by limiting the amount of tension reinforcement to 75% of the amount which will cause the strain in the tension steel to just reach yield strain at crushing strain of the concrete. This strain condition is defined as the "balanced condition," and the amount of reinforcement required to produce a balanced condition at ultimate strength is defined as the "balanced reinforcement ratio ρ_b."

The maximum reinforcement permitted for rectangular section with tension reinforcement only is

$$\rho_{max} = 0.75\ \overline{\rho_b} = 0.75 \left[0.85\beta_1\ \frac{f'_c}{f_y} \times \frac{87{,}000}{87{,}000 + f_y}\right]$$

The maximum reinforcement permitted for flanged section with tension reinforcement only is

$$\rho_{max} = 0.75 \left[\frac{b_w}{b}\left(\overline{\rho_b} + \rho_f\right)\right]$$

The maximum reinforcement permitted for a rectangular section with compression reinforcement is

$$\rho_{max} = 0.75\ \overline{\rho_b} + \rho' \frac{f'_{sb}}{f_y}$$

Note that with compression reinforcement, the portion of ρ_b contributed by compression reinforcement ($\rho' f'_{sb}/f_y$) need not be reduced by the 0.75 factor. For ductile behavior of beams with compression reinforcement, only that portion of the total tension steel balanced by compression in the concrete ($\overline{\rho_b}$) need be limited. See Example 5.2.

It should be realized that the limit on the amount of tension reinforcement for flexural members is a specification-defined limitation for ductile behavior. Tests have shown that beams reinforced with a computed amount of balanced reinforcement actually behave in a ductile manner with gradually increasing deflections and cracking up to ultimate failure. Sudden compression failures do not occur until the amount of reinforcement is considerably higher than the computed balanced amount.

One reason for the increased ductile behavior above the computed condition is the limit on the ultimate concrete strain assumed at $\varepsilon_{cu} = 0.003$ for design. The actual maximum strain based on physical testing may be much higher than this value. The 0.003 value serves as a lower bound on limiting strain. Note discussion under Assumption (10.2.3). Unless unusual amounts of ductility are required, the $0.75\rho_b$ limitation will provide ample ductile behavior for most designs.

10.3.6 - Strength of a member or cross section subject to combined flexure and axial load, M_n and P_n, must satisfy the same two conditions as required for a member subject to flexure only, (1) static equilibrium and (2) compatibility of strains. Equilibrium between the compressive and tensile forces includes the addition of the axial load P_n acting on the cross section. The general form of the stress and strain conditions at nominal strength of a member under combined flexure and axial compression is shown in Fig. 5-13. The tensile or compressive force developed in the reinforcement is determined from the strain condition at the location of the reinforcement.

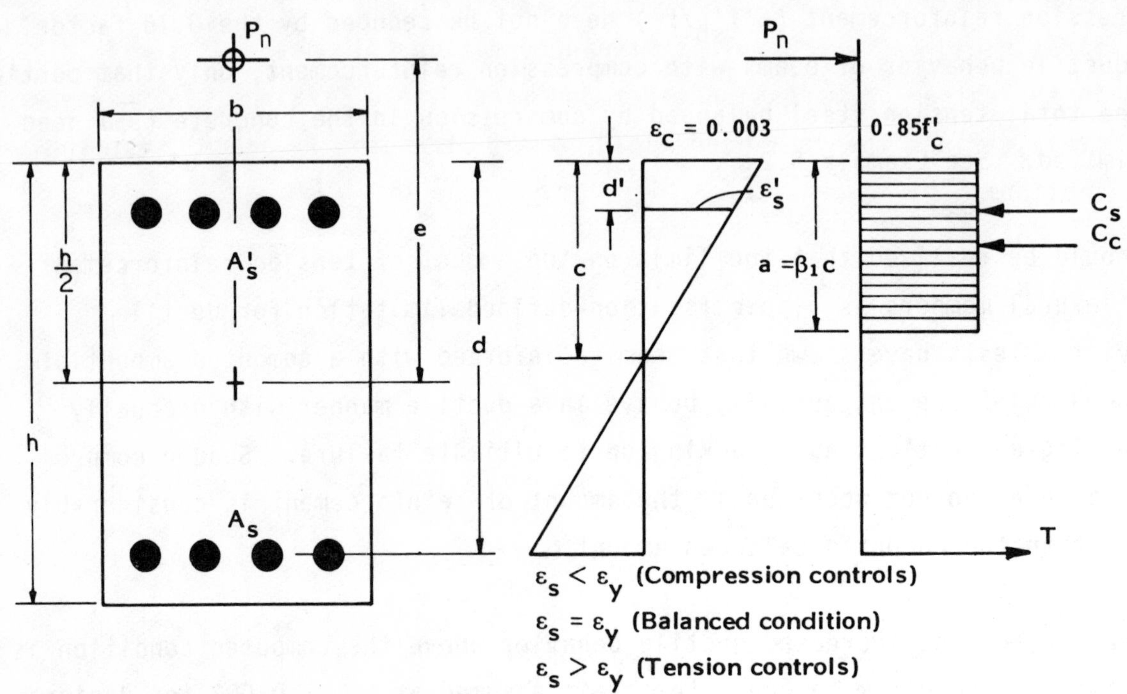

Fig. 5-13 Stress-Strain Conditions for Combined Flexure and Axial Load

Referring to Fig. 5-13,

For A'_s: $C_s = A'_s f'_s = A'_s(E_s \varepsilon'_s)$ when $\varepsilon'_s < \varepsilon_y$ (yield strain)

or $C_s = A'_s f_y$ when $\varepsilon'_s \geq \varepsilon_y$

For A_s: $T = A_s f_s = A_s(E_s \varepsilon_s)$ when $\varepsilon_s < \varepsilon_y$

or $T = A_s f_y$ when $\varepsilon_s \geq \varepsilon_y$

The combined load-moment strength (P_n and M_n) may be computed by equilibrium of forces and moments.

For force equilibrium:

$$P_n = C_c + C_s - T$$

where $C_c = 0.85f'_c ba$

For moment equilibrium about the mid-depth of the section:

$$M_n = P_n e = C_c(\frac{h}{2} - \frac{a}{2}) + C_s(\frac{h}{2} - d') + T(d - \frac{h}{2})$$

For a known strain condition, the corresponding load-monent strength, P_n and M_n, can be computed directly. Assume the strain in the tension steel, A_s, is at first yield ($\varepsilon_s = \varepsilon_y$). This strain condition (simultaneous strain of 0.003 in the extreme compression fiber and first yield strain ε_y in the tension steel) defines the "balanced" load-moment strength, P_b and M_b, for the cross section.

For the linear strain conditions:

$$\frac{c_b}{d} = \frac{\varepsilon_c}{\varepsilon_c + \varepsilon_y}$$

$$= \frac{0.003}{0.003 + f_y/29{,}000{,}000} = \frac{87{,}000}{87{,}000 + f_y}$$

$$a_b = \left(\frac{87{,}000}{87{,}000 + f_y}\right) \beta_1 d$$

and

$$\frac{c_b}{c_b - d'} = \frac{\varepsilon_c}{\varepsilon'_s}$$

$$\varepsilon'_s = 0.003 \left(1 - \frac{d'}{c_b}\right)$$

$$\varepsilon'_s = 0.003 \left[1 - \frac{d'}{d}\left(\frac{87{,}000 + f_y}{87{,}000}\right)\right]$$

$$f'_{sb} = E_s\varepsilon'_s = 87{,}000\left[1 - \frac{d'}{d}\left(\frac{87{,}000 + f_y}{87{,}000}\right)\right]$$ but not greater than f_y

For force equilibrium:

$$P_b = 0.85f'_c ba_b + A'_s f'_{sb} - A_s f_y$$

For moment equilibrium:

$$M_b = P_b e_b = 0.85\ f'_c b a_b \left(\frac{h}{2} - \frac{A_b}{2}\right) +$$

$$A'_s f'_{sb} \left(\frac{h}{2} - d'\right) + A_s f_y \left(d - \frac{h}{2}\right)$$

The "balanced" load-moment strength of a cross section defines only one of many load-moment combinations possible to define the full range of the load-moment interaction relationship for members subject to combined flexure and axial load. The general form of a strength interaction diagram is shown in

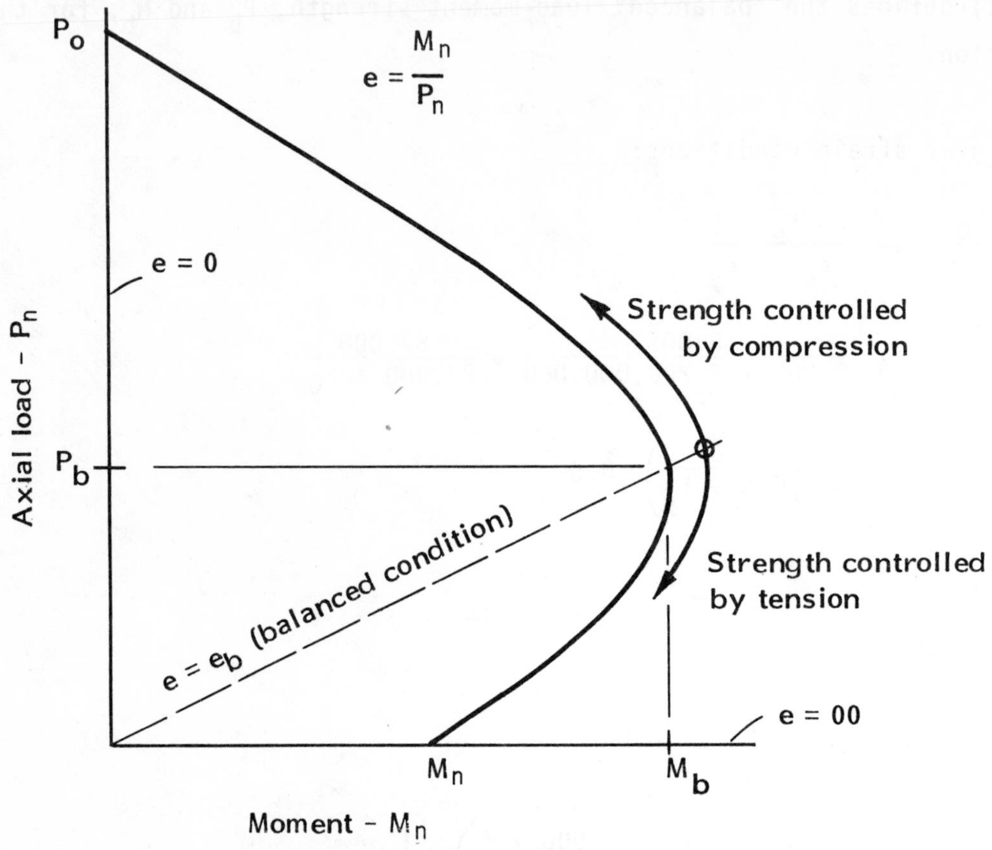

Fig. 5-14 Load-Moment Interaction Strength

Fig. 5-14. The load-moment combination may be such that compression exists over most or all of the section so that the compressive strain in the concrete reaches 0.003 before the tension steel yields ($\varepsilon_s < \varepsilon_y$), known as the "compression controls" segment; or the load combination may be such that tension exists over a large portion of the section so that the strain in the tension steel is greater than the yield strain ($\varepsilon_s > \varepsilon_y$) when the compressive strain in the concrete reaches 0.003, known as the "tension controls" segment. The "balanced" strain condition ($\varepsilon_s = \varepsilon_y$) divides these two segments of the strength curve.

The linear strain variation for the full range of the load-moment interaction strength is illustrated in Fig. 5-15.

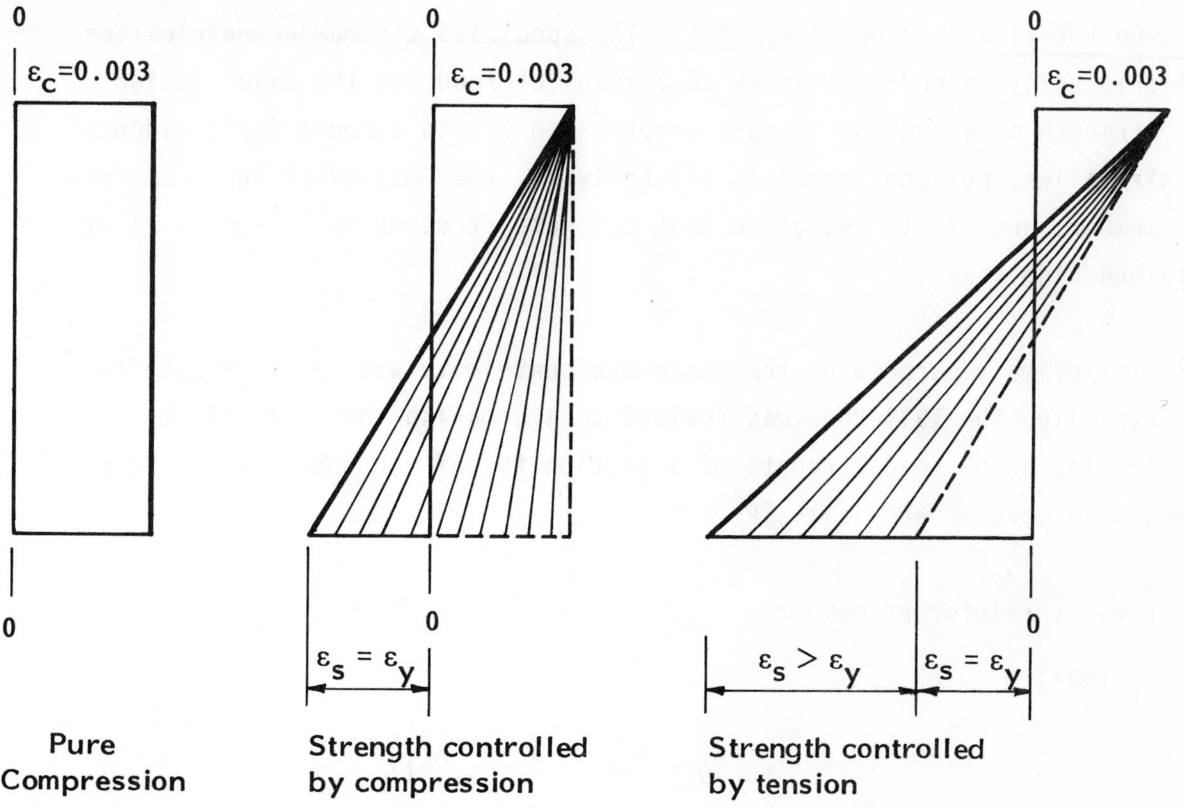

Fig. 5-15 Strain Variation for Full Range of Load-Moment Interaction

At pure compression, the strain is uniform over the entire cross section and equal to 0.003. With increasing load eccentricity (moment), the compressive strain at the "tension face" gradually decreases to zero, then becomes tension, until the tensile strain in the steel most distant from the neutral axis reaches the yield strain ($\varepsilon_s = \varepsilon_y$) at the balanced strain condition. For this range of strain variation, the strength of the section is controlled by compression (0.003 to $\varepsilon_s = \varepsilon_y$). Beyond the balanced strain condition, the steel strain gradually increases ($\varepsilon_s \gg \varepsilon_y$) to the location of pure flexure with infinite load eccentricity ($e = \infty$). For this range of strain variation, strength is controlled by tension ($\varepsilon_s > \varepsilon_y$). With increasing eccentricity, more and more tension exists over the cross section. Each of the many possible strain conditions illustrated in Fig. 5-15 describe a point, P_n and M_n, on the load-moment curve.

10.3.5 - Prior to the 1977 ACI Code, all compression members were required to be designed for a minimum eccentricity of 0.05h for spirally reinforced or 0.10h for tied reinforced members. The specified minimum eccentricities were originally intended to serve as a means of reducing the axial design load strength of a section in pure compression (1) to account for accidental eccentricities, not considered in the analysis, that may exist in a compression member, and (2) to recognize that concrete strength is less than f'_c at sustained high loads.

Since the primary purpose of the minimum eccentricity was to limit pure axial load capacity, the 1977 Code was revised to accomplish this directly by limiting the axial load strength of a section to 85% and 80% of axial load strength at zero eccentricity, P_o.

For spirally reinforced members,

$$P_n\ (\max) = 0.85\ P_o$$

$$= 0.85\ [0.85\ f'_c\ (A_g - A_{st}) + f_y\ A_{st}]$$

For tied reinforced members,

$$P_n\ (\max) = 0.80\ P_o$$

$$= 0.80\ [0.85\ f'_c\ (A_g - A_{st}) + f_y\ A_{st}]$$

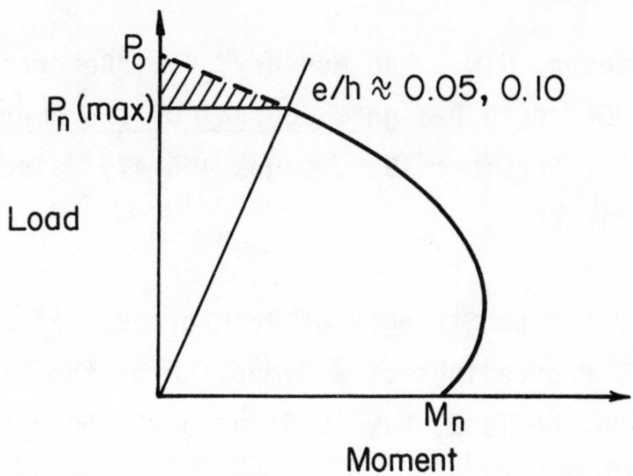

Fig. 5-16 Load-Moment Interaction Strength (Section 10.3.5)

The limiting axial load strength, P_n (max), is illustrated in Fig. 5-16. In essence, design within the cross-hatched portion of the load-moment interaction strength is not permitted. The 85% and 80% values approximate the axial load strengths at e/h ratios of 0.05 and 0.10, respectively. (See Example 5.3.) The designer should note, however, that Code Commentary Section 10.3.5 states that "Design Aids and computer programs based on the minimum eccentricity requirement of the 1963 and 1971 ACI Building Codes may be considered equally applicable for usage."

With deletion of the minimum eccentricity requirement for compression members, a new minimum moment for approximating slenderness effects in compression members was added (Section 10.11.5.4). If factored column moments are very small or zero, the design of slender columns must be based on a minimum moment of $P_u(0.6 + 0.003h)$, where h is the overall thickness of the compression member.

Selected References

5.1 Hognestad, E., Hanson, N.W., and McHenry, D., "Concrete Stress Distribution in Ultimate Strength Design", Journal of the American Concrete Institute, Vol. 52, December 1955, pages 455-479; also PCA Development Department Bulletin D6.

5.2 Hognestad, E., "Ultimate Strength of Reinforced Concrete in American Design Practice", Proceedings of a Symposium on the Strength of Concrete Structures, London, England, May 1955; also PCA Development Department Bulletin D12.

5.3 Hognestad, E., "Confirmation of Inelastic Stress Distribution in Concrete", Journal of Structural Division, Proceedings of the American Society of Civil Engineers, Vol. 83, Proc. Paper 1189, March 1957; also PCA Development Department Bulletin D15.

5.4 Mattock, A.H., Kriz, L.B., and Hognestad, E., "Rectangular Concrete Stress Distribution in Ultimate Strength Design", Journal of the American Concrete Institute, Vol. 57, February 1961, pages 875-928, also PCA Development Department Bulletin D49.

5.5 Wang, C.K., and Salmon, C.G., Reinforced Concrete Design, Third Edition, Harper & Row Publishers, New York, N.Y. 1979, Chap. 1, 2, and 3.

EXAMPLE 5.1 - Strength Factor β_1 for High-Strength Concretes, $f'_c \geq 8000$ psi

The load moment interaction strength diagram for a 20x20-in. column section with four #18 bars (ρ = 4%)and f'_c = 12,000 psi and f_y = 60,000 psi is shown below.

The interaction strength is plotted using two β_1 strength factors as follows:

(1) $\beta_1 = 0.85 - 0.05\left(\dfrac{f'_c - 4000}{1000}\right)$

for f'_c = 12,000, β_1 = 0.45

(2) $\beta_1 = 0.85 - 0.05\left(\dfrac{f'_c - 4000}{1000}\right)$ <u>but not less than 0.65</u>

for f'_c = 12,000, β_1 = 0.65

The lower limit of 0.65 for β_1 primarily affects members subject to axial load plus bending within the intermediate range of the load-moment interaction strength, where interaction strength is controlled by compression.

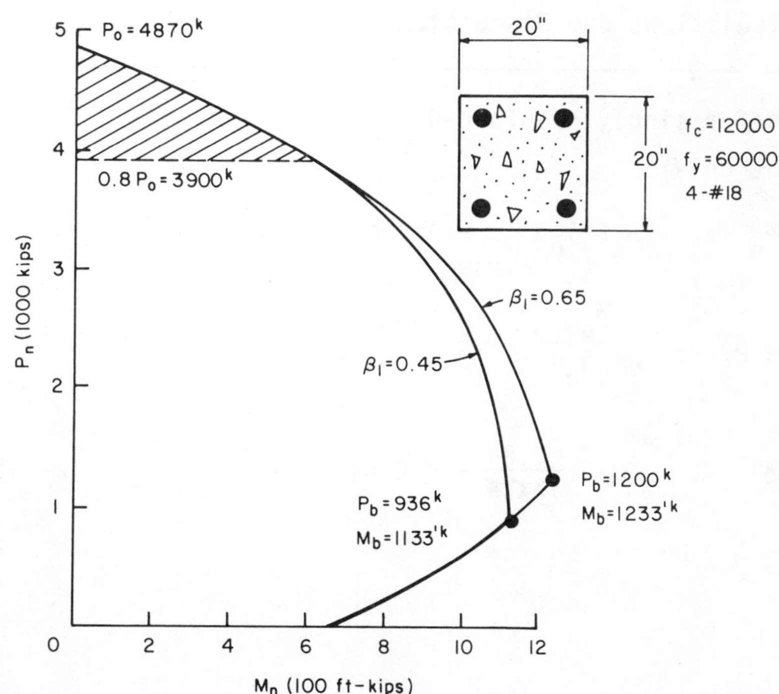

EXAMPLE 5.2 - Definition of ρ_{max} for Beams with Compression Reinforcement

A beam cross section is limited to the size shown. Determine required area of reinforcement for a factored moment M_u = 720 ft kips.

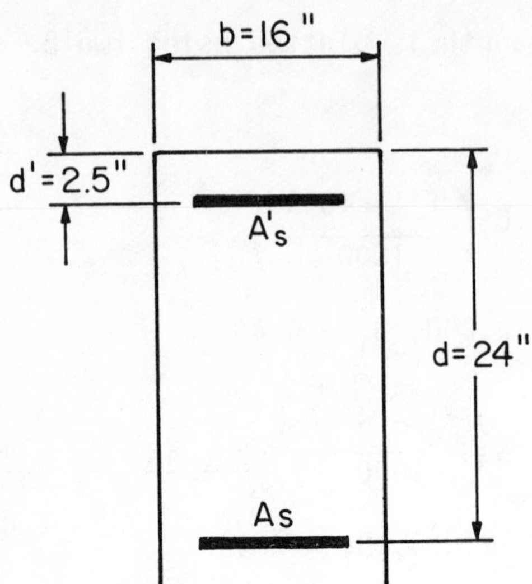

f'_c = 3000 psi
f_y = 60,000 psi

Calculations and Discussion	Code Reference

1. Check design for singly reinforced beam (tension reinforcement only).

$M_u \leq \varphi M_n = \varphi A_s f_y j_u d$ Assume $j_u = 0.85$

$A_s \text{ (req'd)} = \dfrac{M_u}{\varphi f_y j_u d} = \dfrac{720 \times 12}{0.90 \times 60 \times 0.85 \times 24} = 7.84 \text{ in.}^2$

$\rho\text{(req'd)} = \dfrac{A_s}{bd} = \dfrac{7.84}{16 \times 24} = 0.0204$ 10.0

EXAMPLE 5.2 - Continued

Calculations and Discussion	Code Reference

With tension reinforcement only, $\rho_{max} = 0.75\ \overline{\rho_b}$ 10.3.3

$$\overline{\rho_b} = \frac{0.85\ f'_c\ \beta_1}{f_y} \times \frac{87{,}000}{87{,}000 + f_y}$$

$$= \frac{0.85 \times 3 \times 0.85}{60} \times \frac{87{,}000}{87{,}000 + 60{,}000} = 0.0213$$

ρ_{max} = 0.75 (0.0213) = 0.0160

0.0204 > 0.0160 (compression reinforcement required)

Try A_s = 8 #9 bars = 8.00 in.^{2}

ρ = 8.00/16x24 = 0.0208

2. Design as doubly reinforced beam.

With compression reinforcement,

$$\rho_b = \overline{\rho_b} + \rho' \frac{f'_{sb}}{f_y}$$

$$\rho_{max} = 0.75\ \overline{\rho_b} + \rho' \frac{f'_{sb}}{f_y}$$ 10.3.3

Note that for members with compression reinforcement, the portion of ρ_b equalized by compression reinforcement $\rho'\ f'_{sb}/f_y$ need not be reduced by the 0.75 factor.

Solving for $\rho' = \frac{f_y}{f'_{sb}} (\rho_{max} - 0.75\ \overline{\rho_b})$

Assume compression reinforcement yields,

ρ' = 0.0208 - 0.0160 = 0.0048

A'_s = $\rho'bd$ = 0.0048x16x24 = 1.84 in.^{2}

Try A_s = 8 #9 bars = 8.00 in.^{2}

A'_s = 3 #7 bars = 1.80 in.^{2}

EXAMPLE 5.2 - Continued

Calculations and Discussion	Code Reference

3. Check compression reinforcement yield condition

$$\rho - \rho' \geq 0.85\beta_1 \frac{f'_c}{f_y} \frac{d'}{d} \left(\frac{87{,}000}{87{,}000 - f_y} \right)$$

The compression reinforcement yields as assumed; therefore use reinforcement as selected.

EXAMPLE 5.3 - Maximum Axial Strength vs. Minimum Eccentricity

For the tied reinforced column section shown, compare the nominal axial load strength P_n at 0.80 P_o with P_n at 0.1h eccentricity.

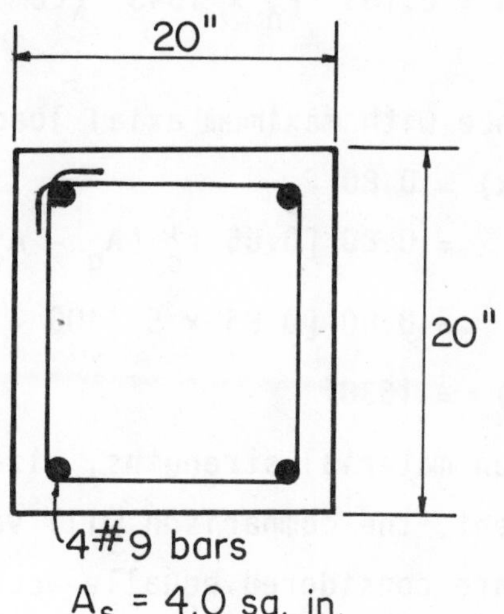

f'_c = 5000 psi

f_y = 60,000 psi

Calculations and Discussion	Code Reference
Prior to ACI 318-77, columns were required to be designed for a minimum eccentricity of 0.1h (tied) and 0.05h (spiral). This required tedious computation to find the axial load strength at these minimum eccentricities. With the 1977 ACI Code, the minimum eccentricity provision was replaced with a maximum axial load strength, 0.80 P_o (tied) and 0.85 P_o (spiral). The 80% and 85% values were chosen to approximate the axial load strengths at e/h ratios of 0.1 and 0.05, respectively.	

EXAMPLE 5.3 - Continued

Calculations and Discussion	Code Reference
1. In accordance with minimum eccentricity criterion:	
At e/h = 0.10: $P_n = 1543^k$ (computer solution)	
2. In accordance with maximum axial load strength criterion:	10.3.5.2
$P_n(max) = 0.80\ P_o$	Eq. (10-2)
$= 0.80\ [0.85\ f'_c\ (A_g - A_{st}) + f_y\ A_{st}]$	
$= 0.80\ [0.85 \times 5\ (400 - 4) + 60 \times 4.0]$	
$P(max) = 1538^k$	
Depending on material strengths, size, and amount of reinforcement, the comparison will vary slightly. Both solutions are considered equally acceptable.	

6

Distribution of Flexural Reinforcement

General Considerations

Provisions of Section 10.6 require proper distribution of tension reinforcement in beams and one-way slabs to control flexural cracking. Structures built in the past using working stress design and reinforcement with a yield strength of 40,000 psi or less had low tensile stresses in the reinforcement at service loads. Laboratory investigations have shown that cracking is generally in proportion to the steel tensile stress. Thus, with low tensile stresses in the reinforcement at service loads, these structures exhibited few flexural cracking problems.

With the advent of high-strength steels with yield stresses of 60,000 or 75,000 psi, and even higher, and the acceptance of strength design concepts where steel reinforcement is stressed to higher proportions of the yield strength, control of flexural cracking by proper reinforcing details is more significant. For example, if a beam were designed using working stress design and a steel yield strength of 40,000 psi, stress in the reinforcement at service loads would be about 20,000 psi. Similarly, using strength design and a steel yield strength of 60,000 psi, stress at service loads could be as high as 36,000 psi. If indeed flexural cracking is proportional to steel tensile stress, then it is quite evident that criteria for crack control must be included in the design process.

From the above, it is apparent that the effective moment of inertia of a section is more dependent upon the cracked section when using strength design concepts. This, of course, is the reason for the emphasis on the cracked section in the Code provisions for calculating deflections in Section 9.5.

Early investigations of crack width in beams and members subject to axial tension indicated that crack width is proportional to steel stress and bar diameter, but inversely proportional to reinforcement percentage. More recent research using modern deformed bars has confirmed that crack width is proportional to steel stress. However, other variables affecting steel detailing were found, such as thickness of concrete cover and the area of concrete in the zone of maximum tension surrounding each individual reinforcing bar. It should be kept in mind that there are large variations in crack widths, even in careful laboratory-controlled work. For this reason, a simple crack control expression is presented in the Code which will give reasonable reinforcing details and still meet the dictates of laboratory work and practical experience.

10.6 Beams and One-Way Slabs

10.6.4 - The Code requires that when the yield strength of the reinforcement exceeds 40,000 psi, detailing of the flexural tension reinforcement should be such that the quantity z given by

$$z = f_s \sqrt[3]{d_c A} \qquad \text{Eq. (10-4)}$$

does not exceed certain specified limits. These limits are 175 for interior exposure and 145 for exterior exposure. For practical considerations, Eq. (10-4) is presented in terms of reinforcing details rather than crack width per se. Eq. (10-4) will provide a distribution of the flexural reinforcement that will assure reasonable control of flexural cracking--a larger number of smaller bars at closer spacing. (See Design Examples 6.1 and 6.2.)

In Eq. (10-4), "A" is defined as the effective tension area of concrete surrounding the flexural tension reinforcement and having the same centroid as the reinforcement, divided by the number of bars and wires. When the flexural reinforcement consists of different bar or wire sizes, the number of bars or wires is computed as the total area of reinforcement divided by the area of the largest bar or wire used. This definition is satisfactory for all reinforcement details except bundled bars. For guidance in determining the number of equivalent bars to use to calculate "A" for bundled bars, refer to the January 1974 ACI Journal article by Leroy A. Lutz entitled, "Crack Control Factor for Bundled Bars and for Bars of Different Sizes."

In the original crack width expression from which Eq. (10-4) is derived, a factor β was included as one of the parameters, where β is the ratio of the distance of the neutral axis from the extreme tension fiber and from the centroid of the reinforcement. To simplify practical design of beams, an approximate value of 1.2 is used for β in Code Eq. (10-4). Since derivation of the original expression, additional tests have indicated that the crack width expression is also applicable to one-way slabs, with a value of β about 1.35. Accordingly, the Commentary to the Code suggests that the maximum z value for one-way slabs be reduced by the ratio of 1.2/1.35, or z = 156 for interior exposure and z = 129 for exterior exposure. Similar adjustments are prudent for other cases where the value of β exceeds 1.2.

As a design aid, Tables 6-1 and 6-2 show maximum bar spacings for Grade 60 reinforcement. The tables are based on a service load stress $f_s = 0.6\ f_y$, as permitted by Section 10.6.4. Computations for f_s would yield about 0.56 f_y for a dead-to-live load ratio of 0.5, and 0.60 f_y for a dead-to-live load ratio of 2. The maximum bar spacings are limited to the condition where reinforcement is in one layer. Review of Tables 6-1 and 6-2 shows that normal spacing of reinforcement generally will satisfy the requirements for 1-1/2 in. or less cover. However, for 2 in. or more cover, the maximum spacing of bars will often be limited by these requirements.

10.6.5 - Data are not available regarding crack width beyond which a corrosion danger exists. Exposure tests indicate that concrete quality, adequate compaction, and ample cover may be of greater importance for corrosion protection than crack width at the concrete surface. The limiting z values of

Table 6-1. **Maximum Bar Spacing in Beams***
(Grade 60 Reinforcement)

Bar Size	Outside Exposure z = 145 Cover-in.			Inside Exposure z = 175 Cover-in.		
	1-1/2	2	3	1-1/2	2	3
# 4	10.7	6.5	3.1	18.8	11.3	5.4
# 5	9.9	6.1	3.0	17.5	10.7	5.2
# 6	9.3	5.8	2.9	16.3	10.2	5.0
# 7	8.7	5.5	2.8	15.3	9.7	4.9
# 8	8.2	5.2	2.7	14.4	9.2	4.7
# 9	7.7	5.0	2.6	13.5	8.7	4.5
#10	7.2	4.7	2.5	12.6	8.3	4.3
#11	6.7	4.5	2.4**	11.8	7.8	4.2

* Values in inches, $f_s = 0.6\ f_y = 36$ ksi, single layer of reinforcement.
** Spacing less than permitted by Section 7.6.1

Table 6-2. **Maximum Bar Spacing In One-Way Slabs***
(Grade 60 Reinforcement)

Bar Size	Outside Exposure z = 129 Cover-in.				Inside Exposure z = 156 Cover-in.			
	3/4	1	1-1/2	2	3/4	1	1-1/2	2
# 4	-	14.7	7.5	4.5	-	-	13.3	8.0
# 5	-	13.4	7.0	4.3	-	-	12.4	7.6
# 6	-	12.2	6.5	4.1	-	-	11.6	7.2
# 7	16.3	11.1	6.1	3.9	-	-	10.8	6.8
# 8	14.7	10.2	5.8	3.7	-	-	10.2	6.5
# 9	13.3	9.4	5.4	3.5	-	16.6	9.6	6.2
#10	12.0	8.6	5.0	3.3	-	15.2	8.9	5.9
#11	10.9	7.9	4.7	3.1	-	14.0	8.4	5.6

* Values in inches, $f_s = 0.6\ f_y = 36$ ksi, single layer of reinforcement. Spacing should not exceed 3 times slab thickness nor more than 18 in. (Section 7.6.5). No value indicates spacing greater than 18 in.

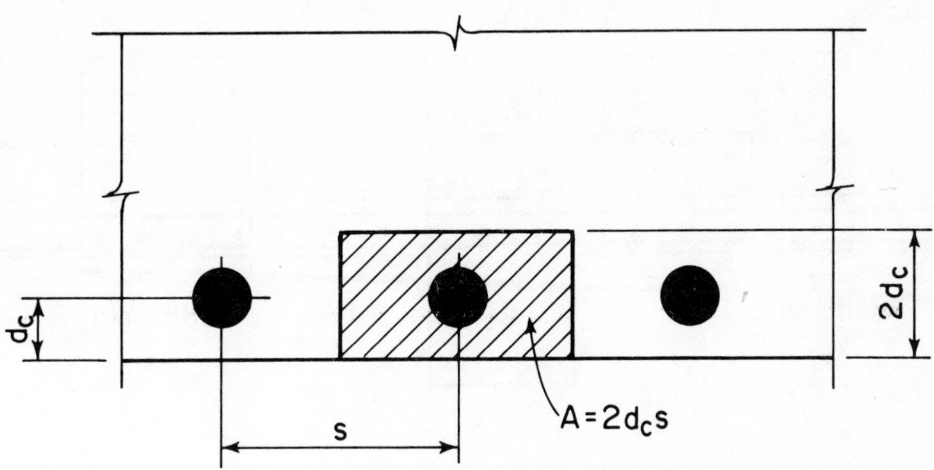

Fig. 6-1. Criteria Used to Develop Tables 6-1 and 6-2

Section 10.6.4 were chosen primarily to give reasonable reinforcing details in terms of practical experience with existing structures. The Code requirements do not apply to structures subject to very aggressive exposure or designed to be watertight; special precautions are required and must be investigated for such cases.

10.6.6 - Tension reinforcement should be well distributed in flexural tension zones. Where flanges are in tension, the primary tension reinforcement should be distributed over a width not exceeding 1/10 the span. If the flange width exceeds 1/10 the span, some additional longitudinal reinforcement as illustrated in Fig. 6-2 must be provided in the outer portions of the flange. (See Design Example 6.2)

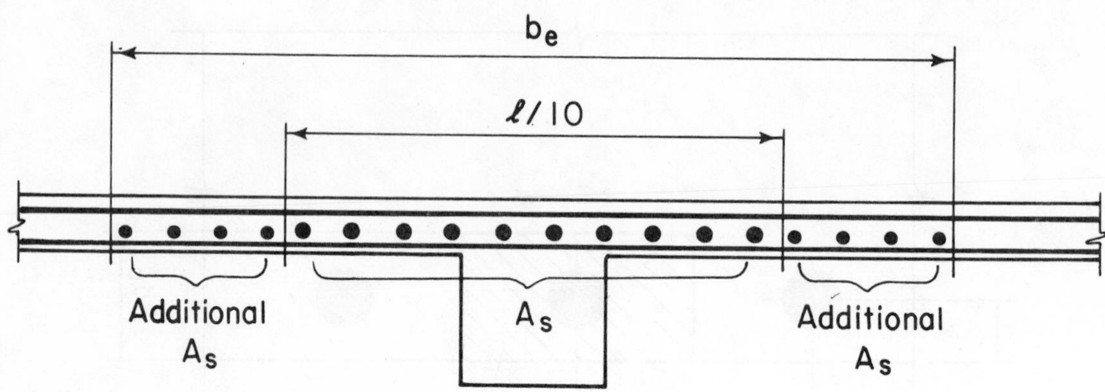

Fig. 6-2 - Negative Moment Reinforcement for Flanged Floor Beams

10.6.7 - In deep flexural members (webs exceeding 3 ft), additional longitudinal reinforcement should be distributed in the flexural tension zone. The total area of this reinforcement should be equal to at least 10% of the area of the primary reinforcement, be placed near the side faces of the beam, and be spaced not more than 12 in. on centers or the web width, whichever is less. The additional crack control reinforcement may be considered in calculating member strength only if strain compatibility is considered. (See Design Example 6.2.)

13.4 Two-Way Slabs

Section 13.4 specifies the maximum bar spacing; details of reinforcement at edge of slab and at beams, walls, and columns; area of reinforcement; special reinforcement in corners of slabs supported on beams; and negative reinforcement at drop panels. In addition, minimum lengths of reinforcement for slabs without beams are given in Code Fig. 13.4.8. Some of the main requirements are as follows:

(1) Spacing of bars in solid slabs should not exceed 2 times the slab thickness.

(2) Where spandrel beams, walls, or columns are provided at the discontinuous edge of an exterior span, all positive moment reinforcement in the slab should be embedded at least 6 in. into these members and all negative moment reinforcement should be bent, hooked, or otherwise anchored into these members.

(3) Area of reinforcement should not be less than required by Section 7.12.

(4) In slabs supported on beams with a flexural stiffness ratio of beam to slab (α) greater than unity, special reinforcement must be provided at the exterior corners. The amount, direction, and location of the special reinforcement are computed in accordance with rules given in Section 13.4.6.

EXAMPLE 6.1 - Distribution of Flexural Reinforcement

Check distribution of reinforcement for beam section using Eq. (10-4). Use z = 145 kips/in. for exterior exposure and Grade 60 reinforcement.

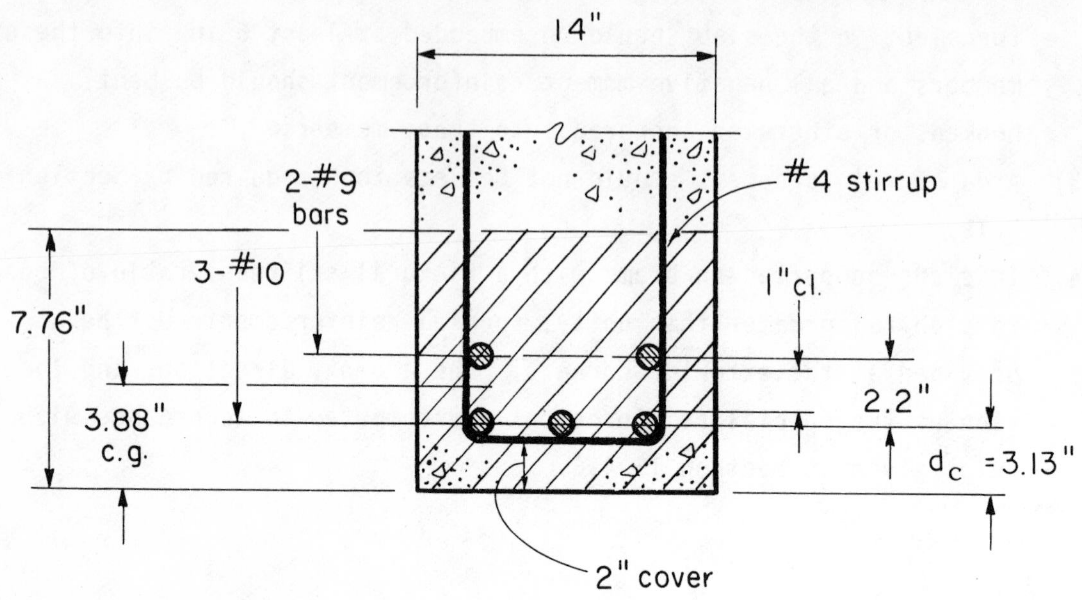

Calculations and Discussion	Code Reference

1. Reinforcement centroid

$$C.G. = \frac{3\ (1.27)\ 3.13 + 2\ (1.0)\ 5.33}{3\ (1.27) + 2\ (1.0)} = 3.88 \text{ in.}$$

2. Effective tension area of concrete

$$= 2 \times 3.88 \times 14 = 108.6 \text{ in.}^2$$

3. Equivalent number of #10 bars

$$= \frac{3\ (1.27) + 2\ (1.0)}{1.27} = 4.57$$

4. $A = \dfrac{\text{effective area}}{\#\text{ bars}} = \dfrac{108.6}{4.57} = 23.8 \text{ in.}^2/\text{bar}$ — 10.0

5. $z = f_s \sqrt[3]{d_c A}$ — Eq. (10-4)

$= 0.6\ (60) \sqrt[3]{3.13 \times 23.8} = 152 > 145$ N.G. — 10.6.4

EXAMPLE 6.1 - Continued

Calculations and Discussion	Code Reference

6. Try 4 - #10 (bottom row) and 2 - #6 (top row)

$$C.G. = \frac{4\ (1.27)\ 3.13 + 2\ (0.44)\ 5.13}{4\ (1.27) + 2\ (0.44)} = 3.43 \text{ in.}$$

$$A = \frac{2 \times 3.43 \times 14}{[4\ (1.27) + 2\ (0.44)]/1.27} = 20.45$$

$$z = 0.6\ (60)\ \sqrt[3]{3.13 \times 20.45} = 144 < 145 \quad \text{O.K.}$$

EXAMPLE 6.2 - Distribution of Reinforcement in Deep Flexural Member with Flanges

Select reinforcement for T-section shown below.

Exposure: Outside f'_c = 4000 psi

Span: 50'-0 continuous f_y = 60,000 psi

Service load moments:

Positive Moment	Negative Moment
$M_d = +265^{'k}$	$M_d = -290^{'k}$
$M_\ell = +690^{'k}$	$M_\ell = -760^{'k}$

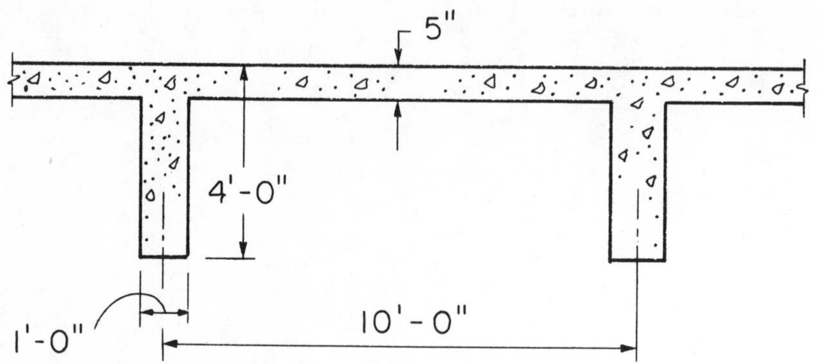

Calculations and Discussion	Code Reference

A. Positive moment reinforcement

1. A_s required = 7.76 in.^{2}

   Try 5 - #11 bars, A_s = 7.80 in.^{2}

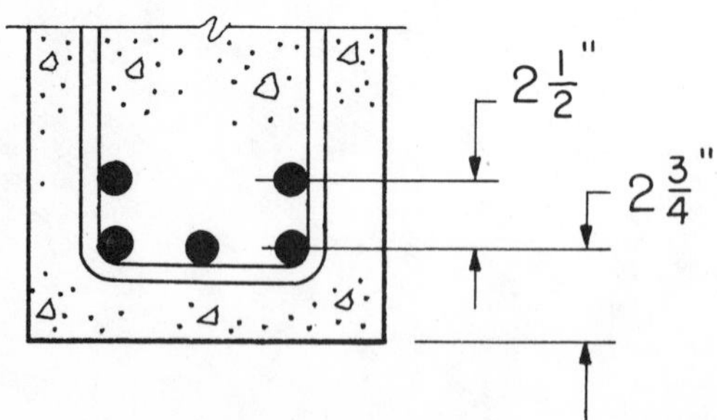

EXAMPLE 6.2 - Continued

Calculations and Discussion	Code Reference

2. C.G. of bar layout

$$= \frac{3\ (1.56)\ 2.75 + 2\ (1.56)\ 5.25}{5\ (1.56)} = 3.75 \text{ in.}$$

Effective concrete area = 2 x 3.75 x 12 = 90 in.^{2}

$A = \frac{90}{5} = 18$ in.^{2}/bar — 10.0

Stress in reinforcement at service load — 10.6.4

$$f_s = \frac{+M}{jdA_s} = \frac{(265 + 690)\ 12}{0.98 \times 44.3 \times 7.80} = 33.9 \text{ ksi}$$

$z = f_s \sqrt[3]{d_c A} = 33.9 \sqrt[3]{2.75 \times 18} = 125 < 145$ O.K. — Eq. (10-4)

3. Web depth exceeds 3 feet — 10.6.7

Crack control A_s = 0.10 x 7.8 = 0.78 in.^{2}

Use 6 - #4 bars (A_s = 1.20 in.^{2}), 3 each face

at 12" max. spacing.

B. <u>Negative moment reinforcement</u>

1. A_s required = 10.0 in.^{2}

Effective flange width = 12 + 2 x 8 x 5 = 92 in. — 8.10.2

Effective width for tension reinforcement — 10.6.6

1/10 x 50 x 12 = 60 in.

Try 10 - #9 bars @ ≃ 6.7 in., A_s = 10.0 in.^{2}

2. Check maximum bar spacing by Table 6-1 for outside exposure. Cover = 2 in., f_y = 60 ksi.

s_{max} = 5.0 in. < 6.7 in. N.G.

3. Try 13 - #8 bars @ 5 in., A_s = 10.27 in.^{2}

From Table 6-1, s_{max} = 5.2 in. O.K.

EXAMPLE 6.2 - Continued

Calculations and Discussion	Code Reference
4. Longitudinal reinforcement in slab outside 60-in. width	
For Grade 60, $A_s = 0.0018 \times 12 \times 5 = 0.108$ in.^{2}/ft	7.12
Use #4 bars @ 18 in., $A_s = 0.13$ in.^{2}/ft	
5. Detail section as shown below.	

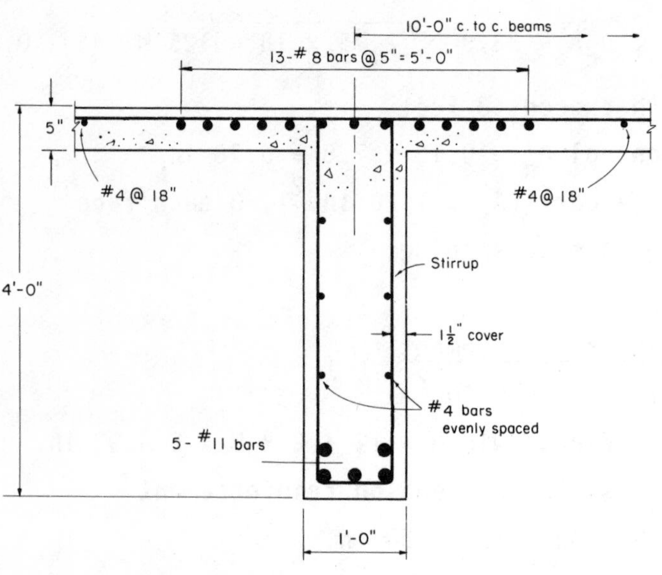

7

Deflections

Introduction

The ACI Code provisions for control of deflections are concerned only with deflections that occur at service load levels under static conditions and may not apply for loads with strong dynamic characteristics such as due to earthquakes, transient winds, and machinery. Because of the variability of concrete structural deformations, in most cases, it is essential that relatively simple procedures be used, so that designers will guard against placing undue reliance on computed or predicted deflection results. An in-depth treatise on the subject of deflection control may be found in References 7.1 and 7.2.

9.5 Control of Deflections

Two methods are given in the Code for controlling deflections of one-way and two-way flexural systems. Deflections may be controlled by means of minimum thickness [Table 9.5(a), or Eqs. (9-11), (9-12), (9-13)] or directly by limiting computed deflections [Table 9.5(b)].

9.5.2 Minimum Thickness for Beams and One-Way Slabs (Nonprestressed)

Deflections of beams and one-way slabs supporting loads commonly found in buildings will normally be satisfactory when the following minimum thickness from Table 9.5(a) of the Code are met or exceeded.

Member	Minimum Thickness, h			
	Simply Supported	One End Continuous	Both Ends Continuous	Cantilever
One-Way Slabs	$\ell/20$	$\ell/24$	$\ell/28$	$\ell/10$
Beams	$\ell/16$	$\ell/18.5$	$\ell/21$	$\ell/8$

Correction Factors for f_y and w_c

f_y, ksi	40	50	60	75	w_c, pcf	90	100	110	120
Cor. Factor	0.80	0.90	1.00	1.15	Cor. Factor	1.20	1.15	1.10	1.09

The designer should especially note that the minimum thickness is intended to apply only for members <u>not</u> supporting or <u>not</u> attached to partitions or other construction likely to be damaged by large deflections . . . otherwise, deflections should be computed.

For shored composite members, the minimum thicknesses in Table 9.5(a) apply as for monolithic T-beams. For unshored construction, if the thickness of a nonprestressed precast member meets the minimum thickness requirements, deflections need not be computed. Section 9.5.5 also states that, if the thickness of an unshored nonprestressed composite member meets the minimum thickness requirements, deflections occurring after the member becomes composite need not be investigated for the magnitude and duration of load prior to the beginning of effective composite action.

9.5.3 Minimum Thickness for Two-Way Slab Systems (Nonprestressed)

Deflections of two-way slab systems with and without beams, drop panels, and column capitals need not be computed when the minimum thickness requirements of Section 9.5.3 are met. The minimum thickness requirements include the effects of panel location (interior, side or corner); panel shape; span ratios; supporting columns and capitals; drop panels; and the yield strength of the reinforcing steel. The minimum thickness equations provide for a transition, Eq. (9-11); from slabs on stiff beams, Eq. (9-12); to slabs without beams, Eq. (9-13).

Typical cases are summarized in Table 7-1. It may be noted in Table 7-1 that the difference between the controlling minimum thickness for square panels and 2- to- 1 rectangular panels in each case is not very large. Also, for two-way slabs without beams, only Eq. (9-13) need be considered.

Minimum Thickness for Prestressed Members

Typical span-depth ratios for general use in design of prestressed members are given in the PCI Design Handbook[7.3], and summarized in Reference 7.2 from several sources.

Table 9.5(b) Maximum Permissible Computed Deflections

The allowable computed deflections specified in Table 9.5(b) apply for both one-way and two-way nonprestressed and prestressed members.

Where excessive deflections may cause damage to nonstructural or other structural elements, only that part of the deflection occurring after the construction of the element needs to be considered. The most stringent deflection limit of $\ell/480$ in Table 9.5(b) is an example of such a case.

Where excessive deflections may result in either esthetic or functional problems, such as objectionable visual sagging, ponding of water, vibration, and improper operation of machinery or sliding doors, the total deflection should be considered. Such examples are not included in Table 9.5(b) and must be dealt with by the designer on a case-by-case basis.

Two-Way Slab System	Minimum h	But Not Less Than	Need Not Exceed
	Eq. (9-11)	Eq. (9-12)	Eq. (9-13)
SQUARE PANELS ($\ell_{n1} = \ell_{n2}$)			
Flat Plate			**
Interior panel — Min. h = 5 in.			$\ell_n/32.7$
Side panel — Min. h = 5 in.			$\ell_n/29.7$
Corner panel — Min. h = 5 in.			$\ell_n/29.7$
Flat Plate with Edge Beams (α = 0.8)	**		
Side panel	$\ell_n/35.2$	$\ell_n/40.7$	$\ell_n/32.7$
Corner panel	$\ell_n/35.0$	$\ell_n/39.0$	$\ell_n/32.7$
Flat Slab with Drop Panels (Length ≧ ℓ/3 and Depth ≧ 1.25 h)			**
Interior panel — Min. h = 4 in.			$\ell_n/36.4$
Side panel — Min. h = 4 in.			$\ell_n/32.7$
Corner panel — Min. h = 4 in.			$\ell_n/32.7$
Flat Slab with Drop Panels (Length ≧ ℓ/3 and Depth ≧ 1.25 h) and Edge Beams (α = 0.8)			**
Side panel	$\ell_n/39.1$	$\ell_n/45.2$	$\ell_n/36.4$
Corner panel	$\ell_n/37.9$	$\ell_n/43.9$	$\ell_n/36.4$
RECTANGULAR PANELS ($\beta = \ell_{n1}/\ell_{n2} = 2$)			
Flat Plate			**
Interior panel — Min. h = 5 in.			$\ell_n/32.7$
Side panel — Min. h = 5 in.			$\ell_n/29.7$
Corner panel — Min. h = 5 in.			$\ell_n/29.7$
Flat Plate with Edge Beams (α = 0.8)	**		
Side panel	$\ell_n/38.3$	$\ell_n/48.6$	$\ell_n/32.7$
Corner panel	$\ell_n/36.6$	$\ell_n/46.4$	$\ell_n/32.7$
Flat Slab with Drop Panels (Length ≧ ℓ/3 and Depth ≧ 1.25 h)			**
Interior panel — Min. h = 4 in.			$\ell_n/36.4$
Side panel — Min. h = 4 in.			$\ell_n/32.7$
Corner panel — Min. h = 4 in.			$\ell_n/32.7$
Flat Slab with Drop Panels (Length ≧ ℓ/3 and Depth ≧ 1.25 h) and Edge Beams (α = 0.8)	**		
Side panel	$\ell_n/42.5$	$\ell_n/54.0$	$\ell_n/36.4$
Corner panel	$\ell_n/40.7$	$\ell_n/51.5$	$\ell_n/36.4$

Table 7-1 – Minimum Thickness For Two-Way Slab Systems (Grade 60 Reinforcement)*

*For f_y other than 60 ksi, muiltiply Table values by 1.05 for f_y = 50 ksi, and 1.10 for f_y = 40 ksi.

**Controlling limit

DEFLECTION OF BEAMS AND ONE-WAY SLABS (Nonprestressed)

Initial or Short-Time Deflection

The effective moment of inertia for cantilevers, simple beams and between inflection points of continuous beams, is given by ACI Eq. (9-7):

$$I_e = (M_{cr}/M_a)^3 \; I_g + [1 - (M_{cr}/M_a)^3] \; I_{cr} \leq I_g \qquad (1)$$

where $M_{cr} = f_r I_g / y_t$

For normal weight concrete, $f_r = 7.5\sqrt{f'_c}$

For lightweight concrete, f_r is modified according to Section 9.5.2.3. See Design Example 7.2.

Values of I_g and I_{cr} may be computed using the equations in Fig. 7-1. M_a in Eq. (1) is the maximum service load moment (unfactored moment) at the stage for which deflections are being considered. For different load levels, the deflection should be computed in each case using Eq. (1) for the total load level being considered, such as dead load or dead plus live load. The incremental deflection, such as for live load, is then computed as the difference in these values.

The effective moment of inertia I_e provides a transition between the well-defined upper and lower bounds of I_g and I_{cr}, as a function of the level of cracking in the form of M_{cr}/M_a. The equation empirically accounts for the effect of tension stiffening -- tensile concrete between cracks and in low tensile stress regions.

For prismatic members (including T-beams with different positive and negative region cracked sections), I_e (and thus M_a) may be determined at the support section for cantilevers and the midspan section for simple and continuous spans. The use of the midspan section properties for continuous prismatic members is considered satisfactory in approximate calculations primarily because the midspan rigidity (including the effect of cracking) has the dominant effect on deflections.

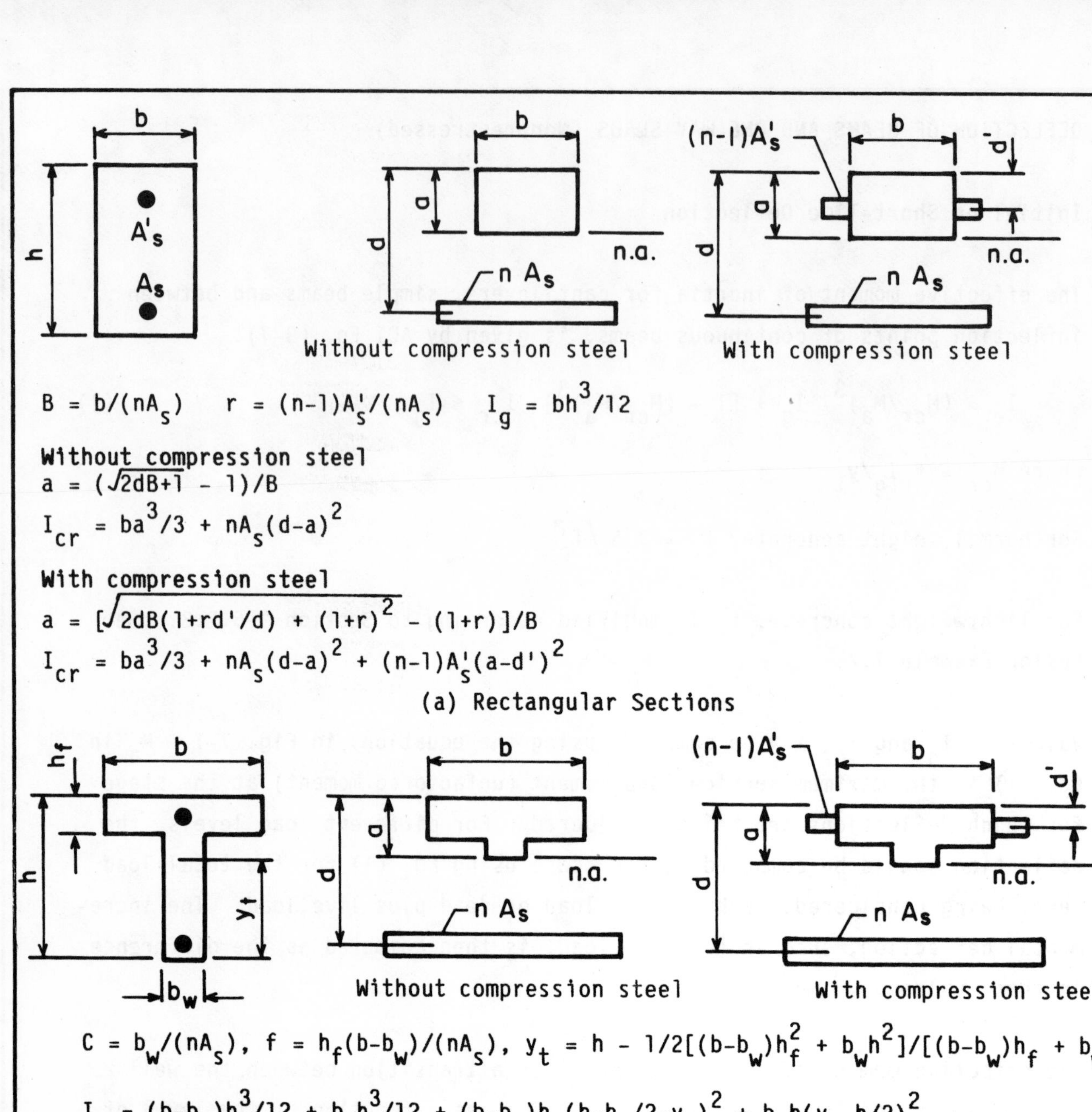

$B = b/(nA_s) \quad r = (n-1)A_s'/(nA_s) \quad I_g = bh^3/12$

Without compression steel

$a = (\sqrt{2dB+1} - 1)/B$

$I_{cr} = ba^3/3 + nA_s(d-a)^2$

With compression steel

$a = [\sqrt{2dB(1+rd'/d) + (1+r)^2} - (1+r)]/B$

$I_{cr} = ba^3/3 + nA_s(d-a)^2 + (n-1)A_s'(a-d')^2$

(a) Rectangular Sections

$C = b_w/(nA_s), \; f = h_f(b-b_w)/(nA_s), \; y_t = h - 1/2[(b-b_w)h_f^2 + b_wh^2]/[(b-b_w)h_f + b_wh]$

$I_g = (b-b_w)h_f^3/12 + b_wh^3/12 + (b-b_w)h_f(h-h_f/2-y_t)^2 + b_wh(y_t-h/2)^2$

Without compression steel

$a = [\sqrt{C(2d+h_ff) + (1+f)^2} - (1+f)]/C$

$I_{cr} = (b-b_w)h_f^3/12 + b_wa^3/3 + (b-b_w)h_f(a-h_f/2)^2 + nA_s(d-a)^2$

With compression steel

$a = [\sqrt{C(2d+h_ff+2rd') + (f+r+1)^2} - (f+r+1)]/C$

$I_{cr} = (b-b_w)h_f^3/12 + b_wa^3/3 + (b-b_w)h_f(a-h_f/2)^2 + nA_s(d-a)^2 + (n-1)A_s'(a-d')^2$

(b) Flanged Sections

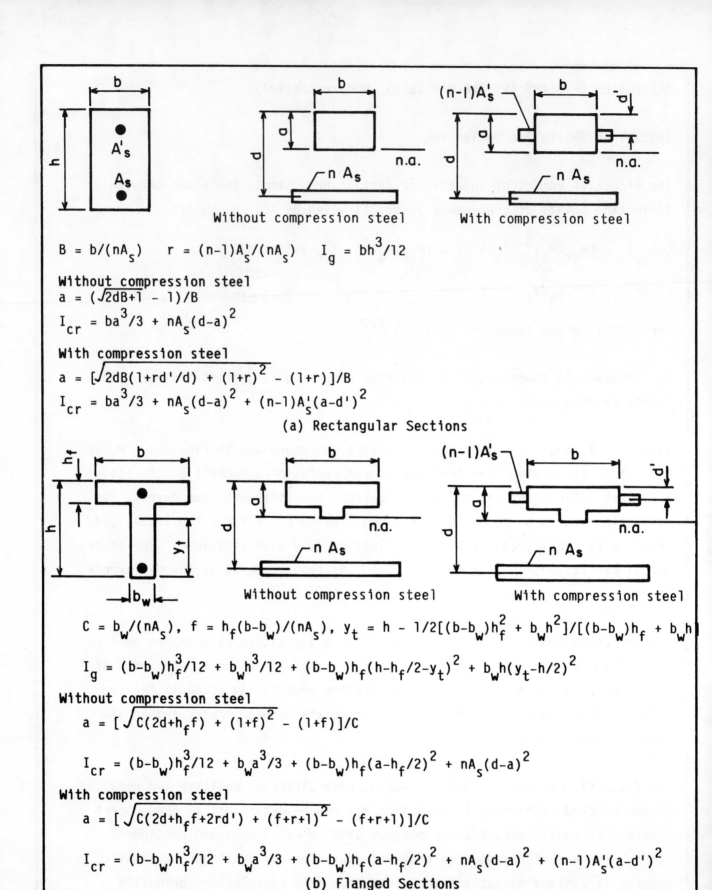

Fig. 7-1 Moment of Inertia of Gross Section, I_g, and Cracked Transformed Section, I_{cr}

Alternatively, for continuous prismatic and nonprismatic members, Section 9.5.2.4 suggests using the average I_e at the critical positive and negative moment sections; and Commentary Section 9.5.2.4 suggests Avg. I_e values for somewhat improved results over the other two methods as follows:

Beams with one end continuous:

$$\text{Avg. } I_e = 0.85\ I_m + 0.15(I_e\text{-Cont. End}) \qquad (2)$$

Beams with both ends continuous:

$$\text{Avg. } I_e = 0.70\ I_m + 0.15(I_{e1} + I_{e2}) \qquad (3)$$

where I_m refers to the midspan section I_e, and I_{e1} and I_{e2} refers to I_e at the respective beam ends. Moment envelopes should be used in computing both positive and negative values of I_e. Moment envelopes based on the approximate moment coefficients of Section 8.3.3 are accurate enough. See Design Example 7.2. For a single heavy concentrated load, only the midspan I_e should be used.

The initial or short-time deflection (a_i) may be computed using the following elastic equation given in Commentary Section 9.5.2.4 for cantilevers, simple and continuous beams. For continuous beams, the midspan deflection may normally be used as an approximation of the maximum deflection.

$$a_i = K\ (5/48)\ M_a\ \ell^2/E_c\ I_e \qquad (4)$$

where M_a is the support moment for cantilevers and the midspan moment (when K is so defined) for simple and continuous beams, and ℓ is the span length (usually the clear span). For uniformly distributed loading w, the theoretical values of the deflection coefficient K are as follows:

Cantilevers (the deflection due to rotation at the supports must be determined in addition)	K = 12/5 = 2.40

Simple Beams $K = 1.00$

Continuous Beams $K = 1.20 - 0.20\ M_o/M_a$

where $M_o = w\,\ell^2/8$ and M_a is the net midspan moment.

Continuous Beams

Fixed-Hinged Beams, Midspan Deflection $K = 0.80$

Fixed-Hinged Beams, Maximum Deflection Using Maximum Moment $K = 0.738$

Fixed-Fixed Beam $K = 0.60$

For other types of loading, K values are given in Reference 7.2.

Since deflections are logically computed for a given continuous span based on the same loading pattern for maximum positive moment and deflections, Eq. (4) is thought to be the most convenient form of a deflection equation. In addition, when using Eq. (4) with only the midspan I_e, the negative moments are not required in the deflection calculation for continous beams, and the use of the ACI approximate moment coefficients for maximum positive moment are considered to be satisfactory in most cases for computing deflections.

Long-Time Deflection

According to Section 9.5.2.5, additional long-time deflections due to the combined effects of creep and shrinkage may be estimated by ACI Eq. (9-10) for sustained loads.

$$\lambda = k_r \xi = \frac{\xi}{1 + 50\,\rho'} \qquad (5)$$

$$a_{(cp + sh)} = \lambda(a_i)_{sus} \quad (6)$$

where ρ' is determined at the support section for cantilevers and the midspan section for simple and continuous spans, and ξ = 2.0 for 5 years or more of sustained loading, 1.4 for 12 months, 1.2 for 6 months, and 1.0 for 3 months. The multiplier λ for additional long-time deflection is new for the 1983 Code. The new multiplier for creep and shrinkage effects is considered to better represent the effect of compression reinforcement by more appropriately relating it to the bulk of the concrete in compression by means of the compression reinforcement ratio ρ', rather than to the ratio of compression to tension reinforcement A'_s/A_s as used in the 1977 Code expression.

Alternatively, creep and shrinkage deflections may be considered separately using the following expressions from References 7.2, 7.4, and 7.5.

$$a_{cp} = \lambda_{cp} (a_i)_{sus} \quad (7)$$

$$a_{sh} = K_{sh}\varphi_{sh}\ell^2 \quad (8)$$

where $$\lambda_{cp} = k_r C_t = \frac{0.85C_t}{1 + 50\rho'}$$

$$\varphi_{sh} = A_{sh}\varepsilon_{sh}/h$$

For average conditions, ultimate values for C_t and ε_{sh} may be taken as C_t = 1.6 and $\varepsilon_{sh} = 400 \times 10^{-6}$. A_{sh} may be taken from Fig. 7-2. Assuming equal positive and negative shrinkage curvatures with an inflection point at the quarter-point of continuous spans (generally satisfactory for this purpose), the following values for the shrinkage deflection coefficient K_{sh} may be used.

Cantilevers $K_{sh} = 1/2 = 0.50$

Simple Spans $K_{sh} = 1/8 = 0.125$

Spans With One End Continuous -- Multi-Span Beams $K_{sh} = 0.09$

Spans With One End Continuous -- Two-Span Beams $K_{sh} = 0.084$

Spans With Both Ends Continuous -- $K_{sh} = 0.065$

The compression reinforcement ratio ρ', and the reinforcement percentages ρ and ρ' used in determining A_{sh} from Fig. 7-2, refer to the support section of cantilevers and the midspan section of simple and continuous beams. For T-beams use $\rho = 100(\rho + \rho_w)/2$ and a similar calculation for any compression steel ρ' in determining A_{sh}, where $\rho_w = A_s/b_w d$. See Design Example 7.2.

With regard to the choice of computing the combined creep and shrinkage deflections by ACI Eq. (9-10) versus the separate creep and shrinkage deflections, the combined calculation is simpler but provides only a rough approximation since shrinkage deflections are only indirectly related to the loading (primarily by means of the steel content). One case in which the separate calculation of creep and shrinkage deflections may be preferable is when part of the live load is considered as a sustained load.

All procedures and properties for computing creep and shrinkage deflections apply equally to normal weight and lightweight concrete.

DEFLECTION OF COMPOSITE FLEXURAL MEMBERS (Nonprestressed)

The ultimate (in time) deflection of unshored and shored composite flexural members may be computed by Eqs. (9) to (12). These equations are derived in detail in References 7.2 and 7.7. Subscripts 1 and 2 are used to refer to the slab (or effect of the slab such as under slab dead load) and precast beam, respectively.

These procedures are described for a composite beam in which both unshored and shored construction are assumed. Design Examples 7.3 and 7.4 demonstrate the beneficial effect of shoring in reducing deflections.

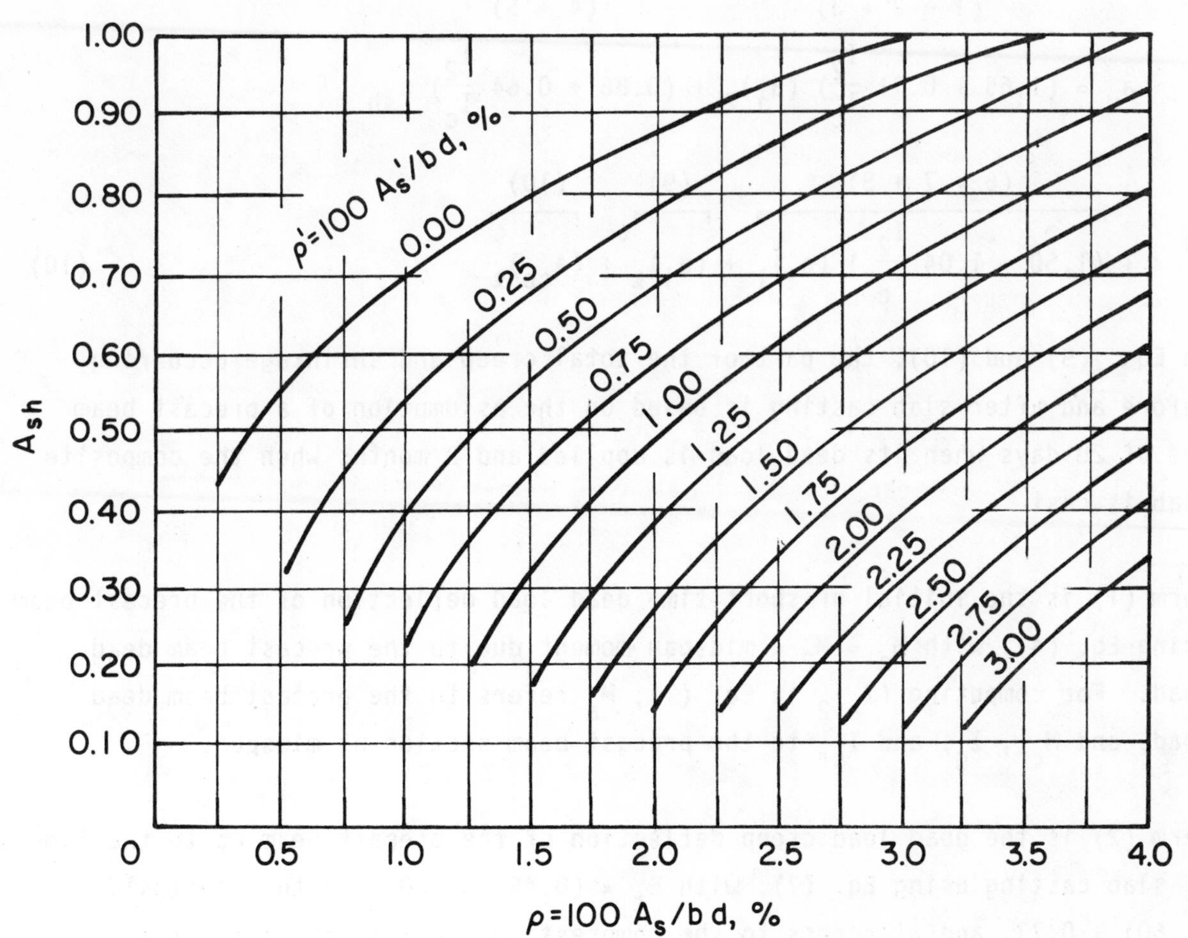

Fig. 7-2 - Values of A_{sh} for Calculating Shrinkage Deflection

Unshored Composite Members

$$a_u = \overbrace{(a_i)_2}^{(1)} + \overbrace{0.77\ k_r\ (a_i)_2}^{(2)} + \overbrace{0.83\ k_r\ (a_i)_2 \frac{I_2}{I_c}}^{(3)} + \overbrace{0.36\ a_{sh}}^{(4)} + \overbrace{0.64\ a_{sh} \frac{I_2}{I_c}}^{(5)}$$

$$+ \overbrace{(a_i)_1}^{(6)} + \overbrace{1.22\ k_r\ (a_i)_1 \frac{I_2}{I_c}}^{(7)} + \overbrace{a_{ds}}^{(8)} + \overbrace{(a_i)_\ell}^{(9)} + \overbrace{(a_{cp})_\ell}^{(10)} \qquad (9)$$

With k_r = 0.85 (no compression steel in the precast beam) and a_{ds} assumed to be equal to 0.50 $(a_i)_1$, Eq. (9) reduces to Eq. (10).

$$a_u = \overbrace{(1.65 + 0.71 \frac{I_2}{I_c}) (a_i)_2}^{(1 + 2 + 3)} + \overbrace{(0.36 + 0.64 \frac{I_2}{I_c}) a_{sh}}^{(4 + 5)}$$

$$+ \overbrace{(1.50 + 1.04 \frac{I_2}{I_c}) (a_i)_1}^{(6 + 7 + 8)} + \overbrace{(a_i)_\ell}^{(9)} + \overbrace{(a_{cp})_\ell}^{(10)} \qquad (10)$$

In Eqs. (9) and (10), the part of the total creep and shrinkage occurring before and after slab casting is based on the assumption of a precast beam age of 20 days when its dead load is applied and 2 months when the composite slab is cast.

Term (1) is the initial or short-time dead load deflection of the precast beam using Eq. (4), with $M_a = M_2$ = midspan moment due to the precast beam dead load. For computing $(I_e)_2$ in Eq. (1), M_a refers to the precast beam dead load, and M_{cr}, I_g, and I_{cr} to the precast beam section at midspan.

Term (2) is the dead load creep deflection of the precast beam up to the time of slab casting using Eq. (7), with C_t = (0.48 for 20 days to 2 months) (1.60) = 0.77, and ρ' refers to the compression steel in the precast beam at midspan when computing k_r.

Term (3) is the creep deflection of the composite beam following slab casting due to the precast beam dead load using Eq. (7), with $C_t = 1.60 - 0.77 = 0.83$. ρ' is the same as in Term (2). The ratio, I_2/I_c, modifies the initial stress (strain) and accounts for the effect of the composite section in restraining additional creep curvature (strain) after the composite section becomes effective. As a simple approximation, $I_2/I_c = [(I_2/I_c)_g + (I_2/I_c)_{cr}]/2$ may be used.

Term (4) is the deflection due to shrinkage warping of the precast beam up to the time of slab casting using Eq. (8), with a 0.36 factor for steam cured concrete (assumed to be the usual case for precast beams) at age 2 months (representing 36% of the total shrinkage). $(\epsilon_{sh})_u = 400 \times 10^{-6}$ in/in.

Term (5) is the shrinkage deflection of the composite beam following slab casting due to the shrinkage of the precast beam concrete using Eq. (8), with a factor of 1.00 - 0.36 = 0.64 or 64% of the total shrinkage. This term does not include the effect of differential shrinkage and creep which is given by Term (8). I_2/I_c is the same as in Term (3).

Term (6) is the initial or short-time deflection of the precast beam under slab dead load using Eq. (4), with the incremental deflection computed as follows: $(a_i)_1 = (a_i)_{1+2} - (a_i)_2$, where $(a_i)_2$ is the same as in Term (1). For computing $(I_e)_{1+2}$ and $(a_i)_{1+2}$ in Eqs. (1) and (4), $M_a = M_1 + M_2$ due to the precast beam plus slab dead load at midspan, and M_{cr}, I_g, and I_{cr} refer to the precast beam section at midspan. When partitions, roofing, etc., are placed at the same time as the slab, or soon thereafter, their dead load should be included in M_1 and M_a.

Term (7) is the creep deflection of the composite beam due to slab dead load using Eq. (7), with (0.76 -- loading age correction factor at age 2 months) (1.60) = 1.22. In this term the initial strains, curvatures and deflections under slab dead load were based on the precast section only. Hence the creep curvatures and deflections refer to the precast beam concrete, although the composite section is restraining the creep curvatures and deflections, as mentioned in Term (3). k_r is the same as in Term (2) and I_2/I_c is the same as in Term (3).

Term (8) is the deflection due to differential shrinkage and creep. As an approximation, $a_{ds} = 0.50(a_i)_1$ may be used.

Term (9) is the initial or short-time live load [plus other loads applied to the composite beam and not included in Term (6)] deflection of the composite beam using Eq. (4), with the incremental deflection estimated as follows: $(a_i)_\ell = (a_i)_{d+\ell} - (a_i)_d$ based on the composite section only. This is thought to be the best and conservative [since the computed $(a_i)_d$ is on the low side and thus the computed $(a_i)_\ell$ is on the high side] approximation, even though the incremental loads are actually resisted by different sections (members). This method is the same as Term (5) of Eq. (11) -- same as for a monolithic beam. Alternatively, Eq. (4) may be used with $M_a = M$ and $I_e = (I_c)_{cr}$ as a simpler rough approximation. The first method is illustrated in Design Example 7.4 and the alternative method in Design Example 7.3.

Term (10) is the partial live load creep deflection for any sustained live load (and other loads) applied to the composite beam using Eq. (7), with $C_u = 1.60$, and ρ' refers to any compression steel in the slab at midspan when computing k_r.

Shored Composite Members

It is assumed in Eqs. (11) and (12) that the composite beam supports all of the dead and live load. The calculation of deflections for shored composite beams is essentially the same as for monolithic beams, except for the deflection due to shrinkage warping of the precast beam which is resisted by the composite section after the slab has hardened, and the deflection due to differential shrinkage and creep of the composite beam. These effects are represented by Terms (3) and (4) in Eq. (11).

$$a_u = \overbrace{(a_i)_{1+2}}^{(1)} + \overbrace{1.80\, k_r\, (a_i)_{1+2}}^{(2)} + \overbrace{a_{sh}\frac{I_2}{I_c}}^{(3)} + \overbrace{a_{ds}}^{(4)} + \overbrace{(a_i)_\ell}^{(5)} + \overbrace{(a_{cp})_\ell}^{(6)} \qquad (11)$$

When $k_r = 0.85$ (neglecting any effect of slab compression steel) and a_{ds} is assumed to be equal to $(a_i)_{1+2}$, Eq. (11) reduces to Eq. (12).

$$a_u = \overbrace{3.53\,(a_i)_{1+2}}^{(1+2+4)} + \overbrace{a_{sh}\frac{I_2}{I_c}}^{(3)} + \overbrace{(a_i)_\ell}^{(5)} + \overbrace{(a_{cp})_\ell}^{(6)} \qquad (12)$$

Term (1) is the initial or short-time deflection of the composite beam due to slab plus precast beam dead load (plus partitions, roofing, etc.) using Eq. (4), with $M_a = M_1 + M_2$ = midspan moment due to slab plus precast beam dead load. For computing $(I_e)_{1+2}$ in Eq. (1), M_a refers to the moment $M_1 + M_2$, and M_{cr}, I_g, and I_{cr} to the composite beam section at midspan.

Term (2) is the creep deflection of the composite beam due to the dead load in Term (1), using Eq. (7). $C_u = 1.80$ (based on the shores being removed at about age 10 days for a moist cured slab), and ρ' refers to any compression steel in the slab at midspan when computing k_r.

Term (3) is the shrinkage deflection of the composite beam after the shores are removed due to the shrinkage of the precast beam concrete, but not including the effect of differential shrinkage and creep which is given by Term (4). Eq. (8) may be used to compute a_{sh}. Assuming the slab is cast at a precast beam concrete (steam-cured) age of 2 months and shores are removed about 10 days later, $(\epsilon_{sh})_u$ = (1 - say 0.37)(400 x 10^{-6}) = 252 x 10^{-6} in/in.

Term (4) is the deflection due to differential shrinkage and creep. As an approximation, $a_{ds} = (a_i)_{1+2}$ may be used.

Term (5) is the initial or short-time live load deflection of the composite beam using Eq. (4). The calculation of the incremental live load deflection follows the same procedure as that of a monolithic beam. This is the same as the first method described in Term (9) of Eq. (9).

Term (6) is the partial live load creep deflection for any sustained live load using Eq. (7). This is the same at Term (10) of Eq. (9).

These procedures suggest using midspan values only, which may normally be satisfactory for both simple composite beams and those with a continuous slab as well. See Reference 7.7 for an example of a continuous slab in composite construction.

DEFLECTION OF NONPRESTRESSED TWO-WAY SLAB SYSTEMS

Initial or Short-Time Deflection

An approximate procedure[7.6] that is compatible with the direct design and equivalent frame methods of Code Chapter 13 will be used to compute the initial or short-time deflection of two-way slab systems. The method is essentially the same for flat plates, flat slabs, and two-way slabs, once the appropriate stiffnesses are computed. In this procedure, the midpanel deflection is computed as the sum of the midspan column strip deflection in one direction, such as a_{cx}, and the midspan middle strip deflection in the other direction, such as a_{my}, as shown in Fig. 7.3.

Under vertical loads, the midspan deflection of an equivalent frame can be considered as the sum of three parts: that of a panel assumed to be fixed at both ends of its span, plus that due to the rotation at each of the two support lines.

Midspan fixed-end deflection of the equivalent frame under uniform loading is given by Eq. (13).

$$\text{Fixed } a_{frame} = w\,\ell^2/384\; E_c\, I_{frame} \qquad (13)$$

where w is the uniformly distributed load per width of frame, and ℓ is the span center-to-center of columns. To include the effect of different positive and negative moment region I values [primarily when using drop panels and/or I_e in Eq. (1)], an average may be used, as given by Eqs. (2) and (3).

Calculation of the midspan fixed-end deflection of the column and middle strips is then based on the M/EI ratio of the strips to the frame.

$$\text{Fixed } a_{c,m} = (LDF)_{c,m}\,(\text{Fixed } a_{frame})\,\frac{(EI)_{frame}}{(EI)_{c,m}} \qquad (14)$$

where $(LDF)_{c,m} = M_{c,m}/M_{frame}$. Typical values of the lateral distribution factor, LDF, are shown in Table 7-2.

If the ends of the columns at the floor above and below are assumed to be fixed (usual case in the equivalent frame analysis), or ideally pinned, the rotation of the column at the floor in question is equal to the net applied moment divided by the stiffness of the equivalent column. This is given by Eq. (15) for the frame, column strip, and middle strip.

$$\text{End } \theta_c = \text{End } \theta_m = \text{End } \theta_{frame} = \text{End } \theta = (M_{net})_{frame}/K_{ec} \qquad (15)$$

where K_{ec} is the gross-section flexural stiffness of the equivalent column. See Equivalent Columns: Commentary Section 13.7.4.

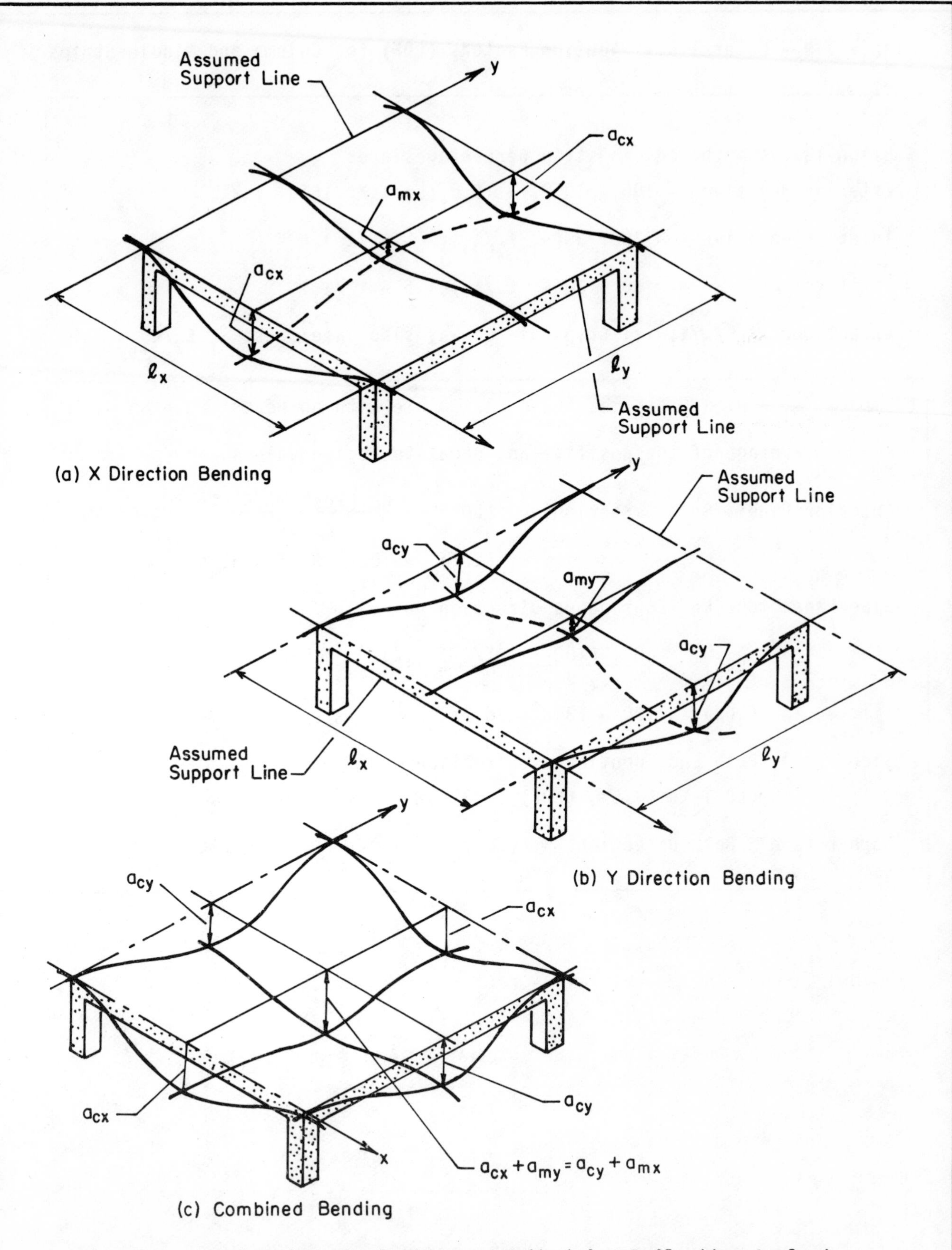

Fig. 7-3 - Basis of Equivalent Frame Method for Deflection Analysis of Two-way Slab Systems, with or without Beams

Table 7-2 - Lateral Distribution Factors (LDF) for Column and Middle Strips

Section 13.6.4 - the column strip percentages are:

Exterior Negative -- $100 - 10\,\beta_t + 12\,\beta_t(\alpha_1\,\ell_2/\ell_1)(1 - \ell_2/\ell_1)$

Interior Negative -- $75 + 30(\alpha_1\,\ell_2/\ell_1)(1 - \ell_2/\ell_1)$

Positive -- $60 + 30(\alpha_1\,\ell_2/\ell_1)(1.5 - \ell_2/\ell_1)$

except when $\alpha_1\,\ell_2/\ell_1 > 1$ (typical two-way slab case), use $\alpha_1\,\ell_2/\ell_1 = 1$

Example: $\alpha_1 = \beta_t = 0$ assumed for a flat plate with no beams. Use an average of the positive and negative region values.

Interior Panel, Both Directions -- $(LDF)_c = \dfrac{60\text{ (Pos)} + 75\text{ (Neg)}}{2} = 67.5\%$

$(LDF)_m = 100 - 67.5 = 32.5\%$

Side Panel, One End Continuous Direction --

$$(LDF)_c = \frac{60 + (100 + 75)/2}{2} = 73.8\%$$

$$(LDF)_m = 100 - 73.8 = 26.2\%$$

Side Panel, Both Ends Continuous Direction --

$(LDF)_c = 67.5\%$, $(LDF)_m = 32.5\%$

Corner Panel, Both Directions -- $(LDF)_c = 73.8\%$, $(LDF)_m = 26.2\%$

In the direct design method, moments are based on the clear span and should theoretically be adjusted to obtain moments and rotations at the column centerlines. However, the use of moments in Eq. (15) at the face of the columns should cause little error. Particularly in the case of flat-plates and flat-slabs, the span center-to-center of columns is thought to be more appropriate in deflection calculations than the clear span.

For practical application, only the exterior column rotation need be considered in most cases when using the direct design moment coefficients with equal spans. When the live load is large compared with the dead load (frequently not the case), the end rotations may be computed by a simple moment-area procedure in which the effect of pattern loading may be included.

Midspan deflection of a member having an end rotation of θ radians, with the far end fixed, is computed by Eq. (16).

$$a_\theta = (\text{End } \theta)\, \ell/8 \tag{16}$$

Because (End θ) is based on the gross-section properties in Eq. (15), when the deflection calculations are based on I_e, Eq. (17) may be used instead of Eq. (16) for consistency.

$$a_\theta = (\text{End } \theta)(\ell/8)(I_g/I_e)_{frame} \tag{17}$$

The combined midspan deflection of a column or middle strip is the sum of the three parts, as computed by Eq. (18).

$$a_{c,m} = \text{Fixed } a_{c,m} + (a_{\theta 1})_{c,m} + (a_{\theta 2})_{c,m} \tag{18}$$

where $a_{\theta 1}$ and $a_{\theta 2}$ refer to the midspan deflections due to rotations at both ends. As shown in Fig. 7-3, the total midpanel deflection is given by Eq. (19).

$$a = a_{cx} + a_{my} = a_{cy} + a_{mx} \tag{19}$$

For other than square symmetrical panels, Eq. (20) may be used.

$$a = [(a_{cx} + a_{my}) + (a_{cy} + a_{mx})]/2 \tag{20}$$

Effective Moment of Inertia

The effective moment of inertia by Eq. (1) is recommended for computing deflections of partially cracked two-way construction. An average I_e of the positive and negative regions in Eqs. (2) and (3) may also be used. The following typical cracking locations have been found empirically, and the corresponding values of I_e have been shown to apply in most cases.[7.2]

Slabs without beams (flat plates, flat slabs)

- All dead load deflections -- I_g
- Dead-plus-live load deflections;
  - For the column strips in both directions -- I_e
  - For the middle strips in both directions -- I_g

These conditions are demonstrated in Design Example 7.5.

Slabs with beams (two-way slabs)

- All dead load deflections -- I_g
- Dead-plus-live load deflections;
  - For the column strips in both directions -- I_g
  - For the middle strips in both directions -- I_e

The I_e of the equivalent frame in each direction is then taken as the sum of the column and middle strip I_e values.

Long-Time Deflection

Since the available data on long-time deflections of two-way construction is too limited to justify more elaborate procedures, the same procedures as those used for one-way members are recommended. ACI Eq. (9-10), with ξ_u = 2.5. See Design Example 7.5.

DEFLECTION OF NONCOMPOSITE PRESTRESSED MEMBERS

The ultimate (in time) camber and deflection of prestressed members may be computed by Eq. (21). The a_u expression is based on a procedure described in Reference 7.2. The procedure includes the use of the I_e method for partially prestressed members as presented in the PCI Design Handbook 7.3 -- as a suggested method to satisfy Code Section 18.4.2(c) for deflection analysis when the computed tensile stress exceeds the modulus of rupture but does not exceed $12\sqrt{f'_c}$. For more information on cracked prestressed deflections and on composite prestressed beam deflections, see Reference 7.2 and 7.7.

$$a_u = \overbrace{-\,a_{po}}^{(1)} + \overbrace{a_o}^{(2)} - \overbrace{\left[-\frac{\Delta P_u}{P_o} + (k_r\ C_u)(1 - \frac{\Delta P_u}{2P_o})\right] a_{po}}^{(3)} + \overbrace{(k_r\ C_u)\ a_o}^{(4)}$$

$$+ \overbrace{a_s}^{(5)} + \overbrace{(\beta_s\ k_r\ C_u)\ a_s}^{(6)} + \overbrace{a_\ell}^{(7)} + \overbrace{(a_{cp})_\ell}^{(8)} \qquad (21)$$

Term (1) is the initial camber due to the initial prestress moment after elastic loss, P_o e. For example, $a_{po} = P_o\ e\ \ell^2/8\ E_{ci}\ I_g$ for a straight tendon.

Term (2) is the initial self-weight deflection of the beam -- such as a_o = $5\ M_o\ell^2/48\ E_{ci}\ I_g$ for a simple beam, where M_o = midspan self-weight moment.

Term (3) is the creep (time-dependent) camber of the beam due to the prestress moment. This term includes the effects of creep and loss of prestress; that is, the creep effect under variable stress. Average values of the prestress-loss ratio after transfer (excluding elastic loss), $\Delta P_u/P_o$, are about 0.18, 0.21, and 0.23 for normal, sand, and all-lightweight concrete, respectively. An average value of C_u = 2.0 might be reasonable for the ultimate prestress-force and self-weight creep factor. The k_r factor takes into account the effect of any nonprestressed tension steel in reducing time-dependent camber, using Eq. (22). This is also used in the PCI Design Handbook[7.3] in a slightly different form.

$$k_r = 1/[1 + (A_s/A_{ps})], \quad \text{for } A_s/A_{ps} < 2 \qquad (22)$$

When k_r = 1 and $\Delta P_u = P_o - P_e$, Terms (1) + (3) can be combined as:

$$-\,a_{po} - \left[-\,a_{po} + a_{pe} + C_u(\frac{a_{po} + a_{pe}}{2})\right] = -\,a_{pe} - C_u(\frac{a_{po} + a_{pe}}{2}).$$

Term (4) is the self-weight creep deflection of the beam. Use the same value of C_u as in Term (3). Since creep under prestress and self-weight takes place under their combined stress, the effect of any nonprestressed tension steel in reducing the creep deformation is included in both the camber Term (3) and the deflection Term (4).

Term (5) is the initial deflection of the beam under a superimposed dead load -- such as $a_s = 5 M_s \ell^2/48 E_c I_g$, where M_s = midspan superimposed dead load moment (uniformly distributed).

Term (6) is the creep deflection of the beam under a superimposed dead load. k_r is the same as in Terms (3) and (4), and is included in this deflection term for the same reason as in Term (4). An average value of $C_u = 1.6$ is recommended, as in Eq. (7) for nonprestressed members. β_s is the creep correction factor for the age of the beam concrete when the superimposed dead load is applied (same values apply for different weight concrete): β_s = 0.85 for age 3 weeks, 0.83 for age 1 month, 0.76 for age 2 months, 0.74 for age 3 months, and 0.71 for age 4 months.

Term (7) is the initial live load deflection of the beam -- such as $a_\ell = 5 M_\ell \ell^2/48 E_c I_e$ for uniformly distributed live load, where M_ℓ = midspan live load moment. For uncracked members, $I_e = I_g$. For partially cracked noncomposite and composite members, see References 7.2 and 7.3. See also Design Example 7.7 for a partially cracked case.

Term (8) is the live load creep deflection of the beam. This deflection increment may be computed as $(a_{cp})\ell = (M_s/M_\ell) C_u a_\ell$, where M_s is the sustained portion of the live load moment and $C_u = 1.6$ as in Term (6).

Summary

In the Design Examples that follow, all 4 of the allowable deflection categories in ACI Table 9.5(b) are satisfied. Only h_{min} is checked for the two-way slab in Design Example 7.6, as required.

Selected References

7.1 American Concrete Institute, Deflections of Concrete Structures, ACI Publication SP 43, 1974, 637 pp.

7.2 Branson, D. E., Deformation of Concrete Structures, McGraw-Hill Advanced Book Program, 1977, 546 pp.

7.3 Prestressed Concrete Institute, PCI Design Handbook, 2nd Ed., 1978, pp. 8-28.

7.4 American Concrete Institute, Designing for Creep and Shrinkage in Concrete Structures, ACI Publication SP 76, 1982, 484 pp.

7.5 American Concrete Institute, Designing for Effects of Creep, Shrinkage, and Temperature in Concrete Structures, ACI Publication SP 27, 1971, 430 pp.

7.6 Nilson, A. H., and Walters, D. B., "Deflection of Two-Way Floor Systems by the Equivalent Frame Method," ACI Journal, Proceedings V. 72, No. 5, May 1975, pp. 210-218.

7.7 Branson, D. E., Chapter 4 -- "Reinforced Concrete Composite Flexural Members," pp. 97-147; and Chapter 5 -- "Prestressed Concrete Composite Flexural Members," pp. 148-210; Handbook of Composite Construction Engineering, Van Nostrand Reinhold Co., Editor, G. M. Sabnis, 1979, 380 pp.

EXAMPLE 7.1 - Simple-Span Nonprestressed Rectangular Beam

Required: Analysis of short-time deflections, and long-time deflections at ages 3 months and 5 years (ultimate values)

Data: f'_c = 3,000 psi (normal wt. concrete)

f_y = 40,000 psi

A_s = 3 # 7 = 1.80 in.^{2}

$\rho = A_s/bd = 0.0077$

A'_s = 3 #4 = 0.60 in.^{2}

$\rho' = A'_s/bd = 0.0026$

(A'_s not required for strength)

Superimposed dead load (not including beam wt.) = 120 lb/ft

Live load = 300 lb/ft (50% sustained)

Span = 25 ft

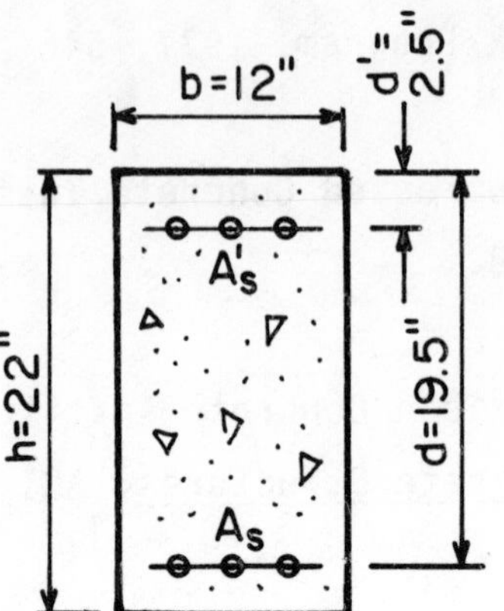

Calculations and Discussion	Code Reference

1. Minimum thickness:

$h_{min} = (\ell/16)(0.80 \text{ for } f_y) = (300/16)(0.80)$ Table 9.5(a)

$= 15.0 \text{ in.} < h = 22 \text{ in.}$

This, the condition for nonstructural elements in Table 9.5(a) is satisfied. This agrees with the deflection checks.

2. Moments:

$w_d = 0.120 + (12)(22)(0.150)/144 = 0.395 \text{ k/ft}$

EXAMPLE 7.1 - Continued

Calculations and Discussion	Code Reference

$M_d = w_d\ell^2/8 = (0.395)(25)^2/8 = 30.9$ ft-k

$M_\ell = w_\ell\ell^2/8 = (0.300)(25)^2/8 = 23.4$ ft-k, $M_{d+\ell} = 54.3$ ft-k

$M_{sus} = M_d + 0.50M_\ell = 30.9 + (0.50)(23.4) = 42.6$ ft-k

3. Modulus of rupture, modulus of elasticity, modular ratio:

$f_r = 7.5\sqrt{f'_c} = 7.5\sqrt{3000} = 411$ psi — Eq. (9-9)

$E_c = 33\sqrt{w_c^3 f'_c} = 33\sqrt{(145)^3(3000)} = 3.16 \times 10^6$ psi — 8.5.1

$n_s = E_c/E = 29/3.16 = 9.2$

4. Gross and cracked section moments of inertia, using Fig. 7-1:

$I_g = b h^3/12 = (12)(22)^3/12 = 10{,}650$ in.^{4}

$B = b/(n A_s) = 12/(9.2)(1.80) = 0.725$ 1/in.

$r = (n - 1)A'_s/(n A_s) = (8.2)(0.60)/(9.2)(1.80) = 0.297$

$a = [\sqrt{2dB(1+rd'/d) + (1+r)^2} - (1+r)]/B$

$= [\sqrt{(2)(19.5)(0.725)(1 + \frac{0.297 \times 2.5}{19.5}) + 1.297^2} - 1.297]/0.725 = 5.89$ in.

$I_{cr} = b a^3/3 + n A_s(d - a)^2 + (n - 1)A'_s(a - d')^2$

$= (12)(5.89)^3/3 + (9.2)(1.80)(19.5 - 5.89)^2 + (8.2)(0.60)(5.89 - 2.5)^2$

$= 3{,}940$ in.^{4}, $I_g/I_{cr} = 2.7$

5. Effective moments of inertia, using Eq. (1):

$M_{cr} = f_r I_g/y_t = (411)(10{,}650)/(11)(12{,}000) = 33.2$ ft-k — Eq. (9-8)

$M_{cr}/M_d = 33.2/30.9 > 1$. Hence $(I_e)_d = I_g = 10{,}650$ in.^{4}

$M_{cr}/M_{sus})^3 = (33.2/42.6)^3 = 0.473$

$(I_e)_{sus} = (M_{cr}/M_a)^3 I_g + [1 - (M_{cr}/M_a)^3] I_{cr} \leq I_g$ — Eq. (9-7)

EXAMPLE 7.1 - Continued

Calculations and Discussion	Code Reference

$= (0.473)(10{,}650) + (1 - 0.473)(3{,}940) = 7{,}110 \text{ in.}^4$

$(M_{cr}/M_{d+\ell})^3 = (33.2/54.3)^3 = 0.229$

$(I_e)_{d+\ell} = (0.229)(10{,}650) + (1 - 0.229)(3{,}940) = 5{,}480 \text{ in.}^4$

6. Initial or short-time deflections, using Eq. (4): 9.5.2.2 9.5.2.3

$$(a_i)d = \frac{K\,(5/48)M_d\,\ell^2}{E_c\,(I_e)_d} = \frac{(1)\,(5/48)(30.9)(25)^2(12)^3}{(3160)(10{,}650)} = 0.103 \text{ in.}$$

$$(a_i)_{sus} = \frac{K\,(5/48)\,M_{sus}\,\ell^2}{E_c\,\,(I_e)_{sus}} = \frac{(1)(5/48)(42.6)(25)^2(12)^3}{(3160)(7{,}110)} = 0.213 \text{ in.}$$

$$(a_i)_{d+\ell} = \frac{K\,(5/48)\,M_{d+\ell}\,\ell^2}{E_c\,\,(I_e)_{d+\ell}} = \frac{(1)(5/48)(54.3)(25)^2(12)^3}{(3160)(5{,}480)} = 0.353 \text{ in.}$$

$(a_i)_\ell = (a_i)_{d+\ell} - (a_i)_d = 0.353 - 0.103 = 0.250$ in.

versus the following allowable deflections from Table 9.5(b):

Flat roofs not supporting and not attached to nonstructural elements likely to be damaged by large deflections --

$(a_i)_\ell \leq \ell/180 = 300/180 = 1.67$ in. OK

Floors not supporting and not attached to nonstructural elements likely to be damaged by large deflections --

$(a_i)_\ell \leq /360 = 300/360 = 0.83$ in. OK

7. Long-time deflections at ages 3 mos and 5 yrs (ult. values):

Combined creep and shrinkage deflections, using Eqs. (5) and (6):

$$\lambda = \frac{\xi_u}{1 + 50\,\rho'} = \frac{2.0}{1 + (50)(0.0026)} = 1.77$$ 9.5.2.5 Eq. (9-10)

$a_{(cp+sh)} = \lambda(a_i)_{sus} = (1.77)(0.213) = 0.377$ in.

$a_{(cp+sh)} + (a_i)_\ell = 0.377 + 0.250 =$ <u>0.63 in.</u> -- ult. value

EXAMPLE 7.1 - Continued

Calculations and Discussion	Code Reference

versus 0.44 in. at age 1 year using $\xi = 1.0$ instead of 2.0 in Eq. (5).

Separate creep and shrinkage deflections, using Eqs. (7) and (8):

$$\lambda_{cp} = \frac{0.85\ C_u}{1+50\ \rho'} = \frac{(0.85)(1.60)}{1+(50)(0.0026)} = 1.20$$

$a_{cp} = \lambda_{cp}\ (a_i)_{sus} = (1.20)(0.213) = 0.256$ in.

A_{sh} (from Fig. 7-2) = 0.455

$\varphi_{sh} = A_{sh}\ (\varepsilon_{sh})_u/h = (0.455)(400 \times 10^{-6})/22 = 8.27 \times 10^{-6}$ 1/in.

$a_{sh} = K_{sh}\ \varphi_{sh}\ \ell^2 = (1/8)(8.27 \times 10^{-6})(25)^2(12)^2 = 0.093$ in.

$a_{cp} + a_{sh} + (a_i)_\ell = 0.256 + 0.093 + 0.250 =$ <u>0.60 in.</u> -- ult. value

versus 0.45 in. at age 3 months using $C_t = (0.56)(1.60) = 0.90$ and

$\varepsilon_{sh} = (0.60)(400 \times 10^{-6}) = 240 \times 10^{-6}$.

These computed ultimate deflections of <u>0.60 in.</u> and <u>0.63 in.</u> are compared with the allowable deflections in Table 9.5(b) as follows: Roof or floor construction supporting or attached to nonstructural elements <u>likely</u> to be damaged by large deflections (very stringent limitation) --

$a_{cp} + a_{sh} + (a_i)_\ell \leq \ell/480 = 300/480 = 0.63$ in. OK by Both Methods

Roof or floor construction supporting or attached to nonstructural elements <u>not likely</u> to be damaged by large deflections --

$a_{cp} + a_{sh} + (a_i)_\ell \leq \ell/240 = 300/240 = 1.25$ in. OK by Both Methods

EXAMPLE 7.2 - Continuous Nonprestressed T-Beam

Required: Analysis of short-time and ultimate long-time deflections of end-span of multi-span beam shown below.

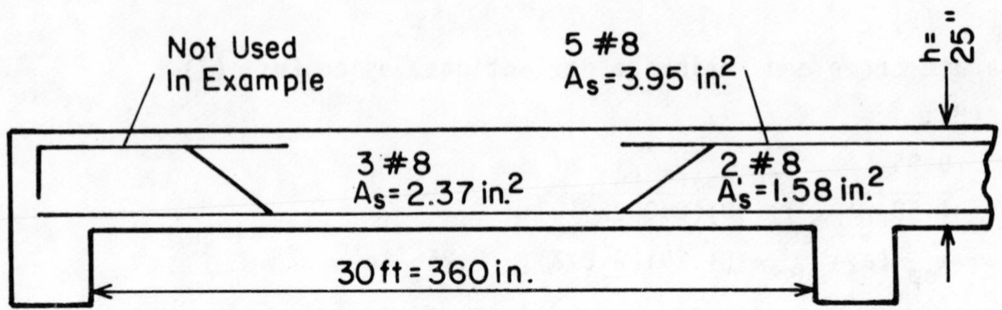

Beam Spacing = 10', b = 360/4 = 90" or 120" or 16(5) + 12 = 92". Use 90"

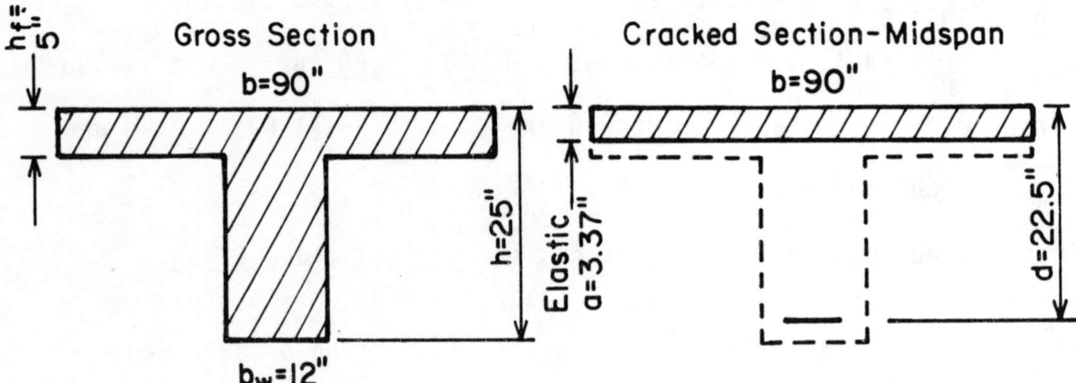

$n\,A_s = (11.3)(2.37) = 26.8 \text{ in.}^2$

$\rho = 2.37/(90)(22.5) = 0.00117$

$\rho_w = 2.37/(12)(22.5) = 0.00878$

$\rho' = 0$

Cracked Section-Interior Support

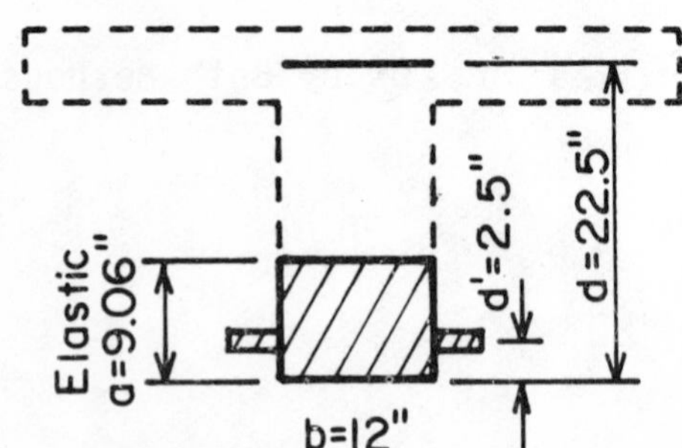

EXAMPLE 7.2 - Continued

Data: f'_c = 4,000 psi (sand-lightweight concrete)

f_y = 50,000 psi

w_c = 115 pcf

Superimposed Dead Load (not including beam weight) = 20 psf

Live Load = 100 psf (30% sustained)

Beam will be assumed to be continuous at one end only for h_{min} in Table 9.5(a), for Avg. I_e in Eq. (2), and for K_{sh} in Eq. (8), since the exterior end is supported by a spandrel beam. The end span might be assumed to be continuous at both ends when supported by an exterior column.

(A'_s not required for strength)

Calculations and Discussion	Code Reference

1. Minimum thickness:

$h_{min} = (\ell/18.5)(0.90 \text{ for } f_y)(1.09 \text{ for } w_c)$ — Table 9.5(a)

$= (360/18.5)(0.90)(1.09) = 19.1 \text{ in.} < h = 25 \text{ in.}$

Thus, the condition for nonstructural elements in Table 9.5(a) is satisfied. This agrees with the deflection checks.

2. Loads and moments:

$w_d = (20 \times 10) + (120 \text{ pcf})(12 \times 20 + 120 \times 5)/144 = 900 \text{ lb/ft}$

$w_\ell = (100 \times 10) = 1000 \text{ lb/ft}$

In lieu of a moment analysis, the ACI approximate moment coefficients may be used as follows: Pos. $M = w\,\ell_h^2/14$ for positive I_e and maximum deflection, and Neg. $M = w\,\ell_h^2/10$ for negative I_e. — 8.3.3

Pos. $M_d = w_d\,\ell_h^2/14 = (0.900)(30)^2/14 = 57.9$ ft-k

Pos. $M_\ell = (1.000)(30)^2/14 = 64.3$ ft-k, Pos. $M_{d+\ell} = 122.2$ ft-k

EXAMPLE 7.2 - Continued

Calculations and Discussion	Code Reference

Pos. $M_{sus} = M_d + 0.30\ M_\ell = 57.9 + (0.30)(64.3) = 77.2$ ft-k

Neg. $M_d = w_d\ \ell_h^2/10 = (0.900)(30)^2/10 = 81.0$ ft-k

Neg. $M_\ell = (1.000)(30)^2/10 = 90.0$ ft-k, Neg. $M_{d+\ell} = 171.0$ ft-k

Neg. $M_{sus} = M_d + 0.30\ M_\ell = 81.0 + (0.30)(90.0) = 108.0$ ft-k

3. Modulus of rupture, modulus of elasticity, modular ratio:

$f_r = (0.85)(7.5)\sqrt{f'_c} = 6.38\sqrt{4000} = 404$ psi — Eq. (9-9), 9.5.2.3(b), 8.5.1

$E_c = 33\sqrt{w_c^3\ f'_c} = 33\sqrt{(115)^3(4000)} = 2.57 \times 10^6$ psi

$n = E_s/E_c = 29/2.57 = 11.3$

4. Gross and cracked section moments of inertia, using Fig. 7-1:

Positive moment section

$y_t = h - (1/2)[(b - b_w)h_f^2 + b_w\ h^2]/[(b - b_w)h_f + b_w\ h]$

$= 25 - (1/2)[(78)(5)^2 + (12)(25)^2]/[(78)(5) + (12)(25)] = 18.15$ in.

$I_g = (b - b_w)h_f^3/12 + b_w\ h^3/12 + (b - b_w)h_f(h - h_f/2 - y_t)^2 + b_w\ h(y_t - h/2)^2$

$= (78)(5)^3/12 + (12)(25)^3/12 + (78)(5)(25 - 2.5 - 18.15)^2$

$+ (12)(25)(18.15 - 12.5)^2 = 33{,}390\ \text{in.}^4$

$B = b/(n\ A_s) = 90/(11.3)(2.37) = 3.361$ 1/in.

$a = (\sqrt{2\ d\ B + 1} - 1)/B = [\sqrt{(2)(22.5)(3.361) + 1} - 1]/3.361 = 3.37$ in.

$< h_f = 5$ in. Hence, treat as a rectangular compression area.

$I_{cr} = b\ a^3/3 + n\ A_s(d - a)^2 = (90)(3.37)^3/3 + (11.3)(2.37)(22.5 - 3.37)^2$

$= 10{,}950\ \text{in.}^4 \qquad I_g/I_{cr} = 3.1$

Negative moment section

$I_g = 33{,}390\ \text{in.}^4$ (same as Pos. Sec.). $I_{cr} = 11{,}740\ \text{in.}^4$ (similar to

EXAMPLE 7.2 - Continued

Calculations and Discussion	Code Reference

Example 7.1, for b = 12 in., d = 22.5 in., d' = 2.5 in., A_s = 3.95 in.^{2}, A'_s = 1.58 in.^{2}).

5. Effective moments of inertia, using Eqs. (1) and (2):

Positive moment section

$M_{cr} = f_r\, I_g/y_t$ = (404)(33,390)/(18.15)(12,000) = 61.9 ft-k   Eq. (9-8)

M_{cr}/M_d = 61.9/57.9 > 1. Hence $(I_e)_d = I_g$ = 33,390 in.^{4}

$(M_{cr}/M_{sus})^3 = (61.9/77.2)^3 = 0.515$

$(I_e)_{sus} = (M_{cr}/M_a)^3\, I_g + [1 - (M_{cr}/M_a)^3]\, I_{cr} \le I_g$ Eq. (9-7)

= (0.515)(33,390) + (1 - 0.515)(10,950) = 22,510 in.^{4}

$(M_{cr}/M_{d+\ell})^3 = (61.9/122.2)^3 = 0.130$

$(I_e)_{d+\ell}$ = (0.130)(33,390) + (1 - 0.130)(10,950) = 13,870 in.^{4}

Negative moment section

M_{cr} = (404)(33,390)/(25 - 18.15)(12,000) = 164.1 ft-k   Eq. (9-8)

M_{cr}/M_d = 164.1/81.0 > 1. Hence $(I_e)_d = I_g$ = 33,390 in.^{4}

M_{cr}/M_{sus} = 164.1/108.0 > 1. Hence $(I_e)_{sus} = I_g$ = 33,390 in.^{4}

$M_{cr}/M_{d+\ell})^3 = (164.1/171.0)^3 = 0.884$

$(I_e)_{d+\ell}$ = (0.884)(33,390)+(1 - 0.884)(11,740) = 30,800 in.^{4} Eq. (9-7)

Average section values

Avg. $(I_e)_d = I_g$ = 33,390 in.^{4}

Avg. $(I_e)_{sus} = 0.85\, I_m + 0.15\, (I_e$ -Cont. End)   Commentary 9.5.2.4

= (0.85)(22,510) + (0.15)(33,390) = 24,140 in.^{4}

EXAMPLE 7.2 - Continued

Calculations and Discussion	Code Reference

Avg. $(I_e)_{d+\ell}$ = (0.85)(13,870) + (0.15)(30,880) = 16,420 in.^{4}

6. Initial or short-time deflections, using Eq. (4) with Midspan I_e and with Avg. I_e: 9.5.2.4

$K = 1.20 - 0.20\, M_o/M_a = 1.20 - (0.20)(w\ell_n^2/8)/(w\ell_n^2/14) = 0.850$

$$(a_i)_d = \frac{K\,(5/48)\,M_d\,\ell^2}{E_c\,(I_e)_d} = \frac{(0.850)(5/48)(57.9)(30)^2(12)^3}{(2570)(33{,}390)} = 0.093 \text{ in.}$$

= 0.093 in. (same) using Avg. I_e in Eq. (2) = 33,390 in.^{4}

$$(a_i)_{sus} = \frac{K\,(5/48)\,M_{sus}\,\ell^2}{E_c\,(I_e)_{sus}} = \frac{(0.850)(5/48)(77.2)(30)^2(12)^3}{(2570)(22{,}510)} = 0.184 \text{ in.}$$

= 0.171 in. using Avg. I_e in Eq. (2) = 24,140 in.^{4}

$$(a_i)_{d+\ell} = \frac{K\,(5/48)\,M_{d+\ell}\,\ell^2}{E_c\,(I_e)_{d+\ell}} = \frac{(0.850)(5/48)(122.2)(30)^2(12)^3}{(2570)(13{,}870)} = 0.472 \text{ in.}$$

= 0.399 in. using Avg. I_e in Eq. (2) = 16,420 in.^{4}

$(a_i)_\ell = (a_i)_{d+\ell} - (a_i)_d = 0.472 - 0.093 =$ <u>0.379 in.</u>

= <u>0.306 in.</u> using Avg. I_e in Eq. (2).

versus the following allowable deflections from Table 9.5(b):

<u>Flat roofs</u> not supporting and not attached to nonstructural elements likely to be damaged by large deflections -- $(a_i)_\ell \leq \ell/180 = 2.00$ in. OK

<u>Floors</u> not supporting and not attached to nonstructural elements likely to be damaged by large deflections -- $(a_i)_\ell \leq \ell/360 = 360/360 = 1.00$ in. OK

7. Ultimate long-time deflections:

Combined creep and shrinkage deflections, using Eqs. (5) and (6):

Pos. $\lambda = \frac{\xi_u}{1 + 50\,\rho'} = \frac{2.0}{1 + 0} = 2.0$ 9.5.2.5 Eq. (9-10)

$a_{(cp + sh)} = \lambda\,(a_i)_{sus} = (2.0)(0.184) = 0.368$ in.

EXAMPLE 7.2 - Continued

Calculations and Discussion	Code Reference

$a_{(cp + sh)} + (a_i)_\ell = 0.368 + 0.379 = \underline{0.75 \text{ in.}}$

$= \underline{0.65 \text{ in.}}$ using Avg. I_e in Eq. (2).

Separate creep and shrinkage deflections, using Eqs. (7) and (8):

Pos. $\lambda_{cp} = \dfrac{0.85\, C_u}{1 + 50\, \rho'} = \dfrac{(0.85)(1.60)}{1 + 0} = 1.36$

$a_{cp} = \lambda_{cp}\, (a_i)_{sus} = (1.36)(0.184) = 0.250$ in.

$= 0.233$ in. using Avg. I_e in Eq. (2).

Pos. $\rho = 100(\rho + \rho_w)/2 = 100(0.00117 + 0.00878)/2 = 0.498$

A_{sh} (from Fig. 7-2) = 0.555

$\varphi_{sh} = A_{sh}\, (\varepsilon_{sh})_u/h = (0.555)(400 \times 10^{-6})/25 = 8.88 \times 10^{-6}$ 1/in.

$a_{sh} = K_{sh}\, \varphi_{sh}\, \ell^2 = (0.090)(8.88 \times 10^{-6})(30)^2(12)^2 = 0.104$ in.

$a_{cp} + a_{sh} + (a_i)_\ell = 0.250 + 0.104 + 0.379 = \underline{0.73 \text{ in.}}$

$= \underline{0.64 \text{ in.}}$ using Avg. I_e in Eq. (2).

These computed ultimate deflections of <u>0.64 in.</u> to <u>0.75 in.</u> are compared with the allowable deflections in Table 9.5(b) as follows: Roof or floor construction supporting or attached to nonstructural elements <u>likely</u> to be damaged by large deflections (very stringent limitation) --

$a_{cp} + a_{sh} + (a_i)_\ell \leq \ell/480 = 360/480 = 0.75$ in. All Results OK

Roof or floor construction supporting or attached to nonstructural elements <u>not likely</u> to be damaged by large deflections --

$a_{cp} + a_{sh} + (a_i)_\ell \leq \ell/240 = 360/240 = 1.50$ in. All Results OK

EXAMPLE 7.3 - Unshored Nonprestressed Composite Beam

Required: Analysis of short-time and ultimate long-time deflections.

Data: Normal wt. concrete

slab f'_c = 3000 psi

precast beam f'_c = 4000 psi

f_y = 40,000 psi

A_s = 3 #9 = 3.00 in.^{2}

Superimposed Dead Load (not including beam and slab weight) = 10 psf

Live Load = 75 psf (20% Sustained)

b_e = 312/4 = 78.0 in. or Spacing = 96.0 in.

or 16(4) + 12 = 76.0

Simple span = 26 ft = 312 in.

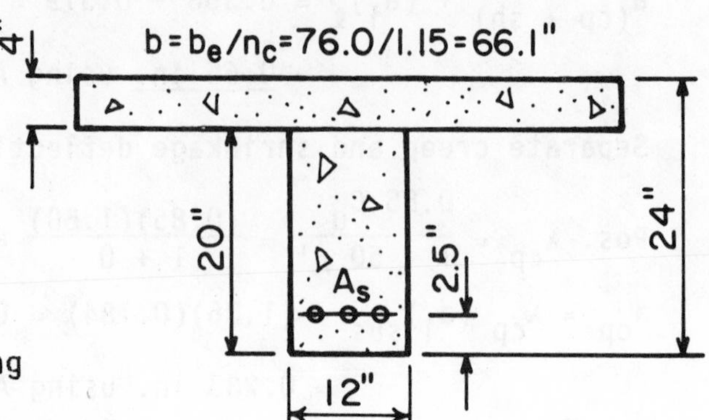

Calculations and Discussion	Code Reference

1. Minimum thickness:

h_{min} = (ℓ/16)(0.80 for f_y) = (312/16)(0.80) — Table 9.5(a)

= 15.6 in. < h = 20 in. or 24 in.

Thus the condition for nonstructural elements in Table 9.5(a) is satisfied. This agrees with the deflection checks.

2. Loads and moments:

w_1 = (10 psf)(8) + (150 pcf)(96)(4)/144 = 480 lb/ft

w_2 = (150 pcf)(12)(20)/144 = 250 lb/ft

w_ℓ = (75 psf)(8) = 600 lb/ft

$M_1 = w_1 \ell^2/8 = (0.480)(26)^2/8 = 40.6$ ft-k

$M_2 = w_2 \ell^2/8 = (0.250)(26)^2/8 = 21.1$ ft-k

$M_\ell = w_\ell \ell^2/8 = (0.600)(26)^2/8 = 50.7$ ft-k

EXAMPLE 7.3 - Continued

Calculations and Discussion	Code Reference

3. Modulus of rupture, modulus of elasticity, modular ratio:

$(E_c)_1 = 33\sqrt{w_c^3 f'_c} = 33\sqrt{(145)^3(3000)} = 3.16 \times 10^6$ psi — 8.5.1

$(f_r)_2 = 7.5\sqrt{f'_c} = 7.5\sqrt{4000} = 474$ psi — Eq. (9-9)

$(E_c)_2 = 33\sqrt{(145)^3(4000)} = 3.64 \times 10^6$ psi — 8.5.1

$n_c = (E_c)_2/(E_c)_1 = 3.64/3.16 = 1.15, \quad n = E_s/(E_c)_2 = 29/3.64 = 8.0$

4. Gross and cracked section moments of inertia, using Fig. 7-1:

Precast Section

$I_g = (12)(20)^3/12 = 8{,}000\ \text{in.}^4 \quad B = b/(n\ A_s) = 12/(8.0)(3.00) = 0.5000\ 1/\text{in.}$

$a = (\sqrt{2\ d\ B + 1} - 1)/B = [\sqrt{(2)(17.5)(0.5000) + 1} - 1]/0.5000 = 6.60$ in.

$I_{cr} = b\ a^3/3 + n\ A_s(d - a)^2$

$= (12)(6.60)^3/3 + (8.0)(3.00)(17.5 - 6.60)^2 = 4{,}000\ \text{in.}^4$

Composite Section

$y_t = h - (1/2)[(b - b_w)h_f^2 + b_w\ h^2]/[(b - b_w)h_f + b_w\ h]$

$= 24 - (1/2)[(54.1)(4)^2 + (12)(24)^2]/[(54.1)(4) + (12)(24)] = 16.29$ in.

$I_g = (b - b_w)h_f^3/12 + b_w\ h^3/12 + (b - b_w)h_f(h - h_f/2 - y_t)^2 + b_w\ h(y_t - h/2)^2$

$= (54.1)(4)^3/12 + (12)(24)^3/12 + (54.1)(4)(24 - 2 - 16.29)^2$

$+ (12)(24)(16.29 - 12)^2 = 26{,}470\ \text{in.}^4$

$B = b/(n\ A_s) = 66.1/(8.0)(3.00) = 2.754$

$a = (\sqrt{2\ d\ B + 1} - 1)/B = [\sqrt{(2)(21.5)(2.754) + 1} - 1]/2.754 = 3.60$ in.

$< h_f = 4$ in. Hence, treat as a rectangular compression area.

$I_{cr} = b\ a^3/3 + n\ A_s(d - a)^2$

$= (66.1)(3.60)^3/3 + (8.0)(3.00)(21.5 - 3.60)^2 = 8{,}720\ \text{in.}^4$

EXAMPLE 7.3 - Continued

Calculations and Discussion	Code Reference

$I_2/I_c = [(I_2/I_c)_g+(I_2/I_c)_{cr}]/2 = [(8,000/26,470)+(4,000/8,720)]/2 = 0.380$

5. Effective moments of inertia, using Eq. (1):

In Term (1), Eq. (9) -- Precast Section,

$M_{cr} = f_r\ I_g/y_t = (474)(8,000)/(10)(12,000) = 31.6$ ft-k — Eq. (9-8)

$M_{cr}/M_2 = 31.6/21.1 > 1$. Hence $(I_e)_2 = I_g = 8,000$ in.^{4}

In Term (6), Eq. (9) -- Precast Section,

$[M_{cr}/(M_1 + M_2)]^3 = [31.6/(40.6 + 21.1)]^3 = 0.134$

$(I_e)_{1+2} = (M_{cr}/M_a)^3\ I_g + [1 - (M_{cr}/M_a)^3]\ I_{cr} \leq I_g$ — Eq. (9-7)

$= (0.134)(8,000) + (1 - 0.134)(4,000) = 4,540$ in.^{4}

6. Deflections, using Eqs. (9) and (10):

Term (1) -- $(a_i)_2 = \dfrac{K\ (5/48)\ M_2\ \ell^2}{(E_c)_2\ (I_e)_2} = \dfrac{(1)(5/48)(21.1)(26)^2(12)^3}{(3640)(8,000)} = 0.088$ in.

Term (2) -- $k_r = 0.85$ (no precast beam compression steel),

$0.77\ k_r\ (a_i)_2 = (0.77)(0.85)(0.088) = 0.058$ in.

Term (3) -- $0.83\ k_r\ (a_i)_2\ \dfrac{I_2}{I_c} = (0.83)(0.85)(0.088)(0.380) = 0.024$ in.

Term (4) -- $K_{sh} = 1/8$. Precast Sec.: $\rho = (100)(3.00)/(12)(17.5) = 1.43\%$

From Fig. 7-2, $A_{sh} = 0.789$

$\varphi_{sh} = A_{sh}\ (\epsilon_{sh})_u/h = (0.789)(400\text{x}10^{-6})/20 = 15.78\text{x}10^{-6}$ 1/in.

$a_{sh} = K_{sh}\ \varphi_{sh}\ \ell^2 = (1/8)(15.78\text{x}10^{-6})(26)^2(12)^2 = 0.192$ in.

$0.36\ a_{sh} = (0.36)(0.192) = 0.069$ in.

Term (5) -- $0.64\ a_{sh}\ \dfrac{I_2}{I_c} = (0.64)(0.192)(0.380) = 0.047$ in.

EXAMPLE 7.3 - Continued

Calculations and Discussion	Code Reference

Term (6) -- $(a_i)_1 = \dfrac{K\,(5/48)(M_1 + M_2)\,\ell^2}{(E_c)_2\,(I_e)_{1+2}} - (a_i)_2$

$= \dfrac{(1)(5/48)(40.6 + 21.1)(26)^2(12)^3}{(3640)(4{,}540)} - 0.088 = 0.366$ in.

Term (7) -- $1.22\ k_r\ (a_i)_1\ \dfrac{I_2}{I_c} = (1.22)(0.85)(0.366)(0.380) = 0.144$ in.

Term (8) -- $a_{ds} = 0.50(a_i)_1 = (0.50)(0.366) = 0.183$ in. (rough estimate)

Term (9) -- Using the alternative method,

$(a_i)_\ell = \dfrac{K\,(5/48)\,M_\ell\,\ell^2}{(E_c)_2\,(I_c)_{cr}} = \dfrac{(1)(5/48)(50.7)(26)^2(12)^3}{(3640)(8{,}720)} = \underline{0.194\text{ in.}}$

Term (10) -- $k_r = 0.85$ (neglecting the effect of any slab compression steel)

$(a_{cp})_\ell = k_r\ C_u\ [0.20\ (a_i)_\ell]$

$= (0.85)(1.60)(0.20 \times 0.194) = 0.053$ in.

In Eq. (9), $a_u = 0.088 + 0.058 + 0.024 + 0.069 + 0.047 + 0.366 + 0.144$

$+ 0.183 + 0.194 + 0.053 = \underline{1.23\text{ in.}}$

Checking Eq. (10) (same solution),

$a_u = (1.65 + 0.71\ \dfrac{I_2}{I_c})(a_i)_2 + (0.36 + 0.64\ \dfrac{I_2}{I_c}\ )\ a_{sh} + (1.50 + 1.04\ \dfrac{I_2}{I_c}\ )(a_i)_1$

$+ (a_i)_\ell + (a_{cp})_\ell$

$= (1.65 + 0.71 \times 0.380)(0.088) + (0.36 + 0.64 \times 0.380)(0.192)$

$+ (1.50 + 1.04 \times 0.380)(0.366) + 0.194 + 0.053 = 1.23$ in. (Same)

Assuming nonstructural elements are installed after the composite slab has hardened, $a_{cp} + a_{sh} + (a_i)_\ell$

= Terms (3) + (5) + (7) + (8) + (9) + (10) -- Excluding (1), (2), (4), and (6)

$= 0.024 + 0.047 + 0.144 + 0.183 + 0.194 + 0.053 = \underline{0.65\text{ in.}}$

Comparisons with the allowable deflections in Table 9.5(b) are shown at the end of Design Example 7.4.

EXAMPLE 7.4 - Shored Nonprestressed Composite Beam

Data: Same as Example 7.3, except using shored construction

Required: Analysis of short-time and ultimate long-time deflections, to show the beneficial effect of shoring in reducing deflections.

Calculations and Discussion	Code Reference

1. Effective moments of inertia, using Eq. (1):

$M_{cr} = f_r I_g/y_t = (474)(26,470)/(16.29)(12,000) = 64.2$ ft-k   Eq. (9-8)

$M_{cr}/(M_1+M_2) = [64.2/(40.6+21.1)]=1.04 > 1.$ Hence $(I_e)_{1+2} = I_g = 26,470$ in.^{4}

In Term (5), Eq. (11) -- Composite Section,

$[M_{cr}/(M_1 + M_2 + M_\ell)]^3 = [64.2/(40.6 + 21.1 + 50.7)]^3 = 0.186$

$(I_e)_{d+\ell} = (M_{cr}/M_a)^3 I_g + [1 - (M_{cr}/M_a)^3] I_{cr} \leq I_g$ Eq. (9-7)

$= (0.186)(26,470) + (1 - 0.186)(8,720) = 12,020$ in.^{4}

versus the alternative method of Example 7.3 using $I_e = (I_c)_{cr} = 8,720$ in.^{4} with the live load moment directly.

2. Deflections, using Eqs. (11) and (12):

Term (1) -- $(a_i)_{1+2} = \dfrac{K(5/48)(M_1 + M_2)\ell^2}{(E_c)_2 (I_e)_{1+2}} = \dfrac{(1)(5/48)(40.6 + 21.1)(26)^2(12)^3}{(3640)(26,470)}$

$= 0.078$ in.

Term (2) -- $k_r = 0.85$ (neglecting the effect of any slab compression steel).

$1.80 k_r (a_i)_{1+2} = (1.80)(0.85)(0.078) = 0.119$ in.

Term (3) -- From Term (4) of Example 7.3, and using $(\varepsilon_{sh})_u =$ 252×10^{-6} in./in.,

$a_{sh} \dfrac{I_2}{I_c} = (252/400)(0.192)(0.380) = 0.046$ in.

Term (4) -- $a_{ds} = (a_i)_{1+2} = 0.078$ in.   (rough estimate)

EXAMPLE 7.4 - Continued

Calculations and Discussion	Code Reference

Term (5) -- $(a_i)_\ell = \dfrac{K\,(5/48)(M_1 + M_2 + M_\ell)\,\ell^2}{(E_c)_2\,(I_e)_{d+\ell}} - (a_i)_{1+2}$

$= \dfrac{(1)(5/48)(40.6+21.1+50.7)(26)^2(12)^3}{(3640)(12,020)} - 0.078 = \underline{0.235 \text{ in.}}$

Term (6) -- $k_r = 0.85$ (neglecting the effect of any slab compression steel),

$(a_{cp})_\ell = k_r C_u[0.20(a_i)_\ell] = (0.85)(1.60)(0.20\text{x}0.235) = 0.064$ in.

In Eq. (11), $a_u = 0.078 + 0.119 + 0.046 + 0.078 + 0.235 + 0.064 = \underline{0.62 \text{ in.}}$ versus $\underline{1.23 \text{ in.}}$ in Example 7.3 with unshored construction. This shows the beneficial effect of shoring in reducing the total deflection.

Checking Eq. (12) (same solution),

$a_u = 3.53\,(a_i)_{1+2} + a_{sh}\dfrac{I_2}{I_c} + (a_i)_\ell + (a_{cp})_\ell$

$= (3.53)(0.078) + 0.046 + 0.235 + 0.064 = \underline{0.62 \text{ in.}}$ (Same)

Assuming nonstructural elements are installed after the shores are removed,

$a_{cp} + a_{sh} + (a_i)_\ell = a_u - (a_i)_{1+2} = 0.62 - 0.08 = \underline{0.54 \text{ in.}}$

The computed deflections of $(a_i)_\ell = \underline{0.19 \text{ in.}}$ in Example 7.3 and $\underline{0.24 \text{ in.}}$ in Example 7.4; and $a_{cp} + a_{sh} + (a_i)_\ell = \underline{0.65 \text{ in.}}$ in Example 7.3 and $\underline{0.54 \text{ in.}}$ in Example 7.4 are compared with the allowable deflections in Table 9.5(b) as follows:

Flat roofs not supporting and not attached to nonstructural elements likely to be damaged by large deflections --

$(a_i)_\ell \leq \ell/180 = 312/180 = 1.73$ in. OK

Floors not supporting and not attached to nonstructural elements likely to be damaged by large deflections --

$(a_i)_\ell \leq \ell/360 = 312/360 = 0.87$ in. OK

EXAMPLE 7.4 - Continued

Calculations and Discussion	Code Reference

Roof or floor construction supporting or attached to nonstructural elements likely to be damaged by large deflections (very stringent limitation) --

$a_{cp} + a_{sh} + (a_i)_\ell \leq \ell/480 = 312/480 = 0.65$ in. OK

Roof or floor construction supporting or attached to nonstructural elements not likely to be damaged by large deflections --

$a_{cp} + a_{sh} (a_i)_\ell \leq \ell/240 = 312/240 = 1.30$ in. OK

All computed deflections are found to be satisfactory in all four categories.

EXAMPLE 7.5 - Slab System Without Beams (Flat Plate)

Required: Analysis of short-time and ultimate long-time deflections of a corner panel

Data: Flat plate with no spandrel beams, designed by direct design method

Slab f'_c = 3 ksi, Col. f'_c = 5 ksi, (Normal Weight Concrete)

f_y = 40 ksi

Square panels -- 15 ft x 15 ft c.c. of columns

Square columns -- 14 in. x 14 in., Clear span = ℓ_n = 15 - 1.17 = 13.83 ft

Story height = 10 ft, Slab thickness = h = 6 in.

Col. Strip -- Pos. A_s = 4 - #5 = 1.24 in.^{2}, Neg. A_s = 6 - #5 = 1.86 in.^{2}

Pos. d = 6.0 - 0.75 - 0.31 (1 #5) = 4.94 in.

Neg. d = 6.0 - 0.75 - 0.63 (2 #5) = 4.62 in.

Middle Strip reinforcement and d values are not required for deflection computations, since the slab remains uncracked in the middle strips.

Superimposed Dead Load = 10 psf

Live Load = 50 psf, Check for 0% and 40% Sustained Live Load

Calculations and Discussion	Code Reference
1. Minimum thickness, using Table 7-1:	9.5.3

From Table 7-1 for either square or rectangular panels, and using the factor 1.10 for f_y = 40 ksi,

Interior panel h_{min} = (ℓ_n = 13.83 x 12)/(32.7 x 1.100) = 4.61 in.

Side, corner panel h_{min} = (ℓ_n = 13.83 x 12)/(29.7 x 1.100) = 5.08 in.

To verify by the governing Code equation,

$$h_{min} = \frac{\ell_n(800 + 0.005\,f_y)}{36{,}000}\ \text{(1.10 when no spandrel beams are used} \qquad \text{Eq. (9-13)}$$

$$= \frac{(13.83 \times 12)[800 + (0.005)(40{,}000)]}{36{,}000}(1.10) = 5.07 \text{ in.} \quad \text{Checks}$$

Since the actual slab thickness is 6 in., deflection calculations are not required; however, as an illustration, deflections will be

EXAMPLE 7.5 - Continued

Calculations and Discussion	Code Reference

checked for a corner panel, where it is found that all allowable deflections per Table 9.5(b) are satisfied.

2. Comment on trial design with regard to deflections:

Based on the minimum thickness limitations versus the actual slab thickness, it appears likely that computed deflections will meet most or all of the Code deflection limitations. It turns out that all are met.

3. Modulus of rupture, modulus of elasticity, modular ratio:

$f_r = 7.5\sqrt{f'_c} = 7.5\sqrt{3000} = 411$ psi — Eq. (9-9)

$E_{cs} = 33\sqrt{w_c^3 f'_c} = 33\sqrt{(145)^3(3000)} = 3.16 \times 10^6$ psi — 8.5.1

$E_{cc} = 33\sqrt{(145)^3(5000)} = 4.07 \times 10^6$ psi, $n = E_s/E_{cs} = 29/3.16 = 9.2$

4. Service load moments and cracking moment:

$w_d = 10 + (150 \text{ pcf})(6.0)/12 = 85.0$ psf

$(M_o)_d = w_d \ell_n^2/8 = (85.0)(15)(13.83)^2/8{,}000 = 30.48$ ft-k

$(M_o)_{d+\ell} = w_{d+\ell}\, \ell_2\, \ell_n^2/8 = (85.0 + 50.0)(15)(13.83)^2/8{,}000 = 48.41$ ft-k

See the following table for the half-column strip and half-middle strip moments:

$(M_{cr})_{c/2} = (M_{cr})_{m/2} = f_r\, I_g/y_t = (411)(15 \times 12)(6.0)^3/$
$(4)(12)(3.0)(12{,}000)$

$= 9.25$ ft-k > All $(M_d)_{c/2}$, $(M_d)_{m/2}$, and $(M_{d+\ell})_{m/2}$ on Lines 5, 8, and 9 of the following table. Hence $I_e = I_g$ for <u>all</u> dead load and all middle strip dead-plus-live load deflections. The $(I_e)_{d+\ell}$ calculations for the half-column strips are shown in the following table.

EXAMPLE 7.5 - Continued

Calculations and Discussion	Code Reference

Table -- Moments (ft-k) and Moments of Inertia (in.^{4})[a]

	Ext. Eq. Fr. $1/\alpha_{ec} = 0.898$		Int. Eq. Fr. $1/\alpha_{ec} = 1.029$	
	Pos.	Int. Neg.	Pos.	Int. Neg.
1. Moment Ratios[b]	0.482	0.697	0.492	0.701
2. Panel M_d = (Line 1)$(M_o)_d$	14.69	21.24	15.00	21.37
3. Panel $M_{d+\ell}$ = (Line 1)$(M_o)_{d+\ell}$	23.33	33.74	23.82	33.94
4. $(LDF)_c$ -- From Table 7-2 with $\alpha_1 = \beta_t = 0$	0.60	0.75	0.60	0.75
5. $(M_d)_{c/2}$ = (Line 2)(Line 4)/2	4.41	7.97	4.50	8.01
6. $(M_{d+\ell})_{c/2}$ = (Line 3)(Line 4)/2	7.00	12.65	7.15	12.73
7. $(M_\ell)_{c/2}$ = Line 6 - Line 5	2.59	4.68	2.65	4.72
8. $(M_d)_{m/2}$ = (Line 2)/2 - Line 5	2.94	2.65	3.00	22.68
9. $(M_{d+\ell})_{m/2}$ = (Line 3)/2 - Line 6	4.67	4.22	4.76	4.24
10. $[(M_{cr})_{c/2}/(\text{Line 6})]^3$	> 1	0.391	> 1	0.384
<u>For Half-Column Strips</u>				
11. No. of #5 Bars	4	6	4	6
12. A_s = (Line 11)(0.31)...in.^{2}	1.24	1.86	1.24	1.86
13. I_{cr} (Separate Calculation)...in.^{4}	--	217	--	217
14. [c]$(I_e)_{d+\ell}$ = (Line 10)I_g + [1 - (Line 10)I_{cr}]...in.^{4}	810	449	810	445

[a]Ext. Neg. Moments and Ext. Neg. I_e are not required in Eq. (2).

[b]Using the Modified Stiffness Method of Commentary Section 13.6.3.3:

$$+M = 0.63 - \frac{0.28}{1+1/\alpha_{ec}}, \quad -M = 0.75 - \frac{0.10}{1+1/\alpha_{ec}}$$

[c]$(I_e)_{d+} = (0.391)(810) + (1 - 0.391)(217) = 449$ in.^{4}...Eq. (1)

where $I_g = (15/4)(12)(6.0)^3/12 = 810$ in.^{4}

EXAMPLE 7.5 - Continued

Calculations and Discussion	Code Reference

5. Flexural stiffness (K_{ec} and α_{ec}) of an exterior equivalent column:

$K_b = 0$ (no beams) — Commentary 13.7.4

$I_s = (I_g)_{frame} = \ell_2 h^3/12 = (15 \times 12)(6.0)^3/12 = 3{,}240 \text{ in.}^4$

$K_s = 4 E_{cs} I_s/\ell_1 = 4 E_{cs} (3{,}240)/(15)(12) = 72.0 E_{cs}$

For Ext. Frame, $K_s = 72.0 E_{cs}/2 = 36.0 E_{cs}$

$K_c = 4 E_{cc} I_c/\ell_c = 4 E_{cc}(14)^4/(12)(10)(12) = 106.7 E_{cc}$

$\Sigma K_c = 2 K_c = (2)(106.7 E_{cc}) = 213.4 E_{cc}$

$$C = (1 - 0.63 x/y)(x^3y/3) = (1 - 0.63 \frac{6.0}{14})(\frac{6.0^3 \times 14}{3}) = 735.8 \text{ in.}^4$$

$$K_t = \frac{\Sigma 9 E_{cs} C}{\ell_2(1 - c_2/\ell_2)^3} = \frac{(2)(9)E_{cs}(735.8)}{(15)(12)(1 - \frac{14}{15 \times 12})^3} = 93.9 E_{cs}$$

For Ext. Frame, $K_t = 93.9 E_{cs}/2 = 47.0 E_{cs}$, $E_{cc} = (4.07/3.16)E_{cs} = 1.288 E_{cs}$

$$K_{ec} = \frac{1}{1/\Sigma K_c + 1/K_t} = \frac{E_{cs}}{[1/(213.4)(1.288)] + (1/93.9)} = 70.0 E_{cs}$$

$$\text{For Ext. Frame, } K_{ec} = \frac{E_{cs}}{[1/(213.4)(1.288)] + (1/47.0)} = 40.1 E_{cs}$$

To use in Eq. (15), Avg. $K_{ec} = (70.0 + 40.1)(3.16 \times 10^6)/(2)(12{,}000)$

$= 14{,}500$ ft-k

$\alpha_{ec} = K_{ec}/\Sigma(K_s + K_b) = 70.0 E_{cs}/72.0 E_{cs} = 0.972$, $1/\alpha_{ec} = 1.029$

For Ext. Frame, $\alpha_{ec} = 40.1 E_{cs}/36.0 E_{cs} = 1.114$, $1/\alpha_{ec} = 0.898$

To compute M_{net} for use in Eq. (15), Avg. $\alpha_{ec} = 1.043$, 1/Avg. $\alpha_{ec} = 0.959$

6. Summary of effective moments of inertia:

Middle Strips -- $(I_e)_d = (I_g)_{d+\ell} = I_g = (7.5)(12)(6.0)^3/12 = 1{,}620 \text{ in.}^4$

Column Strips -- $(I_e)_d = I_g = 1{,}620 \text{ in.}^4$

EXAMPLE 7.5 - Continued

Calculations and Discussion	Code Reference

-- Using the half-column strip values of $(I_e)_{d+\ell}$ from Line 14 of the preceding Table,

Avg. $(I_e)_{d+\ell} = 0.85\ I_m + 0.15(I_e\text{-Cont. End})$ — Eq. (2)

$= (0.85)(810 + 810) + (0.15)(449 + 445) = 1,510\text{ in.}^4$

Equivalent Frame -- $(I_e)_d = (I_g)_c + (I_g)_m = (I_g)_{frame} = (2)(1,620)$

$= 3,240\text{ in.}^4$

-- $(I_e)_{d+\ell} = (I_e)_c + (I_g)_m = 1,510 + 1,620 = 3,130\text{ in.}^4$

7. Deflections, using Eqs. (13) to (19):

Fixed $a_{frame} = w\ \ell^4/384\ E_{cs}\ I_{frame}$ — Eq. (13)

$$(\text{Fixed } a_{frame})_{d,d+\ell} = \frac{(85.0 \text{ or } 135.0)(15)^5(12)^3}{(384)(3.16 \times 10^6)(3,240 \text{ or } 3,130)}$$

$= 0.028\text{ in.}, 0.047\text{ in.}$

Fixed $a_{c,m} = (LDF)_{c,m}(\text{Fixed } a_{frame})(I_{frame}/I_{c,m})$ — Eq. (14)

Pos. and Neg. Avg. $(LDF)_c = 0.738$, $(LDF)_m = 0.262$ — Table 7-2

$(\text{Fixed } a_c)_d = (0.738)(0.028)(2) = 0.041\text{ in.}$

$(\text{Fixed } a_c)_{d+\ell} = (0.738)(0.047)(3,130/1,510) = 0.072\text{ in.}$

$(\text{Fixed } a_c)_\ell = 0.072 - 0.041 = 0.031\text{ in.}$

$(\text{Fixed } a_m)_d = (0.262)(0.028)(2) = 0.015\text{ in.}$

$(\text{Fixed } a_m)_{d+\ell} = (0.262)(0.047)(3,130/1,620) = 0.024\text{ in.}$

$(\text{Fixed } a_m)_\ell = 0.024 - 0.015 = 0.009\text{ in.}$

$(M_{net})_{d+\ell} = 0.65\ (M_o)_{d+\ell}/(1 + 1/\text{Avg. } \alpha_{ec})$

$= (0.65)(48.41)/(1 + 0.959) = 16.06\text{ ft-k}$

$(M_{net})_d = (30.48/48.41)(16.06) = 10.11\text{ ft-k}$

EXAMPLE 7.5 - Continued

Calculations and Discussion	Code Reference

For both column and middle strips,

End θ_d $= (M_{net})_d/\text{Avg } K_{ec} = 10.11/14{,}500 = 0.000697$ rad — Eq. (15)

End $\theta_{d+\ell}$ $= 16.06/14{,}500 = 0.001108$ rad

$a_\theta = (\text{End } \theta)(\ell/8)(I_g/I_e)_{frame}$ — Eq. (17)

$(a_\theta)_d = (0.000697)(15)(12)(1)/8 = 0.016$ in.

$(a_\theta)_{d+\ell} = (0.001108)(15)(12)(3{,}240/3{,}130)/8 = 0.026$ in.

$(a_\theta)_\ell = 0.026 - 0.016 = 0.010$ in.

$a_{c,m} = \text{Fixed } a_{c,m} + (a_{\theta 1})_{c,m} + (a_{\theta 2})_{c,m}$ — Eq. (18)

$a_c)_d = 0.041 + 0.016 + 0 = 0.057$ in.

$(a_m)_d = 0.015 + 0.016 + 0 = 0.031$ in.

$(a_c)_\ell = 0.031 + 0.010 + 0 = 0.041$ in.

$(a_m)_\ell = 0.009 + 0.010 + 0 = 0.019$ in.

$a = a_{cx} + a_{my}$ = midpanel deflection of corner panel — Eq. (19)

$(a_i)_d = 0.057 + 0.031 = 0.088$ in.

$(a_i)_\ell = 0.041 + 0.019 = \underline{0.060 \text{ in.}}$

Using Eqs. (5) with $\xi_u = 2.5$,

$$a_{(cp\ +\ sh)} = \lambda\,(a_i)_{sus} = \frac{\xi_u = 2.5}{1 + (50)(\rho' = 0)}\,[(a_i)_d + 0.40(a_i)_\ell]$$

$= (2.5)[0.088 + (0.40)(0.060)] = 0.280$ in.
(with 40% Sustained LL)

$= (2.5)(0.088) = 0.220$ in. (with 0% Sustained LL)

$a_{(cp\ +\ sh)} + (a_i)_\ell = 0.280 + 0.060 = \underline{0.34 \text{ in.}}$ (with 40% Sustained LL)

$= 0.220 + 0.060 = \underline{0.28 \text{ in.}}$ (with 0% Sustained LL)

These computed deflections are compared with the Code allowable deflections in Table 9.5(b) as follows:

<u>Flat roofs</u> not supporting and not attached to nonstructural elements likely to be damaged by large deflections --

$(a_i)_\ell \le (\ell_n \text{ or } \ell)/180$

EXAMPLE 7.5 - Continued

Calculations and Discussion	Code Reference

= (13.83 or 15)(12)/180 = 0.92 in. or 1.00 in., versus 0.06 in. OK

Floors not supporting and not attached to nonstructural elements likely to be damaged by large deflections --

$(a_i)_\ell \leq (\ell_n$ or $\ell)/360$ = 0.46 in. or 0.50 in., versus 0.06 in. OK

Roof or floor construction supporting or attached to nonstructural elements likely to be damaged by large deflections --

$a_{(cp + sh)} + (a_i)_\ell \leq (\ell_n$ or $\ell)/480$ = 0.35 in.

or 0.38 in., versus 0.28 in. and 0.34 in. OK

Roof or floor construction supporting or attached to nonstructural elements not likely to be damaged by large deflections --

$a_{(cp + sh)} + (a_i)_\ell \leq (\ell_n$ or $\ell)/240$

= 0.69 in. or 0.75 in., versus 0.28 in. and 0.34 in. OK

All computed deflections are found to be satisfactory in all four catagories.

EXAMPLE 7.6 - Slab System with Beams (Two-Way Slab)

Required: Minimum thickness analysis for deflection control

Data:

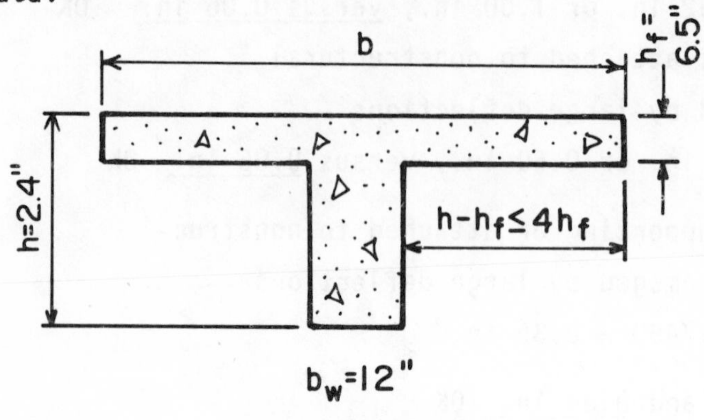

Interior Beam

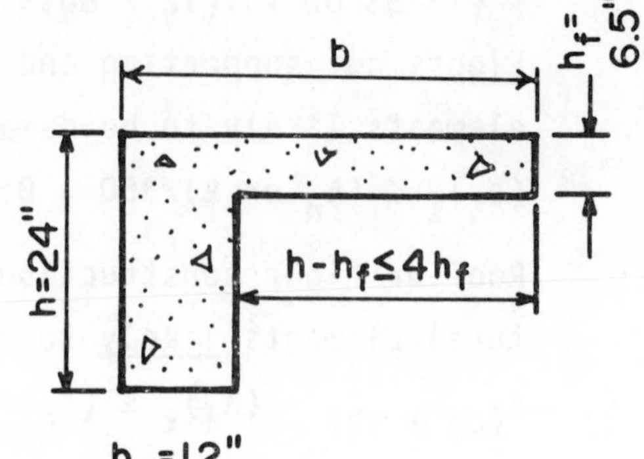

Spandrel Beam

f_y = 60 ksi, Slab thickness -- h_f = 6.5 in.

Square panels -- 22 ft x 22 ft c.c. of columns

All beams -- b_w = 12 in. and h = 24 in. $\ell_n = 22 - 1 = 21$ ft

It is noted that f'_c and the loading are not required in this analysis.

Calculations and Discussion	Code Reference

1. Effective width b and section properties, using Fig. 7-1:

Interior Beam

$I_s = (22)(12)(6.5)^3/12 = 6{,}040 \text{ in.}^4$

$h - h_f = 24 - 6.5 = 17.5 \text{ in.} \le 4\,h_f = (4)(6.5) = 26 \text{ in.}$ OK

Hence, $b = 12 + (2)(17.5) = 47$ in.

$y_t = h - (1/2)[(b - b_w)h_f^2 + b_w\,h^2]/[(b - b_w)h_f + b_w\,h]$

$= 24 - (1/2)[(35)(6.5)^2 + (12)(24)^2]/[(35)(6.5) + (12)(24)]$

$= 15.86$ in.

$I_b = (b - b_w)h_f^3/12 + b_w\,h^3/12 + (b - b_w)h_f(h - h_f/2 - y_t)^2$

EXAMPLE 7.6 - Continued

Calculations and Discussion	Code Reference

$$+ b_w h(y_t - h/2)^2$$

$$= (35)(6.5)^3/12 + (12)(24)^3/12 + (35)(6.5)(24 - 3.25 - 15.86)^2$$

$$+ (12)(24)(15.86 - 12)^2 = 24,360 \text{ in.}^4$$

$$\alpha = E_{cb} I_b / E_{cs} I_s = I_b/I_s = 24,360/6,040 = 4.03$$

Spandrel Beam

$$I_s = (11)(12)(6.5)^3/12 = 3,020 \text{ in.}^4$$

$$b = 12 + (24 - 6.5) = 29.5 \text{ in.}$$

$$y_t = 24 - (1/2)[(17.5)(6.5)^2 + (12)(24)^2]/[(17.5)(6.5) + (12)(24)]$$

$$= 14.48 \text{ in.}$$

$$I_b = (17.5)(6.5)^3/12 + (12)(24)^3/12 + (17.5)(6.5)(24 - 3.25 - 14.48)^2$$

$$+ (12)(24)(14.48 - 12)^2 = 20,470 \text{ in.}^4$$

$$\alpha = I_b/I_s = 20,470/3,020 = 6.78$$

α_m, β_s, and β values

α_m (average value of α for all beams on the edges of a panel), and β_s(ratio of length of continuous edges to total perimeter of a slab panel):

Interior panel -- $\alpha_m = 4.03$, $\beta_s = 1$

Side panel -- $\alpha_m = [(3)(4.03) + 6.78]/4 = 4.72$, $\beta_s = 3/4$

Corner panel -- $\alpha_m = [(2)(4.03) + (2)(6.78)]/4 = 5.41$, $\beta_s = 1/2$

For square panels, β = ratio of clear spans in the two directions = 1

EXAMPLE 7.6 - Continued

Calculations and Discussion	Code Reference

2. Minimum thickness: 9.5.3

$$h_{min} = \frac{\ell_n(800 + 0.005 f_y)}{36,000 + 5000\beta[\alpha_m - 0.5(1 - \beta_s)(1 + 1/\beta)]}$$ Eq. (9-11)

$$h_{min} = \frac{(21)(12)[800 + (0.005)(60,000)]}{36,000 + (5000)(1)[4.03 - (0.5)(1 - 1)(1 + 1)]} = 4.94 \text{ in.}$$

-- Interior panel

Also, h_{min} = 4.75 in. -- Side panel, and h_{min} = 4.58 in. -- Corner panel

But not less than,

$$h_{min} = \frac{\ell_n(800 + 0.005 f_y)}{36,000 + 5000 (1 + \beta_s)}$$ Eq. (9-12)

$$= \frac{(21)(12)[800 + (0.005)(60,000)]}{36,000 + (5000)(1)(1 + 1)} = 6.03 \text{ in.}$$ -- Interior panel

Also, h_{min} = 6.19 in. -- Side panel, and h_{min} = 6.37 in. -- Corner panel

And need not be more than,

$$h_{min} = \frac{\ell_n(800 + 0.005 f_y)}{36,000} = \frac{(21)(12)[800 + (0.005)(60,000)]}{36,000}$$ Eq. (9-13)

= 7.70 in. -- All panels

Hence, the slab thickness of 6.5 in. > 6.37 in. is satisfactory for all panels, and deflections need not be checked.

EXAMPLE 7.7 - Simple-Span Prestressed Single T Beam

Required: Analysis of short-time and ultimate long-time camber and deflection

Data: 8ST36 (Design Details from PCI Handbook)

Span = 80 ft, Beam is Partially Cracked

f'_{ci} = 3.5 ksi, f'_c = ksi (Normal Weight Concrete)

f_{pu} = 270 ksi

14 - 1/2 in. Depressed (1 Pt.) Strands

4 - 1/2 in. Nonprestressed Strands

(Assume same centroid when computing I_{cr})

P_i = (0.7)(14)(0.153)(270) = 404.8 k

P_o = (0.90)(404.8) = 364 k

P_e = (0.78)(404.8) = 316 k

e_e = 11.15 in., e_c = 22.51 in.

y_t = 26.01 in., A_g = 570 $in.^2$, I_g = 68,920 $in.^4$

Self Weight = w_o = 594 lb/ft

Superimposed DL = w_s = (8)(10 psf) = 80 lb/ft

applied at age 2 mos (β_s = 0.76 in Term 6 of Eq. (21)

Live Load = w_ℓ = (8)(51 psf) = 408 lb/ft

Capacity governed by flexural strength

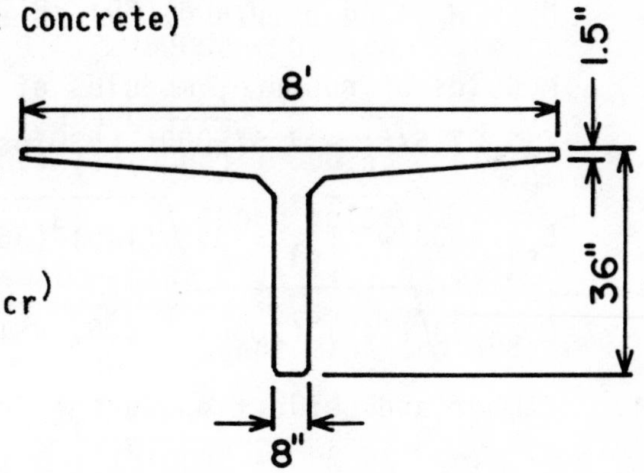

8'

1.5"

36"

8"

Calculations and Discussion	Code Reference

1. Span-depth ratios (using PCI Handbook):
   Typical span-depth ratios for single T beams are 25 to 35 for floors and 35 to 40 for roofs; versus (80)(12)/36 = 27, which is relatively small. It turns out that all allowable deflections in Table 9.5(b) are satisfied.
2. Moments for computing deflections:
   $M_o = w_o \ell^2/8 = (0.594)(80)^2/8 = 475$ ft-k

   (x 0.96 = 456 ft-k at 0.4ℓ for computing stresses and I_e -- 1 Pt. Dep.)

EXAMPLE 7.7 - Continued

Calculations and Discussion	Code Reference

$M_s = w_s\ \ell^2/8 = (0.080)(80)^2/8 = 64$ ft-k (61 ft-k at 0.4ℓ)

$M_\ell = w_\ell\ \ell^2/8 = (0.408)(80)^2/8 = 326$ ft-k (313 ft-k at 0.4ℓ)

3. Modulus of rupture, modulus of elasticity:

$f_r = 7.5\sqrt{f'_c} = 7.5\sqrt{5000} = 530$ psi — Eq. (9-9)

$E_{ci} = 33\sqrt{w^3\ f'_{ci}} = 33\sqrt{(145)^3(3500)} = 3.41 \times 10^6$ psi — 8.5.1

$E_c = 33\sqrt{(145)^3(5000)} = 4.07 \times 10^6$ psi, $n = E_p/E_c = 27/4.07 = 6.6$

4. Camber and deflection, using Eq. (21):

Term (1) -- $a_{po} = \dfrac{P_o(e_c - e_e)\ \ell^2}{12\ E_{ci}\ I_g} + \dfrac{P_o\ e_e\ \ell^2}{8\ E_{ci}\ I_g}$

$= \dfrac{(364)(22.51 - 11.15)(80)^2(12)^2}{(12)(3,410)(68,920)} + \dfrac{(364)(11.15)(80)^2(12)^2}{(8)(3,410)(68,920)}$

$= -\ 3.34$ in.

Term (2) -- $a_o = \dfrac{5\ M_o\ \ell^2}{48\ E_{ci}\ I_g} = \dfrac{(5)(475)(80)^2(12)^2}{(48)(3,410)(68,920)} = 2.33$ in.

Term (3) -- $k_r = 1/[1 + (A_s/A_{ps})] = 1/[1 + (4/14)] = 0.78$

$\left[-\dfrac{\Delta P_u}{P_o} + (k_r\ C_u)(1 - \dfrac{\Delta P_u}{2P_o})\right] a_{po}$

$= [\ -\ 0.18 + (0.78 \times 2.0)(1 - 0.09)](3.34) = -\ 4.14$ in.

Term (4) -- $(k_r\ C_u)\ a_o = (0.78)(2.0)(2.33) = 3.63$ in.

Term (5) -- $a_s = \dfrac{5\ M_s\ \ell^2}{48\ E_c\ I_g} = \dfrac{(5)(64)(80)^2(12)^3}{(48)(4,070)(68,920)} = 0.26$ in.

Term (6) -- $(\beta_s\ k_r\ C_u)\ a_s = (0.76)(0.78)(1.6)(0.26) = 0.25$ in.

Term (7) -- Determination of I_e at 0.4ℓ for 1 Pt. Dep.

EXAMPLE 7.7 - Continued

Calculations and Discussion	Code Reference

Method 1, (Ref. 7.2)

$$(M_\ell)_{cr} = \frac{P_e I_g}{A_g\, y_t} + P_e e - M_{o+s} + \frac{f_r I_g}{y_t} = \frac{(316)(68{,}920)}{(570)(26.01)(12)}$$

$$+ \frac{(316)(20.24)}{12} - 517 + \frac{(0.530)(68{,}920)}{(26.01)(12)} = 255 \text{ ft-k}$$

$(M_\ell)_{cr} / M_\ell = 255/313 = 0.815$ at 0.4ℓ

Method 2, (Ref. 7.2)

$$f_{pe} = \frac{P_e}{A_g} + \frac{P_e e}{S} = \frac{316}{570} + \frac{(316)(20.24)}{2650} = 2.968 \text{ ksi}$$

$$f_{o+s} = \frac{M_{o+s}}{S} = \frac{(517)(12)}{2650} = 2.341 \text{ ksi}$$

$$f_{to} = -f_{pe} + \frac{M_{to}}{S} = -2.968 + \frac{(456 + 61 + 313)(12)}{2650} = 0.790 \text{ ksi}$$

versus $f_r = 0.530$ ksi and $12\sqrt{5000} = 0.849$ ksi. Hence use I_e.

$$\frac{(M_\ell)_{cr}}{M_\ell} = \frac{f_{pe} - f_{o+s} + f_r}{f_{pe} - f_{o+s} + f_{to}} = \frac{2.968 - 2.341 - 0.530}{2.968 - 2.341 - 0.790}$$

$$= 0.817 \text{ at } 0.4\ell$$

PCI Handbook Method,

$f_{t\ell} = f_{to}$ (above) $= 0.790$ ksi

$$f_\ell = \frac{M_\ell}{S} = \frac{(313)(12)}{2650} = 1.417 \text{ ksi}$$

$$\frac{(M_\ell)_{cr}}{M_\ell} = 1 - \frac{f_{t\ell} - f_r}{f_\ell} = 1 - \frac{790 - 530}{1417} = 0.817 \text{ at } 0.4\ell$$

All 3 methods yield the same results.

Neglecting the taper (compression area = 96 in. by 1.5 in.),

$(96)(1.5)(x - 0.75) = (6.6)(18 \times 0.153)(30.23 - x)$

$x = 4.05$ in.

$$I_{cr} = \frac{(96)(1.5)^3}{12} + (96)(1.5)(4.05 - 0.75)^2$$

$$+ (6.6)(18 \times 0.153)(30.23 - 4.05)^2 = 14{,}050 \text{ in.}^4$$

at 0.4ℓ

EXAMPLE 7.7 - Continued

Calculations and Discussion	Code Reference

PCI Handbook Method,

$I_{cr} = n\,A_{st}\,d^2\,(1 - \sqrt{\rho}) = (6.6)(18 \times 0.153)(30.23)^2(1 - \sqrt{0.000949})$

$= 16{,}100 \text{ in.}^4$ at 0.4ℓ USE

$0.817^3 = 0.545$

$(I_e)_\ell = (M_{cr}/M_\ell)^3 \quad I_g + [1 - (M_{cr}/M_\ell)^3]\,I_{cr}$ Eq. (9-7)

$= (0.545)(68{,}920) + (1 - 0.545)(16{,}100) = \underline{44{,}890 \text{ in.}^4}$
at 0.4ℓ

It is noted that for other prestress profiles (straight tendons, 2 Pt. depression, draped, etc.) the stresses and I_e are computed at the midspan instead of 0.4ℓ.

Check using PCI Handbook Chart, p. 3-68,

$f_e = f_{t\ell} - f_r = 790 - 530 = 260$ psi

$f_e/f_\ell = 260/1417 = 0.18, \quad I_{cr}/I_g = 16{,}100/68{,}920 = 0.23$

From chart, $I_e/I_g = 0.66$, $I_e = (0.66)(68{,}920) = 45{,}490 \text{ in.}^4$ OK

<u>Live load deflection at midspan</u>

$$a_\ell = \frac{5\,M_\ell\,\ell^2}{48\,E_c(I_e)_\ell} = \frac{(5)(326)(80)^2(12)^3}{(48)(4{,}070)(44{,}890)} = \underline{2.06 \text{ in.}\downarrow}$$

Combined Results and Comparisons with Code Limitations

(1) (2) (3) (4) (5) (6) (7)

$a_u = -3.34 + 2.33 - 4.14 + 3.63 + 0.26 + 0.25 + 2.06 = 1.05 \text{ in.}\downarrow$ Eq. (21)

Initial Camber = $a_{po} - a_o = 3.34 - 2.33 = 1.01$ in. ↑ versus 1.6 in. at erection in PCI Handbook

Residual Camber = $a_\ell - a_u = 2.06 - 1.05 = 1.01$ in. ↑ versus 1.1 in. in PCI Handbook with no nonprestressed steel

The Residual Camber happened to be the same as the Initial Camber, due to the effect of the Superimposed Dead Load.

EXAMPLE 7.7 - Continued

Calculations and Discussion	Code Reference

Time-Dependent Plus Superimposed Dead Load and Live Load Deflection

$= -4.14 + 3.63 + 0.26 + 0.25 + 2.06 = 2.06$ in. or

$= a_u - (a_o - a_{po}) = 1.05 - (-1.01) =$ <u>2.06 in.</u> ↓ (Coincidentally same as a_ℓ)

These computed deflections are compared with the allowable deflections in Table 9.5(b) as follows:

$\ell/180 = (80)(12)/180 = 5.33$ in. versus $a_\ell = 2.06$ in. OK

$\ell/360 = (80)(12)/360 = 2.67$ in. versus $a_\ell = 2.06$ in. OK

$\ell/480 = (80)(12)/480 = 2.00$ in. versus Time-Dep. etc. = 2.06 in. Say OK*

$\ell/240 = (80)(912)/240 = 4.00$ in. versus Time-Dep. etc. = 2.06 in. OK

*This might be accepted as close enough. It is noted that the nonprestressed steel was used to reduce camber, which has the effect of increasing the downward deflection. In this problem, without compression steel, the time-dep. plus superimposed dead load and live load deflection is computed as 1.89 in. instead of 2.06 in.

8

Moment Redistribution

8.4 Redistribution of Negative Moments in Continuous Nonprestressed Flexural Members

Section 8.4 permits a redistribution of negative moments in continuous flexural members if reinforcement percentages do not exceed a specified amount. This provision recognizes the inelastic behavior of concrete structures and constitutes a move toward "limit design."

A maximum 10 percent adjustment of negative moments was first permitted in the 1963 ACI Code. Experience with the use of that provision was satisfactory but found to be conservative. The 1971 Code increased the maximum adjustment percentage as shown in Fig. 8-1. The increase was justified by additional knowledge of ultimate and service load behavior as demonstrated in tests and analytical studies. The 1983 Code retains the same adjustment percentage criteria.

Application of Section 8.4 will permit, in many cases, substantial reduction in total reinforcement required without reduction in safety, and allow a relief of reinforcement required in congested negative moment areas.

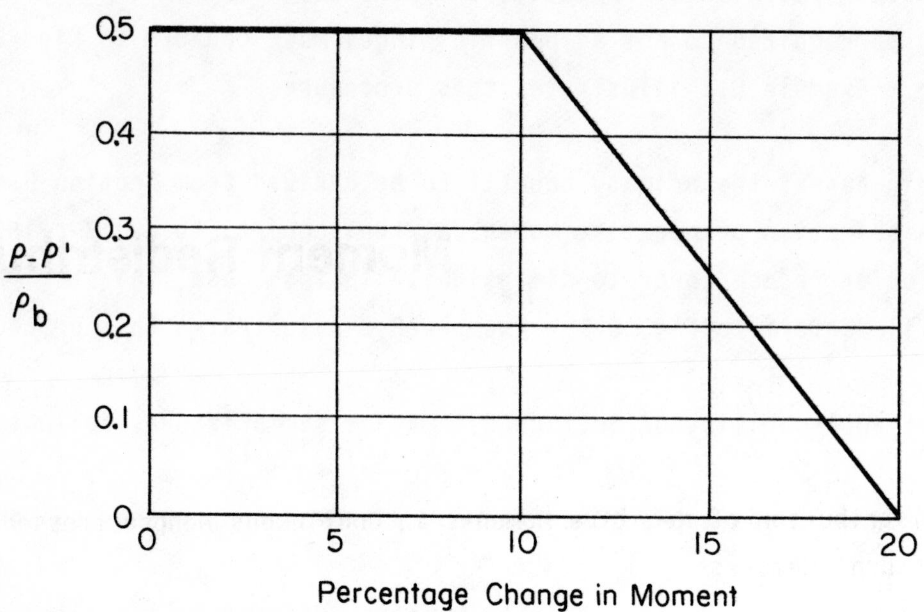

Fig. 8-1 - Permissible Moment Redistribution For Minimum Rotation Capacity

According to Section 8.9, continuous members must resist more than one condition of live load. An elastic analysis is performed for each loading condition, and a moment envelope is obtained for the design of each section. Thus, for any of the loading conditions considered, certain sections in a given span will reach the ultimate moment while others will have reserve capacity. Tests have shown that the structure can continue to carry additional loads if the sections which already have reached the ultimate moment can continue to rotate (rotational capacity) as a hinge, i.e., a "plastic hinge." This allows the moments at other sections to be increased (moment redistribution) until a collapse mechanism is formed.

Recognition of this additional load capacity beyond the intended original design suggests the possibility of redesign with less material. Section 8.4 allows a redesign by decreasing or increasing the elastic negative moments for each loading condition (with the corresponding changes in positive moment required by statics). These moment changes may be such as to reduce both the maximum positive and negative moments in the final moment envelope. Also, to insure proper rotation at capacity, the percentage of steel in the sections that may be required to act as plastic hinges must conform to Fig. 8-1 after redesign. Example 8.1 illustrates this procedure.

In certain cases, the primary benefit to be derived from Section 8.4 will be simply a reduction of negative moment at the supports to avoid reinforcement congestion or reduce concrete dimensions. In this case, the steel percentage must still conform to Fig. 8-1. Example 8.2 illustrates this procedure.

Limits of applicability of Section 8.4 may be summarized as follows:

(1) Provisions apply to continuous flexural nonprestressed members.

(2) Provisions do not apply to members designed by the alternate design method of Appendix B or to slab systems designed by the Direct Design Method, see Section 13.6.1.7.

(3) Bending moments must be determined by analytical methods, such as moment distribution, slope deflection, etc. Approximate methods to determine the original bending moments cannot be used.

(4) Reinforcement ratio at a cross section where moment is to be adjusted must not exceed one-half of the balanced steel ratio, ρ_b, as defined by Eq. (8-1). Solution of Eq. (8-1) is given in Table 8-1.

(5) Maximum allowable percentage increase or decrease of negative moment is defined by the expression, $20(1- \frac{\rho - \rho'}{\rho_b})$.

(6) Adjustment of negative moments is made for each loading condition considered. Members are then proportioned for the maximum adjusted moments resulting from all loading conditions.

(7) Adjusted negative support moments for any span require adjustment of positive moment in the same span. A decrease of a negative support moment requires a corresponding increase in the positive moment for the same span, or an increase in the negative moment requires a decrease in the positive moment.

(8) Static equilibrium must be maintained at all joints before and after moment redistribution.

(9) In the case of unequal negative moments on each side of a fixed support (i.e., where unequal spans are adjacent), the difference between these two moments is taken into the support(s). Should either or both of these negative moments be adjusted, the resulting difference between the adjusted moments is taken into the supports(s).

(10) Moment redistribution may be carried out for as many cycles as deemed practical, provided that, after each cycle of redistribution, a new allowable percentage increase or decrease in negative moment is calculated, based on the final steel ratios provided for the adjusted support moments from the previous cycle.

(11) After the design is completed and the reinforcement selected, the actual steel ratios provided must comply with Fig. 8-1 for the percent moment redistribution taken, to ensure that the requirements of Section 8.4 are met.

Table 8-1. Solution Of Eq. (8-1) With Varying Values Of f_y And f'_c

f_y \ f'_c		3,000	3,500	4,000	4,500	5,000	6,000
40,000	ρ_b	0.0371	0.0433	0.0495	0.0540	0.0582	0.0655
	0.75 ρ_b	0.0278	0.0325	0.0371	0.0405	0.0437	0.0491
	0.50 ρ_b	0.0186	0.0217	0.0247	0.0270	0.0291	0.0328
60,000	ρ_b	0.0214	0.0249	0.0285	0.0311	0.0335	0.0377
	0.75 ρ_b	0.0161	0.0187	0.0214	0.0233	0.0252	0.0283
	0.50 ρ_b	0.0107	0.0125	0.0143	0.0156	0.0168	0.0189

$$\rho_b = \frac{0.85\beta_1 f'_c}{f_y} \times \frac{87,000}{87,000 + f_y} \qquad \text{Eq. (8-1)}$$

where β_1 is a factor defined in Sec. 10.2.7.3:

β_1 = 0.85 for $f'_c \leq 4000$ psi

$= 0.85 - 0.05\left(\frac{f'_c - 4000}{1000}\right)$ for 4000 psi $< f'_c <$ 8000 psi

$= 0.65$ for $f'_c \geq 8000$ psi

EXAMPLE 8.1 - Moment Redistribution

Determine required reinforcement for the one-way joist floor shown using moment redistribution to reduce total reinforcement required.

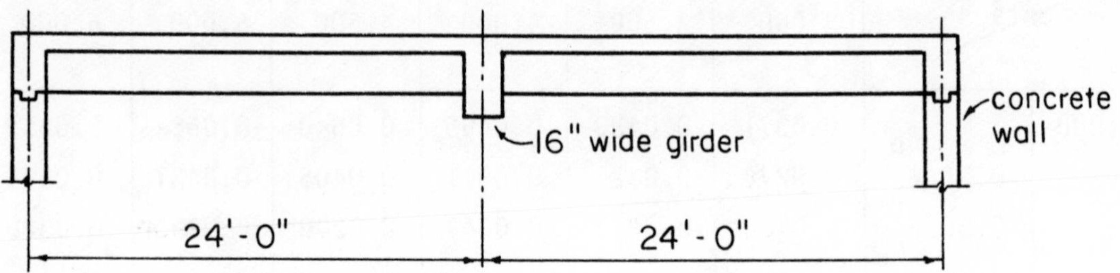

5-in. joist @ 25 in. o.c. (10-in. forms + 2-1/2-in. slab.

f'_c = 4000 psi

f_y = 60,000 psi

D = 80 psf

L = 100 psf

For simplicity, continuity at concrete walls is not considered.

Calculations and Discussion	Code Reference
1. Compute factored loads and balanced reinforcement ratio ρ_b	
ρ_b = 0.0285 (from Table 8-1)	8.4
w_d = 1.4 x 0.08 x 2.08 = 0.234	
w_ℓ = 1.7 x 0.10 x 2.08 = <u>0.354</u>	9.2
Total = 0.588 klf per joist	

EXAMPLE 8.1 - Continued

Calculations and Discussion	Code Reference

2. Obtain moment diagrams by elastic analysis (moments shown in ft. kips)

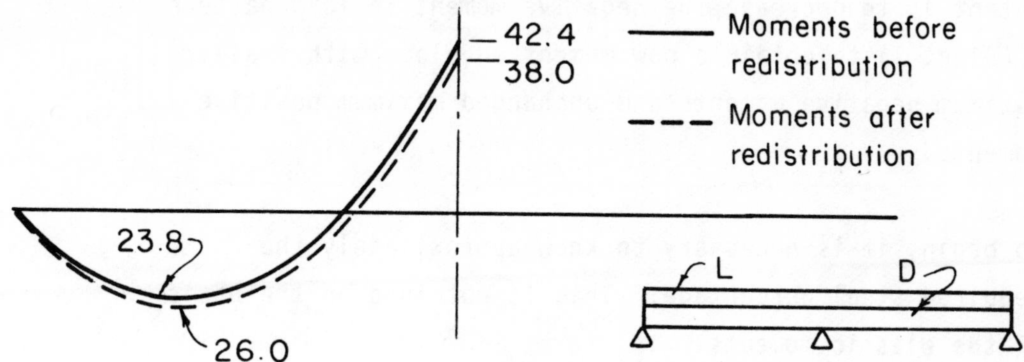

Diag. 1 LOAD PATTERN I

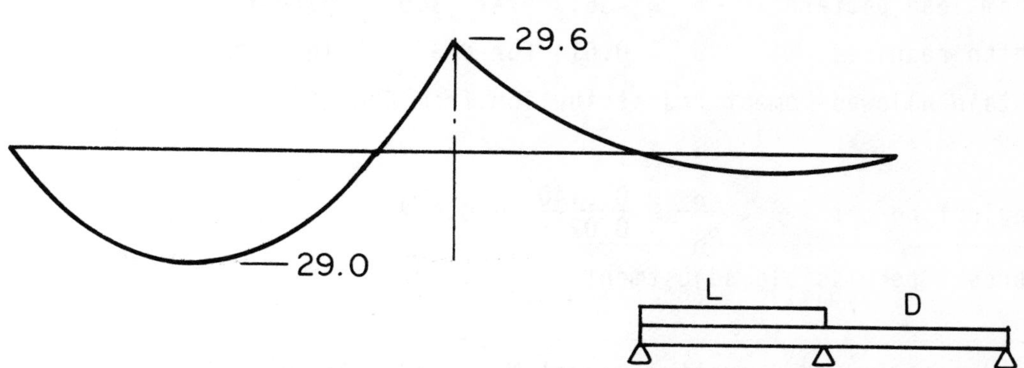

Diag. 2 LOAD PATTERNS II & III (Reverse of II)

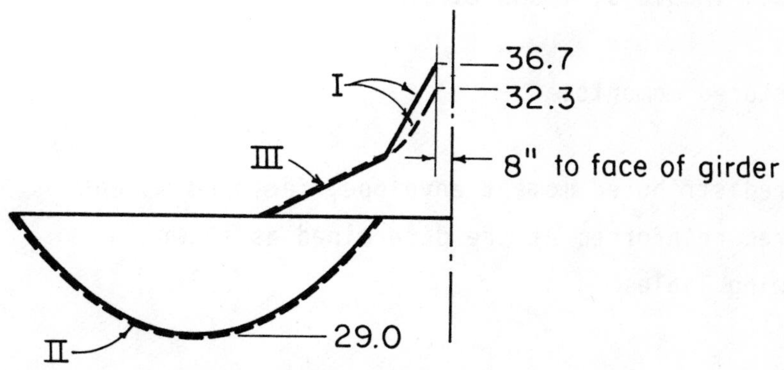

Diag. 3 FACTORED MOMENT ENVELOPE

EXAMPLE 8.1 - Continued

Calculations and Discussion	Code Reference

3. Redistribution of negative moments.

Intent is to decrease the negative moment in load pattern I (Diag. 1) to obtain a new moment envelope with smaller maximum negative moments and unchanged maximum positive moments.

To begin, it is necessary to know approximately the required steel percentage. This is obtained on the basis of the elastic moments.

From load pattern I: $M_u = -36.7^{'k}$ at face of girder
which requires: $\rho = 0.014$ for $d = 11.5$ in.
Obtain allowed moment redistribution from Fig. 8-1.

Neglecting ρ': $\frac{\rho}{\rho_b} = \frac{0.0140}{0.0285} = 0.492$
Percent permissible adjustment $= 10.2\%$

By decreasing the negative moment $M_u = -42.4$ in Diag. 1 by 10%, redistributed moment diagrams are obtained as shown by dashed lines in Diags. 1 and 3.

4. Design factored moments.

From the redistributed moment envelope, factored moments and required reinforcement are determined as shown in the following Table.

EXAMPLE 8.1 - Continued

Calculations and Discussion	Code Reference

SUMMARY OF FINAL DESIGN

Section	Load Pattern		A_s Required		A_s Provided		Redistribution, percent	Permitted Max. ρ (Fig.8-1)
	I	II	in.^{2}	ρ	in.^{2}	ρ		
Support Moment*	$-32.3^{'k}$	-	0.70	0.0122	2-#4 1-#5 (0.71)	0.0124 (b=5")	10.2	0.014
Midspan Moment	-	$29.0^{'k}$	0.57	0.0021	1-#4 1-#6 (0.64)	0.0023** (b=25")	-	-

* calculated at face of support

** check $\rho_{min} = \dfrac{200}{60000} = 0.0033 < \dfrac{A_s}{b_d} = \dfrac{0.64}{5 \times 11.5}$ 10.5.1

$= 0.0111$ provided

Since the provided steel ratios ρ are smaller than the maximum permitted by Section 8.4, the design is O.K.

Final Note:

Moment redistribution has permitted a saving of 10.2% in the negative reinforcement. Since the $29^{'k}$ positive factored moment remained unchanged after redistribution, a <u>total steel saving</u> was obtained without reduction in safety.

EXAMPLE 8.2 - Moment Redistribution

Determine required reinforcement for the spandrel beam at an intermediate floor level as shown using moment redistribution to reduce total reinforcement required.

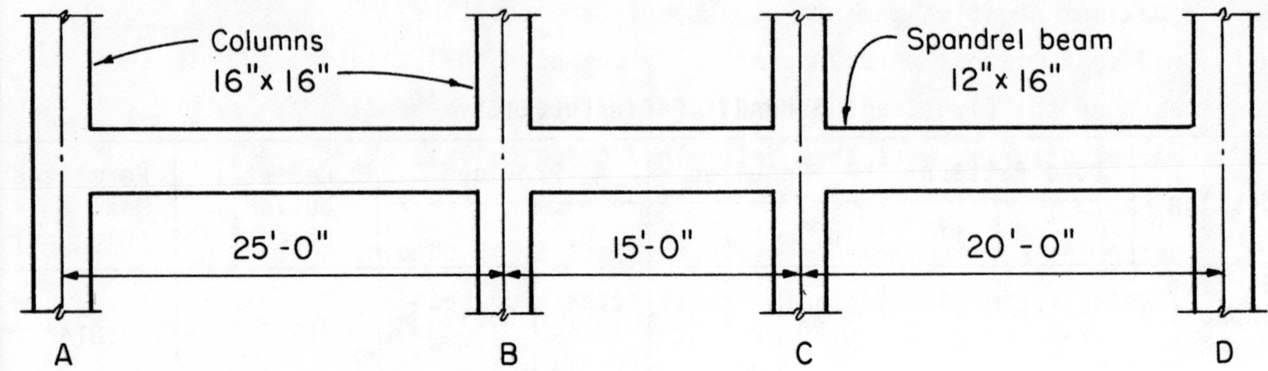

All columns = 16 in. x 16 in.

Story height = 10 ft. 0 in.

Spandrel beam = 12 in. x 16 in.

f'_c = 4,000 psi

f_y = 60,000 psi

D on beam = 1,000 plf

L on beam = 424 plf

Calculations and Discussion	Code Reference
1. Apply load factors to service loads to obtain factored loads.	9.2
$U = 1.4D + 1.7L$	Eq. (9-1)
$w_d = 1.4 \times 1 = 1.4$ klf	
$w_\ell = 1.7 \times 0.424 = 0.72$ klf	

Calculations and Discussion	Code Reference

2. Determine bending moment diagrams for the five load patterns as shown in Diags. 1 to 5. (Maximum negative moments at supports and positive mid-span moments were determined by computer analysis for each of the five loading conditions. Respective bending moment diagrams were then determined graphically). — 8.9.2

3. Determine maximum moment envelope (Diag. 6) to obtain elastic factored moments for cross-section of beam.

4. Determine maximum allowable percentage increase or decrease in negative moments:

d = 13.5 in.; cover = 2 in. — 7.7.1

From Table 8-1:

$\rho_b = 0.0285$

$\rho_{max} = 0.5\ \rho_b = 0.0143$ — 8.4.3

Determine preliminary ρ for maximum elastic negative moments at centerlines of supports:

Support	Factored Moment M_u, ft. kips	Steel Ratios ρ for strength design
A	-103.7	0.0118
B	-109.5	0.0125
C	- 71.0	0.0078
D	- 65.2	0.0071

Obtain percent allowable increase or decrease in negative moments from Fig. 8-1:

EXAMPLE 8.2 - Continued

Calculations and Discussion	Code Reference

Support	$\frac{\rho - \rho'}{\rho_b}$ *	Percent Allowable Adjustment
A	0.414	11.7
B	0.439	11.2
C	0.274	14.5
D	0.249	15.0

$*\rho' = 0$

5. Adjustment of moments.

<u>Note</u>: Adjustment of negative moments, either increase or decrease, is a decision to be made by the engineer. In this example, it was decided to reduce the negative moments at supports B and C and accept the increase in the corresponding positive moments, and not to adjust the negative moments at the exterior supports, A and D.

Referring to Diag. 1 through Diag. 5, the following adjustment in moments is made.

<u>Load Pattern I - Diag. 1</u>

Adjust $M_B = 109.5^{'k}$ (adjustment = 11.2%)

Reduction to $M_B = -109.5 \times 0.1120 = -12.2^{'k}$

Adjusted $M_B = 109.5 - (-12.2) = -97.3^{'k}$

<u>Increase in positive moment in span A-B</u>

$M_A = -103.2^{'k}$

Adjusted $M_B = -97.3^{'k}$

Mid-span ordinate on line M_A to $M_B = \frac{-103.2 + (-97.3)}{2}$

$= -100.3^{'k}$

EXAMPLE 8.2 - Continued

Calculations and Discussion	Code Reference

Moment due to uniform load = $\frac{w\ell^2}{8} = \frac{2.12 \times 25^2}{8} = 166.0'^k$

Adjusted positive moment at mid-span = 166.0 + (-100.3)

$= +65.7'^k$

The adjusted moment is approximately equal to the maximum positive moment. Similar calculations are made to determine the adjusted support and mid-span moments for other loading cases if deemed beneficial.

Results of the additional calculations made for this design indicating the adjusted moments for the various load cases are shown in the following table.

MAXIMUM MOMENTS FOR ONE CYCLE OF REDISTRIBUTION* (moments in ft kips)

Location	Load pattern I		Load pattern II		Load pattern III		Load pattern V	
	M_u	M_{adj}	M_u	M_{adj}	M_u	M_{adj}	M_u	M_{adj}
A	-103.2	-	-103.7	-	- 67.7	-	-103.2	-
Mid-span A-B	+ 59.6	+65.7	+ 60.0	+66.1	+ 38.8	-	+ 59.7	+65.8
B	-109.5	-97.3	-108.2	-96.0	- 73.2	-	-109.3	-97.2
Mid-span B-C	+ 16.8	-	+ 8.3	-	+ 17.3	-	+ 16.0	-
C	- 47.8	-	- 69.4	-59.6	- 71.0	-60.9	- 70.8	-60.7
Mid-Span C-D	+ 25.0	-	+ 38.7	+43.6	+ 38.2	+43.3	+ 38.3	+43.3
D	- 42.2	-	- 65.2	-	- 64.5	-	- 64.6	-

* No adjustment is made to the elastic support moments obtained from load pattern IV. The negative support moments are less

Calculations and Discussion	Code Reference

than the maximum adjusted moments from the other loading patterns and, therefore, would not control the maximum adjusted moment envelope.

6. After the adjusted moments have been determined analytically, the adjusted bending moment diagrams for each loading pattern can be determined. For the solution of this problem, the adjusted moment curves were determined graphically and are indicated by the dashed lines in Diags. 1 to 5.

7. An adjusted maximum moment envelope can now be obtained from the adjusted moment curves as shown in Diag. 6 by dashed lines.

   Since the provided steel ratios ρ are smaller or equal to the maximum permitted by Section 8.4, the design is O.K.

8. Final steel ratios ρ can now be obtained on the basis of the adjusted moments. A final check of these ratios must be made for compliance with Fig. 8-1.

   From the redistributed moment envelopes of Diag. 6, the design factored moments and required reinforcement are obtained as shown in the following table.

EXAMPLE 8.2 - Continued

Calculations and Discussion	Code Reference

SUMMARY OF FINAL DESIGN

Location	Moment, ft.kips	Load Case	A_s Required		A_s Provided		Redistribution, percent	Permitted Max. ρ (Fig. 8-1)
			$in.^2$	ρ	$in.^2$	ρ		
Mid-Span A-B	+66.1	II	1.16	0.0072	1-#5 2-#6 (1.19)	0.0073**	-	-
Support B*	-97.3	I	1.76	0.0108	1-#6 1-#7 1-#8 (1.83)	0.0112	11.2	0.0125
Support C*	-60.9	III	1.06	0.0066	2-#5 1-#6 (1.06)	0.0065	14.5	0.008
Mid-Span C-D	+43.6	II	0.75	0.0046	3-#5 (0.93)	0.0057**	-	-

* centerline moments. Also may use those at face of support

** check ρ min = $\frac{200}{f_y} = \frac{200}{60000} = 0.0033$ 10.5.1

9. Second cycle of moment redistribution.

Should the engineer decide additional adjustment of the moments is warranted, further cycles of redistribution may be made. The procedure to be followed is similar to that used for the first cycle <u>except</u> the allowable percentage redistributions are obtained by using the final steel ratios

Calculations and Discussion	Code Reference

from the previous cycle. To illustrate this, the percentage allowable adjustments for a second cycle redistribution are calculated for supports B and C, and shown in the following Table.

SECOND CYCLE OF REDISTRIBUTION

Support	Factored Moment M_u, ft. kips	Steel Ratio from 1st Cycle	ρ/ρ_b (ρ_b = 0.0285)	Allowable Redistribution (Fig. 8-1)
B	-97.3	0.0112	0.386	12.3%
C	-60.9	0.0065	0.228	15.4%

10. Summary.

To continue with the second cycle of redistribution, the percentage adjustments calculated are then applied to the original elastic moment curves for each of the loading patterns and a second adjusted maximum moment envelope obtained. The moments from this second adjusted maximum moment envelope are then used to determine the required reinforcement. The actual steel ratio provided must comply with Fig. 8-1 for the percent moment redistribution taken. This is an important final step to assure that the design meets the requirements of Section 8.4.

EXAMPLE 8.2 - Continued

Calculations and Discussion	Code Reference

The second cycle distribution at support B results in an adjusted moment of -109.5 (1-0.125) = $-95.8^{'k}$. This is only a reduction of 1.5% from the first cycle redistribution.

For this design example, the design moments were not reduced to the face of the column because the intent was to illustrate moment redistribution alone. In Example 8.1, both moment redistribution and moments at the face of the support were used in design.

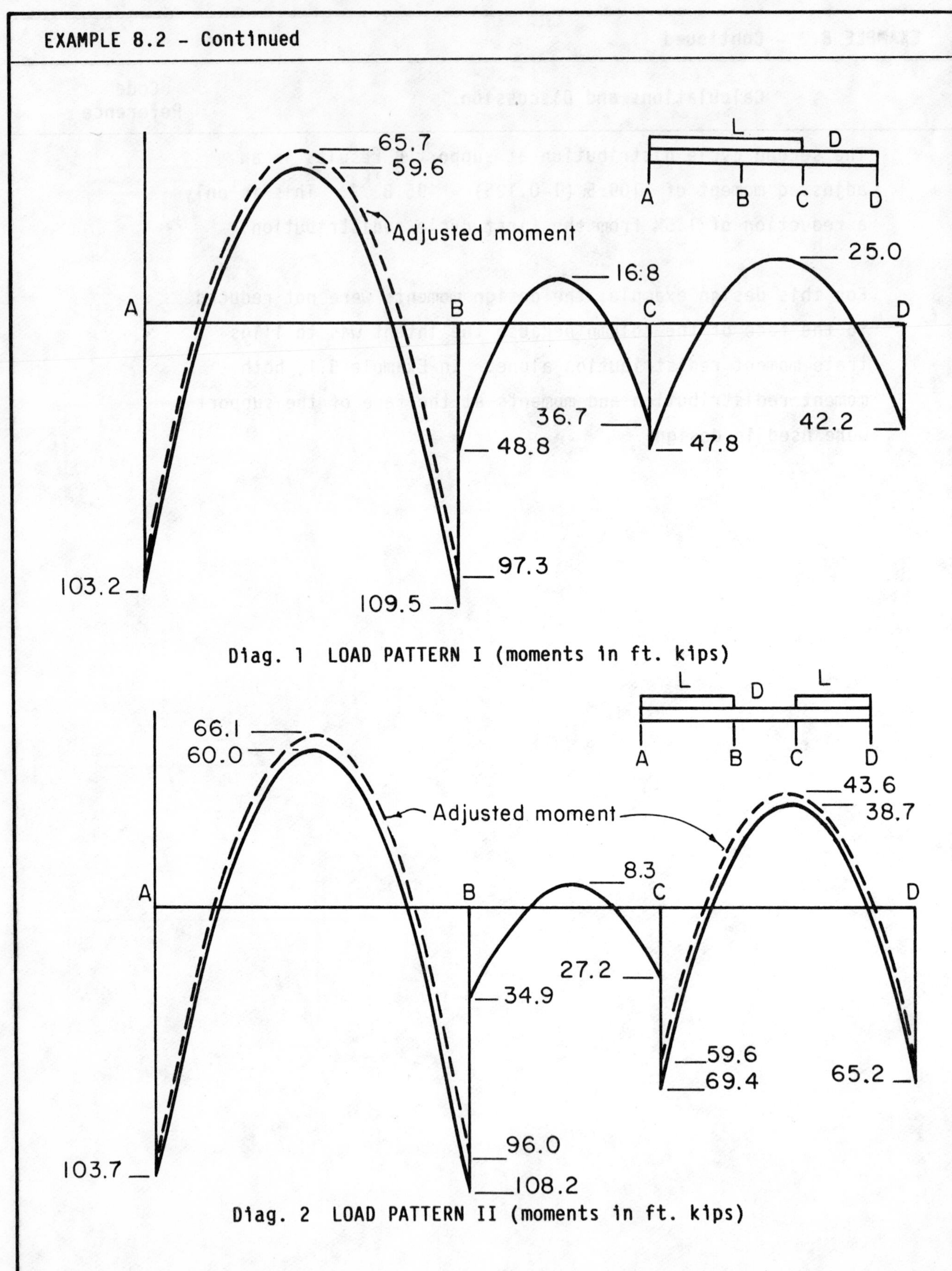

Diag. 1 LOAD PATTERN I (moments in ft. kips)

Diag. 2 LOAD PATTERN II (moments in ft. kips)

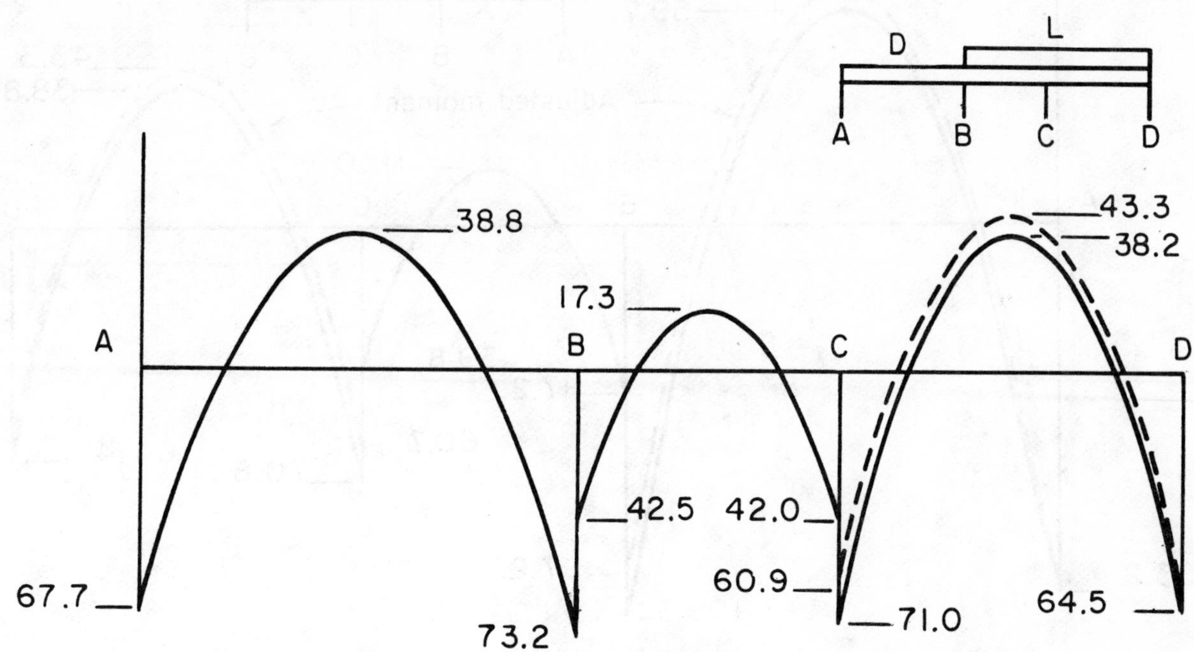

Diag. 3 LOAD PATTERN III (moments in ft. kips)

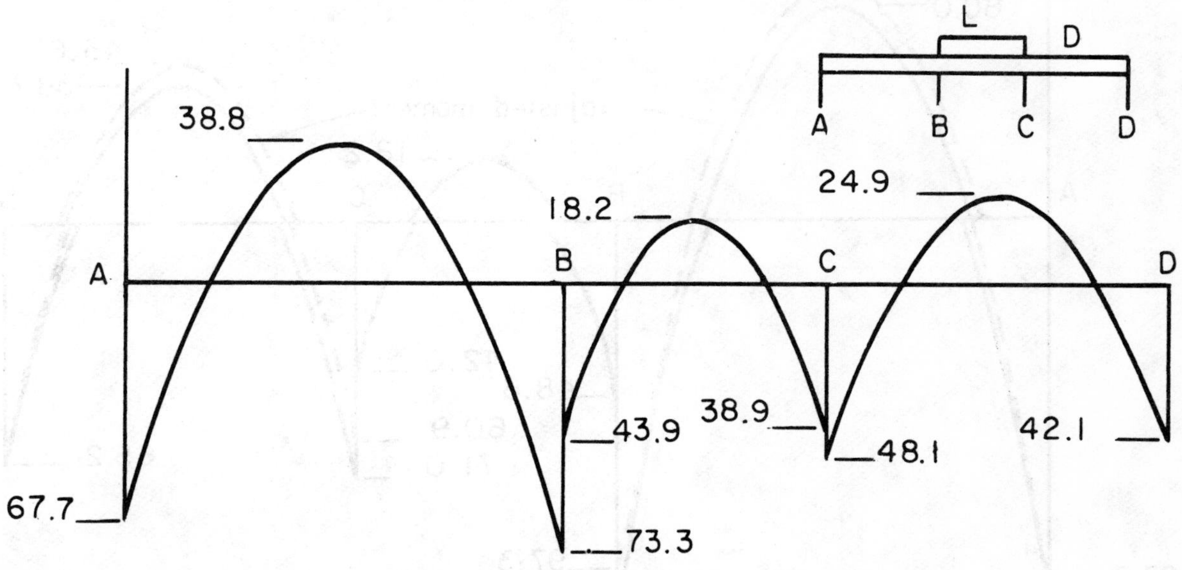

Diag. 4 LOAD PATTERN IV (moments in ft. kips)

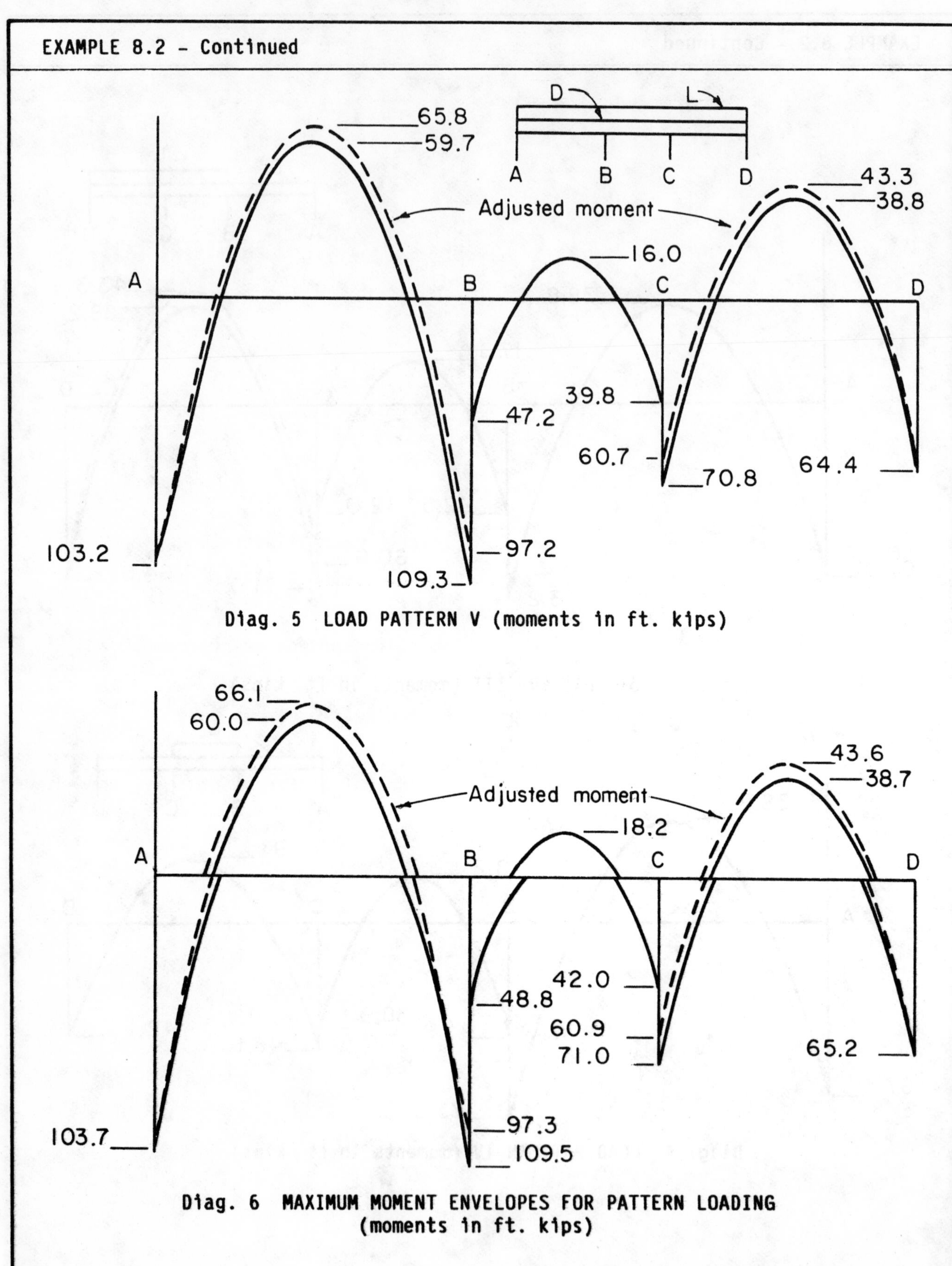

Diag. 5 LOAD PATTERN V (moments in ft. kips)

Diag. 6 MAXIMUM MOMENT ENVELOPES FOR PATTERN LOADING (moments in ft. kips)

9

Design for Flexure

Part 9 presents six design examples for flexure, each illustrating proper application of the various code provisions that govern design of members subject to flexure only. The design examples are prefaced by a step-by-step procedure for design of rectangular sections with tension reinforcement only.

Design Data - Rectangular Sections with Tension Reinforcement Only[9.1]

In the design of rectangular sections with tension reinforcement only, the two conditions of equilibrium are:

$$C = T \tag{1}$$

and

$$M_n = (C \text{ or } T)\left(d - \frac{a}{2}\right) \tag{2}$$

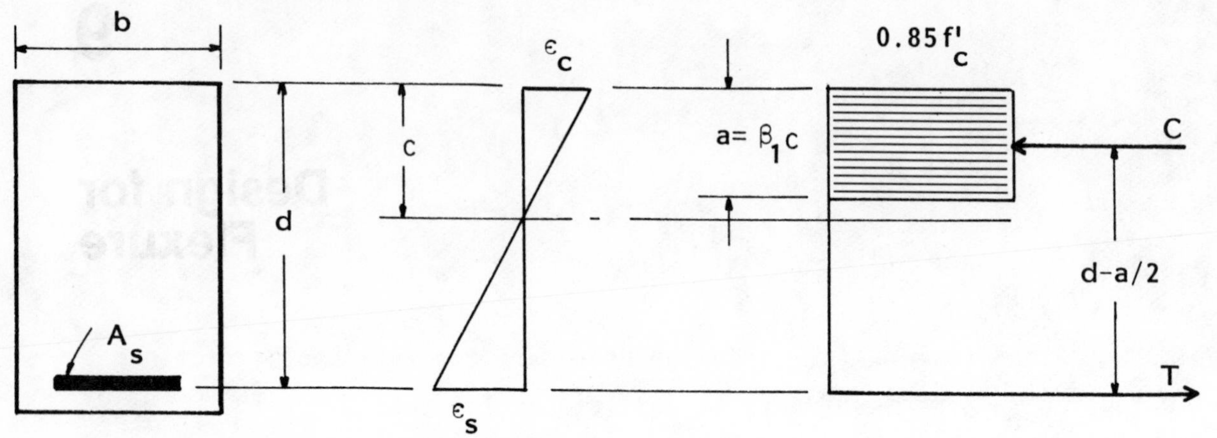

When the reinforcement ratio $\rho = \frac{A_s}{bd}$ is preset, from Equation (1):

$$0.85f'_c ba = \rho bdf_y$$

$$a = \frac{\rho d}{0.85} \frac{f_y}{f'_c}$$

from Equation (2):

$$M_n = \rho bdf_y \left[d - 0.5 \frac{\rho d}{0.85} \frac{f_y}{f'_c} \right]$$

A nominal strength coefficient of resistance R_n is obtained by dividing by bd^2.

$$R_n = \frac{M_n}{bd^2} = \rho f_y \left(1 - 0.5 \frac{\rho f_y}{0.85f'_c}\right) \qquad (3)$$

When b and d are preset, ρ is obtained by solving the quadratic equation for R_n.

$$\rho = \frac{0.85f'_c}{f_y}\left(1 - \sqrt{1 - \frac{2R_n}{0.85f'_c}}\right) \qquad (4)$$

The relationship between ρ and R_n for Grade 60 reinforcement and various values of f'_c is shown in Fig. 9-1.

Using Equations (3) and (4), a design procedure for rectangular sections with tension reinforcement only is outlined as follows:

Step 1 Select an approximate value of tension reinforcement ratio ρ equal to or less than 0.75 ρ_b (Section 10.3.3), but greater than the minimum (Section 10.5.1), where the balanced reinforcement ratio ρ_b is given by:

$$\rho_b = \frac{0.85\beta_1 f'_c}{f_y}\left(\frac{87{,}000}{87{,}000 + f_y}\right)$$

and

$$\beta_1 = 0.85 \text{ for } f'_c \leq 4000 \text{ psi}$$

$$= 0.85 - 0.05\left(\frac{f'_c - 4000}{1000}\right) \text{ for } f'_c > 4000 \text{ psi} < 8000 \text{ psi}$$

$$= 0.65 \text{ for } f'_c \geq 8000 \text{ psi}$$

Values of ρ_b and 0.75 ρ_b are given in Table 9-1.

TABLE 9-1. Balanced Reinforcement Ratio ρ_b (and 0.75 ρ_b) for Rectangular Sections with Tension Reinforcement Only

f_y		$f'_c = 3000$ $\beta_1 = 0.85$	$f'_c = 4000$ $\beta_1 = 0.85$	$f'_c = 5000$ $\beta_1 = 0.80$	$f'_c = 6000$ $\beta_1 = 0.75$
40,000	ρ_b	0.0371	0.0495	0.0582	0.0655
	0.75 ρ_b	0.0278	0.0371	0.0437	0.0491
60,000	ρ_b	0.0214	0.0285	0.0335	0.0377
	0.75 ρ_b	0.0160	0.0214	0.0252	0.0283

Step 2 With ρ preset ($\frac{200}{f_y} \leq \rho < 0.75\rho_b$) compute bd^2 required.

$$bd^2 \text{ (req.)} = \frac{M_u}{\varphi R_n}$$

where $R_n = \rho f_y \left(1 - 0.5 \frac{\rho f_y}{0.85 f'_c}\right)$

φ = 0.90 for flexure

M_u = applied factored moment

Step 3 Size the member so that the value of bd^2 provided is approximately equal to the value of bd^2 required.

Step 4 Compute a revised value of ρ by one of the following methods:

(1) by formula (exact method)

$$\rho = \frac{0.85f'_c}{f_y}\left(1 - \sqrt{1 - \frac{2R_n}{0.85f'_c}}\right)$$

$$\text{where } R_n = \frac{M_u}{\varphi(bd^2 \text{ prov.})}$$

(2) by strength curves such as shown in Fig. 9-1. Values of ρ are given in terms of $R_n = M_u/\varphi bd^2$ for Grade 60 reinforcement.

(3) by moment strength tables such as shown in Table 9-2. Values of $\omega = \rho f_y/f'_c$ are given in terms of moment strength $M_u/\varphi f'_c bd^2$.

(4) by approximate proportion

$$\rho \simeq (\text{original } \rho)\,\frac{(\text{revised } R_n)}{(\text{original } R_n)}$$

Note: From Fig. 9-1, the relationship between R_n and ρ is approximately linear.

Step 5 Compute A_s required

$$A_s = (\text{revised } \rho)\ (\text{bd prov.})$$

When b and d are preset, the A_s required is computed directly from

$$A_s = \rho(\text{bd prov.})$$

where ρ is computed using one of the methods outlined in Step 4.

Selected References

9.1 C.K. Wang and C.G. Salmon, Reinforced Concrete Design, Third Edition, Harper & Row Publishers, New York, N.Y., 1979, Chapter 3.

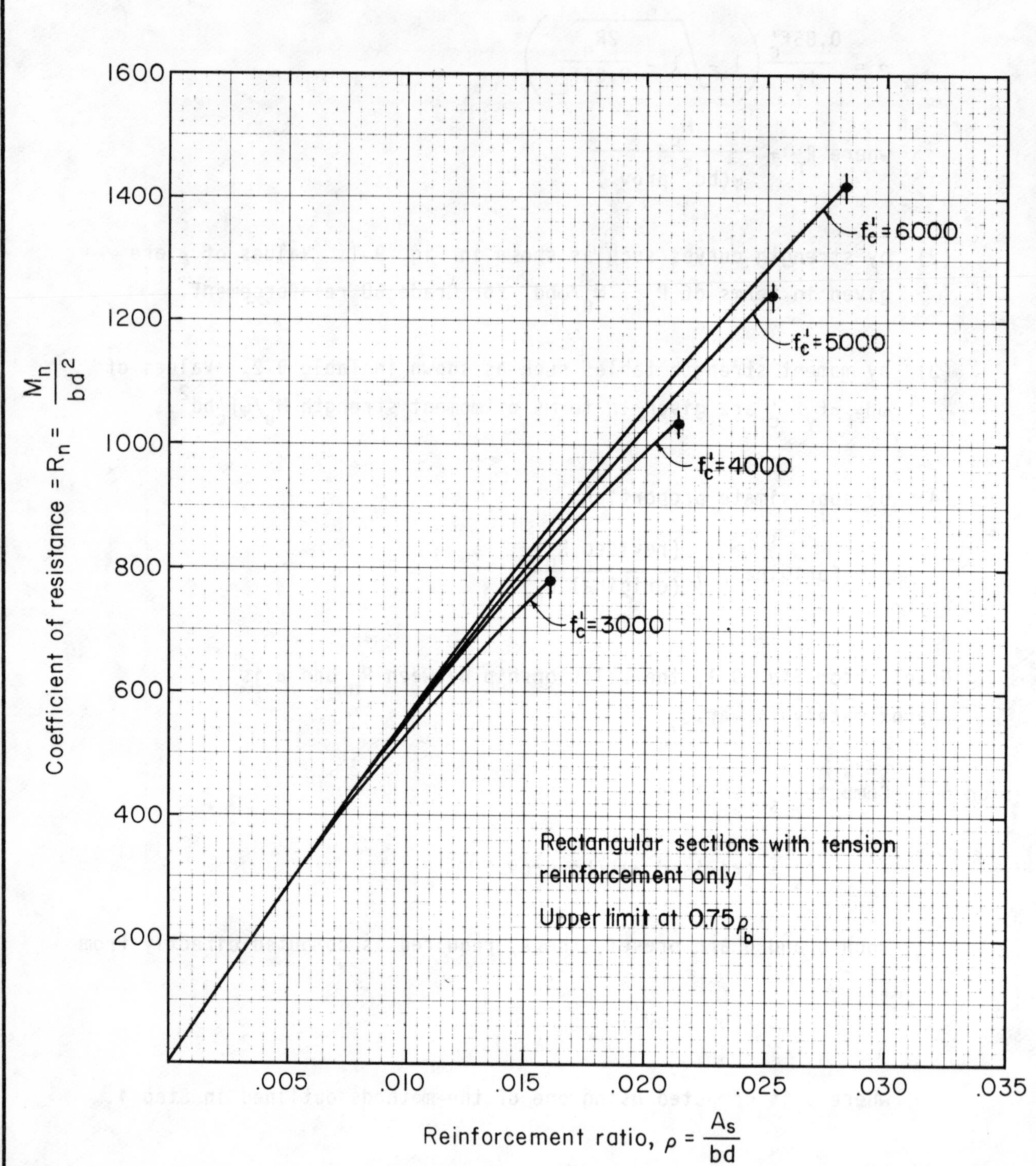

Fig. 9-1 - Strength Curves (R_n vs ρ) for Grade 60 Reinforcement

TABLE 9-2. Moment Strength $M_u/\varphi f'_c bd^2$ or $M_n/f'_c bd^2$ of Rectangular Sections with Tension Reinforcement Only*

ω	.000	.001	.002	.003	.004	.005	.006	.007	.008	.009
0.0	0	.0010	.0020	.0030	.0040	.0050	.0060	.0070	.0080	.0090
0.01	.0099	.0109	.0119	.0129	.0139	.0149	.0159	.0168	.0178	.0188
0.02	.0197	.0207	.0217	.0226	.0236	.0246	.0256	.0266	.0275	.0285
0.03	.0295	.0304	.0314	.0324	.0333	.0343	.0352	.0362	.0372	.0381
0.04	.0391	.0400	.0410	.0420	.0429	.0438	.0448	.0457	.0467	.0476
0.05	.0485	.0495	.0504	.0513	.0523	.0532	.0541	.0551	.0560	.0569
0.06	.0579	.0588	.0597	.0607	.0616	.0625	.0634	.0643	.0653	.0662
0.07	.0671	.0680	.0689	.0699	.0708	.0717	.0726	.0735	.0744	.0753
0.08	.0762	.0771	.0780	.0789	.0798	.0807	.0816	.0825	.0834	.0843
0.09	.0852	.0861	.0870	.0879	.0888	.0897	.0906	.0915	.0923	.0932
0.10	.0941	.0950	.0959	.0967	.0976	.0985	.0994	.1002	.1011	.1020
0.11	.1029	.1037	.1046	.1055	.1063	.1072	.1081	.1089	.1098	.1106
0.12	.1115	.1124	.1133	.1141	.1149	.1158	.1166	.1175	.1183	.1192
0.13	.1200	.1209	.1217	.1226	.1234	.1243	.1251	.1259	.1268	.1276
0.14	.1284	.1293	.1301	.1309	.1318	.1326	.1334	.1342	.1351	.1359
0.15	.1367	.1375	.1384	.1392	.1400	.1408	.1416	.1425	.1433	.1441
0.16	.1449	.1457	.1465	.1473	.1481	.1489	.1497	.1506	.1514	.1522
0.17	.1529	.1537	.1545	.1553	.1561	.1569	.1577	.1585	.1593	.1601
0.18	.1609	.1617	.1624	.1632	.1640	.1648	.1656	.1664	.1671	.1679
0.19	.1687	.1695	.1703	.1710	.1718	.1726	.1733	.1741	.1749	.1756
0.20	.1764	.1772	.1779	.1787	.1794	.1802	.1810	.1817	.1825	.1832
0.21	.1840	.1847	.1855	.1862	.1870	.1877	.1885	.1892	.1900	.1907
0.22	.1914	.1922	.1929	.1937	.1944	.1951	.1959	.1966	.1973	.1981
0.23	.1988	.1995	.2002	.2010	.2017	.2024	.2031	.2039	.2046	.2053
0.24	.2060	.2067	.2075	.2082	.2089	.2096	.2103	.2110	.2117	.2124
0.25	.2131	.2138	.2145	.2152	.2159	.2166	.2173	.2180	.2187	.2194
0.26	.2201	.2208	.2215	.2222	.2229	.2236	.2243	.2249	.2256	.2263
0.27	.2270	.2277	.2284	.2290	.2297	.2304	.2311	.2317	.2324	.2331
0.28	.2337	.2344	.2351	.2357	.2364	.2371	.2377	.2384	.2391	.2397
0.29	.2404	.2410	.2417	.2423	.2430	.2437	.2443	.2450	.2456	.2463
0.30	.2469	.2475	.2482	.2488	.2495	.2501	.2508	.2514	.2520	.2527
0.31	.2533	.2539	.2546	.2552	.2558	.2565	.2571	.2577	.2583	.2590
0.32	.2596	.2602	.2608	.2614	.2621	.2627	.2633	.2639	.2645	.2651
0.33	.2657	.2664	.2670	.2676	.2682	.2688	.2694	.2700	.2706	.2712
0.34	.2718	.2724	.2730	.2736	.2742	.2748	.2754	.2760	.2766	.2771
0.35	.2777	.2783	.2789	.2795	.2801	.2807	.2812	.2818	.2824	.2830
0.36	.2835	.2841	.2847	.2853	.2858	.2864	.2870	.2875	.2881	.2887
0.37	.2892	.2898	.2904	.2909	.2915	.2920	.2926	.2931	.2937	.2943
0.38	.2948	.2954	.2959	.2965	.2970	.2975	.2981	.2986	.2992	.2997
0.39	.3003	.3008	.3013	.3019	.3024	.3029	.3035	.3040	.3045	.3051

*$M_n/f'_c bd^2 = A_s f_y(d-a/2)f'_c bd^2 = \omega(1-0.59\omega)$, where $\omega = \rho f_y/f'_c$ and $a = A_s f_y/0.85f'_c b$.

Design: Using factored moment M_u enter table with $M_u/\varphi f'_c bd^2$; find ω and compute steel percentage ρ from $\rho = \omega f'_c/f'_y$.

Investigation: Enter table with ω from $\omega = \rho f_y/f'_c$; find value of $M_n/f'_c bd^2$ and solve for nominal moment strength, M_n.

EXAMPLE 9.1 - Design of Rectangular Beam with Tension Reinforcement Only

Select a rectangular beam size and required reinforcement A_s to carry service load moments of: M_d = 55 ft kips and M_ℓ = 36 ft kips. Select reinforcement to control flexural cracking for exterior exposure.

Use f'_c = 4000 psi

f_y = 60,000 psi

z = 145 (exterior exposure)

Calculations & Discussion	Code Reference
1. To illustrate a complete design procedure for rectangular sections with tension reinforcement only, a minimum beam depth will be computed using the maximum reinforcement permitted for flexural members, $0.75\rho_b$. The design procedure will follow the method outlined on pages 9-3 through 9-5.	10.3.3
Step 1. Compute maximum reinforcement ratio for material strengths f'_c = 4000 and f_y = 60,000.*	
$$\rho_b = \frac{0.85\beta_1 f'_c}{f_y}\ \frac{87000}{87000 + f_y} = 0.0285$$	
$\rho_{max} = 0.75\rho_b = 0.75(0.0285) = 0.0214$	10.3.3
Step 2. Compute bd^2 required.	
Required moment strength:	
$U = 1.4\ D + 1.7\ L$	Eq. (9-1)
$M_u = 1.4x55 + 1.7x36$	
M_u = 138 ft kips	

*Values of ρ_b and 0.75 ρ_b can also be obtained directly from Table 9-1.

EXAMPLE 9.1 - Continued

Calculations & Discussion	Code Reference

$$R_n = \rho f_y\left(1 - 0.5\,\frac{\rho f_y}{0.85f'_c}\right)$$

$$= 0.0214 \times 60{,}000\left(1 - \frac{0.5 \times 0.0214 \times 60{,}000}{0.85 \times 4000}\right)$$

R_n = 1042 psi

φ = 0.90 for flexure — 9.3.2.1

$$bd^2\ (\text{req'd}) = \frac{M_u}{\varphi R_n} = \frac{138 \times 12 \times 1000}{0.90 \times 1042} = 1766\ \text{in.}^3$$

Step 3. Size member so that:

bd^2 (required) $\leq$ bd^2 (provided)

Set b = 10 in. (column width)

$$d = \sqrt{\frac{1766}{10}} = 13.3\ \text{in.}$$

Minimum beam depth $\simeq$ 13.3 + 2.5 = 15.8 in.

For moment strength, a 10x16-in. beam size is adequate. However, it should be noted that the 16-in. beam depth is somewhat less than traditional designs using the older working stress design procedures. Deflection thus becomes an essential consideration in designing beams by the strength design method. Control of deflection is discussed in Part 7.

Step 4. Using the 16-in. beam depth, compute a revised value of ρ. For illustration, ρ will be computed by the four methods outlined on page 9-5.

d = 16 - 2.5 = 13.5 in.

EXAMPLE 9.1 - Continued

Calculations & Discussion	Code Reference

(1) by formula (exact method):

$$R_n = \frac{M_u}{\varphi(bd^2 \text{ prov.})} = \frac{138 \times 12 \times 1000}{0.90(10 \times 13.5^2)} = 1010 \text{ psi}$$

$$\rho = \frac{0.85f'_c}{f_y}\left(1 - \sqrt{1 - \frac{2R_n}{0.85f'_c}}\right)$$

$$= \frac{0.85 \times 4}{60}\left(1 - \sqrt{1 - \frac{2 \times 1010}{0.85 \times 4000}}\right) = 0.0206$$

(2) by strength curves such as that shown in Fig. 9-1:

for $R_n = 1010$ psi, $\rho \simeq 0.0205$

(3) by strength tables such as that shown in Table 9-2:

$$\text{for } \frac{M_u}{\varphi f'_c bd^2} = \frac{138 \times 12 \times 1000}{0.90 \times 4000 \times 10 \times 13.5^2} = 0.252$$

$\omega = 0.308$

$\rho = \omega f'_c/f_y = 0.308 \times 4/60 = 0.0205$

(4) by approximate proportion:

$$\rho \simeq (\text{original } \rho)\ \frac{(\text{revised } R_n)}{(\text{original } R_n)}$$

$$\rho \simeq 0.0214 \times \frac{1010}{1042} = 0.0207$$

Step 5. Compute A_s required.

A_s = (revised ρ) (bd prov.)

$A_s = 0.0206 \times 10 \times 13.5 = 2.78$ in.^{2}

2. A review of the correctness of the computations can be made by considering the simple statics shown on page 9-2.

EXAMPLE 9.1 - Continued

Calculations & Discussion	Code Reference

$T = \rho bdf_y = A_sf_y = 2.78 \times 60 = 166.8$ kips

$$a = \frac{C \text{ or } T}{0.85f'_cb} = \frac{166.8}{0.85 \times 4 \times 10} = 4.91 \text{ in.}$$

Design moment strength:

$$\varphi M_n = \varphi\left[A_sf_y(d - \frac{a}{2})\right] = \left[0.9\ 166.8(13.5 - 4.91/2)\right]$$

$$M_n = 1658.1 \text{ in. kips} = 138.2 \text{ ft kips}$$

Required moment strength ≤ design moment strength

$M_u \leq \varphi M_n$

$138 \leq 138.2$ OK

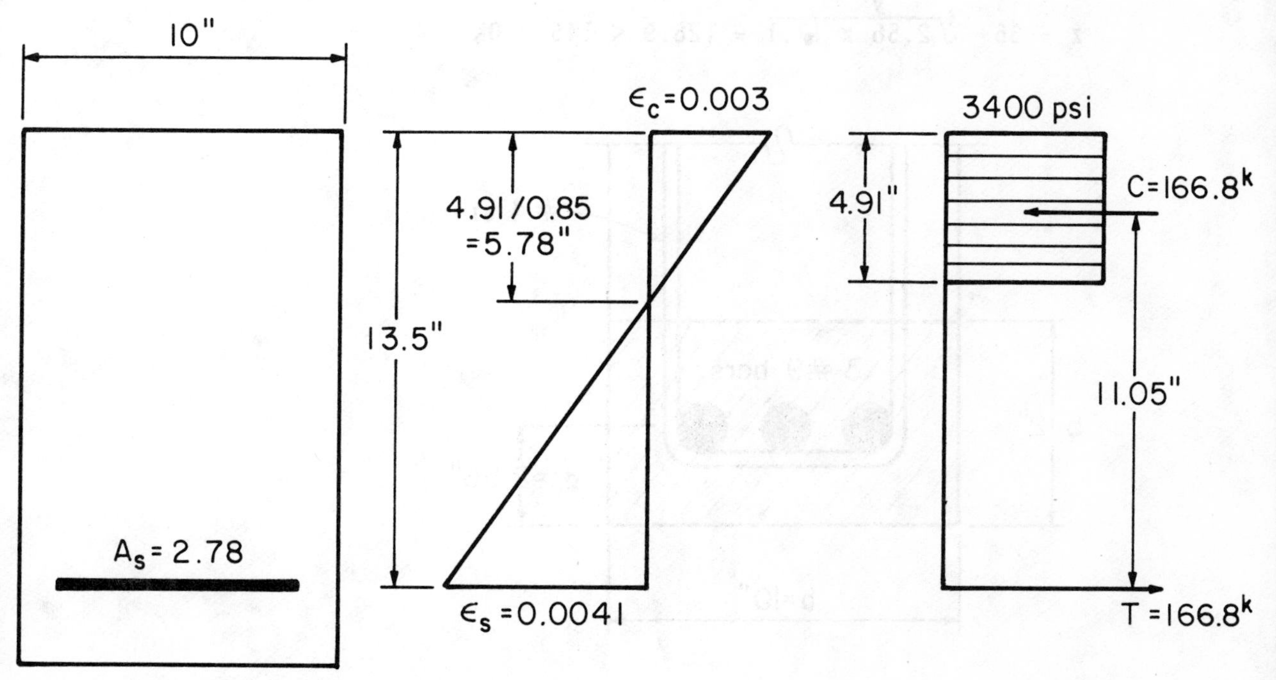

EXAMPLE 9.1 - Continued

Calculations & Discussion	Code Reference
3. Select reinforcement to satisfy distribution of flexural reinforcement requirements of Section 10.6. Use z = 145 for exterior exposure.	10.6
A_s (req'd) = 2.78 in.^{2}	
Select 3 - #9 bars A_s = 3.00 in.^{2}	
$z = f_s \sqrt[3]{d_c A}$	Eq. (10-4)
d_c = cover req'd + 1/2 bar dia. + stirrup dia.	10.0
= 1.5 + 0.56 + 0.5 = 2.56 in.	
(cover to #9 bars = 1.5 + 0.5 = 2.0 in.)	7.7.1
(exposed to weather)	
$A = 2d_c b$/no. of bars	10.0
= 2 x 2.56 x 10/3 = 17.1 in.^{2}/bar	
Use $f_s = 0.6f_y$ = 0.6 x 60 = 36 ksi	10.6.4
$z = 36 \sqrt[3]{2.56 \times 17.1}$ = 126.9 < 145 OK	

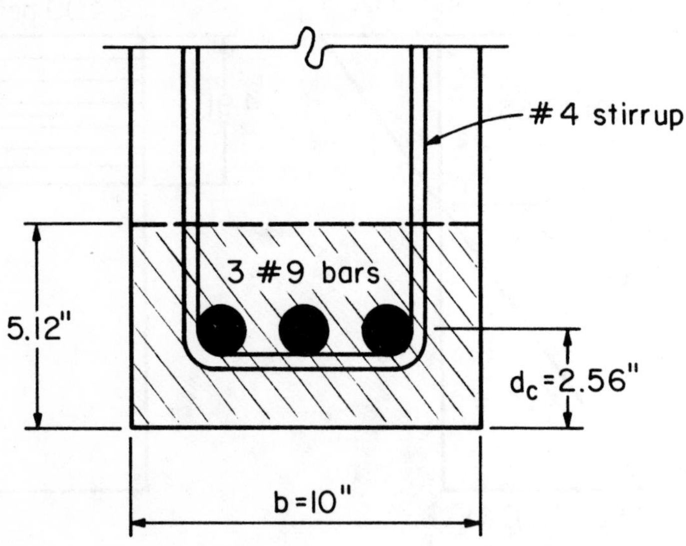

4. Check beam width:	
b ≥ 2 x cover + 3 x 1.128 + 2 x 1.128	7.6.1
= 2 x 2 + 5.64 = 9.64 < 10 in. (prov'd) OK	7.7.1

EXAMPLE 9.2 - Design of Rectangular Beam with Compression Reinforcement

A beam cross section is limited to the size shown. Determine required area of reinforcement for a factored moment M_u = 900 ft kips.

f'_c = 4000 psi

f_y = 60,000 psi

Exterior exposure

z = 145

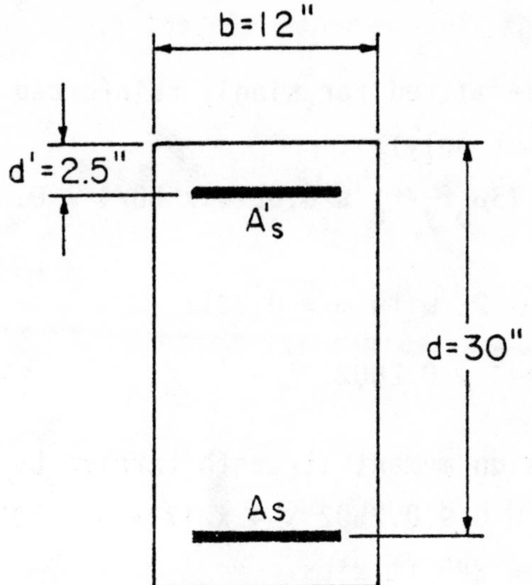

Calculations & Discussion	Code Reference

1. Check design for tension reinforcement only.

Compute required tension reinforcement using strength Table 9-2:

$$\frac{M_u}{\varphi f'_c bd^2} = \frac{900 \times 12 \times 1000}{0.90 \times 4000 \times 12 \times 30^2} = 0.2777$$

From Table 9-2, ω = 0.350

Ratio of tension reinforcement required:

$$\rho = \omega f'_c/f_y = 0.350 \times 4/60 = 0.0233$$

With tension reinforcement only:

$\rho_{max} = 0.75\rho_b$ — 10.3.3

From Table 9-1, with f'_c = 4000 and f_y = 60,000:

ρ_{max} = 0.0214

0.0233 > 0.0214 ∴ compression reinforcement required

EXAMPLE 9.2 - Continued

Calculations & Discussion	Code Reference

2. Compute reinforcement required, A_s and A'_s:

Maximum ω permitted for singly reinforced beam (tension reinforcement only):

$$\omega \leq 0.75\rho_b f_y/f'_c \leq 0.0214 \times 60/4 = 0.321$$

From Table 9-2, with $\omega = 0.321$:

$$M_n/f'_c bd^2 = 0.2602$$

Maximum design moment strength carried by concrete:

$$\varphi M_{nc} = 0.9(0.2602 \times 4 \times 12 \times 30^2/12)$$
$$= 843 \text{ ft kips}$$

Required moment strength to be carried by compression reinforcement:

$$M'_u = 900 - 843 = 57 \text{ ft kips}$$

Assume compression reinforcement yields, $f'_s = f_y$

$$\rho' = \frac{A'_s}{bd} = \frac{M'_u}{\varphi f_y(d - d')bd}$$

$$\rho' = \frac{57 \times 12 \times 1000}{0.9 \times 60{,}000(30 - 2.5)12 \times 30} = 0.00128$$

$$\rho = 0.75\rho_b + \rho = 0.0214 + 0.00128 = 0.0227$$

Note: For members with compression reinforcement, the portion of ρ_b contributed by compression reinforcement need not be reduced by the 0.75 factor. See Code Commentary Table 10.3.2. 10.3.3

$$A'_s = \rho' bd = 0.00128 \times 12 \times 30 = 0.46 \text{ in.}^2$$

$$A_s = \rho bd = 0.0227 \times 12 \times 30 = 8.17 \text{ in.}^2$$

Calculations & Discussion	Code Reference

Check compression reinforcement yield condition:

$$\frac{A_s - A_s'}{bd} \geq \frac{0.85\beta_1 f_c' d'}{f_y d}\left(\frac{87000}{87000 - f_y}\right)$$

$$0.0227 - 0.00128 \geq \frac{0.85 \times 0.85 \times 4 \times 2.5}{60 \times 30}\left(\frac{87000}{87000 - 60000}\right)$$

$$0.0214 \geq 0.0129$$

∴ compression reinforcement yields as assumed OK.

3. A review of the correctness of the computations may be made by using the strength equations given in Code Commentary Section 10.3(A)(3). When the compression reinforcement yields:

$$\varphi M_n = \varphi\left[(A_s - A_s')f_y(d - \frac{a}{2}) + A_s' f_y(d - d')\right]$$

$$= 0.9\left[7.71 \times 60(30 - \frac{11.34}{2}) + 0.46 \times 60(30 - 2.5)\right]/12$$

$$= 901 \text{ ft kips}$$

where $a = \frac{(A_s - A_s')f_y}{0.85 f_c' b} = \frac{7.71 \times 60}{0.85 \times 4 \times 12} = 11.34$ in.

4. Select reinforcement to satisfy control of flexural cracking criteria of Section 10.6 for exterior exposure.

Compression reinforcement:

Select 2 - #5 bars (A_s' = 0.62 in.^{2} > 0.46in.^{2})

Tension reinforcement:

Select 8 - #9 bars (A_s = 8.00 in.^{2} ≃ 8.17 in.^{2})

(2% less than req'd...say OK)

EXAMPLE 9.2 - Continued

Calculations & Discussion	Code Reference

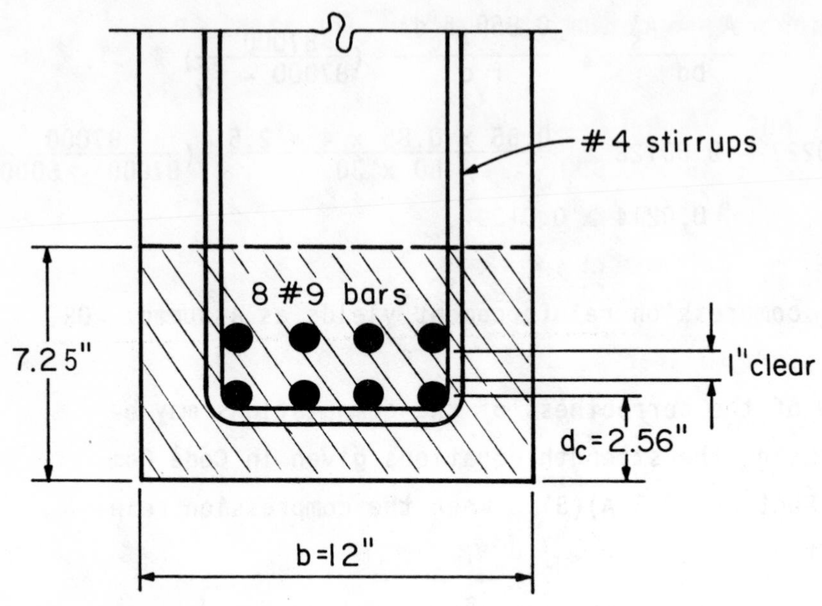

$z = f_s \sqrt[3]{d_c A}$ — Eq. (10-4)

d_c = cover + 1/2 bar dia. + stirrup dia. — 10.0

= 1.5 + 0.56 + 0.5 = 2.56

(cover to #9 bars = 1.5 + 0.5 = 2.0 in.) — 7.7.1

(exposed to weather)

A = 7.25 x 12/8 = 10.9 in.^{2}/bar — 10.0

Use $f_s = 0.6 f_y$ = 36 ksi — 10.6.4

$z = 36 \sqrt[3]{2.56 \times 10.9} = 109 < 145$

5. Check beam width:

b = 2 x cover + 4 x 1.128 + 3 x 1.128 — 7.6.1

= 2 x 2 + 4.51 + 3.38 = 11.9 < 12 in. (prov'd) OK — 7.7.1

EXAMPLE 9.2 - Continued

	Calculations & Discussion	Code Reference
6.	Stirrups or ties are required throughout distance where compression reinforcement is required for strength.	7.11.1
	Max. spacing: 16 x 0.625 = 10 in. 48 x 0.900 = 24 in. least dimension of member = 12 in.	7.10.5.2
	Use s_{max} = 10 in. for #3 stirrups	

EXAMPLE 9.3 - Design of One-Way Solid Slab

Determine required thickness and reinforcement for a one-way slab continuous over two or more equal spans. Clear span, ℓ_n = 18'-0". Use f'_c = 4000 psi and f_y = 60,000 psi.

Service loads: w_ℓ = 50 psf
w_d = 75 psf (assume 6-in. slab)

Calculations & Discussion	Code Reference

1. Compute required moment strengths using approximate moment analysis permitted by Section 8.3.3.* Design to be based on end span.

Factored load:

$U = 1.4\ D + 1.7\ L$ — Eq. (9-1)

$w_u = 1.4 \times 75 + 1.7 \times 50 = 190$ psf

Positive moment:

$+M_u = w_u \ell_n^2/14$ — 8.3.3

$= 0.190 \times 18^2/14 = 4.40$ ft kips/ft of width

Negative moment at exterior face of first interior support:

$-M_u = w_u \ell_n^2/10$ — 8.3.3

$= 0.190 \times 18^2/10 = 6.16$ ft kips/ft of width

2. Compute required slab thickness.

Choose a reinforcement percentage ρ equal to about $0.375\rho_b$, or one-half the maximum permitted, to have reasonable deflection control. — 10.3.3

From Table 9-1, for f'_c = 4000 and f_y = 60,000,

---

*With the 1983 Code, an additional condition "members must be prismatic" was added to Section 8.3.3 to clarify that the approximate moments and shears are not applicable for members with haunches and varying cross sections.

EXAMPLE 9.3 - Continued

Calculations & Discussion	Code Reference

$\rho_b = 0.0285$

Set $\rho = 0.375(0.0285) = 0.0107$

Design procedure to follow method outlined on pages 9-3 through 9-5:

$$R_n = \rho f_y \left(1 - 0.5 \frac{\rho f_y}{0.85 f'_c}\right)$$

$$= 0.0107 \times 60{,}000 \left(1 - \frac{0.5 \times 0.0107 \times 60{,}000}{0.85 \times 4000}\right)$$

$$R_n = 581 \text{ psi}$$

$$\text{required } d = \sqrt{\frac{M_u}{\varphi R_n b}} = \sqrt{\frac{6.16 \times 12{,}000}{0.90 \times 581 \times 12}} = 3.43 \text{ in.}$$ 9.3.2.1

Assume #5 bars.

required h = 3.43 + 0.31 + 0.75 = 4.49 in.

The above design indicates a slab thickness of 4-1/2 in. is adequate. However, Table 9.5(a) of the Code requires a minimum thickness of $\ell/28$ = 7-3/4 in. unless deflections are computed. Also note that Table 9.5(a) is applicable only for "members not supporting or attached to partitions or other construction likely to be damaged by large deflections" . . . otherwise deflections must be computed. Compute required reinforcement for h = 4-1/2 in., d = 3.43 in.

EXAMPLE 9.3 - Continued

Calculations & Discussion	Code Reference

3. Compute required reinforcement.

$-A_s$ (req'd) = $\rho bd = 0.0107 \times 12 \times 3.43 = 0.440$ in.^{2}/ft

<u>Use #5 @ 8 in</u>. ($A_s = 0.47$ in.^{2}/ft)

<u>or #6 @ 12 in</u>. ($A_s = 0.44$ in.^{2}/ft)

For positive moment, using strength Table 9-2:

$$\frac{M_u}{\varphi f'_c bd^2} = \frac{4.40 \times 12000}{0.9 \times 4000 \times 12 \times 3.43^2} = 0.1039$$

From Table 9-2, $\omega = 0.111$

$\rho = \omega f'_c/f_y = 0.111 \times 4/60 = 0.0074$

$+A_s$ (req'd) = $\rho bd = 0.0074 \times 12 \times 3.43 = 0.305$ in.^{2}/ft

<u>Use #4 @ 8 in.</u> ($A_s = 0.30$ in.^{2}/ft)

<u>or #5 @ 12 in.</u> ($A_s = 0.31$ in.^{2}/ft)

EXAMPLE 9.4 - Design of Flanged Section with Tension Reinforcement Only

Select reinforcement for the "T" section shown to carry service dead and live load moments of M_d = 72 ft kips and M_ℓ = 88 ft kips.

f'_c = 4000 psi

f_y = 60,000 psi

Exterior exposure

(z = 145)

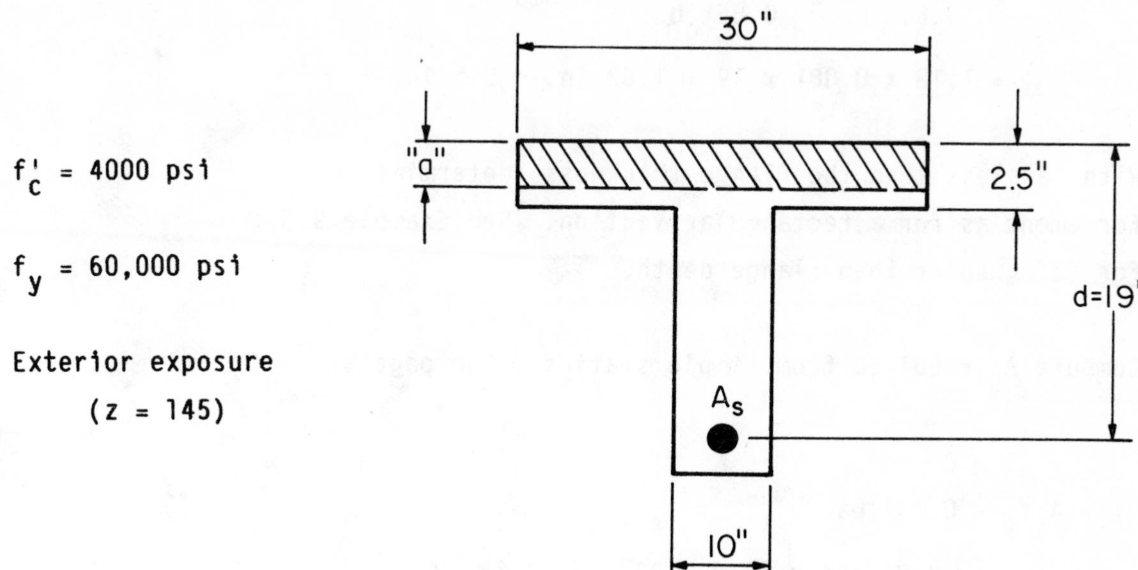

Calculations & Discussion	Code Reference
1. Determine required moment strength (factored load moment).	
$M_u = 1.4\ M_d + 1.7\ M_\ell$	Eq. (9-1)
$= 1.4 \times 72 + 1.7 \times 88$	
$= 250$ ft kips	
2. Using Table 9-2, determine depth of equivalent stress block "a" as for a rectangular section.	
For $\dfrac{M_u}{\varphi f'_c bd^2} = \dfrac{250 \times 12}{0.9 \times 4 \times 30 \times 19^2} = 0.077$	9.3.2.1

EXAMPLE 9.4 - Continued

Calculations & Discussion	Code Reference

From Table 9-2, $\omega = \rho f_y/f'_c = 0.081$

$$a = \frac{A_s f_y}{0.85f'_c b} = \frac{\rho d f_y}{0.85f'_c b} = 1.18\omega d$$

$$= 1.18 \times 0.081 \times 19 = 1.82 \text{ in.} < 2.5 \text{ in.}$$

With "a" less than the flange thickness, determine reinforcement as for a rectangular section. See Example 9.5 for "a" greater than flange depth.

3. Compute A_s required from simple statics. See page 9-2.

$$T = C$$

$$A_s f_y = 0.85f'_c ba$$

$$A_s = \frac{0.85 \times 4 \times 30 \times 1.82}{60} = 3.09 \text{ in.}^2$$

Try 2 - #11 bars (A_s = 3.12 in.^{2})

4. Check minimum required reinforcement. 10.5

$\rho_{min} = 200/f_y = 200/60{,}000 = 0.0033$ Eq. (10-3)

$\rho = A_s/b_w d = 3.12/10 \times 19 = 0.0164$

0.0164 > 0.0033 OK

5. Check distribution of reinforcement for exterior exposure (z = 145). 10.6

$z = f_s \sqrt[3]{d_c A}$ Eq. (10-4)

d_c = cover + 1/2 bar diameter

= 2 + 0.71 = 2.71 in.

$A = 2d_c b_w$/no. of bars

= 2x2.71 x 10/2 = 27.1 in.^{2}/bar

$z = 0.6 \times 60 \sqrt[3]{2.71 \times 27.1} = 150.8 > 145$ NG 10.6.4

EXAMPLE 9.4 - Continued

Calculations & Discussion	Code Reference
Since the limiting value of z is exceeded for exterior exposure, unacceptable tensile cracking is indicated. Smaller bar sizes must be used.	
Try 3 - #9 bars (A_s = 3.00 in.^{2}) (3% less than req'd...say OK)	
d_c = 2 + 0.56 = 2.56 in.	
A = 2 x 2.56 x 10/3 = 17.1 in.^{2}/bar	
$z = 0.6 \times 60 \sqrt[3]{2.56 \times 17.1} = 127 < 145$ OK	
6. Check minimum web width.	
$b_w \geq 2 \times \text{cover} + 3 \times 1.128 + 2 \times 1.128$	7.6.1
$= 2 \times 2 + 5.64 = 9.64 < 10$ in. (prov'd) OK	7.7.1

EXAMPLE 9.5 - Design of Flanged Section with Tension Reinforcement Only

Select reinforcement for the "T" section shown to carry a factored moment of M_u = 400 ft kips.

f'_c = 4000 psi

f_y = 60,000 psi

Exterior exposure
(z = 145)

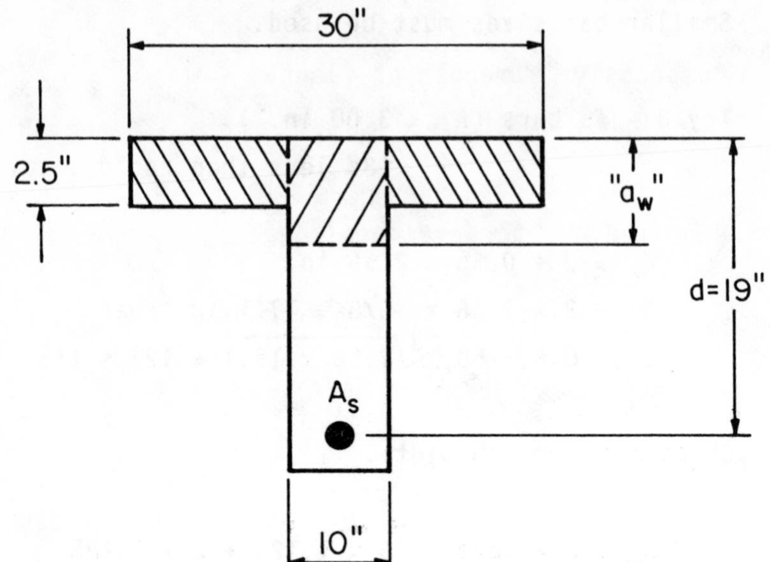

Calculations & Discussion	Code Reference
1. Using Table 9-2, determine depth of equivalent stress block "a" as for a rectangular section.	
For $\frac{M_u}{\varphi f'_c bd^2} = \frac{400 \times 12}{0.9 \times 4 \times 30 \times 19^2} = 0.123$	9.3.2.1
From Table 9-2, $\omega = \rho\, f_y/f'_c = 0.133$	
"a" = 1.18 ωd	
= 1.18 x 0.133 x 19 = 2.98 > 2.5 in.	
Since the required value of "a" as a rectangular section exceeds the flange thickness, the equivalent stress block is not rectangular and the design must be based on a "T" section. See Example 9.4 for "a" less than flange depth.	

EXAMPLE 9.5 - Continued

Calculations & Discussion	Code Reference

2. Compute required reinforcement A_{sf} and moment strength φM_{nf} to develop beam flange.

Compressive strength of flange:

$$C_f = 0.85f'_c(b - b_w)h_f$$

$$= 0.85 \times 4(30 - 10)2.5 = 170^k$$

Required A_{sf} to develop flange:

$$A_{sf} = \frac{C_f}{f_y} = \frac{170}{60} = 2.83 \text{ in.}^2$$

Design moment strength of flange:

$$\varphi M_{nf} = \varphi[A_{sf}f_y(d - 0.5h_f)]$$

$$= 0.9[2.83 \times 60(19 - 1.25)]/12 = 226 \text{ ft kips}$$

Required moment strength to be carried by beam web:

$$M_{uw} = M_u - \varphi M_{nf} = 400 - 226 = 174 \text{ ft kips}$$

3. Using Table 9-2, compute reinforcement A_{sw} required to develop moment strength to be carried by web.

$$\text{For } \frac{M_{uw}}{\varphi f'_c bd^2} = \frac{174 \times 12}{0.9 \times 4 \times 10 \times 19^2} = 0.161$$

From Table 9-2, $\omega = 0.180$

$$a_w = 1.18\,\omega d = 1.18 \times 0.180 \times 19 = 4.04 \text{ in.}$$

$$A_{sw} = \frac{0.85f'_c b_w a_w}{f_y} = \frac{0.85 \times 4 \times 10 \times 4.04}{60} = 2.29 \text{ in.}^2$$

Alternatively, A_{sw} can be computed directly from:

$$A_{sw} = \frac{\omega f'_c b_w d}{f_y} = \frac{0.180 \times 4 \times 10 \times 19}{60} = 2.28 \text{ in.}^2$$

EXAMPLE 9.5 - Continued

Calculations & Discussion	Code Reference

4. Total reinforcement required to carry factored moment M_u = 400 ft kips:

$$A_s = A_{sf} + A_{sw} = 2.83 + 2.29 = 5.12 \text{ in.}^2$$

5. Check maximum tension reinforcement permitted according to Section 10.3.3. See Code Commentary Fig. 10.3.2(c) and Table 10.3.2: — 10.3.3

(2) For flanged section with tension reinforcement only:

$$\rho_{max} = 0.75 \left[\frac{b_w}{b} (\bar{\rho}_b + \rho_f)\right]$$

$$\rho_f = 0.85 \frac{f'_c}{f_y} (b - b_w) h_f/b_w d$$

$$\rho_f = 0.85 \frac{4}{60} (30 - 10)2.5/10 \times 19 = 0.0149$$

From Table 9-1, $\bar{\rho}_b = 0.0285$

$$\rho_{max} = 0.75 \left[\frac{10}{30} (0.0285 + 0.0149)\right] = 0.0109$$

$$A_{s(max)} = 0.0109 \times 30 \times 19 = 6.21 \text{ in.}^2 > 5.12 \quad \text{OK}$$

6. Select reinforcement to satisfy crack control criteria for exterior exposure (z = 145). — 10.6

Try 4 - #9 and 2 - #7 bars, (A_s = 5.20 in.^{2})

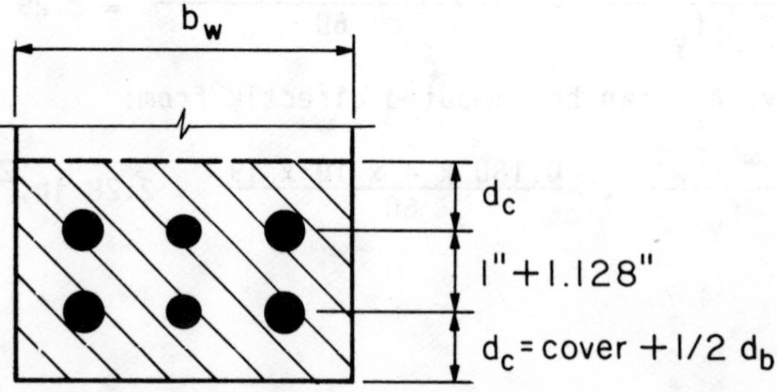

EXAMPLE 9.5 - Continued

Calculations & Discussion	Code Reference
For exterior exposure:	
$d_c = 2 + 0.564 = 2.564$ in.	10.0
Effective tension area of concrete:	
$A = (2d_c + 1 + 1.128)b_w$ /equiv. no. of #9 bars	10.0
$= 7.256 \times 10/(5.20/1.0) = 13.9$ in.^{2}	
$z = f_s \sqrt[3]{d_c A} = 0.6 \times 60 \sqrt[3]{2.564 \times 13.9}$	Eq. (10-4)
$= 118 < 145$ OK	
7. Check required web width.	7.6.1
Req'd $b_w = 2 \times \text{cover} + 2d_{b1} + 2d_{b1}^* + d_{b2}$	7.7.1
$= 2 \times 2 + 4 \times 1.128 + 0.875$	
$= 9.39$ in. < 10 in. OK	
*Clear distance between bars not less than d_b or 1 in.	

EXAMPLE 9.6 - Design of One-Way Joist

Compute required d and reinforcement for one-way joist system continuous over two or more equal spans. Assume 10 + 3 joists 5 in. wide @ 35 in. cts. shown below. Clear span, ℓ_n = 20'-0". Use f'_c = 4000 psi and f_y = 60,000 psi.

Service loads: w_ℓ = 100 psf

w_d = 58 psf

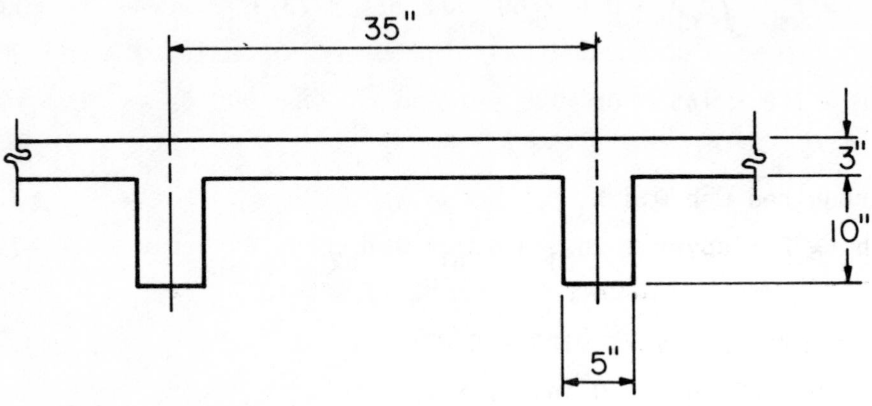

Calculations & Discussion	Code Reference
1. Compute required moment strengths using approximate moment coefficients permitted by Code Section 8.3.3.* Design to be based on end span.	
Factored load:	
$w_u = 1.4 \times 58 + 1.7 \times 100 = 251$ psf	Eq. (9-1)
Positive moment:	
$+M_u = w_u \ell_n^2/14$	8.3.3
$= 0.251 \times 20^2/14 = 7.17$ ft kips/ft	
$+M_u$(per joist) = 7.17 x 35/12 = 20.9 ft kips	

*With the 1983 Code, an additional condition "member must be prismatic" was added to Section 8.3.3 to clarify that the approximate moments and shears are not applicable for members with haunches and varying cross sections.

EXAMPLE 9.6 - Continued

Calculations & Discussion	Code Reference

Negative moment at exterior face of first interior support:

$-M_u = w_u \ell_n^2/10$ 8.8.3

$= 0.251 \times 20^2/10 = 10.04$ ft kips/ft

$-M_u$ (per joist) $= 10.04 \times 35/12 = 29.3$ ft kips

2. Compute required joist depth.

For reasonable deflection control, choose a reinforcement percentage ρ about one-half the maximum permitted.

From Table 9-1, for $f'_c = 4000$ psi and $f_y = 60{,}000$ psi, 10.3.3

$\rho_{max} = 0.75\rho_b = 0.0214$.

$0.5(0.0214) = 0.0107$ Set $\rho = 0.012$

Note: The design meets the minimum thickness of Table 9.5(a). $h = \ell/18.5 = 12 \times 20/18.5 = 12.9 < 13$

Using Table 9-2 as a design aid:

$\omega = \rho f_y/f'_c = 0.012 \times 60/4 = 0.18$

From Table 9-2, $M_u/\varphi f'_c bd^2 = 0.1609$

For negative moment section, $M_u = 29.3$ ft kips

$$d = \sqrt{\frac{M_u}{\varphi f'_c b(0.1609)}} = \sqrt{\frac{29.3 \times 12}{0.9 \times 4 \times 5(0.1609)}} = 11.0 \text{ in.}$$

(10 + 3 joist OK)

3. Compute required reinforcement.

$-A_s$ (req'd) $= \rho bd = 0.0120 \times 5 \times 11.0 = 0.66$ in.^{2}

Use 1 - #6 trussed bar = 0.44

1 - #5 top bar = 0.31

0.75 in.^{2}

For positive moment, $M_u = 20.9$ ft kips

$$\text{Using Table 9-2: } \frac{M_u}{\varphi f'_c bd^2} = \frac{20.9 \times 12}{0.9 \times 4 \times 35 \times 11.75^2} = 0.0144$$

$\omega = 0.0145$

EXAMPLE 9.6 - Continued

Calculations & Discussion	Code Reference
$\rho = \omega f'_c/f_y = 0.0145 \times 4/60 = 0.00097$	
$+A_s$ (req'd) $= 0.00097 \times 35 \times 11.75 = 0.398$ in.^{2}	
Use 1 - #6 bar $= 0.44$ in.^{2}	
4. Shear strength needs to be checked at supports.	

10

Design for Flexure and Axial Load

General Considerations

Design or investigation of a short compression member is based primarily on the strength of its cross section. Strength of a cross section under combined flexure and axial load must satisfy both stress and strain compatibility. The combined nominal axial load and moment strength (P_n, M_n) is then multiplied by the appropriate strength reduction factor φ to obtain the design strength (φP_n, φM_n) of the section. The value of φ may be increased linearly from the value for compression members (φ = 0.75 or 0.70) to the value for flexure (φ = 0.90) as the design axial load strength φP_n decreases from 0.10 $f'_c Ag$ or φP_b, whichever is smaller, to zero. A "strength interaction diagram" can then be generated between the design axial load strength φP_n and the design moment strength about an axis φM_n; this diagram defines the "usable" strength of a section. A typical schematic strength interaction diagram is shown in Fig. 10-1, illustrating the various strength curves for design.

Maximum strain at the extreme concrete compression fiber is always assumed as 0.003. Tensile strength of the concrete is neglected in strength computations. Note that the engineer has the option to use any compressive stress-strain relationship for concrete, provided it is in agreement with the results of comprehensive tests. The equivalent rectangular concrete stress block can be used in lieu of other complex stress-strain relationships for concrete. The user should recognize, however, that the use of different stress-strain relationships for concrete may result in slightly different values for the strength of the section. All PCA published strength design aids and computer programs are based on a parabolic stress-strain relationship (See Fig. 5-8, page 5-13). Note that the required strength (P_u, M_u) must be at least equal to the structural effects of the load groups which represent various combinations of loads and forces to which a structure may be subjected.

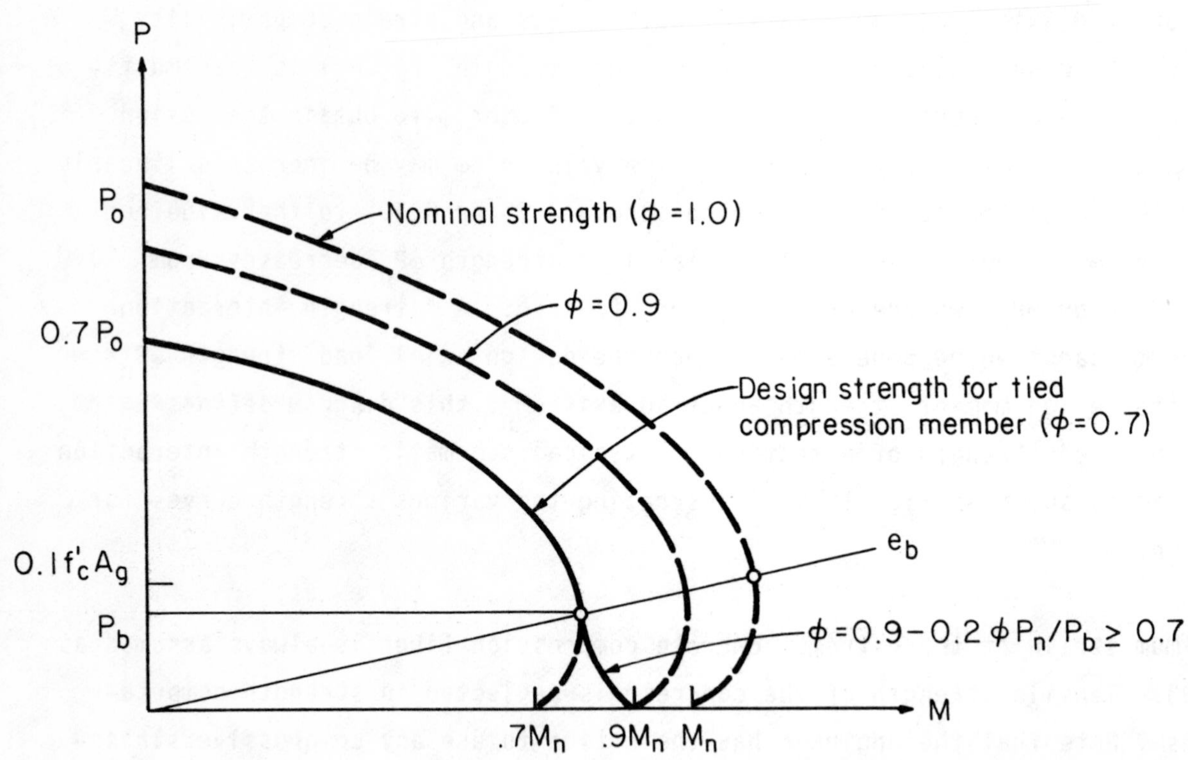

Fig. 10-1 φ Factor Increase for Compression Members

The minimum design eccentricities included in the 1971 ACI Code were deleted from the 1977 Code, except for consideration of slenderness effects in compression members with small or zero computed end moments. The specified minimum eccentricities were originally intended to serve as a means of reducing the axial design load strength of a section in pure compression to account for accidental eccentricities (moments) not considered in the analysis, and to recognize that concrete strength is less under sustained high loads. The primary purpose was to limit the axial load strength for design of compression members with small or zero computed end moments. Beginning with the 1977 Code edition, this design condition is accomplished directly in Section 10.3.5 by limiting the axial load strength of a section in pure compression to 85 or 80 percent of the pure axial load strength.

For spirally reinforced members:

$$P_n\,(max) = 0.85\,(P_o) \qquad \text{Eq. (10-1)}$$

For tied reinforced members:

$$P_n\,(max) = 0.80\,(P_o) \qquad \text{Eq. (10-2)}$$

The percentage values approximate the axial load strengths at e/h ratios of 0.05 and 0.10 specified in the 1971 Code for the spirally reinforced and tied reinforced members, respectively. (See Fig. 10-2). The same axial load limitation applies to both cast-in-place and precast members. For prestressed compression members, the P_o strength includes the effect of the prestressing force. See Design Example 10.1.

The current provisions (1977 and 1983 Codes) for maximum axial load strength eliminated the concerns expressed by engineers about the excessively high minimum design moment required for large column sections and the often asked question as to whether the minimum moments were required to be transferred to other interconnecting members (beams, footings, etc.) To permit existing design aids and computer programs based on the minimum eccentricity requirements of the 1971 Code edition to be used for design, the Code Commentary states that... "designs based on the minimum eccentricity criteria are equally applicable."

Four points along the load-moment strength interaction diagram are significant to define the behavior of members subject to combined axial load and flexure. Referring to Fig. 10-2, (1) pure compression...P_o, (2) maximum axial load strength permitted by the Code...P_n (max), (3) balanced conditions...P_b, M_b, and (4) pure flexure...M_n. For values of axial load strength greater than balanced conditions $P_n > P_b$, compression in the concrete controls the strength and, for values of axial load strength less than balanced conditions $P_n < P_b$, tension in the reinforcement controls the strength. Fig. 10-2 illustrates the compression and tension design regions.

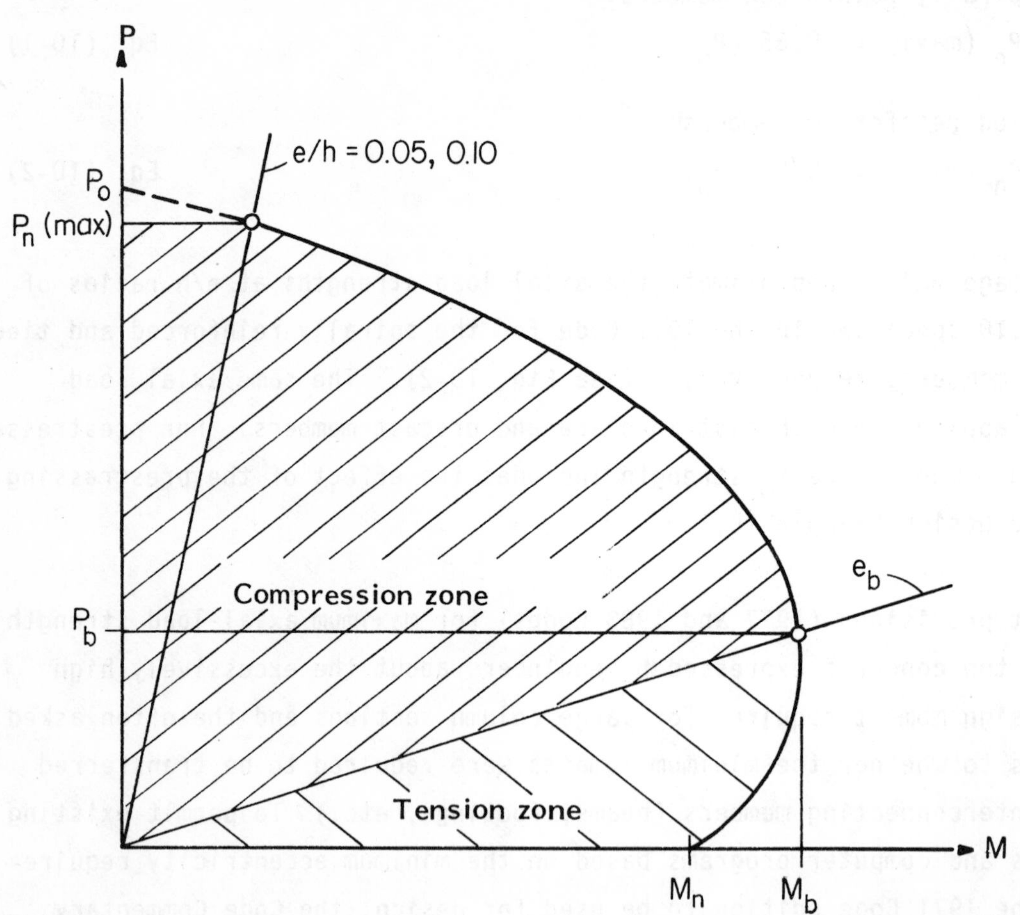

Fig. 10-2 General Form of Load-Moment Interaction Strength

Design Examples 10.1 through 10.6 consider combined axial loading and uniaxial bending for rectangular and circular sections. For circular sections biaxial bending is considered. The examples illustrate the use of the selected design aids that follow. For a complete treatise on biaxial loading, including Design Examples, see Part 11.

Selected Design Aids

The design aids may be obtained from the sponsoring organization.

EB9 - **Strength Design of Reinforced Concrete Columns**, Portland Cement Association, Skokie, EB009.02D, 1978, 48 pp. Provides design tables of column strength in terms of load in kips versus moment in ft-kips for concrete strength of 5000 psi and Grade 60 reinforcement. The design tables do not include the strength reduction factor φ in the tabulated values. Tabulated strength values are nominal strengths (P_n and M_n). When designing with this aid, the factored load and moment (required strength) values must be divided by the φ factor to enter the strength tables directly,

$$\frac{P_u}{\varphi} \leq P_n \text{ (tabulated axial load strength)}$$

$$\frac{M_u}{\varphi} \leq M_n \text{ (tabulated moment strength)}$$

For members with small or zero computed end moments, a design based on the maximum axial load strength of 0.85 or 0.80 of the axial load strength at zero eccentricity is required. The design data for this condition can be found under the "special conditions" P_o. To enter the strength tables directly,

For spirally reinforced members:

$$\frac{P_u}{(0.85)(0.75)} \leq P_o \text{ (tabulated axial load strength)}$$

For tied reinforced members:

$$\frac{P_u}{(0.80)(0.70)} \leq P_o \text{ (tabulated axial load strength)}$$

SP17A - **Design Handbook, Volume 2, Columns**, American Concrete Institute, Detroit SP-17A (78), 1978, 228 pp. Provides design aids (tables and graphs) for use in the engineering design and analysis of reinforced concrete columns by the strength design method. Design examples illustrating the use of the design aids are included.

SR14 - **Strength Design of Reinforced Concrete Column Sections**, Computer Program, Portland Cement Association, Skokie, SR014D. Program designs reinforced concrete columns to resist a given combination of loads, or investigates the adequacy of a given cross section to resist a similar set of loadings. Each loading case consists of an axial compressive force combined with uniaxial and biaxial bending. Method of solution is based on the ultimate strength theories for reinforced concrete design and, where applicable, assumptions and limits conform to ACI 318.

EXAMPLE 10.1 - Design for Pure Compression

Design a concentrically loaded square column with tied reinforcement. Use Design Aid EB9.

Service dead load = 320^k

Service live load = 190^k

Unsupported length = 8'6"

Braced against sidesway

f'_c = 4,000 psi

f_y = 60,000 psi

Calculations and Discussion	Code Reference
1. Determine required strength.	
$P_u = 1.4D + 1.7L$	Eq. (9-1)
$= 1.4\ (320) + 1.7\ (190) = 771^k$	
2. Check column slenderness. Assume 18 in. square column.	
$k = 1.0$ for braced frame	10.11.2.1
$r = 0.3 \times 18 = 5.4$	10.11.3
$k\ell_u/r = 1.0 \times 8.5 \times 12/5.4 = 18.9$	
$k\ell_u/r < 34 - 12\ (M_1/M_2)$	10.11.4.1
$18.9 < 22$ Slenderness may be neglected.	
3. Using Design Aid EB9:	
(a) Required $P_n = P_u/\varphi = 771/0.70 = 1101.5^k$	9.3.2.2(b)
Adjust required strength P_n for concrete strength lower than tabulated values.	
$1101.5\ (5000/4000) = 1376.9^k$	

EXAMPLE 10.1 - Continued

Calculations and Discussion	Code Reference

(b) For columns with zero computed end moments, design is based on the limiting axial load strength of 0.80 P_o. To enter the tables directly, — 10.3.5

Required P_o = 1376.9/0.80 = 1721^k

From strength table for 18 in. x 18 in. column size, (see Table 10-1), read PCNT = 1.93 for $P_o = 1725^k$.

Adjust PCNT for 4,000 psi concrete.

Required ρ = 1.93 (4000/5000) = 1.544%

Use 4-#10 bars (ρ = 1.57%)

4. Select column ties.

Use #3 lateral ties — 7.10.5.1

Spacing not less than: 16 x 1.27 = 20.3 in. — 7.10.5.2

48 x 0.375 = 18 in.

Column Size = 18 in.

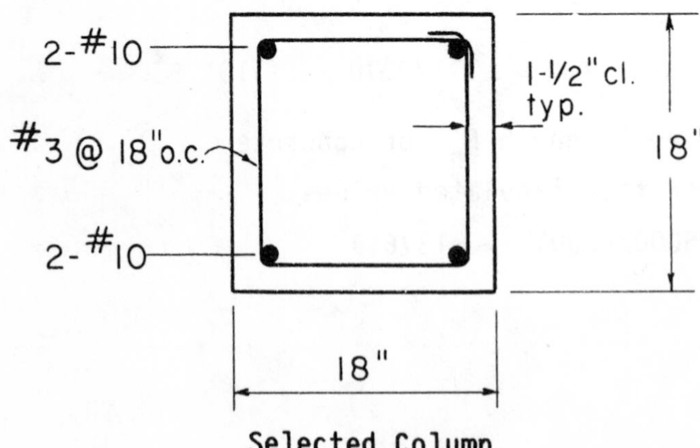

Selected Column

EXAMPLE 10.1 - Continued

Calculations and Discussion	Code Reference

PCA–LOAD AND MOMENT STRENGTH TABLES FOR CONCRETE COLUMNS

No A1 — No. A2 — 1 1/2" clear — Axis of bending

COLUMN SIZE =18X18 INCHES
FPC=5,000 FY=60,000

BARS				SPECIAL CONDITIONS			AXIAL LOAD STRENGTH P_n (KIPS)																	
NO	NO	SIZE	PCNT	0.	BALANCE		0	40	80	120	160	260	360	460	560	660	760	860	960	1060	1160	1260	1360	1560
A1	A2	NO		PO	PB	MB	MOMENT STRENGTH M_n (FT-KIPS)																	
8	0	6	1.09	1573	540	355	133	156	179	202	223	272	312	341	353	342	327	307	282	249	208			
6	0	7	1.11	1578	538	356	135	159	181	203	225	273	313	342	354	343	328	309	283	250	210			
4	0	9	1.23	1600	533	366	148	171	193	215	236	283	323	353	363	351	336	316	291	259	219			
10	0	6	1.36	1622	538	383	163	187	209	231	252	300	341	370	380	366	350	329	303	271	232			
6	0	8	1.46	1641	533	391	174	196	218	240	261	309	348	378	387	373	356	335	310	278	240	194		
8	0	7	1.48	1645	535	395	176	199	221	243	264	312	352	382	391	376	359	338	312	281	242	197		
4	0	10	1.57	1660	528	398	183	206	227	249	269	316	355	386	393	378	361	341	316	285	247	202		
6	0	9	1.85	1711	528	429	215	237	259	280	300	347	387	417	424	406	387	365	339	309	272	229		
4	0	11	1.93	1725	523	432	220	242	263	284	304	350	389	420	425	408	389	367	342	312	276	234		
8	0	8	1.95	1729	530	441	226	249	270	292	312	359	399	429	436	417	397	375	348	318	282	239		
6	0	10	2.35	1802	523	477	266	288	309	330	350	396	435	467	469	449	427	403	377	347	312	272	226	
8	0	9	2.47	1823	524	492	281	302	324	344	364	411	451	481	484	462	440	415	388	358	323	284	238	
4	0	14	2.78	1879	512	510	305	326	346	366	385	430	468	499	500	477	454	430	404	375	342	304	261	
6	0	11	2.89	1899	516	528	321	342	362	383	402	447	487	518	518	494	470	445	418	388	354	316	273	
4	0	18	4.94	2269	468	689	506	525	543	562	579	620	657	687	671	648	619	590	561	530	499	465	430	349
0	0	0	6.00	2461	439	769	599	617	635	652	669	709	745	765	744	723	699	668	637	606	574	541	506	432
0	0	0	7.00	2641	410	840	683	701	718	735	752	790	825	829	808	787	766	740	708	676	644	611	577	505
0	0	0	8.00	2822	377	908	765	782	799	816	832	869	903	890	869	847	826	804	779	746	713	680	646	575
6	2	6	1.09	1573	544	326	134	157	179	200	221	263	292	314	325	318	307	291	268	237	199			
8	4	5	1.15	1584	547	326	141	164	186	207	225	264	297	315	326	319	308	293	271	241	203			
6	2	7	1.48	1645	541	356	177	199	220	241	260	297	324	345	354	344	332	315	293	264	229			
10	6	5	1.53	1654	547	353	183	205	224	242	260	297	327	344	352	343	331	315	294	267	232	190		
8	4	6	1.63	1671	543	361	194	215	236	253	269	306	335	352	360	350	337	321	300	273	239	198		
6	2	8	1.95	1729	537	391	226	248	268	288	306	336	361	381	388	375	361	343	321	295	262	223		
10	6	6	2.17	1769	543	398	249	267	285	302	319	351	378	391	396	383	369	352	332	307	277	241		
8	4	7	2.22	1778	539	403	256	275	291	306	321	357	381	395	401	387	373	356	335	310	280	243		
6	2	9	2.47	1823	532	429	280	300	320	338	353	378	401	420	425	409	393	374	353	327	297	262	220	
8	4	8	2.93	1906	533	453	322	337	353	367	382	416	434	446	448	432	416	397	377	353	325	293	255	
6	2	10	3.14	1943	526	477	346	365	384	396	406	429	452	470	471	453	435	415	393	368	340	308	271	
8	4	9	3.70	2046	527	506	389	404	419	433	447	474	490	501	500	482	463	444	422	399	373	343	310	
6	2	11	3.85	2073	520	528	415	432	442	451	460	483	504	522	520	500	480	458	436	412	385	354	321	240
4	4	6	1.09	1573	547	304	134	157	178	197	212	249	278	294	304	300	292	279	258	230	193			
4	2	7	1.11	1578	543	317	136	159	181	202	221	259	285	305	316	311	301	286	264	235	197			
4	6	6	1.36	1622	549	312	164	182	199	217	233	265	291	304	312	308	300	288	269	244	211			
4	2	8	1.46	1641	540	340	174	196	217	237	255	287	311	330	339	331	320	305	284	257	222			
4	4	7	1.48	1645	544	326	177	198	215	230	245	279	303	317	325	319	311	297	278	253	220			
4	2	9	1.85	1711	536	366	215	236	256	274	292	316	338	356	364	353	341	326	306	280	249	210		
4	4	8	1.95	1729	540	352	224	239	254	268	282	314	333	345	350	342	332	319	301	278	249	213		
4	2	10	2.35	1802	531	398	265	285	304	320	329	351	372	390	394	382	368	352	333	309	280	245	204	
4	4	9	2.47	1823	535	380	267	281	294	308	321	350	363	374	378	368	356	343	326	305	279	247	208	
4	2	11	2.89	1899	526	432	317	336	350	359	367	388	408	425	427	413	397	380	361	338	312	280	243	
10	2	5	1.15	1584	545	342	141	164	187	208	229	275	306	329	341	332	319	301	278	246	207			
8	2	6	1.36	1622	543	355	164	186	208	229	250	291	320	343	353	343	330	312	289	259	222			
10	2	6	1.63	1671	541	383	194	216	238	258	278	320	349	372	381	368	353	334	310	281	246	203		
8	2	7	1.85	1711	538	395	217	239	260	280	299	336	363	384	392	378	363	344	321	293	260	219		
8	2	8	2.44	1817	533	441	278	299	319	339	357	386	412	432	437	420	402	382	360	333	302	265	222	
8	2	9	3.09	1934	528	492	344	364	384	402	414	440	465	484	486	466	446	425	402	375	346	312	274	
10	4	5	1.34	1619	546	347	162	185	207	228	245	285	317	336	346	337	324	308	286	256	220			
6	4	6	1.36	1622	545	332	164	186	207	225	241	278	306	323	332	325	314	299	279	251	216			
6	4	7	1.85	1711	541	365	216	237	253	268	283	318	342	356	363	353	341	326	306	281	250	212		
10	4	6	1.90	1720	541	390	223	245	264	281	297	335	364	381	387	375	360	343	321	294	261	222		
6	4	8	2.44	1817	537	402	273	288	303	318	332	365	383	396	399	387	373	358	338	315	287	253	213	
6	4	9	3.09	1934	531	443	328	342	356	370	384	412	427	438	439	424	409	392	373	351	326	296	260	
6	6	5	1.15	1584	550	312	141	164	183	201	219	257	285	302	312	307	299	285	264	236	200			
8	6	5	1.34	1619	548	332	162	184	203	221	239	277	306	323	332	325	314	300	279	251	216			
6	6	6	1.63	1671	547	341	192	210	228	245	262	293	320	333	340	333	323	309	290	265	233	193		
8	6	6	1.90	1720	545	369	220	238	256	273	290	322	349	362	368	358	346	330	311	286	255	217		

Table 10-1 - Reproduced from Page 12 of EB9

EXAMPLE 10.2 - Design for Small Axial Load

Design a square tied reinforced column for service dead and live loads of 20 and 15 kips respectively. Service dead and live load moments at each end have values of 90 and 70 ft kips respectively. Column has an unsupported height of 12 ft and is bent in double curvature.

Use f'_c = 5,000 psi and f_y = 60,000 psi. Column size is set at 18 in. minimum by architectural considerations and is braced against sidesway. Use Design Aid EB9.

Calculations and Discussion	Code Reference
1. Determine required strength.	
$P_u = 1.4\ (20) + 1.7\ (15) = 53.5^k$	Eq. (9-1)
$M_u = 1.4\ (90) + 1.7\ (70) = 245^{'k}$	
2. Check column slenderness. Assume 18 in. x 18 in. column section.	
$k = 1.0,\quad \ell_u = 12$ ft	10.11.2.1
$r = 0.3\ (18) = 5.4$	10.11.3
$k\ell_u/r = 1.0 \times 12 \times 12/5.4 = 30$	
$k\ell_u/r < 34 - 12\ (-245/245) = 46$	10.11.4.1
$30 < 46$ Slenderness may be neglected.	
3. Since column section is large and required strength $P_u = 53.5^k$ is relatively small, check whether the strength reduction factor φ may be increased ($\varphi = 0.70 \rightarrow 0.90$) in accordance with Code Section 9.3.2.2(b):	
$(h - d' - d_s)/h = (18 - 2.5 - 2.5)/18 = 0.72 > 0.70$ (OK)	
$0.10\ f'_c\ A_g = 0.10 \times 5\ (18 \times 18) = 162^k > 53.5^k$ (OK)	

Calculations and Discussion	Code Reference

Increase φ using straight line interpolation:

$$\varphi = 0.90 - 0.2\ (P_u/0.10\ f'_c\ A_g) \geq 0.70$$

$$= 0.9 - 0.2\ (53.5/162) = 0.83$$

4. <u>Using Design Aid EB9</u>:

$$\text{Required } P_n = P_u/\varphi = 53.5/0.83 = 64.5^k$$

$$M_n = M_u/\varphi = 245/0.83 = 296^{'k}$$

From strength table for 18 x 18 column size, f'_c = 5,000 and f_y = 60,000 (see Table 10-2), <u>select 6 #10 bars</u> (PCNT = 2.35) to provide a load moment strength of P_n = 65 and M_n = 301.

5. Select lateral ties.

<u>Use #3 ties</u> with #10 longitudinal bars. 7.10.5.1

Spacing not less than: 16 x 1.27 = 20.3 in. 7.10.5.2

48 x 0.375 = 18 in.

Column size = 18 in.

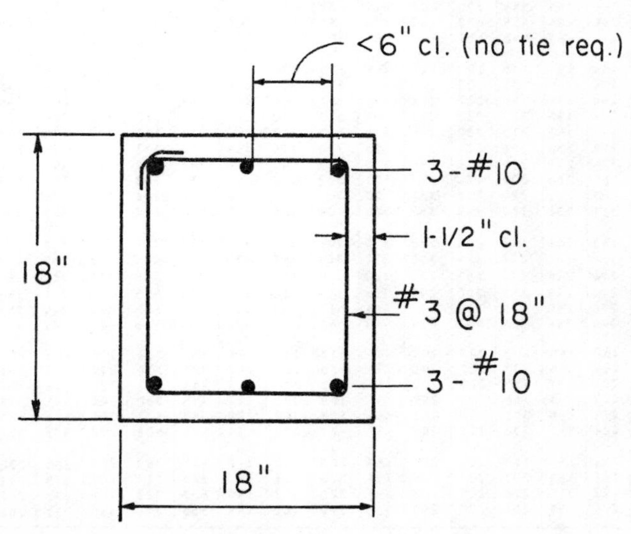

EXAMPLE 10.2 - Continued

Calculations and Discussion	Code Reference

PCA–LOAD AND MOMENT STRENGTH TABLES FOR CONCRETE COLUMNS

1 1/2" clear

No A1   No.A2   Axis of bending

COLUMN SIZE =18X18 INCHES
FPC=5.000 FY=60.000

BARS				SPECIAL CONDITIONS			AXIAL LOAD STRENGTH P_n (KIPS)																	
NO	NO	SIZE	PCNT	0.	BALANCE		0	40	80	120	160	260	360	460	560	660	760	860	960	1060	1160	1260	1360	1560
A1	A2	NO		PO	PB	MB	MOMENT STRENGTH M_n (FT-KIPS)																	
8	0	6	1.09	1573	540	355	133	156	179	202	223	272	312	341	353	342	327	307	282	249	208			
6	0	7	1.11	1578	538	356	135	159	181	203	225	273	313	342	354	343	328	309	283	250	210			
4	0	9	1.23	1600	533	366	148	171	193	215	236	283	323	353	363	351	336	316	291	259	219			
10	0	6	1.36	1622	538	383	163	187	209	231	252	300	341	370	380	366	350	329	303	271	232			
6	0	8	1.46	1641	533	391	174	196	218	240	261	309	348	378	387	373	356	335	310	278	240	194		
8	0	7	1.48	1645	535	395	176	199	221	243	264	312	352	382	391	376	359	338	312	281	242	197		
4	0	10	1.57	1660	528	398	183	206	227	249	269	316	355	386	393	378	361	341	316	285	247	202		
6	0	9	1.85	1711	528	429	215	237	259	280	300	347	387	417	424	406	387	365	339	309	272	229		
4	0	11	1.93	1725	523	432	220	242	263	284	304	350	389	420	425	408	389	367	342	312	276	234		
8	0	8	1.95	1729	530	441	226	249	270	292	312	359	399	429	436	417	397	375	348	318	282	239		
6	0	10	2.35	1802	523	477	266	288	309	330	350	396	435	467	469	449	427	403	377	347	312	272	226	
8	0	9	2.47	1823	524	492	281	302	324	344	364	411	451	481	484	462	440	415	388	358	323	284	238	
4	0	14	2.78	1879	512	510	305	326	346	366	385	430	468	499	500	477	454	430	404	375	342	304	261	
6	0	11	2.89	1899	516	528	321	342	362	383	402	447	487	518	518	494	470	445	418	388	354	316	273	
4	0	18	4.94	2269	468	689	506	525	543	562	579	620	657	687	671	648	619	590	561	530	499	465	430	349
0	0	0	6.00	2461	439	769	599	617	635	652	669	709	745	765	744	723	699	668	637	606	574	541	506	432
0	0	0	7.00	2641	410	840	683	701	718	735	752	790	825	829	808	787	766	740	708	676	644	611	577	505
0	0	0	8.00	2822	377	908	765	782	799	816	832	869	903	890	869	847	826	804	779	746	713	680	646	575
6	2	6	1.09	1573	544	326	134	157	179	200	221	263	292	314	325	318	307	291	268	237	199			
8	4	5	1.15	1584	547	326	141	164	186	207	225	264	297	315	326	319	308	293	271	241	203			
6	2	7	1.48	1645	541	356	177	199	220	241	260	297	324	345	354	344	332	315	293	264	229			
10	6	5	1.53	1654	547	353	183	205	224	242	260	297	327	344	352	343	331	315	294	267	232	190		
8	4	6	1.63	1671	543	361	194	215	236	253	269	306	335	352	360	350	337	321	300	273	239	198		
6	2	8	1.95	1729	537	391	226	248	268	288	306	336	361	381	388	375	361	343	321	295	262	223		
10	6	6	2.17	1769	543	398	249	267	285	302	319	351	378	391	396	383	369	352	332	307	277	241		
8	4	7	2.22	1778	539	403	256	275	291	306	321	357	381	395	401	387	373	356	335	310	280	243		
6	2	9	2.47	1823	532	429	280	300	320	338	353	378	401	420	425	409	393	374	353	327	297	262	220	
8	4	8	2.93	1906	533	453	322	337	353	367	382	416	434	446	448	432	416	397	377	353	325	293	255	
6	2	10	3.14	1943	526	477	346	365	384	396	406	429	452	470	471	453	435	415	393	368	340	308	271	
8	4	9	3.70	2046	527	506	389	404	419	433	447	474	490	501	500	482	463	444	422	399	373	343	310	
6	2	11	3.85	2073	520	528	415	432	442	451	460	483	504	522	520	500	480	458	436	412	385	354	321	240
4	4	6	1.09	1573	547	304	134	157	178	197	212	249	278	294	304	300	292	279	258	230	193			
4	2	7	1.11	1578	543	317	136	159	181	202	221	259	285	305	316	311	301	286	264	235	197			
4	6	6	1.36	1622	549	312	164	182	199	217	233	265	291	304	312	308	300	288	269	244	211			
4	2	8	1.46	1641	540	340	174	196	217	237	255	287	311	330	339	331	320	305	284	257	222			
4	4	7	1.48	1645	544	326	177	198	215	230	245	279	303	317	325	319	311	297	278	253	220			
4	2	9	1.85	1711	536	366	215	236	256	274	292	316	338	356	364	353	341	326	306	280	249	210		
4	4	8	1.95	1729	540	352	224	239	254	268	282	314	333	345	350	342	332	319	301	278	249	213		
4	2	10	2.35	1802	531	398	265	285	304	320	329	351	372	390	394	382	368	352	333	309	280	245	204	
4	4	9	2.47	1823	535	380	267	281	294	308	321	350	363	374	378	368	356	343	326	305	279	247	208	
4	2	11	2.89	1899	526	432	317	336	350	359	367	388	408	425	427	413	397	380	361	338	312	280	243	
10	2	5	1.15	1584	545	342	141	164	187	208	229	275	306	329	341	332	319	301	278	246	207			
8	2	6	1.36	1622	543	355	164	186	208	229	250	291	320	343	353	343	330	312	289	259	222			
10	2	6	1.63	1671	541	383	194	216	238	258	278	320	349	372	381	368	353	334	310	281	246	203		
8	2	7	1.85	1711	538	395	217	239	260	280	299	336	363	384	392	378	363	344	321	293	260	219		
8	2	8	2.44	1817	533	441	278	299	319	339	357	386	412	432	437	420	402	382	360	333	302	265	222	
8	2	9	3.09	1934	528	492	344	364	384	402	414	440	465	484	486	466	446	425	402	375	346	312	274	
10	4	5	1.34	1619	546	347	162	185	207	228	245	285	317	336	346	337	324	308	286	256	220			
6	4	6	1.36	1622	545	332	164	186	207	225	241	278	306	323	332	325	314	299	279	251	216			
6	4	7	1.85	1711	541	365	216	237	253	268	283	318	342	356	363	353	341	326	306	281	250	212		
10	4	6	1.90	1720	541	390	223	245	264	281	297	335	364	381	387	375	360	343	321	294	261	222		
6	4	8	2.44	1817	537	402	273	288	303	318	332	365	383	396	399	387	373	358	338	315	287	253	213	
6	4	9	3.09	1934	531	443	328	342	356	370	384	412	427	438	439	424	409	392	373	351	326	296	260	
6	6	5	1.15	1584	550	312	141	164	183	201	219	257	285	302	312	307	299	285	264	236	200			
8	6	5	1.34	1619	548	332	162	184	203	221	239	277	306	323	332	325	314	300	279	251	216			
6	6	6	1.63	1671	547	341	192	210	228	245	262	293	320	333	340	333	323	309	290	265	233	193		
8	6	6	1.90	1720	545	369	220	238	256	273	290	322	349	362	368	358	346	330	311	286	255	217		

Table 10-2 - Reproduced from Page 12 of EB9

EXAMPLE 10.3 - Design for General Loading (Rectangular Section)

Design a rectangular tied reinforced column for the service dead and live loads and moments shown below. Architectural considerations limit the width of column to 14 in. Column is braced against sidesway with an unsupported height of 7 ft-6 in. Assume moments at the bottom of the column are 1/2 those at the top, and the column is bent in double curvature about the strong axis and single curvature about the weak axis. Use Design Aids EB9 and SP17A.

$P_d = 350^k$ $M_d = 100^{'k}$) Normal to strong axis.

$P_\ell = 240^k$ $M_\ell = 80^{'k}$) Moments are negligible about weak axis.

$f'_c = 5{,}000$ psi

$f_y = 60{,}000$ psi

Calculations and Discussion	Code Reference

1. Determine required strength.

$P_u = 1.4\ (350) + 1.7\ (240) = 898^k$ Eq. (9-1)

$M_u = 1.4\ (100) + 1.7\ (80) = 276^{'k}$

2. Check column slenderness. Assume 14 in. x 22 in. column size.

(a) Slenderness about weak axis (14 in. width):

$k = 1.0$ 10.11.2.1

$r = 0.3 \times 14 = 4.2$ 10.11.3

$k\ell_u/r = 1.0 \times 7.5 \times 12/4.2 = 21.4$

with negligible moments about weak axis assume $M_1/M_2 = 1.0$ for slenderness consideration.

$k\ell_u/r < 34 - 12\ (M_1/M_2) = 22$ 10.11.4.1

$21.4 < 22$ Slenderness may be neglected about weak axis.

EXAMPLE 10.3 - Continued

Calculations and Discussion	Code Reference

(b) Slenderness about strong axis (22 in. width):

$k = 1.0$

$r = 0.3 \times 22 = 6.6$

$k\ell_u/r = 1.0 \times 7.5 \times 12/6.6 = 13.6$

$k\ell_u/r < 34 - 12\ (-\frac{1}{2}) = 40$ 10.11.4.1

$13.6 < 40$ Slenderness may be neglected about strong axis.

Using Design Aid EB9:

3. Required $P_n = P_u/\varphi = 898/0.70 = 1282^k$ 9.3.2.2(b)

$M_n = M_u/\varphi = 276/0.70 = 395^{'k}$

4. Since strength tables in EB9 are given for square sections only, systematic interpolation is required. Select smallest square column capable of supporting required strength, then a series of larger "rectangular" sections to suit required loading. From strength tables, 18 in. x 18 in. is the smallest tabulated size to support $P_n = 1282^k$. Adjust for smaller width as follows:

Column Size	Ratio of Actual to Square Width	Adjusted Required Strength	
		P_n/Width ratio	M_n/Width ratio
18 x 18	14/18 = 0.78	1645	506
20 x 20	14/20 = 0.70	1835	564
22 x 22	14/22 = 0.64	2005	617
24 x 24	14/24 = 0.58	2215	681

EXAMPLE 10.3 - Continued

Calculations and Discussion	Code Reference

5. An 18 x 18 section (see Table 10-2 of Design Example 10.2) is not adequate to support the adjusted required strength $P_n = 1645^k$.

   A 20 x 20 section (see Table 10-3) with 6% reinforcement has a combined load-moment strength of 1835^k and $642^{'k}$ which is adequate to support the adjusted strengths (1835^k and $564^{'k}$). For a 14 x 20 section, the required area of reinforcement, $A_{st} = 0.06\ (14 \times 20) = 16.8\ in.^2$

   A 22 x 22 section (see Table 10-4) with 4% reinforcement has a load-moment strength of 2005^k and $689^{'k}$, which is adequate for the adjusted strengths (2005^k and $617^{'k}$). For a 14 x 22 section, $A_{st} = 0.04\ (14 \times 22) = 12.3\ in.^2$

   A 24 x 24 section (see Table 10-5) with 2.34% reinforcement has a load-moment strength of 2215^k and $675^{'k}$ (Say OK.). For a 14 x 24 section, $A_{st} = 0.0234\ (14 \times 24) = 7.86\ in.^2$

   Assume no architectural limitation on the long dimension. To allow room for lap splicing of the bars, <u>select 14 x 24 section with 4-#14 bars</u> ($A_{st} = 9.0\ in.^2$).

6. Select Column Ties

   <u>Use #4 lateral ties</u> — 7.10.5.1

   Spacing not less than: 16 x 1.693 = 27 in. — 7.10.5.2

   48 x 0.50 = 24 in.

   Column Size = 14 in.

Calculations and Discussion	Code Reference

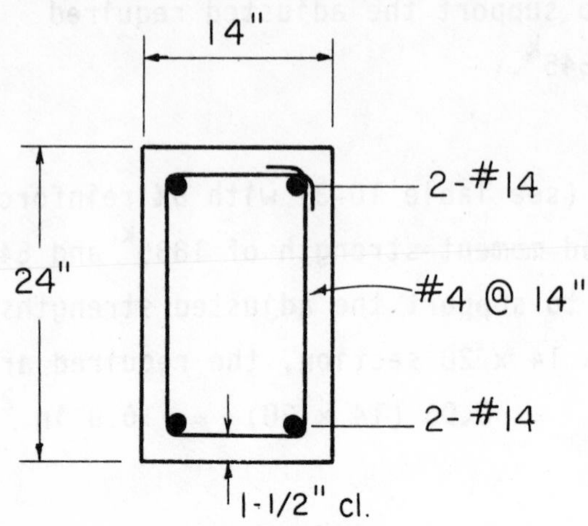

Selected Design

7. Check limiting axial load strength. For the 14 x 24 column with 4 #14 bars:

$$0.8\ \varphi P_o = 0.8\ \varphi[0.85\ f'_c\ (A_g - A_{st}) + f_y\ A_{st}]$$ Eq. (10-2)

$$= 0.8\ (0.70)\ [0.85 \times 5\ (336 - 9) + 60 \times 9]$$

$$= 1081^k$$

$898^k < 1081^k$ The column selected is adequate for required strength about both axis.

Note: Design for rectangular columns by Design Aid EB9 is an approximate trial procedure. A more exact and direct procedure is available using Design Aid SP17A. See below and also Design Example 10.5.

Using Design Aid SP17A:

8. When the column section is preset, selection of reinforcement by Design Aid SP17A is simple and direct. Select reinforcement for 14 x 24 column section ($A_g = 336$ in.^{2}).

EXAMPLE 10.3 - Continued

Calculations and Discussion	Code Reference

Compute $\dfrac{P_u}{A_g} = \dfrac{898}{336} = 2.67$

Compute $\dfrac{M_u}{A_g h} = \dfrac{276 \times 12}{336 \times 24} = 0.41$

Estimate $\gamma \simeq \dfrac{h - 5.5}{h} = \dfrac{24 - 5.5^*}{24} = 0.77$

*2 (cover + tie diameter + 1/2 bar diameter).

9. For rectangular section with bars along two faces, $f'_c = 5$ ksi, $f_y = 60$ ksi, and $\gamma = 0.77$, use column chart 7.12.3 for E5 - 60.75 columns (see Fig. 10-3). Read $\rho_g = 0.02$.

$$A_{st} = \rho_g A_g = 0.02\,(336) = 6.72 \text{ in.}^2$$

As shown above, the approximate interpolation procedure required for rectangular sections by Design Aid EB9 yields somewhat conservative results for this design.

Calculations and Discussion	Code Reference

PCA–LOAD AND MOMENT STRENGTH TABLES FOR CONCRETE COLUMNS

No A1 — No.A2 — 1 1/2" clear — Axis of bending

COLUMN SIZE =20X20 INCHES
FPC=5,000 FY=60,000

BARS				SPECIAL CONDITIONS			AXIAL LOAD STRENGTH P_n (KIPS)																	
NO	NO	SIZE	PCNT	0.	BALANCE		0	40	140	240	340	440	540	640	740	840	940	1040	1140	1240	1340	1540	1740	1940
A1	A2	NO		PO	PB	MB	MOMENT STRENGTH M_n (FT-KIPS)																	
4	0	9	1.00	1923	670	475	170	196	260	318	371	415	449	471	468	456	440	419	392	359	319			
10	0	6	1.10	1945	677	495	187	213	278	337	390	435	468	489	487	473	455	433	406	373	333			
6	0	8	1.19	1964	671	504	199	225	289	347	399	444	478	499	495	480	462	440	413	380	341			
8	0	7	1.20	1968	673	508	202	228	292	351	403	448	481	503	499	484	466	443	416	383	344			
4	0	10	1.27	1983	665	513	211	236	299	357	408	452	487	509	502	487	469	447	420	388	349			
10	0	7	1.50	2034	671	553	248	274	337	396	448	493	526	548	541	522	502	478	450	417	379	283		
6	0	9	1.50	2034	666	548	247	272	334	392	444	489	523	544	536	518	498	475	447	415	377	282		
4	0	11	1.56	2048	660	552	253	279	340	397	448	492	527	549	539	521	501	478	450	419	381	288		
8	0	8	1.58	2052	668	562	260	286	348	406	458	503	537	558	549	530	509	485	458	425	387	293		
6	0	10	1.90	2125	660	604	307	332	393	450	501	545	580	601	588	566	544	519	491	458	422	332		
10	0	8	1.98	2140	664	620	320	346	407	465	517	562	596	617	604	581	557	531	503	470	433	343		
8	0	9	2.00	2146	662	621	323	348	409	466	518	563	597	618	604	582	558	532	503	471	434	345		
4	0	14	2.25	2202	649	644	353	377	436	491	541	585	621	643	623	600	576	550	522	491	455	370		
6	0	11	2.34	2222	653	664	370	395	454	510	561	605	641	662	643	618	593	566	537	505	469	384		
8	0	10	2.54	2266	654	696	402	427	486	543	594	638	673	694	674	648	621	593	563	530	494	410	306	
6	0	14	3.38	2453	639	802	516	540	597	651	701	744	782	801	771	742	712	681	649	616	581	501	406	
4	0	18	4.00	2592	623	866	589	612	667	718	765	808	844	861	829	798	767	736	704	671	636	560	473	370
0	0	0	5.00	2815	606	985	718	739	792	842	888	929	965	977	943	909	875	841	807	773	738	663	581	487
0	0	0	6.00	3038	577	1094	841	862	913	962	1007	1047	1083	1079	1053	1017	981	946	910	875	838	764	684	597
0	0	0	7.00	3261	545	1198	961	982	1031	1078	1122	1162	1197	1174	1149	1123	1086	1049	1012	976	939	863	784	701
0	0	0	8.00	3484	509	1297	1078	1098	1146	1191	1234	1273	1289	1264	1238	1212	1187	1151	1113	1075	1038	961	883	801
6	2	7	1.20	1968	680	464	202	228	289	344	386	417	440	458	459	448	435	416	393	363	326			
10	6	5	1.24	1977	687	460	210	235	292	343	385	418	441	455	456	446	433	416	394	365	330			
8	4	6	1.32	1994	683	470	222	247	307	353	395	430	450	465	465	454	441	423	401	372	337			
12	8	5	1.55	2046	688	491	257	280	335	382	423	453	475	487	485	474	460	442	420	393	360	272		
6	2	8	1.58	2052	676	504	260	285	344	397	432	461	483	500	497	484	468	449	425	397	363	276		
10	6	6	1.76	2092	684	512	289	311	363	411	447	478	496	508	506	493	478	459	437	411	379	297		
8	4	7	1.80	2101	679	518	294	318	370	414	454	484	501	514	511	497	482	463	441	414	382	300		
6	2	9	2.00	2146	672	548	322	346	403	452	481	508	529	544	538	522	505	485	462	434	402	321		
8	4	8	2.37	2229	675	576	377	398	442	484	522	546	561	572	565	549	531	511	489	463	433	359		
10	6	7	2.40	2235	680	576	377	397	446	489	522	550	563	573	566	550	533	513	491	466	436	363		
6	2	10	2.54	2266	666	604	400	423	477	514	542	568	587	601	591	573	553	532	508	481	451	377		
8	4	9	3.00	2369	669	638	460	477	519	559	596	613	625	635	625	605	586	564	541	516	487	419	333	
6	2	11	3.12	2396	660	664	481	503	552	579	605	630	648	661	647	626	605	582	557	531	501	432	345	
10	6	8	3.16	2405	675	651	477	496	543	577	608	631	641	648	638	619	599	578	555	530	502	436	352	
8	4	10	3.81	2550	662	717	559	575	615	654	682	697	707	715	700	678	656	633	609	584	556	493	416	324
6	2	14	4.50	2703	646	802	666	677	702	726	750	773	790	801	778	753	728	702	676	649	620	556	483	395
4	6	6	1.10	1945	690	413	188	212	264	311	348	378	397	409	412	408	399	385	365	339	305			
4	2	8	1.19	1964	679	446	200	225	285	338	374	402	424	441	442	434	422	405	383	354	319			
4	4	7	1.20	1968	684	429	203	227	282	325	364	394	411	425	427	421	411	396	376	349	315			
4	6	7	1.50	2034	688	442	245	265	313	356	387	414	429	439	440	433	424	410	392	368	337			
4	2	9	1.50	2034	676	475	247	272	329	380	408	434	455	471	470	459	446	429	407	380	347	260		
4	4	8	1.58	2052	681	459	259	282	327	367	405	429	444	456	455	447	436	421	402	377	347	265		
4	2	10	1.90	2125	672	513	306	330	385	425	450	475	494	509	505	492	477	460	438	412	382	303		
4	6	8	1.98	2140	685	476	304	323	368	403	431	455	465	473	472	464	453	439	422	400	373	302		
4	4	9	2.00	2146	678	492	318	335	375	413	448	466	479	489	486	476	464	449	430	407	379	306		
4	2	11	2.34	2222	667	552	367	390	442	470	494	517	535	549	542	527	511	492	471	446	417	345		
4	4	10	2.54	2266	673	533	379	395	433	470	498	512	522	531	526	514	500	485	466	445	419	354		
4	2	14	3.38	2453	656	644	507	527	549	571	593	614	631	642	629	610	591	571	549	525	499	436	357	
12	2	5	1.08	1942	683	471	185	212	275	333	383	418	445	465	466	454	439	419	393	361	323			
8	2	6	1.10	1945	682	462	187	214	276	333	379	412	438	456	457	447	432	413	389	358	320			
10	2	6	1.32	1994	680	495	222	248	310	366	412	446	471	490	488	475	458	438	413	383	346			
8	2	7	1.50	2034	677	508	249	274	335	389	430	462	485	503	500	486	470	450	426	396	361	270		
10	2	7	1.80	2101	675	553	295	320	380	434	475	507	530	548	542	525	506	484	459	430	395	308		
8	2	8	1.98	2140	673	562	320	345	403	456	490	520	542	558	551	534	515	494	469	441	407	324		
10	2	8	2.37	2229	669	620	380	404	462	514	548	579	600	617	606	585	564	540	514	485	452	372		
8	2	9	2.50	2257	667	621	397	421	477	523	554	582	602	618	607	587	566	543	518	489	457	381		
8	2	10	3.17	2408	661	696	493	516	570	605	633	661	679	694	678	655	631	606	580	552	521	449	360	

Table 10-3 - Reproduced From Page 13 of EB9

EXAMPLE 10.3 - Continued

Calculations and Discussion | **Code Reference**

PCA–LOAD AND MOMENT STRENGTH TABLES FOR CONCRETE COLUMNS

1 1/2" clear

No A1 — No. A2 — Axis of bending

COLUMN SIZE =22X22 INCHES
FPC=5.000 FY=60.000

BARS				SPECIAL CONDITIONS			AXIAL LOAD STRENGTH P_n (KIPS)																	
NO	NO	SIZE	PCNT	0.	BALANCE		0	100	200	300	400	500	600	700	800	900	1000	1200	1400	1600	1800	2000	2200	2400
A1	A2	NO		PO	PB	MB	MOMENT STRENGTH M_n (FT-KIPS)																	
4	0	10	1.05	2340	818	648	238	310	379	444	502	552	594	624	645	637	622	583	525	444	338			
12	0	6	1.09	2351	828	664	249	323	393	459	518	569	609	639	660	654	637	596	536	455	348			
10	0	7	1.24	2391	824	692	280	353	423	488	547	598	638	668	689	680	662	618	558	477	373			
6	0	9	1.24	2391	819	688	278	350	420	484	542	593	634	664	685	675	657	614	555	475	371			
4	0	11	1.29	2405	812	693	287	358	426	490	547	598	640	670	691	679	661	618	559	481	377			
8	0	8	1.31	2409	821	703	293	365	435	499	558	609	650	680	700	690	671	626	567	487	383			
12	0	7	1.49	2458	822	742	332	405	474	539	598	649	689	719	739	727	706	658	596	517	415			
6	0	10	1.57	2482	812	752	347	417	486	549	607	658	700	730	750	734	713	665	604	526	427			
10	0	8	1.63	2497	817	769	362	433	502	566	625	676	717	746	767	751	729	679	617	538	439			
8	0	9	1.65	2503	815	770	364	435	504	568	626	677	719	748	768	752	730	680	618	540	441			
4	0	14	1.86	2559	800	798	400	469	535	597	654	704	748	778	798	775	752	701	641	564	470			
6	0	11	1.93	2579	806	819	419	488	555	618	676	727	769	799	818	797	772	719	657	580	485	370		
10	0	9	2.07	2614	810	853	450	520	588	652	710	762	803	832	851	830	803	747	683	605	510	396		
8	0	10	2.10	2623	807	856	455	524	592	655	712	764	806	835	855	831	805	749	685	608	514	400		
8	0	11	2.58	2753	799	946	550	618	684	747	804	856	898	927	946	916	886	824	757	680	589	481		
6	0	14	2.79	2810	791	977	586	653	718	779	836	887	931	960	974	944	913	850	783	706	617	513		
4	0	18	3.31	2949	774	1053	673	737	799	858	913	962	1006	1039	1045	1012	980	914	846	770	685	588	474	
0	0	0	4.00	3136	762	1172	798	861	921	979	1032	1081	1124	1160	1158	1122	1087	1016	944	868	784	692	585	467
0	0	0	5.00	3406	745	1338	973	1034	1092	1148	1200	1248	1291	1328	1316	1277	1239	1163	1086	1007	924	834	737	627
0	0	0	6.00	3676	729	1499	1142	1201	1257	1312	1362	1409	1452	1489	1469	1429	1388	1307	1227	1145	1061	974	880	779
0	0	0	7.00	3946	697	1645	1306	1363	1418	1471	1521	1567	1609	1644	1615	1576	1534	1450	1366	1283	1197	1110	1019	923
0	0	0	8.00	4216	659	1784	1465	1521	1575	1626	1675	1720	1761	1771	1741	1712	1676	1590	1504	1419	1333	1245	1154	1061
10	6	5	1.02	2334	843	587	236	308	370	427	477	515	547	568	583	584	575	545	496	422				
8	4	6	1.09	2351	839	598	250	321	387	439	487	529	557	578	594	594	584	554	504	430	331			
12	8	5	1.28	2403	844	622	291	357	418	472	519	556	585	605	618	617	606	575	526	456	359			
6	2	8	1.31	2409	830	638	294	364	429	488	533	566	594	616	634	630	617	583	532	460	363			
10	6	6	1.45	2449	840	647	326	392	450	504	548	584	613	630	643	640	627	594	546	478	386			
8	4	7	1.49	2458	835	654	332	401	459	509	554	594	617	636	650	646	633	599	550	482	390			
14	10	5	1.54	2472	845	657	342	407	465	517	562	597	623	643	654	651	638	605	557	488	397			
6	2	9	1.65	2503	826	688	364	433	496	553	590	622	647	668	684	677	662	624	573	504	414			
12	8	6	1.82	2548	841	696	398	459	517	565	609	640	665	682	692	687	673	637	589	524	437			
8	4	8	1.96	2586	831	719	428	492	541	589	633	668	687	703	716	708	691	653	604	540	456			
10	6	7	1.98	2592	837	719	432	490	545	596	634	666	691	705	716	709	693	655	607	543	461			
6	2	10	2.10	2623	821	752	453	519	580	630	661	691	714	734	749	738	719	677	626	560	476	372		
12	8	7	2.48	2726	838	785	523	581	631	676	714	740	763	775	783	774	755	714	665	604	526	428		
8	4	9	2.48	2726	826	789	530	582	630	675	718	746	762	776	787	776	756	714	664	602	525	429		
6	2	11	2.58	2753	816	819	547	610	668	705	736	764	785	803	817	802	780	735	682	618	539	442		
10	6	8	2.61	2762	833	804	547	602	655	698	733	762	781	792	801	791	771	729	679	619	543	450		
8	4	10	3.15	2907	820	880	647	694	740	784	825	843	856	868	878	862	840	793	741	681	608	522	419	
10	6	9	3.31	2949	828	896	670	723	771	806	839	866	878	887	895	880	858	811	759	700	630	546	443	
6	2	14	3.72	3060	803	977	761	819	849	877	905	931	949	964	977	952	926	872	816	753	681	598	500	
8	4	11	3.87	3101	814	975	766	812	856	898	931	944	956	966	974	954	929	878	824	764	695	615	522	
4	8	6	1.09	2351	848	547	249	310	368	415	457	489	515	532	543	546	542	521	480	413				
4	6	7	1.24	2391	844	568	279	339	393	444	481	513	539	553	564	566	560	537	496	432	340			
4	2	9	1.24	2391	831	605	279	348	412	469	507	538	563	585	601	600	590	560	512	442	346			
4	4	8	1.31	2409	838	587	293	359	411	457	499	535	554	571	583	583	576	550	507	442	351			
4	8	7	1.49	2458	848	584	321	379	429	472	511	536	560	573	581	582	575	553	515	456	370			
4	2	10	1.57	2482	827	648	347	414	475	527	557	586	609	629	644	640	627	595	548	482	393			
4	6	8	1.63	2497	843	607	351	405	456	499	533	562	582	594	603	603	595	570	531	473	392			
4	4	9	1.65	2503	835	624	362	419	465	509	550	579	595	610	621	619	609	582	540	480	397			
4	2	11	1.93	2579	822	693	417	482	540	580	609	636	657	675	690	682	667	632	586	523	441			
4	6	9	2.07	2614	841	649	423	475	524	557	588	615	629	639	646	643	633	607	569	516	444			
4	4	10	2.10	2623	831	672	444	489	532	574	613	633	647	659	669	664	652	622	582	526	451			
4	4	11	2.58	2753	827	722	517	560	602	642	676	689	701	711	720	712	698	666	626	573	505	418		
4	2	14	2.79	2810	813	798	579	637	674	700	725	749	768	783	796	782	763	722	674	616	544	455		
14	2	5	1.02	2334	836	626	236	310	379	443	499	541	575	602	621	619	606	569	514	435	329			
10	2	6	1.09	2351	834	627	249	322	391	453	508	546	578	603	622	620	607	571	517	440	337			
8	2	7	1.24	2391	832	642	280	352	419	480	530	566	595	620	638	634	621	585	532	457	357			
10	2	7	1.49	2458	829	692	333	404	470	531	581	617	646	670	688	681	665	624	569	496	399			
8	2	8	1.63	2497	827	703	362	431	496	555	599	633	660	683	700	692	675	635	581	510	417			
10	2	8	1.96	2586	824	769	429	498	563	622	666	700	727	749	766	754	733	687	631	560	471			
8	2	9	2.07	2614	822	770	449	517	580	637	673	705	730	751	768	755	735	690	636	567	480	373		
10	2	9	2.48	2726	818	853	534	601	663	720	757	789	814	835	851	833	809	758	700	630	546	443		
8	2	10	2.62	2765	816	856	560	625	686	733	766	796	819	839	854	836	812	762	706	639	558	459		
8	2	11	3.22	2927	809	946	676	739	797	831	863	892	913	931	945	922	895	841	782	715	637	546	439	

Table 10-4 - Reproduced From Page 14 of EB9

Calculations and Discussion

Code Reference

PCA—LOAD AND MOMENT STRENGTH TABLES FOR CONCRETE COLUMNS

No A1
No. A2
1 1/2" clear
Axis of bending

COLUMN SIZE =24X24 INCHES
FPC=5.000 FY=60.000

BARS				SPECIAL CONDITIONS			AXIAL LOAD STRENGTH P_n (KIPS)																	
NO A1	NO A2	SIZE NO	PCNT	0. PO	BALANCE PB	MB	0	100	200	300	400	500	600	700	900	1100	1300	1500	1700	1900	2100	2300	2500	2700
							MOMENT STRENGTH M_n (FT-KIPS)																	
10	0	7	1.04	2782	993	854	312	393	472	547	616	678	733	776	837	837	799	746	676	584	469			
6	0	9	1.04	2782	987	850	310	391	469	543	612	674	729	773	834	832	795	743	673	582	467			
14	0	6	1.07	2791	996	864	320	402	482	556	626	688	742	786	847	847	807	754	683	591	476			
4	0	11	1.08	2796	980	855	320	399	477	550	618	680	735	779	841	837	799	748	679	589	474			
8	0	8	1.10	2800	989	867	326	407	486	560	629	691	746	789	850	849	809	756	686	595	480			
12	0	7	1.25	2849	990	910	370	451	530	604	673	736	790	833	894	889	845	790	719	628	516			
6	0	10	1.32	2873	980	921	387	466	544	616	685	747	802	846	906	898	854	798	728	639	529			
10	0	8	1.37	2888	986	940	403	483	561	635	703	766	821	864	925	916	870	813	742	652	542			
8	0	9	1.39	2894	983	942	406	486	563	637	705	768	823	866	927	917	871	814	743	654	545			
4	0	14	1.56	2950	968	974	447	525	600	672	739	800	855	901	961	945	896	840	770	683	577			
6	0	11	1.63	2970	973	997	468	546	622	694	762	824	879	924	984	967	917	858	787	699	593	468		
12	0	8	1.65	2977	983	1013	479	559	636	709	778	841	896	939	999	984	932	871	798	710	603	477		
10	0	9	1.74	3005	979	1034	502	581	658	731	799	862	918	960	1020	1003	949	887	815	727	621	496		
8	0	10	1.76	3014	975	1037	508	586	662	735	803	865	921	964	1024	1005	951	890	818	730	625	501		
8	0	11	2.17	3144	967	1139	615	692	767	839	906	968	1024	1068	1127	1099	1038	972	897	811	709	590		
6	0	14	2.34	3201	958	1175	656	732	806	877	943	1004	1060	1106	1164	1131	1068	1001	926	841	742	626		
4	0	18	2.78	3340	940	1263	756	829	900	969	1033	1093	1147	1196	1256	1210	1143	1073	998	914	819	710	585	
0	0	0	4.00	3732	914	1542	1051	1121	1189	1254	1316	1374	1428	1477	1539	1468	1389	1309	1228	1142	1050	950	840	715
0	0	0	5.00	4054	895	1762	1282	1350	1416	1479	1540	1597	1650	1698	1759	1673	1587	1502	1415	1327	1235	1138	1034	921
0	0	0	6.00	4375	876	1976	1507	1573	1636	1698	1757	1813	1865	1913	1964	1873	1782	1692	1602	1511	1418	1322	1220	1114
0	0	0	7.00	4696	857	2184	1724	1789	1851	1911	1968	2023	2074	2121	2163	2068	1974	1881	1788	1694	1599	1503	1403	1299
0	0	0	8.00	5017	826	2378	1936	1999	2059	2118	2174	2227	2278	2324	2352	2259	2163	2067	1971	1875	1779	1681	1582	1480
12	8	5	1.08	2794	1015	776	324	401	471	537	594	644	685	718	762	769	743	703	644	561	453			
6	2	8	1.10	2800	1000	793	327	406	482	551	614	662	697	728	777	783	755	711	650	566	458			
10	6	6	1.22	2840	1011	803	364	441	508	571	630	677	716	750	790	794	766	725	666	586	483			
8	4	7	1.25	2849	1006	811	370	448	521	580	635	685	727	756	797	801	772	730	670	590	487			
14	10	5	1.29	2863	1016	815	383	456	526	588	644	691	731	761	803	806	778	736	677	597	495			
6	2	9	1.39	2894	996	850	407	484	557	625	685	726	759	788	835	835	802	757	696	616	514			
12	8	6	1.53	2939	1013	858	447	517	584	645	697	743	779	809	847	846	814	772	714	638	540			
8	4	8	1.65	2977	1002	884	478	552	618	672	724	772	812	835	871	869	834	790	732	656	561			
10	6	7	1.67	2983	1008	884	484	554	617	676	732	772	808	840	872	870	836	792	735	661	567			
6	2	10	1.76	3014	991	921	507	582	653	718	771	807	837	864	907	902	863	815	755	679	584	468		
14	10	6	1.83	3037	1015	913	525	595	658	715	767	807	843	869	904	900	864	819	762	688	595	482		
12	8	7	2.08	3117	1010	958	588	655	719	772	821	862	892	919	949	942	903	857	800	730	640	531		
8	4	9	2.08	3117	998	963	593	664	719	771	821	867	900	920	952	944	904	857	799	727	639	531		
6	2	11	2.17	3144	986	997	612	685	753	816	858	892	919	944	985	974	929	879	818	744	655	547		
10	6	8	2.19	3153	1005	979	618	681	742	800	847	883	916	943	969	961	920	873	816	746	660	554		
8	4	10	2.65	3298	993	1064	738	793	845	895	944	986	1010	1027	1055	1041	994	943	885	816	734	636	522	
12	8	8	2.74	3329	1007	1076	751	815	868	918	964	996	1023	1047	1069	1056	1009	959	902	834	753	655	540	
10	6	9	2.78	3340	1001	1083	758	819	877	930	969	1002	1033	1053	1075	1061	1014	963	905	838	757	663	548	
6	2	14	3.13	3451	974	1175	856	923	986	1023	1054	1085	1109	1130	1165	1141	1087	1030	967	897	815	722	612	
8	4	11	3.25	3492	987	1172	875	927	977	1026	1072	1110	1125	1140	1164	1143	1091	1037	977	909	831	742	637	
4	6	7	1.04	2782	1016	716	312	386	448	506	561	601	638	670	703	713	697	664	612	535	432			
4	2	9	1.04	2782	1000	757	311	390	464	531	592	633	665	695	742	750	726	687	628	546	439			
4	10	6	1.07	2791	1024	707	316	386	450	505	557	597	634	660	696	705	691	661	609	533	431			
4	4	8	1.10	2800	1009	737	327	403	472	525	576	622	663	687	724	732	713	678	624	546	444			
4	8	7	1.25	2849	1020	734	364	429	492	545	592	635	665	692	724	731	714	684	635	562	464			
4	2	10	1.32	2873	997	805	387	463	535	600	655	689	719	747	790	794	767	726	669	591	491			
4	6	8	1.37	2888	1015	759	399	461	520	576	623	660	693	721	748	754	735	702	653	583	490			
4	4	9	1.39	2894	1006	779	405	477	535	586	634	679	714	734	767	771	749	714	662	590	495			
4	2	11	1.63	2970	993	855	467	541	610	672	716	748	776	801	842	841	810	768	712	638	544			
4	8	8	1.65	2977	1021	783	456	519	574	620	665	699	726	751	775	777	758	727	682	617	529			
4	6	9	1.74	3005	1014	806	482	541	598	651	688	722	753	775	797	799	776	743	697	633	549			
4	4	10	1.76	3014	1003	832	502	563	613	661	707	749	776	793	822	822	795	759	710	643	557			
4	4	11	2.17	3144	1000	888	597	645	693	739	784	823	840	855	879	875	845	808	760	697	618	520		
4	2	14	2.34	3201	984	974	651	719	782	823	853	882	906	927	962	952	913	867	812	744	662	562		
8	2	7	1.04	2782	1001	798	312	393	469	540	605	658	696	730	781	787	757	712	649	562	452			
14	2	6	1.22	2840	1000	864	364	445	521	594	659	718	758	794	847	848	810	760	694	607	498			
10	2	7	1.25	2849	999	854	370	450	526	597	662	715	753	787	838	839	803	755	690	605	498			
8	2	8	1.37	2888	997	867	403	482	556	625	688	736	771	803	851	851	815	767	704	622	518			
10	2	8	1.65	2977	993	940	479	557	631	700	763	811	846	877	925	919	876	824	759	678	578			
8	2	9	1.74	3005	992	942	502	578	651	718	779	820	853	882	928	921	879	828	765	686	589	471		
10	2	9	2.08	3117	988	1034	597	673	745	812	873	914	946	975	1021	1007	958	901	836	756	661	547		
8	2	10	2.20	3156	986	1037	626	701	771	836	888	925	955	982	1025	1011	962	907	843	767	675	565		
8	2	11	2.71	3318	979	1139	757	830	897	960	1000	1036	1063	1088	1127	1107	1052	993	928	852	765	663	544	

Table 10-5 - Reproduced From Page 15 of EB9

Calculations and Discussion	Code Reference

COLUMNS 7.12.3—Load-moment strength interaction diagram for E5-60.75 columns

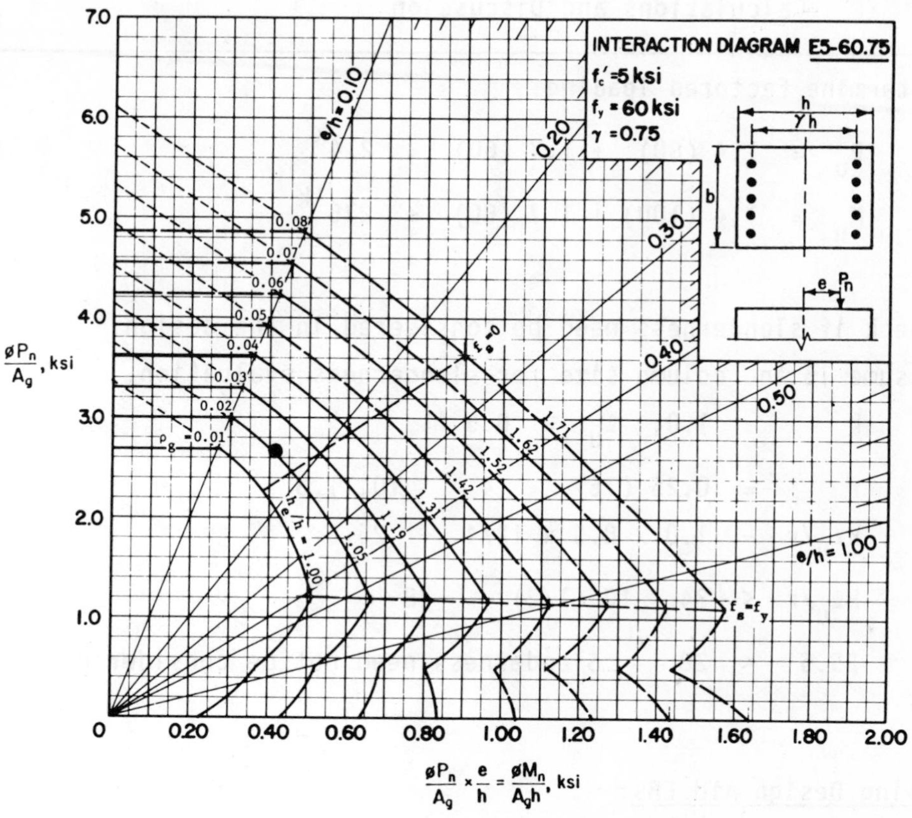

Fig. 10-3 - Reproduced From Page 93 of SP17A
Courtesy of American Concrete Institute

EXAMPLE 10.4 - Design for General Loading (Circular Section)

Design a circular spirally reinforced column with a minimum diameter of 16 in. Column has an unsupported height of 8 ft- 6 in. (braced against sidesway). Use f'_c = 5,000 psi and f_y = 60,000 psi. Required service load strength: $P_d = 80^k$, $P_\ell = 60^k$, $M_d = 100^{'k}$, and $M_\ell = 80^{'k}$. Moments at one end of the column are half those of the other end (single curvature bending). Use Design Aid EB9 and check result by Design Aid SP17A.

Calculations and Discussion	Code Reference
1. Determine factored loading.	
$P_u = 1.4\ (80) + 1.7\ (60) = 214^k$	Eq. (9-1)
$M_u = 1.4\ (100) + 1.7\ (80) = 276^{'k}$	
2. Check if slenderness need be considered in the design.	
Assume 16 in. column size for slenderness evaluation.	
$k = 1.0;\ \ell_u = 8.5$ ft	10.11.2.1
$r = 0.25\ (16) = 4$ 10.11.3	
$k\ell_u/r = 1.0 \times 8.5 \times 12/4 = 25.5$	
$k\ell_u/r < 34 - 12\ (1/2) = 28$	10.11.4.1
$25.5 < 28$ Slenderness need not be considered.	
Using Design Aid EB9:	
3. Required $P_n = P_u/\varphi = 214/0.75 = 285^k$	9.3.2.2(b)
$M_n = M_u/\varphi = 276/0.75 = 368^{'k}$	

Calculations and Discussion	Code Reference

A review of the strength tables indicates that the smallest column section that will support the required load-moment strength is a 20 in. diameter section with ≃ 3 percent reinforcement (See Table 10-6).

Possible Reinforcement Selections

Bars	PCNT	P_n	M_n
12-#8	3.02	285	374
10-#9	3.18	285	382
8-#10	3.23	285	386
6-#11	2.98	285	368

Use 20 in. diameter with 6-#11 bars.

Using Design Aid SP17A, check required reinforcement for 20 in. diameter column (A_g = 314 in.^{2}).

4. Compute $\dfrac{P_u}{A_g} = \dfrac{214}{314} = 0.68$

Compute $\dfrac{M_u}{A_g h} = \dfrac{276 \times 12}{314 \times 20} = 0.53$

Estimate $\gamma \simeq \dfrac{h - 5}{h} = \dfrac{20 - 5}{20} = 0.75$

5. For circular section, f'_c = 5 ksi, f_y = 60 ksi, and γ = 0.75, use column chart 7.23.3 for C5-60.75 columns. (See Fig. 10-4). Read ρ_g = 0.03. (Check OK)

EXAMPLE 10.4 - Continued

Calculations and Discussion	Code Reference

6. Select Spiral Reinforcement.

$\rho_s = 0.45\ (A_g/A_c - 1)\ f'_c/f_y$ — Eq. (10-5)

$A_g = 20^2 \times \pi/4 = 314 \text{ in.}^2$

$A_c = (20 - 3)^2 \pi/4 = 227 \text{ in.}^2$

$\rho_s = 0.45\ (314/227-1)\ 5/60 = 0.0144$

$$\rho_s = \frac{\text{Volume of spiral in one hoop}}{\text{Volume of core in one pitch}}$$

$$= \frac{A_s\ \pi(D_c - D_s)}{A_c s}$$

where s = spiral pitch

D_c = core diameter

D_s = spiral diameter

A_s = spiral area

Minimum spiral diameter = 3/8 in. — 7.10.4.2

Solving for maximum spiral pitch,

$$s = \frac{0.11\ \pi(17 - 0.375)}{227\ (0.0144)} = 1.8 \text{ in.}$$

Clear spacing limitations: s > 1 in. — 7.10.4.3

< 3 in.

> 4/3 (max. aggr.) — 3.3.3

Use #3 Spiral @ 1-3/4 in. pitch

EXAMPLE 10.4 - Continued

Calculations and Discussion	Code Reference

PCA–LOAD AND MOMENT STRENGTH TABLES FOR CONCRETE COLUMNS

1 1/2" clear

COLUMN SIZE =20X20 INCHES
FPC=5,000 FY=60,000

BARS			SPECIAL CONDITIONS			AXIAL LOAD STRENGTH P_n (KIPS)																	
NO	SIZE	PCNT	0.	BALANCE		0	20	120	220	320	420	520	620	720	820	920	1020	1120	1220	1320	1420	1520	1620
TOT	NO		PO	PB	MB	MOMENT STRENGTH M_n (FT-KIPS)																	
11	5	1.09	1525	493	278	131	140	186	225	252	271	279	277	271	258	240	214	181	141				
12	5	1.18	1543	511	285	139	149	196	232	261	277	285	282	276	263	245	220	188	149				
13	5	1.28	1560	499	291	149	159	204	240	266	285	291	288	280	268	251	226	195	156	113			
14	5	1.38	1577	511	296	159	169	212	248	274	290	296	293	286	274	256	232	201	163	121			
15	5	1.48	1594	502	302	170	178	221	255	281	297	302	298	291	279	262	238	207	170	128			
16	5	1.58	1612	512	308	178	187	230	263	288	303	308	304	296	284	267	243	213	177	136			
17	5	1.68	1629	504	314	187	197	238	271	295	310	314	309	301	289	272	249	219	184	143			
18	5	1.78	1646	512	320	197	207	246	278	302	316	320	314	306	294	277	254	226	191	150			
19	5	1.87	1664	506	326	206	215	255	286	309	322	326	320	312	299	282	260	232	198	157			
8	6	1.12	1531	506	279	132	142	191	226	254	273	279	278	272	261	242	216	183	143				
9	6	1.26	1556	481	288	147	158	201	237	266	281	288	286	279	268	249	224	192	153	111			
10	6	1.40	1580	508	296	162	171	213	250	273	291	296	293	287	274	256	232	201	163	122			
11	6	1.54	1605	490	305	173	183	226	259	284	299	305	301	293	281	263	240	210	174	133			
12	6	1.68	1630	510	313	186	196	238	269	294	309	313	308	300	288	271	248	219	184	143			
13	6	1.82	1654	496	322	201	210	248	282	303	318	322	316	307	295	278	256	228	194	153			
14	6	1.96	1679	510	330	213	221	260	291	314	326	329	323	314	302	286	264	237	203	163	121		
15	6	2.10	1703	499	339	225	233	272	301	323	337	338	331	322	310	293	272	245	212	173	132		
16	6	2.24	1728	511	346	238	247	282	313	332	344	346	339	329	317	301	280	253	221	183	143		
17	6	2.38	1752	502	355	251	259	294	322	343	354	354	346	337	324	308	287	261	230	193	153		
6	7	1.15	1536	502	279	135	144	188	228	259	272	279	277	271	260	242	218	186	147				
7	7	1.34	1569	460	290	152	161	208	245	267	285	290	288	282	270	253	230	198	160	117			
8	7	1.53	1603	504	302	172	182	226	256	282	298	302	298	292	280	264	240	209	172	131			
9	7	1.72	1636	476	314	192	202	238	271	296	309	313	309	302	290	273	250	221	185	145			
10	7	1.91	1670	506	325	207	215	254	287	307	323	325	320	312	299	282	260	232	198	159			
11	7	2.10	1703	486	337	223	232	273	299	322	333	336	330	321	309	292	270	244	211	173	133		
12	7	2.29	1737	508	348	242	251	285	314	334	346	347	341	331	318	302	281	255	224	187	146		
13	7	2.48	1770	492	361	258	265	300	328	346	358	359	351	340	328	312	291	267	237	200	159		
14	7	2.67	1803	509	371	273	281	317	341	361	370	370	361	350	338	322	302	278	249	213	173	132	
15	7	2.86	1837	496	383	290	298	329	355	372	382	381	371	360	347	332	313	289	260	226	187	146	
6	8	1.51	1599	499	299	167	176	218	256	283	293	298	295	288	277	260	238	209	173	132			
7	8	1.76	1643	453	315	191	200	244	275	294	311	313	309	302	290	274	253	225	190	148			
8	8	2.01	1688	502	329	217	227	262	290	314	326	328	323	315	304	288	267	239	205	165	124		
9	8	2.26	1732	470	345	239	245	280	311	331	341	343	337	328	317	301	279	253	221	183	142		
10	8	2.51	1776	503	359	257	265	302	328	346	359	358	351	342	329	313	292	267	236	201	161		
11	8	2.77	1820	481	375	280	288	321	345	366	372	372	365	355	342	325	305	281	252	218	180	140	
12	8	3.02	1864	506	389	302	308	338	365	379	390	387	379	368	354	338	318	295	268	235	198	157	
13	8	3.27	1908	488	405	320	327	359	380	396	404	402	392	380	367	351	332	309	283	252	215	175	
14	8	3.52	1952	507	418	341	348	376	398	413	420	417	406	394	380	364	345	324	298	268	232	192	152
6	9	1.91	1670	495	320	201	210	250	287	308	316	319	314	306	295	280	259	232	200	162			
7	9	2.23	1725	445	342	232	241	283	306	324	338	338	332	323	312	297	277	253	221	182	140		
8	9	2.55	1781	499	358	266	272	300	326	349	357	357	349	340	329	314	296	270	239	203	162		
9	9	2.86	1837	463	379	285	292	324	354	367	376	375	367	357	346	331	311	286	257	223	184	144	
10	9	3.18	1893	500	395	311	318	353	372	388	397	393	385	375	363	346	326	303	276	244	207	168	
11	9	3.50	1948	474	416	341	347	371	394	410	414	412	403	392	378	361	342	320	294	264	229	191	152
12	9	3.82	2004	502	432	361	367	395	418	428	436	431	421	408	394	377	359	337	312	284	251	214	173
6	10	2.43	1760	491	348	243	252	290	325	338	345	346	339	330	319	304	285	261	232	197	157		
7	10	2.83	1831	434	375	284	293	326	344	360	374	369	361	352	340	325	307	285	259	224	183	142	
8	10	3.23	1902	495	395	316	321	347	372	393	395	392	383	373	361	347	329	308	280	247	210	169	
9	10	3.64	1972	453	422	343	349	379	404	413	420	416	406	395	382	368	350	328	301	271	236	198	158
10	10	4.04	2043	495	441	377	384	409	425	439	444	439	428	417	404	389	370	348	323	295	262	226	187
6	11	2.98	1857	486	377	287	295	332	363	370	375	374	365	355	343	329	311	289	263	232	197	156	
7	11	3.48	1944	422	411	338	346	366	383	398	410	402	392	382	370	355	338	318	294	265	229	188	146
8	11	3.97	2031	489	433	366	371	396	419	435	434	430	420	408	396	381	364	345	321	292	258	220	180
9	11	4.47	2118	442	467	402	408	437	452	460	466	458	447	434	421	407	391	371	346	319	288	253	215
6	14	4.30	2088	472	444	387	394	428	437	441	443	438	427	415	402	387	371	352	331	306	277	245	210
7	14	5.01	2213	392	493	440	443	457	472	485	489	477	465	453	439	425	408	390	370	347	322	292	254

Table 10-6 - Reproduced From Page 39 of EB9

EXAMPLE 10.4 - Continued

Calculations and Discussion	Code Reference

COLUMNS 7.23.3—Load-moment strength interaction diagram for C5-60.75 columns

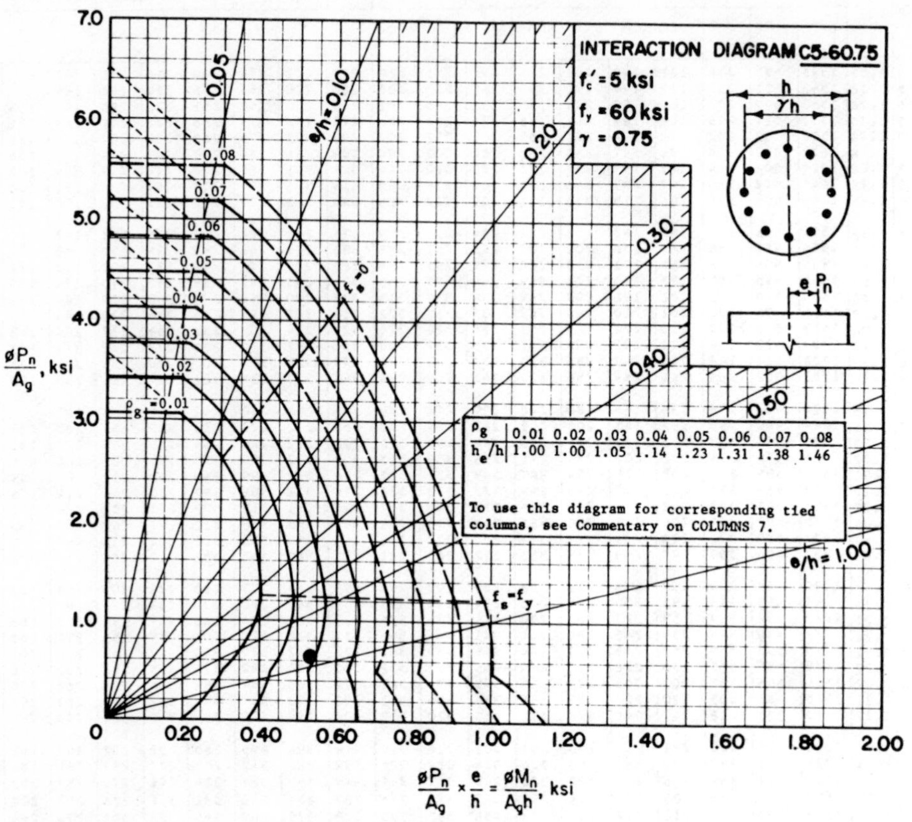

Fig. 10-4 - Reproduced From Page 115 of SP17A

Courtesy of American Concrete Institute

EXAMPLE 10.5 - Design for Reinforcement with Column Size Preset

Select reinforcement for a 10 in. x 20 in. tied reinforced column section using Design Aid SP17A. Use f'_c = 5,000 psi and f_y = 60,000 psi. Assume slenderness may be neglected. Required strength: $P_u = 190^k$ and M_u (about strong axis) = $235^{'k}$.

Calculations and Discussion	Code Reference

1. Using Design Aid SP17A:

   Compute $\frac{P_u}{A_g} = \frac{190}{200} = 0.95$

   Compute $\frac{M_u}{A_g h} = \frac{235 \times 12}{200 \times 20} = 0.71$

   Estimate $\gamma = \frac{h - 5}{h} = \frac{20 - 5}{20} = 0.75$

2. For rectangular section with bars along two faces, f'_c = 5 ksi, f_y = 60 ksi, and γ = 0.75, use column chart 7.12.3 for E5-60.75 columns. (See Fig. 10-5). Read ρ_g = 0.025.

   $A_{st} = \rho_g A_g = 0.025\ (10 \times 20) = 5.00\ \text{in.}^2$

   Select 4 #10 bars (A_s = 5.08 in.^{2})

3. Select lateral ties.

   Use #3 ties with #10 longitudinal bars. 7.10.5.1

   Spacing not less than: 16 x 1.27 = 20.3 in.

   48 x 0.375 = 18 in.

   Column Size = 10 in. (governs)

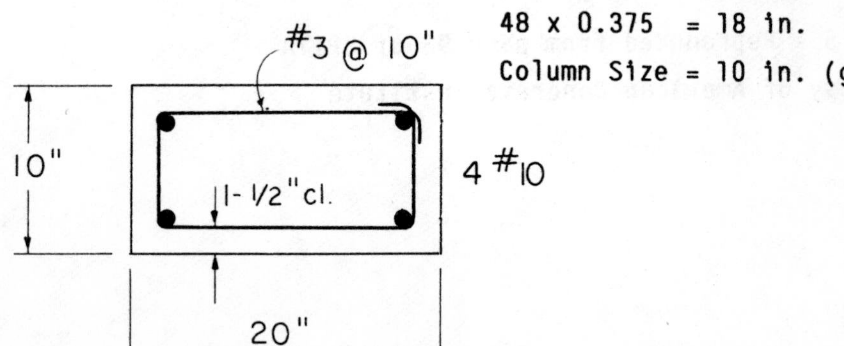

EXAMPLE 10.5 - Continued

Calculations and Discussion	Code Reference

COLUMNS 7.12.3—Load-moment strength interaction diagram for E5-60.75 columns

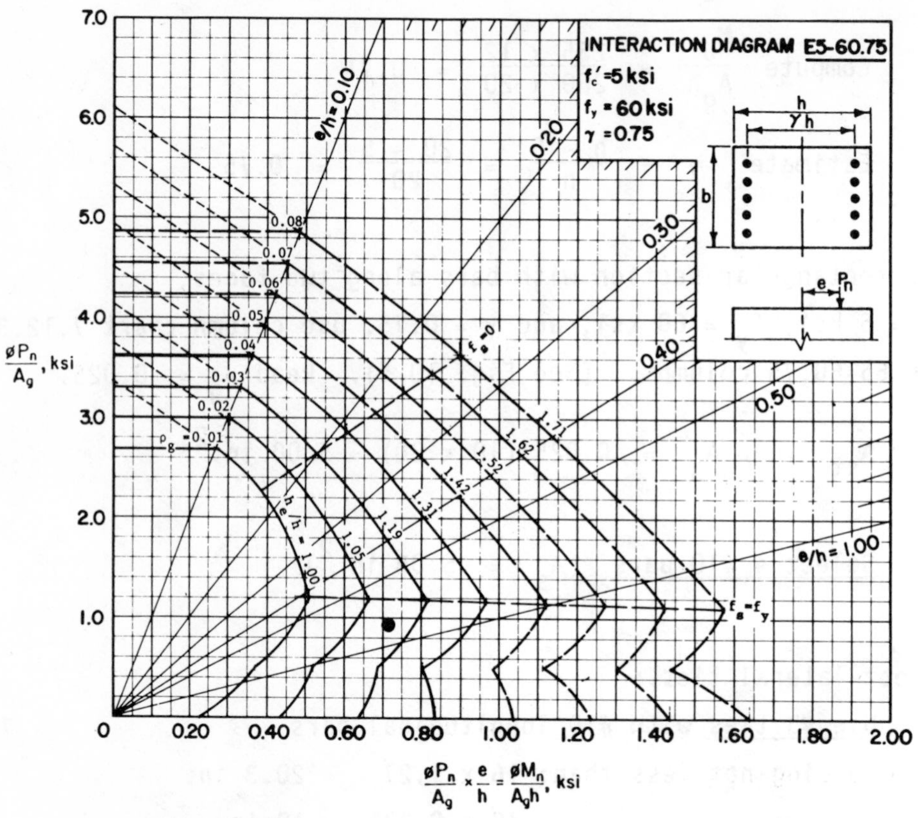

Fig. 10-5 - Reproduced from page 93 of SP17A
Courtesy of American Concrete Institute

EXAMPLE 10.6 - Design for Biaxial Loading (Circular Section)

Design a circular spirally reinforced column for the service load conditions given below. Column has an unsupported height of 7 ft and is braced against sidesway. Use f'_c = 5,000 psi and f_y = 60,000 psi. Column size is limited to 16 in. diameter by architectural considerations. Use Design Aid EB9.

	Dead	Live
Axial Loads	80^k	73^k
EW Moments	$40^{'k}$	$30^{'k}$
NS Moments	$52^{'k}$	$42^{'k}$

Calculations and Discussion	Code Reference

1. Required strength.

$$P_u = 1.4\ (80) + 1.7\ (73) = 236^k$$ Eq. (9-1)

$$M_u\ (EW) = 1.4\ (40) + 1.7\ (30) = 107^{'k}$$

$$M_u\ (NS) = 1.4\ (52) + 1.7\ (42) = 144^{'k}$$

Since this is a circular column, moments about both axes can be combined into a resultant moment and the column can be treated as one subject to uniaxial loading.

$$M_u\ (\text{resultant}) = \sqrt{107^2 + 144^2} = 179^{'k}$$

2. Check if slenderness need be considered in the design.
Assume 16 in. column size for slenderness evaluation.

$k = 1.0;\quad \ell_u = 7.0$ 10.11.2.1

$r = 0.25\ (16) = 4$ 10.11.3

$k\ell_u/r = 1.0 \times 7 \times 12/4 = 21$

$k\ell_u/r < 34 - 12^* = 22$ 10.11.4.1

21 < 22 slenderness neet not be considered

*ratio of end moments assumed equal to 1.0

EXAMPLE 10.6 - Continued

Calculations and Discussion	Code Reference
3. Using Design Aid EB9:	
Required $P_n = P_u/\varphi = 236/0.75 = 315^k$	9.3.2.2(b)
$M_n = M_u/\varphi = 179/0.75 = 238'^k$	
From strength table for 16-in.-diameter column (see Table 10-7), 7 #11 bars (PCNT = 5.43) are required, with a strength provided of $P_n = 315^k$ and $M_n = 239'^k$.	
4. Select spiral reinforcement as illustrated in Design Example 10.4	

Calculations and Discussion	Code Reference

PCA–LOAD AND MOMENT STRENGTH TABLES FOR CONCRETE COLUMNS

1 1/2" clear

COLUMN SIZE =16X16 INCHES
FPC=5,000 FY=60,000

BARS			SPECIAL CONDITIONS			AXIAL LOAD STRENGTH P_n (KIPS)																	
NO	SIZE	PCNT	0.	BALANCE		0	20	40	60	80	100	120	160	200	240	340	440	540	640	740	840	940	1040
TOT	NO		PO	PB	MB	MOMENT STRENGTH M_n (FT-KIPS)																	
7	5	1.08	975	283	137	64	72	79	87	94	101	107	117	124	131	138	136	128	111	85			
8	5	1.23	993	308	142	72	80	87	95	102	107	112	121	130	137	142	140	132	115	89	56		
9	5	1.39	1010	291	146	80	88	95	100	106	112	117	128	135	141	146	144	135	119	94	62		
10	5	1.54	1027	307	150	87	94	100	106	113	119	124	133	140	146	150	147	139	123	99	68		
11	5	1.70	1045	295	155	93	100	107	113	120	125	129	138	145	150	154	151	142	127	104	74		
12	5	1.85	1062	307	159	100	107	114	120	125	130	135	144	150	155	158	155	146	131	109	80		
13	5	2.00	1079	297	163	108	114	120	125	131	136	141	149	155	160	163	159	150	135	114	85		
14	5	2.16	1096	306	168	114	120	126	132	137	142	146	154	161	164	167	162	153	139	119	90		
6	6	1.31	1002	304	143	75	82	89	96	102	108	114	126	133	138	143	141	132	117	92	60		
7	6	1.53	1026	277	149	85	92	99	106	113	119	124	131	138	144	149	146	138	123	100	68		
8	6	1.75	1051	304	155	96	103	111	116	121	125	130	139	146	152	155	151	143	129	106	75		
9	6	1.97	1075	285	161	106	112	117	123	128	133	138	148	153	157	160	156	148	134	112	83		
10	6	2.19	1100	302	167	114	120	126	132	138	143	147	153	160	165	166	162	154	140	119	91		
11	6	2.41	1124	289	174	123	130	136	141	145	149	153	161	168	171	172	167	159	145	125	98	67	
12	6	2.63	1149	302	179	133	138	143	148	153	157	162	169	173	177	178	172	164	150	131	106	75	
6	7	1.79	1055	300	155	95	102	108	114	121	127	132	143	147	151	155	151	143	129	107	78		
7	7	2.09	1089	268	164	109	116	123	129	136	139	142	149	155	160	162	158	150	136	117	88		
8	7	2.39	1122	299	171	124	129	134	138	143	147	151	159	166	169	170	165	157	144	125	97	65	
9	7	2.69	1156	277	180	133	138	144	149	154	159	163	170	174	177	178	172	164	152	133	107	76	
10	7	2.98	1189	297	187	145	151	156	162	166	169	172	178	183	187	185	179	171	159	140	116	86	
11	7	3.28	1222	280	196	158	163	166	170	174	178	181	188	192	194	193	187	178	166	148	125	97	
6	8	2.36	1119	295	169	118	124	130	136	142	148	153	160	163	166	168	163	155	142	123	97	65	
7	8	2.75	1163	258	181	137	143	150	153	156	159	162	168	174	179	178	172	164	152	134	110	78	
8	8	3.14	1207	294	190	151	155	159	163	167	171	175	182	188	189	188	182	173	161	145	122	91	
9	8	3.54	1251	266	202	164	168	173	178	183	187	190	193	197	200	198	191	182	171	155	133	105	73
10	8	3.93	1295	290	211	180	185	189	192	194	197	200	205	210	211	208	200	192	181	164	144	117	87
6	9	2.98	1189	288	184	142	147	153	159	165	170	174	177	179	182	182	176	167	156	139	116	88	
7	9	3.48	1245	245	199	166	170	172	175	178	181	183	189	194	198	194	188	179	168	152	132	104	72
8	9	3.98	1301	286	210	178	182	186	189	193	197	200	207	209	210	207	200	191	179	165	146	119	88
9	9	4.48	1356	253	225	196	200	205	209	212	214	216	218	221	224	219	211	202	191	178	158	134	105
6	10	3.79	1279	279	203	171	177	182	187	193	194	195	197	199	201	200	193	184	172	158	138	113	85
7	10	4.42	1350	229	221	194	197	199	202	204	207	209	214	218	221	215	208	198	187	173	156	134	105
6	11	4.66	1376	268	222	201	207	212	214	215	216	217	218	220	221	218	210	201	190	176	159	138	112
7	11	5.43	1463	209	245	222	224	226	229	231	233	235	240	244	243	236	228	219	208	195	179	160	138

Table 10-7 - Reproduced From Page 37 of EB9

EXAMPLE 10.7 - Design and Investigation of Members for Flexure and Axial Load Using Computer Solution (Design Aid SR14)

Design a 38 in. square tied reinforced column for a service dead load of 3916 kips and a service live load of 1017 kips. Dead and live load service moments are 750 and 720 ft kips, respectively. Use f'_c = 9,000 psi* and f_y = 75,000 psi. Investigate the 38 x 38 section selected for load-moment strength interaction data. Use Design Aid SR14.

*The use of high-strength concrete in columns of high-rise buildings reduces the column size to an acceptable limit for architectural considerations. In the use of reinforcement with a specified yield strength exceeding 60,000 psi (Grade 60), note Code Section 3.5.3.2.

Calculations and Discussion	Code Reference
1. Design Option	
For the design option of computer program SR14, the input/output sheets are reproduced on pages 10-35 and 10-36. Assuming slenderness need not be considered, the required strength is	
P_u = 1.4 (3196) + 1.7 (1017) = 7211^k	Eq. (9-1)
M_u = 1.4 (750) + 1.7 (720) = $2274^{'k}$	
This is part of the input data for design. From the output data sheet, 28 #11 bars, AST = 43.68 (PCT = 3.02) are required, with a design load-moment interaction strength of $P_n = 7232^k$ and $M_n = 2273^{'k}$.	
#4 lateral ties are required with #11 longitudinal	7.10.5.1
bars, spaced not less than: 16 x 1.41 = 22.6 in.(governs)	7.10.5.2
48 x 0.50 = 24.0 in.	
Column Size = 38.0 in.	

EXAMPLE 10.7 - Continued

Calculations and Discussion	Code Reference
To provide lateral support to every alternate longitudinal bar at each tie location, the multiple cross ties shown are required. Bundled bar arrangement may also be used to reduce the number of cross ties.	7.10.5.3

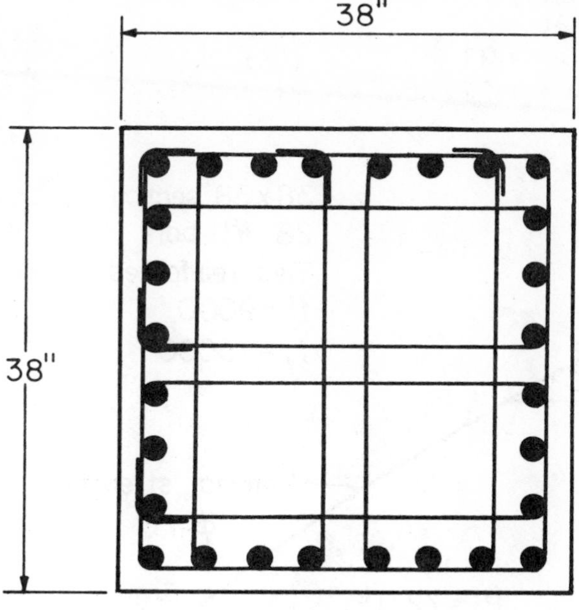

Selected design

2. Investigation Option

Under the investigation option of the computer program, the complete load-moment interaction strength data is obtained for the 38 x 38 section with the selected 28 #11 bars. The input/output sheets are reproduced on pages 10-37 and 10-38. For completeness, the strength interaction diagram is plotted for nominal strength (φ = 1.0) and design strength (φ = 0.7), with the required load-moment values, $P_u = 7211^k$ and $M_u = 2274'^k$ indicated.

EXAMPLE 10.7 - Continued

Calculations and Discussion	Code Reference
The design axial load strength φP_n of compression members must not be taken greater than,	10.3.5
$\varphi P_{n\,(max)} = 0.8\,\varphi[0.85\,f'_c\,(A_g - A_{st}) + f_y\,A_{st}]$	Eq. (10-2)
$= 0.80\,(0.70)\,[0.85 \times 9\,(1444 - 43.7) + 75\,(43.7)]$	
$= 7834^k > 7211^{1k}$ (OK)	

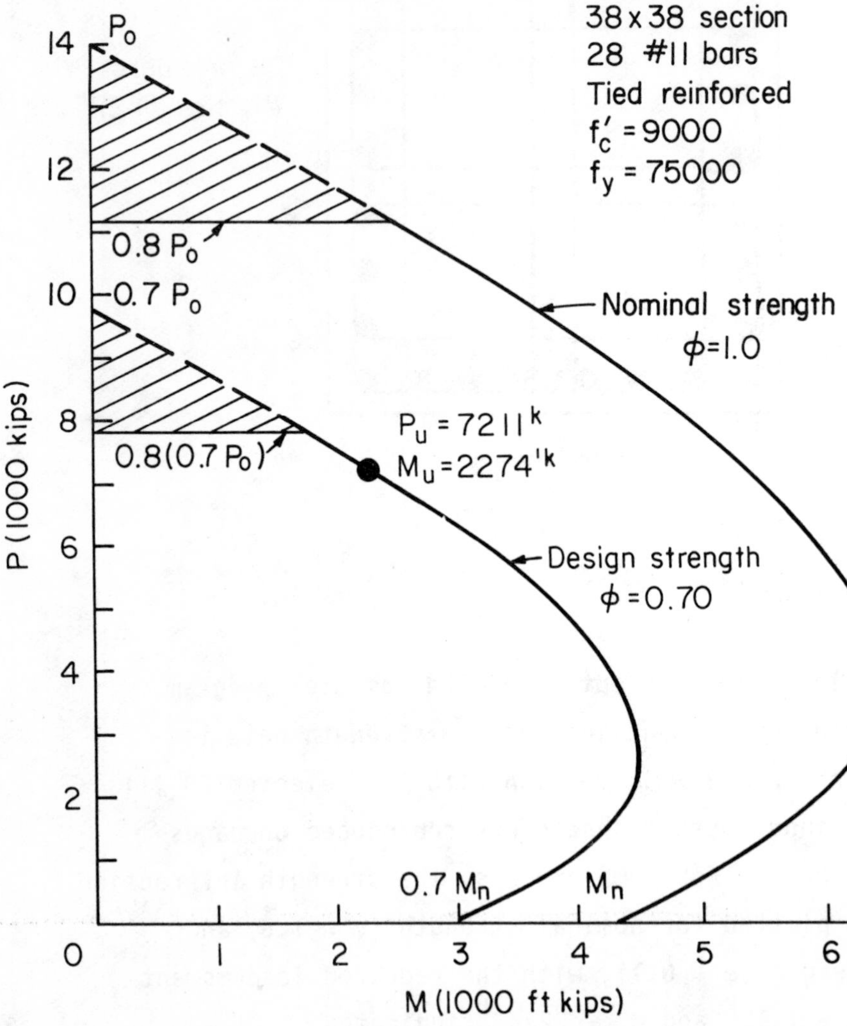

38 x 38 section
28 #11 bars
Tied reinforced
$f'_c = 9000$
$f_y = 75000$

P_o
$0.8\,P_o$
$0.7\,P_o$
$0.8(0.7\,P_o)$
Nominal strength $\phi = 1.0$
Design strength $\phi = 0.70$
$P_u = 7211^k$
$M_u = 2274^{'k}$
$0.7\,M_n$
M_n
P (1000 kips)
M (1000 ft kips)
0 1 2 3 4 5 6 7
2 4 6 8 10 12 14

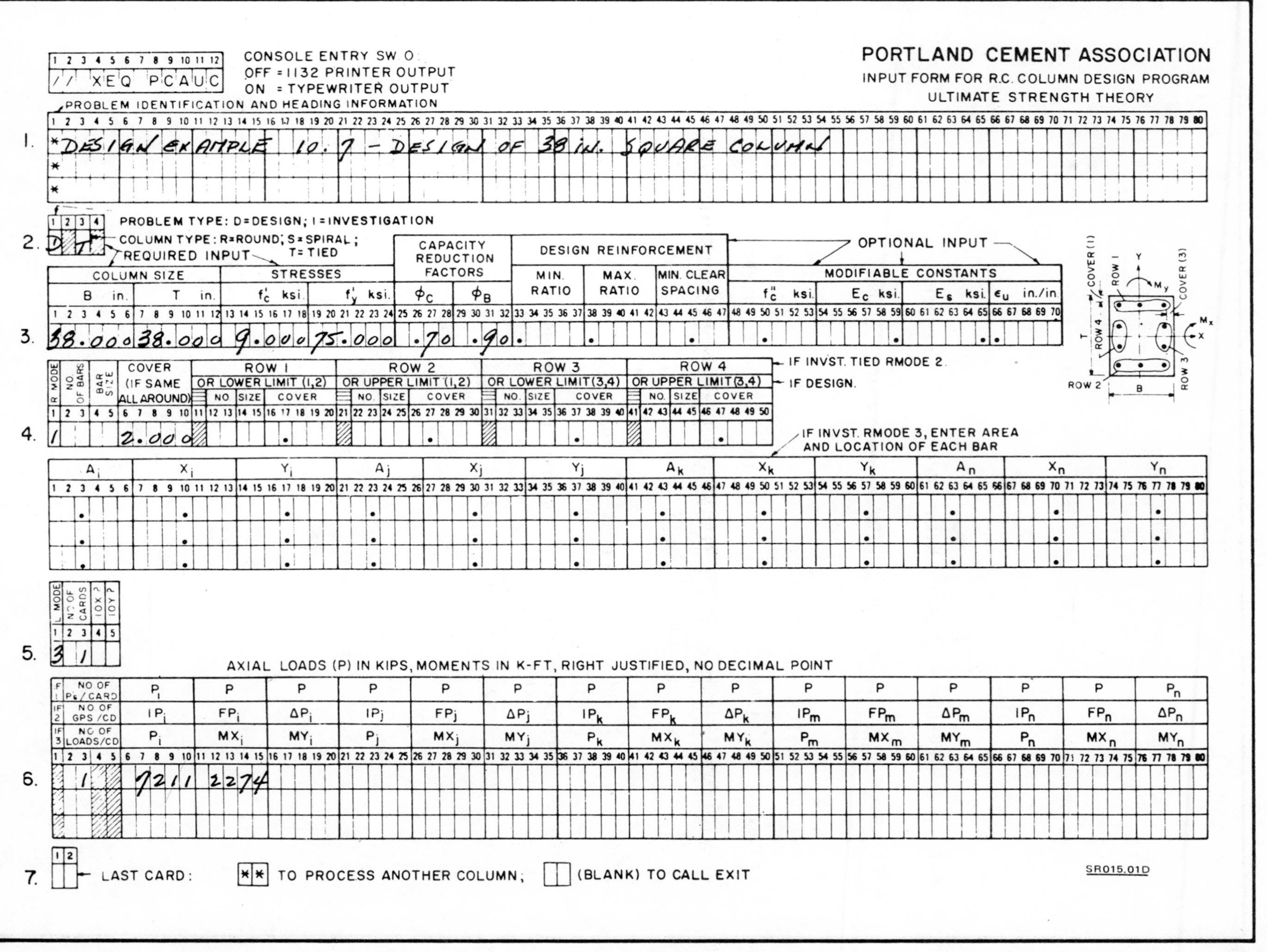

PORTLAND CEMENT ASSOCIATION

INPUT FORM FOR R.C. COLUMN DESIGN PROGRAM

ULTIMATE STRENGTH THEORY

1 2 3 4 5 6 7 8 9 10 11 12
// XEQ PCAUC

CONSOLE ENTRY SW. O:
OFF = 1132 PRINTER OUTPUT
ON = TYPEWRITER OUTPUT

PROBLEM IDENTIFICATION AND HEADING INFORMATION

1. *DESIGN EXAMPLE 10.7 – DESIGN OF 38 in. SQUARE COLUMN
*
*

2. | 1 | 2 | 3 | 4 |
|---|---|---|---|
| D | T | | |

PROBLEM TYPE: D = DESIGN; I = INVESTIGATION
COLUMN TYPE: R = ROUND; S = SPIRAL; T = TIED
REQUIRED INPUT

OPTIONAL INPUT

3.

COLUMN SIZE B in. (1–6)	COLUMN SIZE T in. (7–12)	STRESSES f'_c ksi (13–18)	STRESSES f'_y ksi (19–24)	CAPACITY REDUCTION FACTORS ϕ_C (25–28)	ϕ_B (29–32)	DESIGN REINFORCEMENT MIN. RATIO (33–37)	MAX. RATIO (38–42)	MIN. CLEAR SPACING (43–47)	MODIFIABLE CONSTANTS f''_c ksi (48–53)	E_c ksi (54–59)	E_s ksi (60–65)	ϵ_u in./in (66–70)
38.000	38.000	9.000	75.000	.70	.90	.	.	.	.	.	.	.

4.

R MODE (1)	NO. OF BARS (2–3)	BAR SIZE (4–5)	COVER (IF SAME ALL AROUND) (6–10)	ROW 1 OR LOWER LIMIT (1,2) NO (12–13)	SIZE (14–15)	COVER (16–20)	ROW 2 OR UPPER LIMIT (1,2) NO (22–23)	SIZE (24–25)	COVER (26–30)	ROW 3 OR LOWER LIMIT (3,4) NO (32–33)	SIZE (34–35)	COVER (36–40)	ROW 4 OR UPPER LIMIT (3,4) NO (42–43)	SIZE (44–45)	COVER (46–50)
1			2.000			.			.			.			.

← IF INVST. TIED RMODE 2.
← IF DESIGN.

IF INVST. RMODE 3, ENTER AREA AND LOCATION OF EACH BAR

A_i (1–6)	X_i (7–13)	Y_i (14–20)	A_j (21–26)	X_j (27–33)	Y_j (34–40)	A_k (41–46)	X_k (47–53)	Y_k (54–60)	A_n (61–66)	X_n (67–73)	Y_n (74–80)
.	.	.	.	.	.	.	.	.	.	.	.
.	.	.	.	.	.	.	.	.	.	.	.
.	.	.	.	.	.	.	.	.	.	.	.

5.

L MODE (1)	NO. OF CARDS (2–3)	IOX ? (4)	IOY ? (5)
3	1		

AXIAL LOADS (P) IN KIPS, MOMENTS IN K-FT, RIGHT JUSTIFIED, NO DECIMAL POINT

6.

(1–5)	6–10	11–15	16–20	21–25	26–30	31–35	36–40	41–45	46–50	51–55	56–60	61–65	66–70	71–75	76–80
IF 1: NO OF Ps/CARD	P_i	P	P	P	P	P	P	P	P	P	P	P	P	P	P_n
IF 2: NO OF GPS/CD	IP_i	FP_i	ΔP_i	IP_j	FP_j	ΔP_j	IP_k	FP_k	ΔP_k	IP_m	FP_m	ΔP_m	IP_n	FP_n	ΔP_n
IF 3: NO OF LOADS/CD	P_i	MX_i	MY_i	P_j	MX_j	MY_j	P_k	MX_k	MY_k	P_m	MX_m	MY_m	P_n	MX_n	MY_n
1	7211	2274													

7. LAST CARD: ** TO PROCESS ANOTHER COLUMN; (BLANK) TO CALL EXIT

SR015.01D

```
P.C.A. - STRENGTH DESIGN OF R.C. COMPRESSION MEMBERS

DESIGN EXAMPLE  10.7 - DESIGN OF 38IN. SQUARE COLUMN

DESIGN OF
TIED COMPRESSION MEMBERS

B= 38.00  T= 38.00  FC= 9.000  FY= 75.000  PHIC= 0.700  PHIB= 0.900

USE- 28 NO.11 BARS.  AST = 43.68 SQ.IN. = 3.02 PCT.   COVER = 2.000 IN.

              ROW 1   ROW 2   ROW 3   ROW 4

NO. OF BARS     8       8       6       6
COVER         2.000   2.000   2.000   2.000

 LOAD       APPLIED  LOADS          COMPUTED  STRENGTH
 CASE       AP      AMX     AMY        UP      UMX     UMY        UP/AP

   1      7211.   2274.      0.     7232.   2273.      0.       1.003
```

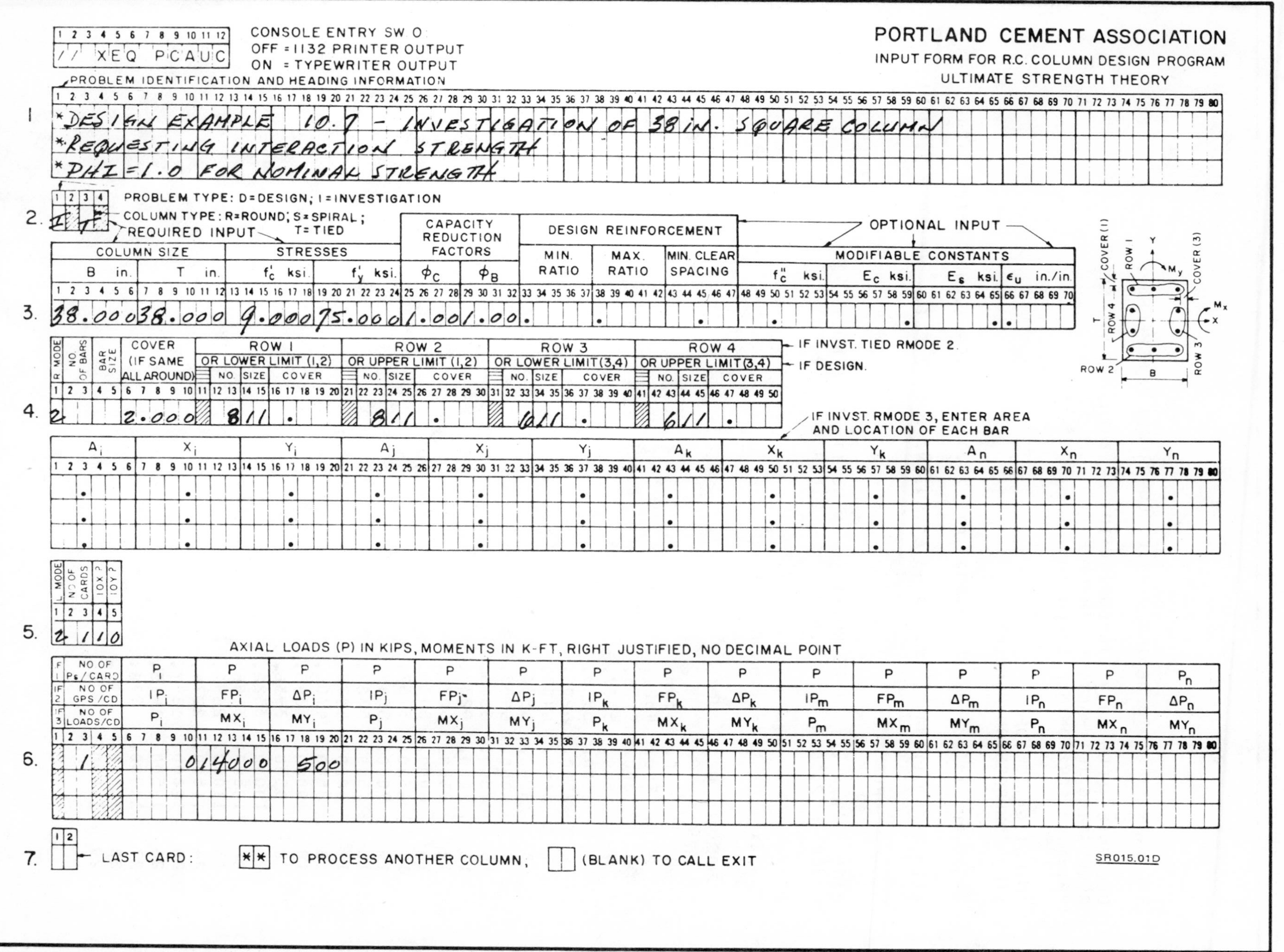

PORTLAND CEMENT ASSOCIATION
INPUT FORM FOR R.C. COLUMN DESIGN PROGRAM
ULTIMATE STRENGTH THEORY

1 2 3 4 5 6 7 8 9 10 11 12
// XEQ PCAUC

CONSOLE ENTRY SW O
OFF = 1132 PRINTER OUTPUT
ON = TYPEWRITER OUTPUT

PROBLEM IDENTIFICATION AND HEADING INFORMATION

1. *DESIGN EXAMPLE 10.7 - INVESTIGATION OF 38 in. SQUARE COLUMN
*REQUESTING INTERACTION STRENGTH
*PHI=1.0 FOR NOMINAL STRENGTH

2. I T F
PROBLEM TYPE: D=DESIGN; I=INVESTIGATION
COLUMN TYPE: R=ROUND; S=SPIRAL; T=TIED
REQUIRED INPUT

COLUMN SIZE B in.	COLUMN SIZE T in.	STRESSES f'_c ksi	STRESSES f'_y ksi	CAPACITY REDUCTION FACTORS ϕ_C	ϕ_B	DESIGN REINFORCEMENT MIN. RATIO	MAX. RATIO	MIN. CLEAR SPACING	MODIFIABLE CONSTANTS f''_c ksi	E_c ksi	E_s ksi	ϵ_u in./in.
1–6	7–12	13–18	19–24	25–28	29–32	33–37	38–42	43–47	48–53	54–59	60–65	66–70
3. 38.000	38.000	9.000	75.000	1.00	1.00	.	.	.	.	.	.	.

OPTIONAL INPUT

R MODE	NO. OF BARS	BAR SIZE	COVER (IF SAME ALL AROUND)	ROW 1 OR LOWER LIMIT (1,2) NO.	SIZE	COVER	ROW 2 OR UPPER LIMIT (1,2) NO.	SIZE	COVER	ROW 3 OR LOWER LIMIT (3,4) NO.	SIZE	COVER	ROW 4 OR UPPER LIMIT (3,4) NO.	SIZE	COVER
1	2 3	4 5	6–10	12 13	14 15	16–20	22 23	24 25	26–30	32 33	34 35	36–40	42 43	44 45	46–50
4. 2			2.000	8	11	.	8	11	.	6	11	.	6	11	.

← IF INVST. TIED RMODE 2.
← IF DESIGN.

IF INVST. RMODE 3, ENTER AREA AND LOCATION OF EACH BAR

A_i	X_i	Y_i	A_j	X_j	Y_j	A_k	X_k	Y_k	A_n	X_n	Y_n
1–6	7–13	14–20	21–26	27–33	34–40	41–46	47–53	54–60	61–66	67–73	74–80
.	.	.	.	.	.	.	.	.	.	.	.
.	.	.	.	.	.	.	.	.	.	.	.
.	.	.	.	.	.	.	.	.	.	.	.
.	.	.	.	.	.	.	.	.	.	.	.

L MODE	NO OF CARDS	IOX?	IOY?
1	2 3	4	5
5. 2	1	1	0

AXIAL LOADS (P) IN KIPS, MOMENTS IN K-FT, RIGHT JUSTIFIED, NO DECIMAL POINT

	1–5	6–10	11–15	16–20	21–25	26–30	31–35	36–40	41–45	46–50	51–55	56–60	61–65	66–70	71–75	76–80
IF 1	NO OF Ps/CARD	P_i	P	P	P	P	P	P	P	P	P	P	P	P	P	P_n
IF 2	NO OF GPS/CD	IP_i	FP_i	ΔP_i	IP_j	FP_j	ΔP_j	IP_k	FP_k	ΔP_k	IP_m	FP_m	ΔP_m	IP_n	FP_n	ΔP_n
IF 3	NO OF LOADS/CD	P_i	MX_i	MY_i	P_j	MX_j	MY_j	P_k	MX_k	MY_k	P_m	MX_m	MY_m	P_n	MX_n	MY_n
6.	1	0	14000	500												

7. LAST CARD: ** TO PROCESS ANOTHER COLUMN, (BLANK) TO CALL EXIT

SR015.01D

P.C.A. - STRENGTH DESIGN OF R.C. COMPRESSION MEMBERS

DESIGN EXAMPLE 10.7 -INVESTIGATION OF 38IN. SQUARE COLUMN
REQUESTING INTERACTION STRENGTH
PHI=1.0 FOR NOMINAL STRENGTH

INVESTIGATION OF
TIED COMPRESSION MEMBERS

B= 38.00 T= 38.00 FC= 9.000 FY= 75.000 PHIC= 1.000 PHIB= 1.000

WITH 28 NO.11 BARS. AST = 43.67 SQ.IN. = 3.02 PCT. COVER = 2.000 IN.

	ROW 1	ROW 2	ROW 3	ROW 4
BARS	8 NO.11	8 NO.11	6 NO.11	6 NO.11
COVER	2.000	2.000	2.000	2.000

UNIAXIAL INTERACTION REQUESTED

LOAD	UP	UMX
1	0.	4308.
2	500.	4822.
3	1000.	5253.
4	1500.	5637.
5	2000.	5958.
6	2500.	6206.
7	3000.	6403.
8	3500.	6530.
9	4000.	6488.
10	4500.	6366.
11	5000.	6237.
12	5500.	6096.
13	6000.	5938.
14	6500.	5757.
15	7000.	5548.
16	7500.	5307.
17	8000.	5028.
18	8500.	4706.
19	9000.	4343.
20	9500.	3936.
21	10000.	3522.
22	10500.	3108.
23	11000.	2692.
24	11500.	2274.
25	12000.	1847.
26	12500.	1415.
27	13000.	978.
28	13500.	517.
29	14000.	0.

11

Biaxial Loading

General Considerations

A uniaxial interaction diagram defines the load-moment strength in a single plane of a section under an axial load P and a uniaxial moment M. The biaxial bending resistance of an axially loaded column can be represented schematically (see Fig. 11-1) as a surface formed by a series of uniaxial interaction curves drawn radially from the P axis. Data for these intermediate curves are obtained by varying the angle of the neutral axis (in the assumed strain configurations) with respect to the major axes (see Fig. 11-2).

The difficulty associated with the determination of the strength of reinforced columns subject to combined axial load and biaxial bending is primarily an arithmetic one. The bending resistance of an axially loaded column about a particular skewed axis is determined through iteration of simple, but lengthy calculations. These extensive calculations are compounded when optimization of the reinforcement or cross section is sought.

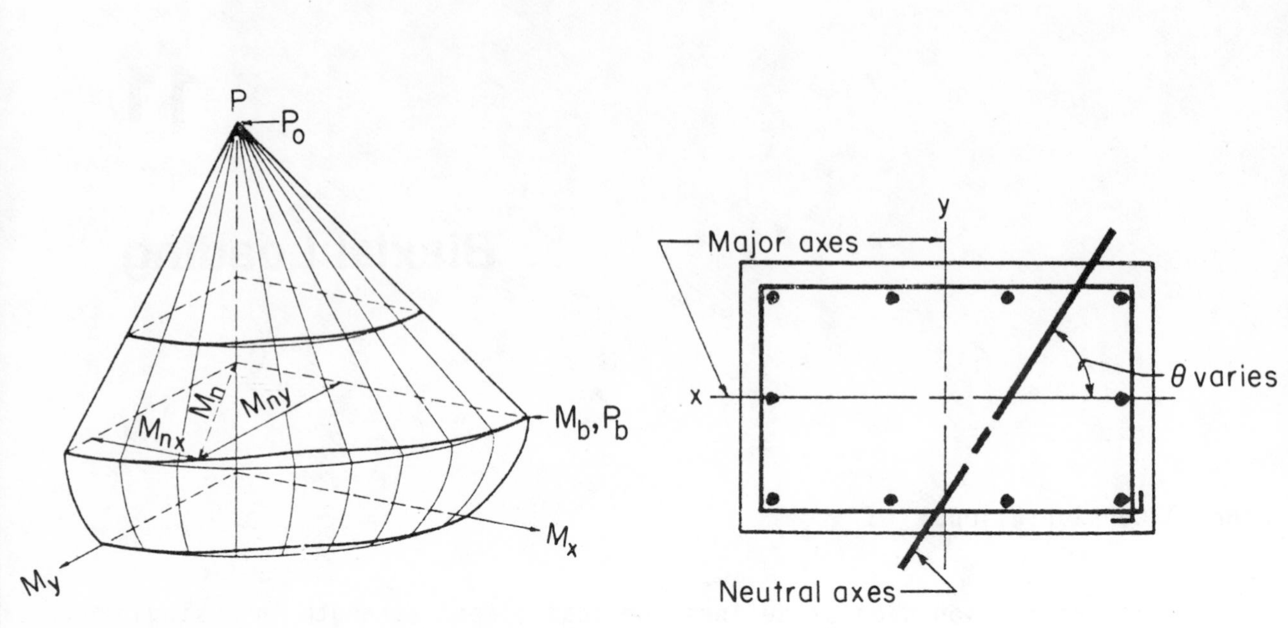

Fig. 11-1
Biaxial Interaction Surface

Fig. 11-2
Neutral Axis at Angle to Major Axis

For uniaxial bending, it is customary to utilize design aids in the form of interaction curves or tables. However, for biaxial bending, because of the voluminous nature of the data and the difficulty in multiple interpolations, the development of interaction curves or tables for the various ratios of bending moments about each axis is impractical. Instead, several approaches (based on acceptable approximations) have been developed that relate the response of a column in biaxial bending to its uniaxial resistance about each major axis.

Failure Surfaces

The nominal strength of a section under biaxial bending and compression is a function of three variables P_n, M_{nx} and M_{ny} which may be expressed in terms of an axial load acting at eccentricities $e_x = \frac{M_{ny}}{P_n}$ and $e_y = \frac{M_{nx}}{P_n}$ as shown in Fig. 11-3. A failure surface may be described as a surface produced by plotting the failure load P_n as a function of its eccentricities e_x and e_y or of its associated bending moments M_{ny} and M_{nx}.

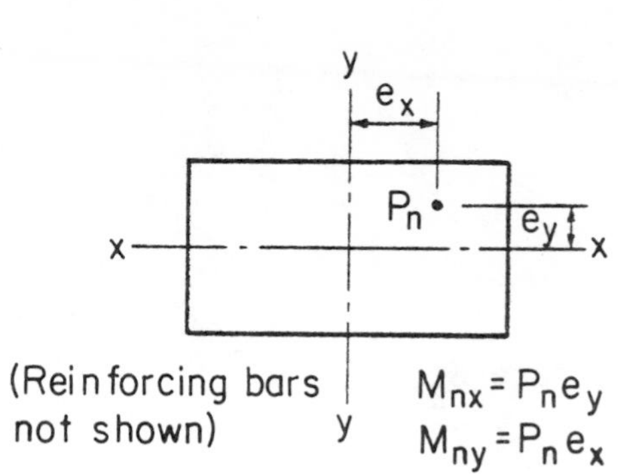

Fig. 11-3 - Notation for Biaxial Loading

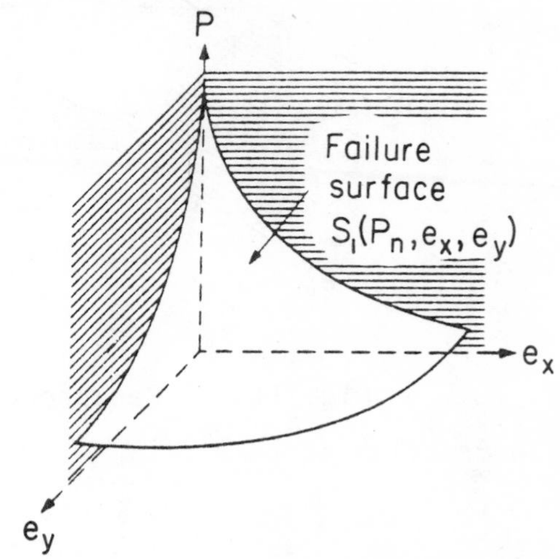

Fig. 11-4 - Failure Surface S_1

Three types of failure surfaces have been defined.[11.2, 11.3, 11.4] The basic surface S_1 is defined by a function which is dependent upon the variables P_n, e_x and e_y, as shown in Fig. 11-4. A reciprocal surface can be derived from S_1 in which the reciprocal of the nominal axial load P_n is employed to produce the surface S_2 ($1/P_n$, e_x, e_y) as illustrated in Fig. 11-5. The third type of failure surface, shown in Fig. 11-6, is obtained by relating the nominal axial load P_n to moments M_{nx} and M_{ny} to produce surface S_3 (P_n, M_{nx}, M_{ny}). Failure surface S_3 is the three dimensional extension of the uniaxial interaction diagram previously described.

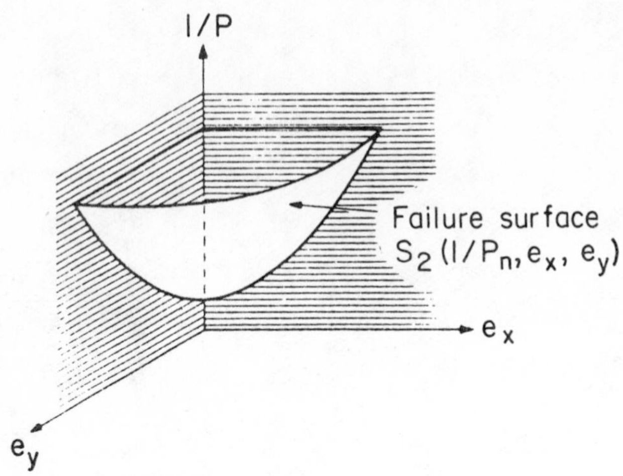

Fig. 11-5 - Reciprocal Failure Surface S_2

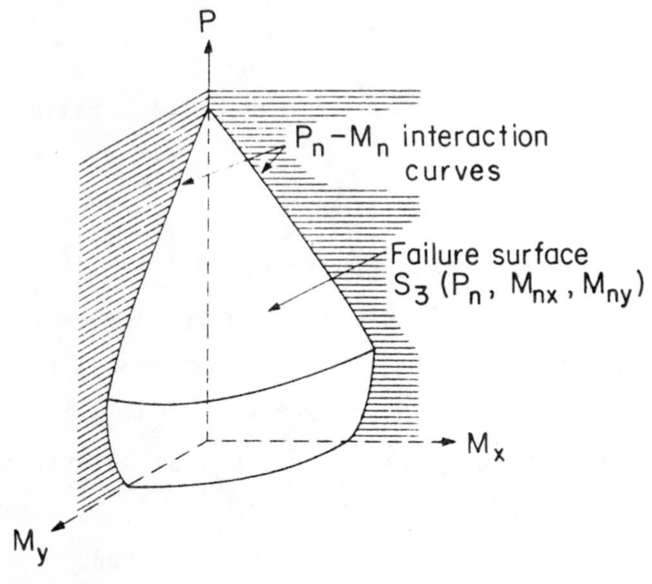

Fig. 11-6 - Failure Surface S_3

A number of investigators have made approximations for both the S_2 and S_3 failure surfaces for use in design and analysis.[11.4-11.10] An explanation of those methods used in current practice along with design examples of each follows.

A. Bresler Reciprocal Load Method

This method approximates the ordinate $1/P_n$, on the surface S_2 $(1/P_n, e_x, e_y)$ by a corresponding ordinate $1/P'_n$ on the plane S'_2 $(1/P'_n, e_x, e_y)$, which is defined by the characteristic points A, B and C as indicated in Fig. 11-7. For any particular cross section, the value P_o (corresponding to point C) is the load strength under pure axial compression; P_{ox} (corresponding to point B) and P_{oy} (corresponding to point A) are the load strengths under uniaxial eccentricities e_y and e_x, respectively. Each point on the true surface is approximated by a different plane; therefore, the entire surface is approximated using an infinite number of planes.

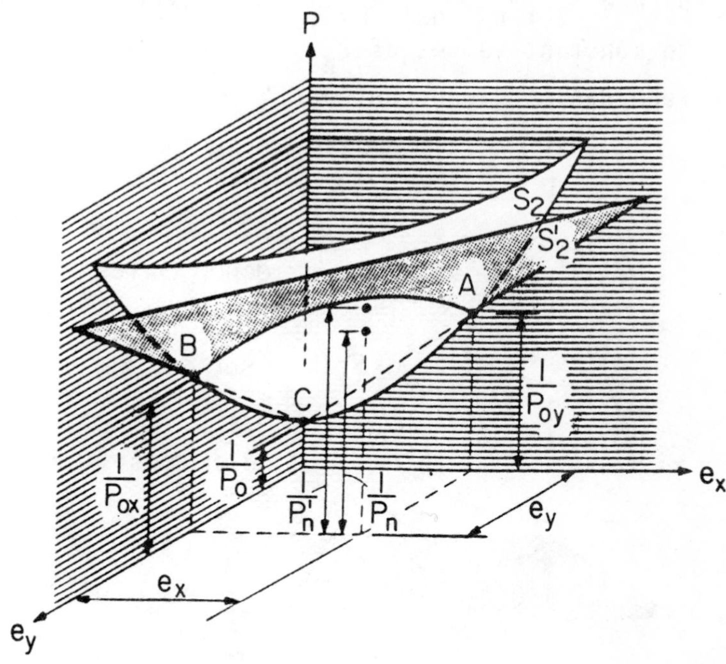

Fig. 11-7 - Reciprocal Load Method

The general expression for any values of e_x and e_y when derived[11.4] yields the following equation:

$$\frac{1}{P_n} \simeq \frac{1}{P'_n} = \frac{1}{P_{ox}} + \frac{1}{P_{oy}} - \frac{1}{P_o}$$

Rearranging variables yields:

$$P_n \approx \frac{1}{(1/P_{ox}) + (1/P_{oy}) - (1/P_o)} \qquad \text{Eq. (1)}$$

This equation is simple in form and the variables are easily determined. Axial load strengths P_o, P_{ox} and P_{oy} are determined using any of the methods presented in Part 10. Experimental test results have shown the equation to be reasonably accurate when flexure does not govern design. The equation should only be used when:

$$P_n \geq 0.1 f'_c A_g \qquad \text{Eq. (2)}$$

B. Bresler Load Contour Method

In this method, the surface S_3 (P_n, M_{nx}, M_{ny}) is approximated by a family of curves corresponding to constant values of P_n. These curves, as illustrated in Fig. 11-8, may be regarded as "load contours."

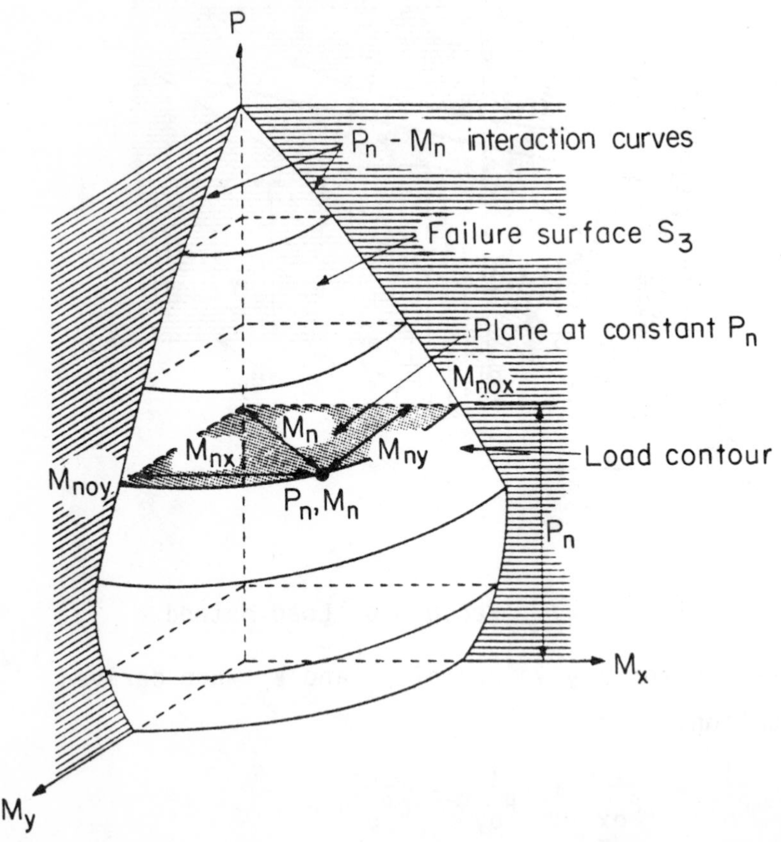

Fig. 11-8 - Load Contours for Constant P_n on Failure Surface S_3

The general expression for these curves can be approximated[11.4] by a non-dimensional interaction equation of the form

$$\left(\frac{M_{nx}}{M_{nox}}\right)^{\alpha} + \left(\frac{M_{ny}}{M_{noy}}\right)^{\beta} = 1.0 \qquad \text{Eq. (3)}$$

In the above equation, M_{nx} and M_{ny} are the nominal biaxial moment strengths in the direction of the x and y axis respectively. M_{nx} and M_{ny} are the vectorial equivalent of the nominal uniaxial moment strength M_n. M_{nox} and M_{noy} are the nominal uniaxial moment strengths with bending considered in the direction of the x and y axis separately. The values of the exponents α and β are a function of the amount, distribution and location of reinforcement, the dimensions of the column, and the strength and elastic properties of the steel and concrete. Bresler[11.4] indicates that it is reasonably accurate to assume that $\alpha = \beta$; therefore, Eq. (3) becomes

$$\left(\frac{M_{nx}}{M_{nox}}\right)^{\alpha} + \left(\frac{M_{ny}}{M_{noy}}\right)^{\beta} = 1.0 \qquad \text{Eq. (4)}$$

which is shown graphically in Fig. 11-9.

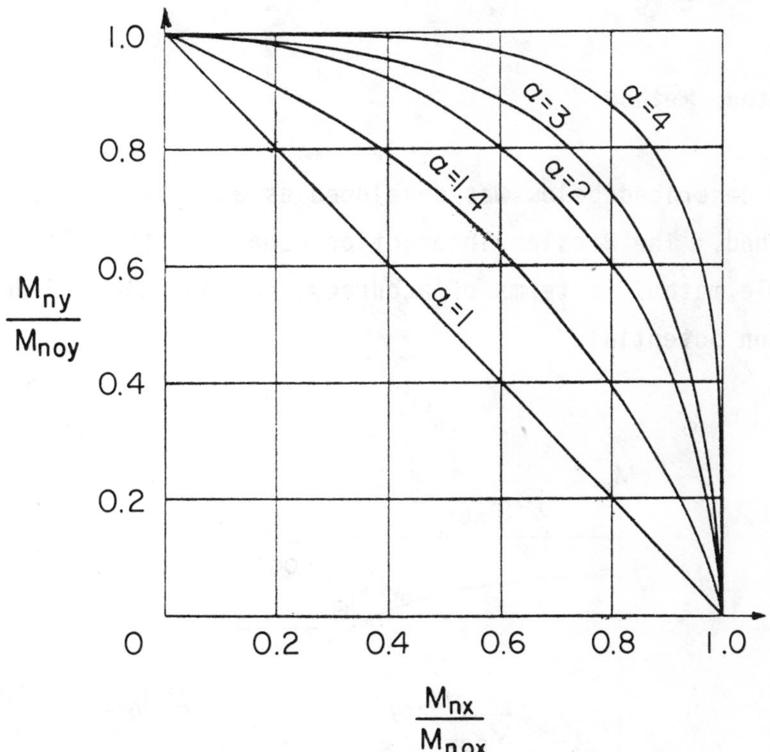

Fig. 11-9 - Interaction Curves for Eq. (4)

When using Eq. (4) or Fig. 11-9, it is still necessary to determine the α value for the cross section being designed. Bresler indicated that, typically, α varied from 1.15 to 1.55, with a value of 1.5 being reasonably accurate for most square and rectangular sections with uniformly distributed reinforcement.

With α set at unity, the equation becomes that of a straight line, as shown in Fig. 11-9, and will always yield conservative results:

$$\frac{M_{nx}}{M_{nox}} + \frac{M_{ny}}{M_{noy}} = 1.0 \qquad \text{Eq. (5)}$$

The use of Eq. (5) becomes overly conservative for high axial loads or low percentages of reinforcement. It should only be used when

$$P_n < 0.1\, f'_c\, A_g \qquad \text{Eq. (6)}$$

C. PCA Load Contour Method

The PCA approach described below was developed as an extension of the Bresler Load Contour Method. The Bresler interaction equation [Eq. (4)] was chosen as the most viable method in terms of accuracy, condensation of design aids and simplification potential.

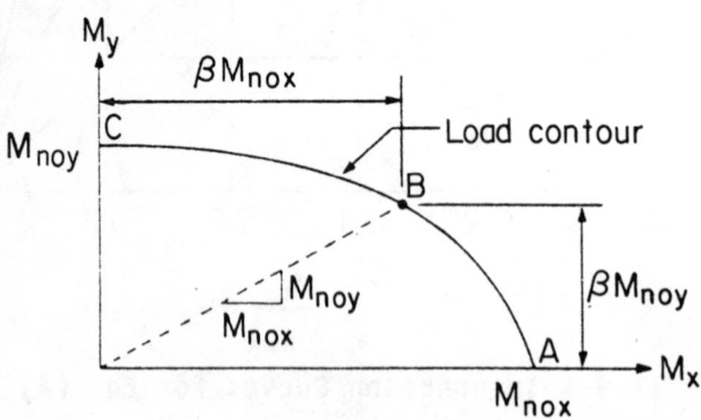

Fig. 11-10 - Load Contour at Plane of Constant P_n of Failure Surface S_3

A typical Bresler load contour is shown in Fig. 11-10. In the PCA method,[11.9] a point B is defined such that the nominal biaxial moment strengths M_{nx} and M_{ny} at this point are in the same ratio as the uniaxial moment strengths M_{nox} and M_{noy}; therefore, at point B

$$\frac{M_{nx}}{M_{ny}} = \frac{M_{nox}}{M_{noy}} \qquad \text{Eq. (7)}$$

When the load contour of Fig. 11-10 is nondimensionalized, it takes the form shown in Fig. 11-11, and the point B will have x and y coordinates of β. When the bending resistance is plotted in terms of the dimensionless parameters P_n/P_o, M_{nx}/M_{nox} and M_{ny}/M_{noy} designated as the relative moments, the failure surface S_4 (P_n/P_o, M_{nx}/M_{nox}, M_{ny}/M_{noy}) generated assumes the typical shape shown in Fig. 11-12. The advantage of expressing the behavior in relative terms is that the contours of the surface (Fig. 11-11) - i.e., the intersection formed by planes of constant P_n/P_o and the surface - can be considered for design purposes to be symmetrical about the vertical plane bisecting the two coordinate planes. Even for sections that are rectangular or have unequal reinforcement in the two adjacent faces, this approximation yields values sufficiently accurate for design.

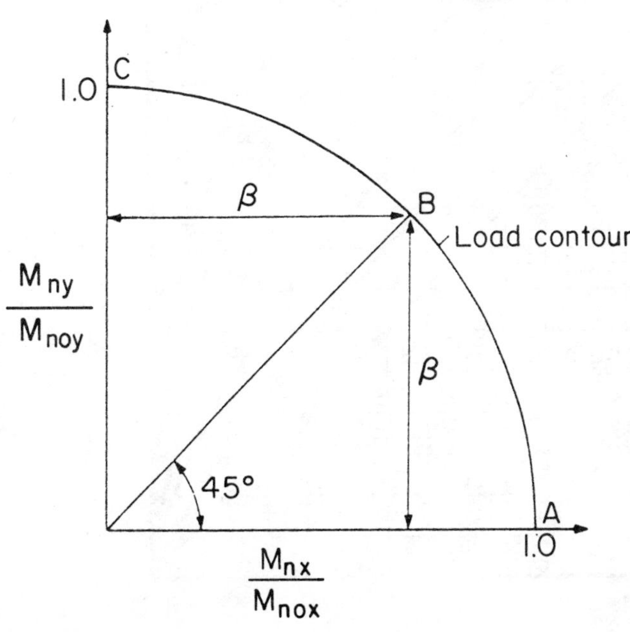

Fig. 11-11 - Nondimensional Load Contour at Constant P_n

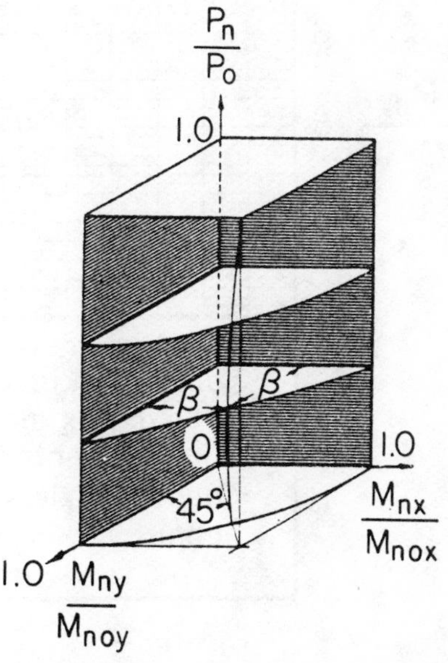

Fig. 11-12 - Failure Surface S_4 (P_n/P_o, M_{nx}/M_{nox}, M_{ny}/M_{noy})

The relationship of α from Eq. (4) and β is obtained by substituting the coordinates of point B from Fig. 11-10 into Eq. (4) and solving for α in terms of β. This yields:

$$\alpha = \frac{\log 0.5}{\log \beta}$$

Thus Eq. (4) may be written:

$$\left(\frac{M_{nx}}{M_{nox}}\right)^{\log 0.5/\log\beta} + \left(\frac{M_{ny}}{M_{noy}}\right)^{\log 0.5/\log\beta} = 1 \qquad \text{Eq. (8)}$$

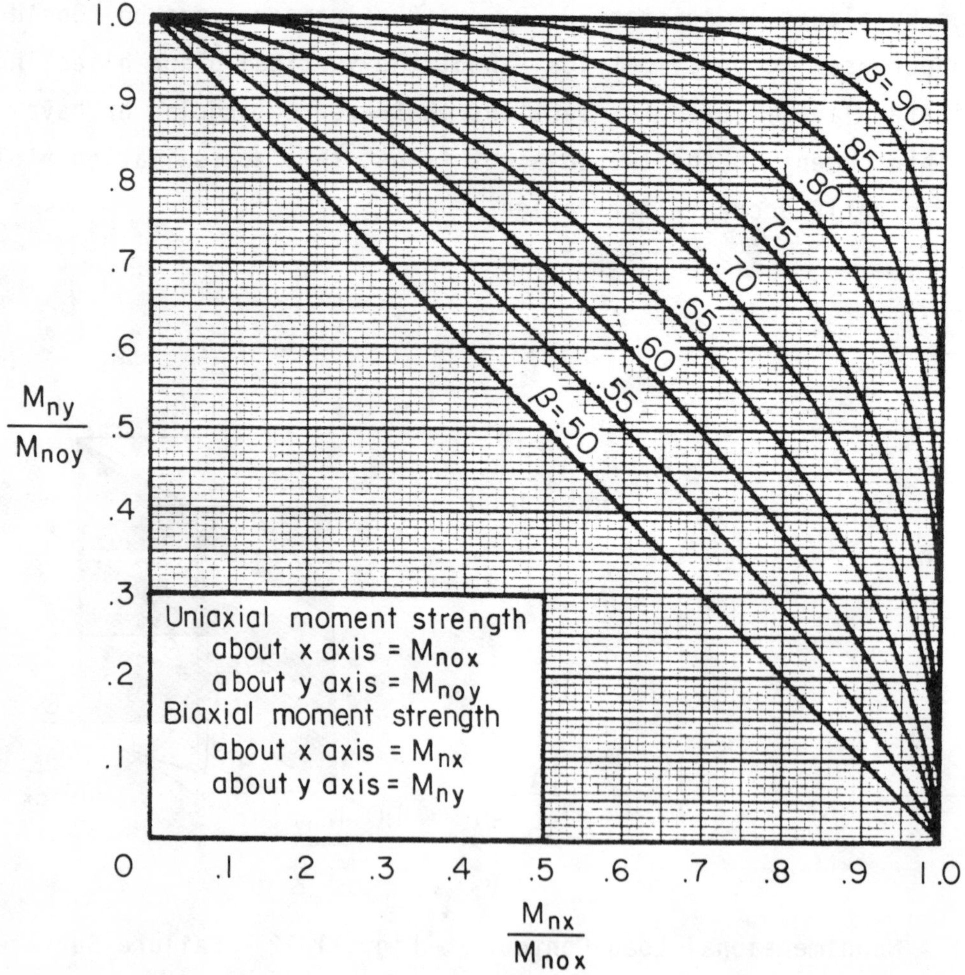

Fig. 11-13 - Biaxial Moment Strength Relationship

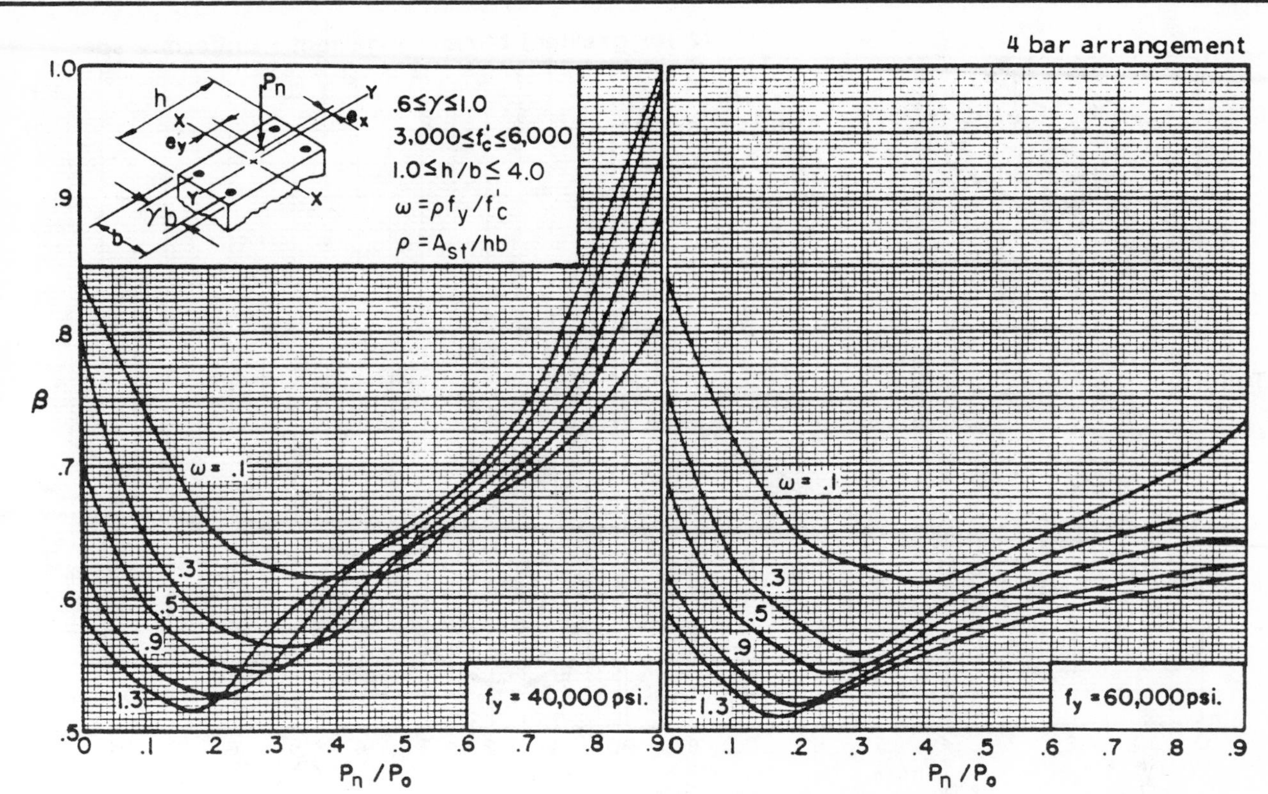

Fig. 11-14 - Biaxial Bending Design Constants

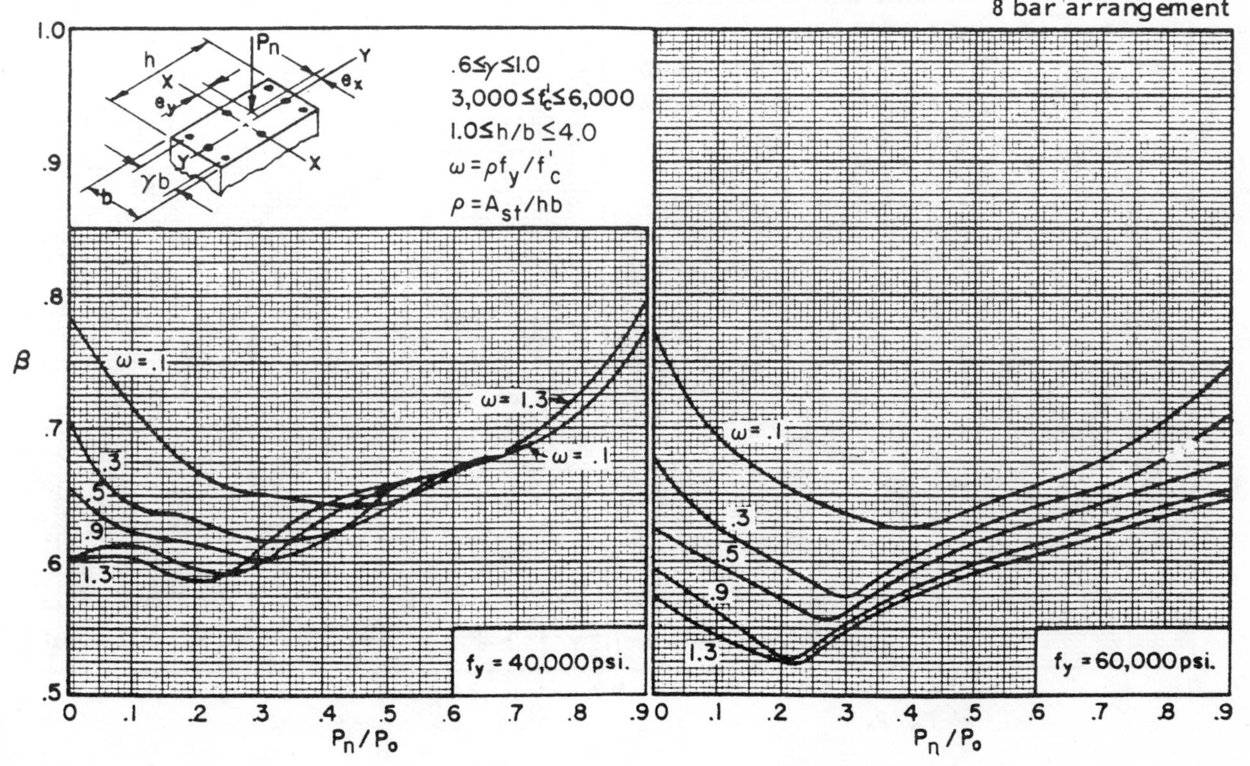

Fig. 11-15 - Biaxial Bending Design Constants

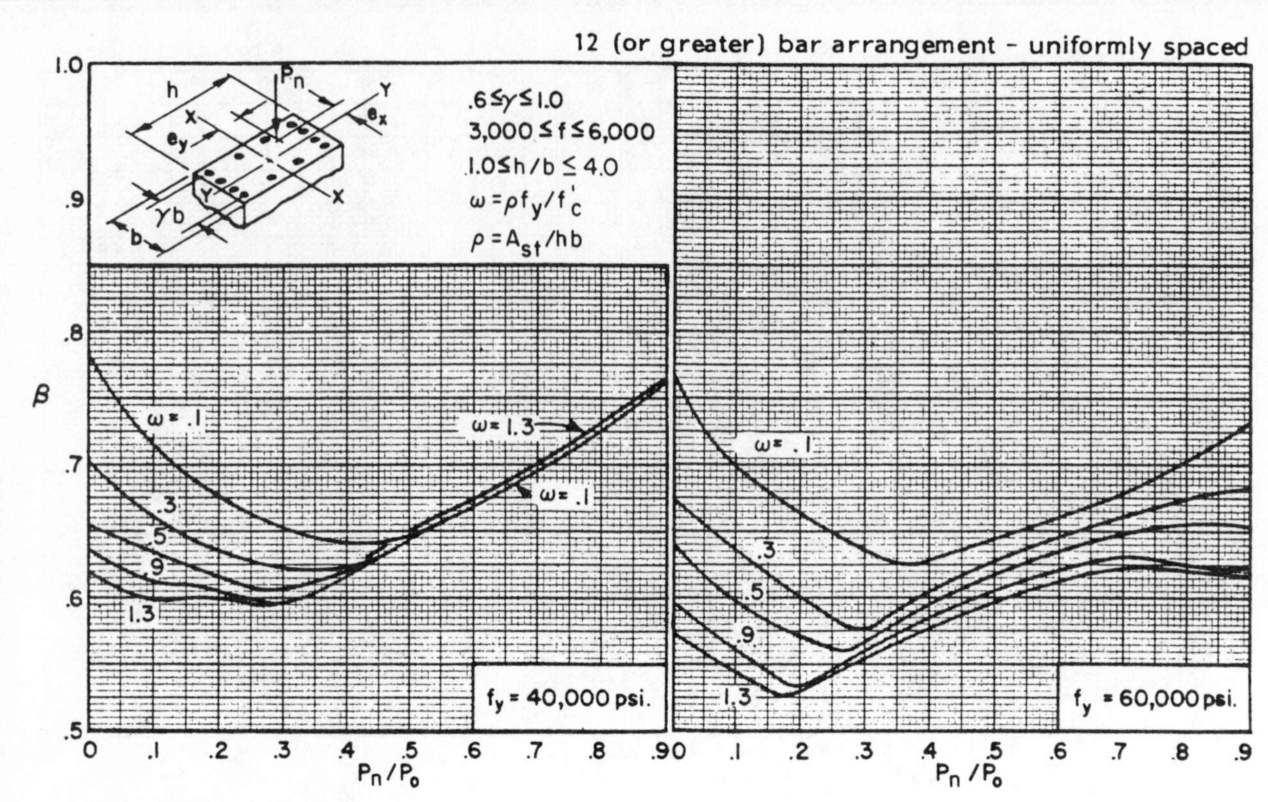

Fig. 11-16 - Biaxial Bending Design Constants

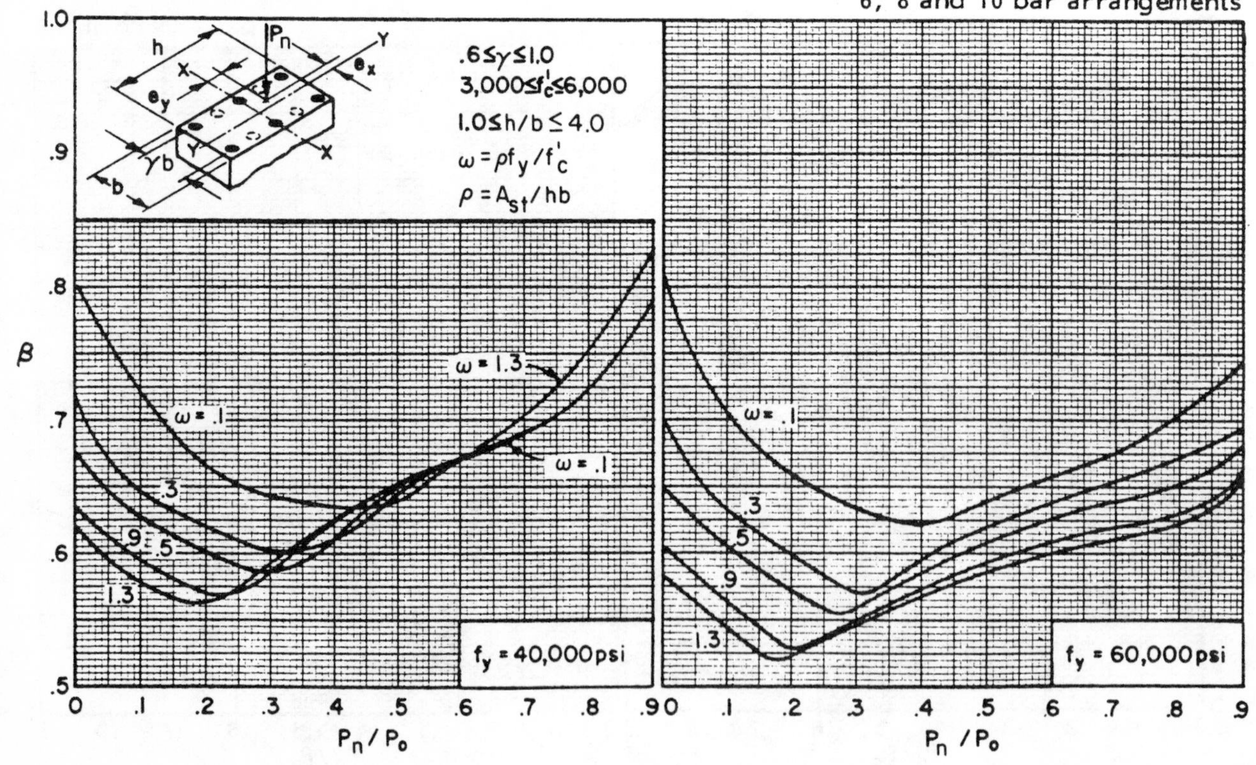

Fig. 11-17 - Biaxial Bending Design Constants

For design convenience, a plot of the curves generated by nine values of β are given in Fig. 11-13. Note that when $\beta = 0.5$, its lower limit, Eq. (8) describes a straight line joining the points at which the relative moments equal 1.0 at the coordinate planes. When $\beta = 1.0$, its upper limit, Eq. (8) describes two lines, each of which is parallel to one of the coordinate planes.

Data for β were prepared on the basis of Section 10.2 of the Code, utilizing a rectangular stress block and basic principles of equilibrium. It was found that the parameters γ, b/h, and f'_c had minor effects. The maximum difference in β amounted to about 5% for a given value of P_n/P_o ranging from 0.1 to 0.9. The bulk of the values, especially those in the most frequently used range of P_n/P_o, did not differ by more than 3%. In view of these small differences, only envelopes of the lowest β values were developed for two values of f_y and different bar arrangements. These are shown in Figs. 11-14 to 11-17.

As can be seen from an inspection of these four figures, β is dependent primarily on the ratio P_n/P_o and to a lesser, though still significant extent, on the bar arrangement, the reinforcement index ω and the strength of the reinforcement.

Fig. 11-13, in combination with Figs. 11-14 to 11-17, furnish a convenient and direct means of determining the biaxial moment strength of a given cross section subject to an axial load, since the values of P_o, M_{nox} and M_{noy} can be readily obtained by methods described here and in Part 10.

While analysis of a particular section has been simplified, the determination of a section which will satisfy the strength requirements imposed by a load eccentric about both axes can only be achieved by successive analyses of assumed sections. Rapid and easy convergence to a satisfactory section can be achieved by approximating the curves in Fig. 11-13 by two straight lines intersecting at the 45 degree line as shown in Fig. 11-18.

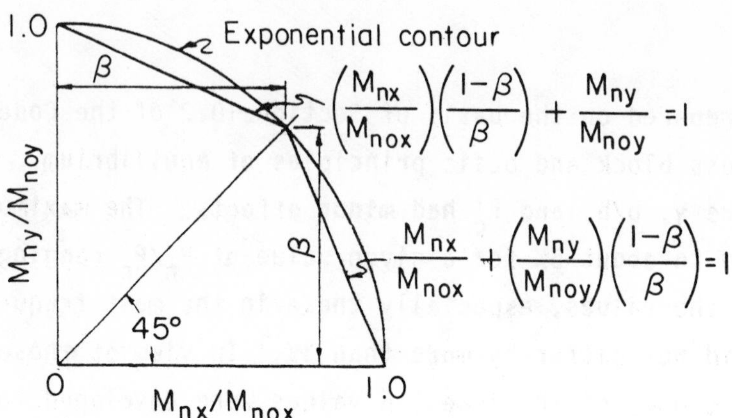

Fig. 11-18 - Bilinear Approximation of Nondimensionalized Load Contour

By simple geometry, it can be shown that the equation of the lines is:

when $M_{ny}/M_{nx} > M_{noy}/M_{nox}$

$$\frac{M_{ny}}{M_{noy}} + \frac{M_{nx}}{M_{nox}} \frac{(1-\beta)}{\beta} = 1 \qquad \text{Eq. (9)}$$

which can be restated for design convenience

$$M_{ny} + M_{nx} \frac{M_{noy}}{M_{nox}} \frac{(1-\beta)}{\beta} = M_{noy} \qquad \text{Eq. (10)}$$

For rectangular sections with reinforcement equally distributed on all faces, Eq. (10) can be approximated by:

$$M_{ny} + M_{nx} \frac{b}{h} \frac{(1-\beta)}{\beta} \simeq M_{noy} \qquad \text{Eq. (11)}$$

and similarly, when $M_{ny}/M_{nx} < M_{noy}/M_{nox}$

$$\frac{M_{ny}}{M_{noy}} \frac{(1 - \beta)}{\beta} + \frac{M_{nx}}{M_{nox}} = 1 \qquad \text{Eq. (12)}$$

$$M_{ny} \frac{M_{nox}}{M_{noy}} \frac{(1 - \beta)}{\beta} + M_{nx} = M_{nox} \qquad \text{Eq. (13)}$$

and

$$M_{ny} \frac{h}{b} \frac{(1 - \beta)}{\beta} + M_{nx} \simeq M_{nox} \qquad \text{Eq. (14)}$$

In design Eqs. (11) and (14), the ratio b/h or h/b must be chosen and the value of β must be assumed. For lightly loaded columns, β will generally vary from 0.55 to about 0.70. Hence, a value of 0.65 for β is generally a good initial choice in a biaxial bending analysis.

Manual Design Procedure

To aid the engineer in designing for biaxial bending, a procedure, in outline form, for manual design is presented below:

1. Choose the value of β at 0.65 or use Figs. 11-14 to 11-17 to make an estimate.

2. If M_{ny}/M_{nx} is greater than b/h, use Eq. (11) to calculate an approximate equivalent uniaxial moment strength M_{noy}. If M_{ny}/M_{nx} is less than b/h, use Eq. (14) to calculate an approximate equivalent uniaxial moment strength M_{nox}.

3. Design the section using any of the methods presented in Part 10 (i.e., uniaxial bending with axial load) to provide an axial load strength P_n and equivalent uniaxial moment strength M_{noy} or M_{nox}.

4. Verify the section chosen by any one of the following three methods:

A. **Bresler Reciprocal Load Method:**

$$P_n \leq \frac{1}{(1/P_{ox}) + (1/P_{oy}) - (1/P_o)} \qquad \text{Eq. (1)}$$

B. **Bresler Load Contour Method:**

$$\frac{M_{nx}}{M_{nox}} + \frac{M_{ny}}{M_{noy}} \leq 1.0 \qquad \text{Eq. (5)}$$

C. **PCA Load Contour Method:** Use Eq. (8) or,

$$\frac{M_{ny}}{M_{noy}} + \frac{M_{nx}}{M_{nox}} \frac{(1-\beta)}{\beta} \leq 1.0 \left[\text{for } \frac{M_{ny}}{M_{nx}} > \frac{M_{noy}}{M_{nox}}\right] \qquad \text{Eq. (9)}$$

$$\frac{M_{ny}}{M_{noy}} \frac{(1-\beta)}{\beta} + \frac{M_{nx}}{M_{nox}} \leq 1.0 \left[\text{for } \frac{M_{ny}}{M_{nx}} < \frac{M_{noy}}{M_{nox}}\right] \qquad \text{Eq. (12)}$$

Selected References

11.1 General Reference: Wang, C. K., and Salmon, C. G., Reinforced Concrete Design, Third Edition, Harper & Row Publishers, New York, 1979, Section 13.23, Biaxial Bending and Compression.

11.2 Pannell, F. N., "The Design of Biaxially Loaded Columns by Ultimate Load Methods," Magazine of Concrete Research, London, July 1960, pp. 103-104.

11.3 Pannell, F. N., "Failure Surfaces for Members in Compression and Biaxial Bending," ACI Journal, Proceedings Vol. 60, January 1963, pp. 129-140.

11.4 Bresler, Boris, "Design Criteria for Reinforced Columns under Axial Load and Biaxial Bending," ACI Journal, Proceedings Vol. 57, November 1960, pp. 481-490, discussion pp. 1621-1638.

11.5 Furlong, Richard W., "Ultimate Strength of Square Columns under Biaxially Eccentric Loads," _ACI Journal_, _Proceedings_ Vol. 57, March 1961, pp. 1129-1140.

11.6 Meek, J. L., "Ultimate Strength of Columns with Biaxially Eccentric Loads," _ACI Journal_, _Proceedings_ Vol. 60, August 1963, pp. 1053-1064.

11.7 Aas-Jakobsen, A., "Biaxial Eccentricities in Ultimate Load Design," _ACI Journal_, _Proceedings_ Vol. 61, March 1964, pp. 293-315.

11.8 Ramamurthy, L. N., "Investigation of the Ultimate Strength of Square and Rectangular Columns under Biaxially Eccentric Loads," Symposium on Reinforced Concrete Columns, American Concrete Institute, Detroit, 1966, pp. 263-298.

11.9 _Capacity of Reinforced Rectangular Columns Subject to Biaxial Bending_, EB011D, PCA, 1966.

11.10 _Biaxial and Uniaxial Capacity of Rectangular Columns_, EB031D, PCA, 1967.

EXAMPLE 11.1 - Design of a Square Column for Biaxial Loading

Determine the required square column size and reinforcing for the factored load and moments given. Assume the reinforcement is equally distributed in all faces.

P_u = 1200 kips
M_{ux} = 1800 ft-kips
M_{uy} = 750 ft-kips
f_y = 60 ksi
f'_c = 4000 psi

Calculations and Discussion	Code Reference

1. Determine required nominal load-moment strengths. For tied reinforced column $\varphi = 0.70$

$$P_n = P_u/\varphi = 1200/0.7 = 1714^k$$

$$M_{nx} = M_{ux}/\varphi = 1800/0.7 = 2571'^k$$

$$M_{ny} = M_{uy}/\varphi = 750/0.7 = 1071'^k$$

2. Assume $\beta = 0.65$

3. Determine an equivalent uniaxial moment strength M_{nox} or M_{noy}.

$$\frac{M_{ny}}{M_{nx}} = \frac{1071}{2571} = 0.42 \text{ is less than } b/h = 1.0 \text{ (square column)}$$

Therefore, using Eq. (14)

$$M_{nox} \simeq M_{ny} \frac{h}{b} \frac{(1-\beta)}{\beta} + M_{nx}$$ Eq. (14)

$$= 1071 \; (1.0) \; \frac{(1-0.65)}{0.65} + 2571$$

$$= 3148'^k$$

EXAMPLE 11.1 - Continued

Calculations and Discussion	Code Reference

4. Assuming a 36-in.-square column, determine reinforcement required toprovide an axial load strength P_n = 1714 kips and an equivalent uniaxial moment strength M_{nox} = 3148 ft. kips.

   Using Design Aid SP17A (See Part 10, Page 10-6)

   For square section with bars uniformly distributed on all sides, f'_c = 4,000 psi, f_y = 60,000 psi, and $\gamma \simeq 0.9$, use column chart 7.4.4 for R4-60.90 columns (See Fig. 11-19). Note that the strength reduction factor φ is already included in the load-moment strength curves.

   Compute $\dfrac{\varphi P_n}{A_g} = \dfrac{0.7\ (1714)}{1296} = 0.93$

   Compute $\dfrac{\varphi M_{nox}}{A_g h} = \dfrac{0.7\ (3148)\ 12}{(1296)\ 36} = 0.57$

   From Fig. 11-19, read $\rho_g = 0.022$

   $A_{st} = \rho_g A_g = 0.022\ (1296) = 28.51\ \text{in.}^2$

   Select 24 #10 bars ($A_s = 30.40\ \text{in.}^2$)

   ρ_g (actual) = 0.023

5. Selected section will now be checked for biaxial loading strength by each of the three methods presented in the discussion.

   A. <u>Bresler Reciprocal Load Method</u>

   Check $P_n \geqq 0.1\ f'_c\ A_g$ — Eq. (2)

   $1714^k > 0.1\ (4)\ (1296) = 518^k$

EXAMPLE 11.1 - Continued

Calculations and Discussion	Code Reference

To employ this method, P_o, P_{ox}, and P_{oy} must be determined.

$$P_o = 0.85 f'_c (A_g - A_{st}) + A_{st} f_y$$

$$= 0.85 (4) (1296 - 30.4) + 30.4 (60)$$

$$= 6127^k$$

Knowing ρ_g (actual) and required nominal load-moment strengths (P_n, M_{nx} and M_{ny}), P_{ox} and P_{oy} can be determined.

<u>x-axis</u>

$$P_n = 1714^k, M_{nx} = 2571'^k$$

$$e_y = \frac{M_{nx}}{P_n} = \frac{2571 (12)}{1714} = 18 \text{ in.}$$

$$\frac{e_y}{h} = \frac{18}{36} = 0.5$$

From Fig. 11-19, with $e_y/h = 0.5$ and $\rho_g = 0.023$,

read $\frac{\varphi P_{ox}}{A_g} = 1.16$

Therefore, $P_{ox} = \frac{1.16 (1296)}{0.70} = 2148^k$

<u>y-axis</u>

$$P_n = 1714^k, M_{ny} = 1071'^k$$

$$e_x = \frac{M_{ny}}{P_n} = \frac{1071 (12)}{1714} = 7.5 \text{ in.}$$

$$\frac{e_x}{h} = \frac{7.5}{36} = 0.21$$

From Fig. 11-19, with $e_x/h = 0.21$ and $\rho_g = 0.023$

EXAMPLE 11.1 - Continued

Calculations and Discussion	Code Reference

read $\dfrac{\varphi P_{oy}}{A_g} = 2.05$

Therefore, $P_{oy} = \dfrac{2.05\ (1296)}{0.70} = 3795^k$

Using the above values, Eq. (1) can now be evaluated.

$$P_n \leq \frac{1}{(1/P_{ox}) + (1/P_{oy}) + (1/P_o)} \qquad \text{Eq. (1)}$$

$$\leq \frac{1}{(1/2148) + (1/3795) - (1/6127)} = 1767^k$$

Required $P_n = 1714^k < 1767^k$ (OK)

B. Bresler Load Contour Method

Due to lack of available data, a conservative α value of 1.0 is chosen; i.e., Eq. (5). Although $P_u > 0.1 f'_c A_g$, the necessary calculations will be carried out for example purposes. Since the section is symmetrical, M_{nox} is equal to M_{noy}. Knowing ρ_g (actual) and required axial load strength P_n, M_{nox} is determined as the nominal moment strength with bending considered in the direction of the x axis only.

$$P_n = 1714^k, \quad \frac{\varphi P_n}{A_g} = \frac{0.7\ (1714)}{1296} = 0.93$$

From Fig. 11-19, with $\varphi P_n/A_g = 0.93$ and $\rho_g = 0.023$,

read $\dfrac{\varphi M_{nox}}{A_g h} = 0.59$

Therefore, $M_{nox} = \dfrac{0.59\ (1296)\ 36/12}{0.70} = 3277^{'k}$

Using the above value, Eq. (5) can now be evaluated.

$$\frac{M_{nx}}{M_{nox}} + \frac{M_{ny}}{M_{noy}} \leq 1.0 \qquad \text{Eq. (5)}$$

EXAMPLE 11.1 - Continued

Calculations and Discussion	Code Reference

$$\frac{2571}{3277} + \frac{1071}{3277} = 0.785 + 0.327 = 1.112 > 1.0 \quad \text{(NG)}$$

Due to the inherent conservation when using $\alpha = 1.0$, the section is inadequate when checked by this method.

C. PCA Load Contour Method

To employ this method, P_o, M_{nox}, M_{noy} and the true value of β must first be found.

$$P_o = 0.85\, f'_c\, (A_g - A_{st}) + A_{st}\, f_y$$

$$= 0.85\,(4)\,(1296 - 30.4) + 30.4\,(60) = 6127^k$$

Since the section is symmetrical, M_{nox} and M_{noy} are equal. Knowing ρ_g (actual) and required axial load strength P_n, M_{nox} is determined as follows:

$$P_n = 1714^k, \quad \frac{\varphi P_n}{A_g} = \frac{0.7\,(1714)}{1296} = 0.93$$

From Fig. 11-19, with $P_n/A_g = 0.93$ and $\rho_g = 0.023$

read $\frac{\varphi M_{nox}}{A_g h} = 0.59$

Therefore, $M_{nox} = \frac{0.59\,(1296)\,36/12}{0.70} = 3277'^k$

Having found P_o and using ρ_g (actual), the true β value is determined as follows:

$$P_n/P_o = 1714/6127 = 0.28$$

$$\omega = \rho_g f_y/f'_c = 0.023\,(60)/4 = 0.345$$

From Fig. 11-16, read $\beta = 0.573$

Calculations and Discussion	Code Reference

Using the above values, Eq. (8) can now be evaluated.

$$\left(\frac{M_{nx}}{M_{nox}}\right)^{\log 0.5/\log\beta} + \left(\frac{M_{ny}}{M_{noy}}\right)^{\log 0.5/\log\beta} \leq 1.0 \qquad \text{Eq. (8)}$$

$$\log 0.5 = -0.3$$

$$\log \beta = \log 0.573 = -0.242$$

$$\log 0.5/\log\beta = 1.24$$

$$\left(\frac{2571}{3277}\right)^{1.24} + \left(\frac{1071}{3277}\right)^{1.24} = 0.740 + 0.250$$

$$= 0.990 < 1.0 \qquad \text{(OK)}$$

The section can also be checked using the bilinear approximation. Since $M_{ny}/M_{nx} < M_{noy}/M_{nox}$, Eq. (12) should be used.

$$\frac{M_{ny}}{M_{noy}} \frac{(1 - \beta)}{\beta} + \frac{M_{nx}}{M_{nox}} \leq 1.0 \qquad \text{Eq. (12)}$$

$$\frac{1071}{3277} \frac{(1 - 0.573)}{0.573} + \frac{2571}{3277} = 0.244 + 0.785 = 1.029 \simeq 1.0 \qquad \text{(OK)}$$

COLUMNS 7.4.4—Load-moment strength interaction diagram for R4-60.90 columns

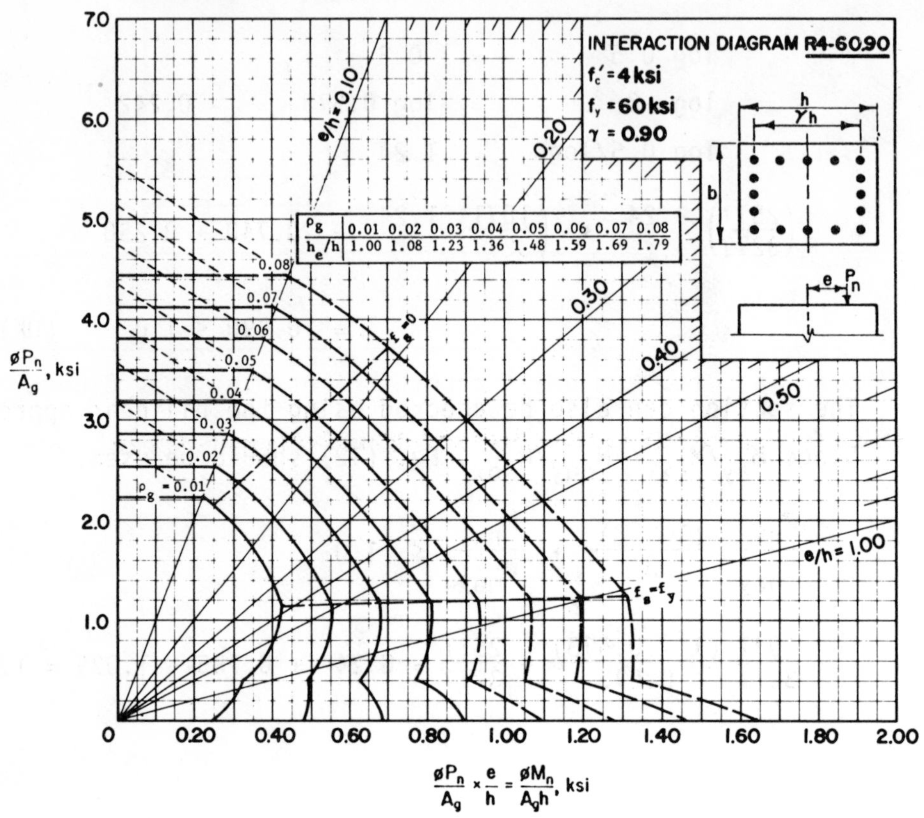

Fig. 11-19 - Reproduced from page 77 of SP17A.
Courtesy of American Concrete Institute

EXAMPLE 11.2 - Design of A Rectangular Column for Biaxial Loading

Determine required rectangular column size and reinforcement for the load and moments given. Assume $\frac{h}{b} = 1.5$, and that the reinforcement is equally distributed on all faces.

$P_u = 1700$ kips

$M_{ux} = 2900$ ft-kips

$M_{uy} = 1200$ ft-kips

$f_y = 60$ ksi

$f'_c = 4000$ psi

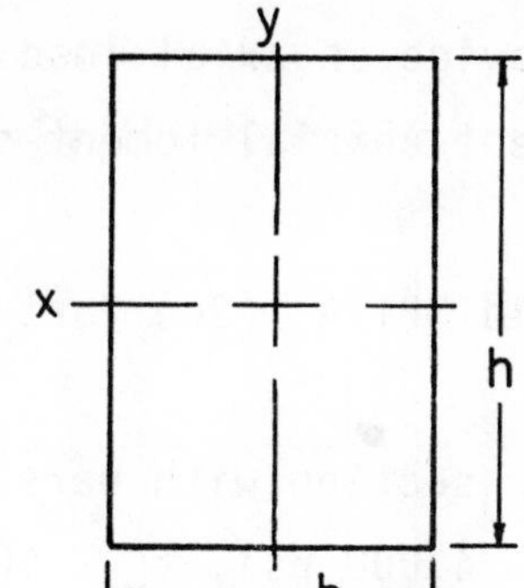

h/b = 1.5

Calculations and Discussion	Code Reference

1. Determine required nominal strengths. For tied reinforced column $\varphi = 0.70$

$$P_n = P_u/\varphi = 1700/0.7 = 2429^k$$

$$M_{nx} = M_{ux}/\varphi = 2900/0.7 = 4143^{'k}$$

$$M_{ny} = M_{uy}/\varphi = 1200/0.7 = 1714^{'k}$$

2. Assume $\beta = 0.65$

3. Determine an equivalent uniaxial moment strength M_{nox} or M_{noy}.

$$\frac{M_{ny}}{M_{nx}} = \frac{1714}{4143} = 0.41 \text{ is less than } b/h = 0.67$$

Therefore, using Eq. (14)

$$M_{nox} \simeq M_{ny} \frac{h}{b} \frac{(1 - \beta)}{\beta} + M_{nx}$$ Eq. (14)

$$= 1714\,(1.5)\,\frac{(1 - 0.65)}{0.65} + 4143$$

$$= 5527^{'k}$$

EXAMPLE 11.2 - Continued

Calculations and Discussion	Code Reference

4. Assuming b = 32 in. and h = 48 in., determine reinforcement required to provide an axial load strength P_n = 2429 kips and an equivalent unaxial moment strength M_{nox} = 5527 ft. kips.

Using Design Aid SP17A (See Part 10, page 10-6)

For rectangular section with bars uniformly distributed on all sides, f'_c = 4,000 psi, f_y = 60,000 psi, and $\gamma \simeq 0.90$, use column chart 7.4.4 for R4 - 60.90 columns (See Fig. 11-19). Note that the strength reduction factor φ is already included in the strength curves.

Compute $\dfrac{\varphi P_n}{A_g} = \dfrac{0.7\ (2429)}{32 \times 48} = 1.11$

Compute $\dfrac{\varphi M_{nox}}{A_g h} = \dfrac{0.7\ (5527)12}{(32 \times 48)\ 48} = 0.63$

From Fig. 11-19, read $\rho_g = 0.027$

$A_{st} = \rho_g A_g = 0.027\ (32 \times 48) = 41.47\ \text{in.}^2$

<u>Select 30 #11 bars</u> (A_s = 46.8 in.^{2})

ρ_g (actual) = 46.8/32 x 48 = 0.030

5. Selected section will now be checked for biaxial loading strength by the first and third methods previously described. The Bresler load contour method will not be used due to the conservatism involved in the procedure, as noted in Example 11.1.

A. Bresler Reciprocal Load Method

Check $P_n \geq 0.1\ f'_c\ A_g$ — Eq. (2)

$2429^k > 0.1\ (4)\ (32 \times 48) = 614^k$

Calculations and Discussion	Code Reference

To employ this method, P_o, P_{ox}, and P_{oy} must be determined.

$$P_o = 0.85\, f'_c\, (A_g - A_{st}) + A_{st}\, f_y$$

$$= 0.85\,(4)\,(1536 - 46.8) + 46.8\,(60)$$

$$= 7871^k$$

Knowing ρ_g and required nominal load-moment strengths (P_n, M_{nx}, and M_{ny}), P_{ox} and P_{oy} can be determined.

<u>x-axis</u>

$$P_n = 2429^k, \quad M_{nx} = 4143^{'k}$$

$$e_y = \frac{M_{nx}}{P_n} = \frac{4143\,(12)}{2429} = 20.47 \text{ in.}$$

$$\frac{e_y}{h} = \frac{20.47}{48} = 0.43$$

From Fig. 11-19, with $e_y/h = 0.43$ and $\rho_g = 0.030$,

read $\dfrac{\varphi P_{ox}}{A_g} = 1.50$

Therefore, $P_{ox} = \dfrac{1.50\,(32 \times 48)}{0.70} = 3291^k$

<u>y-axis</u>

$$P_n = 2429^k, \quad M_{ny} = 1714^{'k}$$

$$e_x = \frac{M_{ny}}{P_n} = \frac{1714\,(12)}{2429} = 8.47 \text{ in.}$$

$$\frac{e_x}{b} = \frac{8.47}{32} = 0.26$$

From Fig. 11-19, with $e_x/b = 0.26$ and $\rho_g = 0.030$,

read $\dfrac{\varphi P_{oy}}{A_g} = 2.05$

EXAMPLE 11.2 - Continued

Calculations and Discussion	Code Reference

Therefore, $P_{oy} = \dfrac{2.05\ (32 \times 48)}{0.70} = 4498^k$

Using the above values, Eq. (1) can now be evaluated.

$$P_n \le \frac{1}{(1/P_{ox}) + (1/P_{oy}) - (1/P_o)} \qquad \text{Eq. (1)}$$

$$\le \frac{1}{(1/3291) + (1/4498) - (1/7871)} = 2505^k$$

Required $P_n = 2429^k < 2505^k$ (OK)

B. PCA Load Contour Method

To employ this method, P_o, M_{nox}, M_{noy}, and the true value of β must be found.

$$P_o = 0.85\ f'_c\ (A_g - A_{st}) + A_{st}\ f_y$$

$$= 0.85\ (4)\ (1536 - 46.8) + 46.8\ (60) = 7871^k$$

Knowing ρ_g and required axial load strength P_n, M_{nox} and M_{noy} are determined as follows:

$$P_n = 2429^k, \quad \frac{\varphi P_n}{A_g} = \frac{0.7\ (2429)}{32 \times 48} = 1.11$$

From Fig. 11-19, with $\varphi P_n/A_g = 1.11$ and $\rho_g = 0.030$

read $\dfrac{\varphi M_n}{A_g h} = 0.68$

Therefore, $M_{nox} = \dfrac{0.68\ A_g h}{\varphi} = \dfrac{0.68\ (32 \times 48)\ 48/12}{0.70} = 5968'^k$

$M_{noy} = \dfrac{0.68\ A_g b}{\varphi} = \dfrac{0.68\ (32 \times 48)\ 32/12}{0.70} = 3979'k$

EXAMPLE 11.2 - Continued

Calculations and Discussion	Code Reference

Having found P_o and using ρ_g, the true value of β is determined as follows:

$$P_n/P_o = 2429/7871 = 0.309$$

$$\omega = \rho_g f_y/f'_c = 0.030\ (60)/4 = 0.45$$

From Fig. 11-16, read $\beta = 0.57$

Using the above values, Eq. (8) can now be evaluated.

$$\left(\frac{M_{nx}}{M_{nox}}\right)^{\log 0.5/\log\beta} + \left(\frac{M_{ny}}{M_{noy}}\right)^{\log 0.5/\log\beta} \leq 1.0 \qquad \text{Eq. (8)}$$

$$\log 0.5 = -0.3$$

$$\log\ \beta = \log 0.57 = -0.244$$

$$\log 0.5/\log\beta = 1.23$$

$$\left(\frac{4143}{5968}\right)^{1.23} + \left(\frac{1714}{3979}\right)^{1.23} = 0.638 + 0.355$$

$$= 0.993 \quad 1.0 < \text{(OK)}$$

The section can also be checked using the bilinear approximation. Since $M_{ny}/M_{nx} < M_{noy}/M_{nox}$, Eq. (12) should be used.

$$\frac{M_{ny}}{M_{noy}}\ \frac{(1-\beta)}{\beta} + \frac{M_{nx}}{M_{nox}} \leq 1.0 \qquad \text{Eq. (12)}$$

$$\frac{1714}{3979}\ \frac{(1-0.57)}{0.57} + \frac{4143}{5968} = 0.325 + 0.694$$

$$= 1.019 \simeq 1.0 \quad \text{(OK)}$$

12

Slenderness Effects

Update for '83 Code

The moment magnification method for consideration of column slenderness effects was first introduced in the 1971 edition of the ACI Code. The design provisions remained essentially the same in the 1977 Code edition. With publication of ACI 318-83, application of the moment magnification procedure for unbraced frames has been significantly clarified by a new Eq. (10-6). For an unbraced frame, the new magnification equation expresses the column secondary moments separately as the sum of (1) the magnification due to non-sway moments (gravity load effects) plus (2) the magnification due to sway moments (lateral load effects). The differentiation between the nonsway and sway magnification effects is supported by frame tests. The new procedures will generally result in substantially lessened column design moments for unbraced frames, since gravity load moments will be magnified by a braced frame magnification factor. The new provisions will lead to economy in both columns and floor systems where unbraced frames are used. For an in-depth discussion addressing new Eq. (10-6) the reader is referred to Code Commentary Section 10.11.5.1.

10.10 Slenderness Effects in Compression Members

The basic design provisions for column slenderness consideration call for the use of improved structural analysis procedures wherever possible or practical. With greater use of computers, more accurate procedures become feasible. If such an analysis is not possible or practical, an approximate moment magnification procedure is permitted by Section 10.11.

10.11 Approximate Evaluation of Slenderness Effects

The approximate moment magnification procedure is similar to the method used for structural steel design. The moment magnifier δ is a function of the ratio of the axial load to the critical or buckling load of the column, the ratio of the moments at the ends of the column, and the deflected shape of the column.

The objective of column design is the selection of a cross section with reinforcement for a specified combination of factored axial load P_u and factored moment M_u. A column is said to be slender if its cross-section dimensions are small in comparison to its length. The degree of slenderness is expressed in terms of the slenderness ratio $k\ell_u/r$, where k is the effective length factor, which is dependent on end conditions of the column and bracing against sidesway, and r is the radius of gyration of the column cross-section. Con cepts of three ranges of slenderness ratios are given along with column design methods proposed for each range.

Slenderness limits are given in Section 10.11.4, below which secondary moments can be disregarded and only the axial load and primary moment used to select the column cross-section and reinforcement. More than 90 percent of the columns in braced frames and 40 percent in unbraced frames fall into this classification.

Within moderate slenderness limits, the approximate analysis based on a moment magnifier is suggested. Whenever the slenderness of a column exceeds moderate slenderness, a more rational second-order analysis is required (Section 10.10.1). No upper limits for slenderness are given. When high slenderness ratios are encountered, the analysis must take into account the influence of axial loads and variable moment of inertia on column stiffness and forces, and the effects of duration of the loads. Slenderness effects are considered for both braced and unbraced frames.

10.11.1-10.11.3 Unsupported and Effective Lengths

The unsupported length ℓ_u of a column is described in Section 10.11.1. This is to be taken as the clear distance between lateral support as shown in Fig. 12-1. It is also to be understood that the length ℓ_u may be different in each of the principal axes of the column cross-section. The radius of gyration may be taken as 0.3 of the overall dimension of a rectangular section and 0.25 of the diameter of a circular section, as shown in Fig. 12-2. For other shapes, the radius of gyration is to be computed.

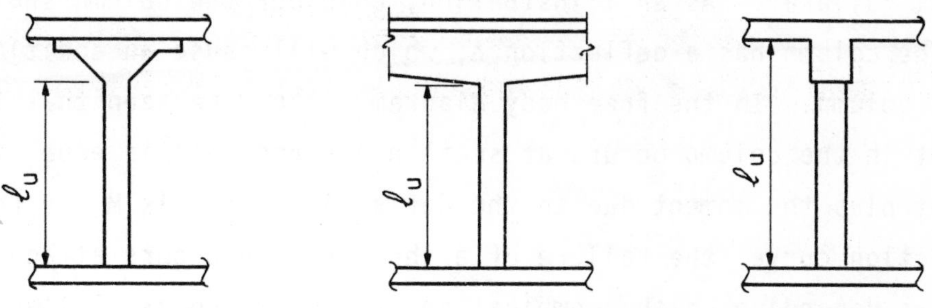

Fig. 12-1 - Unsupported Length (ℓ_u)

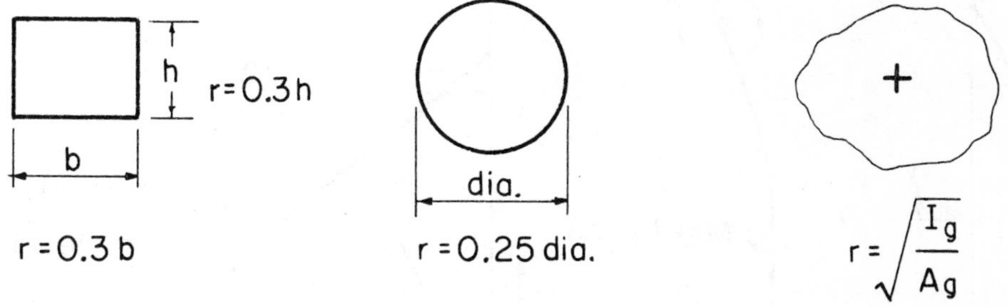

Fig. 12-2 - Radius of Gyration (r)

It must be understood that the term "short column" is used to denote a column that has a strength equal to or greater than that computed for the cross-section using the forces and moments obtained from an analysis for combined bending and axial load. A "slender column" is defined as a column whose strength is reduced by second order deformations. By these definitions, a column with a given slenderness ratio may be a short column under another combination of restraints. With the use of high-strength steels and concretes, and with more accurate design methods, it is possible, for a given axial load with or without simultaneous bending, to design a much smaller cross-section. This results in a more slender member, and because of this, reliable and rational design procedures for slender columns are more important.

A short column may fail due to a combination of moment and axial load that may exceed the strength of the cross-section. This type of failure is known as a "material failure". As an illustration, consider the column shown in Fig. 12-3. The column has a deflection Δ, which will cause an additional moment in the column. In the free body diagram, it can be seen that the maximum moment in the column occurs at section A-A and this is equal to the applied moment plus the moment due to the deflection, that is $M = P(e + \Delta)$. In the interaction curve, the failure of a short column occurs at any point along the curve depending on the combination of moment and axial load applied. As mentioned above, some deflections would occur and a "material failure" would result when the load P and $M = P(e + \Delta)$ combination

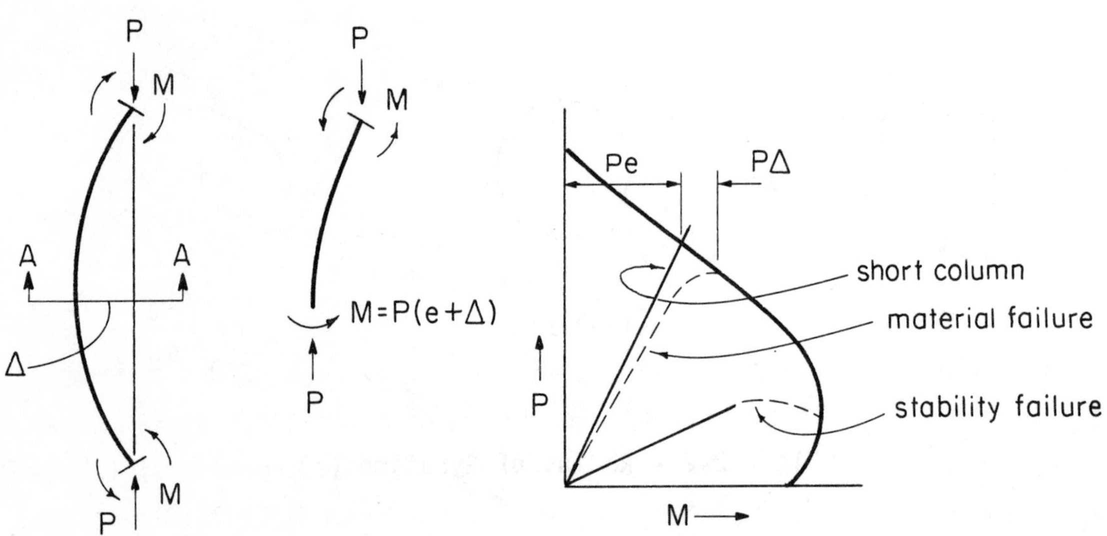

Fig. 12-3 - Interaction in Slender Columns

intersects the particular cross-section interaction curve. If the column is very slender, it may reach a deflection due to the axial force P and the moment Pe, such that deflections can increase indefinitely with small increases in load P. The change in moment occurs without any increase in load. This type of failure is known as a "stability failure" and may occur in a slender column.

The basic information on the behavior of straight, concentrically loaded slender columns was developed by Euler more than 200 years ago. It states that a member will fail by buckling at the critical load $P_c = \pi^2 EI/(\ell_e)^2$, where ℓ_e is the effective length $k\ell_u$. For a very stocky column, the value of the buckling load calculated from this equation exceeds the direct crushing strength. For more slender members (larger $k\ell_u/r$ values), the failure occurs by buckling, with the buckling load decreasing for greater slenderness (see Fig. 12-4).

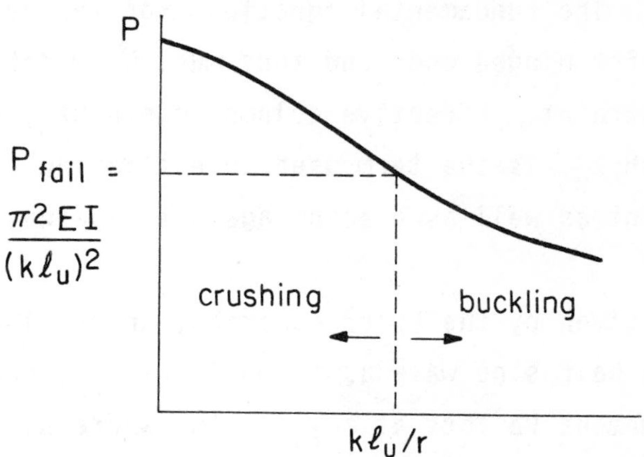

Fig. 12-4 - Column Curve

As shown above, it is possible to depict the slenderness effect and amplified moment on a typical interaction curve. Hence, a family of slender column interaction diagrams for members of varying slenderness ratios can be developed as shown in Fig. 12-5. The interaction diagram for $k\ell_u/r = 0$ is that which corresponds to the combination of moment and loads for a particular section with reinforcement as in a short column. The shape of the interaction curves for higher $k\ell_u/r$ values is dependent on the moments applied to the column.

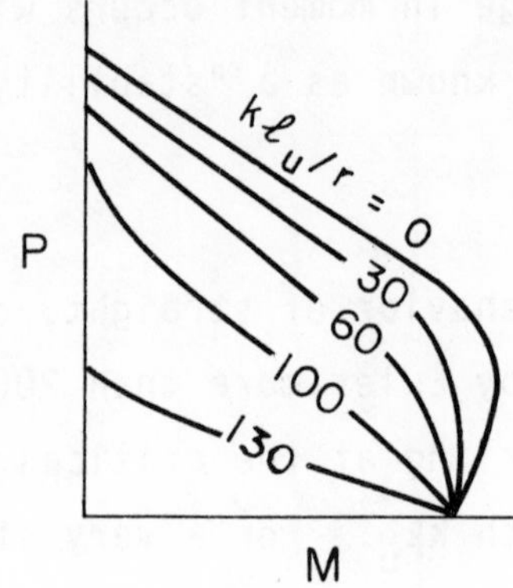

Fig. 12-5 - Slender Column Interaction Diagrams

In most cases, the designer can determine the effective length of a column by analyzing the moments acting on the column by means of a free body diagram, and experience. The fundamental equations for the design of slender columns were derived for hinged ends and thus must be modified to account for the effect of end restraint. Effective column length $k\ell_u$, as contrasted to actual unbraced length ℓ_u, is the term used in estimating column strength and considers end restraint as well as bracing against sidesway.

In the critical load given by the Euler equation, an originally straight member buckles into a half sine wave as shown in Fig. 12-6(a). In this configuration, bending moment $P\Delta$ acts at any section where Δ is the deflection at that point. This deflection continues to increase until the bending stress caused by the increasing moment, together with the original compression stress, exceeds the compressive strength and the member fails. The effective length ℓ_e $(k\ell_u)$ is between pinned ends, zero moments or inflection points, and in this case is equal to the unsupported length ℓ_u. If the member is fixed against rotation at both ends, as shown in Fig. 12-6(b), it will buckle in the shape shown. Inflection points will occur as shown and the effective length ℓ_e $(k\ell_u)$ will be one half of the unsupported length. When Euler's equation is applied to this column, the column will carry four times as much load as when ends are hinged. Rarely are columns in actual structures either hinged or fixed, rather they are partially restrained against rotation by abutting members and thus the effective length will occur between

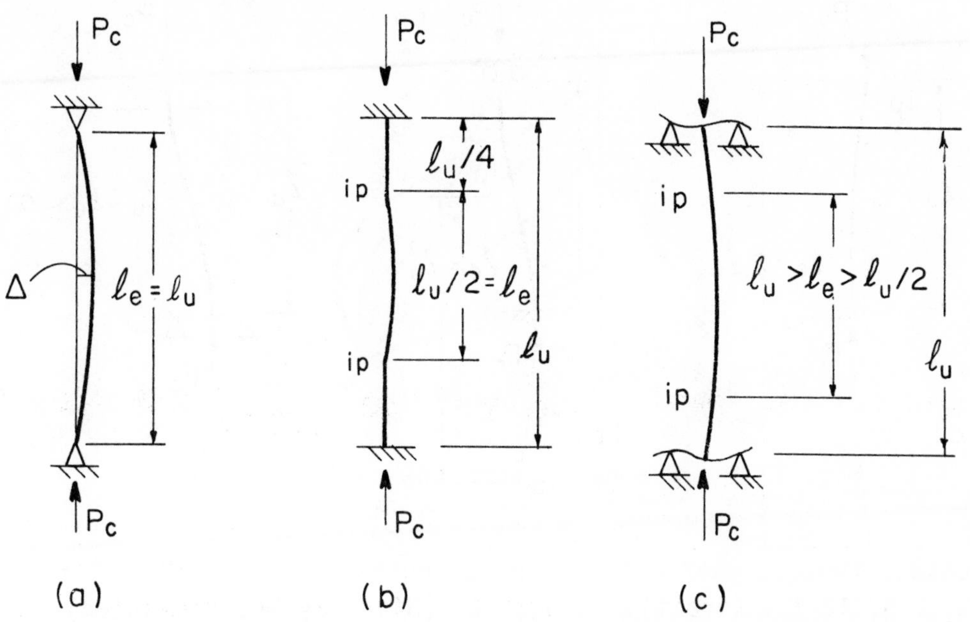

Fig. 12-6 - Effective Length ℓ_e (Sidesway Prevented)

$\ell_u/2$ and ℓ_u as shown in Fig. 12-6(c). The precise value will depend on the rigidity of the members abutting the column.

A column that is fixed at one end and entirely free at the other end would buckle as shown in Fig. 12-7(a). The upper end would move laterally in respect to the lower. This is known as sidesway. The inflection points would occur at the upper end of the member and thus would be similar to the upper end of the sine curve. The effective length would be twice the height. If the column is fixed against rotation at both ends but one end can move laterally, it will buckle as shown in Fig. 12-7(b). The effective length would be equal to the height with an inflection point occurring as shown. If the buckling load of the column in Fig. 12-7(b) were compared to that of the column in Fig. 12-6(b), which is braced against sidesway, it would be only one quarter that if sidesway is permitted. Again, rarely are the ends of columns either hinged or fixed, but rather they are partially restrained against rotation by abutting members and thus the effective length, where

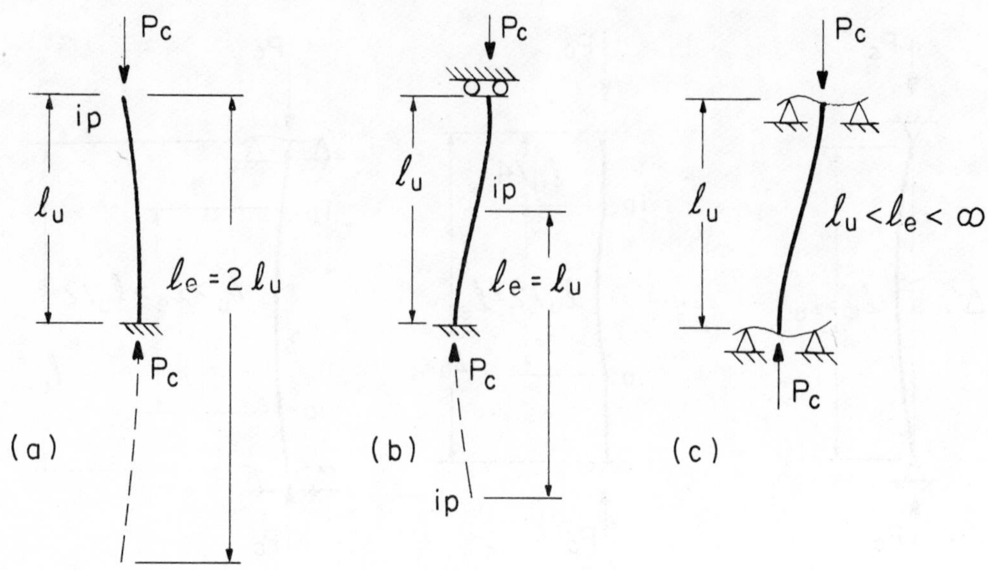

Fig. 12-7 - Effective Length ℓ_e (Sidesway Not Prevented)

sidesway is not prevented, will vary between ℓ_u and ∞, as shown in Fig. 12-7(c). If the beams are very rigid as compared to the column, the case in Fig. 12-7(b) is approached. If, on the other hand, the beams are fairly flexible, a hinged condition is approached at both ends and the structure would not be very stable.

In typical reinforced concrete structures, the designer rarely is concerned with single members, but rather with rigid frames of various types. The buckling behavior of a frame that is not braced against sidesway can be illustrated by a simple portal frame as shown in Fig. 12-8. Without any lateral restraint at the upper end, the entire (unbraced) frame is free to move sideways. The bottom end may be pin-ended or partially restrained against rotation as indicated. It can be seen that the effective length ℓ_e may exceed $2\ell_u$ and would be dependent on the degree of rotational restraint at each end.

In summary, the following comments can be made:

1. For columns braced against sidesway, the effective length falls between $\ell_u/2$ and ℓ_u, where ℓ_u is the actual unsupported length of the column.

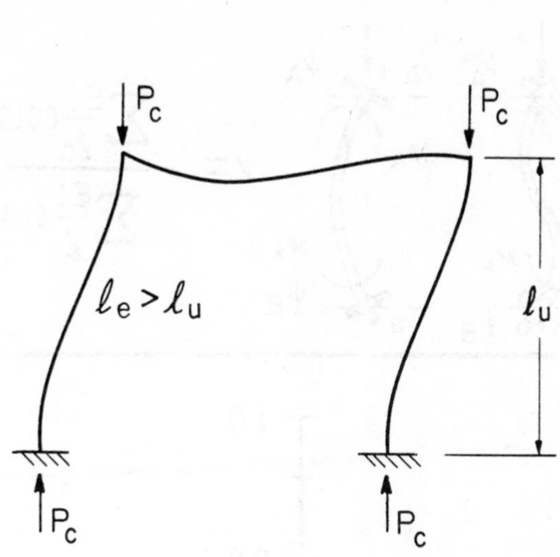

Fig. 12-8 - Rigid Frame (Sidesway Not Prevented)

2. For columns not braced against sidesway, the effective length is always longer than the actual length of the column ℓ_u and may be more like $2\ell_u$ and higher. A value of ℓ_e or $k\ell_u$ less than 1.2 for columns not braced against sidesway normally would not be realistic.

3. Use of the alignment charts shown in Figs. 12-9 and 12-10 allows graphical determination of the effective length factors for both braced and unbraced frames. If both ends have very little stiffness or approach $\psi = \infty$, then $k = 1.0$. If both ends have or approach full fixity, $\psi = 0$, then $k = 0.5$.

In determining the effective length factor k in Figs. 12-9 and 12-10, the rigidity (EI) may be calculated on the basis of the moment of inertia of the cracked transformed section and the rigidity of the columns on the basis of EI from Equation (10-10) with $\beta_d = 0$. Alternately, using a value of $0.5I_g$ for the beams (to account for the effect of cracking and reinforcement on relative stiffness) and I_g for the columns when computing ψ will usually result in reasonable member sizes for columns with $k\ell_u/r$ less than 60.

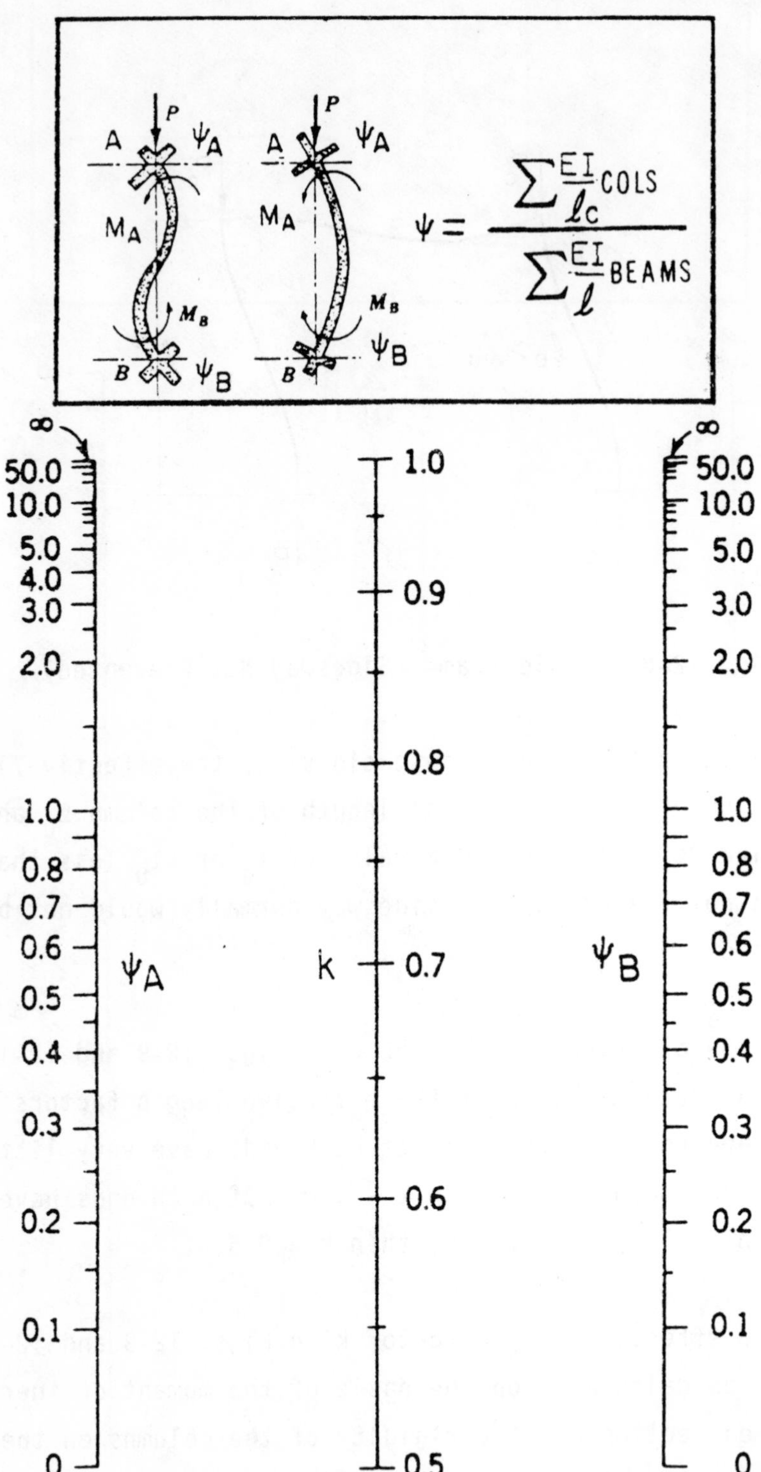

Fig. 12-9 - Effective Length Factors for Braced Columns

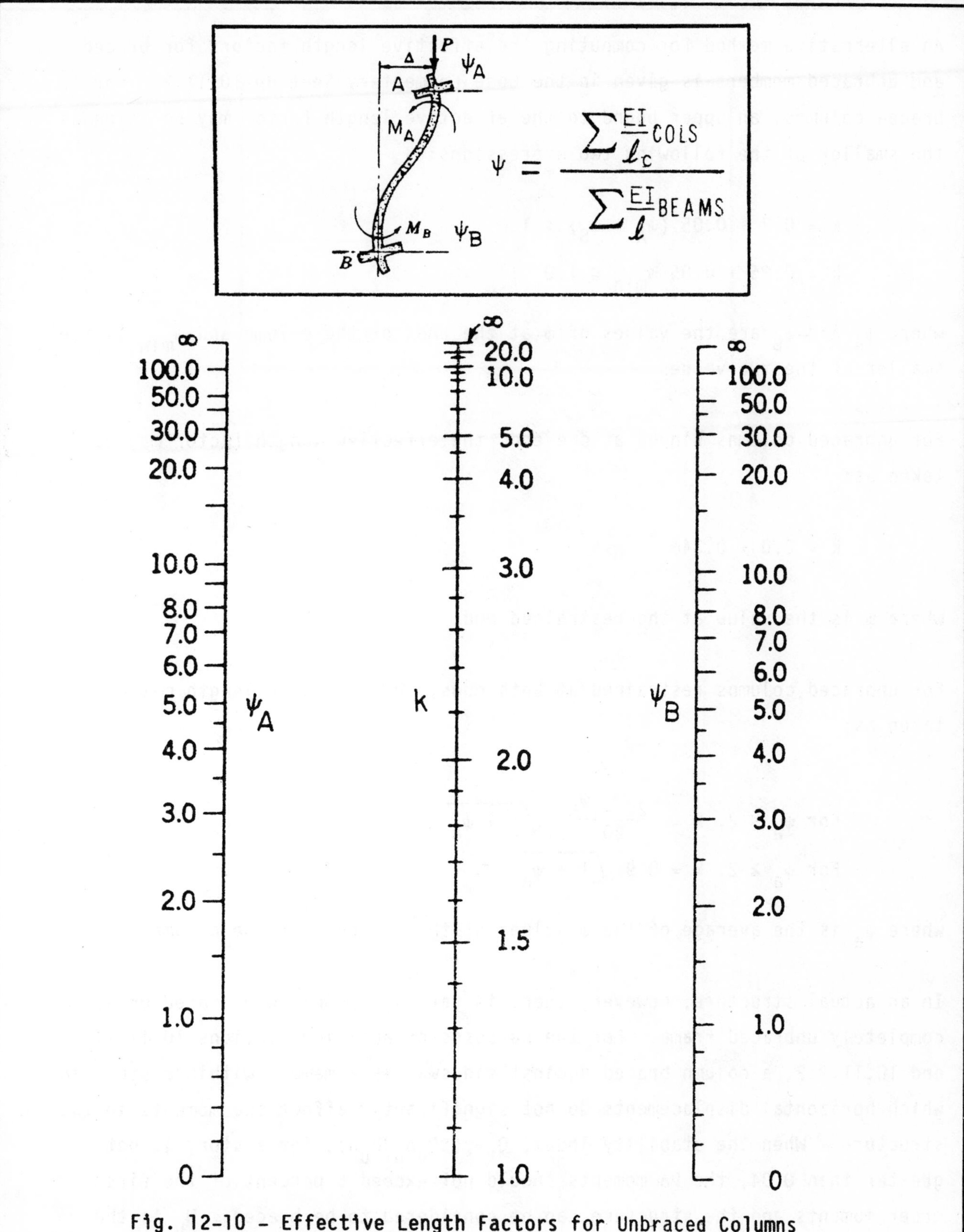

Fig. 12-10 - Effective Length Factors for Unbraced Columns

An alternative method for computing the effective length factors for braced and unbraced members is given in the Code Commentary Section 10.11.2. For braced columns, an upper bound to the effective length factor may be taken as the smaller of the following two expressions:

$$k = 0.7 + 0.05\ (\psi_A + \psi_B) \leq 1.0$$

$$k = 0.85 + 0.05\ \psi_{min} \leq 1.0$$

where ψ_A and ψ_B are the values of ψ at the ends of the column and ψ_{min} is the smaller of the two values.

For unbraced columns hinged at one end, the effective length factor may be taken as:

$$k = 2.0 + 0.34\psi$$

where ψ is the value at the restrained end.

For unbraced columns restrained at both ends, the effective length may be taken as:

$$\text{For } \psi_a < 2,\ k = \frac{20 - \psi_a}{20} \sqrt{1 + \psi_a}$$

$$\text{For } \psi_a \geq 2,\ k = 0.9 \sqrt{1 + \psi_a}$$

where ψ_a is the average of the ψ values at the two ends of the column.

In an actual structure, however, there is rarely a completely braced or a completely unbraced frame. For the purposes of applying Sections 10.11.2.1 and 10.11.2.2, a column braced against sidesway is a member within a story in which horizontal displacements do not significantly affect the moments in the structure. When the stability index, $Q = (\Sigma P_u \Delta_u / H_u h)$, for a story is not greater than 0.04, the $P\Delta$ moments should not exceed 5 percent of the first-order moments and the structure can be considered to be braced. H_u is the total design shear force acting within the story and Δ_u is the elastically-computed first-order lateral deflection due to H_u (neglecting $P\Delta$ effects) at the top of the story relative to the bottom of the story. In many cases, one will be able to tell by inspection whether a story is braced or unbraced.

Alternatively, a more approximate procedure can be used to determine if a story is braced or unbraced. A column may be assumed braced if located in a story in which the bracing elements (structural walls, or other lateral bracing elements) have a total stiffness, resisting lateral movement of the story, at least six (6) times the sum of the stiffnesses of all columns within the story. With this amount of lateral stiffness, lateral deflections of the story will not be large enough to affect the column strength significantly. What constitutes adequate bracing in a given case must be left to the judgment of the engineer, depending on the layout and arrangement of the structural framing members. A value of k less than 1.2 for columns not braced against sidesway normally would not be realistic.

In determining the effective length factor k, the stiffness of the beams may be calculated on the basis of the moment of inertia of the cracked transformed section and the stiffness of the columns by using the EI from Eq. (10-10), or conservatively from Eq. (10-11), with $\beta_d = 0$. For preliminary design, using $0.5I_g$ for the beams and I_g for the columns will usually result in a reasonable estimate of member sizes.

10.11.4 Consideration of Slenderness Effects

This section provides the lower and upper slenderness ratio limits for use with the moment magnification method. For columns braced against sidesway, the effects of slenderness may be neglected when $k\ell_u/r$ is less than 34-12 M_{1b}/M_{2b}. M_{2b} is the larger end moment on the braced column, obtained by elastic frame analysis and M_{1b} is the smaller end moment. The ratio M_{1b}/M_{2b} is positive if the column is in single curvature, negative if in double curvature. (M_{1b} and M_{2b} are factored end moments.) For columns not braced against sidesway, the effects of slenderness may be neglected when $k\ell_u/r$ is less than 22.

The upper limit for columns that may be designed by the approximate method is $k\ell_u/r$ equal to 100. When $k\ell_u/r$ is greater than 100, an analysis as defined in Section 10.10.1 must be used which takes into account the influence of axial loads and variable moment of inertia on member stiffness and fixed-end moments, effect of deflections on the moments and forces, and the effects of the duration of the loads.

The lower slenderness ratio limits will allow a larger number of columns to be excluded from slenderness consideration. Considering the slenderness ratio $k\ell_u/r$ in terms of ℓ_u/h for rectangular columns, the effects of slenderness may be neglected in design when ℓ_u/h is less than 10 for a column braced against sidesway with zero restraint at the ends. This lower limit increases to 18 for a braced column in double curvature with equal end moments and a ratio of column-to-beam stiffness equal to one at each end. In most braced frames, it is sufficiently accurate to evaluate the limit on slenderness by using estimated values of the effective length factor k. For columns hinged at both ends, k of 1.0 should be used. For stocky columns restrained by flat slab floors, k ranges from about 0.95 to 1.0 and conservatively can be estimated as 1.0 for a preliminary slenderness evaluation. For columns in beam-column frames, k ranges from about 0.75 to 0.9, and conservatively can be estimated as 0.90. If the initial computation of slenderness ratio based on estimated values of k indicates that effects of slenderness must be considered in the design, a more accurate value of k should be calculated and slenderness re-evaluated. For an unbraced column with a column-to-beam stiffness ratio equal to one at both ends, the effects of slenderness may be neglected when ℓ_u/h is less than 5. This value reduces to 3 if the beam stiffness is reduced to one-fifth of the column stiffness at each end of the column. Thus, beam stiffnesses at the top and bottom of a column of a high rise structure where the sidesway is not prevented become very important on the slenderness effects.

The upper limit on the slenderness ratio of $k\ell_u/r$ equal to 100 corresponds to an ℓ_u/h equal to 30 for a column braced against sidesway with zero restraint at the ends. This ℓ_u/h limit increases to 39 with a ratio of column-to-beam stiffness equal to one at each end.

10.11.5 Moment Magnification

The slender column approximate design equations of Section 10.11.5 are based on the concept of a moment magnifier δ which amplifies the column moments to account for the effect of axial load on the column moments. The column is then designed for the axial load and the magnified moment. In application, δ is a function of the ratio of the axial load on the column to the critical "buckling" load of the column (P_c), and the ratio of end moments and deflected shape of the column (C_m). Code provisions provide for different methods of calculating δ for the braced and unbraced conditions. For a braced condition, the magnifier δ is based on an effective length factor of 1.0 or less. For an unbraced condition, the magnifier δ is based on an effective length factor greater than 1.0. C_m factors for braced range from 0.4 to 1.0; for unbraced, a value of 1.0 is used. Thus, critical loads P_c are higher and magnification factors are smaller for the braced condition.

For ACI 318-83, application of the moment magnifier procedure for unbraced frames has been clarified by a new Eq. (10-6). For an unbraced frame, the new magnifier equation expresses the column secondary moments separately as the sum of (1) the magnification due to essentially nonsway moments (gravity load effects) plus (2) the magnification due to sway moments (lateral load effects):

$$M_c = \delta_b M_{2b} + \delta_s M_{2s} \qquad \text{Eq. (10-6)}$$

The first term is the magnified gravity load moments due to effects of member curvature only; where δ_b is a braced frame magnifier,

$$\delta_b = \frac{C_m}{1 - P_u/\varphi P_{cb}} \qquad \text{Eq. (10-7)}$$

$$P_{cb} = \frac{\pi^2 EI}{(k_b \ell_u)^2} \qquad \text{Eq. (10-9)}$$

and $C_m = 0.6 + 0.4\,(M_{1b}/M_{2b})$

where P_{cb} is the critical load computed for a braced condition using an effective length factor k_b of 1.0 or less according to Section 10.11.2.1.

The second term of Eq. (10-6) is the magnified lateral load moments due to effects of lateral drift; where δ_s is a sway frame magnifier,

$$\delta_s = \frac{1}{1 - \Sigma P_u / \varphi\, \Sigma P_{cs}} \qquad \text{Eq. (10-8)}$$

$$P_{cs} = \frac{\pi^2 EI}{(k_s \ell_u)^2} \qquad \text{Eq. (10-9)}$$

where P_{cs} is the critical load computed for an unbraced condition using an effective length factor k_s greater than 1.0 according to Section 10.11.2.2. ΣP_u and ΣP_{cs} are the summations for all columns within a story.

When Eq. (10-6) is applied for design of the columns of a moment resisting frame (beam or slab and column framing) that is not braced by structural walls or other bracing elements, both terms of Eq. (10-6) must be evaluated. M_{2b} is the larger end moment due to gravity load (dead and live load). M_{2s} is the moment on the column resulting from lateral load effects of wind or earthquake. Both moments, M_{2b} and M_{2s}, are computed using a conventional (first order) frame analysis. The member stiffnesses used in the analysis for the lateral load effects should allow, at least approximately, for cracking of the flexural members (beams and slabs).

When Eq. (10-6) is applied for design of the columns of a moment resisting frame that is effectively braced against sidesway by structural walls or other bracing elements, the δ_s term becomes 1.0 since the drift of a braced frame is small under lateral loads. When shear walls are tall and slender, or flexible, the δ_s term may need to be evaluated, as the moment resisting frame may not be effectively braced by the shearwalls. Designer's judgment combined with appropriate analysis are essential elements in determining the effectiveness of such "braced" framing systems.

To illustrate proper application of new Eq. (10-6) for slender column design, the following summary of design equations, and how they are applied in conjunction with Eq. (10-6), may be helpful. Both Design Examples 12.1 and 12.2 illustrate application of Eq. (10-6) for column design of an unbraced frame, and braced frame, respectively.

For the general case of an unbraced frame resisting gravity loads (dead and live load) plus wind loads:

(1) Factored load combinations to be considered:

(a) Gravity loads

$P_u = 1.4D + 1.7L$ Eq. (9-1)

$M_{1b} = 1.4D + 1.7L$ (smaller end moment)

$M_{2b} = 1.4D + 1.7L$ (larger end moment)

(b) Gravity plus wind loads

$P_u = 0.75(1.4D + 1.7L + 1.7W)$ Eq. (9-2)

$M_{2b} = 0.75(1.4D + 1.7L)$

$M_{2s} = 0.75(1.7W)$

or $P_u = 0.9D + 1.3W$ Eq. (9-3)

$M_{2b} = 0.9D$

$M_{2s} = 1.3W$

(2) For gravity load combination:

$M_c = \delta_b M_{2b}$ Eq. (10-6)

where $M_{2b} = 1.4D + 1.7L$

but not less than P_u (0.6 + 0.03h) 10.11.5.4

$$\delta_b = \frac{C_m}{1 - P_u/\varphi P_{cb}}$$ Eq. (10-7)

with $C_m = 0.6 + 0.4 (M_{1b}/M_{2b}) \leq 1.0$ Eq. (10-12)

and $$P_{cb} = \frac{\pi^2 EI}{(k_b \ell_u)^2}$$ Eq. (10-9)

k_b is evaluated as for a braced condition according to Section 10.11.2.1.

(3) For gravity plus wind load combination:

$M_c = \delta_b M_{2b} + \delta_s M_{2s}$ Eq. (10-6)

where $M_{2b} = 0.75(1.4D + 1.7L)$

$$\delta_b = \frac{C_m}{1 - P_u/\varphi P_{cb}}$$ Eq. (10-7)

with $$P_{cb} = \frac{\pi^2 EI}{(k_b \ell_u)^2}$$ Eq. (10-9)

C_m, P_{cb}, and k_b are the same values as for the gravity load combination

where $M_{2s} = 0.75(1.7W)$
but not less than $P_u(0.6 + 0.03h)$ 10.11.5.5

$$\delta_s = \frac{1.0}{1 - \Sigma P_u/\varphi \Sigma P_{cs}}$$ Eq. (10-8)

with $$P_{cs} = \frac{\pi^2 EI}{(k_s \ell_u)^2}$$ Eq. (10-9)

k_s is evaluated as for an unbraced condition according to Section 10.11.2.2. ΣP_u and ΣP_{cs} are the summations for all columns within the story under consideration. Note that in the calculation of column stiffness EI by Eq. (10-10) or (10-11), if the sway moment M_{2s} is entirely caused by wind, $\beta_d = 0$.

The same general procedure is applied for load combination Eq. (9-3), using the appropriate P_u, M_{2b}, and M_{2s} values.

(4) Final selection of column size and reinforcement is the more severe of the three load combinations considered, either Eq. (9-1), Eq. (9-2) or Eq. (9-3).

(5) For a frame effectively braced by structural walls and frame-shearwall interaction considered in the lateral load analysis, design of the frame columns is somewhat simplified by taking δ_s equal to 1.0 in the gravity plus wind load combination. For this condition, the M_{2s} column moments are directly additive to the magnified gravity load moments. Thus, Eq. (10-6) reduces to

$$M_c = \delta_b M_{2b} + M_{2s}$$ Eq. (10-6)

(6) For a braced frame with columns resisting gravity loads only (lateral load effects assumed in the analysis to the resisted by shearwalls), design of the columns reduces to Step (2) only. Thus, Eq. (10-6) reduces to

$$M_c = \delta_b M_{2b} \qquad \text{Eq. (10-6)}$$

Since the moment magnification method of Section 10.11 is an approximate procedure, some judgment must be used in its application. The application of gravity loads to an unbraced frame in an unsymmetrical loading pattern, or unsymmetrical unbraced frame, will result in some calculated sidesway of the frame. Unless the lateral deflection due to gravity loading is appreciable ($\Delta/\ell_u > 1/1500$), the minor effect of this sway component can be neglected and corresponding moments can be considered as nonsway moments, magnified by the braced frame magnifier. The definitions of M_{2b} and M_{2s} both contain the terminology "appreciable sway." For use in the approximate technique, a deflection ratio of $\Delta > \ell_u/1500$ is a reasonable upper limit above which sway deflections become "appreciable." In computing Δ for this purpose, only the nonlateral loads should be considered.

In the case of columns subjected to transverse loading between supports, it is possible that maximum moment will occur at a section away from the end of the member. If this occurs, the value of the largest calculated moment occurring anywhere along the member should be used for the value of M_{2b} in Eq. (10-6). In accordance with Section 10.11.5.3, C_m must be taken as 1.0 for this case.

In Eq. (10-12), the ratio of the end moments indicates the shape of the deformed column. The positive algebraic sign should be used for moments if each moment induces compression on the same face of the column. For example, if end moments are equal, C_m is equal to unity, but if one end of a column is restrained, and a carryover moment is taken as minus one-half the top moment, C_m is equal to 0.4 (see Fig. 12-11).

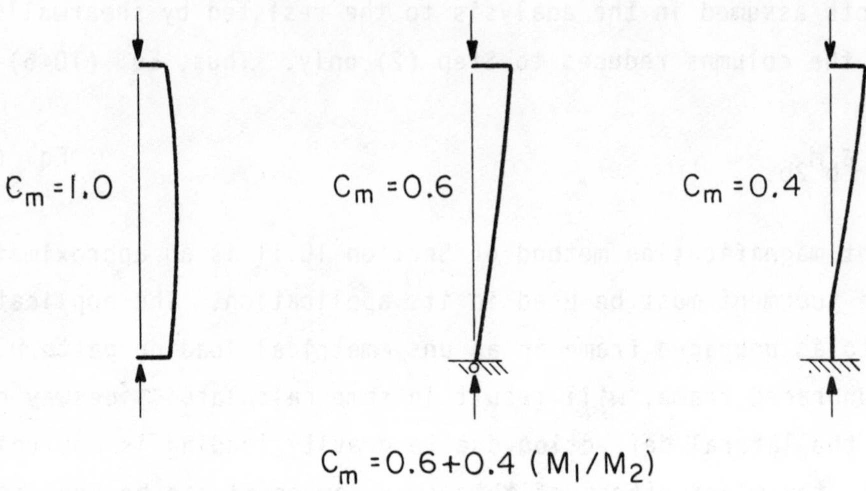

Fig. 12-11 - Moment Factor

10.11.5.2 Column Stiffness EI

In defining the critical column load, the biggest problem is the choice of a stiffness parameter EI in Eq. (10-9) which reasonably approximates the stiffness variations due to cracking, creep, and the nonlinearity of the concrete stress-strain curve. When more precise values are not available, it is recommended that EI be defined by Eqs. (10-10) and (10-11):

$$EI = [(E_c I_g/5) + E_s I_{se}]/(1 + \beta_d) \qquad \text{Eq. (10-10)}$$

$$EI = (E_c I_g/2.5)/(1 + \beta_d) \qquad \text{Eq. (10-11)}$$

The Code equations for EI approximate the lower limits of EI for practical cross sections and thus are conservative for secondary moment calculations. The approximate nature of the EI equations is shown in Fig. 12-12 where they are compared with values derived from load-moment-curvature diagrams for the case of no sustained load ($\beta_d = 0$).

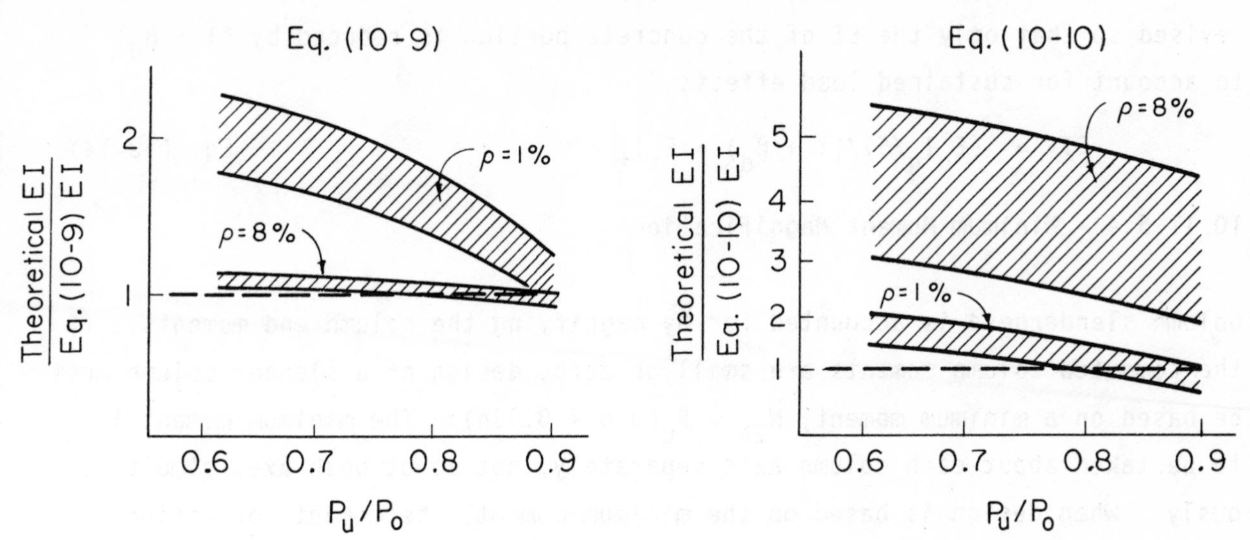

Fig. 12-12 - Comparison of Equations for EI

Eq. (10-10) represents the lower limit of the practical range of stiffness values. This is especially true for heavily reinforced columns. Eq. (10-11) is simpler to use but greatly underestimates the effect of reinforcement in heavily reinforced columns (see Fig. 12-12). They were derived for small e/h values and high P_u/P_o values, where the effect of axial load is most pronounced. P_o is the nominal axial load strength at zero eccentricity.

For reinforced concrete columns subject to sustained loads, creep of concrete transfers some of the load from the concrete to the steel, thus increasing steel stresses. For lightly reinforced columns, this load transfer may cause the compression steel to yield prematurely, resulting in a loss in the effective value of EI. This is taken into account by dividing the EI term by $(1 + \beta_d)$ where β_d is the ratio of factored dead load moment to factored total load moment on the member. Note that β_d is based on absolute values of the moments, irrespective of sign. The β_d factor gives a correct trend when compared to analysis and tests of columns under sustained loads.

For composite columns in which a steel pipe or structural steel shape makes up a larger percentage of the total column cross section, load transfer due to creep is not significant. Accordingly, with the 1983 Code, Eq. (10-14) is revised so that only the EI of the concrete portion is reduced by $(1 + \beta_d)$ to account for sustained load effects.

$$EI = (E_c I_g/5)/(1 + \beta_d) + E_s I_t \qquad \text{Eq. (10-14)}$$

10.11.5.4 - Minimum Moment Magnification

Column slenderness is accounted for by magnifying the column end moments. If the computed column moments are small or zero, design of a slender column must be based on a minimum moment, $M_{2b} = P_u(0.6 + 0.03h)$. The minimum moment is to be taken about each column axis separately, not about both axes simultaneously. When design is based on the minimum moment, the moment correction factor C_m is, however, evaluated using the computed column moments for M_{1b} and M_{2b} in Eq. (10-12). If computations show that there is no moment at both ends of a column due to greater relative flexibility of the restraining members at the column ends, the ratio M_{1b}/M_{2b} is taken equal to 1.0.

10.11.6 - Moment Magnification for Flexural Members

The strength of a laterally unbraced frame is governed by the stability of the columns and by the degree of end restraint provided by the beams in the frame. If plastic hinges form in the restraining beam, the structure approaches a mechanism and its axial load capacity is drastically reduced. Section 10.11.6 requires that the designer make certain that the restraining flexural members (beams or slabs) have the capacity to resist the magnified column moments. The ability of the moment magnification method to provide a good approximation of the actual magnified moments at the member ends in an unbraced frame is a significant improvement over the reduction factor method prescribed in earlier ACI Codes to account for member slenderness in design.

10.11.7 Moment Magnifier δ for Biaxial Bending

When biaxial bending occurs in a column, the computed moments about each of the principal axes must be magnified. The magnification factors δ_b and

δ_s are computed considering the buckling load P_c about each axis separately, based on the appropriate effective lengths and the related stiffness ratios of columns to beams in each direction. Thus, different buckling capacities about the two axes are reflected in different magnification factors. The moments about each of the two axes are magnified separately, and the cross section is then proportioned for an axial load P_u and magnified biaxial moments.

Modified R Method

Prior to adoption of the moment magnification procedure for column slenderness, the ACI Code used a column reduction factor R and an effective length h' for slenderness evaluation of unbraced columns. The modified R method given in the Code Commentary, within the limits noted, leads to an accuracy equal to that of the "moment magnification" method of Section 10.11.5. Hence, the R method may be used an an alternative within the stated limits. It is to be noted that both the axial load and the moment must be divided by the appropriate factor R.

10.10.1 Second-Order Frame Analysis

The Code encourages the use of a second-order frame analysis for consideration of slenderness effects in compression members, which takes into account effects of sway deflections on the axial loads and moments in the frame. Since first introduction of the current code provisions for column slenderness (ACI 318-71), extensive studies have been made and it is now feasible for a designer to use second-order analyses in the design of reinforced concrete buildings. Generally, the moments from a second-order analysis provide a better approximation of the real moments than those obtained from Section 10.11. For sway frames or lightly braced frames, economies can be achieved by the use of second-order analyses. Procedures for carrying out a second-order analysis are given in Code Commentary References 10.21, 10.22, and 10.23. The reader is referred to Code Commentary Section 10.10.1, which discusses minimum requirements for an adequate second order analysis under Section 10.10.1.

EXAMPLE 12.1 - Slenderness Effects by Approximate Design Method (Section 10.11) Columns not Braced Against Sidesway

Design columns for the first story of a 10-story office Building. Columns are not braced against sidesway. Clear height of first floor is 18'0". Clear height above first floor and below first floor (basement) is 11'0". Building layout is 7x3 bays.

f'_c = 5000 psi

f_y = 60,000 psi

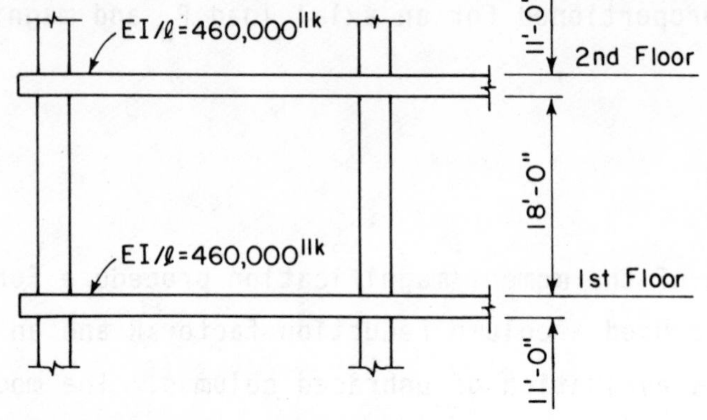

Calculations and Discussion	Code Reference

1. Load data from frame analysis.

	Exterior Columns	Interior Columns
Service loads	D = 282^k L = 64^k W = 4^k	D = 363^k L = 128^k W = 0^k
Service moments	Top: D = 35$^{'k}$ L = 20$^{'k}$ W = 25$^{'k}$	Top: D = 16.5$^{'k}$ L = 13.5$^{'k}$ W = 50$^{'k}$
	Btm: D = 58$^{'k}$ L = 30$^{'k}$ W = 25$^{'k}$	Btm: D = 43$^{'k}$ L = 34$^{'k}$ W = 50$^{'k}$

2. Factored load combinations to be considered. 9.2

For application in Eq. (10-6), it is helpful to distinguish between the factored "nonsway" moments due to gravity loads (M_{2b}) and the factored "sway" moments due to lateral loads (M_{2s}).

EXAMPLE 12.1 - Continued

Calculations and Discussion	Code Reference

Interior Columns

(a) gravity loads

$U = 1.4D + 1.7L$ — Eq. (9-1)

$P_u = 1.4(363) + 1.7(128) = 726^k$

$M_{2b} = 1.4(43) + 1.7(34) = 118'^k$

(b) gravity + wind loads

$U = 0.75(1.4D + 1.7L + 1.7W)$ — Eq. (9-2)

$P_u = 0.75(1.4 \times 363 + 1.7 \times 128) = 544^k$

$M_{2b} = 0.75(1.4 \times 43 + 1.7 \times 34) = 89'^k$

$M_{2s} = 0.75(1.7 \times 50) = 64'^k$

or $U = 0.9D + 1.3W$ — Eq. (9-3)

$P_u = 0.9(363) = 327^k$

$M_{2b} = 0.9(43) = 39'^k$

$M_{2s} = 1.3(50) = 65'^k$

Exterior Columns

(a) gravity loads

$U = 1.4D + 1.7L$ — Eq. (9-1)

$P_u = 1.4(282) + 1.7(64) = 504^k$

$M_{2b} = 1.4(58) + 1.7(30) = 132'^k$

(b) gravity + wind loads

$U = 0.75(1.4D + 1.7L + 1.7W)$ — Eq. (9-2)

$P_u = 0.75(1.4 \times 282 + 1.7 \times 64 + 1.7 \times 4) = 383^k$

$M_{2b} = 0.75(1.4 \times 58 + 1.7 \times 30) = 99'^k$

$M_{2s} = 0.75(1.7 \times 25) = 32'^k$

EXAMPLE 12.1 - Continued

Calculations and Discussion	Code Reference

or $U = 0.9D + 1.3W$ — Eq. (9-3)

$P_u = 0.9(282) + 1.3(4) = 259^k$

$M_{2b} = 0.9(58) = 52^{'k}$

$M_{2s} = 1.3(25) = 33^{'k}$

3. Preliminary selection of column size and reinforcement.

Use Design Aid EB9. (See Part 10, page 10-5). Note that the EB9 Design Tables do not include the strength reduction factor φ in the tabulated values; P_u/φ and M_u/φ must be used when designing with this aid.

Base preliminary selection on gravity load combination, excluding slenderness effects.

Interior Columns

required $P_n = P_u/\varphi = 726/0.7 = 1037^k$

$M_n = M_u/\varphi = 118/0.7 = 169^{'k}$

Exterior Columns

required $P_n = P_u/\varphi = 504/0.7 = 720^k$

$M_n = M_u/\varphi = 132/0.7 = 189^{'k}$

Try 20 x 20 Column Size See Table 12-1 (reproduced from EB9)

To account for slenderness effects, try 4 #9 bars for both interior and exterior columns. $A_s = 4.00$ in.^{2}, $\rho = 1.00\%$.

For $P_n = 1037$, $M_n \simeq 420$

$P_n = 720$, $M_n \simeq 469$

Calculations and Discussion	Code Reference

PCA–LOAD AND MOMENT STRENGTH TABLES FOR CONCRETE COLUMNS

1 1/2" clear — No A1 — No. A2 — Axis of bending

COLUMN SIZE =20X20 INCHES
FPC=5,000 FY=60,000

BARS				SPECIAL CONDITIONS			AXIAL LOAD STRENGTH P_n (KIPS)																	
NO	NO	SIZE	PCNT	0.	BALANCE		0	40	140	240	340	440	540	640	740	840	940	1040	1140	1240	1340	1540	1740	1940
A1	A2	NO		PO	PB	MB	MOMENT STRENGTH M_n (FT-KIPS)																	
4	0	9	1.00	1923	670	475	170	196	260	318	371	415	449	471	468	456	440	419	392	359	319			
10	0	6	1.10	1945	677	495	187	213	278	337	390	435	468	489	487	473	455	433	406	373	333			
6	0	8	1.19	1964	671	504	199	225	289	347	399	444	478	499	495	480	462	440	413	380	341			
8	0	7	1.20	1968	673	508	202	228	292	351	403	448	481	503	499	484	466	443	416	383	344			
4	0	10	1.27	1983	665	513	211	236	299	357	408	452	487	509	502	487	469	447	420	388	349			
10	0	7	1.50	2034	671	553	248	274	337	396	448	493	526	548	541	522	502	478	450	417	379	283		
6	0	9	1.50	2034	666	548	247	272	334	392	444	489	523	544	536	518	498	475	447	415	377	282		
4	0	11	1.56	2048	660	552	253	279	340	397	448	492	527	549	539	521	501	478	450	419	381	288		
8	0	8	1.58	2052	668	562	260	286	348	406	458	503	537	558	549	530	509	485	458	425	387	293		
6	0	10	1.90	2125	660	604	307	332	393	450	501	545	580	601	588	566	544	519	491	458	422	332		
10	0	8	1.98	2140	664	620	320	346	407	465	517	562	596	617	604	581	557	531	503	470	433	343		
8	0	9	2.00	2146	662	621	323	348	409	466	518	563	597	618	604	582	558	532	503	471	434	345		
4	0	14	2.25	2202	649	644	353	377	436	491	541	585	621	643	623	600	576	550	522	491	455	370		
6	0	11	2.34	2222	653	664	370	395	454	510	561	605	641	662	643	618	593	566	537	505	469	384		
8	0	10	2.54	2266	654	696	402	427	486	543	594	638	673	694	674	648	621	593	563	530	494	410	306	
6	0	14	3.38	2453	639	802	516	540	597	651	701	744	782	801	771	742	712	681	649	616	581	501	406	
4	0	18	4.00	2592	623	866	589	612	667	718	765	808	844	861	829	798	767	736	704	671	636	560	473	370
0	0	0	5.00	2815	606	985	718	739	792	842	888	929	965	977	943	909	875	841	807	773	738	663	581	487
0	0	0	6.00	3038	577	1094	841	862	913	962	1007	1047	1083	1079	1053	1017	981	946	910	875	838	764	684	597
0	0	0	7.00	3261	545	1198	961	982	1031	1078	1122	1162	1197	1174	1149	1123	1086	1049	1012	976	939	863	784	701
0	0	0	8.00	3484	509	1297	1078	1098	1146	1191	1234	1273	1289	1264	1238	1212	1187	1151	1113	1075	1038	961	883	801
6	2	7	1.20	1968	680	464	202	228	289	344	386	417	440	458	459	448	435	416	393	363	326			
10	6	5	1.24	1977	687	460	210	235	292	343	385	418	441	455	456	446	433	416	394	365	330			
8	4	6	1.32	1994	683	470	222	247	307	353	395	430	450	465	465	454	441	423	401	372	337			
12	8	5	1.55	2046	688	491	257	280	335	382	423	453	475	487	485	474	460	442	420	393	360	272		
6	2	8	1.58	2052	676	504	260	285	344	397	432	461	483	500	497	484	468	449	425	397	363	276		
10	6	6	1.76	2092	684	512	289	311	363	411	447	478	496	508	506	493	478	459	437	411	379	297		
8	4	7	1.80	2101	679	518	294	318	370	414	454	484	501	514	511	497	482	463	441	414	382	300		
6	2	9	2.00	2146	672	548	322	346	403	452	481	508	529	544	538	522	505	485	462	434	402	321		
8	4	8	2.37	2229	675	576	377	398	442	484	522	546	561	572	565	549	531	511	489	463	433	359		
10	6	7	2.40	2235	680	576	377	397	446	489	522	550	563	573	566	550	533	513	491	466	436	363		
6	2	10	2.54	2266	666	604	400	423	477	514	542	568	587	601	591	573	553	532	508	481	451	377		
8	4	9	3.00	2369	669	638	460	477	519	559	596	613	625	635	625	605	586	564	541	516	487	419	333	
6	2	11	3.12	2396	660	664	481	503	552	579	605	630	648	661	647	626	605	582	557	531	501	432	345	
10	6	8	3.16	2405	675	651	477	496	543	577	608	631	641	648	638	619	599	578	555	530	502	436	352	
8	4	10	3.81	2550	662	717	559	575	615	654	682	697	707	715	700	678	656	633	609	584	556	493	416	324
6	2	14	4.50	2703	646	802	666	677	702	726	750	773	790	801	778	753	728	702	676	649	620	556	483	395
4	6	6	1.10	1945	690	413	188	212	264	311	348	378	397	409	412	408	399	385	365	339	305			
4	2	8	1.19	1964	679	446	200	225	285	338	374	402	424	441	442	434	422	405	383	354	319			
4	4	7	1.20	1968	684	429	203	227	282	325	364	394	411	425	427	421	411	396	376	349	315			
4	6	7	1.50	2034	688	442	245	265	313	356	387	414	429	439	440	433	424	410	392	368	337			
4	2	9	1.50	2034	676	475	247	272	329	380	408	434	455	471	470	459	446	429	407	380	347	260		
4	4	8	1.58	2052	681	459	259	282	327	367	405	429	444	456	455	447	436	421	402	377	347	265		
4	2	10	1.90	2125	672	513	306	330	385	425	450	475	494	509	505	492	477	460	438	412	382	303		
4	6	8	1.98	2140	685	476	304	323	368	403	431	455	465	473	472	464	453	439	422	400	373	302		
4	4	9	2.00	2146	678	492	318	335	375	413	448	466	479	489	486	476	464	449	430	407	379	306		
4	2	11	2.34	2222	667	552	367	390	442	470	494	517	535	549	542	527	511	492	471	446	417	345		
4	4	10	2.54	2266	673	533	379	395	433	470	498	512	522	531	526	514	500	485	466	445	419	354		
4	2	14	3.38	2453	656	644	507	527	549	571	593	614	631	642	629	610	591	571	549	525	499	436	357	
12	2	5	1.08	1942	683	471	185	212	275	333	383	418	445	465	466	454	439	419	393	361	323			
8	2	6	1.10	1945	682	462	187	214	276	333	379	412	438	456	457	447	432	413	389	358	320			
10	2	6	1.32	1994	680	495	222	248	310	366	412	446	471	490	488	475	458	438	413	383	346			
8	2	7	1.50	2034	677	508	249	274	335	389	430	462	485	503	500	486	470	450	426	396	361	270		
10	2	7	1.80	2101	675	553	295	320	380	434	475	507	530	548	542	525	506	484	459	430	395	308		
8	2	8	1.98	2140	673	562	320	345	403	456	490	520	542	558	551	534	515	494	469	441	407	324		
10	2	8	2.37	2229	669	620	380	404	462	514	548	579	600	617	606	585	564	540	514	485	452	372		
8	2	9	2.50	2257	667	621	397	421	477	523	554	582	602	618	607	587	566	543	518	489	457	381		
8	2	10	3.17	2408	661	696	493	516	570	605	633	661	679	694	678	655	631	606	580	552	521	449	360	

Table 12-1 - Reproduced from page 13 of EB9

EXAMPLE 12.1 - Continued

Calculations and Discussion	Code Reference
Estimate $k\ell_u/r \simeq 1.2(18 \times 12)/0.3 \times 20 = 43$	10.11.3
$22 < (k\ell_u/r = 43) < 100$	10.11.4
Column slenderness effects must be included, with the approximate method an acceptable procedure.	10.11.4.3

4. Compute properties of 20 x 20 Section with 4 #9 bars for slenderness evaluation.

(a) Effective length factors (k) — 10.11.2

k for braced condition must be taken as 1.0, unless an analysis shows that a lower value may be used. k for unbraced condition must take into account effect of cracking and reinforcement on relative stiffness, and must be greater than 1.0. With $k\ell_u/r < 60$, use $0.5EI_g$ for beams (to account for effect of cracking and reinforcement on relative stiffness) and EI_g for columns to evaluate k factors. Determine k factors from Figs. 12-9 and 12-10.

$$I_g = 20^4/12 = 13{,}333 \text{ in.}^4$$

$$E_c = 57000\sqrt{5000} = 4.03 \times 10^3 \text{ ksi}$$

For 18'-0 columns $\dfrac{EI_g}{\ell_c} = \dfrac{4.03 \times 10^3 \times 13{,}333}{18 \times 12} = 249 \times 10^{3}\text{"k}$

For 11'-0 columns $\dfrac{EI_g}{\ell_c} = \dfrac{4.03 \times 10^3 \times 13{,}333}{11 \times 12} = 407 \times 10^{3}\text{"k}$

EXAMPLE 12.1 - Continued

Calculations and Discussion	Code Reference

Interior Columns

$$(\text{top \& btm})\ \psi'_a = \psi'_b = \frac{\Sigma(EI/\ell_c)}{\Sigma(EI/\ell)} = \frac{249 + 407}{2(0.5 \times 460)} = 1.43$$

From Fig. 12-9 (braced k factor) k = 0.81

From Fig. 12-10 (unbraced k factor) k = 1.40

Exterior Columns

$$(\text{top \& btm})\ \psi_A = \psi_B = \frac{249 + 407}{0.5 \times 460} = 2.85$$

From Fig. 12-9 (braced k factor) k = 0.88

From Fig. 12-10 (unbraced k factor) k = 1.80

(b) Critical load $P_c = \pi^2 EI/(k\ell_u)^2$ Eq. (10-9)

Compute EI from Eq. (10-10)

$I_{se} = 4.00\ (7.5)^2 = 225\ \text{in.}^4$

$E_s = 29 \times 10^3$ ksi 8.5.2

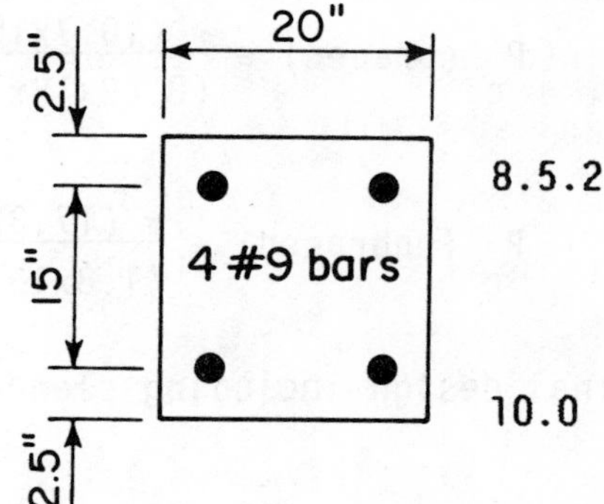

For calculation of δ_b:

$$\beta_d = \frac{1.4D}{1.4D+1.7L}$$ 10.0

$$\beta_d\ (\text{interior}) = \frac{1.4(43)}{118} = 0.51$$

$$\beta_d\ (\text{exterior}) = \frac{1.4(58)}{132} = 0.62$$

$$EI = \frac{(E_c I_g/5) + E_s I_{se}}{1 + \beta_d}$$ Eq. (10-10)

$$EI_{(int)} = \frac{(4.03 \times 10^3 \times 13{,}333/5) + 29 \times 10^3 \times 225}{1 + 0.51}$$

$$= \frac{17.3 \times 10^6}{1.51} = 11.4 \times 10^6$$

EXAMPLE 12.1 - Continued

Calculations and Discussion	Code Reference

$$EI_{(ext)} = \frac{17.3 \times 10^6}{1 + 0.62} = 10.7 \times 10^6$$

For calculation of δ_s; $\beta_d = 0$:

$$EI_{(int\ \&\ ext)} = \frac{17.3 \times 10^6}{1.0} = 17.3 \times 10^6$$

Interior Columns

$$P_c\ (\text{braced}) = \frac{\pi^2(11.4 \times 10^6)}{(0.81 \times 18 \times 12)^2} = 3676^k$$

$$P_c\ (\text{unbraced}) = \frac{\pi^2(17.3 \times 10^6)}{(1.4 \times 18 \times 12)^2} = 1867^k$$

Exterior Columns

$$P_c\ (\text{braced}) = \frac{\pi^2(10.7 \times 10^6)}{(0.88 \times 18 \times 12)^2} = 2923^k$$

$$P_c\ (\text{unbraced}) = \frac{\pi^2(17.3 \times 10^6)}{(1.8 \times 18 \times 12)^2} = 1130^k$$

5. Final design including slenderness effects

Interior Columns

(a) For gravity load combination — Eq. (9-1)

$P_u = 726^k$
$M_{2b} = 118^{'k}$
$M_{1b} = 1.4(16.5)+1.7(13.5) = 46^{'k}$

C_m may be taken as 1.0, unless computed by

EXAMPLE 12.1 - Continued

Calculations and Discussion	Code Reference

$C_m = 0.6+0.4(M_{1b}/M_{2b}) = 0.6+0.4(46/118) = 0.76$ — Eq. (10-12)

$$\delta_b = \frac{C_m}{1-(P_u/\varphi P_c)} = \frac{0.76}{1-(726/0.7x3676)} = 1.06$$ Eq. (10-7)

Minimum moment for slenderness effects — 10.11.5.4

$M_{2b} \geq P_u(0.6+0.03h)$

$= 726(0.6+0.03x20)/12 = 73'^{k} < 118$

$M_c = \delta_b M_{2b} = 1.06(118) = 125'^{k}$ — Eq. (10-6)

required $P_n = P_u/\varphi = 726/0.7 = 1037^k$

$M_n = M_u/\varphi = 125/0.7 = 179'^{k}$

Review of Table 12-1 indicates that the 4 #9 bars are more than adequate for gravity loading.

(b) For gravity + wind load combination — Eq. (9-2)

$P_u = 544^k$
$M_{2b} = 89'^{k}$
$M_{2s} = 64'^{k}$

$$\delta_b = \frac{0.76}{1-(544/0.7x3676)} = 0.96 < 1.0 \quad \text{use } 1.0$$ Eq. (10-7)

$$\delta_s = \frac{1.0}{1-\Sigma P_u/\varphi \Sigma P_c}$$ Eq. (10-8)

For the 7x3 bay building, axial loads P_u and P_c must be summed for the 20 exterior columns and 12 interior columns. Assume corner columns have 1/2 load of edge columns; use 18 for the exterior summation of P_u.

EXAMPLE 12.1 - Continued

Calculations and Discussion	Code Reference

$\Sigma P_u = 18(383)+12(544) = 6894+6528 = 13422^k$

$\Sigma P_c = 20(1130)+12(1867) = 22600+22404 = 45004^k$

$$\delta_s = \frac{1.0}{1-(13422/0.7x45004)} = 1.74$$

Minimum moment for slenderness effects — 10.11.5.5

$M_{2s} \geq P_u(0.6+0.03h)$

$= 544(0.6+0.03x18)/12 = 52^{'k} < 64$

$M_c = \delta_b M_{2b} + \delta_s M_{2s} = 1.0(89)+1.74(64) = 200^{'k}$ — Eq. (10-6)

required $P_n = 544/0.7 = 777^k$

$M_n = 200/0.7 = 286^{'k}$

Review of Table 12-1 indicates that the 4 #9 bars are more than adequate for gravity plus wind loading.

4 #9 bars

P_n	740	777	840
M_n	468	464	456

By inspection, Eq. (9-3) load combination does not govern.

EXAMPLE 12.1 - Continued

Calculations and Discussion	Code Reference

Exterior Columns

(a) For gravity load combination — Eq. (9-1)

$P_u = 504^k$

$M_{2b} = 132^{'k}$

$M_{1b} = 1.4(35)+1.7(20) = 83^{'k}$

$C_m = 0.6+0.4(83/132) = 0.85$ — Eq. (10-12)

$\delta_b = \dfrac{0.85}{1-(504/0.7x2923)} = 1.13$ — Eq. (10-7)

Minimum moment for slenderness effects — 10.11.5.4

$M_{2b} \geq P_u(0.6+0.03h)$

$= 504(0.6+0.03x20)/12 = 50^{'k} < 132$

$M_c = 1.13(132) = 149^{'k}$

required $P_n = 504/0.7 = 720^k$

$M_n = 149/0.7 = 213^{'k}$

Review of Table 12-1 indicates that the 4 #9 bars are more than adequate for gravity loading.

EXAMPLE 12.1 - Continued

Calculations and Discussion	Code Reference

(b) For gravity plus wind load combination — Eq. (9-2)

$P_u = 383^k$
$M_{2b} = 99^{'k}$
$M_{2s} = 32^{'k}$

$$\delta_b = \frac{0.85}{1-(383/0.7x2923)} = 1.05$$ Eq. (10-7)

$$\delta_s = \frac{1.0}{1-\Sigma P_u/\varphi\Sigma P_c} = 1.74 \text{ (Same as for interior columns)}$$ Eq. (10-8)

Minimum moment for slenderness effects — 10.11.5.5

$$M_{2s} \geq P_u(0.6+0.03h)$$
$$= 383(0.6+0.03x20)/12 = 38^{'k} > 32$$

$$M_c = 1.05(99)+1.74(38) = 170^{'k}$$ Eq. (10-6)

required $P_n = 383/0.7 = 547^k$
$M_n = 170/0.7 = 243^{'k}$

Review of Table 12-1 indicates that the 4 #9 bars are more than adequate for gravity plus wind loading. For the exterior columns somewhat less reinforcement could be used. Such a redesign is left to the reader.

<u>Use 20 x 20 column size with 4 #9 bars for all columns of first story</u>.

Calculations and Discussion	Code Reference

6. For illustration, compare results with ACI 318-77 procedures for the interior columns.

(a) For gravity load combination

$P_u = 726^k$

$M_2 = 118^{'k}$

Evaluate 20 x 20 section with 4 #9 bars:

k(unbraced) = 1.40

$\beta_d = 0.51$

$EI = 11.4 \times 10^6$

$$P_c(\text{unbraced}) = \frac{\pi^2(11.4 \times 10^6)}{(1.4 \times 18 \times 12)^2} = 1230^k$$

$$\delta = \frac{C_m}{1 - (P_u/\varphi P_c)} = \frac{1.0}{1 - (726/0.7 \times 1230)} = 6.4$$

Note: With this high a moment magnification, a more exact method for slenderness evaluation would be required to give reasonable results using ACI 318-77 procedures.

(b) For gravity plus wind load combination

$P_u = 544^k$

$M_2 = 89 + 64 = 153^{'k}$

$k = 1.40$

$$\beta_d = \frac{0.75(1.4 \times 43)}{153} = 0.30$$

$$EI = \frac{17.3 \times 10^6}{1 + 0.30} = 13.3 \times 10^6$$

$$P_c = \frac{\pi^2(13.3 \times 10^6)}{(1.4 \times 18 \times 12)^2} = 1435^k$$

EXAMPLE 12.1 - Continued

Calculations and Discussion	Code Reference

$$\delta = \frac{1.0}{1 - (544/0.7 \times 1435)} = 2.18$$

$$\delta M_2 = 2.18(153) = 334^{'k}$$

$$\text{required } P_n = 544/0.7 = 777^k$$

$$M_n = 334/0.7 = 477^{'k} > 464 \quad \text{N.G.}$$

For this load combination, the 20 x 20 section with 4 #9 bars is not quite adequate using ACI 318-77 procedures.

With gravity load moments now magnified by a braced frame magnifier, new Eq. (10-6), substantially lessened column design moments will generally result. The new slenderness procedures will lead to more economical column design for unbraced frames.

Note: See Example 12.2 for design of columns for same building frame with lateral load effects of wind resisted by shearwalls and columns designed as braced members.

EXAMPLE 12.2 - Slenderness Effects by Approximate Design Method (Section 10.11) Columns Braced Against Sidesway

For Design Example 12.1, assume lateral load effects of wind are resisted by shearwalls with columns resisting gravity load effects of dead and live load only. Assume shearwall bracing is adequate to classify the building frame as braced against sidesway for column design.

Calculations and Discussion	Code Reference

1. Factored gravity loads (from Example 12.1).

Interior columns

$P_u = 726^k$
$M_{2b} = 118^{'k}$
$M_{1b} = 46^{'k}$

Exterior columns

$P_u = 504^k$
$M_{2b} = 132^{'k}$
$M_{1b} = 83^{'k}$

2. Preliminary selection of column size and reinforcement.

Base preliminary selection on load combinations excluding slenderness effects (if any).

EXAMPLE 12.2 - Continued

Calculations and Discussion	Code Reference

Interior columns

required $P_n = P_u/\varphi = 726/0.7 = 1037^k$ 9.3.2.2

$M_n = M_u/\varphi = 118/0.7 = 169^{'k}$

Exterior columns

required $P_n = P_u/\varphi = 504/0.7 = 720^k$ 9.3.2.2

$M_n = M_u/\varphi = 132/0.7 = 189^{'k}$

Using Design Aid EB9, try 18x18 column size
(See Table 12.2 reproduced from EB9).

To account for column slenderness, try 4 #9 bars for both interior and exterior columns. $A_s = 4.00$ in.^{2} $\rho = 1.23\%$

For $P_n = 1037$, $M_n \simeq 266$

$P_n = 720$, $M_n \simeq 342$

Check if slenderness effects need be considered with the 18x18 column size. 10.11.4

Estimate $k\ell_u/r \simeq 1.0(18\text{x}12)/0.3\text{x}18 = 40$ 10.11.3

For braced condition; $k\ell_u/r$ must be less than:

$34 - 12(M_{1b}/M_{2b}) = 34 - 12(46/118) = 29.3$ 10.11.4.1

$29 < (k\ell_u/r = 40) < 100$

Column slenderness must be considered, with the approximate method an acceptable procedure. 10.11.4.3

EXAMPLE 12.2 - Continued

Calculations and Discussion	Code Reference

PCA–LOAD AND MOMENT STRENGTH TABLES FOR CONCRETE COLUMNS

No A1 — No. A2 — 1 1/2" clear — Axis of bending

COLUMN SIZE =18X18 INCHES
FPC=5,000 FY=60,000

BARS NO A1	NO A2	SIZE NO	PCNT	SPECIAL CONDITIONS 0. PO	BALANCE PB	MB	P_n (KIPS) 0	40	80	120	160	260	360	460	560	660	760	860	960	1060	1160	1260	1360	1560
							MOMENT STRENGTH M_n (FT-KIPS)																	
8	0	6	1.09	1573	540	355	133	156	179	202	223	272	312	341	353	342	327	307	282	249	208			
6	0	7	1.11	1578	538	356	135	159	181	203	225	273	313	342	354	343	328	309	283	250	210			
4	0	9	1.23	1600	533	366	148	171	193	215	236	283	323	353	363	351	336	316	291	259	219			
10	0	6	1.36	1622	538	383	163	187	209	231	252	300	341	370	380	366	350	329	303	271	232			
6	0	8	1.46	1641	533	391	174	196	218	240	261	309	348	378	387	373	356	335	310	278	240	194		
8	0	7	1.48	1645	535	395	176	199	221	243	264	312	352	382	391	376	359	338	312	281	242	197		
4	0	10	1.57	1660	528	398	183	206	227	249	269	316	355	386	393	378	361	341	316	285	247	202		
6	0	9	1.85	1711	528	429	215	237	259	280	300	347	387	417	424	406	387	365	339	309	272	229		
4	0	11	1.93	1725	523	432	220	242	263	284	304	350	389	420	425	408	389	367	342	312	276	234		
8	0	8	1.95	1729	530	441	226	249	270	292	312	359	399	429	436	417	397	375	348	318	282	239		
6	0	10	2.35	1802	523	477	266	288	309	330	350	396	435	467	469	449	427	403	377	347	312	272	226	
8	0	9	2.47	1823	524	492	281	302	324	344	364	411	451	481	484	462	440	415	388	358	323	284	238	
4	0	14	2.78	1879	512	510	305	326	346	366	385	430	468	499	500	477	454	430	404	375	342	304	261	
6	0	11	2.89	1899	516	528	321	342	362	383	402	447	487	518	518	494	470	445	418	388	354	316	273	
4	0	18	4.94	2269	468	689	506	525	543	562	579	620	657	687	671	648	619	590	561	530	499	465	430	349
0	0	0	6.00	2461	439	769	599	617	635	652	669	709	745	765	744	723	699	668	637	606	574	541	506	432
0	0	0	7.00	2641	410	840	683	701	718	735	752	790	825	829	808	787	766	740	708	676	644	611	577	505
0	0	0	8.00	2822	377	908	765	782	799	816	832	869	903	890	869	847	826	804	779	746	713	680	646	575
6	2	6	1.09	1573	544	326	134	157	179	200	221	263	292	314	325	318	307	291	268	237	199			
8	4	5	1.15	1584	547	326	141	164	186	207	225	264	297	315	326	319	308	293	271	241	203			
6	2	7	1.48	1645	541	356	177	199	220	241	260	297	324	345	354	344	332	315	293	264	229			
10	6	5	1.53	1654	547	353	183	205	224	242	260	297	327	344	352	343	331	315	294	267	232	190		
8	4	6	1.63	1671	543	361	194	215	236	253	269	306	335	352	360	350	337	321	300	273	239	198		
6	2	8	1.95	1729	537	391	226	248	268	288	306	336	361	381	388	375	361	343	321	295	262	223		
10	6	6	2.17	1769	543	398	249	267	285	302	319	351	378	391	396	383	369	352	332	307	277	241		
8	4	7	2.22	1778	539	403	256	275	291	306	321	357	381	395	401	387	373	356	335	310	280	243		
6	2	9	2.47	1823	532	429	280	300	320	338	353	378	401	420	425	409	393	374	353	327	297	262	220	
8	4	8	2.93	1906	533	453	322	337	353	367	382	416	434	446	448	432	416	397	377	353	325	293	255	
6	2	10	3.14	1943	526	477	346	365	384	396	406	429	452	470	471	453	435	415	393	368	340	308	271	
8	4	9	3.70	2046	527	506	389	404	419	433	447	474	490	501	500	482	463	444	422	399	373	343	310	
6	2	11	3.85	2073	520	528	415	432	442	451	460	483	504	522	520	500	480	458	436	412	385	354	321	240
4	4	6	1.09	1573	547	304	134	157	178	197	212	249	278	294	304	300	292	279	258	230	193			
4	2	7	1.11	1578	543	317	136	159	181	202	221	259	285	305	316	311	301	286	264	235	197			
4	6	6	1.36	1622	549	312	164	182	199	217	233	265	291	304	312	308	300	288	269	244	211			
4	2	8	1.46	1641	540	340	174	196	217	237	255	287	311	330	339	331	320	305	284	257	222			
4	4	7	1.48	1645	544	326	177	198	215	230	245	279	303	317	325	319	311	297	278	253	220			
4	2	9	1.85	1711	536	366	215	236	256	274	292	316	338	356	364	353	341	326	306	280	249	210		
4	4	8	1.95	1729	540	352	224	239	254	268	282	314	333	345	350	342	332	319	301	278	249	213		
4	2	10	2.35	1802	531	398	265	285	304	320	329	351	372	390	394	382	368	352	333	309	280	245	204	
4	4	9	2.47	1823	535	380	267	281	294	308	321	350	363	374	378	368	356	343	326	305	279	247	208	
4	2	11	2.89	1899	526	432	317	336	350	359	367	388	408	425	427	413	397	380	361	338	312	280	243	
10	2	5	1.15	1584	545	342	141	164	187	208	229	275	306	329	341	332	319	301	278	246	207			
8	2	6	1.36	1622	543	355	164	186	208	229	250	291	320	343	353	343	330	312	289	259	222			
10	2	6	1.63	1671	541	383	194	216	238	258	278	320	349	372	381	368	353	334	310	281	246	203		
8	2	7	1.85	1711	538	395	217	239	260	280	299	336	363	384	392	378	363	344	321	293	260	219		
8	2	8	2.44	1817	533	441	278	299	319	339	357	386	412	432	437	420	402	382	360	333	302	265	222	
8	2	9	3.09	1934	528	492	344	364	384	402	414	440	465	484	486	466	446	425	402	375	346	312	274	
10	4	5	1.34	1619	546	347	162	185	207	228	245	285	317	336	346	337	324	308	286	256	220			
6	4	6	1.36	1622	545	332	164	186	207	225	241	278	306	323	332	325	314	299	279	251	216			
6	4	7	1.85	1711	541	365	216	237	253	268	283	318	342	356	363	353	341	326	306	281	250	212		
10	4	6	1.90	1720	541	390	223	245	264	281	297	335	364	381	387	375	360	343	321	294	261	222		
6	4	8	2.44	1817	537	402	273	288	303	318	332	365	383	396	399	387	373	358	338	315	287	253	213	
6	4	9	3.09	1934	531	443	328	342	356	370	384	412	427	438	439	424	409	392	373	351	326	296	260	
6	6	5	1.15	1584	550	312	141	164	183	201	219	257	285	302	312	307	299	285	264	236	200			
8	6	5	1.34	1619	548	332	162	184	203	221	239	277	306	323	332	325	314	300	279	251	216			
6	6	6	1.63	1671	547	341	192	210	228	245	262	293	320	333	340	333	323	309	290	265	233	193		
8	6	6	1.90	1720	545	369	220	238	256	273	290	322	349	362	368	358	346	330	311	286	255	217		

Table 12-2 - Reproduced from Page 12 of EB9

EXAMPLE 12.2 - Continued

Calculations and Discussion	Code Reference

3. Compute properties of 18x18 section with 4 #9 bars for column slenderness evaluation.

(a) Effective length factors (k) — 10.11.2

k for braced condition must be taken as 1.0, unless an analysis shows that a lower value may be used. Determine k factors from Fig. 12-9. With $k\ell_u/r <$ 60, use $0.5EI_g$ for beams (to account for effect of cracking and reinforcement), and EI_g for columns to evaluate k factors.

$$I_g = 18^4/12 = 8748 \text{ in.}^4$$

$$E_c = 57000\sqrt{5000} = 4.03 \times 10^3 \text{ ksi}$$

For 18'-0 columns $\dfrac{EI_g}{\ell_c} = \dfrac{4.03\text{x}10^3\text{x}8748}{18\text{x}12} = 163 \times 10^3 \text{"k}$

For 11'-0 columns $\dfrac{EI_g}{\ell_c} = \dfrac{4.03\text{x}10^3\text{x}8748}{11\text{x}12} = 267 \times 10^3 \text{"k}$

Interior columns

$$\psi'_A = \psi'_B = \frac{\Sigma EI/\ell_c}{\Sigma EI/\ell} = \frac{163 + 267}{2(0.5\text{x}460)} = 0.93$$

From Fig. 12-9, k =0.76

Recheck slenderness consideration: — 10.11.4

$$k\ell_u/r = 0.76(18\text{x}12)/0.3\text{x}18 = 30.4 \simeq 29.3$$

Considering the approximations in the evaluation procedure, the above "numbers" are reasonably close to neglect slenderness for the interior columns. However,

Calculations and Discussion — Code Reference

to illustrate design procedure including slenderness, design of the interior columns will include slenderness effects.

Exterior columns

$$\psi'_A = \psi'_B = \frac{163 + 267}{0.5 \times 460} = 1.87$$

From Fig. 12-9, k = 0.85

recheck slenderness consideration: — 10.11.4

$$k\ell_u/r = 0.85(18x12)/0.3x18 = 34$$

$$34 - 12(83/132) = 26.5 < 34$$ — 10.11.4.1

Slenderness must be considered for the exterior columns.

(b) Critical load $P_c = \pi^2 EI/(k\ell_u)^2$ — Eq. (10-9)

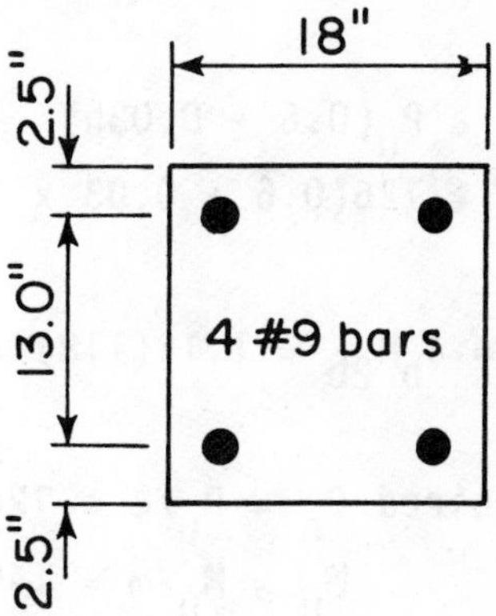

18"

2.5"

13.0"

4 #9 bars

2.5"

Compute EI from Eq. (10-10)

$$I_{se} = 4.0(6.5)^2 = 169 \text{ in.}^4$$

$$E_s = 29 \times 10^3 \text{ ksi}$$

$$\beta_d = \frac{1.4D}{1.4D+1.7L}$$ — 10.0

$$\beta_d \text{ (interior)} = 1.4(43)/118 = 0.51$$

$$\beta_d \text{ (exterior)} = 1.4(58)/132 = 0.62$$

$$EI = \frac{(E_c I_g/5) + E_s I_{se}}{1 + \beta_d}$$ — Eq. (10-10)

EXAMPLE 12.2 - Continued

Calculations and Discussion	Code Reference

$$EI_{(int)} = \frac{(4.03x10^3x8478/5) + 29x10^3x169}{1 + 0.51} = 7.8 \times 10^6$$

$$EI_{(ext)} = \frac{(4.03x10^3x8478/5) + 29x10^3x169}{1 + 0.62} = 7.2 \times 10^6$$

$$P_{c(int)} = \frac{\pi^2(7.8x10^6)}{(0.76x18x12)} = 2857^k$$

$$P_{c(ext)} = \frac{\pi^2(7.2x10^6)}{(0.85x18x12)^2} = 2108^k$$

4. Final design including slenderness effects.

(a) Interior Columns

$$C_m = 0.6+0.4(M_{1b}/M_{2b}) = 0.6+0.4(46/118) = 0.76$$ Eq. (10-12)

$$\delta_b = \frac{C_m}{1-(P_u/\varphi P_c)} = \frac{0.76}{1-(726/0.7x2857)} = 1.19$$ Eq. (10-7)

Minimum moment for slenderness effects 10.11.5.4

$$M_{2b} \geq P_u(0.6 + 0.03h)$$
$$\geq 726(0.6 + 0.03 \times 18)/12 = 69^{'k} < 118$$

$$M_c = \delta_b M_{2b} = 1.19(118) = 140^{'k}$$ Eq. (10-6)

required $P_n = P_u/\varphi = 726/0.7 = 1037^k$
$M_n = M_u/\varphi = 140/0.7 = 200^{'k}$

Review of Table 12-2 indicates that the 4 #9 bars are more than adequate for the interior columns.

4 #9 bars

960	1037	1060
291	266	259

Calculations and Discussion	Code Reference
(b) Exterior Columns	
$C_m = 0.6+0.4(83/132) = 0.85$	Eq. (10-12)
$\delta_b = \dfrac{0.85}{1-(504/0.7\text{x}2108)} = 1.29$	Eq. (10-7)
Minimum moment for slenderness effects	10.11.5.4
$M_{2b} \geq 504(0.6+0.03\text{x}18)/12 = 48^{'k} < 132$	
$M_c = 1.29(132) = 170^{'k}$	
required $P_n = P_u/\varphi = 504/0.7 = 720^k$ $M_n = M_u/\varphi = 170/0.7 = 243^{'k}$	
Review of Table 12-2 indicates that the 4 #9 bars are more than adequate for the exterior columns.	
<u>Use 18x18 column size with 4 #9 bars for all columns of first story</u>	

13

Shear

Introduction

Code provisions for shear are presented in terms of shear forces (rather than stresses) to be compatible with the other design conditions for the strength design method, which are expressed in terms of loads, moments, and forces.

Accordingly, shear is expressed in terms of the factored shear force V_u directly, using the basic shear strength equality:

$$\text{required shear strength} \leq \text{design shear strength}$$

$$V_u \leq \varphi V_n \qquad \text{Eq. (11-1)}$$

$$\leq \varphi V_c + \varphi V_s \qquad \text{Eq. (11-2)}$$

where the design shear strength φV_n is simply the sum of shear strength provided by concrete φV_c plus shear strength provided by shear reinforcement φV_s.

The Code equations expressing area of shear reinforcement A_v are presented in terms of shear strength V_s for direct application into Eqs. (11-1) and (11-2), rather than A_v directly. As an aid to the designer, and to assure correct application of the strength reduction factor φ, equations for computing required area of shear reinforcement A_v are developed in Code Commentary Section 11.5.6.

Only nonprestressed members having clear-span-to-effective-depth ratios greater than 5 are considered in Part 13. Shear design for deep flexural members is presented in Part 18. Shear design for prestressed members is illustrated in Part 26.

General Considerations

The relatively abrupt nature of a "shear" failure, as compared to a ductile flexural failure, makes it desirable to design members so that their strength in shear is equal to, or greater than, their strength in flexure. To ensure a ductile flexural failure, the Code (1) limits the maximum amount of longitudinal reinforcement and (2) except for certain types of construction (Section 11.5.5.1), requires a minimum amount of shear reinforcement in all flexural members.

The determination of the amount of shear reinforcement is based on a modified form of the truss analogy. The truss analogy assumes that shear reinforcement resists the total transverse shear. Considerable research has indicated that shear strength provided by concrete V_c can be assumed equal to the shear causing inclined cracking; therefore, shear reinforcement need be designed to carry only the excess shear.

Briefly, the design of a member for shear involves the following steps:

(1) Determine factored shear forces V_u at critical sections along the length of the member.

(2) For a given section, calculate the shear strength provided by concrete φV_c.

(3) If $(V_u - \varphi V_c)$ is greater than $\varphi 8\sqrt{f'_c}\, b_w d$, increase the cross-section (Section 11.5.6.8). Otherwise, proportion shear reinforcement to carry the excess shear $(V_u - \varphi V_c)$. If V_u is greater than 1/2 φV_c, provide minimum required shear reinforcement (Section 11.5.5).

Only a relatively few controlling sections along a member's length need be considered; the required spacing of shear reinforcement at intermediate points is usually evident from the computed values at controlling points.

11.1 Shear Strength

Shear strength at any section is computed using Eqs. (11-1) and (11-2), where the factored shear force V_u is obtained by applying the load factors specified in Section 9.2. For gravity loads, $V_u = 1.4\ V_d + 1.7\ V_\ell$. The strength reduction factor $\varphi = 0.85$ is specified in Section 9.3.2.3.

The shear strength of a section is increased if a reaction produces compression in the end region of a member. For this condition, the provisions of Section 11.1.2.1 allow sections between the support and a distance "d" from the face of the support to be designed for the same shear force V_u as that computed at a distance "d". Typical support conditions where the factored shear force V_u at a distance "d" from the support can be used include: members supported by bearing at the bottom of the member, as shown in Fig. 13-1(a); and members framing monolithically into another member, as illustrated in Figs. 13-1(b) and (c).

Support conditions where this provision can not be applied include members framing into a supporting member in tension, as shown in Fig. 13-1(d). In this case, the critical section for shear must be taken at the face of the support and the shear within the connection should also be investigated. Also, Section 11.1.2.1 does not apply for shear in columns as shown in Fig. 13-1(e). Although the shear would generally be the same throughout the length of a column, the moment M_u at the face of support must be used if the strength provided by concrete V_c is computed from Eqs. (11-6) and (11-7).

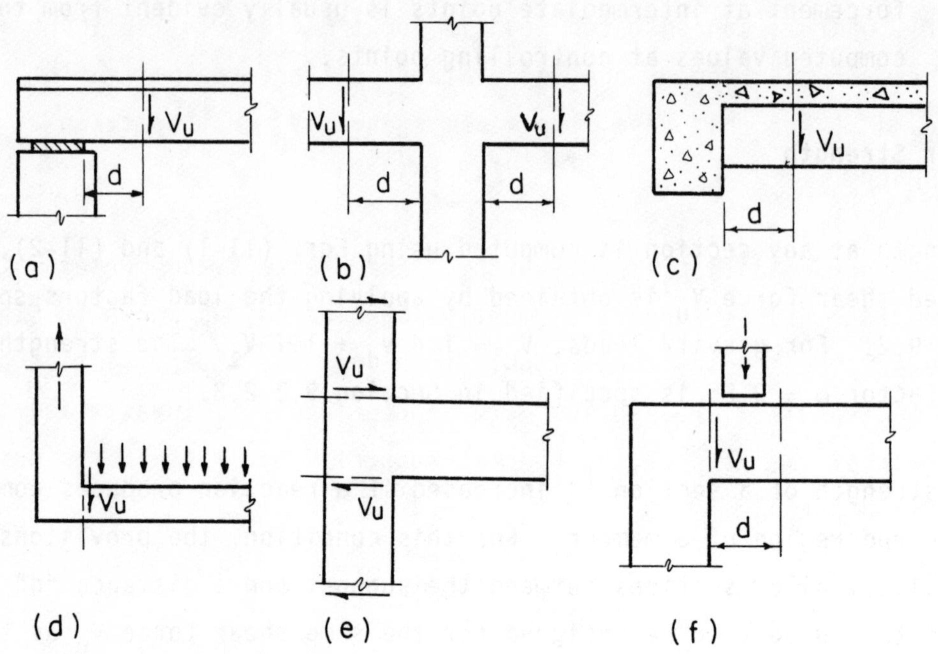

Fig. 13-1 - Typical Support Conditions for Locating Factored Shear Force V_u

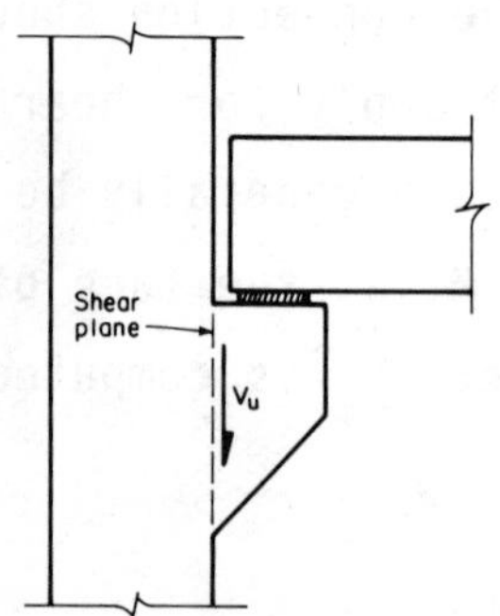

Fig. 13-2 - Critical Shear Plane for Brackets

With the 1983 Code, application of Section 11.1.2.1 was further restricted to cases where an abrupt change in shear does not occur between the face of support and distance "d". An example of such a condition is illustrated in Fig. 13-1(f), where a concentrated load is located close to the support. The shear between the support and "d" distance differs radically from that of distance "d". For this case, the maximum shear V_u must be taken at the face of support.

One other support condition is noteworthy. For brackets and corbels, the shear at the face of the support V_u must be considered, as shown in Fig. 13.2. However, the loading condition is such that the shear-friction provisions of Section 11.7 are applied more appropriately to investigate the shear strength at the face of the bracket support. See Part 16 for design of brackets and corbels.

11.1.1.1 **Web Openings**

Often it is necessary to integrate mechanical and electrical service systems into the structural components of buildings. Passing these services through openings in the webs of the floor beams within the floor-ceiling sandwich eliminates a significant amount of dead space and results in a more economical design. However, the effect of the openings on the shear strength of the floor beams must be considered, especially when such openings are located in regions of high shear near supports. With the 1983 Code a new provision is added to alert the designer to the importance of considering the effect of openings on the shear strength of members. Because of the many variables such as opening shape, size, and location along the span, specific design rules are not stated. Code Commentary references are, however, given for design guidance. Generally it is desirable to provide additional vertical stirrups adjacent to both sides of a web opening, except for small isolated openings. The additional shear reinforcement can be proportioned to carry the total shear force at the section where an opening is located. Design Example 13.5 illustrates application of a design method recommended in Code Commentary Reference 11.8.

11.2 Lightweight Concrete

Since the shear strength of lightweight aggregate concrete may be less than that of normal weight concrete with equal compressive strength, adjustments in the value of V_c, as computed for normal weight concrete, are necessary.

When average splitting tensile strength f_{ct} is specified, $f_{ct}/6.7$ is substituted for $\sqrt{f'_c}$ in all equations for V_c; however, the value of $f_{ct}/6.7$ cannot be taken greater than $\sqrt{f'_c}$. When f_{ct} is not specified, $\sqrt{f'_c}$ is reduced by a factor of 0.75 for all-lightweight concrete or 0.85 for sand-lightweight concrete, with linear interpolation allowed when partial sand replacement is used.

11.3 Shear Strength Provided by Concrete for Nonprestressed Members

When computing the shear strength provided by concrete for members subject to shear and flexure only, designers have the option of using either $V_c = 2\sqrt{f'_c}\, b_w d$ or the more elaborate expression given by Eq. (11-6). In computing V_c from Eq. (11-6), it should be noted that V_u and M_u are the values which occur simultaneously at the section considered. A maximum value of 1.0 is allowed for the ratio $V_u d/M_u$ for members not subject to axial compression to limit V_c near points of contraflexure (at these points M_u is zero or very small).

For members subject to shear and flexure with axial compression, axial tension, or torsion, simplified V_c expressions are given in Section 11.3.1, with optional more elaborate expressions for V_c available in Section 11.3.2.

Fig. 13-3 shows the variation of shear strength provided by concrete V_c with the ratio given by $V_u d/M_u$ for two values of concrete strength f'_c and reinforcement ratio ρ_w = 0.5, 1 and 2 percent.

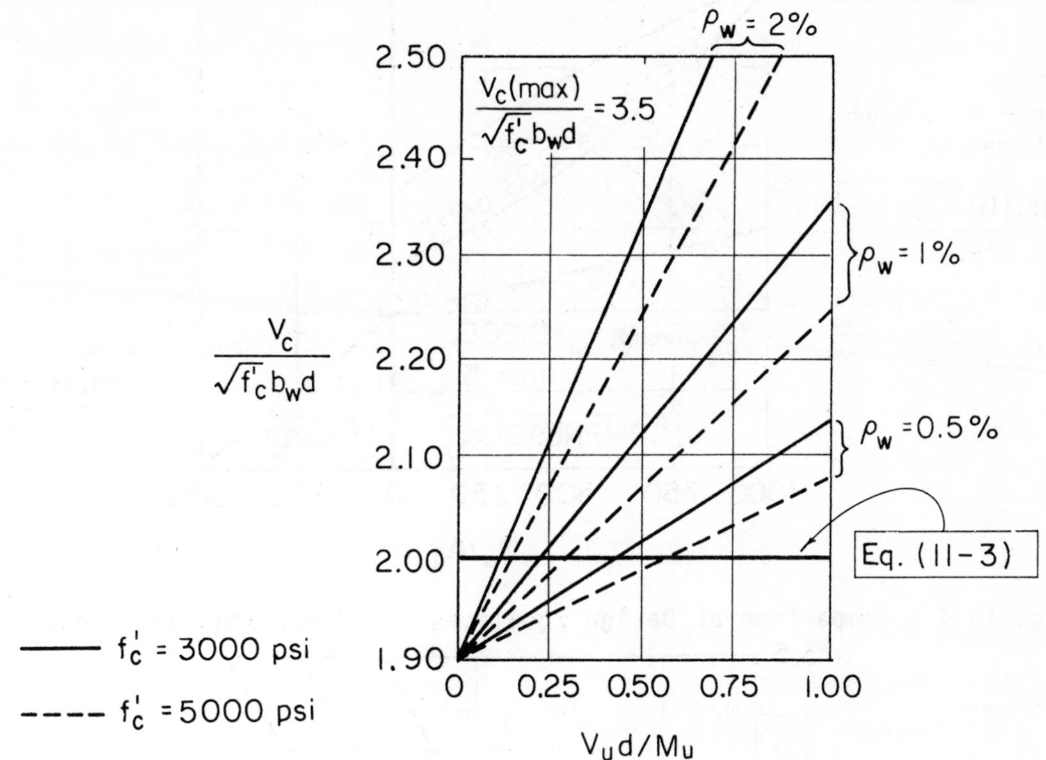

Fig. 13-3 - Variation of $V_c/\sqrt{f'_c}\, b_w d$ with f'_c, ρ_w and Ratio $V_u d/M_u$, as per Eq. (11-6)

Fig. 13-4 shows the approximate range of values of V_c for sections under axial compression, as obtained from Eqs. (11-6) and (11-7). Values are based on a 6x12 in. beam section with an effective depth of 10.8 in. The curves corresponding to the alternate expressions for V_c given by Eqs. (11-4) and (11-8), as well as that corresponding to Eq. (11-9) for members subject to axial tension, are also indicated.

Fig. 13-5 shows the variation of V_c with N_u/A_g and f'_c for sections subject to axial compression, based on Eq. (11-4). For the range of N_u/A_g values shown, V_c varies from about 49% to 57% of the value of V_c as defined by Eq. (11-8).

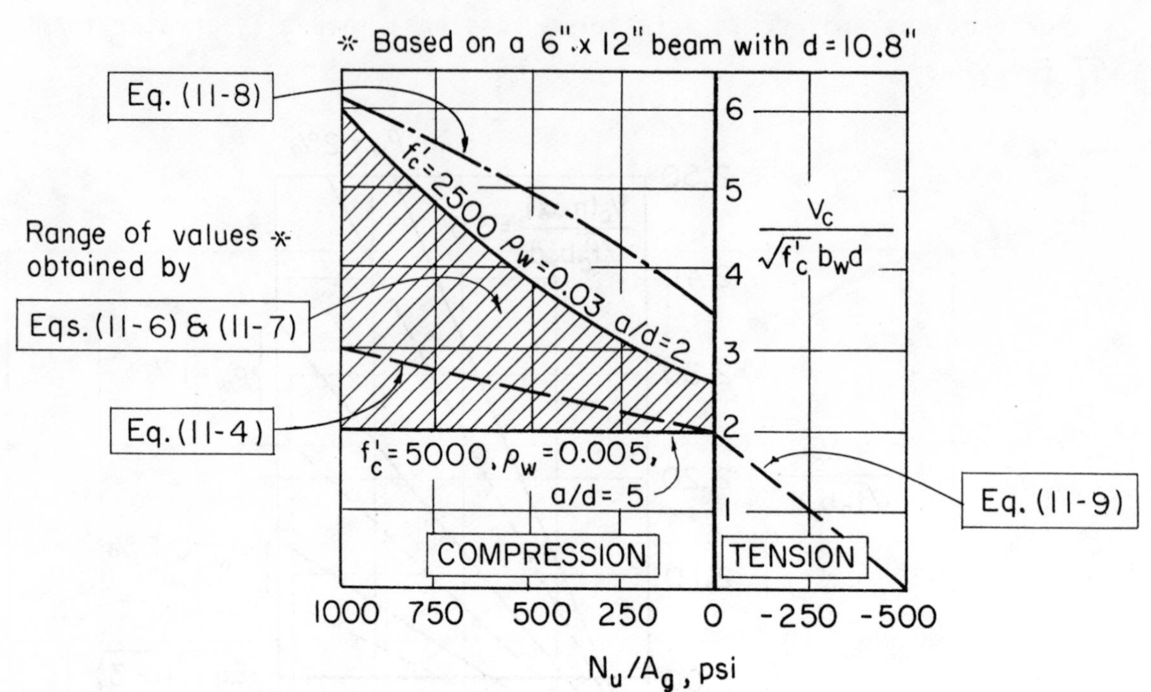

Fig. 13-4 - Comparison of Design Equations for Shear and Axial Load

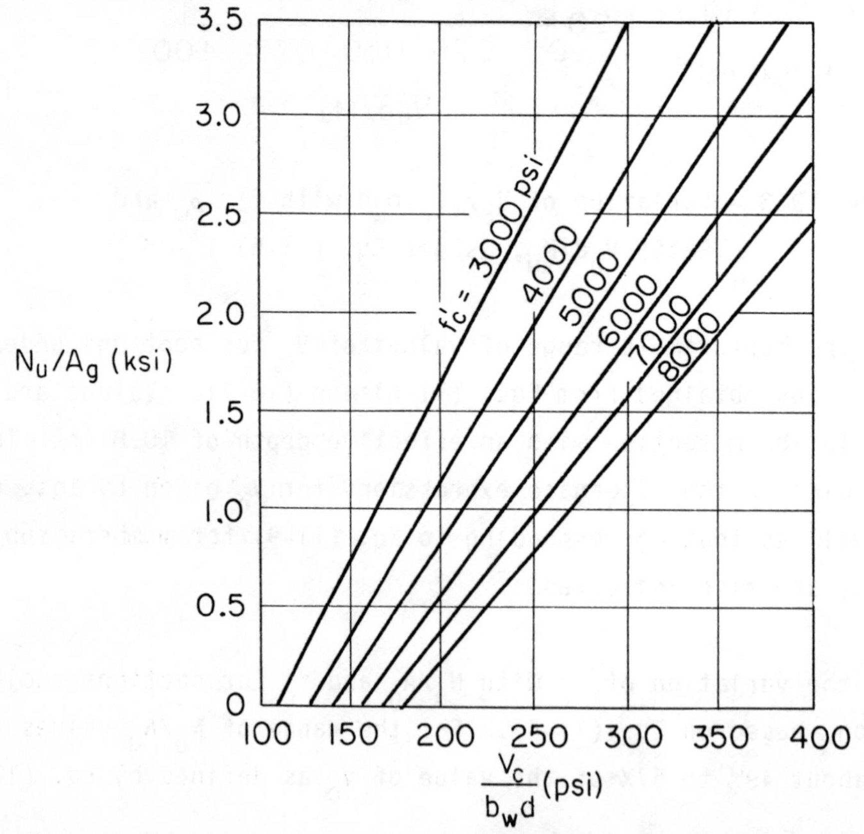

Fig. 13-5 - Variation of V_c/b_wd with f'_c and Ratio N_u/A_g, as per Eq. (11-4)

Fig. 13-6 is a plot of Eq. (11-5) giving the variations of V_c with the ratio $C_t\, T_u/V_u$ for sections subject to a factored torsional moment T_u greater than $\varphi(0.5\sqrt{f'_c}\,\Sigma x^2 y)$.

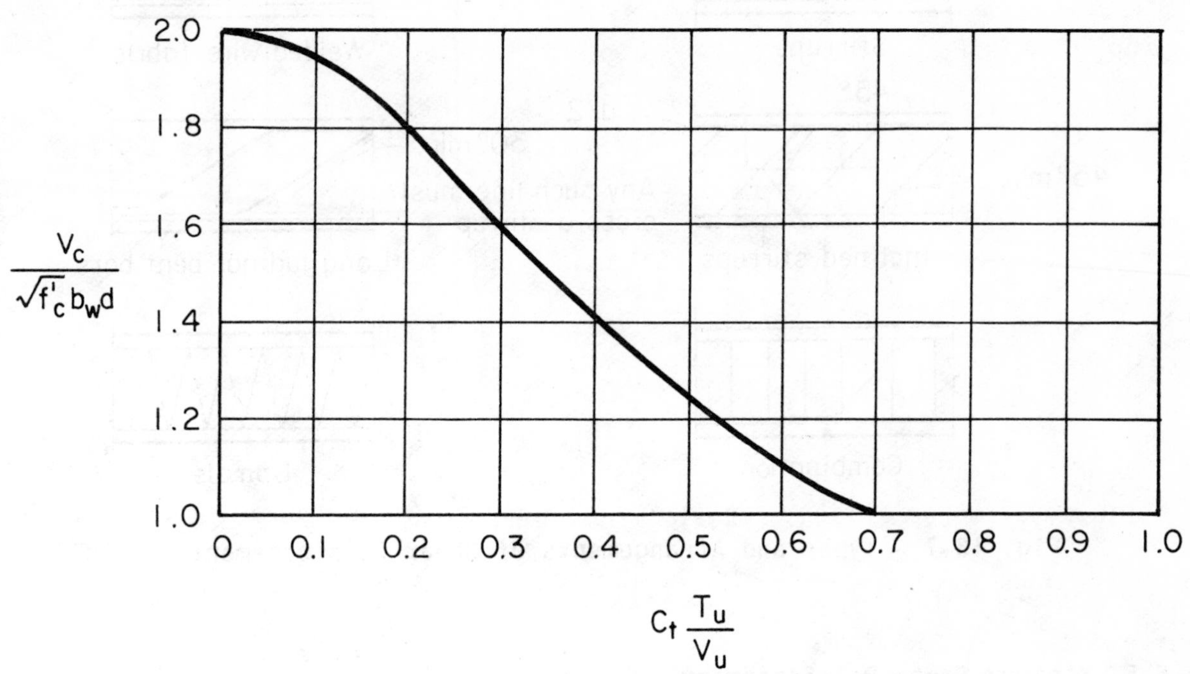

Fig. 13-6 - Variation of $V_c/\sqrt{f'_c}\, b_w d$ with Ratio $C_t\, T_u/V_u$, as per Eq. (11-5)

11.5 Shear Strength Provided by Shear Reinforcement

When the factored shear force V_u exceeds shear strength φV_c, shear reinforcement must be provided to carry the excess shear. Where shear reinforcement perpendicular to the axis of the member is used, required area of shear reinforcement A_v within a distance s is computed from

$$A_v = \frac{(V_u - \varphi V_c)s}{\varphi f_y d}$$

Several types of arrangements of shear reinforcement, as permitted by Sections 11.5.1.1 and 11.5.1.2, are illustrated in Fig. 13-7.

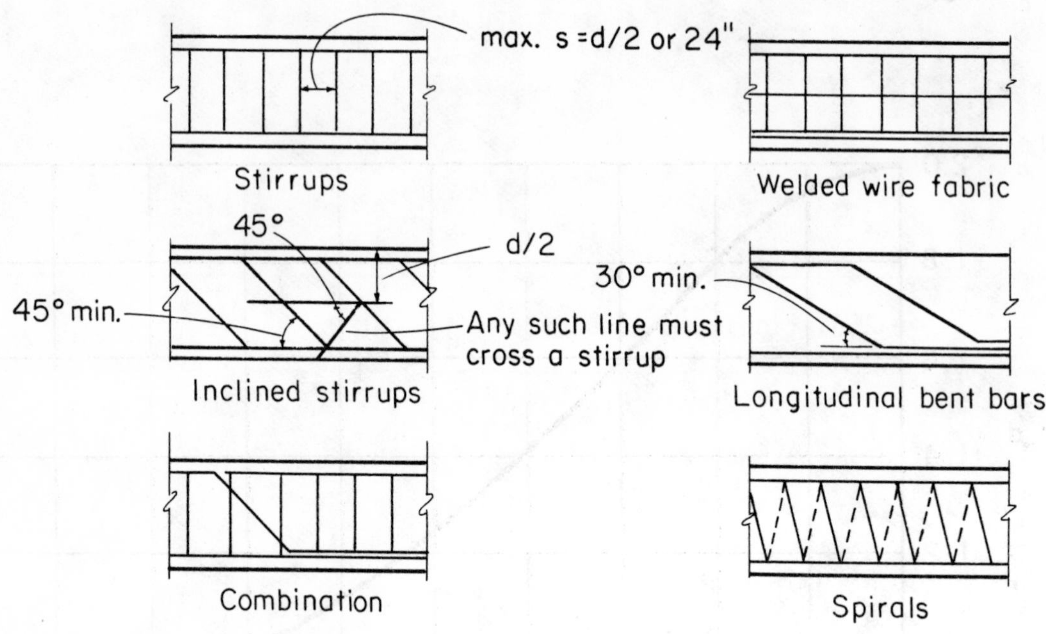

Fig. 13-7 - Types and Arrangements of Shear Reinforcement

11.5.5 Minimum Shear Reinforcement

In general, all concrete flexural members must be provided with a minimum amount of shear reinforcement, except for slabs and footings, floor joists and wide, shallow beams (Section 11.5.5.1). For nonprestressed members, the required minimum shear reinforcement is computed from

$$A_v = 50 \frac{b_w s}{f_y} \qquad \text{Eq. (11-14)}$$

The essence of Eq. (11-14) is that, when minimum shear reinforcement is used, the total shear strength of a section is $V_c + 50\ b_w d$.

EXAMPLE 13.1 - Design for Shear - Members Subject to Shear and Flexure Only

Determine required size and spacing of vertical U-stirrups for a 30-foot span, simply supported beam.

b_w = 13 in. d = 20 in.

f'_c = 3,000 psi f_y = 40,000 psi

w_u = 4.5 klf

Calculations and Discussion	Code Reference
For the purpose of this example, the live load will be assumed fixed, so that design shear at centerline of span is zero. (A design shear greater than zero is obtained by considering partial live loading of the span.)	
1. Determine factored shear forces	
@ support: $V_u = 4.5\ (15) = 67.5$ kips	
@ distance d from support:	
$V_u = 67.5 - 4.5\ (20/12) = 60$ kips	11.1.2.1
2. Determine shear strength provided by concrete	
$\varphi V_c = \varphi 2 \sqrt{f'_c}\, b_w d$	Eq. (11-3)
$= 0.85\ (2)\sqrt{3{,}000}\ (13)\ 20 = 24.2$ kips	
3. Determine distance x_c from support beyond which concrete can carry total shear	
From the sketch, $(15 - x_c)/15 = \varphi V_c/V_u$ @ support	
$x_c = 15\ [1 - (\varphi V_c/V_u)] = 15\ [1 - (24.2/67.5)] = 9.6$ ft	

EXAMPLE 13.1 - Continued

Calculations and Discussion	Code Reference

4. Determine distance x_m from support over which minimum shear reinforcement must be provided (i.e., up to where $V_u = \varphi V_c/2$). 11.5.5.1

$$x_m = 15\,[(67.5 - 12.1)/67.5] = 12.3 \text{ ft}$$

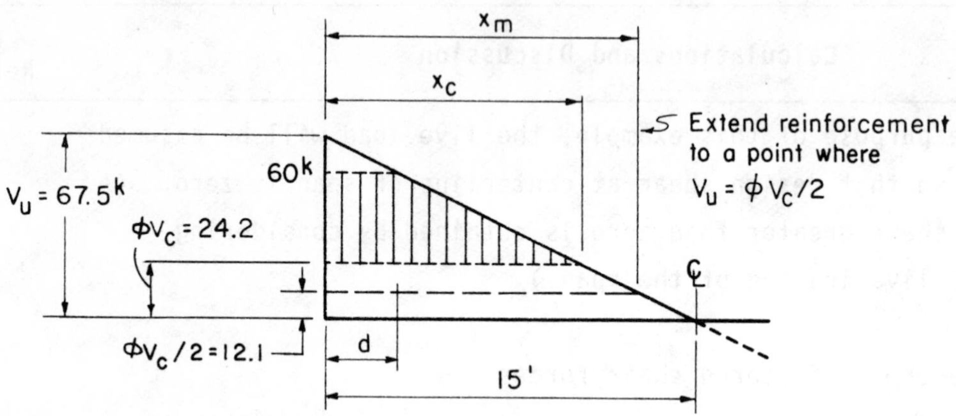

5. Determine required spacing of U-stirrups. 11.5.6.2

$$s\ (\text{req'd}) = \varphi A_v f_y d/(V_u - \varphi V_c)$$ Commentary Section 11.5.6

Assuming #4 U-stirrups (A_v = 0.40 in.^{2}),

@ distance d from support:

$$s\ (\text{req'd}) = 0.85\,(0.40)\,40\,(20)/(60 - 24.2) = 7.6 \text{ in.}$$

Since $(V_u - \varphi V_c)$ varies linearly between x = d and x = x_c and required spacing varies inversely with $(V_u - \varphi V_c)$, required spacing at any section between these points can be obtained directly from the value of s (req'd) corresponding to x = d.

EXAMPLE 13.1 - Continued

Calculations and Discussion	Code Reference

For instance, at a section $x = d + [(x_c - d)/2] = 5.63$ ft from support

s (req'd) = 7.6/1/2 = 15.2 in.

6. Check maximum permissible spacing of stirrups.

s (max) of vertical stirrups $\leq d/2 = 10$ in. 11.5.4.1
or ≤ 24 in.

s (max) of #4 U-stirrups corresponding to minimum reinforcement area requirements

s (max) = $A_v f_y/50 b_w$ = 0.40 (40,000)/50 (13) Eq. (11-14)
= 24.6 in.

<u>s (max) = 10 in.</u>

Summary:

Stirrup spacing using #4 U-stirrups:

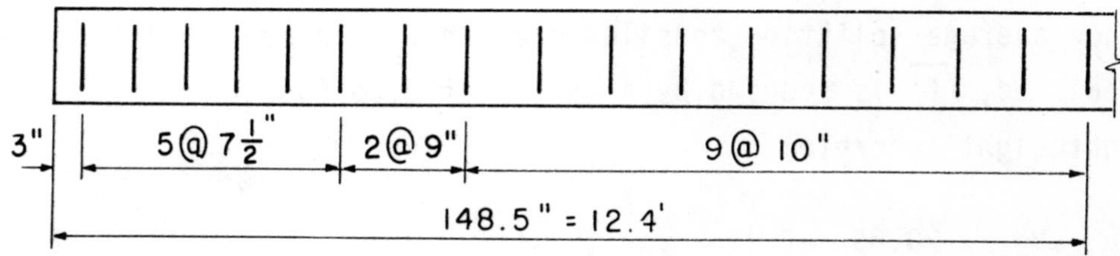

6 stirrups @ 7.5 in.

2 stirrups @ 9. in.

9 stirrups @ 10 in.

EXAMPLE 13.2 - Design for Shear - with Axial Tension

Determine required spacing of vertical U-stirrups for a beam subject to axial tension.

f'_c = 3,600 psi (sand-lightweight concrete, f_{ct} not specified)
f_y = 40,000 psi
M_d = 43.5 ft-kips
M_ℓ = 32.0 ft-kips
V_d = 12.8 kips
V_ℓ = 9.0 kips
N_d = - 2.0 kips (tension)
N_ℓ = -15.7 kips (tension)

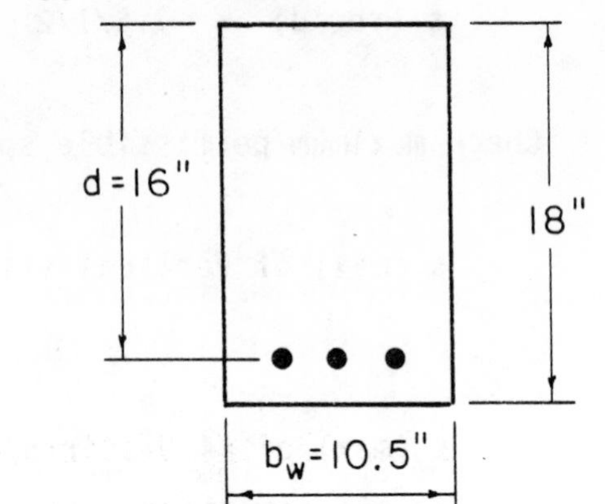

Calculations and Discussion	Code Reference
1. Determine factored loads	Eq. (9-1)

$$M_u = 1.4\ (43.5) + 1.7\ (32.0) = 115.3 \text{ ft-kips}$$

$$V_u = 1.4\ (12.8) + 1.7\ (9.0) = 33.2 \text{ kips}$$

$$N_u = 1.4\ (-2.0) + 1.7\ (-15.2) = -28.6 \text{ kips (tension)}$$

2. Determine shear strength provided by concrete

Since average splitting tensile strength f_{ct} is not specified, $\sqrt{f'_c}$ is reduced by a factor of 0.85 (sand-lightweight concrete). 11.2.1.2

$$\varphi V_c = 0.85 \quad \varphi 2 \left[1 + \frac{N_u}{500\ A_g}\right]\sqrt{f'_c}\ b_w d \qquad \text{Eq. (11-9)}$$

$$= 0.85\ (0.85)\ 2 \left[1 + \frac{(-28,600)}{500\ (18)\ 10.5}\right]\sqrt{3,600}\ (10.5)\ 16$$

$$= 10.2 \text{ kips}$$

EXAMPLE 13.2 - Continued

Calculations and Discussion	Code Reference

3. Check adequacy of cross-section.

$(V_u - \varphi V_c) \leq 0.85\left(\varphi 8\sqrt{f'_c}\ b_w d\right)$ — 11.5.6.8

23.0 kips < 58.3 kips OK

4. Determine required spacing of U-stirrups

s (req'd) = $\varphi A_v f_y d/(V_u - \varphi V_c)$ — Commentary Section 11.5.6

Assuming #3 U-stirrups (A_v = 0.22 in.^{2}),

s (req'd) = 0.85 (0.22) 40 (16)/23.0 = 5.2 in.

5. Check maximum permissible spacing of stirrups

$(V_u - \varphi V_c) \leq 0.85\left(\varphi 4\sqrt{f'_c}\ b_w d\right)$ — 11.5.4.3

23.0 kips < 29.1 kips OK

Provisions of Section 11.5.4.1 apply

s (max) of vertical stirrups ≤ d/2 = 8 in. — 11.5.4.1
or ≤ 24 in.

s (max) of #3 U-stirrups corresponding to minimum reinforcement area requirements

s (max) = $A_v f_y/50\ b_w$ = 0.22 (40,000)/50 (10.5) = 16.8 in.

<u>s (max) = 8 in.</u>

Summary:

<u>Use #3 vertical stirrups @ 5.0 in. spacing.</u>

EXAMPLE 13.3 - Design for Shear - with Axial Compression

A tied compression member has been designed for the given load conditions. However, the original design did not take into account the fact that, under a reversal in the direction of lateral load (wind), the axial load, due to the combined effects of gravity and lateral loads, becomes P_u = 10 kips, with essentially no change in the values of M_u and V_u. Check shear reinforcement requirements for the column under (1) original design loads and (2) reduced axial load.

M_u = 86 ft-kips
P_u = 160 kips
V_u = 20 kips
f'_c = 3,000 psi
f_y = 40,000 psi

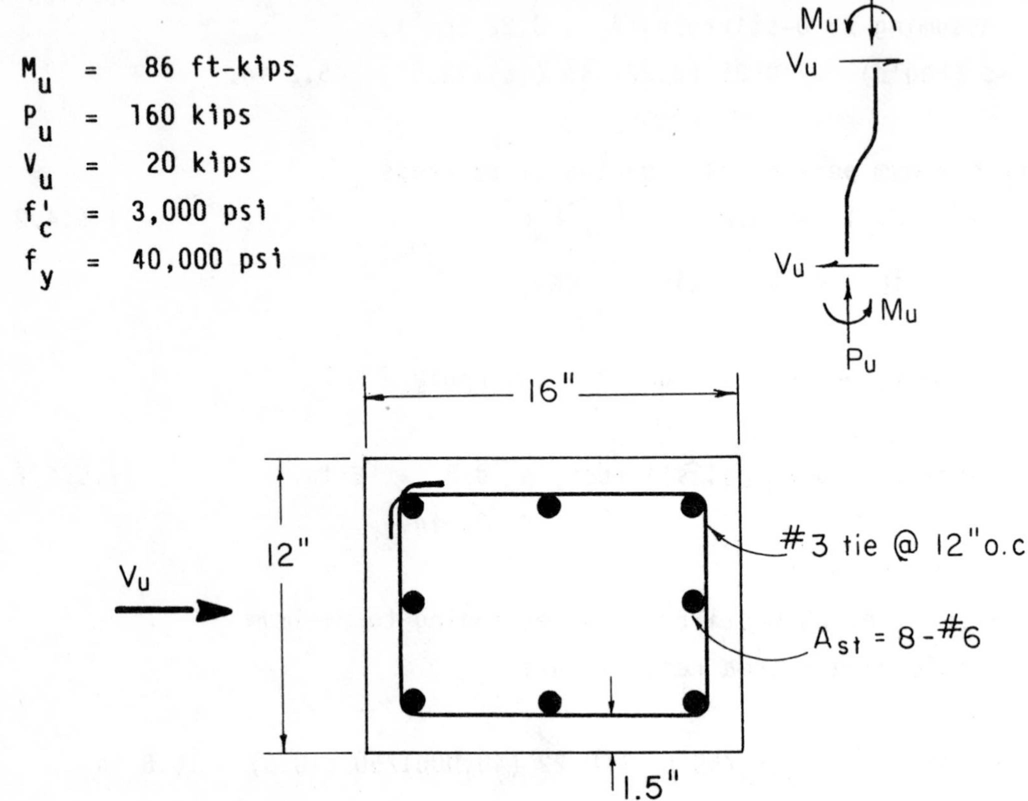

Calculations and Discussion	Code Reference

Condition 1: $P_u = N_u$ = 160 kips

1. Determine shear strength provided by concrete

$$d = 16 - [1.5 + 0.375 + (0.750/2)] = 13.75 \text{ in.}$$

$$\varphi V_c = \varphi 2 \left[1 + \frac{N_u}{2{,}000\ A_g}\right] \sqrt{f'_c}\ b_w d$$ Eq. (11-4)

EXAMPLE 13.3 - Continued

Calculations and Discussion	Code Reference

$$\varphi V_c = 0.85\ (2) \left[1 + \frac{160{,}000}{2{,}000\ (16)\ 12}\right] \sqrt{3{,}000}\ (12)\ 13.75$$

$$= 21.8 \text{ kips} > 20 \text{ kips}$$

Condition 2: $P_u = N_u = 10$ kips

1. Determine shear strength provided by concrete.

$$\varphi V_c = 0.85\ (2) \left[1 + \frac{10{,}000}{2{,}000\ (16)\ 12}\right] \times$$ Eq. (11-14)

$$\times[\sqrt{3{,}000}\ (12)\ 13.75]$$

$$= 15.8 \text{ kips} < 20 \text{ kips}$$

Shear reinforcement must be provided to carry excess shear.

2. Determine maximum permissible spacing of #3 ties

$$s\ (\max) = d/2 = 13.75/2$$ 11.5.4.1

$$= 6.9 \text{ in.} < 12 \text{ in. (provided)}$$ NG

<u>reduce spacing of #3 ties from 12 in. to 6.75 in. o.c.</u>

3. Check total shear strength with #3 @ 6.75 in.

$$\varphi V_s = \varphi A_v f_y \frac{d}{s} = 0.85\ (0.22)\ 40\ (13.75)/6.75 = 15.2 \text{ kips}$$

$$\varphi V_c + \varphi V_s = 15.8 + 15.2 = 31.0 \text{ kips} > 20 \text{ kips}$$

EXAMPLE 13.4 - Design for Shear - Concrete Floor Joist

Check shear requirements in the uniformly loaded floor joist shown below.

f'_c = 3,600 psi $\qquad f_y$ = 40,000 psi

w_d = 58 psf $\qquad w_\ell$ = 120 psf

Assumed longitudinal reinforcement:

One #5 bottom bar

One #6 trussed bar

One #5 top bar

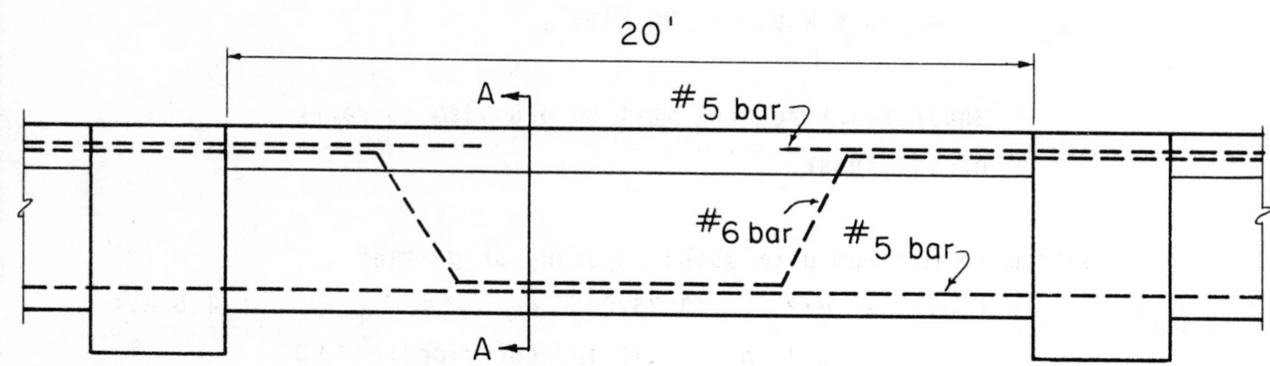

Joist Elevation

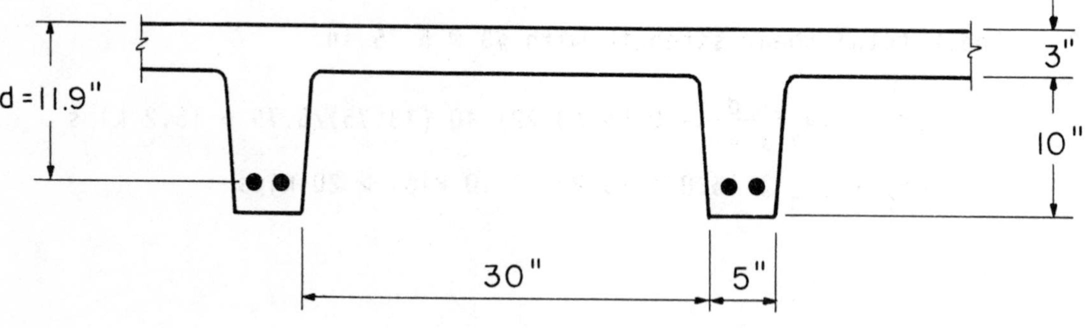

Section A-A

EXAMPLE 13.4 - Continued

Calculations and Discussion	Code Reference

1. Determine factored load.

w_u = [1.4 (58) + 1.7 (120)] 35/12 = 832 plf — Eq. (9-1)

2. Determine factored shear force.

@ distance d from support: — 11.1.2.1

V_u = 0.832 (10) - 0.832 (11.9/12) = 7.5 kips — 8.3.3

3. Determine shear strength provided by concrete.

According to the provisions of Section 8.11.8, V_c may be increased by 10 percent.

$\varphi V_c = 1.1\ \varphi 2 \sqrt{f'_c}\ b_w d$ — Eq. (11-3)

$= 1.1\ (0.85)\ 2 \sqrt{3{,}600}\ (5)\ 11.9$

$= 6.7$ kips < 7.5 kips NG

Calculate V_c using Eq. (11-6)

Compute ρ_w and $V_u\ d/M_u$ at distance d from support:

$\rho_w = A_s/b_w d = (0.31 + 0.44)/5\ (11.9) = 0.0126$

$M_u = -w_u \ell_n^2/11 + w_u \ell_n\ d/2 - w_u d^2/2$ — 8.3.3

$= -0.832\ (20)^2/11 + 0.832\ (20)\ (11.9/12)/2 - (0.832/2)\ (11.9/12)^2$

$= -30.3 + 8.3 - 0.4 = -22.4$ ft-kips

$V_u d/M_u = 7.5\ (11.9)/22.4\ (12) = 0.33 < 1$ — 11.3.2.1

$\varphi V_c = 1.1\ (0.85)\ [1.9 \sqrt{3{,}600} + 2{,}500\ (0.0126)\ 0.33] \times [5\ (11.9)]$

$= 6.9$ kips < 7.5 kips NG

EXAMPLE 13.4 - Continued

Calculations and Discussion	Code Reference

In accordance with the provisions of Section 8.11.8, the shear strength of concrete joist floor construction may be increased by use of shear reinforcement or by widening the end of ribs. Therefore, an increase in the joist section near the supports will be considered as follows.

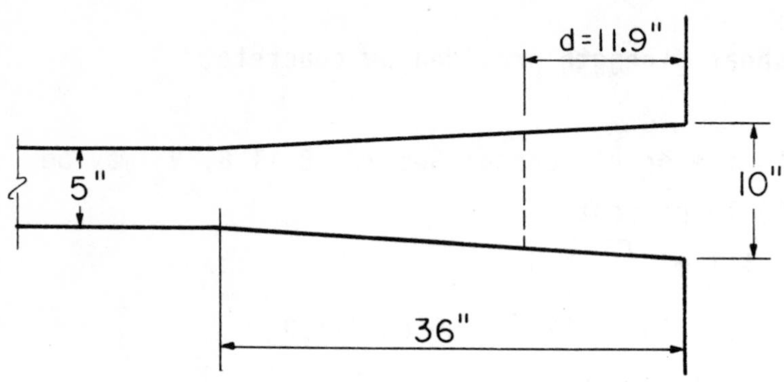

Plan View of End of Joist

Compute b_w at distance d from face of support

$$b_w = 5 + 5\,[(36 - 11.9)/36] = 8.3 \text{ in.}$$

shear strength provided by concrete at distance d from support

$$\varphi V_c = 1.1(0.85)2\sqrt{3,600}(8.3)11.9 \qquad \text{Eq. (11-3)}$$

$$= 11.1 \text{ kips} > 7.5 \text{ kips} \quad \text{OK}$$

EXAMPLE 13.5 - Design for Shear - Shear Strength at Web Openings

The simply supported prestressed double tee beam shown below has been designed without web openings to carry a live load of 50 psf (w_u = 1520 lb/ft). Two 10-in.-deep by 36-in.-long web openings are required for passage of mechanical and electrical services. Investigate the shear strength of the beam at web opening A.

This design example is based on an experimental and analytical investigation reported in "Behavior and Design of Prestressed Concrete Beams with Large Web Openings," Research and Development Bulletin RD054D, Portland Cement Association, Skokie, Ill. Also Code Commentary Reference 11.8.

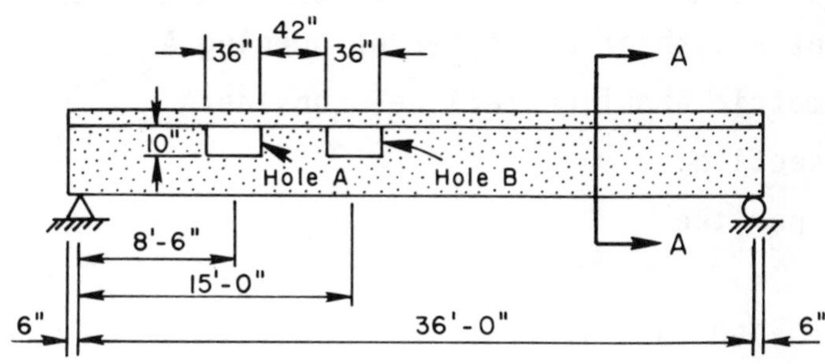

ELEVATION

Beam f'_c = 6,000 psi

Topping f'_c = 3,000 psi

f_{pu} = 270,000 psi

f_y = 60,000 psi

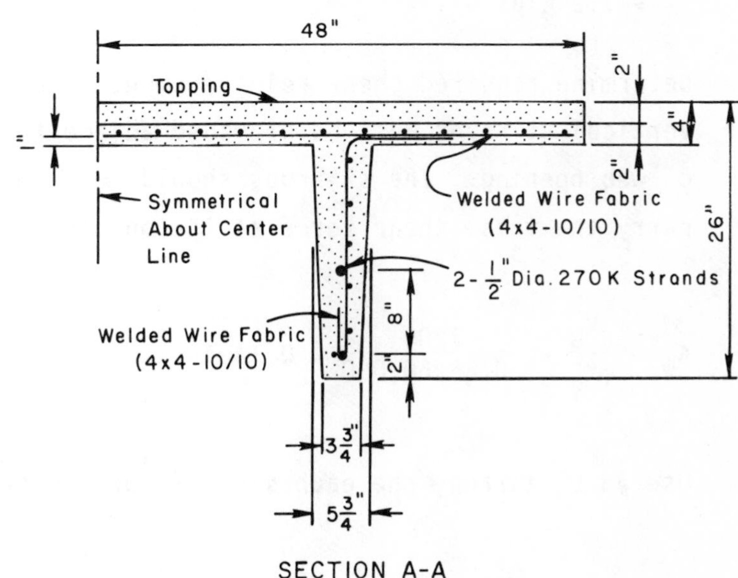

SECTION A-A

EXAMPLE 13.5 - Continued

This example treats only the shear strength considerations for the web opening. Other strength considerations need to be investigated, such as: to avoid slip of the prestressing strand, openings must be located outside the required strand development length, and strength of the struts to resist flexure and axial loads must be checked. The reader is referred to the complete design example in RD054D for such calculations. The design example in RD054D also illustrates procedures for checking service load stresses and deflections around the openings.

Calculations and Discussion	Code Reference

1. Determine factored moment and shear at center of opening A. Since double tee is symmetric about centerline, consider one-half of double tee section.

w_u = 1520/2 = 760 lb/ft per tee.

$M_u = 0.760\ (36/2)\ 8.5 - 0.760\ (8.5^2)/2$

$= 1066$ ft-kips

$V_u = 0.760\ (36/2) - 0.760\ (8.5)$

$= 7.2$ kips

2. Determine required shear reinforcement adjacent to opening. Vertical stirrups must be provided adjacent to both sides of web opening. The stirrups should be proportioned to carry the total shear force at the opening.

$$A_v = \frac{V_u}{\varphi f_y} = \frac{7200}{0.85 \times 60{,}000} = 0.14 \text{ in.}^2$$

Use #3 U-stirrup, one each side of opening (A_v = 0.22 in.^{2})

EXAMPLE 13.5 - Continued

Calculations and Discussions	Code Reference

3. Using a simplified analytical procedure developed in reference RD054D, the axial and shear forces acting on the "struts" above and below opening A are calculated. Results are shown on the diagram below. The reader is referred to the complete design example in RD054D for the actual force calculations. Axial forces should be accounted for in the shear design of the struts.

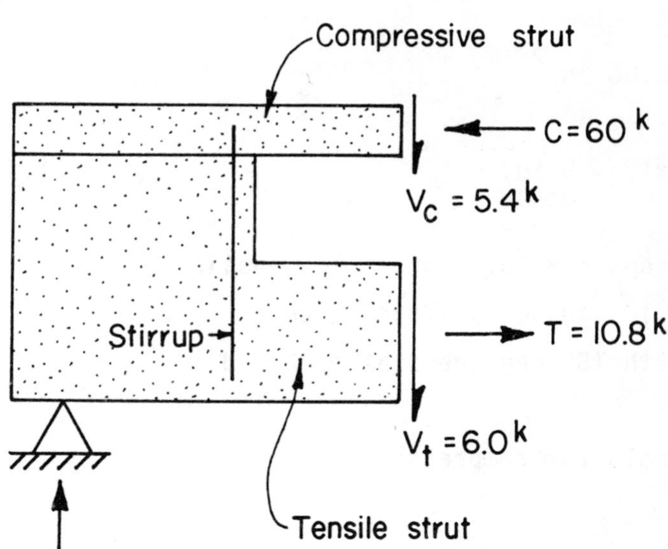

4. Investigate shear strength for tensile strut.

V_u = 6.0 kips

N_u = -10.8 kips — 11.0

d = 0.8h = 0.8(12) = 9.6 in. — 11.0

b_w = minimum width of tensile strut = 3.75 in.

Calculations and Discussions	Code Reference

$$V_c = 2\left(1 + \frac{N_u}{500A_g}\right)\sqrt{f'_c}\, b_w d$$ (Eq. 11-9)

$$= 2\left(1 - \frac{10,800}{500\text{x}53.2}\right)\sqrt{6000}\,(3.75)(9.6)$$

$$= 3.31 \text{ kips}$$

$$\varphi V_c = 0.85\,(3.31) = 2.82 \text{ kips}$$ 9.3.2.3

$$V_u > \varphi V_c$$

6.0 > 2.82 (shear reinforcement required in tensile strut)

$$A_v = \frac{(V_u - \varphi V_c)s}{\varphi f_y d}$$

$$= \frac{(6.0 - 2.82)9}{0.85\text{x}60\text{x}9.6} = 0.06 \text{ in.}^2$$

where s = 0.75h = 0.75x12 = 9 in. 11.5.4.1

Use #3 single leg stirrups @ 9-in. centers in tensile strut. (A_v = 0.11 in.^{2}). Anchor stirrups around prestressing strands with 180 deg. bend at each end.

5. Investigate shear strength for compressive strut.

V_u = 5.4 kips
N_u = 60 kips
d = 0.8h = 0.8(4) = 3.2 in.
b_w = 48 in.

$$V_c = 2\left(1 + \frac{N_u}{2000A_g}\right)\sqrt{f'_c}\, b_w d$$ (Eq. 11-4)

$$= 2\left(1 + \frac{60,000}{2000\text{x}192}\right)\sqrt{3000}\,(48)(3.2)$$

$$= 19.5 \text{ kips}$$

Example 13.5 - Continued

Calculations and Discussions	Code Reference

$\varphi V_c = 0.85\ (19.5) = 16.5$ kips

$V_u < \varphi V_c$

$5.4 < 16.5$ (shear reinforcement not required in compressive strut.)

6. Design Summary - See reinforcement details below.

(a) Use U-shaped #3 stirrup adjacent to both edges of opening to contain cracking within the struts.

(b) Use single-leg #3 stirrups at 9-in. centers as additional reinforcement in the tensile strut.

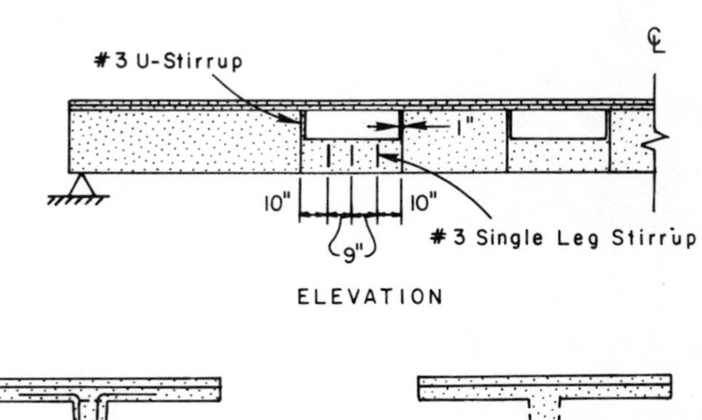

ELEVATION

U-SHAPED STIRRUP

SINGLE-LEG STIRRUP

DETAILS OF ADDITIONAL REINFORCEMENT

A similar design procedure is required for opening B.

14

Torsion

Introduction

The code provisions for torsion are presented in the same format as that used for shear, with the code requirements expressed in terms of the factored torsional moment T_u directly, using the basic torsional moment strength equality:

required torsional moment strength ≤ design torsional moment strength

$$T_u \leq \varphi T_n \qquad \text{Eq. (11-20)}$$

$$\leq \varphi T_c + \varphi T_s \qquad \text{Eq. (11-21)}$$

where the design torsional moment strength φT_n is simply the sum of torsional moment strength provided by concrete φT_c plus torsional moment strength provided by torsion reinforcement φT_s.

Similar to shear, the code equations expressing area of torsion reinforcement A_t are presented in terms of torsional moment strength T_s for direct application into Eqs. (11-20) and (11-21), rather than A_t directly. As an aid to the designer, and to assure correct application of the strength reduction factor φ, equations for computing required area of torsional reinforcement A_t are developed in Code Commentary Section 11.6.9.

Design for torsion is analagous to that for shear. The factored torsional moment T_u is first computed; then the torsional moment strength T_c provided by the concrete is determined. If this value, modified by the strength reduction factor φ, is less than the factored torsional moment, torsion reinforcement must be provided to carry the difference. Torsion reinforcement, when required, must consist of both closed stirrups and longitudinal bars. Shear reinforcement requirements are added to the torsion requirements to determine size and spacing of combined closed stirrups. Similarly, longitudinal bars for torsion are added to those for flexure and axial force.

The interaction and design requirements for shear and torsion can be represented by elliptical interaction curves such as that shown in Fig. 14-1. Definitions of the Zones of the interaction diagram are given in Decision Table 14-1, which also give the governing code sections and applicable code equations. In Fig. 14-1, the inner curve represents the strength provided by the concrete to resist combined shear and torsion. The outer curve is the maximum combination of shear and torsion strength permitted by the Code.

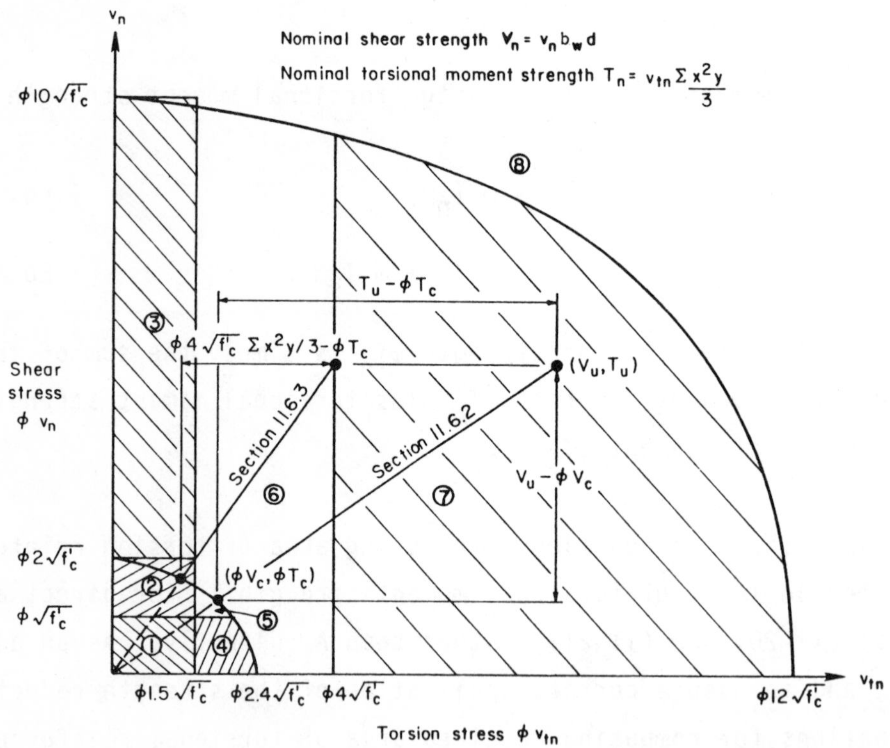

Fig. 14-1 - Interaction of Shear and Torsion

TABLE 14-1 DECISION TABLE FOR COMBINED SHEAR AND TORSION

Interaction Zone Fig. 14-1	Design Conditions	Code Reference	Required Reinforcement
(1)	1. $T_u < \varphi(0.5\sqrt{f'_c}\Sigma x^2 y)$ Torsion may be neglected 2. $V_u < \varphi V_c/2$	11.6.1 11.5.5.1	None
(2)	1. $T_u < \varphi(0.5\sqrt{f'_c}\Sigma x^2 y)$ Torsion may be neglected 2. $\varphi V_c > V_u > \varphi V_c/2$	11.6.1 11.5.5.3	Minimum Shear Only $A_v = \dfrac{50\ b_w s}{f_y}$ (11-14)
(3)	1. $T_u < \varphi(0.5\sqrt{f'_c}\Sigma x^2 y)$ Torsion may be neglected 2. $V_u > \varphi V_c$	11.6.1 11.5.6.1	Calculate shear only $A_v = \dfrac{(V_u - \varphi V_c)s}{\varphi f_y d}$ (11-17)
(4)	1. $T_u > \varphi(0.5\sqrt{f'_c}\Sigma x^2 y)$ 2. $V_u < \varphi V_c/2$	11.6.1 11.5.5.1	Minimum torsion only $2\ A_t = \dfrac{50\ b_w s}{f_y}$ (11-16) A_ℓ = Eq. (11-24) or (11-25)
(5)	1. $T_u > \varphi(0.5\sqrt{f'_c}\Sigma x^2 y)$ 2. $\varphi V_c > V_u > \varphi V_c/2$	11.6.1 11.5.5.5	Minimum combined shear and torsion $A_v + 2\ A_t = \dfrac{50\ b_w s}{f_y}$ (11-16) A_ℓ = Eq. (11-24) or (11-25)
(6) or (7)	1. $T_u > \varphi T_c$ 2. Torsional moment required for equilibrium 3. Design for T_u	11.6.9.1 11.6.2	Calculate combined shear and torsion $A_t = \dfrac{(T_u - \varphi T_c)s}{\varphi f_y \alpha_t x_1 y_1}$ (11-23) A_ℓ = Eq. (11-24) or (11-25)
(6)	1. $T_u > \varphi T_c$ 2. Uncracked section analysis for torsional moment T_u 3. Design for T_u or over design for cracking torque $T_u = \varphi(4\sqrt{f'_c}\Sigma x^2 y/3)$	11.6.9.1	Calculate combined shear and torsion $A_t = \dfrac{(T_u - \varphi T_c)s}{\varphi f_y \alpha_t x_1 y_1}$ (11-23) A_ℓ = Eq. (11-24) or (11-25)
(7)	1. $T_u > \varphi T_c$ 2. Redistribution of torsional moment after cracking 3. Design for cracking torque $T_u = \varphi(4\sqrt{f'_c}\Sigma x^2 y/3)$	11.6.9.1 11.6.3	Calculate combined shear and torsion $A_t = \dfrac{(T_u - \varphi T_c)s}{\varphi f_y \alpha_t x_1 y_1}$ (11-23) A_ℓ = Eq. (11-24) or (11-25)
(8)	1. $T_u > \varphi\ 5\ T_c$	11.6.9.4	Increase member section

11.6.1 When Torsion Must Be Considered

Torsion effects may be neglected when the factored torsional moment T_u is equal to or less than $\varphi(0.5\sqrt{f'_c}\ \Sigma x^2 y)$, corresponding to a torsion stress of $1.5\sqrt{f'_c}$. This will include Zones (1), (2), & (3) in Fig. 14-1. See Fig. 14-2 for evaluation of $\Sigma x^2 y$ for typical member cross sections.

11.6.2 Torsional Moment to Maintain Equilibrium

As discussed in Code Commentary Sections 11.6.2 and 11.6.3, there are structural framing conditions in which the torsional moment is introduced by statically determinate loads. In such cases the member must be designed for the full torsional moment, as shown in either Zone (6) or (7) of Fig. 14-1.

11.6.3 Reduction of Torsional Moment

Where torsional moments occur in statically indeterminate framing conditions, the magnitude of the torsional moment will depend on the redistribution of loads between the member under consideration and the interacting structure. If the torsional moment (before redistribution) is greater than $\varphi(4\sqrt{f'_c}\ \Sigma x^2 y/3)$, torsional cracking is assumed. This will permit a large twist in the member for load redistribution while limiting the torsional moment to the cracking value. Thus, for Zones (6) and (7) in Fig. 14-1, the torsional moment for design may be taken conservatively as $\varphi(4\sqrt{f'_c}\ \Sigma x^2 y/3)$, corresponding to a torsional stress of $4\sqrt{f'_c}$. The reduced torsional moment, $\varphi(4\sqrt{f'_c}\ \Sigma x^2 y/3)$, is then used to determine adjusted shears and moment in the adjoining structural members. Note the Commentary caution concerning unusual framing conditions where cracking redistribution may not be realized.

If it is expected that the factored torsional moment will be less than the cracking torque, the structure may be analyzed using equilibrium and compatibility conditions (uncracked section analysis) to determine the torsional moment. If the torsional moment, so determined, is less than the cracking value, $\varphi(4\sqrt{f'_c}\ \Sigma x^2 y/3)$, the calculated value can be used to determine torsion reinforcement requirements. Such cases would fall in Zone (6) in Fig. 14-1.

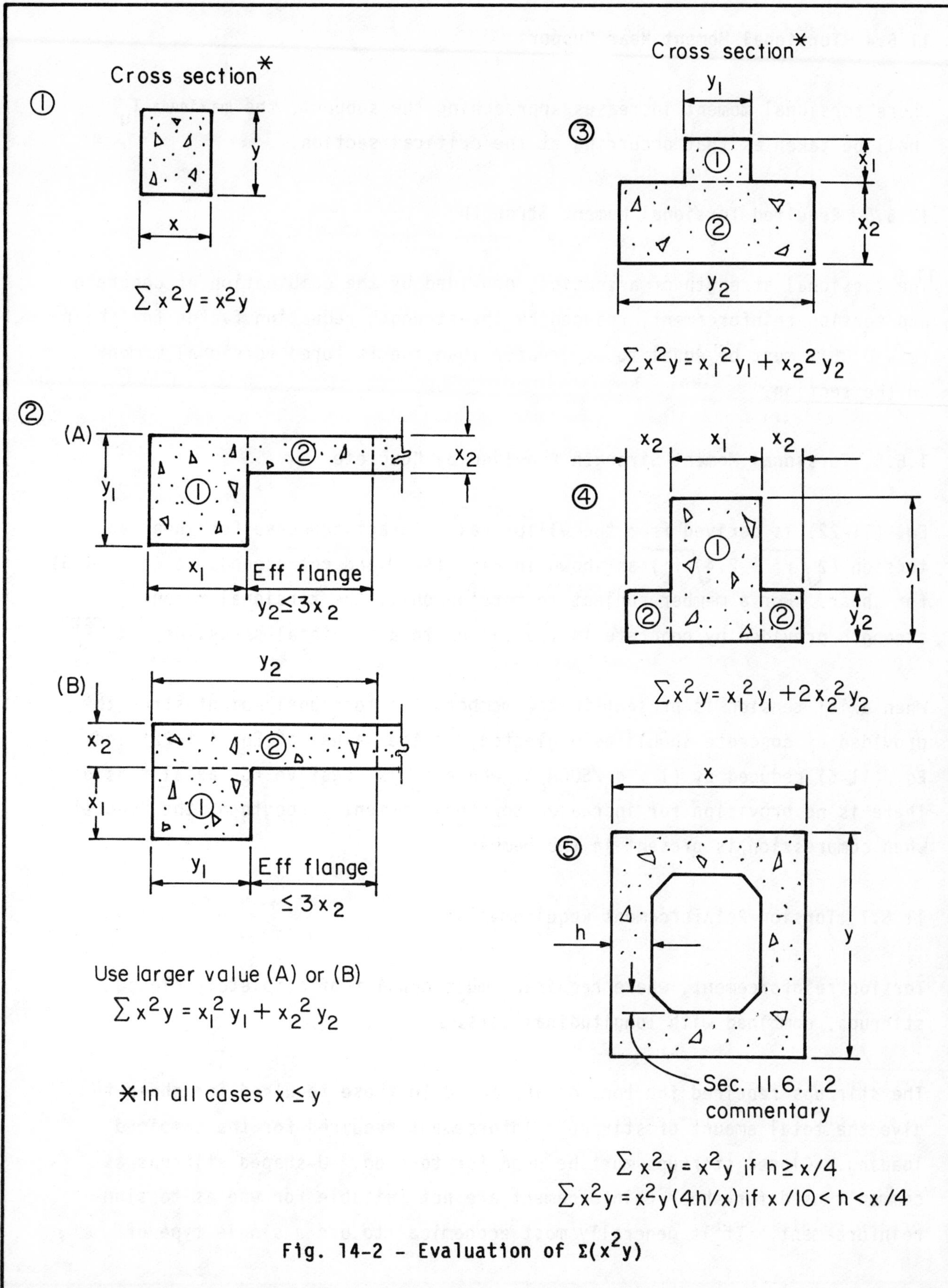

Fig. 14-2 - Evaluation of $\Sigma(x^2y)$

11.6.4 Torsional Moment Near Support

Where torsional moment increases approaching the support, the maximum T_u shall be taken as that occurring at the critical section.

11.6.5 Required Torsional Moment Strength

The torsional strength of a section, provided by the combination of concrete and torsion reinforcement, reduced by the strength reduction factor for shear (φ = 0.85), must be equal to or greater than the factored torsional moment on the section.

1.6.6 Torsional Moment Strength Provided by Concrete

Eq. (11-22) is derived from the elliptical interaction curve for shear and torsion ($2\sqrt{f'_c} + 2.4\sqrt{f'_c}$) as shown in Fig. 14-1, and corresponds to Eq. (11-5) for shear. For a member subject to torsion only, the torsional moment strength provided by concrete is equivalent to a torsional stress of $2.4\sqrt{f'_c}$.

When axial tension is present in the member, the torsional moment strength provided by concrete should be neglected, or the values of Eq. (11-22) and Eq. (11-5) reduced by $(1 + N_u/500A_g)$, where N_u is negative for axial tension. There is no provision for increased torsional moment strength of the concrete when compression is present in the member.

11.6.7 Torsion Reinforcement Requirements

Torsion reinforcement, where required, must consist of completely closed stirrups, combined with longitudinal bars.

The stirrups required for torsion are added to those required for shear to give the total amount of stirrup reinforcement required for the combined loading. Closed stirrups must be used for torsion. U-shaped stirrups as commonly used for shear reinforcement are not suitable for use as torsion reinforcement. It is generally most economical to use a single type of stirrup in a particular beam and to combine the shear and torsion reinforcement. In this case all the stirrups must be closed stirrups. When the reinforcement is combined in this way, the most restrictive requirements for spacing and placement of both torsion and shear reinforcement must be met. In large beams, subject to heavy shear loads it may be convenient to provide multiple-legged stirrups to carry the shear and closed perimeter stirrups to carry the torsion. Detailing for closed stirrups used as torsion reinforcement is discussed on page 2-13, with recommended two-piece closed stirrup details illustrated in Figs. 2-9 and 2-10 on page 2-14.

The longitudinal bars required for torsion must be distributed around the perimeter of the closed stirrups. At least one longitudinal bar must be placed in each corner of the closed stirrups. The longitudinal torsion bars required near the flexural tension and compression faces of a beam may be combined with the areas of flexural reinforcement required when the reinforcement in these locations is detailed. When detailing the longitudinal torsion bars, due care should be taken to ensure that it is effectively anchored so that its yield strength can be developed as assumed in design. This requires that the longitudinal bars have full development length provided beyond the $(d + b_t)$ distance called for in Section 11.6.7.6. Practically, this will usually mean continuous longitudinal reinforcement for the full length of the member. Note: With the 1980 Supplemental revisions, the expression defining the distance torsion reinforcement must be extended was clarified to be equal to the effective depth d plus the width of that part of the cross section containing the closed stirrups resisting torsion b_t.

11.6.8 Spacing Limits for Torsion Reinforcement

These provisions insure proper performance of the torsion reinforcement cage and simplify reinforcement fabrication. Where combined shear and torsion stirrups are used, the spacing may be controlled by the shear requirement of d/2, as provided in Section 11.5.4.1.

11.6.9 Design of Torsion Reinforcement

When the concrete section alone is not adequate for torsion, reinforcement must be provided in accordance with Eqs. (11-23), (11-24) and (11-25). For the usual case of combined shear and torsion, Eq. (11-23) will be more convenient for design in the form:

$$\frac{A_t}{s} = \frac{T_s}{\varphi f_y \alpha_t x_1 y_1} = \frac{(T_u - \varphi T_c)}{\varphi f_y \alpha_t x_1 y_1}$$

which gives the required area of <u>one leg</u> of a closed stirrup to resist torsion within a distance s. This area can then be combined with the similar shear reinforcement requirement ($A_t/s + A_v/2s$) to give the required area of one leg of a combined stirrup per unit length of member. For a given stirrup size, the required stirrup spacing can be determined directly.

For flanged members in which closed stirrups are placed in more than one component of the section, Eq. (11-23) in combination with Eqs. (11-20) and (11-21) can be written:

$$\frac{A_t}{s} = \frac{(T_u - \varphi T_c)}{\varphi f_y \Sigma\alpha_t x_1 y_1}$$

assuming that all stirrups will be of the same bar size and spacing. The procedure for combining with shear reinforcement requirements would then be the same as above.

Minimum closed stirrup requirements for combined shear and torsion are set in Section 11.5.5.5, and will include Zone (5) in the interaction diagram of Fig. 14-1. When shear force is negligible ($V_u \leq \varphi V_c/2$), minimum area of closed stirrups for torsion can be determined from Eq. (11-16):

$$2A_t = \frac{50 b_w s}{f_y}$$

which applies to Zone (4) in Fig. 14-1. In more convenient form this becomes:

$$\frac{A_t}{s} = \frac{25 b_w}{f_y}$$

These minimum values may also apply in the lower strength levels of Zones (6) and (7). Stirrups (and longitudinal reinforcement to resist torsion) are not required in Zone (1), where both shear ($V_u \leq \varphi V_c/2$) and torsion ($T_u \leq \varphi 0.5\sqrt{f'_c}\ \Sigma x^2 y$) are minimal.

The contribution of the concrete to combined shear and torsion strength of a beam with web reinforcement varies between 40 and 100 percent of the torque at diagonal tension cracking. Therefore, in order that the strength of a cracked beam will be slightly greater than the cracking strength, a minimum amount of shear and torsion reinforcement must be provided to prevent failure at cracking.

In the case of shear without torsion, it has been found that a minimum web reinforcement area equal to 50 $b_w s/f_y$ will prevent a failure at shear cracking. For pure torsion, the contribution of the concrete to strength after cracking is much less than the cracking torque. Because of this, four times as much minimum web reinforcement is necessary to avoid failure at cracking in the case of pure torsion than in the case of shear, if equal volumes of web and longitudinal reinforcement are provided.

For the small quantities of reinforcement corresponding to minimum reinforcement, the contribution of the torsion reinforcement to ultimate strength is proportional to the total volume of longitudinal and web reinforcement, and is essentially independent of the ratio of web reinforcement to longitudinal reinforcement. Because of this, the Code is able to specify the same minimum web reinforcement for any combination of torsion and shear; that is, $A_v + 2A_t = 50\ b_w s/f_y$. But this requires the use of more than an equal volume of longitudinal reinforcement for torsion, so that the total volume of stirrups plus longitudinal bars will be sufficient to ensure a ductile failure. This is achieved by use of Eq. (11-25), which also reduces the minimum amount of longitudinal torsion reinforcement as the ratio of torsion to shear decreases. Eq. (11-25) will govern for A_ℓ rather than Eq. (11-24) if A_t is less than $\frac{100xs}{f_y}\left(\frac{T_u}{T_u + \frac{V_u}{3C_t}}\right)$. In Eq. (11-25), the x term is the length of the shorter side of the component rectangle of the cross-section which contains the torsion reinforcement. In a conventionally proportioned beam, x is the width of the web. In a wide, shallow beam, x is the overall depth of the beam.

If different yield strengths for the longitudinal bars and the closed stirrups are used, Eqs (11-24) and (11-25) are modified as follows:

$$A_\ell = 2A_t \left(\frac{x_1 + y_1}{s} \right) f_{vy}/f_y$$

and

$$A_\ell = \left[\frac{400xs}{f_{vy}} \left(\frac{T_u}{T_u + \frac{V_u}{3C_t}} \right) - 2A_t\ f_{vy}/f_y \right] \left(\frac{x_1 + y_1}{s} \right)$$

where f_{vy} = yield strength of closed stirrups

f_y = yield strength of longitudinal bars

The validity of the provisions for design of torsion reinforcement depends on the development of the yield strength of the reinforcement at ultimate load. To ensure that yielding will occur, the maximum amount of reinforcement is set by an upper limit on the torsional moment strength of $5T_c$, or

$$\frac{12\sqrt{f'_c}\ \Sigma x^2 y}{3\sqrt{1 + \left(\frac{0.4\ V_u}{C_t\ T_u} \right)^2}}$$

The corresponding upper limit on shear strength is

$$\frac{10\sqrt{f'_c}\ b_w d}{\sqrt{1 + \left(2.5\ C_t \frac{T_u}{V_u} \right)^2}}$$

This interaction relationship, shown as the outer curve in Fig. 14-1, limits the shear stress to $10\sqrt{f'_c}$ when torsion is zero, and the torsion stress to $12\sqrt{f'_c}$ when shear is zero.

EXAMPLE 14.1 - SPANDREL BEAM DESIGN FOR COMBINED SHEAR AND TORSION

Design spandrel beam CI in parking garage second level for combined shear and torsion, assuming sections noted to be adequately sized for flexure and shear. The omission of columns in parking structures in many cases introduces appreciable torsion into spandrel members.

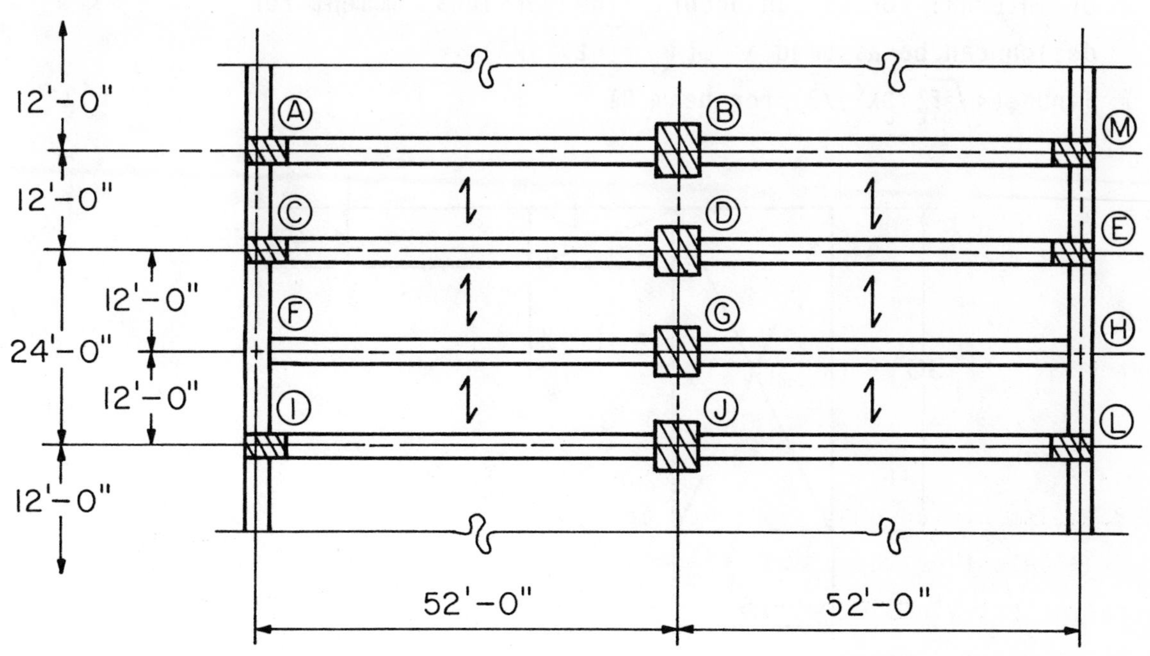

Partial Plan of Parking Garage

Note: Columns omitted at (F) and (H) for entry and exit

Design Criteria:

Typical bay = 12' x 52'

Slab thickness = 4-1/2"

Live load = 50 psf uniform, or 2^k concentrated

f'_c = 4,000 psi (normal-weight concrete)

f_y = 60,000 psi

Height = 10' (floor to floor)

All beams = 15" x 30"

Exterior columns = 15" x 24"

Interior columns = 24" x 24"

EXAMPLE 14.1 - Continued

Calculations and Discussion	Code Reference

1. The new provisions of Section 11.6.3 greatly simplify the determination of the torsional moment in beam CI, since it is part of an indeterminate framing system in which redistribution of internal forces can occur. The torsional moment for design can be assumed as $\varphi(4\sqrt{f'_c}\ \Sigma x^2y/3)$.

   Find $\varphi(4\sqrt{f'_c}\ \Sigma x^2y/3)$ for beam CI — 11.6.3

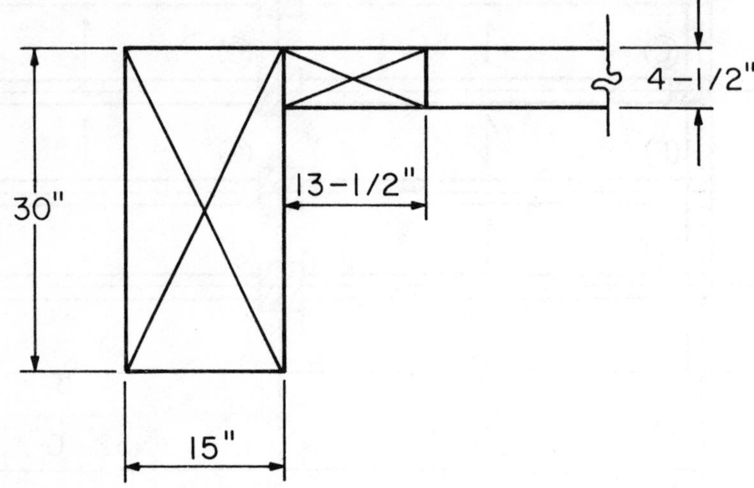

For slab portion

$y = 3 \times 4.5 = 13.5$ — 11.6.1.1

$\Sigma x^2y = 15^2 \times 30 + 4.5^2 \times 13.5 = 7023 \text{ in.}^3$

$\varphi = 0.85$ — 9.3.2.3

$$\varphi(4\sqrt{f'_c}\ \Sigma x^2y/3) = 0.85\ (4\sqrt{4{,}000} \times \frac{7023}{3}) = 503{,}397 \text{ in. lb}$$
$$= 42^{'k}$$

This value must be used in determining the redistribution of moment in beam FG. The resulting reaction at (F) will determine the shear in beam CI to be used in combination with the torsion. — 11.6.3.1

2. Determine fixed end moments in beam FG.

$$\text{Service DL} = \left(\frac{4.5}{12} \times 12 + \frac{25.5 \times 15}{144}\right) 150 = 1{,}073 \text{ lb/ft}$$

EXAMPLE 14.1 - Continued

Calculations and Discussion	Code Reference
Service LL = 50x12 = 600 lb/ft	
Factored load, U = 1.4 x 1,073 + 1.7 x 600 = 2,522 lb/ft	9.2.1

$$F.E.M. = \frac{w\ell^2}{12} = \frac{2.522 \times 52^2}{12} = 568^{'k}$$

3. Apply reduced restraining torsional moment from step 1 = $42^{'k}$. Since this torsional moment is applied from both sides of (F), the end moment on FG will be 2x42 = $84^{'k}$.

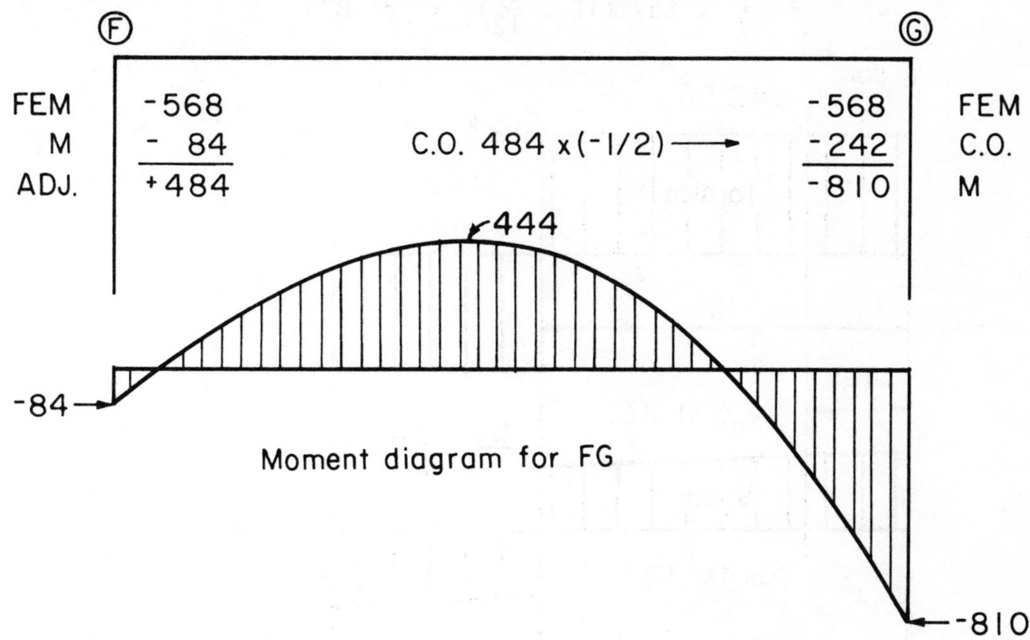

Moment diagram for FG

4. Find beam reaction at (F) and resulting shear in beam CI.

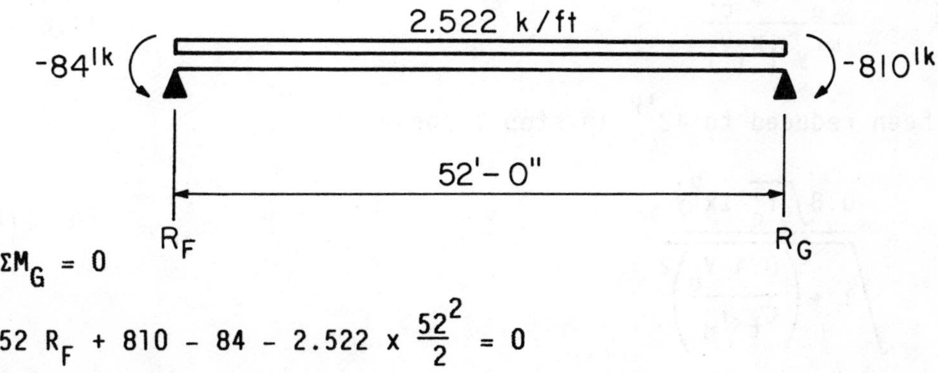

$\Sigma M_G = 0$

$$52\ R_F + 810 - 84 - 2.522 \times \frac{52^2}{2} = 0$$

EXAMPLE 14.1 - Continued

Calculations and Discussion	Code Reference

$$R_F = \frac{-810 + 84 + 3410}{52} = 51.6^k$$

$$\text{DL in CI} = \frac{30 \times 15}{144} 150 = 469 \text{ lb/ft} = 0.469^k\text{/ft}$$

$$U = 1.4 \times 0469 = 0.657^k\text{/ft}$$ 9.2.1

Critical section in CI: (Assume d = 27.5 in.)

$$27.5 + \frac{15}{2} = 35 \text{ in. from col. } ℄$$ 11.6.4

$$V_u = \frac{51.6}{2} + 0.657\left(12 - \frac{35}{12}\right) = 31.8^k$$

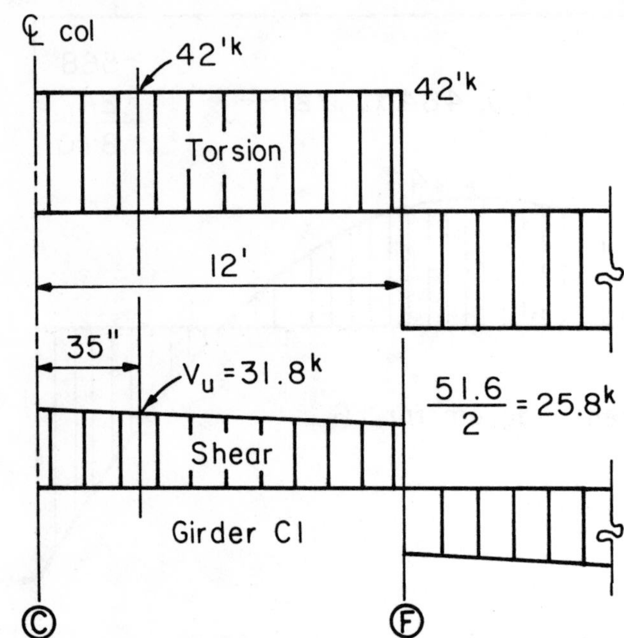

5. Calculate required area of closed stirrups for torsion.

$$\frac{A_t}{s} = \frac{(T_u - \varphi T_c)}{\varphi f_y \alpha_t x_1 y_1}$$ 11.6.5, 11.6.9.1

T_u has been reduced to $42^{'k}$ in step 1 above.

$$T_c = \frac{0.8\sqrt{f'_c}\,\Sigma x^2 y}{\sqrt{1 + \left(\frac{0.4\,V_u}{C_t\,T_u}\right)^2}}$$ Eq. (11-22)

EXAMPLE 14.1 - Continued

Calculations and Discussion	Code Reference

$$C_t = \frac{b_w d}{\Sigma x^2 y} = \frac{15 \times 27.5}{7023} = 0.0587$$ 11.0

$$T_c = \frac{0.8\sqrt{4{,}000} \times 7{,}023}{\sqrt{1 + \left(\frac{0.4 \times 31.8}{0.0587 \times 42 \times 12}\right)^2}}$$

Assuming 1-1/2 in. cover and #4 stirrup size 7.7.1

$$x_1 = 15 - 2\,(1.5 + 0.25) = 11.5 \text{ in.}$$

$$y_1 = 30 - 2\,(1.5 + 0.25) = 26.5 \text{ in.}$$

$$\alpha_t = 0.66 + 0.33\,\frac{26.5}{11.5} = 1.42$$ 11.6.9.1

$$\frac{A_t}{s} = \frac{(42 - 0.85 \times 27.2)\,12}{0.85 \times 60 \times 1.42 \times 11.5 \times 26.5} = 0.0103 \text{ in.}^2\text{/in./leg}$$

6. Calculate required area of stirrups for shear.

$$V_c = \frac{2\sqrt{f'_c}\,b_w d}{\sqrt{\left(1 + 2.5\,C_t\,\frac{T_u}{V_u}\right)^2}} = \frac{2\sqrt{4{,}000} \times 15 \times 27.5}{\sqrt{1 + \left(2.5 \times 0.0587 \times \frac{42 \times 12}{31.8}\right)^2}}$$ Eq. (11-5)

$$V_c = 20{,}610 \text{ lb} = 20.6^k$$

$$V_u = \varphi(V_c + V_s)$$ Eq. (11-1)

$$V_s = \frac{V_u}{\varphi} - V_c = \frac{31.8}{0.85} - 20.6 = 16.8^k$$ Eq. (11-2)

$$\frac{A_v}{s} = \frac{V_s}{f_y d} = \frac{16.8}{60 \times 27.5} = 0.0102 \text{ in.}^2\text{/in.}$$ Eq. (11-17)

7. Determine combined shear and torsion stirrup requirements.

$$\frac{A_t}{s} + \frac{A_v}{2s} = 0.0103 + \frac{0.0102}{2} = 0.0154 \text{ in.}^2\text{/in./leg}$$

Try #3 bar, $A_b = 0.11 \text{ in.}^2$

EXAMPLE 14.1 - Continued

Calculations and Discussion	Code Reference

$$s = \frac{0.11}{0.0154} = 7.14 \text{ in; Space \#3 closed stirrups at 7 in.}$$

8. Check maximum stirrup spacing.

$$\frac{x_1 + y_1}{4} = \frac{11.5 + 26.5}{4} = 9.5 > 7 \quad \text{O.K.}$$ 11.6.8.1

$$\frac{d}{2} = \frac{27.5}{2} = 13.75 > 7 \quad \text{O.K.}$$ 11.5.4.1

9. Check requirements at center of span.

$$V_s = \frac{25.8}{0.85} - 20.6 = 9.75^k$$

$$\frac{A_v}{s} = \frac{9.75}{60 \times 27.5} = 0.0059$$

$$\frac{A_t}{s} + \frac{A_v}{2s} = 0.0103 + \frac{0.0059}{2} = 0.0133 \text{ in.}^2\text{/in./leg}$$

$$s = \frac{0.11}{0.0133} = 8.27 \text{ in.; Use 7-in spacing full length}$$

10. Check minimum stirrup area. 11.6.9.2

$$A_v + 2A_t = \frac{50b_w s}{f_y} = \frac{50 \times 15 \times 7}{60{,}000} = 0.0875 \text{ in.}^2$$ Eq. (11-16)

Area provided $= 2 \times 0.11 = 0.22 \text{ in.}^2$ O.K.

11. Calculate longitudinal torsion reinforcement. 11.6.9.3

$$A_\ell = \frac{2A_t}{s}(x_1 + y_1) = 2 \times 0.0103\,(11.5 + 26.5)$$ Eq. (11-24)

$$= 0.783 \text{ in.}^2$$

$$A_\ell = \left[\frac{400 \times s}{f_y}\left(\frac{T_u}{T_u + \frac{V_u}{3C_t}}\right) - 2A_t\right]\left(\frac{x_1 + y_1}{s}\right)$$

(or substituting $\frac{50b_w s}{f_y}$ for $2A_t$)

$$\frac{50b_w s}{f_y} = 0.0875 < 2A_t = 2 \times 0.0103 \times 7 = 0.1442$$

Use $2A_t$

EXAMPLE 14.1 - Continued

Calculations and Discussion	Code Reference

$$A_{\ell} = \left[\frac{400 \times 15 \times 7}{60{,}000}\left(\frac{42 \times 12}{42 \times 12 + \dfrac{31.8}{3 \times 0.0587}}\right) - 0.1442\right]\left(\frac{11.5 + 26.5}{7}\right)$$

$$= 2.01 \text{ in.}^2$$

Provide A_{ℓ} = 2.01 in.^{2}. Place longitudinal bars around perimeter of the closed stirrups, spaced at not more than 12 in., and locate one longitudinal bar in each corner of the closed stirrups. Longitudinal bars may be combined with the flexural reinforcement.

12. Flexural analysis of beam CI: (Ignoring flange action)

Consider column-beam joint at Ⓒ:

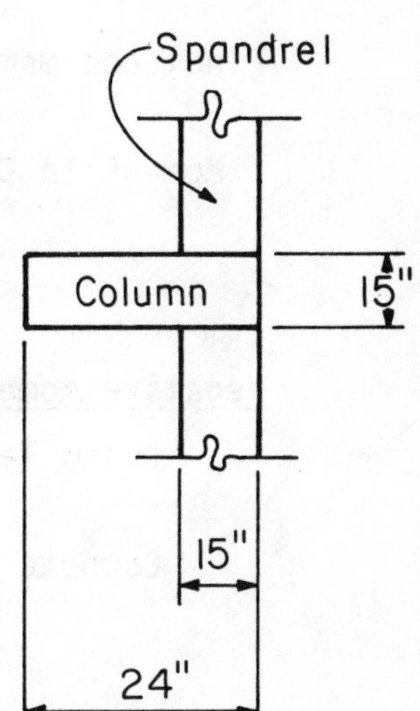

Plan joint Ⓒ

Col. stiffness (above) $= \dfrac{4EI}{L} = \dfrac{4E \times 24 \times 15^3}{120 \times 12} = 225E$

Col. stiffness (below) $= 225E$

Bm. CI stiffness $= \dfrac{4EI}{L} = \dfrac{4E \times 15 \times 30^3}{24 \times 12 \times 12} = 469E$

Bm. AC stiffness $= \dfrac{4EI}{L} = \dfrac{4E \times 15 \times 30^3}{12 \times 12 \times 12} = 938E$

$\Sigma = 1857E$

Dist. Factors at Ⓒ (and at Ⓘ):

Col. (above & below) $= \dfrac{225}{1857} = 0.121$

CA $= \dfrac{938}{1857} = 0.505$

CI $= \dfrac{469}{1857} = 0.253$

F.E.M.:

$$CA = \frac{w\ell^2}{12} = \frac{0.657 \times 12^2}{12} = 8^{'k}$$

$$CI = \frac{P\ell}{8} + \frac{w\ell^2}{12} = \frac{51.6 \times 24}{8} + \frac{0.657 \times 24^2}{12} = 186^{'k}$$

Calculations and Discussion	Code Reference

Two-step Moment Distribution: (Moments distributed to columns above and below not shown)

	Ⓒ		Ⓘ	
D.F.	.505	.253	.253	.505
F.E.M.	− 8	−186	−186	− 8
D.	− 90	+ 45	+ 45	− 90
C.O.	0	− 23	− 23	0
D.	− 12	+ 6	+ 6	− 12
	−110	−158	−158	−110

Final end moments in beam CI = 158'k

$$\text{Moment in CI} = \frac{P\ell}{8} + \frac{w\ell^2}{24} + (186 - 158)$$

$$= \frac{51.6 \times 24}{8} + \frac{0.657 \times 24^2}{24} + 28 = 199'^{k}$$

<u>Negative moment reinforcement</u>

Using Table 9-2, page 9-7:

$$\text{Compute } \frac{M_u}{\varphi f'_c bd^2} = \frac{158 \times 12}{0.9 \times 4 \times 15 \times 27.5^2} = 0.0464$$

From Table 9-2, read $\omega \simeq 0.048$

$$A_s = \rho bd = \frac{\omega f'_c bd}{f_y} = \frac{0.048 \times 4 \times 15 \times 27.5}{60} = 1.31 \text{ in.}^2$$

<u>Positive moment reinforcement</u>

$$\text{Compute } \frac{M_u}{\varphi f'_c\, bd^2} = \frac{199 \times 12}{0.9 \times 4 \times 15 \times 27.5^2} = 0.0585$$

From Table 9-2, read $\omega \simeq 0.061$

$$A_s = \frac{f'_c bd}{f_y} = \frac{0.061 \times 4 \times 15 \times 27.5}{60} = 1.67 \text{ in.}^2$$

EXAMPLE 14.1 - Continued

Calculations and Discussion	Code Reference

13. Size combined longitudinal reinforcement. Eight longitudinal bars are required for torsion reinforcement to meet maximum spacing requirements. 11.6.8.2

Two corner bars (top and bottom) will be combined with the flexural reinforcement.

Positive moment section:

$$\frac{A_\ell}{4} + A_s = \frac{2.01}{4} + 1.67 = 2.17 \text{ in.}^2$$

Use 6 - #6 bars (A_s = 2.64 in.^{2})

Negative moment section:

$$\frac{A_\ell}{4} + A_s = \frac{2.01}{4} + 1.31 = 1.81 \text{ in.}^2$$

Use 6 - #5 bars (A_s = 1.86 in.^{2})

Extended positive moment bars: 12.11.1

$$\frac{A_\ell}{4} + \frac{A_s}{4} = \frac{2.01}{4} + \frac{1.67}{4} = 0.92 \text{ in.}^2$$

Use 3 - #6 (A_s = 1.32 in.^{2})

Torsion bars in side of beam:

$$\frac{A_\ell}{8} = \frac{2.01}{8} = 0.25 \text{ in.}^2$$

Use #5 bar (A_b = 0.31 in.^{2})

Torsion bars in top corners of beam:

Extend two negative moment corner bars full length of beam.

Use #5 bar (A_b = 0.31 in.^{2})

EXAMPLE 14.1 - Continued

Calculations and Discussion	Code Reference

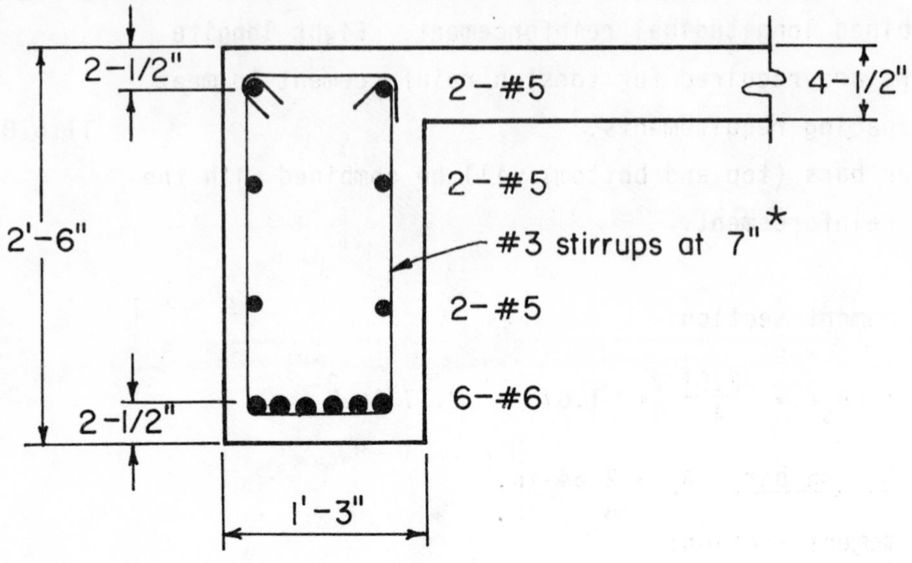

Positive Moment Section

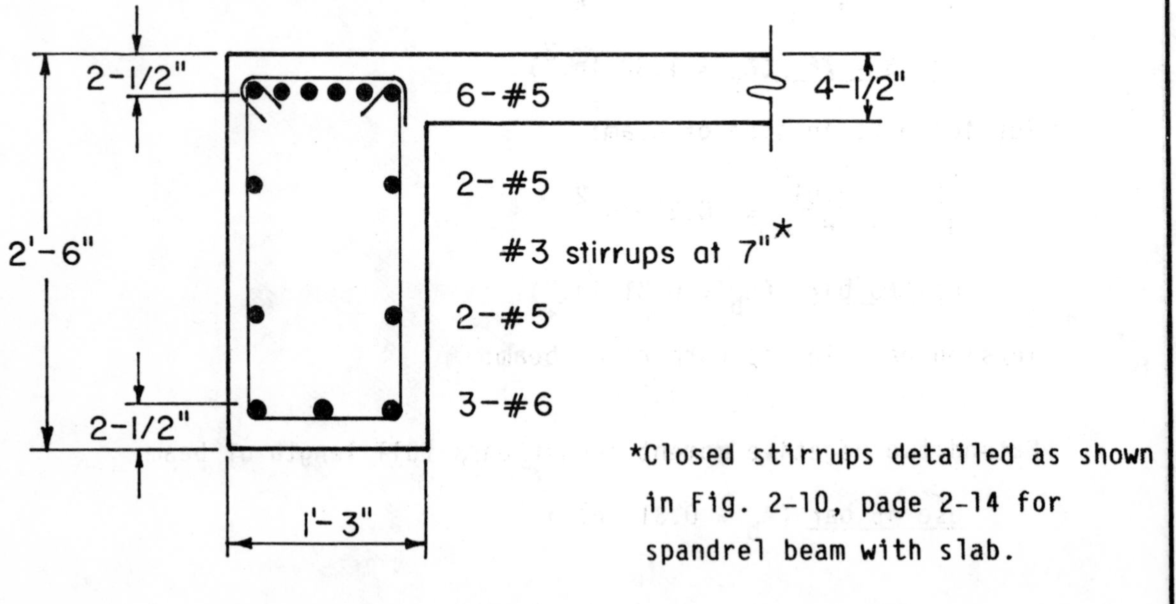

*Closed stirrups detailed as shown in Fig. 2-10, page 2-14 for spandrel beam with slab.

Negative Moment Section

EXAMPLE 14.2 - Precast Spandrel Beam Design for Combined Shear and Torsion

Design a precast reinforced concrete spandrel beam for combined shear and torsion. Roof members are simply supported on spandrel ledge. Spandrel beams are connected to columns to transfer torsion. Continuity between spandrel beams is not provided.

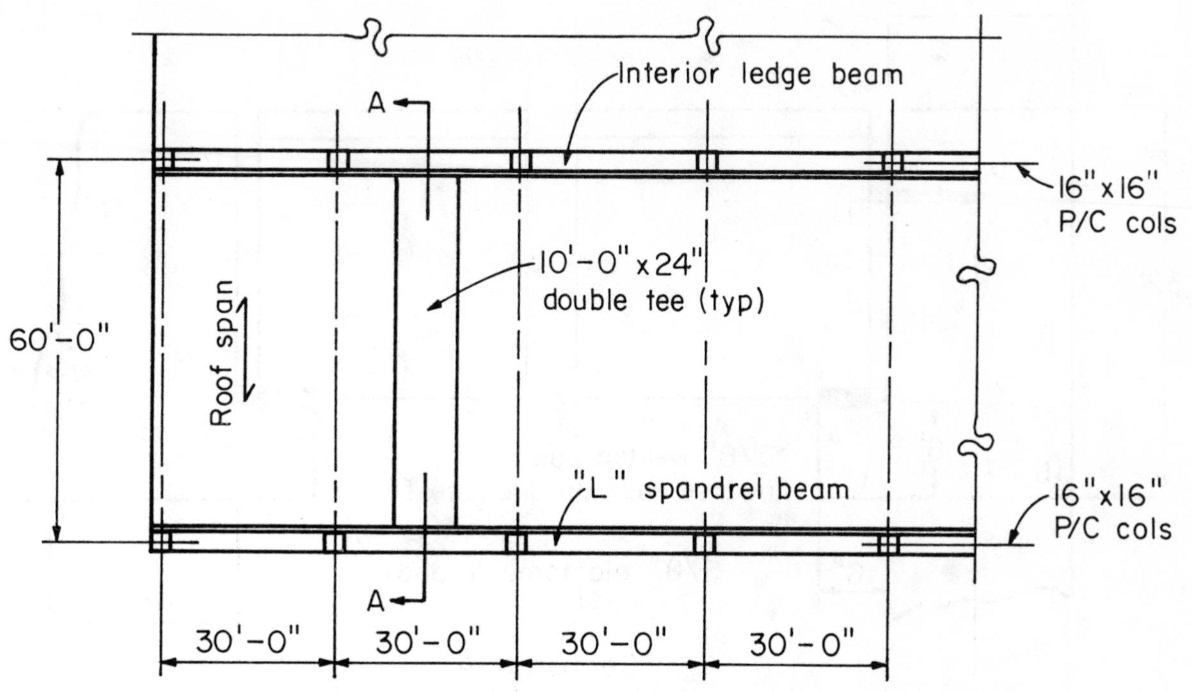

Partial plan of precast roof system

Design Criteria:

Live load = 30 psf

Dead Load = 64 psf (double tee + insulation + roofing)

f'_c = 5,000 psi

f_y = 60,000 psi

EXAMPLE 14.2 - Continued

Roof members are 10-ft-wide double-tee units, 24-in. deep. (Design of these units not included in this design example). For lateral support, alternate ends of roof members are fixed to supporting beams.

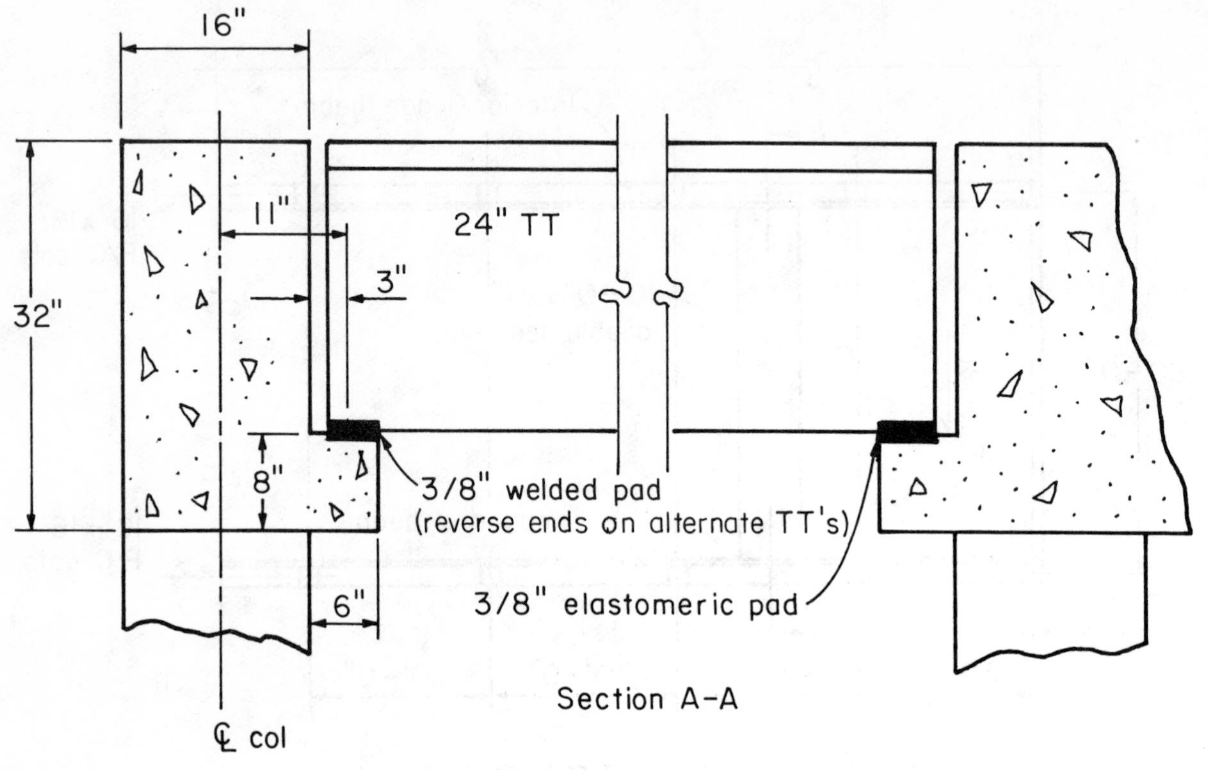

Calculations and Discussion	Code Reference
1. Assume double tee loading on spandrel beam as uniform. Calculate factored loading M_u, V_u, and T_u for spandrel beam.	
Dead Load:	
Superimposed = 0.064 x 60/2 = 1.92	
Spandrel = (1.33x2.67x0.5x0.67) 0.150 = 0.58	
Total = 2.50 k/ft	
Live load = 0.030 x 60/2 = 0.9 k/ft	
Factored load = 1.4 x 2.50 + 1.7 x 0.9 = 5.03 k/ft	9.2.1

EXAMPLE 14.2 - Continued

Calculations and Discussion	Code Reference

At center of span, $M_u = \frac{5.03 \times 30^2}{8} = 566^{'k}$

End shear $V_u = 5.03 \times 30/2 = 75.45^k$

Torsional factored load = $1.4 \times 1.92 + 1.7 \times 0.9 = 4.22$ k/ft

End torsional moment $T_u = 4.22 \times 30/2 \times 11/12 = 58.0^{'k}$.

Critical section is at "d" distance from face of support. 11.6.4

Assume d = 29.5 in.; critical section is at 29.5 + 8 = 37.5 in. from ℄ of column. 11.1.2.1

At critical section: (15.0 - 37.5/12 = 11.88 ft from ℄ of span)

$$V_u = 75.45 \times \frac{11.88}{15} = 59.8^k$$

$$T_u = 58.0 \times \frac{11.88}{15} = 45.9^{'k}$$

The spandrel beam must be designed for the full factored torsional moment since it is required to maintain equilibrium. 11.6.2

2. Determine Σx^2y of spandrel beam section. 11.6.1.1

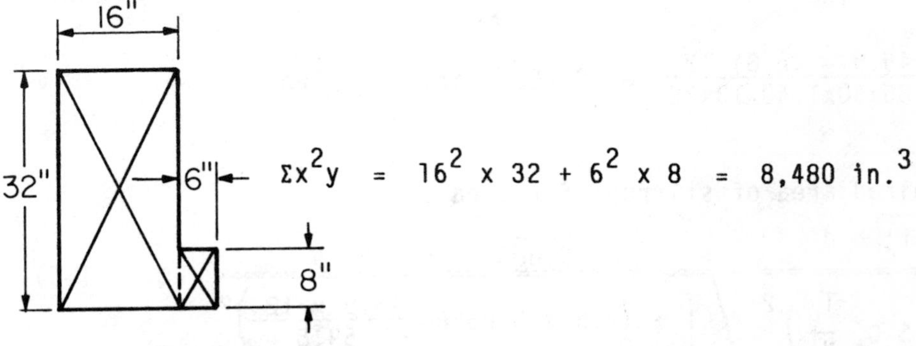

$$\Sigma x^2y = 16^2 \times 32 + 6^2 \times 8 = 8{,}480 \text{ in.}^3$$

3. Check if torsion may be neglected. 11.6.1

$$\varphi(0.5\sqrt{f'_c}\,\Sigma x^2y) = 0.85\,(0.5\sqrt{5{,}000} \times 8{,}480)$$
$$= 254{,}841 \text{ in. lb}$$
$$= 21.2^{'k} < T_u = 45.9^{'k}$$

Torsion must be considered.

EXAMPLE 14.2 - Continued

Calculations and Discussion	Code Reference

4. Calculate torsional moment strength provided by concrete.

$$T_c = \frac{0.8\sqrt{f'_c}\ \Sigma x^2 y}{\sqrt{1 + \left(\frac{0.4\ V_u}{C_t\ T_u}\right)^2}}$$ 11.6.6.1 Eq. (11-22)

$$C_t = \frac{b_w d}{\Sigma x^2 y} = \frac{16 \times 29.5}{8{,}480} = 0.05566$$ 11.0

$$T_c = \frac{0.8\sqrt{5{,}000} \times 8{,}480}{\sqrt{1 + \left(\frac{0.4 \times 59.8}{0.05566 \times 45.9 \times 12}\right)^2}} = 378{,}203 \text{ in. lb} = 31.5^{'k}$$

5. Determine required area of closed stirrups for torsion.

$$\frac{A_t}{s} = \frac{(T_u - \varphi T_c)}{\varphi f_y \alpha_t x_1 y_1}$$ 11.6.5 11.6.9.1

Assuming 1-1/4 in. cover and #4 stirrups for exterior exposure 7.7.2

$x_1 = 16 - 2\ (1.25 + 0.25) = 13$

$y_1 = 32 - 2\ (1.25 + 0.25) = 29$

$\alpha_t = 0.66 + 0.33\ (\frac{29}{13}) = 1.40$

$$\frac{A_t}{s} = \frac{(45.9 - 26.8)\ 12}{0.85 \times 60 \times 1.40 \times 13 \times 29} = 0.00851 \text{ in.}^2\text{/in./leg}$$

6. Calculate required area of stirrups for shear.

$$V_c = \frac{2\sqrt{f'_c}\ b_w d}{\sqrt{1 + \left(2.5\ C_t\ \frac{T_u}{V_u}\right)^2}} = \frac{2\sqrt{5{,}000} \times 16 \times 29.5}{\sqrt{1 + \left(2.5 \times 0.05566\ \frac{45.9 \times 12}{59.8}\right)^2}}$$ Eq. (11-5)

$V_c = 41{,}062 \text{ lb} = 41.1^k$

EXAMPLE 14.2 - Continued

Calculations and Discussion	Code Reference
$V_u \le \varphi(V_c - V_s)$	Eq. (11-1)
$V_s = \dfrac{V_u}{\varphi} - V_c = \dfrac{59.8}{0.85} - 41.1 = 29.3^k$	Eq. (11-2)
$\dfrac{A_v}{s} = \dfrac{V_s}{f_y d} = \dfrac{29.3}{60 \times 29.5} = 0.01655 \text{ in.}^2/\text{in.}$	Eq. (11-17)

7. Determine combined shear and torsion stirrup requirements.

$$\frac{A_t}{s} + \frac{A_v}{2s} = 0.00851 + \frac{0.01655}{2} = 0.01679 \text{ in.}^2/\text{in./leg}$$

Try #3, $A_b = 0.11 \text{ in.}^2$

$$s = \frac{0.11}{0.01679} = 6.55 \text{ in.}$$

8. Check maximum stirrup spacing.

$$\frac{x_1 + y_1}{4} = \frac{13 + 29}{4} = 10.5 \text{ in. or 12 in.}$$ 11.6.8.1

$$\frac{d}{2} = \frac{29.5}{2} = 14.75 \text{ in. or 24 in.}$$ 11.5.4.1

$$4\sqrt{f'_c}\, b_w d = 4 \times \sqrt{5{,}000} \times \frac{16 \times 29.5}{1{,}000} = 133.5^k > V_s = 29.3^k \quad \text{O.K.}$$ 11.5.4.3

Use 6-1/2-in. minimum and 10-1/2-in. maximum spacing.

9. Check minimum stirrup area. 11.6.9.2

$$A_v + 2A_t \ge \frac{50 b_w s}{f_y} = \frac{50 \times 16 \times 6.5}{60{,}000} = 0.087 \text{ in.}^2$$ Eq. (11-16)

Area provided = $2 \times 0.11 = 0.22 \text{ in.}^2$ O.K.

10. Determine stirrup layout.

Since both shear and torsion are zero at the center of span, and are assumed to vary linearly to the maximum value at the

Calculations and Discussion	Code Reference

critical section, the start of maximum stirrup spacing can be determined by simple proportion.

$$\frac{s(\text{critical})}{s(\text{maximum})} \times 11.88 = \frac{6.5}{10.5} \times 11.88 = 7.35 \text{ ft}$$

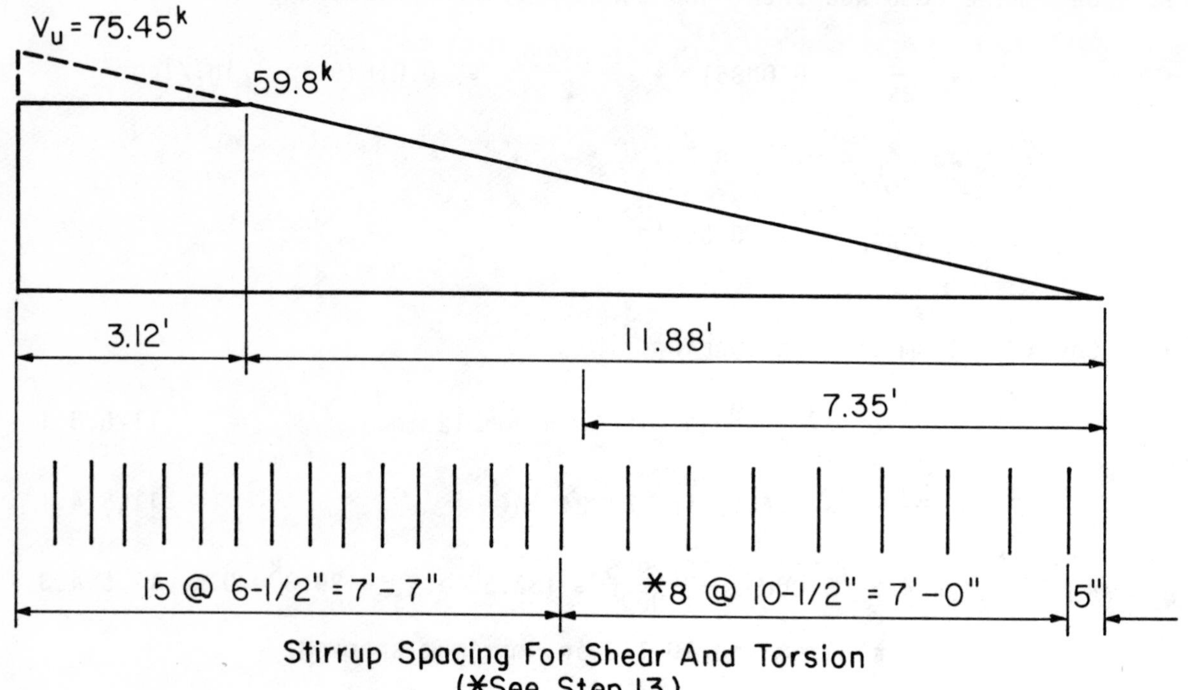

Stirrup Spacing For Shear And Torsion
(*See Step 13)

11. Calculate longitudinal torsion reinforcement. 11.6.9.3

$$A_\ell = 2A_t\left(\frac{x_1 + y_1}{s}\right) = 2 \times 0.00851\ (13 + 29) = 0.715 \text{ in.}^2 \qquad \text{Eq. (11-24)}$$

$$A_\ell = \left[\frac{400xs}{f_y}\left(\frac{T_u}{T_u + \frac{V_u}{3C_t}}\right) - 2A_t\right]\left(\frac{x_1 + y_1}{s}\right) \qquad \text{Eq. (11-25)}$$

(or substituting $\frac{50b_w s}{f_y}$ for $2A_t$)

EXAMPLE 14.2 - Continued

Calculations and Discussion	Code Reference

$$\frac{50b_w s}{f_y} = 0.087 < 2A_t = 2 \times 0.00851 \times 6.5 = 0.1106$$

Use $2A_t$

$$A_\ell = \left[\frac{400 \times 16 \times 6.5}{60{,}000}\left(\frac{45.9 \times 12}{45.9 \times 12 + \frac{59.8}{3 \times 0.05566}}\right) - 0.1106\right]\left(\frac{13 + 29}{6.5}\right) = 2.00 \text{ in.}^2$$

Provide A_ℓ = 2.00 in.^{2}. Place longitudinal bars around perimeter of the closed stirrups, spaced at not more than 12 in., and locate one longitudinal bar in each corner of the closed stirrups. Longitudinal bars may be combined with the flexural reinforcement.

12. Size combined longitudinal reinforcement.

$$\frac{A_\ell}{8} = \frac{2.00}{8} = 0.25 \text{ in.}^2$$

Use #5 bar in sides and top corners of spandrel beam.

$$A_b = 0.31 \text{ in.}^2$$

Using Table 9-2, page 9-7:

$$\frac{M_u}{\varphi f'_c bd^2} = \frac{566 \times 12}{0.9 \times 5 \times 16 \times 29.5^2} = 0.1084$$

From Table 9-2, read ω = 0.1165

$$A_s = \frac{\omega f'_c bd}{f_y} = \frac{0.1165 \times 5 \times 16 \times 29.5}{60} = 4.58 \text{ in.}^2$$

At center of span:

$$\frac{A_\ell}{4} + A_s = \frac{2.00}{4} + 4.58 = 5.08 \text{ in.}^2$$

At end of span (extended reinforcement): 12.11.1

$$\frac{A_\ell}{4} + \frac{A_s}{3} = \frac{2.00}{4} + \frac{4.58}{3} = 2.03 \text{ in.}^2$$

Use 4 - #10 bars (A_s = 5.08 in.^{2})

EXAMPLE 14.2 - Continued

Calculations and Discussion	Code Reference

Extend 2 - #10 bars to end of girder

A_s = 2.54 $in.^2$

13. Check required area of beam stirrups used as "hanger" reinforcement for beam ledge. Sufficient stirrups in beam section must be available to act also as hanger reinforcement for the beam ledge. See Part 16 for design of beam ledges.

Reaction from one double tee stem (5 ft-0 between stems):

$R_u = (1.4 \times 1.92 + 1.7 \times 0.9)5 = 21.1^k\text{/stem}$

$$A_v(\text{one leg}) = \frac{R_u}{\varphi f_y} = \frac{21.1}{0.85 \times 60} = 0.414 \text{ in.}^2\text{/stem}$$

Effective width of ledge over which hanger forces can be distributed may be evaluated from Reference 16.1. For the ledge loading and dimensions of this example, b_e = 26 in.

$A_v/s = 0.414/26 = 0.0160 \text{ in.}^2\text{/in.}$

For #3 stirrups, $s_{max} = 0.11/0.0160 = 6.9$ in.

The #3 stirrups @ 6 1/2 in. must be used for the full span length to act also as hanger reinforcement for the beam ledge.

Calculations and Discussion	Code Reference

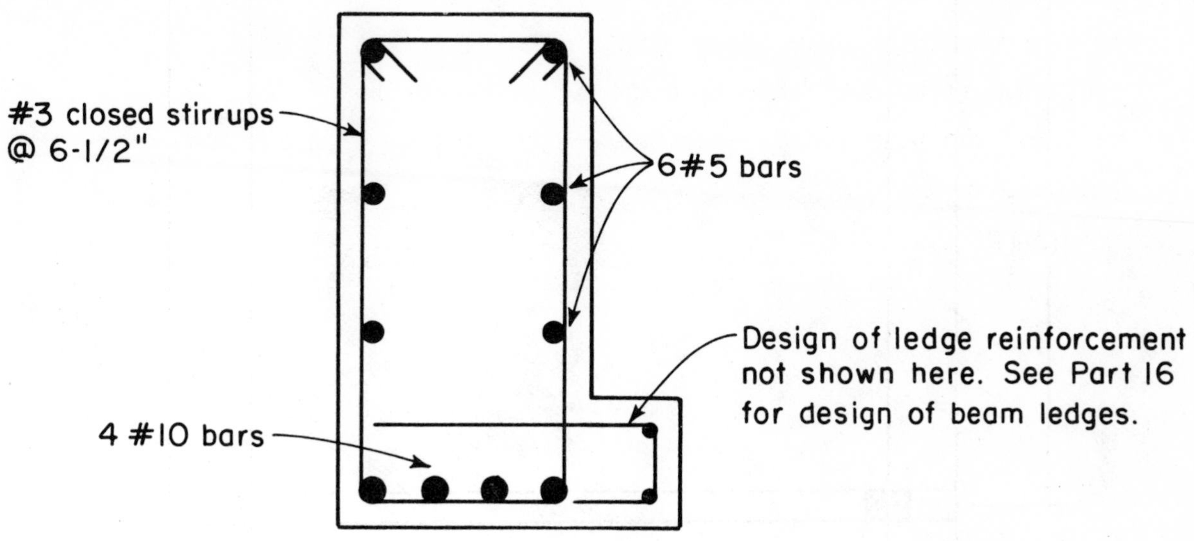

Reinforcement Details

*Closed stirrup detailed as shown in Fig. 2-10, page 2-14 for isolated beam.

EXAMPLE 14.3 - Spandrel Beam Analysis for Combined Shear and Torsion

For the floor slab framing system shown, develop the shear force and torsional moment diagrams for design of the spandrel beams.

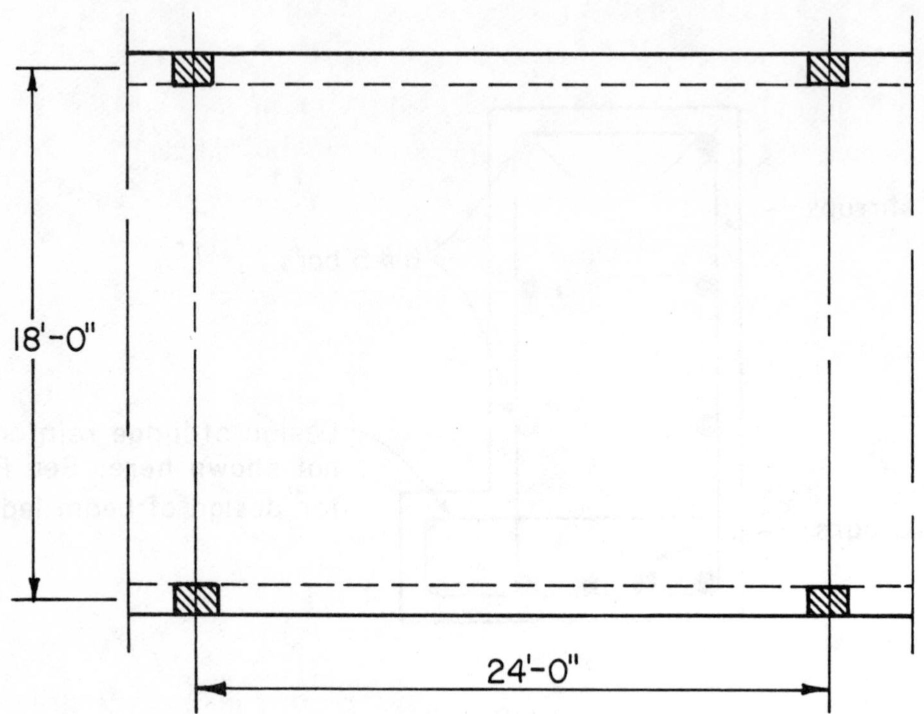

Design Data:

Slab thickness = 7" (87.5 psf)

Spandrel beams = 24" x 8" (200 plf)

Columns = 8" x 12"

f'_c = 3000 psi

Slab effective depth d = 21.5"

Live load = 150 psf uniform

Dead load = 11 psf superimposed

EXAMPLE 14.3 - Continued

Calculations and Discussion	Code Reference

In lieu of computing the torsional stiffness of the spandrel beam and the flexural stiffness of the slab to determine the magnitude of the torque to be applied to the spandrel beam, Section 11.6.3 greatly simplifies the determination of torsional moments in members of statically indeterminate framing systems where reduction of torsional moment in a member can occur due to redistribution of internal forces. A maximum factored torsional moment equal to $T_u = \varphi(4\sqrt{f'_c}\ \Sigma x^2y/3)$ may be assumed at critical sections of such members.

1. For the floor slab system shown, torsional loading from the slab may be assumed uniformly distributed along the spandrel beam (Section 11.6.3.2) with maximum torque ($T_u = \varphi 4\sqrt{f'_c}\ \Sigma x^2y/3$) in the beam taken at distance "d" from face of column support (Section 11.6.4) and decreasing linearly to zero at midspan.

   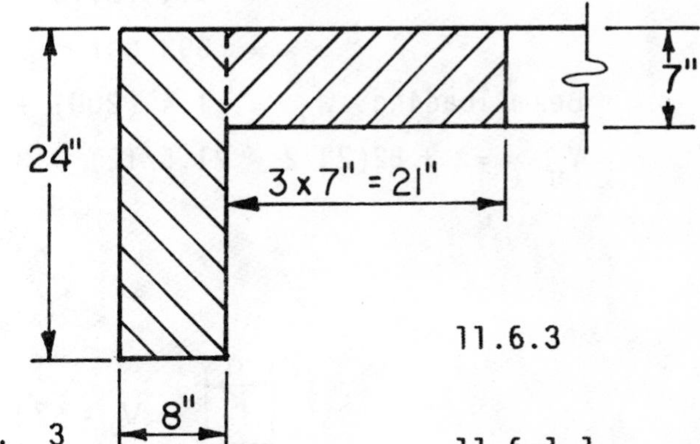

   For the spandrel beam:

   $T_u = \varphi(4\sqrt{f'_c}\ \Sigma x^2y/3)$

   $0.85\ (4\sqrt{3{,}000} \times 2565/3)$

   $= 159.2^{\prime\prime k}$

   $T_u = 13.3^{\prime k}$ 11.6.3

where

$\Sigma x^2y = 8^2 \times 24 + 7^2 \times 21 = 2565 \text{ in.}^3$ 11.6.1.1

EXAMPLE 14.3 - Continued

Calculations and Discussion	Code Reference

The torsional moment diagram is sketched as follows:

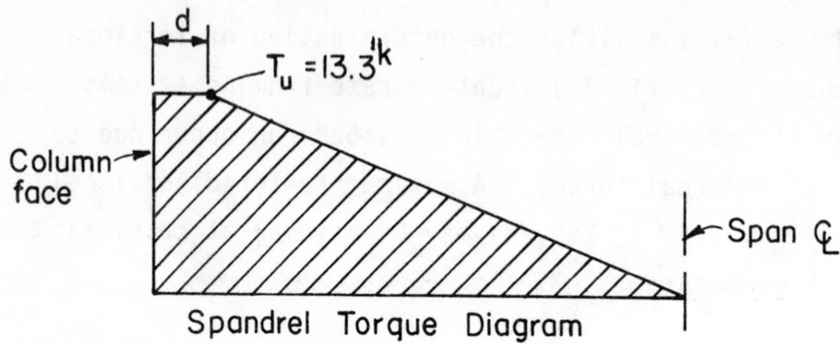

2. The shear force diagram is sketched as follows:

Slab loading w_u = 1.4D + 1.7L

= 1.4 (87.5 + 11) + 1.7 (150)

= 393 psf

Beam loading w_u = 1.4 (200) + 393 x 18/2 = 3.82 k/ft

V_u = 3.82(23/2 - 21.5/12) = 37.1^k @ distance d from column face

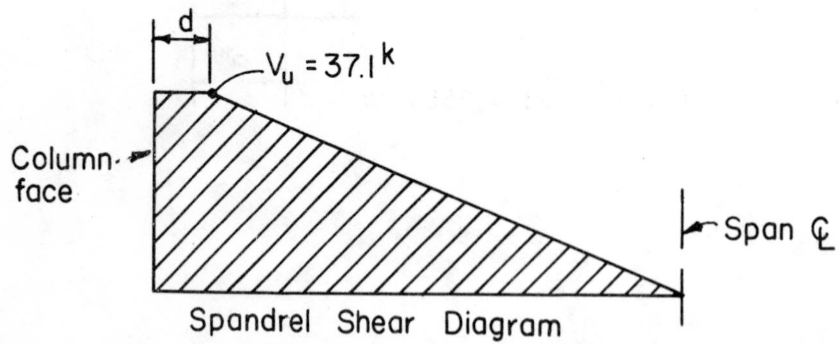

EXAMPLE 14.3 - Continued

Calculations and Discussion	Code Reference

Torsion reinforcement for the spandrel beam is determined directly from the above shear force and torsional moment diagrams. Torsion reinforcement need not be provided where T_u is less than $\varphi(0.5\sqrt{f'_c}\ \Sigma x^2 y)$.

$T_u < 0.85(0.5\sqrt{3000} \times 2565) < 59.7^{\prime\prime k} < 4.98^{\prime k}$

3. For slab design, the slab moments must be adjusted by the assumed torque from the spandrel beam. 11.6.3.1

Torsional loading on spandrel beams, assumed as uniformly distributed, is approximated as $2T_u/\ell = 2 \times 13.3 / 19.4 = 1.37^{\prime k}/ft$

Slab $+ M_u = w_u \ell_n^2/8 - 1.37 = 393 \times 16.67^2/8 - 1.37 = 12.28^{\prime k}/ft$

$- M_u = 1.37^{\prime k}/ft$

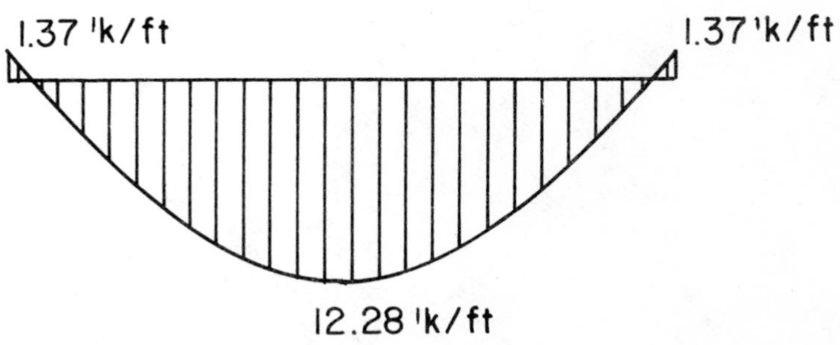

Slab Moment Diagram

Application of Section 11.6.3 for a spandrel beam with torque applied as a single concentration is illustrated in Example 14.1.

15

Shear Friction

11.7 Shear-Friction

The shear-friction concept provides a convenient tool for the design of members for direct shear where it is inappropriate to design for diagonal tension, as in precast connections, brackets, and corbels. The concept is simple to apply and allows the designer to visualize the structural action within the member or joint. The approach is to assume that a crack has appeared at an unwanted, or unexpected location, as illustrated in Fig. 15-1. As slip begins to occur along the crack surface, the roughness of the crack forces the opposing faces of the crack to separate. This separation is resisted by reinforcement (A_{vf}) across the assumed crack (a separation of only 0.01 in. is sufficient to develop the yield strength of Grade 40 bars). The tensile force ($A_{vf}f_y$) developed in the reinforcement by this strain furnishes an equal and opposite normal clamping force, which in turn maintains a frictional force parallel to the crack ($A_{vf}f_y\mu$) to resist further slip.

11.7.1 Applications

Shear-friction design is to be used where direct shear is being transferred across a given plane. Examples of shear friction applications are shown in Fig. 15-2, including potential crack locations. Successful application of the shear-friction method depends on proper selection of the location of the assumed crack. Note that in end or edge bearing applications, the crack

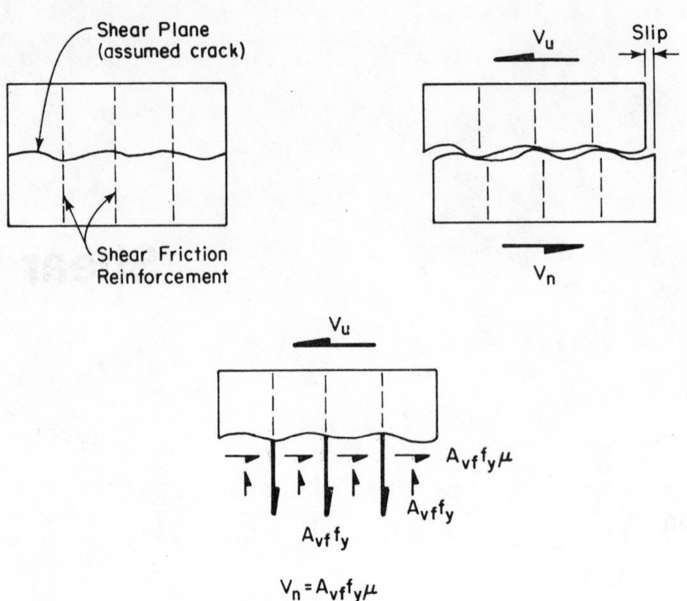

Fig. 15-1 Idealization of Shear-Friction Concept

tends to occur at an angle of about 20 degrees from the direction of the force application.

For the 1983 Code, Section 11.7 has been completely rewritten to expand the shear-friction concept to include (1) applications where the shear-friction reinforcement is placed at an angle other than 90 degrees to the shear plane, (2) applications where concrete is cast against hardened concrete not intentionally roughened, and (3) applications with lightweight concrete. In addition, a performance statement is added to allow "any other shear transfer design method" substantiated by tests. One such method is outlined in Code Commentary Section 11.7.3. Application of the "Modified Shear-Friction Method" is illustrated in Part 16, Example 16.2. Other acceptable methods are presented in publications by the Prestressed Concrete Institute.

Actually, the shear-friction design method presented in the 1971 and 1977 code editions, and for 1983--Section 11.7.4--is based on the simplest model of shear-transfer behavior, and results in a more conservative prediction of shear-transfer strength. Other less simple shear-transfer relationships result in a closer prediction of shear-transfer strength. The new performance statement of Section 11.7.3 now includes the "other methods" within the scope and intent of Section 11.7. However, it should be noted that the provisions of Sections 11.7.5 through 11.7.10 apply for whatever shear-transfer method is used.

It is noteworthy that new Section 11.9 refers to Section 11.7 for the direct shear transfer in brackets and corbels. See Part 16.

11.7.4 Shear-Friction Design Method

As with the other shear design applications, the code provisions for shear-friction are presented in terms of shear transfer strength V_n for direct application in the basic shear strength equality:

required shear transfer strength ≤ design shear-transfer strength

$$V_u \leq \varphi V_n \qquad \text{Eq. (11-1)}$$

and, for shear-friction reinforcement perpendicular to shear plane

$$V_u \leq \varphi A_{vf} f_y \mu \qquad \text{Eq. (11-26)}$$

where required area of shear-friction reinforcement can be computed directly by

$$A_{vf} = \frac{V_u}{\varphi f_y \mu}$$

or, for shear-friction reinforcement inclined to the shear plane; using Eq. (11-27)

$$A_{vf} = \frac{V_u}{\varphi f_y(\mu \mathrm{Sin}\alpha_f + \mathrm{Cos}\alpha_f)}$$

The basic Eq. (11-26) for nominal shear-friction strength $V_n = A_{vf} f_y \mu$ remains

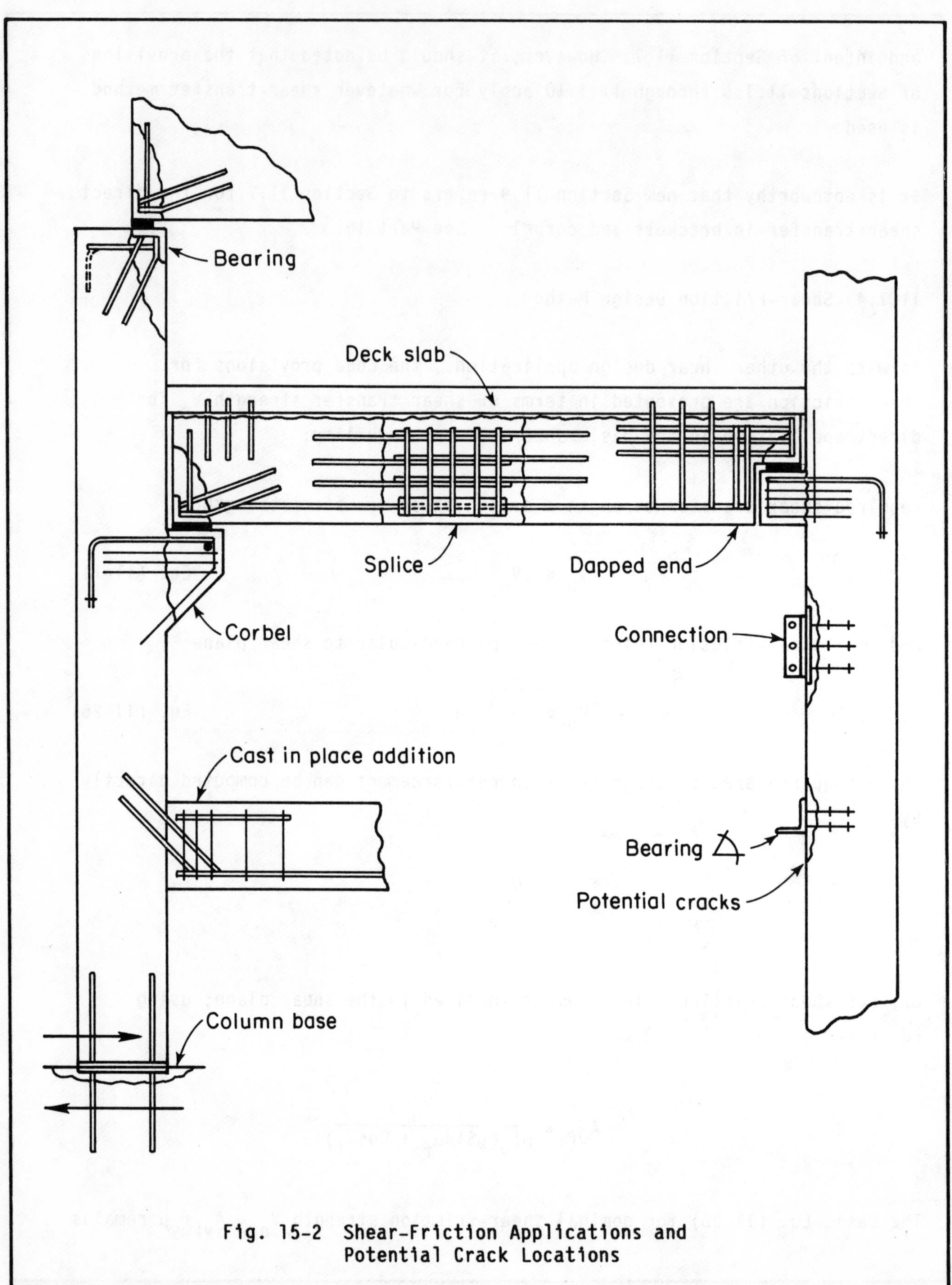

Fig. 15-2 Shear-Friction Applications and Potential Crack Locations

unchanged from the previous Codes, where μ is an "effective" coefficient of friction providing a conservative lower bound to test data (Fig. 15-3). The actual mechanics of resistance to direct shear are more complex than Eq. (11-26) would indicate, since dowel action and the apparent cohesive strength of the concrete both contribute to direct shear strength. The high values for the basic coefficients of friction, μ, in Section 11.7.4.3 partially account for these discrepancies. The modified shear-friction strength given in Commentary Section 11.7.3 more closely approximates the effects of these factors.

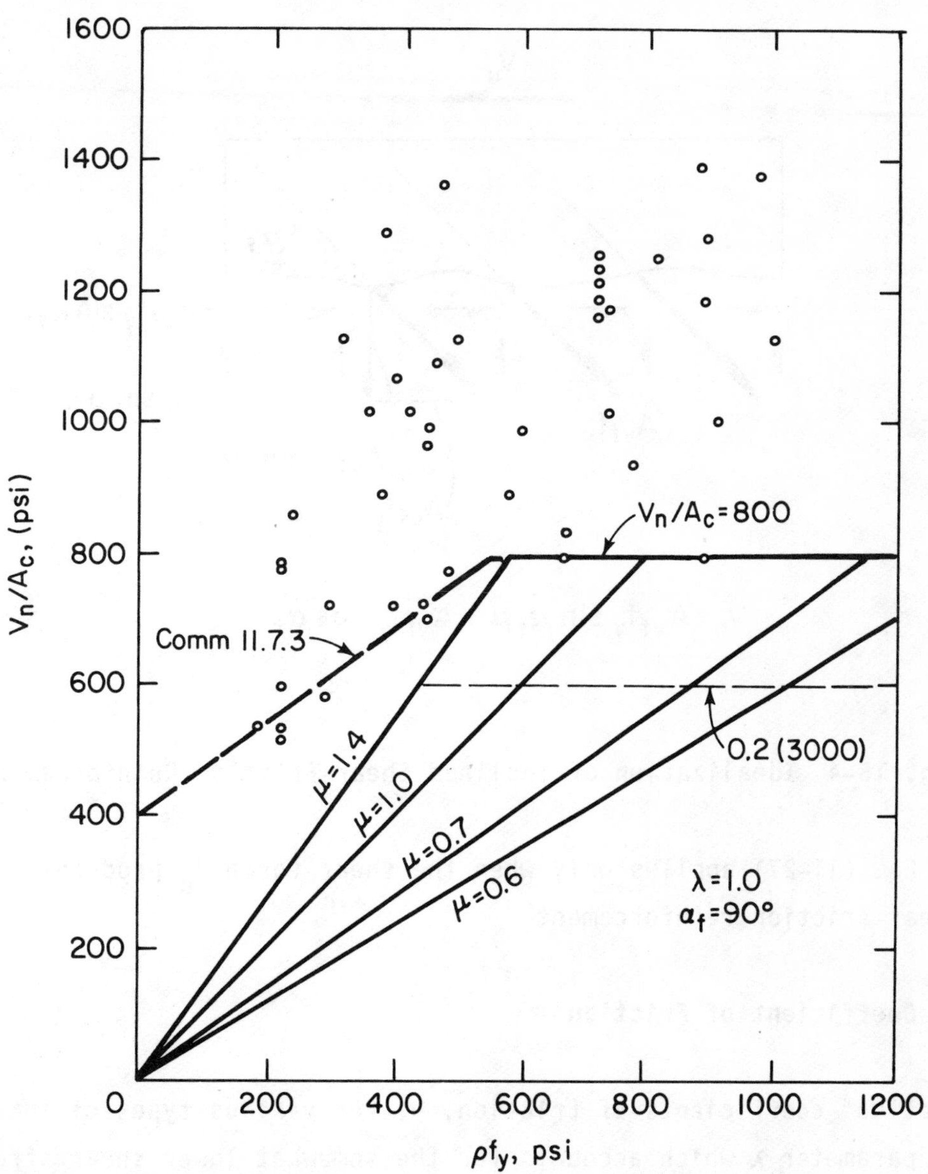

Fig. 15-3 Effect of Shear-Friction Reinforcement on Shear Transfer Strength (dots indicate test results)

With ACI 318-83, an additional Eq. (11-27) is given for design applications where the shear-friction reinforcement crosses the shear-plane at an angle α_f other than 90 degrees, as illustrated in Fig. 15-4. Eq. (11-27) resolves the total tensile force ($A_{vf}f_y$) into two components: (1) a clamping component ($A_{vf}f_y \sin\alpha_f$) with a frictional force ($A_{vf}f_y \sin\alpha_f\mu$) plus, (2) a component contributed directly by the inclination of the shear-friction reinforcement ($A_{vf}f_y \cos\alpha_f$).

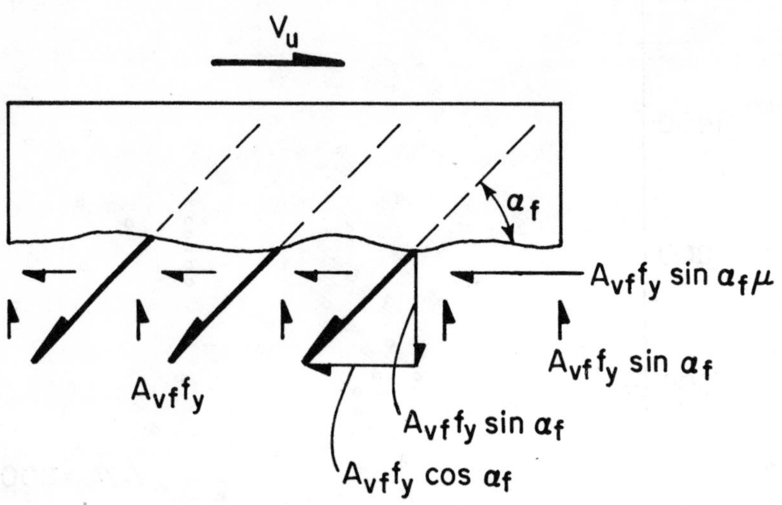

$$V_n = A_{vf}f_y \sin \alpha_f\mu + A_{vf}f_y \cos \alpha_f$$

Fig. 15-4 Idealization of Inclined Shear-Friction Reinforcement

Note that Eq. (11-27) applies only when the shear force V_u produces tension in the shear-friction reinforcement.

11.7.4.3 Coefficient of Friction

The "effective" coefficients of friction, μ, for various types of interfaces include a parameter λ which accounts for the somewhat lower shear strength of all-lightweight and sanded-lightweight concretes; for example, the μ value for all-lightweight concrete placed against hardened concrete not intentionally roughened is 0.6(0.75) = 0.45.

Also, for lightweight, Section 11.9.3.2.2 limits the nominal shear-transfer strength V_n along the shear plane for design applications with low shear-to-depth ratios a/d, as for brackets and corbels. This further restriction in the use of lightweight concrete is applied in Design Example 15.1.

11.7.5 - As in the previous code provisions for shear-friction, the shear-transfer strength V_n cannot be taken greater than $0.2f'_c$ nor 800 psi times the area of concrete section resisting shear transfer. This upper limit on V_n effectively limits the maximum reinforcement as shown by Fig. 15-3.

11.7.7 **Normal Forces**

Eqs. (11-26) and (11-27) assume that there are no additional forces acting on the shear plane. A certain amount of moment is almost always present in brackets and corbels and other connections due to eccentricity of loads or applied moments at connections. Most joints also carry a significant amount of tension due to restrained shrinkage or thermal shortening of the connected members. A direct tensile force of at least $0.2V_u$ is required for design of connections such as brackets or corbels (Section 11.9.3.4), unless the actual force is accurately known. Friction of bearing pads, for example, can cause appreciable tensile forces on a corbel supporting a member subject to shortening. Reinforcement must be provided for direct tension according to Section 11.7.7, using $A_s = N_{uc}/\varphi f_y$, where N_{uc} is the factored tensile force, with φ taken as 0.85 (see Section 11.9.3.1).

Tensile reinforcement required for bending due to eccentricity of the loads, or other causes, is sized in the normal manner. Note that Section 11.9.3.1 specifies a φ of 0.85 for all design conditions. Special consideration should also be given to development of the full strength of the flexural reinforcement.

Since direct tension perpendicular to the shear plane (crack face) detracts from the shear-transfer strength, it follows that compression will add to the strength. This is now provided for in Section 11.7.7 which allows the use of "permanent net compression" as an additive force to the shear-friction clamping force. It is advisable, though not required, to use a load factor of 0.9 with such compressive loads.

11.7.8 - 11.7.10 Details

Section 11.7.8 requires that the shear-friction reinforcement be "appropriately placed" along the shear plane. Where no moment acts on the shear plane, uniform distribution of the bars is proper. Where a moment exists, the reinforcement should be distributed in the flexural tension portion of the shear plane.

Reinforcement should be adequately embedded on both sides of the shear plane to develop the full yield strength of the bars. Since space is limited in thin walls, corbels, and brackets, it is often necessary to use special anchorage details such as welded plates, angles, or cross bars. Reinforcement should be anchored in confined concrete. Confinement may be beam or column ties, "external" concrete, or special added reinforcement. In certain applications, confinement reinforcement both parallel and normal to the primary reinforcement may be required, with the area of bars in each direction equal to $V_u/8f_y$.

In Section 11.7.9, the term "intentionally roughened" concrete at the interface is defined as "roughened to a full amplitude of approximately 1/4 in." This can be accomplished by raking the plastic concrete or by bushhammering or chiseling hardened concrete surfaces.

A final requirement of Section 11.7.10, often overlooked, is that structural steel interfaces must be clean and free of paint. This requirement is based on tests to evaluate the friction coefficient for concrete anchored to unpainted structural steel by studs or rebars ($\mu = 0.7$). Data is not available for painted surfaces. If painted surfaces are to be used, a lower value of μ would be appropriate.

Design Examples

In addition to Design Examples 15.1 and 15.2, shear-friction design is also illustrated for direct shear transfer in brackets and corbels (see Part 16), and horizontal shear transfer at column/footing connections (see Part 23).

EXAMPLE 15.1 - Shear-Friction Design

The tilt-up wall panel is subject to seismic shear forces as shown. The wall also shortens due to temperature and shrinkage changes causing a strain of 0.0005 in./in. Design the shear connectors shown below. The connectors are detailed to minimize build-up of tensile forces in the connection due to wall shortening. Use lightweight concrete, w_c = 95 pcf. f'_c = 4,000 psi and f_y = 60,000 psi. Panel width = 16 ft-0.

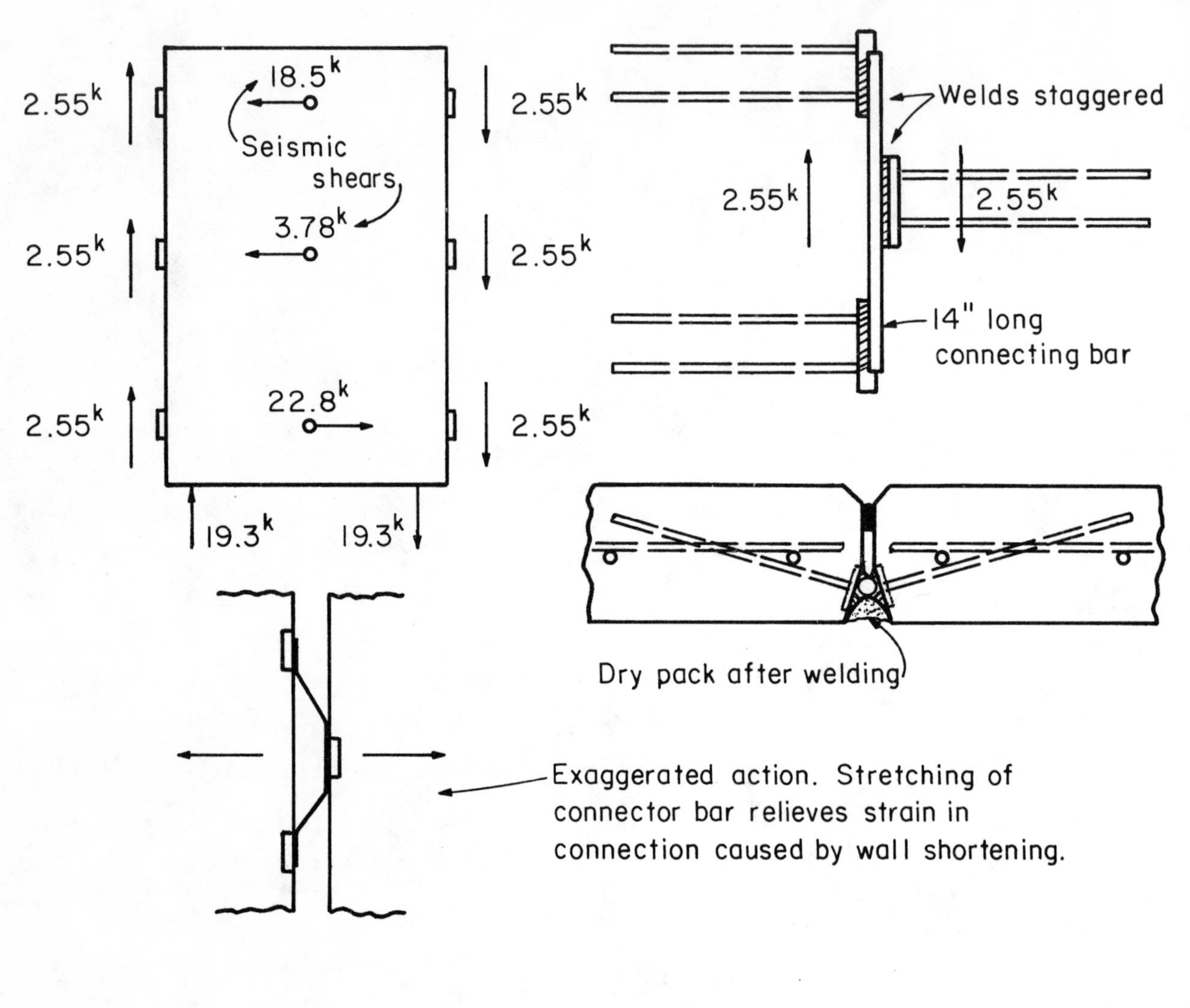

EXAMPLE 15.1 - Continued

Calculations and Discussion	Code Reference

1. Calculate tension in "connecting" bar due to shear forces and wall shortening of 0.0005 in./in. Try #4 connecting bar (A_b = 0.20 in.^{2}).

For 16-ft wide panel:

Total strain at joint = 0.0005 (16)(12) = 0.096 in.

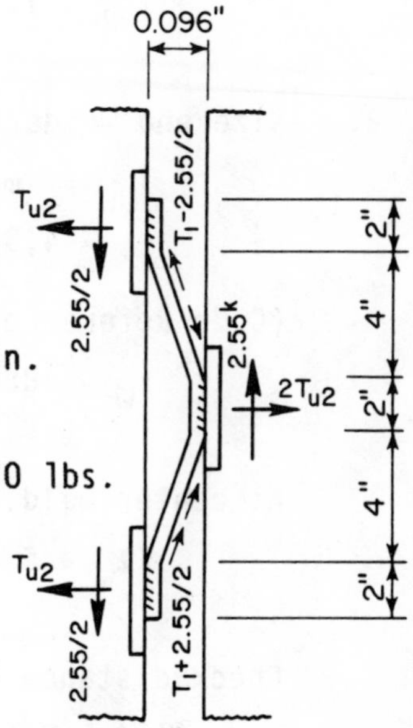

Length of 4-in. portion of bar between welds after

straining $= \sqrt{4^2 + 0.096^2} = 4.00115183$ in.

$\varepsilon = (4.00115183 - 4.0)/4.0 = 0.00028796$ in./in.

$T_1 = E_s \varepsilon_s A_b = 29\text{x}10^6\ (0.00028796)(0.20) = 1{,}670$ lbs.

0.096"
T_{u2}
2.55/2
T_1-2.55/2
2.55^k
$2T_{u2}$
T_{u2}
2.55/2
T_1+2.55/2
2"
4"
2"
4"
2"

The general building code (UBC) requires for seismic design of shear walls that:

$$U = 1.4(D+L) + 2.0\ E$$

Shrinkage and temperature forces should be included as live load due to their unpredictable nature. The total shear force is assumed as divided between both ends of the connector bar.

$T_{u1} = 1.4\ (1670) \pm 2.0\ (2550)/2 = 4{,}888$ lbs. max.

$\varphi T_n = 0.85\ (60{,}000)(0.20) = 10{,}200$ lbs.

$T_{u1} \leq T_n$

EXAMPLE 15.1 - Continued

Calculations and Discussion	Code Reference

4888 < 10200   <u>#4 Connector bar OK</u>

(#3 would also be adequate)

Force in the connection plates normal to the shear plane can be determined from geometry of detail.

$$T_{u2} = 1.4\ (1670)(0.096/4.0) = 56\ \text{lbs. (negligible)}$$

2. Size end welds for T_{u1} = 4888 lbs.

$$T_{uw} = \varphi(25000)\ \ell_w t_w = 0.70(25000)\ell_w(0.25)$$
$$= 4{,}375\ \ell_w$$

(Code Reference 16.3 recommends φ = 0.70 for welds)

ℓ_w = 4888/4375 = 1.12 in.   <u>Use 1-1/4 welds at ends</u>

At center weld, 2.0(2550) = 5,100 lbs must be developed.

ℓ_w = 5,100/4375 = 1.17 in.   <u>Use 1-1/4 in. weld at center</u>

Check distance between welds.

(14 - 2.5 - 1.25 - 1.25)/2 = 4.5 in. > 4.0 in. OK

3. Design anchor plates using shear-friction method.

Center plate most heavily loaded. Try 2 in. x 4 in. x 1/4 in. plate.

V_u = 2.0 E = 2(2550) = 5,100 lbs.

$V_u \leq \varphi V_n$ Eq. (11-1)

$V_u \leq \varphi(A_{vf} f_y \mu)$ Eq. (11-26)

$$\text{Solving for } A_{vf} = \frac{V_u}{\varphi f_y \mu}$$

$$= \frac{5100}{0.85\ (60{,}000)\ 0.525} = 0.19\ \text{in.}^2$$

EXAMPLE 15.1 - Continued

Calculations and Discussion	Code Reference

where, for lightweight concrete (95 pcf),

$\mu = 0.75(0.7) = 0.525$ — 11.7.4.3

Add A_s for direct tensile force.

$2T_{u2} = 2(56) = 112$ lbs. (negligible)

Use 2 #3 bars per plate ($A_{vf} = 0.22$ in.^{2})

Weld bars to plates to develop full f_y.

Check maximum shear transfer strength permitted for connection. For lightweight aggregate concrete: — 11.9.3.2.2

$V_n(max) = [0.2 - 0.7(a/d)]\ f'_cA_c$ or $(800 - 780\ a/d)A_c$

Use $a/d \simeq 0.25/2.5 = 0.1$ $A_c = 2x4 = 8$ in.^{2}

$= [0.2 - 0.7(0.1)]\ 4000(8) = 4{,}160$ lbs.

$\varphi V_n = 0.85\ (4160) = 3536$ lbs.

$V_u \leq \varphi V_n$ — Eq. (11-1)

$5{,}100 > 3536$ NG

Try: 2 in. x 5 in. x 1/4 in. plate ($A_c = 10$ in.^{2} $a/d = 0.07$)

$V_n(max) = [0.2 - 0.7(0.07)]\ 4000(10) = 6040$ lbs (governs)

$= [800 - 780(0.07)]\ 10 = 7454$ lbs

$\varphi V_n = 0.85(6040) = 5134$ lbs

$5100 < 5134$ OK

Use 2 in. x 5 in. x 1/4 in. plates

EXAMPLE 15.2 - Shear-Friction Design (Inclined Shear Plane)

For the pilaster beam support shown, design for shear-transfer across the potential crack plane. In edge bearing applications a crack tends to occur at an angle of about 20 degrees from the direction of force application. Beam reactions are $D = 25^k$, $L = 30^k$, and $T = 20^k$ due to estimate of shrinkage and temperature change effects.

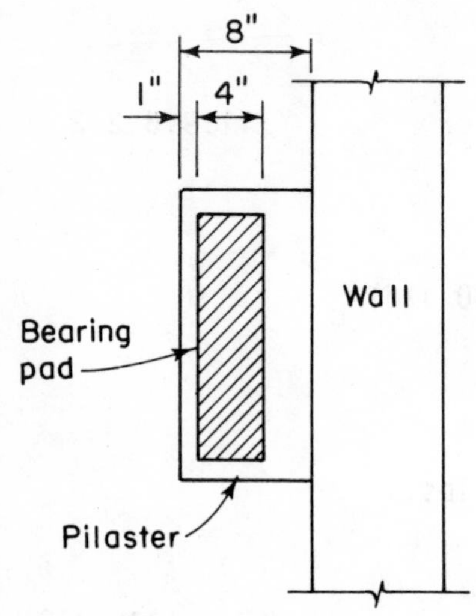

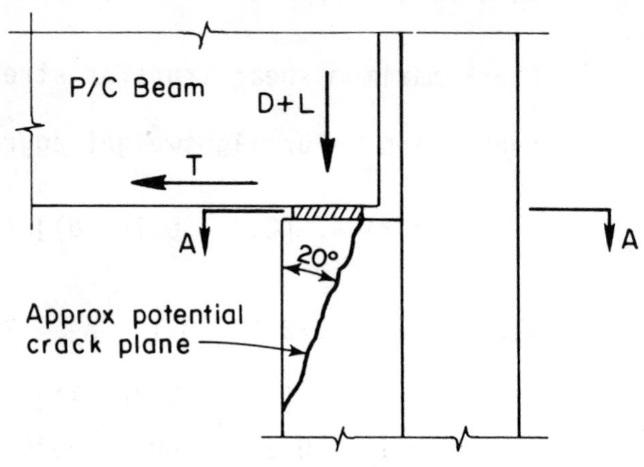

Calculations and Discussion	Code Reference

1. Factored loads to be considered.

Beam reaction $R_u = 1.4D + 1.7L = 1.4(25) + 1.7(30)$

$= 35 + 57 = 86^k$ — Eq. (9-1)

Shrinkage and temperature effects $T_u = 1.7(20) = 34^k$

but not less than $0.2(R_u) = 0.2(86) = 17.2^k$

Calculations and Discussion	Code Reference

Note: For "T" effects, the higher live load factor of 1.7 is used due to the low confidence level of determining shrinkage and temperature effects occurring in service. Also a minimum value of 20 percent of the beam reaction is considered. (See Section 11.9.3.4 for corbel design.)

2. Evaluate force conditions along potential crack plane

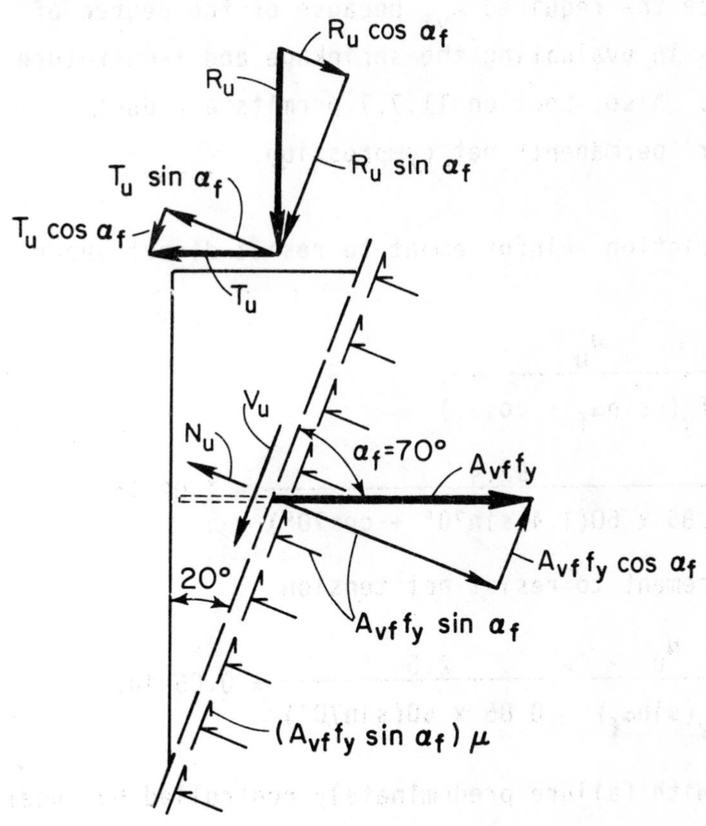

Direct shear-transfer force along shear plane:

$$\begin{aligned} V_u &= R_u \sin\alpha_f + T_u \cos\alpha_f \\ &= 86(\sin 70°) + 34(\cos 70°) \\ &= 80.8 + 11.6 = 92.4^k \end{aligned}$$

Calculations and Discussion	Code Reference

Net tension (or compression) along shear plane:

$$N_u = T_u \sin\alpha_f - R_u \cos\alpha_f$$
$$= 34(\sin 70°) - 86\ (\cos 70°)$$
$$= 31.9 - 29.4 = 2.5^k \text{ (net tension)}$$

Note: If the load conditions were such to result in net compression across the shear plane, it should not be used to reduce the required A_{vf} because of the degree of accuracy in evaluating the shrinkage and temperature effects. Also, Section 11.7.7 permits a reduction in A_{vf} only for "permanent" net compression.

3. Shear-friction reinforcement to resist direct shear transfer

$$A_{vf} = \frac{V_u}{\varphi f_y(\mu \sin\alpha_f + \cos\alpha_f)}$$ Eq. (11-27)

$$= \frac{92.4}{0.85 \times 60(1.4 \sin 70° + \cos 70°)} = 1.09 \text{ in.}^2$$ 11.7.4.3

4. Reinforcement to resist net tension

$$A_n = \frac{N_u}{\varphi f_y(\sin\alpha_f)} = \frac{2.5}{0.85 \times 60(\sin 70°)} = 0.05 \text{ in.}^2$$

Note: With failure predominately controlled by shear, use $\varphi = 0.85$. (See Section 11.9.3.1 for corbel design.)

5. Add A_{vf} and A_n for uniform distribution along potential crack plane.

$$A_s = 1.09 + 0.05 = 1.14 \text{ in.}^2$$

EXAMPLE 15.2 - Continued

Calculations and Discussion	Code Reference

Use #3 closed ties (2 legs per tie)

$$\text{Number required} = \frac{1.14}{2(0.11)} = 5.2$$

Ties should be distributed along length of potential crack plane. Approximate depth at bearing = 5(tan70°) ≃ 14 in.

Use 6-#3 closed ties at ± 3in. spacing at top of pilaster.

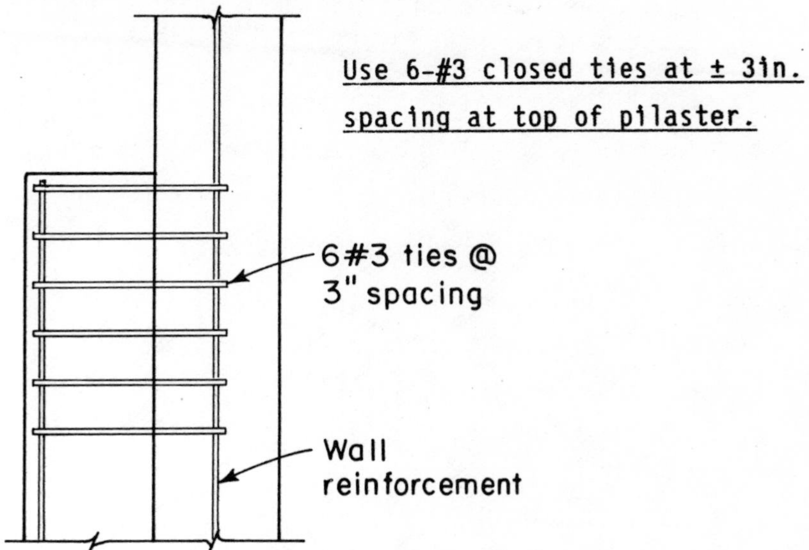

6. Check reinforcement requirements for dead load only plus shrinkage and temperature effects. Use 0.9 load factor for dead load to maximize net tension across shear plane.

$R_u = 0.9D = 0.9(25) = 22.5^k$
$T_u = 34^k$

$V_u = 22.5(\sin 70°) + 34(\cos 70°)$
$= 21.1 + 11.6 = 32.7^k$

$N_u = 34(\sin 70°) - 22.5(\cos 70°)$
$= 31.9 - 7.7 = 24.2^k$

EXAMPLE 15.2 - Continued

Calculations and Discussion	Code Reference

$$A_{vf} = \frac{32.7}{0.85 \times 60\ (1.4 \sin 70° + \cos 70°)} = 0.39 \text{ in.}^2$$

$$A_n = \frac{24.2}{0.85 \times 60 \times \sin 70°} = 0.50 \text{ in.}^2$$

$A_s = 0.39 + 0.50 = 0.89 \text{ in.}^2 < 1.14$ (original design for full dead load + live load governs.)

16

Brackets and Corbels and Beam Ledges

11.9 Brackets and Corbels

Design provisions for brackets and corbels have been completely revised for the 1983 Code, eliminating the empirical equations of the 1977 Code and simplifying design by using the shear-friction method exclusively for shear-transfer strength V_n.

The new design procedure for brackets and corbels recognizes the deep beam or simple truss action of these short shear span members, as illustrated in Fig. 16-1. Four possible failure modes must be controlled: (1) Direct shear failure at the interface between bracket or corbel and supporting member; (2) Yielding of the tension tie due to moment and direct tension; (3) Crushing of the internal compression "strut"; and (4) Localized bearing or shear failure under the loaded area.

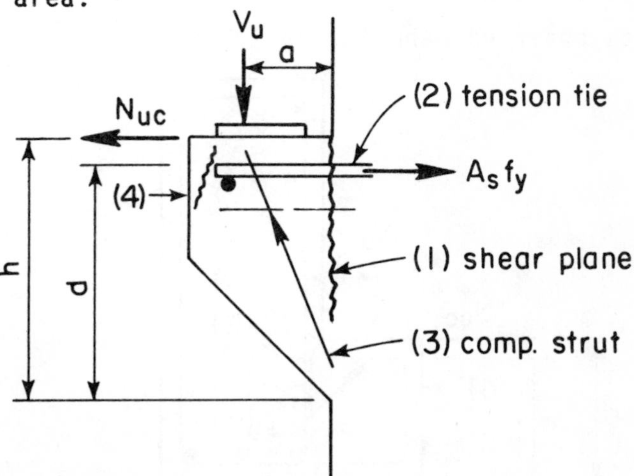

Fig. 16-1 Structural Action of Corbel

The design provisions of Section 11.9 apply only to members having a shear span-to-depth ratio of unity or less ($a/d \leq 1$) since, for longer spans, diagonal tension cracks may form and the use of horizontal shear reinforcement may not suffice. Furthermore, the method has not been validated by tests beyond $a/d = 1$.

A second restriction limits the design method to cases where the factored shear force V_u, exceeds the factored tensile force N_{uc}, again because no test data is available for load conditions with N_{uc} exceeding V_u.

Beam Ledges

Design of beam ledges is similar to that of a bracket or corbel with respect to loading conditions, design considerations, and reinforcing details. Accordingly, even though not specifically addressed by the Code, special design of beam ledges is included in Part 16. The four failure modes discussed above for brackets and corbels are also noted for beam ledges in Fig. 16-2. One additional failure mode needs to be considered for the beam ledge; (5) Separation between ledge and beam web near the top of the ledge in the vicinity of the ledge load. The vertical component of the inclined compression strut must be picked up by the beam stirrups (stirrup legs A_v adjacent to the side face of the beam) acting as "hangar" reinforcement to carry ledge load to top of beam. Note that the critical section for moment is taken at center of beam stirrups not at face of beam. Also, for beam ledges, the internal moment arm should not be taken greater than 0.8 for flexural strength. This reflects observed behavior in tests.[16.1, 16.2].

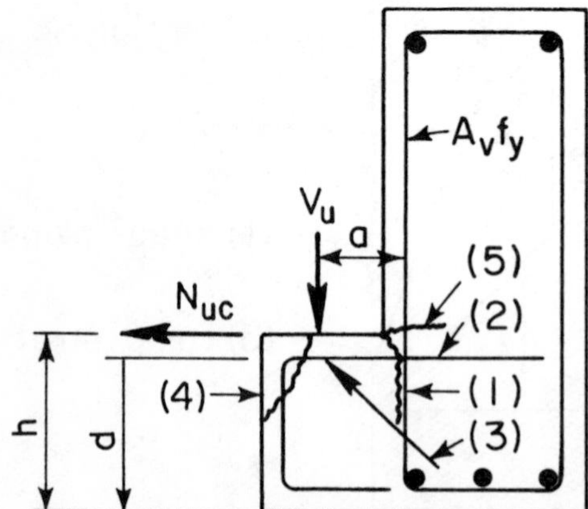

Fig. 16-2 Structural Action of Beam Ledge

Design Procedure

The critical section for design of a bracket or corbel is taken at the face of the support; however, in the case of beam ledges (Fig. 16-2) the critical section for flexural design must be taken at the centerline of the beam stirrup, which acts as a "hanger" to carry vertical loads.

The critical section is "designed to resist simultaneously a shear V_u, a moment $[V_u a + N_{uc}(h-d)]$ and a horizontal tensile force N_{uc}". The value of N_{uc} should be not less than $0.2V_u$, unless some provision is made to avoid tensile forces. Since slip joints and flexible bearings do not always function as designed, good practice dictates a value of at least $0.2V_u$ for N_{uc}, in any case. Since the tensile force N_{uc} is due to indeterminate causes such as restrained shrinkage or temperature stresses, a load factor of 1.7, as for live load, is required.

The critical section must be proportioned such that the required V_n is not greater than $0.2f'_c b_w d$ nor $800b_w d$ (as in shear-friction design); except, for lightweight concrete, V_n is somewhat more restricted by Section 11.9.3.2.2. Tests show that for lightweight concrete, the maximum shear strength is a function of a/d as well as f'_c.

For design purposes, the total reinforcement required is divided into three parts, with each determined separately: (A_{vf}) area of shear-friction reinforcement to resist direct shear V_u; (A_f) area of flexural reinforcement to resist moment $V_u a + N_{uc}(h-d)$; and (A_n) area of tensile reinforcement to resist direct tensile force N_{uc}. With behavior predominantly controlled by shear, a single value of $\varphi = 0.85$ should be used for all design conditions.

Once the separate areas of reinforcement A_{vf}, A_f, and A_n have been determined, the actual reinforcement to be provided, A_s and A_h, may be sized, where A_s will act as the primary tension reinforcement and A_h will act as shear reinforcement.

Since the flexural reinforcement A_f is balanced by a compression area, it also contributes to the direct shear-transfer strength of the section: thus, the total required reinforcement ($A_s + A_h$) will be somewhat less than the total of ($A_{vf} + A_f + A_n$).

The required reinforcement is distributed to conform to the results of tests. The total of ($A_s + A_h$) must be the greater of two amounts: (1) the sum of A_{vf} and A_n or (2) the sum of $3/2A_f$ and A_n, thus a comparison of A_{vf} and $3/2A_f$ (or $2/3A_{vf}$ and A_f) will determine control.

With control determined, primary tension reinforcement A_s may be found from either $A_s = (2/3\ A_{vf} + A_n)$ or $A_s = (A_f + A_n)$, whichever is larger. The required area of shear reinforcement parallel to A_s is then computed as $A_h = 0.5(A_s - A_n)$. This results in the total ($A_s + A_h$) as required above. A_h is to be distributed uniformly within two-thirds of the depth "d" next to A_s. For typical shallow ledge members, distribution of A_h depends on the magnitude of the punching shear strength around the bearing area. If the required V_n is greater than the punching shear strength, A_h should be distributed in the upper third of the ledge depth, otherwise A_h may be added to A_s. See Design Example 16.3.

A minimum amount of primary tension reinforcement $\rho = 0.04(f'_c/f_y)$ is required to assure a ductile failure after cracking under moment and direct tensile force. All reinforcement must be fully developed on both sides of the critical section. Anchorage in the support is usually accomplished by embedment or hooks. Within the bracket or corbel, the distance between load and support face is usually too short, so special anchorage must be provided at the outer ends of both A_s and A_h. Anchorage of A_s is normally provided by welding a cross bar of equal size across the ends of A_s (Fig. 16-3) or welding to an armor angle. In the former case, the cross bar must be located beyond the edge of the loaded area, a change from ACI 318-77. Where anchorage is provided by a hook or a loop in A_s, the load must not project beyond the straight portion of the hook or loop (Fig. 16-4). In beam ledges, anchorage may be provided by a hook or loop, with the same limitation on the load location (Fig. 16-5). Where a corbel or beam ledge is designed to resist specific horizontal forces, the bearing plate should be welded to A_s.

The closed stirrups or ties used for A_h must be similarly anchored, usually by bending around a "framing bar" of the same diameter as the closed stirrups or ties.

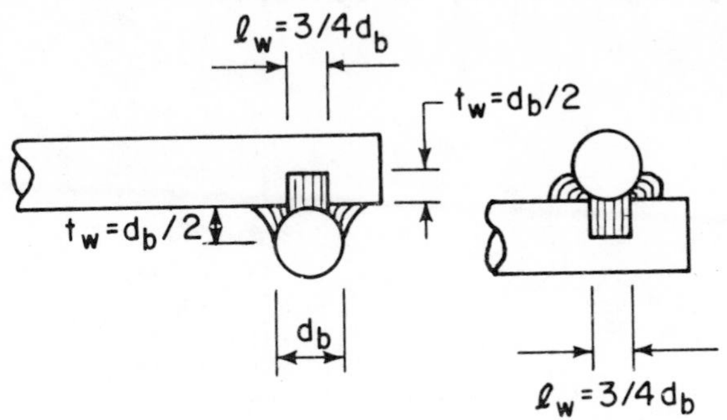

Fig. 16-3 Cross-Bar Weld Details

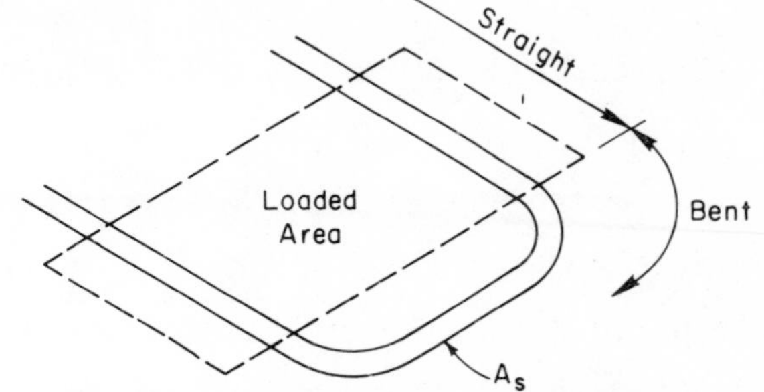

Fig. 16-4 Load Area Limitation with Loop Bar Detail

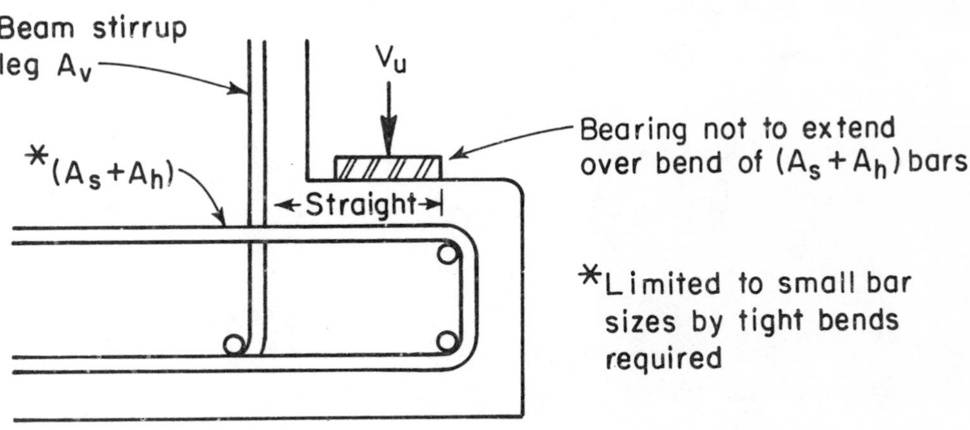

Fig. 16-5 Bar Details for Beam Ledge

Selected References

16.1 Mirza, Sher Ali, and Furlong, Richard W., "Strength Criteria for Concrete Inserted T-Girder," Journal of Structural Engineering V. 109, No. 8, Aug. 1983, pp. 1836-1853.

16.2 Mirza, Sher Ali, and Furlong, Richard W., "Serviceability Behavior and Failure Mechanisms of Concrete Inserted T-Beam Bridge Bent Caps," ACI Journal, Proceedings V. 80, No. 4, July-August 1983, pp 294-304.

EXAMPLE 16.1 - Corbel Design

Design a corbel with minimum dimensions to support a beam as shown below. Corbel to project from a 14-in.-square column. Restrained creep and shrinkage create a horizontal force of 20 kips at the welded bearing.

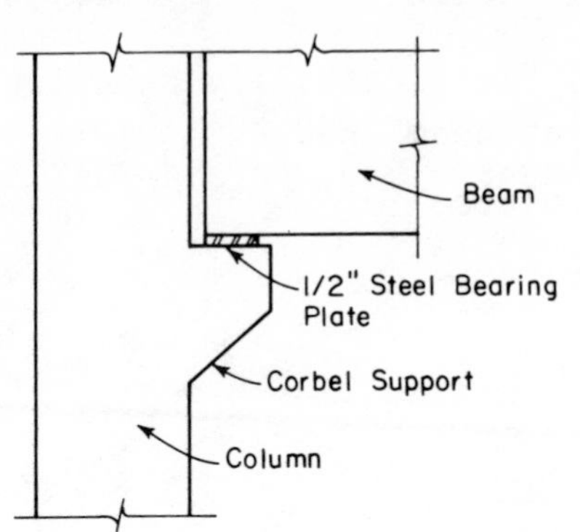

f'_c = 5,000 psi (normal weight)

f_y = 60,000 psi

Beam reactions:

$DL = 24^k$

$LL = 37.5^k$

$T = 20^k$

Calculations and Discussion	Code Reference

1. Size bearing based on bearing strength on concrete according to Section 10.15. Width of bearing = 14 in.

$V_u = 1.4(24) + 1.7(37.5) = 97.4^k$

$V_u \leq \varphi P_{nb} = \varphi\ (0.85 f'_c\ A_1)$

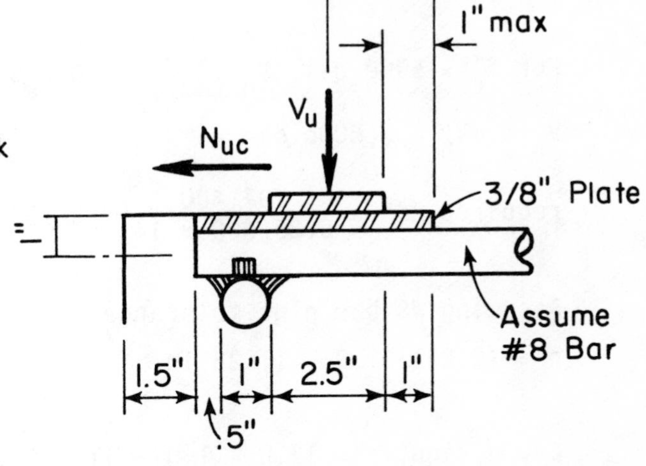

Preliminary sketch of bearing details.

$97.4 = 0.70(0.85 \times 5A_1) = 2.975A_1$ — 9.3.2.4

$A_1 = 97.4/2.975 = 32.74\ \text{in.}^2$

Bearing length = 32.74/14 = 2.34 in.

<u>Use 2.5 in. x 14 in. bearing</u>

EXAMPLE 16.1 - Continued

Calculations and Discussion	Code Reference

$N_{uc} = 1.7(20) = 34^k$ (treated as Live Load) 11.9.3.4

2. Determine "a" with 1 in. max. clearance at beam end. Beam reaction at third point at bearing plate.

$a = 2/3(2.5) + 1.0 = 2.67$ in.

<u>Use a = 3 in. maximum</u>
Detail cross bar just outside outer bearing edge.

3. Determine total depth of corbel based on limiting shear-transfer strength V_n.

For $f'_c = 5000$ psi, $V_{n\,(max)} = 800b_w d$ 11.9.3.2.1

$V_u \leq \varphi V_n = \varphi(800b_w d)$

$$\text{required "d"} = \frac{97,400}{0.85(800 \times 14)} = 10.23 \text{ in.}$$

Assuming #8 bar plus tolerance,
$h = 10.23 + 1.0 = 11.23$ in. <u>Use h = 12 in.</u>

For design, $d = 12.0 - 1.0 = 11.0$ in. $a/d = 0.27$

4. Determine shear-friction reinforcement A_{vf}. 11.9.3.2

$$A_{vf} = \frac{V_u}{\varphi f_y \mu} = \frac{97.4}{0.85(60)(1.4 \times 1)} = 1.36 \text{ in.}^2$$ 11.7.4.1 11.7.4.3

EXAMPLE 16.1 - Continued

Calculations and Discussion	Code Reference

5. Determine moment reinforcement A_f. — 11.9.3.3

$M_u = V_u a + N_{uc}(h-d)$

$= 97.4(3) + 34(12-11) = 326.2^{\text{"k}}$

Find A_f using ordinary flexural design methods or conservatively use $j_u d = 0.9d$

$$A_f = \frac{326.2}{0.85(60)(0.9 \times 11)} = 0.646 \text{ in.}^2$$

Note: For all design calculations, $\varphi = 0.85$ — 11.9.3.1

6. Determine direct tension reinforcement A_n. — 11.9.3.4

$$A_n = \frac{N_{uc}}{\varphi f_y} = \frac{34}{0.85(60)} = 0.667 \text{ in.}^2$$

7. Determine primary tension reinforcement A_s. — 11.9.3.5

$(2/3)A_{vf} = (2/3)1.36 = 0.907 \quad A_f = 0.646$; $(2/3)A_{vf}$ controls design

$A_s = 2/3A_{vf} + A_n = 0.907 + 0.667 = 1.57 \text{ in.}^2$

<u>Use 2 #8 bars</u> ($A_s = 1.58 \text{ in.}^2$) — 11.9.5

Check minimum reinforcement A_s:

$\rho_{min} = 0.04(f'_c/f_y) = 0.04(5/60) = 0.003$

$A_{s(min)} = 0.003(14)(11) = 0.462 \text{ in.}^2 < 1.58$ OK

8. Determine shear reinforcement A_h. — 11.9.4

Calculations and Discussion	Code Reference

$A_h = 0.5(A_s - A_n) = 0.5(1.57 - 0.667) = 0.454$ in.^{2}

<u>Use 3 #3 stirrups</u> (A_h = 0.66 in.^{2})

Distribute stirrups in two-thirds of depth adjacent to A_s.

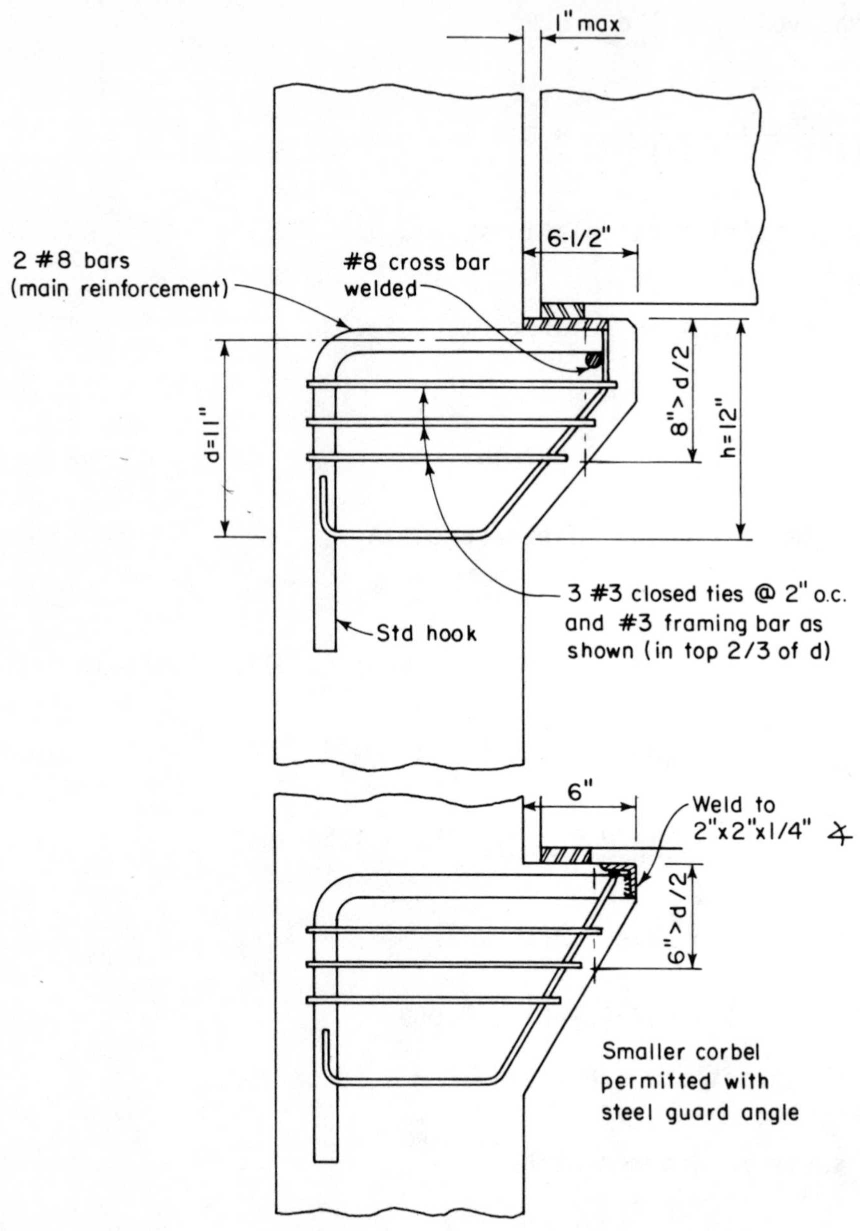

Reinforcement Details for Corbel

EXAMPLE 16.2 - Corbel Design....Using Lightweight Concrete and "Modified Shear-Friction Method."

Design a corbel to project from a 14-in.-square column to support the following beam reactions:

dead load = 32 kips

live load = 30 kips

horizontal force = 24 kips

f'_c = 4000 psi (all lightweight)

f_y = 60,000 psi

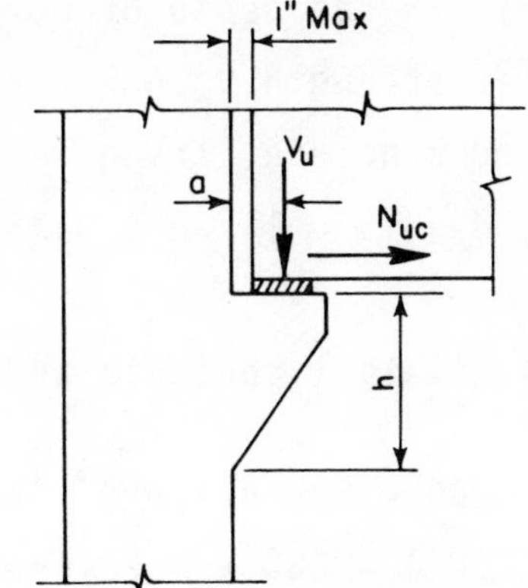

Calculations and Discussion	Code Reference
1. Size bearing plate	
$V_u = 1.4(32) + 1.7(30) = 95.8^k$	Eq.(9-1)
$V_u \leq \varphi P_{nb} = \varphi(0.85f'_c A_1)$	10.15.1
$95.8 = 0.70(0.85 \times 4 \times A_1)$	9.3.2.4
solving $A_1 = 40.3$ in.^{2}	
length of bearing required = 40.3/14 = 2.9 in.	
<u>Use 14 in. x 3 in. plate</u>	
Assume beam reaction to act at outer third point of bearing plate. Assume 1-in. gap between back edge of bearing plate and column face. Hence,	
$a = 1 + 2/3(3) = 3$ in.	

EXAMPLE 16.2 - Continued

	Calculations and Discussion	Code Reference
2	Determine total depth of corbel based on limiting shear-transfer strength V_n. For easier placement of reinforcement and concrete, try h = 15 in. Assuming #8 bar with 1-in. cover, d = 15 - 1 - 0.5 = 13.5 in. a/d = 0.22	
	For lightweight concrete and $f'_c \geq 4000$ psi:	11.9.3.2.2
	$V_n = (800 - 280\ a/d)b_w d$, in pounds $= (800 - 280 \times 0.22)14 \times 13.5/1000 = 139.6^k$ $\varphi V_n = 0.85(139.6) = 118.7 > V_u = 95.8$ OK	
3.	Determine shear-friction reinforcement A_{vf}.	11.9.3.2
	Using a "Modified Shear-Friction Method" as permitted by Section 11.7.3 (See commentary Section 11.7.3):	
	$V_n = 0.8A_{vf}f_y + K_1 b_w d$, with $A_{vf}f_y/b_w d$ not less than 200 psi	
	For all lightweight concrete, $k_1 = 200$ psi	
	$V_u \leq \varphi V_n \leq \varphi(0.8A_{vf}f_y + 0.2b_w d)$	
	Solving for A_{vf}:	
	$A_{vf} = \frac{(V_u - \varphi 0.2b_w d)}{\varphi 0.8f_y}$, but not less than $0.2b_w d/f_y$ $= \frac{(95.8 - 0.85 \times 0.2 \times 14 \times 13.5)}{0.85(0.8)60} = 1.56\ \text{in.}^2$	
	but not less than $0.2 \times 14 \times 13.5/60 = 0.63\ \text{in.}^2$	
	For comparison, compute A_{vf} by Code Eq. (11-26):	
	For all-lightweight concrete, $\mu = 1.4\lambda = 1.4(0.75) = 1.05$	11.7.4.3

EXAMPLE 16.2 - Continued

Calculations and Discussion	Code Reference

$$A_{vf} = \frac{V_u}{\varphi f_y \mu} = \frac{95.8}{0.85 \times 60 \times 1.05} = 1.79 \text{ in.}^2$$

4. Determine moment reinforcement A_f. 11.9.3.3

$$M_u = V_u a + N_{uc}(h-d)$$

$$= 95.8(3) + 40.8(15 - 13.5) = 348.6''^k$$

where $N_{uc} = 1.7(24) = 40.8^k$ 11.9.3.4

Find A_f using ordinary flexural design methods, or conservatively use $j_u d = 0.9d$

$$A_f = \frac{M_u}{\varphi f_y j_u d} = \frac{348.6}{0.85 \times 60 \times 0.9 \times 13.5} = 0.56 \text{ in.}^2$$

Note: For all design calculations, $\varphi = 0.85$ 11.9.3.1

5. Determine direct tension reinforcement A_n. 11.9.3.4

$$A_n = \frac{N_{uc}}{\varphi f_y} = \frac{40.8}{0.85 \times 60} = 0.80 \text{ in.}^2$$

6. Determine primary tension reinforcement A_s. 11.9.3.5

$(2/3)A_{vf} = (2/3)1.56 - 1.04 > A_f = 0.56$; $(2/3)A_{vf}$ controls design

$A_s = (2/3)A_{vf} + A_n = 1.04 + 0.80 = 1.84 \text{ in.}^2$

Use 2 #9 bars ($A_s = 2.00 \text{ in.}^2$)

or 3 #7 bars ($A_s = 1.80 \text{ in.}^2$)

check $A_{s(min)} = 0.04(4/60)\ 14 \times 13.5 = 0.504 \text{ in.}^2 < 1.84$ OK

EXAMPLE 16.2 - Continued

Calculations and Discussion	Code Reference

7. Determine shear reinforcement A_h. 11.9.4

$A_h = 0.5(A_s - A_n) = 0.5\ (1.84 - 0.80) = 0.52\ \text{in.}^2$

<u>Use 3 #3 stirrups</u> ($A_h = 0.66\ \text{in.}^2$)

The shear reinforcement is to be placed within two thirds of the effective depth adjacent to A_s.

$s_{max} = (2/3)\ 13.5/3 = 3\ \text{in.}$ <u>Use 3 in. stirrup spacing</u>

8. Corbel Details

Corbel will project (1 + 3 + 2) = 6 in. from column face.

Use 6-in. depth at outer face of corbel, then depth at outer edge of bearing plate will be greater than d/2. 11.9.2

$6 + 7.5/3 = 8.5\ \text{in.} > 13.5/2$ OK

A_s to be anchored at front face of corbel by welding a #7 bar transversely across ends of A_s bars. 11.9.6

A_s must be anchored within column by standard hook.

Calculations and Discussion	Code Reference

Corbel Details

EXAMPLE 16.3 - Beam Ledge Design

f'_c = 5 ksi (normal weight)

f_y = 60 ksi

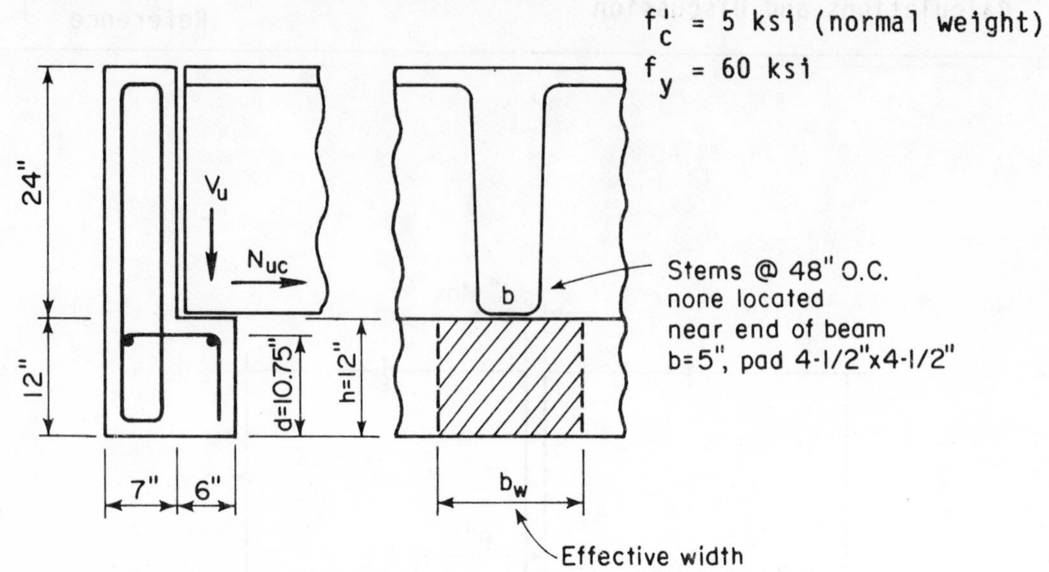

The L-beam shown is to support a double-tee parking deck spanning 64 feet. Maximum service loads per stem are: DL = 11.1 kips; LL = 6.4 kips; total load = 17.5 kips. The loads may occur at any location on the L-beam ledge except near beam ends. The stems of the double-tees rest on 4.5 in. x 4.5 in. x 1/4 in. neoprene bearing pads (1000 psi maximum service load).

Design in accordance with the 1983 Code provisions for brackets and corbels may require a wider ledge than the 6 in. shown. To maintain the 6-in. width, one of the following may be necessary: (1) Use of a higher strength bearing pad (up to 2000 psi); or (2) Anchoring primary ledge reinforcement A_s to an armor angle.

This example will be based on the 6-in. ledge with 4.5-in.-square bearing pad. At the end of the example an alternative design will be shown.

EXAMPLE 16.3 - Continued

Calculations and Discussion	Code Reference

1. Check 4.5x4.5 bearing pad size (1000 psi maximum service load).
   $4.5 \times 4.5(1) - 20.25^k > 17.5$ OK

2. Check concrete bearing strength.

   $V_u = 1.4(11.1) + 1.7(6.4) = 26.42^k$ Eq. (9-1)

   $\varphi P_{nb} = \varphi(0.85f'_c\ A_1)$ 10.15.1

   $= 0.85(0.85 \times 5 \times 4.5 \times 4.5)$

   $= 73.15^k > 26.42$ OK

3. Determine shear-span "a" for both shear and flexure. The reaction is considered to be at the outer third point of the bearing pad.
   For shear, a = 4.5(2/3) + 1.0 = 4 in.

   For flexure, the critical section is at the center line of the hanger reinforcement (A_v). Assume 1-in. cover and #4 bar stirrups.
   a = 4 + 1 + 0.25 = 5.25 in.

4. Determine width of ledge to be considered as effective for each stem reaction. An effective width equal to width of bearing plus 4 times shear-span "a" is suggested.[16.1] (If a stem reaction were located near the end of a beam ledge, the effective width may be further restricted to twice the distance between center of bearing and end of ledge).
   b + 4a = 4.5 + 4(4) = 20.5 in.

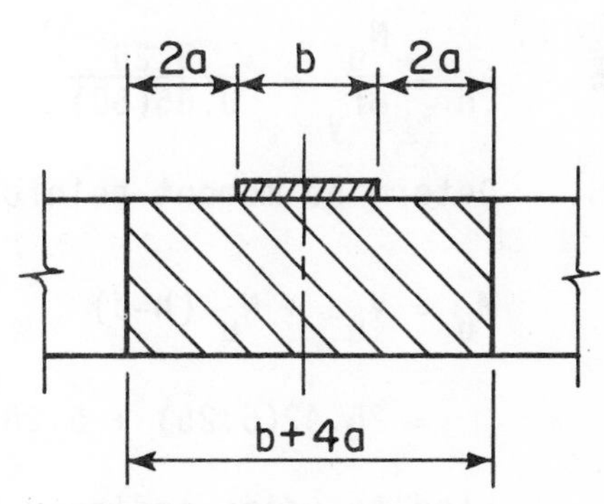

5. Check effective ledge section for shear-transfer strength V_n. 11.9.3.2.1

EXAMPLE 16.3 - Continued

Calculations and Discussion	Code Reference

For f'_c = 5000 psi; $V_{n(max)} = 800\ A_c$, where $A_c = (b + 4a)d$

$$V_n = 800(20.5)(10.75)/1000 = 176.3^k$$

$$\varphi V_n = 0.85(176.3) = 149.8^k > 26.42 \quad \text{OK}$$

6. Determine shear-friction reinforcement A_{vf}. 11.9.3.2

$$A_{vf} = \frac{V_u}{\varphi f_y \mu} = \frac{26.42}{0.85(60)1.4} = 0.37 \text{ in.}^2\text{/stem}$$ 11.7.4.1 11.7.4.3

7. Determine reinforcement to resist direct tension A_n. 11.9.3.4
Unless special provisions are made to reduce direct tension, N_u should be taken not less than $0.2V_u$ to account for unexpected forces due to restrained long-time deformation of the supported member, or other causes. When the beam ledge is designed to resist specific horizontal forces, the bearing plate should be welded to the tension reinforcement A_s.

$$N_u = 0.2V_u = 0.2(26.42) = 5.28^k$$

$$A_n = \frac{N_u}{\varphi f_y} = \frac{5.28}{0.85(60)} = 0.104 \text{ in.}^2\text{/stem}$$

8. Determine moment reinforcement A_f.

$$M_u = V_u a + N_u\,(h-d)$$

$$= 26.42(5.25) + 5.28(12 - 10.75) = 145.3^{\prime\prime k}$$

Find A_f using ordinary flexural design methods. For beam ledges, use $j_u d = 0.8d$. 11.9.3.1
Also, φ should be taken as 0.85

$$A_f = \frac{145.3}{0.85(60)(0.8 \times 10.75)} = 0.331 \text{ in.}^2\text{/stem}$$

EXAMPLE 16.3 - continued

Calculations and Discussion	Code Reference

9. Determine primary tension reinforcement A_s. — 11.9.3.5

$(2/3)A_{vf} = (2/3)0.37 = 0.247 < A_f = 0.435$; A_f controls design

$A_s = A_f + A_n = 0.331 + 0.104 = 0.435$ in.^{2}/stem

Check minimum $A_s = 0.04(f'_c/f_y)\ bd$ — 11.9.5

$= 0.04(5/60)20.5 \times 10.75 = 0.735$ in.^{2}/stem $\gg 0.435$*

*For typical shallow ledge members, minimum A_s by Section 11.9.5 will almost always govern.

10. Determine shear reinforcement A_h. — 11.9.4

$A_h = 0.5(A_s - A_n) = 0.5A_f$

$= 0.5(0.331) = 0.166$ in.^{2}/stem

11. Determine final size and spacing of ledge reinforcement.

Distribution of A_h depends on the magnitude of the punching shear strength around the bearing area:

$V_n = 4\ f'_c b_o d = 4\ 5000 \times 36 \times 10.75/1000 = 109.5^k$

where $b_o = (b + 2w + 2d)$

$= (4.5 + 2 \times 5 + 2 \times 10.75) = 36$ in.

$\varphi V_n = 0.85(109.5) = 93^k > 26.42$

Combine $A_s + A_h$ in one layer.

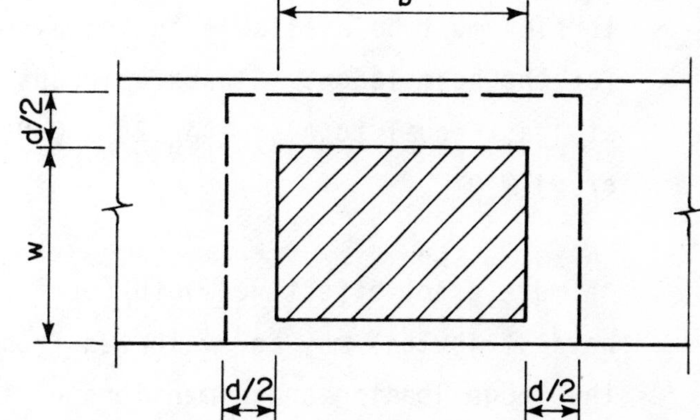

EXAMPLE 16.3 - continued

Calculations and Discussion	Code Reference

Since all required areas of reinforcement are computed for the reaction of one stem of the double-tee, divide areas by effective ledge width (20.5) to obtain area per inch of ledge so that an adequate amount of reinforcement is provided along the effective width no matter where the tee stems are located on the ledge.

$A_s + A_h = (0.735 + 0.166)/20.5 = 0.044$ in.^{2}/in. of ledge

Try #4 bar, $A_b = 0.20$ in.^{2} Try #5 bar, $A_b = 0.31$

$s_{max} = 0.20/0.044 = 4.55$ in. $s_{max} = 0.31/0.044 = 7.05$ in.

Use #5 @ 7 in.

12. Check required area of beam stirrups used as "hangar" reinforcement.

$$A_v \text{ (one leg)} = \frac{V_u}{\varphi f_y} = \frac{26.42}{0.85(60)} = 0.518 \text{ in.}^2\text{/leg}$$

Use 20.5 in. effective ledge width (somewhat conservative) for distribution of hangar reinforcement.*

A_v (one leg) = 0.518/20.5 = 0.0253 in.^{2}/in. of ledge.

Sufficient stirrups in beam section for combined shear and torsion must be available to act also as hanger reinforcement for the beam ledge. Closed stirrups in beam section must be at least equal to $A_s/s + A_v/2s \geq 0.0253$ in.^{2}/in.; i.e, #3 @ 4 or #4 @ 8

*A more exact effective width (over which hanger forces can be distributed) may be evaluated from Reference 16.1. For the ledge loading and dimensions of this example, $b_e = 37.5$ in.

EXAMPLE 16.3 - continued

Calculations and Discussion	Code Reference

A_v (one leg) = 0.518/37.5 = 0.0138 $in.^2/in.$

thus, $A_t/s + A_v/2s \geq 0.0138$ $in.^2/in.$; i.e, #3 @ 8 or #4 @ 14

13. Reinforcement Details

In accordance with Section 11.9.7, bearing area (4.5-in. pad) must not extend beyond straight portion of beam ledge reinforcement, nor beyond inside edge of transverse anchor bar. With a 4.5-in. bearing pad, this requires that the width of ledge be increased to 9 in. as shown below. Alternately a 6-in. ledge with a 3-in. medium strength pad (1500 psi) and the ledge reinforcement welded to an armor angle would satisfy the intent of Section 11.9.7.

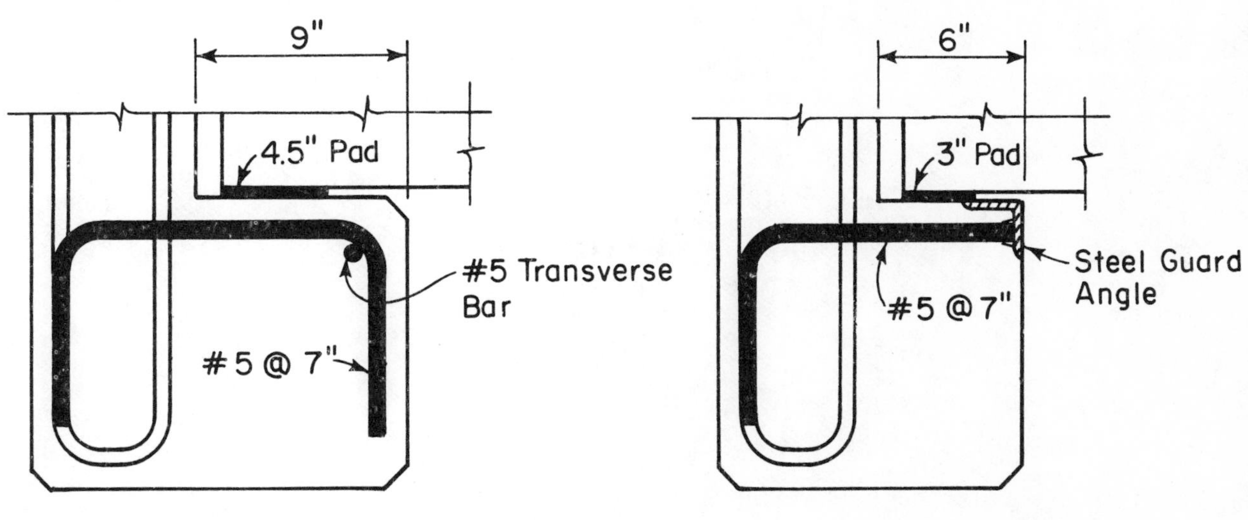

9 in. ledge detail

6 in. ledge detail

17

Shear in Slabs

11.11 Special Provisions for Slabs and Footings

Design of slabs for shear in the region near concentrated loads and reactions is regulated by Section 11.11. For flexural members, such as one-way slabs in which the bending action is primarily in one direction, the requirements of Sections 11.1 through 11.5 must be satisfied. When two-way action is the primary mode of behavior, such as in flat slabs, then the failure mechanism changes to that of "punching," thus necessitating a different design approach. Even though beam action shear rarely controls the strength of two-way slabs, the designer must ensure that shear strength for beam action is not exceeded. Tributary areas and corresponding critical sections for beam action shear and two-way action shear at a slab and column support are illustrated in Fig. 17-1. For design application see Example 20.1.

Two-way action shear strength of slabs (without shear reinforcement) is affected by the following four principal variables:

1. concrete strength,
2. relationship between size of loaded area and slab thickness,
3. shape of loaded area,
4. shear-moment ratio at the critical section.

These variables are taken into account in the formulation of Eq. (11-36). Shear strength affected by the shape of loaded area β_c was first introduced in the 1977 Code edition. The β_c variable provides a transition between two-way action shear ($4\sqrt{f'_c}$) and beam action shear ($2\sqrt{f'_c}$) as the support size becomes more elongated (See Fig. 17-2). For a support size less than 2 to 1 ($1/\beta_c \geq 0.5$), Eq. (11-36) reduces to $V_c = 4\sqrt{f'_c}b_o d$.

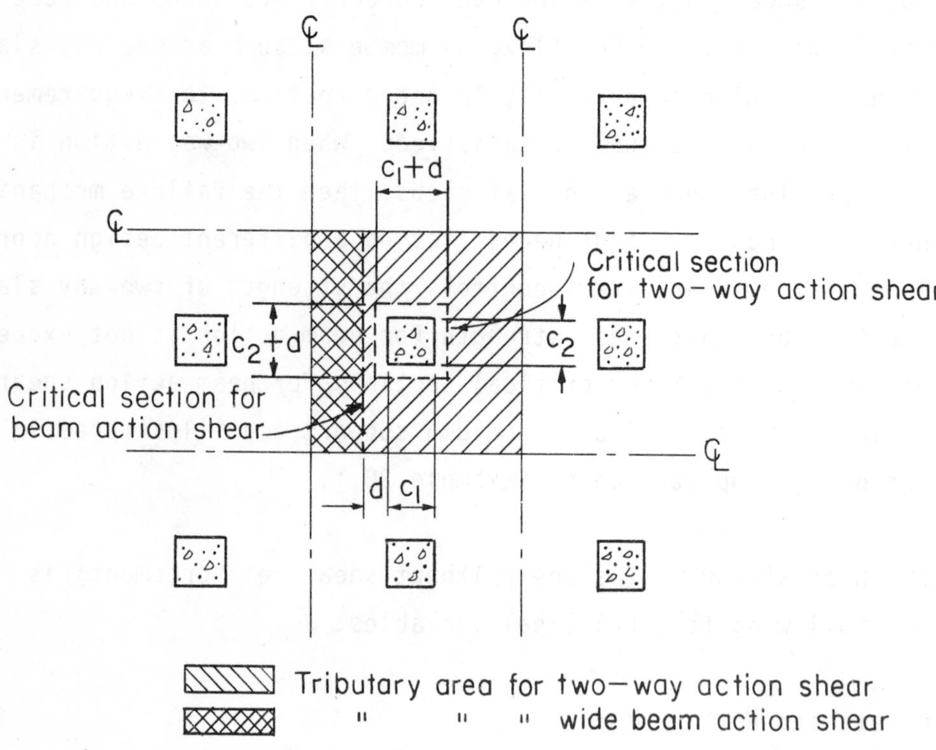

Fig. 17-1 - Tributary Areas and Critical Sections for Slab Shear at a Column Support

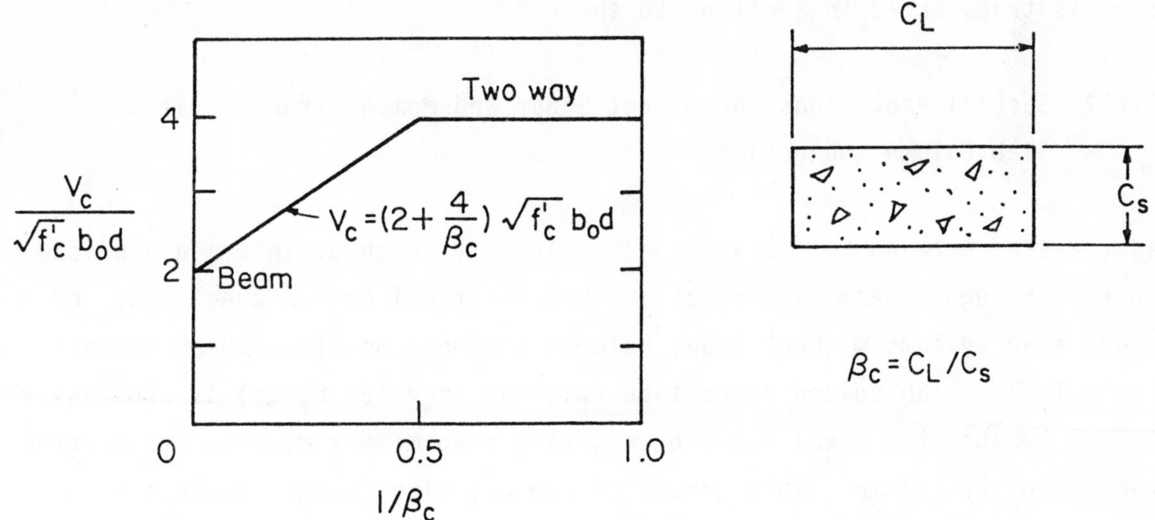

Fig. 17-2 - Shear Strength for Slabs

Under certain design conditions, it may be desirable to increase the shear strength of slabs by using shear reinforcement consisting of bars, wires, or steel I or channel shapes (shearheads). In the design of shearhead reinforcement, three important basic criteria must be considered as follows:

1. To ensure that the required shear strength of the slab is reached before the flexural strength of the shearhead is exceeded, a minimum flexural strength must be provided for the shearhead (Section 11.11.4.6).

2. Shear strength in the slab at the end of the shearhead reinforcement must not be exceeded (Section 11.11.4.8).

3. After the above requirements are satisfied, certain reduction in column strip negative moment reinforcement may be made in proportion to the contribution of the shearhead (Section 11.11.4.9).

Application of these principles is illustrated in Design Example 17.3.

With the 1983 Code, the shearhead reinforcement provisions were extended to include design of shearheads at edge columns and at interior columns when moment is transferred in addition to shear.

11.12.2 Special Provisions for Direct Shear and Moment Transfer at Slab-Column Connections

Load transfer between a slab and column directly, without intermediate load transfer through a beam, is one of the more critical design conditions for two-way slab systems without beams between column supports. Shear strength at an exterior slab-column connection (without spandrel beams) is especially critical because the total exterior negative slab moment must be transferred directly to the column. This aspect of two-way slab design should not be taken lightly by the designer. Two-way slab systems are quite "forgiving" for an error in distribution or even in amount of flexural reinforcement, but there is no forgiveness if a critical error in shear strength is made.

Note that the provisions of Section 11.12.2 (or Section 13.3.3) do not apply to slab systems with beams framing into the column support. With beams, load transfer from slab through the beams to the columns is considerably less critical. Shear strength in slab systems with beams is covered in Section 13.6.8.

The Code specifies that the unbalanced moment between a slab (without beams) and column be transferred by eccentricity of shear (Section 11.12.2) and by flexure (Section 13.3.3) at the slab-column connection. The general mechanism of transfer is illustrated in Fig. 17-3. Shear transfer is assumed to act on a critical section at d/2 from the face of the column (Section 11.12.2.2), while the fraction of unbalanced moment transferred by flexure is resisted by a width of slab equal to the transverse column width c_2, plus 1.5h on each side of the column (Section 13.3.3.2).

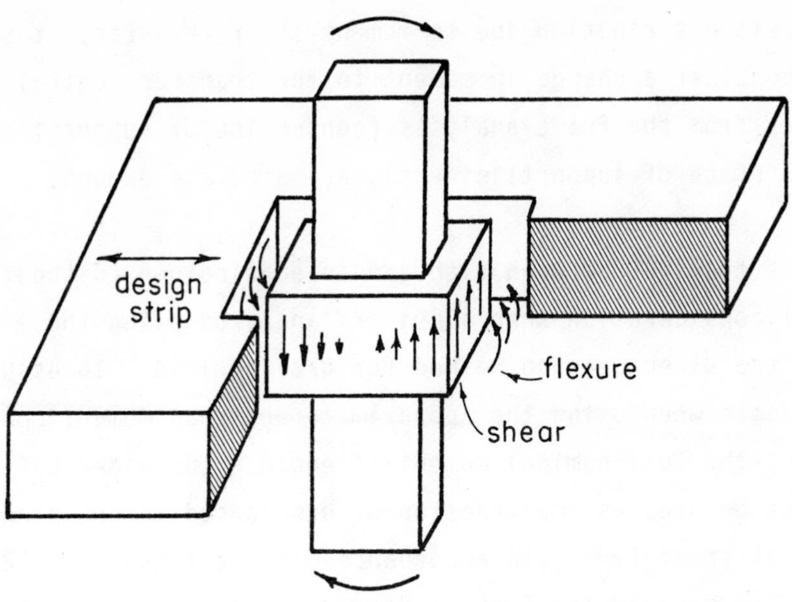

Fig. 17-3 - Direct Shear and Moment Transfer

The fraction of unbalanced moment transferred by eccentricity of shear and by flexure is respectively:

$$\gamma_v = 1 - \frac{1}{1 + 2/3\sqrt{\dfrac{c_1 + d}{c_2 + d}}} \qquad \text{Eq. (11-40)}$$

and

$$\gamma_f = \frac{1}{1 + 2/3\sqrt{\dfrac{c_1 + d}{c_2 + d}}} \qquad \text{Eq. (13-1)}$$

Fig. 17-4 gives a graphical solution to the above equations. The unbalanced moment transferred by eccentricity of shear is $\gamma_v M_u$, where M_u is the unbalanced moment at the centroid of the critical section. The unbalanced moment M_u at an exterior support of an end span will generally not be computed at the centroid of the critical transfer section in the frame analysis.

For the Direct Design Method of Chapter 13, moments are computed at the face of support. Considering the approximate nature of the procedure used to evaluate the stress distribution due to moment-shear transfer, it seems unwarranted to consider a change in moment to the transfer centroid...use of the moment values from the frame analysis (centerline of support) or from Section 13.6.3.3 (face of support) directly are accurate enough.

Unbalanced moment transfer between slab and an edge column (without edge beams) requires special consideration when slabs are analyzed using the moment coefficients of the Direct Design Method for gravity load. To assure adequate shear strength when using the approximate end-span moment coefficient, Section 13.6.3.6, the full nominal moment strength M_n provided by the column strip must be used as the fraction of unbalanced moment transferred by eccentricity of shear ($\gamma_v M_n$) in accordance with Sections 11.12.2.3 and 11.12.2.4 See Part 20 for further discussion of this special shear strength requirement and its application in Design Example 20.1.

For two-way slabs analyzed by the Equivalent Frame Method for gravity load, the computed frame moment at the exterior support for an end span is used directly as the unbalanced transfer moment.

The factored shear stress on the critical transfer section is the sum of direct and unbalanced moment-shear as follows:

$$v_u = V_u/A_c + \gamma_v M_u c/J$$

or

$$v_u = V_u/A_c + \gamma_v M_u c'/J$$

Expressions for A_c, c, c', J/c and J/c' are given in Figs. 17-5 and 17-6. Shear stress v_u must not exceed $\varphi(2 + 4\beta_c)\sqrt{f'_c}$ nor $\varphi 4\sqrt{f'_c}$. Effect of the β_c factor is shown in Fig. 17.2. Design Example 20.1 illustrates shear stress calculations for unbalanced moment-shear transfer between slab and an edge column.

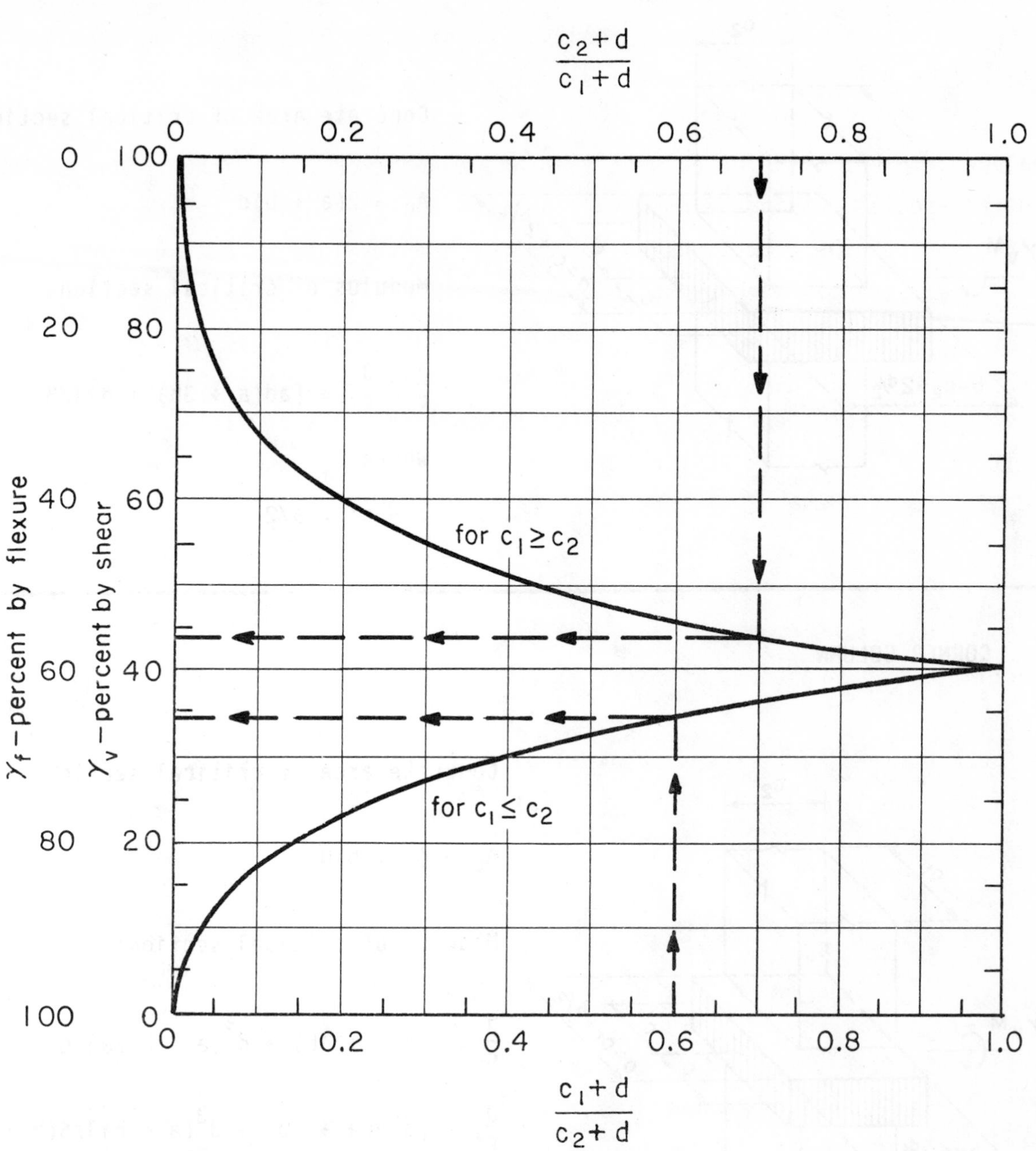

Fig. 17-4 - Solution for Eq. (11-40) and (13-1)

INTERIOR COLUMN

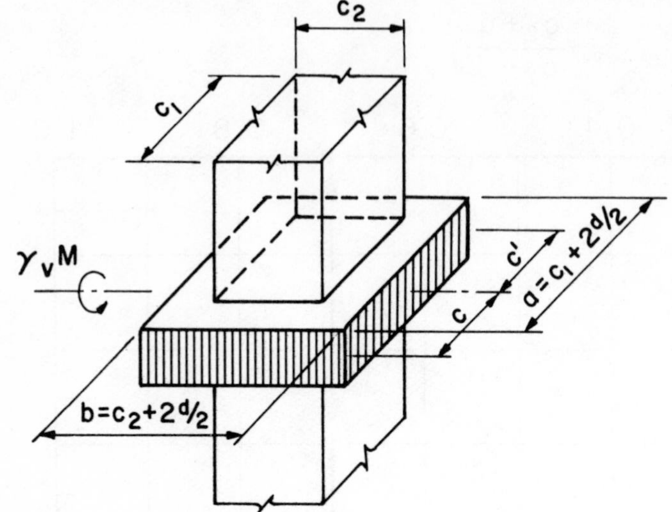

Concrete area of critical section:

$A_c = 2(a + b)d$

Modulus of critical section:

$\frac{J}{c} = \frac{J}{c'} = [ad(a + 3b) + d^3]/3$

where

$c = c' = a/2$

CORNER COLUMN

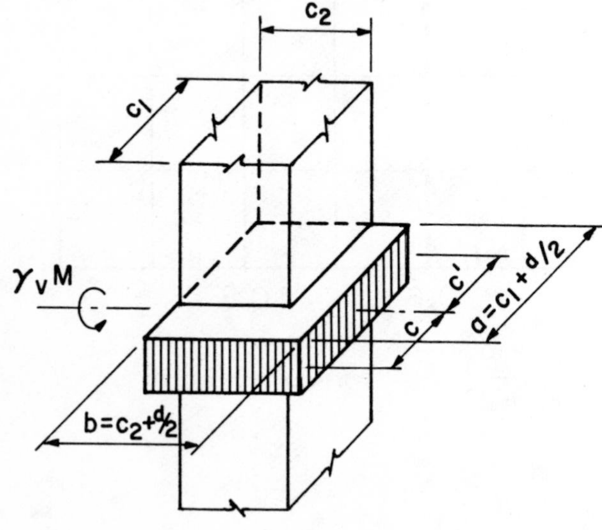

Concrete area of critical section:

$A_c = (a + b)d$

Modulus of critical section:

$\frac{J}{c} = [ad(a + 4b) + d^3(a + b)/a]/6$

$\frac{J}{c'} = [a^2d(a + 4b) + d^3(a + b)]/6(a + 2b)$

where

$c = a^2/2(a + b)$

$c' = a(a + 2b)/2(a + b)$

Fig. 17-5 - Section Properties for Shear Stress Computations

EDGE COLUMN (Bending Parallel to edge)

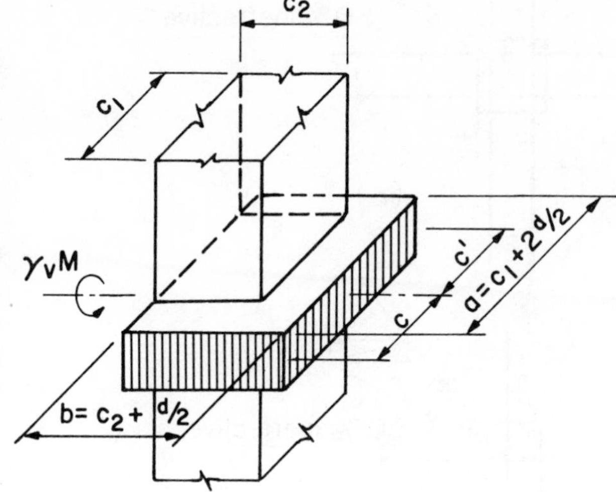

Concrete area of critical section:

$$A_c = (a + 2b)d$$

Modulus of critical section:

$$\frac{J}{c} = \frac{J}{c'} = [ad(a + 6b) + d^3]/6$$

where

$$c = c' = a/2$$

EDGE COLUMN (Bending perpendicular to edge)

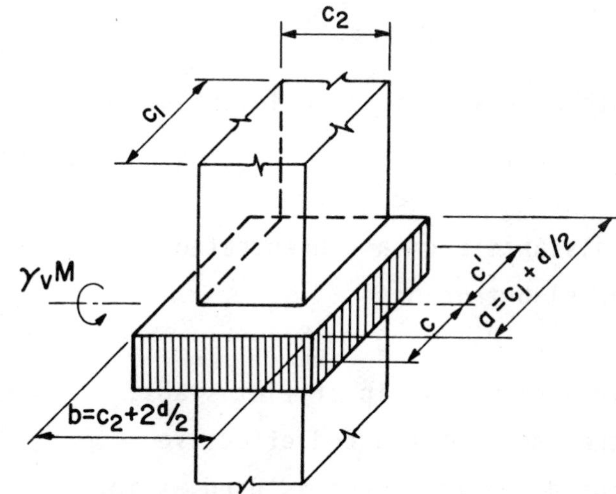

Concrete area of critical section:

$$A_c = (2a + b)d$$

Modulus of critical section:

$$\frac{J}{c} = [2ad(a + 2b) + d^3(2a + b)/a]/6$$

$$\frac{J}{c'} = [2a^2d(a + 2b) + d^3(2a + b)]/6(a + b)$$

where

$$c = a^2/2(a + b)$$

$$c' = a(a + b)/(2a + b)$$

Fig. 17-6 - Section Properties for Shear Stress Computations

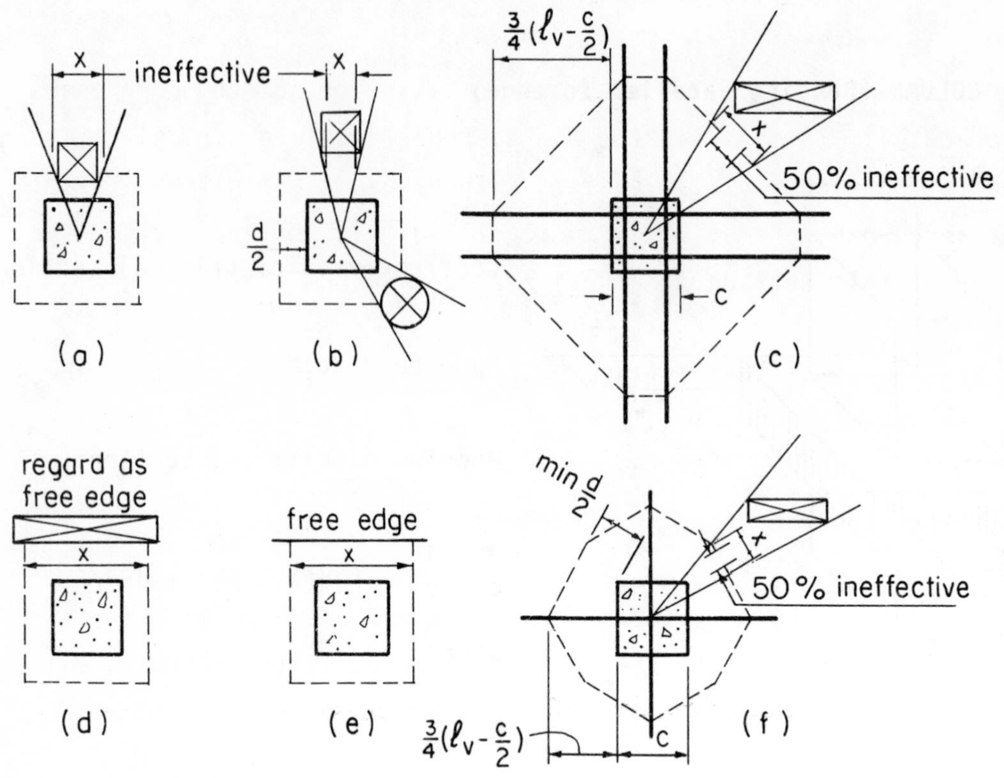

Fig. 17-7 - Effect of Openings in Slabs

11.11.5 Openings in Slabs

The effects of openings on the shear strength of a slab must be investigated in the following circumstances:

1. When the opening is located at a distance from a concentrated load or reaction less than 10 times the slab thickness.

2. When the opening is located within a column strip of flat slabs, the effects are investigated on the basis of reduced effective perimeter as follows: If no shearhead reinforcement is used as in Figs. 17-7(a), (b), (d) and (e), effective periphery = $(b_o - x)$. If shearhead reinforcement is used, as in Figs. 17-7(c) and (f), effective periphery = $(b_o - x/2)$.

13.5 Openings in Slab Systems

The openings shown in Fig. 17-8 are representative of common occurrences found in practice. In this Section of the Code certain empirical limiting dimensions for openings in slab systems without beams are given. These dimensions are shown schematically in Fig. 17-8. To further satisfy the requirements of this Section, the designer must ensure that:

1. Reinforcement interrupted by an opening is added on all sides of the opening.

2. Shear strength is not exceeded.

Openings larger than those shown also can be used, provided an analysis is carried out to determine the effect of the openings on the strength and the serviceability of the slab.

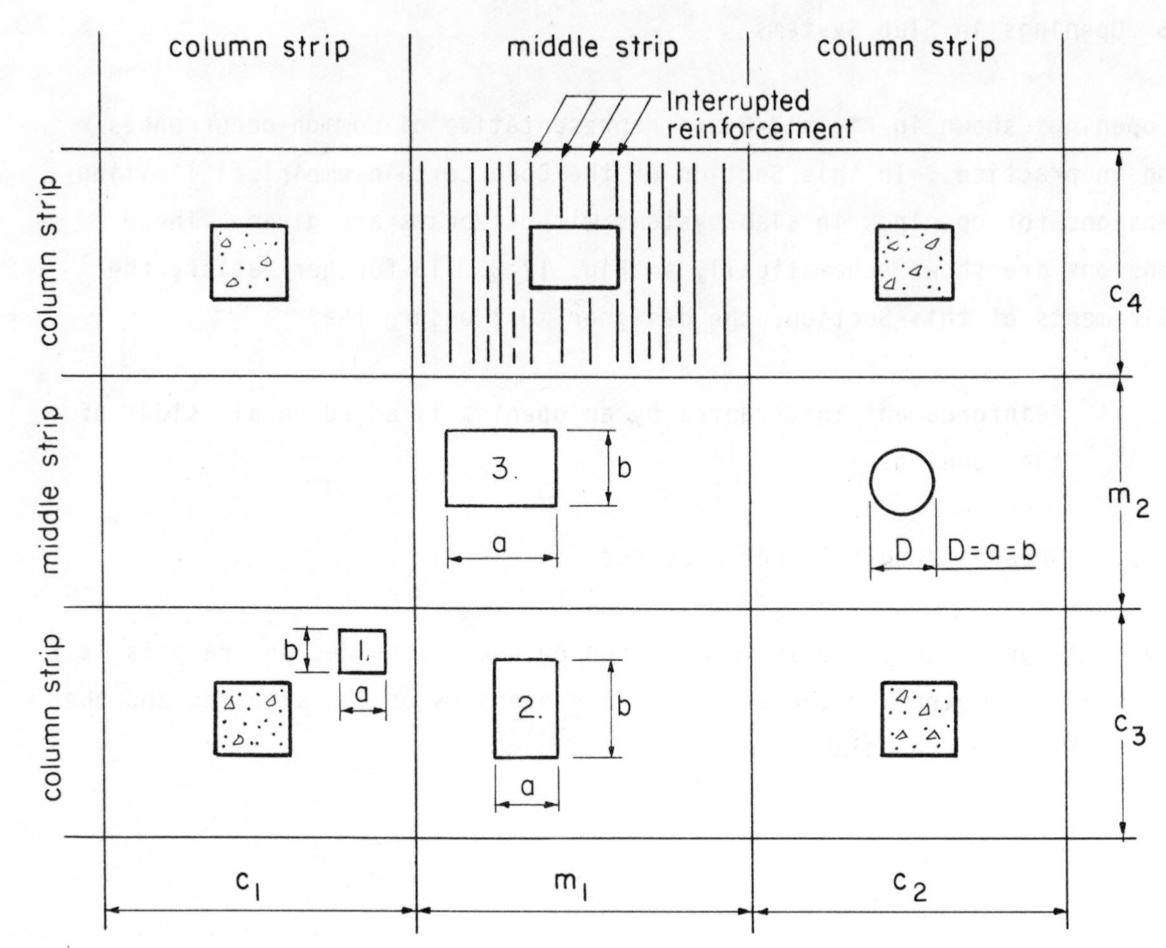

Opening	Limiting Dimension	Code Reference
1	$a \leq \frac{1}{8} c_1$ $b \leq \frac{1}{8} c_3$	13.5.2.2
2	$a \leq \frac{1}{4} m_1$ $b \leq \frac{1}{4} c_3$	13.5.2.3
3	$a \leq m_1$ $b \leq m_2$	13.5.2.1

Fig. 17-8 - Openings in Slab Systems with No Beams

EXAMPLE 17.1 - Shear Strength of Slab at Column Support

Determine two-way action shear strength at an interior column support of a flat plate slab system for the following design conditions.

Column = 48 in. x 8 in.

Slab d = 6.5 in.

f'_c = 3000 psi

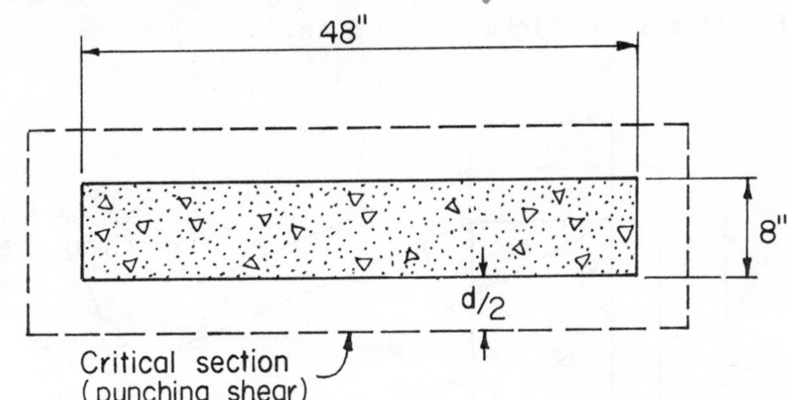

Calculations and Discussion	Code Reference
1. Two-way action shear (punching shear).	
$V_u \leq \varphi V_n$	Eq. (11-1)
$\leq \varphi V_c$	11.11.2
2. Excluding size effect β_c:	
$\varphi V_c = (4\sqrt{f'_c}\, b_o d)$	
$= 0.85\ (4\sqrt{3000} \times 138 \times 6.5) = 167$ kips	
where $b_o = 2(48 + 6.5 + 8 + 6.5) = 138$ in.	11.11.1.2
$\varphi = 0.85$	9.3.2.3
3. Including size effect β_c:	
$\varphi V_c = \varphi(2 + 4/\beta_c)\sqrt{f'_c}\, b_o d$	Eq. (11-36)
$= 0.85\ (2 + 4/6)\sqrt{3000} \times 138 \times 6.5 = 111$ kips	
where $\beta_c = 48/8 = 6$	
4. Summary: Size effect β_c results in a 33% reduction in two-way action shear strength for the 48 x 8 in. column support. Note: the 1971 Code permitted the higher shear strength, excluding size effect β_c.	

EXAMPLE 17.2 - Shear Strength for Non-Rectangular Support

For the L-shaped column support shown, check "punching" shear strength for a factored shear force transfer of $V_u = 105^k$. Use $f'_c = 3,000$ psi. Say effective slab thickness = 5.5 in.

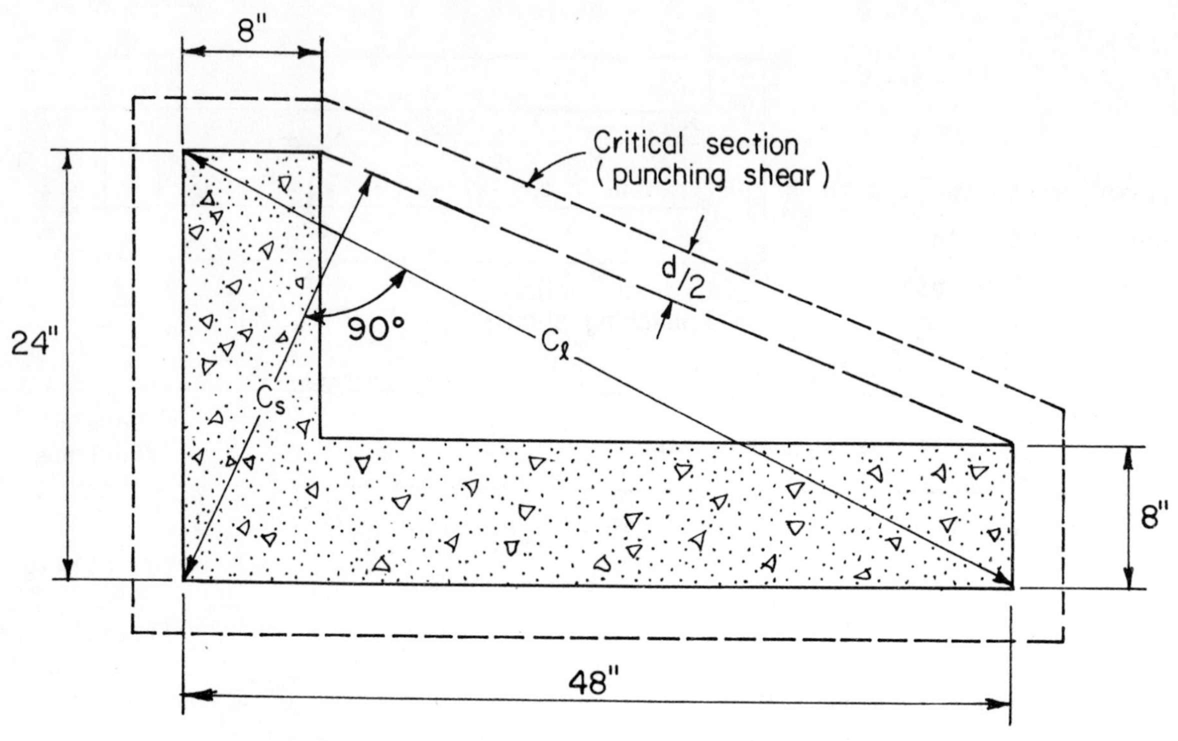

Calculations and Discussion	Code Reference
1. For shapes other than rectangular, the Code Commentary recommends that β_c be taken as the ratio of the longest overall dimension of the effective loaded area c_ℓ to the largest overall dimension of the effective loaded area c_s, measured perpendicular to c_ℓ. As shown above, $\beta_c = c_\ell/c_s \simeq 54/25^* = 2.16$	11.11.2
For the critical section shown, b_o (perimeter of critical section) $\simeq$ 141 in.*	11.11.1.2

*Scaled dimensions are accurate enough.

EXAMPLE 17.2 - Continued

Calculations and Discussion	Code Reference
2. Punching shear strength:	
$V_u \leq \varphi V_n$	Eq. (11-1)
$\leq \varphi V_c$	
$\leq \varphi[(2 + 4/\beta_c)\sqrt{f'_c}\, b_o d]$	Eq. (11-36)
$\leq 0.85\ [(2 + 4/2.16)\sqrt{3000} \times 141 \times 5.5] = 139.1^k$	9.3.2.3
$105^k < 139.1^k$ OK	
Note: The quantity $(2 + 4/\beta_c)$ cannot be taken greater than 4.	
$(2 + 4/2.16) = 3.85 < 4$ OK	

EXAMPLE 17.3 - Shear Strength of Slab with Shear Reinforcement

Consider an interior panel of a flat plate slab system supported by a 10 in. x 10 in. column. Determine shear strength of slab at column support and if not adequate, increase the shear strength by shear reinforcement. Total factored shear force to be transferred from slab to column, V_u = 105 kips. Overall slab thickness, h = 7.5 in. (d ≃ 6 in.).

f'_c = 3,000 psi

f_y = 60,000 psi (reinforcing bars)

f_y = 36,000 psi (structural steel)

Column strip negative moment M_u = 175 ft. kips

Calculations and Discussion	Code Reference
A. Determine shear strength of slab without shear reinforcement.	
1. Assuming two-way action governs, locate the critical section at d/2 = 3 in. from column perimeter. See Fig. 17-1.	11.11.1.2
$b_o = 4(10 + 6) = 64$ in.; $\beta_c = 1.0$	11.11.2
$V_u \leq \varphi V_c$	Eq. (11-1)
$\leq \varphi[(2 + 4/\beta_c)\sqrt{f'_c}\, b_o d]$	Eq. (11-36)
but not greater than $\varphi 4\sqrt{f'_c}\, b_o d$	
$\leq 0.85\,[(2 + 4/1.0)\sqrt{3000} \times 64 \times 6] = 107.3^k$	
$0.85 \times 4\sqrt{3000} \times 64 \times 6 = 71.5^k$ (governs)	
$105^k > 71.5^k$	
Shear strength of slab is not adequate to transfer the factored shear force V_u = 105 kips from slab to column support. Shear strength may be increased by	

EXAMPLE 17.3 - Continued

Calculations and Discussion	Code Reference

(1) increasing concrete strength f'_c

(2) increasing slab thickness at column support by capital and/or drop panel

(3) providing shear reinforcement (bars, wires, or steel I or channel shapes).

B. Increase shear strength by increasing strength of slab concrete.

$V_u \leq \varphi V_c$ Eq. (11-1)

$105{,}000 \leq 0.85 \times 4\sqrt{f'_c} \times 64 \times 6$ Eq. (11-36)

Solving for $f'_c \simeq 6500$ psi

C. Increase shear strength by increasing slab thickness at column support by drop panel. Provide drop panel size in accordance with Sections 9.5.3.2 and 13.4.7 (see Fig. 17-9). Minimum overall slab thickness at drop panel = 1.25(7.5) = 9.375 in. Check 9.5 in. slab thickness; $d \simeq 8.00$ in., $b_o = 4(10 + 8) = 72$ in.

$V_u \leq \varphi V_c = 0.85 \times 4\sqrt{3000} \times 72 \times 8 = 107.3^k$ Eq. (11-36)

$105^k < 107.3^k$ OK

EXAMPLE 17.3 - Continued

Calculations and Discussion	Code Reference

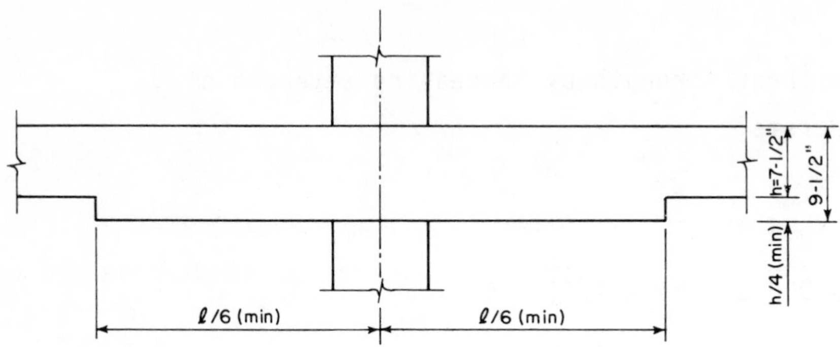

Fig. 17-9 Drop Panel Details

D. Increase shear strength by bar reinforcement. 11.11.3

1. Check maximum shear strength permitted with bars.

$V_u \leq \varphi V_n$ Eq. (11-1)

$\leq \varphi 6\sqrt{f'_c}\, b_o d$ 11.11.3.2

$\leq 0.85 \times 6\sqrt{3000} \times 64 \times 6 = 107.3^k$

$105^k < 107.3^k$ OK

2. Determine shear strength provided by concrete <u>with shear reinforcement</u>.

$V_c = 2\sqrt{f'_c}\, b_o d$ 11.11.3.4

$= 2\sqrt{3000} \times 64 \times 6 = 42^k$

3. Design shear reinforcement in accordance with Section 11.5. Required area of shear reinforcement A_v is computed from Eqs. (11-1), (11-2), and (11-17) as follows: 11.11.3.5

$V_u \leq \varphi V_n$ Eq. (11-1)

EXAMPLE 17.3 - Continued

Calculations and Discussion	Code Reference

$$\leq \varphi V_c + \varphi V_s$$ Eq. (11-2)

$$\leq \varphi V_c + \varphi A_v f_y \frac{d}{s}$$ Eq. (11-17)

Solving for $A_v = \dfrac{(V_u - \varphi V_c)s}{\varphi f_y d}$

Assume s = 3 in. (max. spacing permitted = d/2) 11.5.4.2

$$A_v = \frac{(105 - 0.85 \times 42)\ 3}{0.85 \times 60 \times 6} = 0.68 \text{ in.}^2$$

where A_v is total area of shear reinforcement required on four sides of column support (See Fig 17-10).

A_v (per side) = 0.68/4 = 0.17 in.^{2}

<u>Use #3 stirrups @ 3 in.</u> (A_v = 0.22 in.^{2})

4. Determine distance from column support where stirrups may be terminated. (See Fig. 17-10) 11.11.3.3

$$V_u \leq \varphi V_c$$ Eq. (11-1)

$$\leq \varphi 2\sqrt{f'_c}\, b_o d$$ 11.11.3.4

where $b_o = 4\ (10 + a\sqrt{2})$

$$105{,}000 \leq 0.85 \times 2\sqrt{3{,}000} \times 4\ (10 + a\sqrt{2})\ 6$$

Solving for a = 26.1 in.

Extending the shear reinforcement a distance "d" beyond the theoretical cut-off point, use 11 - #3 (closed stirrups with standard hooks) @ 3 in. along each column line as shown in Fig. 17-10.

EXAMPLE 17.3 - Continued

Calculations and Discussion	Code Reference

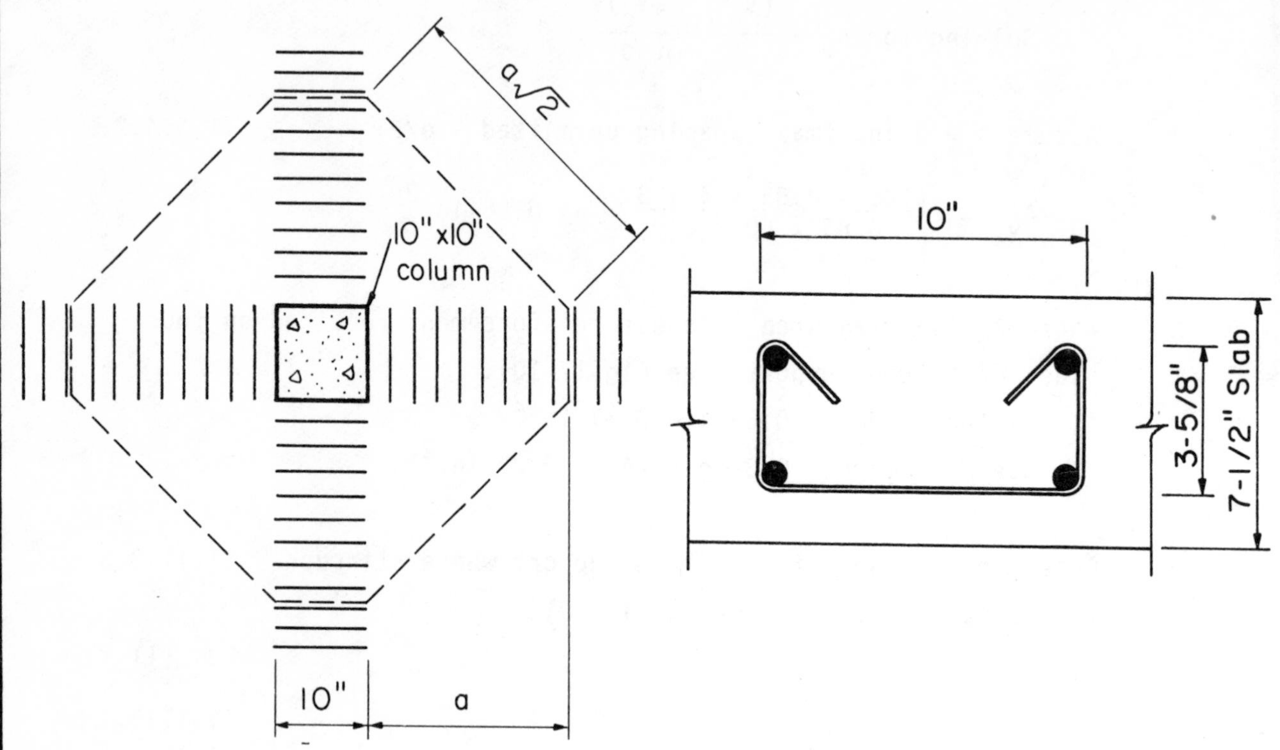

Fig. 17-10 - Bar Reinforcement for Shear in Slabs

E. Increase shear strength by Steel I shapes (shearheads) — 11.11.4

1. Check maximum shear strength permitted with steel shapes as shear reinforcement.

$V_u \leq \varphi V_n$ — Eq. (11-1)

$\leq \varphi 7\sqrt{f'_c}\, b_o d$ — 11.11.4.8

$\leq 0.85 \times 7\sqrt{3000} \times 64 \times 6 = 125^k$

$105^k < 125^k$ OK

EXAMPLE 17.3 - Continued

Calculations and Discussion	Code Reference

2. Determine minimum required perimeter b_o at critical section to limit shear strength V_n equal to $4\sqrt{f'_c}\, b_o d$ — 11.11.4.7, 11.11.4.8

$$V_u \leq \varphi V_n$$ Eq. (11-1)

$$105{,}000 \leq 0.85 \times 4\sqrt{3000}\, b_o \times 6$$ 11.11.4.8

Solving for $b_o = 94$ in. — 11.11.4.7

3. Determine required length of shearhead arm ℓ_v to satisfy $b_o = 94$ in. at 3/4 the distance $(\ell_v - c_1/2)$ — 11.11.4.7

$$b_o \simeq 4\sqrt{2}\left[\frac{c_1}{2} + \frac{3}{4}\left(\ell_v - \frac{c_1}{2}\right)\right]$$

with $b_o = 94$ in. and $c_1 = 10$ in; solve for $\ell_v = 20.5$ in.

4. To ensure that premature flexural failure of shearhead does not occur before shear strength of slab is reached, check plastic moment strength M_p of each shearhead arm.

$$\varphi M_p = \frac{V_u}{2\eta}\left[h_v + \alpha_v\left(\ell_v - c_1/2\right)\right]$$ Eq. (11-38)

For a four (identical) arm shearhead, $\eta = 4$; and assuming $h_v = 4$ in. and $\alpha_v = 0.25$: — 11.11.4.5

$$\varphi M_p = \frac{105}{2(4)}\left[4 + 0.25\,(20.5 - 10/2)\right] = 103.4 \text{ in. kips}$$

required $M_p = 103.4/0.9 = 115$ in. kips

Try S4 x 7.7 (plastic modulus $Z_x = 3.51$ in.^{3})

$$M_p = Z_x f_y = 3.51 \times 36{,}000 = 126 > 115 \quad \text{OK}$$

EXAMPLE 17.3 - Continued

Calculations and Discussion	Code Reference

5. Check depth limitation of S4 x 7.7 shearhead — 11.11.4.2

$70t_w = 70\ (0.193) = 13.51$ in. $>$ 4 in. OK

6. Determine location of compression flange of steel shape with respect to compression surface of slab. — 11.11.4.4

$0.3d = 0.3(6) = 1.8$ in. $< 0.75 + 2(0.625)^* = 2$ in. NG

* 3/4 clear cover + 2 #5 bars.

Therefore, the #5 bars in the lower layer must be cut; see Fig. 17-11.

7. Determine relative stiffness ratio α_v — 11.11.4.5

EI (S4 x 7.7) = 6 x 29,000 = 174,000 kip. in.^{2}
c.g. of S4 x 7.7 from compression face = 2.75 in.
$c_2 + d = 10 + 6 = 16$ in.

A_s provided for $M_u = 175^{'k}$, #5 @ 5 in., $\rho = 0.0103$

EI for cracked section assuming full composite action = 790,000 kip in.^{2}

$$\alpha_v = \frac{174{,}000}{790{,}000} = 0.22 > 0.15 \quad \text{OK}$$

Therefore, S4 x 7.7 section satisfies all code requirements for shearhead reinforcement. See Fig. 17-11 for details.

EXAMPLE 17.3 - Continued

Calculations and Discussion	Code Reference

8. Determine contribution of shearhead to negative moment strength of column strip.

$$M_v = \frac{\varphi\alpha_v V_u}{2\eta}\left(\ell_v - \frac{c_1}{2}\right)$$ Eq. (11-39)

$$= \frac{0.9 \times 0.22 \times 105}{2 \times 4}(24 - 5) = 49 \text{ in. kips}$$

However, M_v must not exceed either $M_p = 115^{"k}$ or $0.3(175)12 = 630^{"k}$, or the change in column strip moment over the length ℓ_v. For this design, approximately 2% of the column strip negative moment may be considered resisted by the shearhead reinforcement.

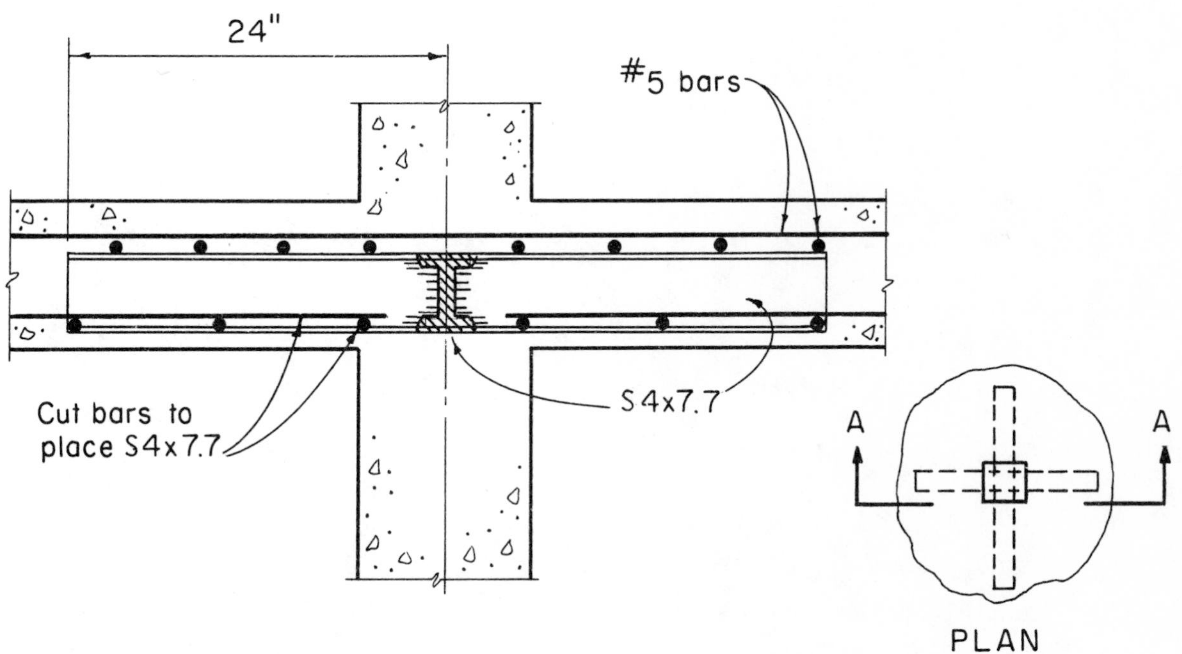

Fig. 17-11 - Details of Shearhead Reinforcement

18

Deep Flexural Members

General Considerations

The Code gives two definitions for "deep" members. For flexure, members with overall depth-to-clear-span ratios greater than 2/5 for continuous spans or 4/5 for simple spans are defined as "deep" (Section 10.7.1). For shear, a "deep" member is one with an overall depth-to-span ratio of 1/5 or greater (Section 11.8).

No specific provisions for designing deep members for flexure are found in the Code, but such members must be designed "taking into account nonlinear distribution of strain and lateral buckling." Appropriate references for the design of deep beams for flexure are given in the Code Commentary and at the end of Part 18.

Information on lateral buckling is more difficult to find. Fortunately, most walls and beams receive lateral support from supported floor or roof members, so lateral buckling of the compression flange is rarely a problem. See Fig. 18-1(a). Some form of lateral support is required at intervals not exceeding 50 times the least width of the compression flange (Section 10.4.1), even if the member is free-standing. See Fig. 18-1(b). For free standing walls, a lateral stability check should be made and an adequate margin of safety against lateral buckling provided.

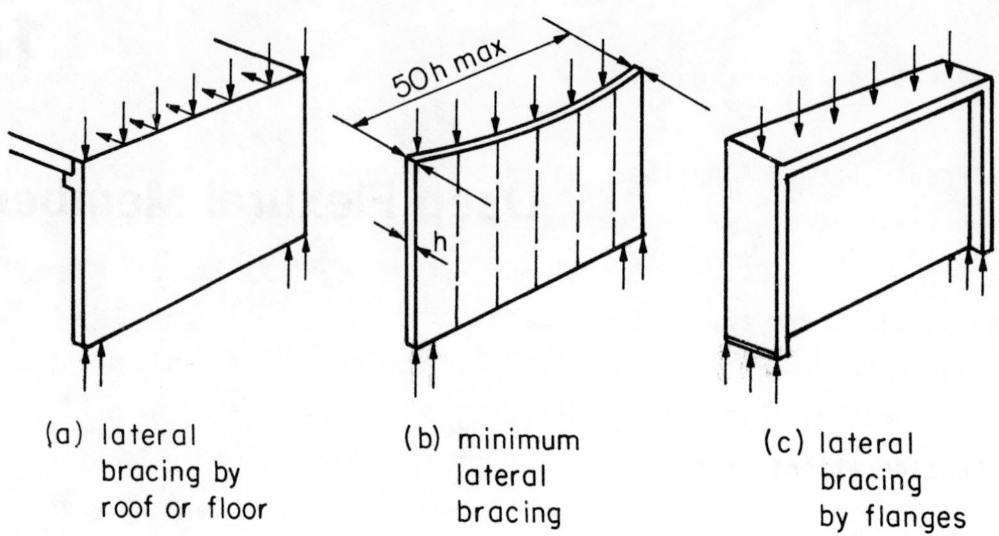

Fig. 18-1 - Lateral Support for Deep Flexural Members

Lateral buckling in a vertical direction (Fig. 18-2), particularly near concentrated loads and at supports, can be checked by the column moment magnifier, or by numerical or energy methods. A simplified procedure for wall-like beams (tilt-up panels) is provided in Reference 18.3. If the height-to-thickness ratio of a member is kept below 25, buckling should not be a problem.

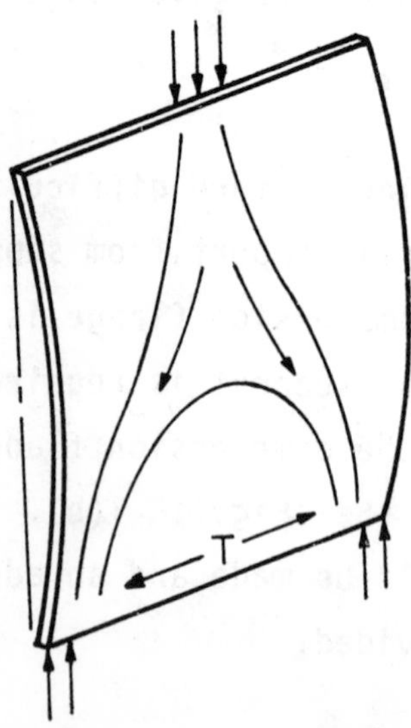

Fig. 18-2 - Lateral Buckling of Deep Flexural Member

Shear and flexure provisions for deep members are discussed in more detail in the following paragraphs.

10.7 Flexural Considerations

The Code requires that "nonlinear distribution of strain" be taken into account in flexural design of deep members. Elastic analysis by Dischinger and others[18.1,18.2] has shown that the shape of the elastic stress curve can be quite different from the linear distribution usually assumed. The neutral axis moves away from the loaded face of the member and, in some cases, the resultant elastic tensile forces can be within a third of the member depth of the extreme compression fiber. Although tensile stresses found by Dischinger are usually less than would be expected from a linear analysis, such stresses can be as much as 31 percent higher.

Nonlinear distribution of strains and stresses assumes an uncracked, homogeneous cross section and, therefore, does not apply to design at the ultimate moment strength (nominal moment strength M_n for design), since cracking usually occurs before the moment strength can be developed. This would imply that the tensile reinforcement required to develop the moment strength M_n could be placed near the extreme tensile fiber as is customary for ordinary flexural members. Reference 18.2, however, recommends that tensile reinforcement be distributed throughout the tensile area and centered at or near the resultant of the tensile forces, so that, when cracking occurs, there will not be a sudden shift in the location of the resultant tensile force. Both methods of sizing and placing reinforcement are demonstrated in Design Example 18.1 and it is left to the judgment of the designer to choose the method he deems appropriate.

Development of horizontal tensile reinforcement in single-span deep members can be a problem. Moments increase rapidly from zero at the face of the support, and the reinforcement may not have sufficient anchorage length to develop the required moment strength near the support. Tensile bars may be anchored by development length (if available), standard hooks, or by special anchorage devices.

The most radical departure from a linear strain and stress distribution is in compression areas at or near supports of continuous members. Compressive forces may be confined to the bottom 5 or 10 percent of the member depth and compressive stresses may be as high as 14 times those indicated by linear strain and stress distribution.[18.1] In these cases, reinforcing details require special consideration. If service load compressive stresses approach about $0.5f'_c$, it may be necessary to treat the compression area as an axially loaded member, using confined reinforcement to carry the compressive forces as the moment strength is approached.

11.8 Shear Considerations

The special shear provisions for deep members apply only to members having an overall depth-to-clear-span ratio of 1/5 or greater. The deep members must be loaded at the top face as shown in Fig. 18-3. When loads are applied through the sides or bottom of the member, ordinary shear design methods should be used (Sections 11.1 through 11.6). Since the principal tensile forces in deep members are primarily horizontal (vertical cracking), horizontal shear reinforcement is most effective in resisting the tensile forces. Truss bars are, therefore, not recommended as shear reinforcement in deep members.

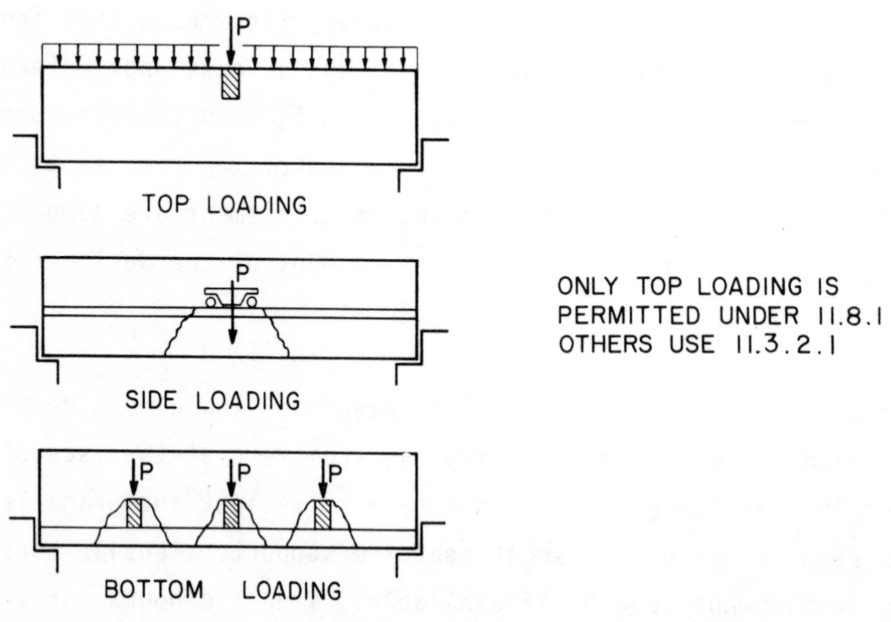

Fig. 18-3 - Loading of Deep Flexural Members

For shear design of deep members, the required shear strength V_u is calculated at a distance from the face of the support defined as 0.15 times the clear span for uniformly loaded beams or 0.50 times the shear span "a" for beams with concentrated loads, but in no case greater than "d" distance from face of support.

The factored shear force V_u must be less than or equal to the shear strength provided by the section $\varphi(V_c + V_s)$, where V_c is the shear strength provided by concrete and V_s is the shear strength provided by shear reinforcement, both horizontal and vertical. V_c may be computed from either the more complex Eq. (11-30), which takes into account the effects of the tensile reinforcement and M_u/V_ud at the critical section, or may be determined from the simpler Eq. (11-29), $V_c = 2\sqrt{f'_c}\, b_wd$. Eq. (11-30) is illustrated in Fig. 18-4.

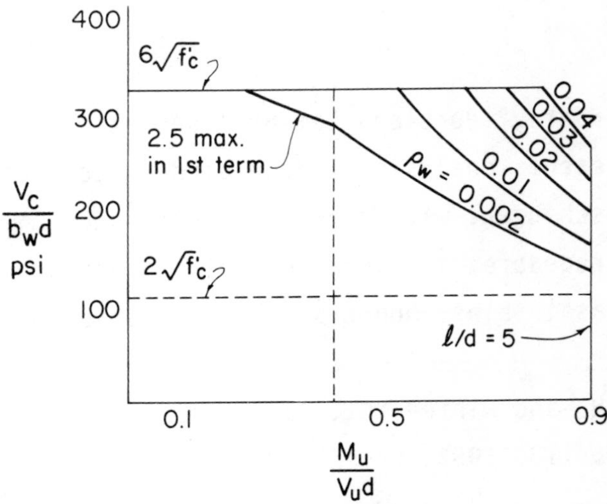

Fig. 18-4 - Shear Strength of Deep Flexural Members

The first step in design is to check if V_u is less than φV_c, with V_c equal to $2\sqrt{f'_c}\, b_w d$. If the shear strength provided by concrete is not adequate to carry the factored shear force V_u, then calculate φV_s for minimum shear reinforcement and add to φV_c. Using the minimum shear reinforcement requirements of Section 11.8.8 ($A_v = 0.0015 b_w s$) and Section 11.8.9 ($A_{vh} = 0.0025 b_w s_2$), shear strength Eq. (11-31) reduces to

$$V_s = [0.029d - 0.001\ell_n]\, b_w f_y/12$$

The shear strength with minimum shear reinforcement becomes

$$V_u \le \varphi(V_c + V_s)$$

$$V_u \le \varphi[2\sqrt{f'_c}\, b_w d + (0.029d - 0.001\ell_n)\, b_w f_y/12]$$

If the shear strength with minimum shear reinforcement is still not adequate, the more complex Eq. (11-30) is used to calculate a higher concrete shear strength, or additional shear reinforcement, A_v and A_{vh}, may be added to increase the shear strength of the section. Shear reinforcement required at the critical section must be provided throughout the span in all cases (Section 11.8.10).

Selected References

18.1 "Design of Deep Girders," Portland Cement Association, Skokie, IL, IS079D, 10 pp. Presents analysis of deep girders according to elastic theory of Franz Dischinger, with special studies and numerical examples added. Data and procedures illustrated apply to design of deep wall-like members such as in bins, hoppers, and foundation walls.

18.2 Chow, Li., Conway H. and Winter, G., "Stresses in Deep Beams," Transactions, ASCE, Vol. 118, 1953, pp. 686-708.

18.3 "Tilt-Up Load-Bearing Walls," A Design Aid, Portland Cement Association, Skokie, IL, EB074D, 1980, 28 pp. A "column model" (a panel considered hinged along loaded edges and free along vertical edges) is used to compute load capacities of reinforced concrete tilt-up wall panels (with a central curtain of reinforcement) that rest on continuous footings. An approximate but rational means of evaluating effects of isolated footings and sustained loads on capacity of these slender walls is included, as well as load-moment interaction charts and tables and design applications.

EXAMPLE 18.1 - Design of Deep Flexural Members

This design example has been adapted from the PCA publication, "Design of Deep Girders," and modified in accordance with the ACI Code and the strength design method. The publication may be used directly to design deep members by the Alternate Design Method, or it may be used to locate the tensile resultants and check cracking under the Strength Design Method of the Code.

An interior span of a continuous deep girder is shown below.

Width of beam and support, b_w = 15 in.

Uniform loads: Live load = 10 klf, Dead load = 10 klf

f'_c = 3,000 psi

f_y = 40,000 psi

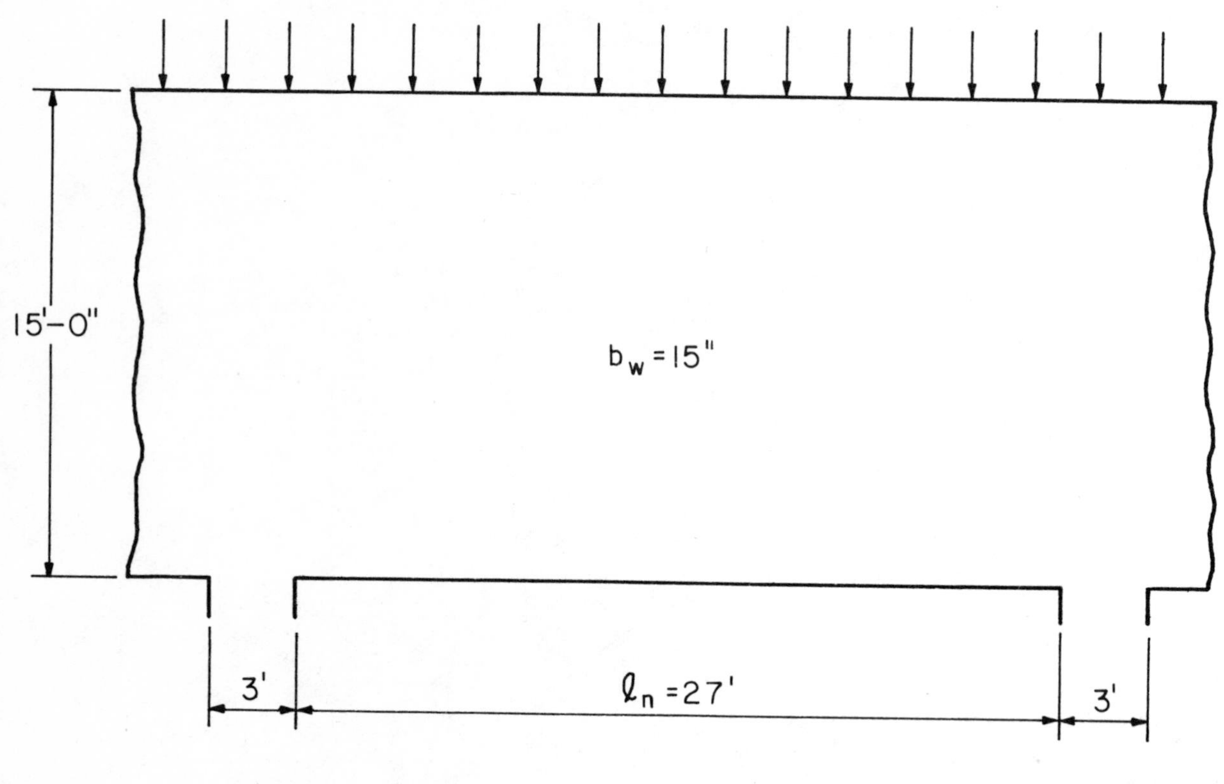

EXAMPLE 18.1 - Continued

Calculations and Discussion	Code Reference

A. Design for Flexure

1. Determine moment stresses (at service loads)

Refer to Reference 18.1, "Design of Deep Girders," for design constants as follows:

$\varepsilon = C/L = 3/30 = 1/10 \quad \beta = H/L = 15/30 = 1/2$

$w = 10 + 10 = 20$ k/ft

$= 20{,}000/12 = 1{,}667$ lbs/in.

From Figs. 2, 3, 4, and 5 (Ref. 18.1), the service load moment stresses at mid-span and support are:

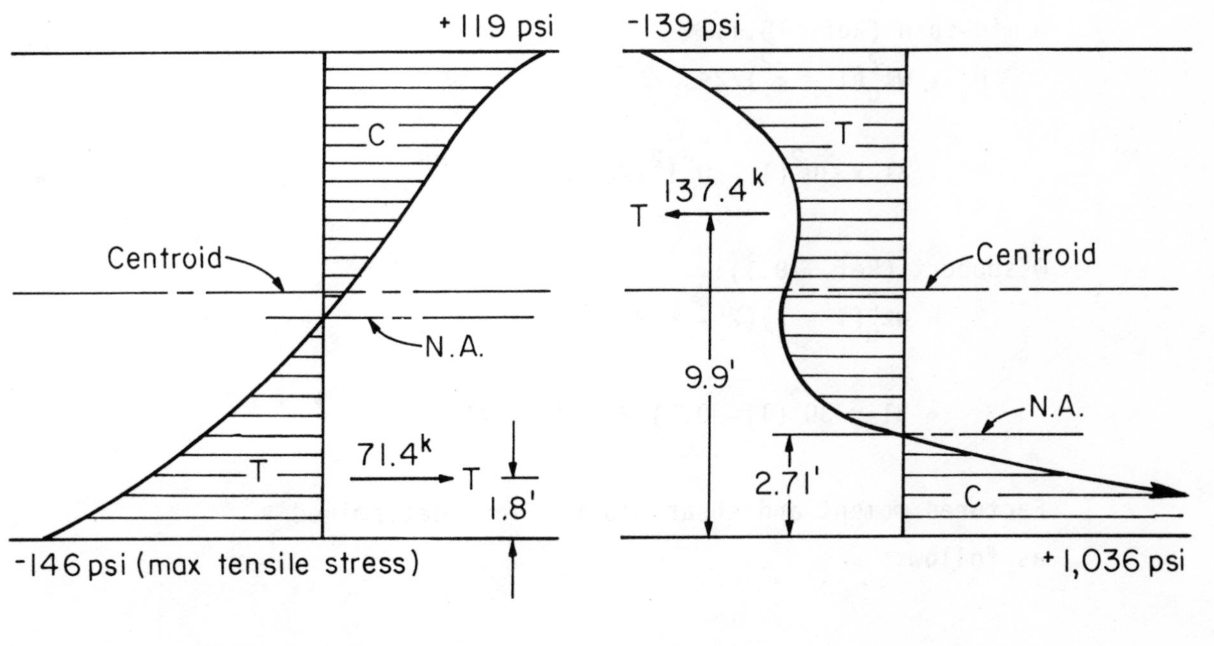

At Mid-span

At Support

EXAMPLE 18.1 - Continued

Calculations and Discussion	Code Reference

To avoid cracking at service loads, tensile stresses should not exceed the modulus of rupture.

$$f_r = 7.5\sqrt{f'_c} = 7.5\sqrt{3{,}000} = 411 \text{ psi} > 146 \quad \text{(OK)}$$ Eq. (9-9)

Designers using the Alternate Design Method of Appendix B may proceed directly with the flexural design as outlined in Reference 18.1, calculating the required reinforcement from the tensile resultants (T) and distributing the reinforcement appropriately. The following procedure is in accordance with the Strength Design Method of the Code.

2. Determine required moment strengths

$$U = 1.4(10) + 1.7(10) = 31.0 \text{ k/ft}$$ Eq. (9-1)

@ mid-span (Ref. 18.1):

$$M_u = w\ell_u^2(1 - \varepsilon^2)/24$$

$$= 31 \times 30^2(1 - 0.1^2)/24 = 1{,}151^{'k}$$

@ support (Ref. 18.1):

$$M_u = w\ell_u^2(1 - \varepsilon)(2 - \varepsilon)/24$$

$$= 31 \times 30^2(1 - 0.1)(2 - 0.1)/24 = 1{,}988^{'k}$$

Factored moment and shear diagrams are determined as follows:

EXAMPLE 18.1 - Continued

Calculations and Discussion	Code Reference

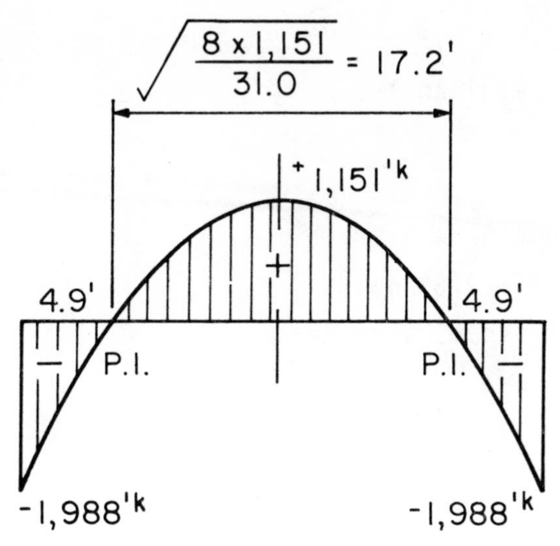

Factored moments

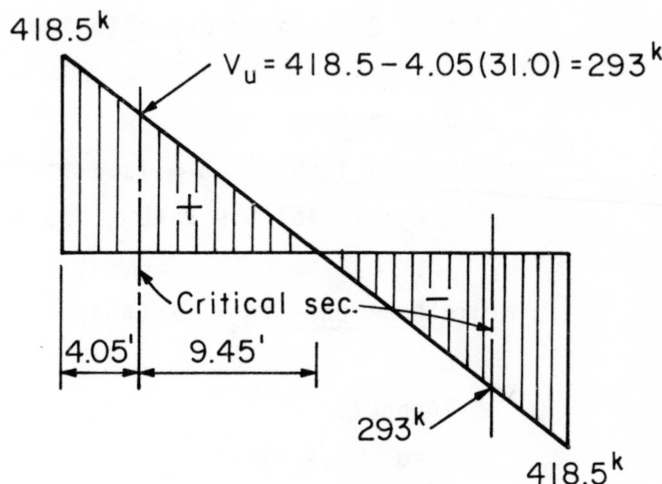

Factored shear force

3. Determine flexural reinforcement

Method 1 - (using full effective depth d)

$d = 15.0 - 2/12 = 14.83$ ft; assume $j_u = 0.9$

$A_s = M_u/\varphi f_y j_u d$

@ mid-span:

$A_s = 1,151/0.9 \times 40 \times 0.9 \times 14.83 = 2.40 \text{ in.}^2$

<u>Use 4 #7 bars</u> ($A_s = 2.40 \text{ in.}^2$)

@ support:

$A_s = 1,988/0.9 \times 40 \times 0.9 \times 14.83 = 4.14 \text{ in.}^2$

<u>Use 2 #10 and 2 #8 bars</u> ($A_s = 4.12 \text{ in.}^2$)

Calculations and Discussion	Code Reference

Locate primary reinforcement A_s (top and bottom) as close to tension face as cover and other reinforcement allow.

Method 2 - (using depth to tensile resultant)

@ mid-span:

$d = 15.0 - 1.8 = 13.2$ ft; assume $j_u = 0.9$

$A_s = 1{,}151/0.9 \times 40 \times 0.9 \times 13.2 = 2.69$ in.^{2}

Use 6 #6 bars ($A_s = 2.64$ in.^{2})

@ support:

$d = 9.9$ ft

$A_s = 1{,}988/0.9 \times 40 \times 0.9 \times 9.9 = 6.20$ in.^{2}

Use 14 #6 bars ($A_s = 6.16$ in.^{2})

Reinforcement determined by this method should be distributed in the total tensile area, approximately centered on the resultant tensile force.

4. Determine minimum horizontal and vertical reinforcement in side faces of girder. The minimum "wall" type reinforcement will be used in addition to the primary tensile reinforcement. 10.7.4

Horizontal reinforcement:

$A_{sh} = 0.0025 b_w s_2$ 14.3.3

$= 0.0025 \times 15 \times 12 = 0.45$ in.^{2}/ft

Use #5 bars @ 16 in. (each face) $A_s = 0.46$ in.^{2}

EXAMPLE 18.1 - Continued

Calculations and Discussion	Code Reference

Vertical reinforcement:

$A_v = 0.0015b_w s_1$ — 14.3.2

$= 0.0015 \times 15 \times 12 = 0.27$ in.^{2}/ft

Use #4 bars @ 17.5 in. (each face) $A_s = 0.27$ in.^{2}

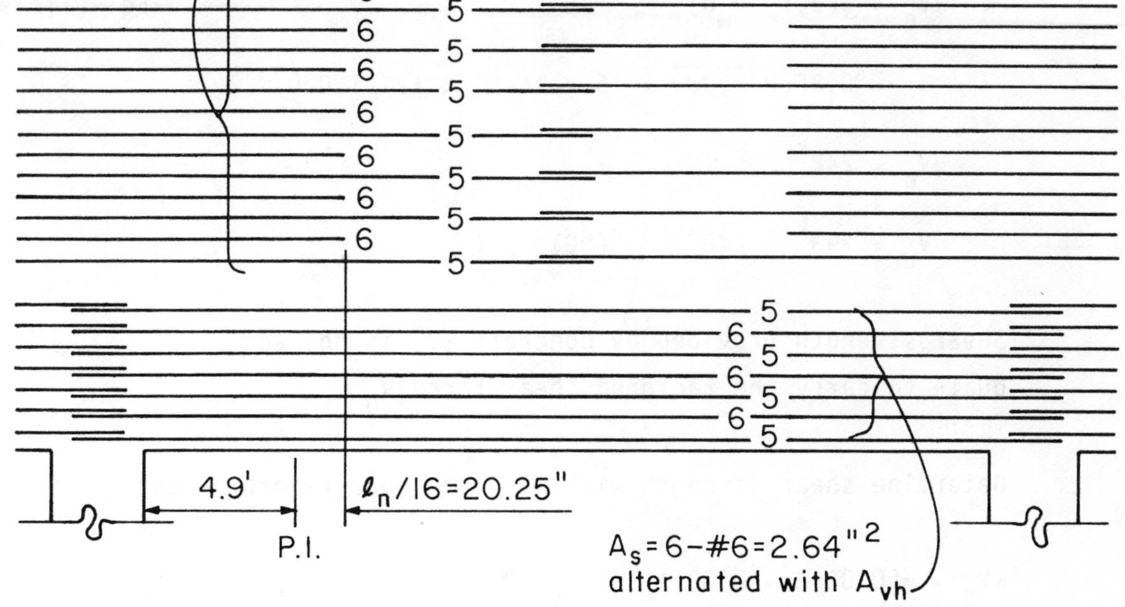

Space all horizontal bars at 8 in. or 16 in. for simplicity. Extension of the negative reinforcement beyond the point of inflection is normally the largest of d, $12d_b$, or $\ell_n/16$. For deep members, this extension requirement (d controls) would not allow any bar cut-off. It is suggested that for deep members with minimum reinforcement A_{sh} provided throughout the span, this excessive d extension is unnecessary. $\ell_n/16 = 20.25$ in. is used in this example.

EXAMPLE 18.1 - Continued

Calculations and Discussion	Code Reference

An alternate distribution of reinforcement might be to use the #6 bars @ 9-1/2 in. throughout, since the area of reinforcement is the same as alternating #5 and #6 bars @ 8 in. as suggested above.

B. Design for Shear

1. Determine shear strength without shear reinforcement

$\varphi V_c = \varphi(2\sqrt{f'_c}\, b_w d)$ Eq. (11-29)

$= 0.85(2\sqrt{3,000} \times 15 \times 14.83 \times 12)/1,000$

$\varphi V_c = 248^k$

$V_u = 293^k > 248^k$ (NG)

Shear strength provided by concrete φV_c is not adequate to carry the factored shear force V_u.

2. Determine shear strength with minimum shear reinforcement

$\varphi V_s = \varphi[0.029d - 0.001\ell_n]\, b_w f_y/12$

$= 0.85\ [0.029 \times 14.83 \times 12 - 0.001 \times 27 \times 12]\ 15 \times 40/12$

$\varphi V_s = 206^k$

$\varphi(V_c + V_s) = 248 + 206 = 454^k > 293^k$ (OK)

Shear strength with minimum shear reinforcement will usually be adequate for deep beam designs. If the more complicated Eq. (11-30) is required to provide higher concrete shear strength, $M_u/V_u d$ will invariably be small ($V_u d/M_u$ large) for uniformly loaded continuous members, and V_c will be limited by $6\sqrt{f'_c}\, b_w d$. Eq. (11-30) may be academic for all but rare cases.

19

Horizontal Shear in Composite Flexural Members

General Considerations

Horizontal shear forces act over the area of contact between the interconnected concrete surface of composite flexural members, where horizontal slip may occur due to flexural strain. Full transfer of these horizontal shear forces at contact surfaces of interconnected elements is required by the Code. Horizontal shear strength must be investigated and provisions must be made to transfer horizontal shear to supporting elements. The Code considers that the strength of a composite member is the same whether or not the first element cast is shored during the casting and curing of the second element.

17.5 - Horizontal Shear Strength

With ACI 318-83, horizontal shear strength must be investigated in all composite flexural members. The simplified alternative (Section 17.5.2 of ACI 318-77) is deleted. Application of Section 17.5.2 did not consider an upper limit on horizontal shear strength, as provided in the other two design procedures presented in Section 17.5. (When the computed horizontal shear exceeds 350 psi, design for horizontal shear must be in accordance with the shear-friction procedures of Section 11.7.) A reported distress in a member for which all the interface conditions of Section 17.5.2 were satisfied but for which the computed horizontal shear was greater than 350 psi, prompted reconsideration of the merits of Section 17.5.2. If the provisions of Section 17.5.2 were retained including an upper limit for application, the procedure would essentially be that provided by Section 17.5.2 (new numbering for '83). Thus the decision to simply delete.

The maximum limits of the horizontal shear force depend upon the contact surface conditions. The nominal horizontal shear strength V_{nh} may be evaluated on the basis of the provisions for intentional roughening or minimum ties, or both. The common requirement for all provisions is that interfacing surfaces must be clean and free of laitance. The horizontal shear strength may not be taken greater than $80b_vd$ for <u>either</u> intentionally roughened surfaces or minimum ties <u>alone</u>, or $350b_vd$ for both roughened surfaces <u>and</u> minimum ties together. Degree of roughness is specified only for the higher permissible shear strength (350 psi). Scoring the surface with a stiff bristled broom is common practice to satisfy the "intentionally roughened" requirement of Section 17.5.2.1. To permit the higher horizontal shear strength, a heavy raking or grooving of the surface is common practice to satisfy the "full 1/4 in. amplitude" requirement of Section 17.5.2.3.

If the horizontal shear exceeds 350 b_vd, the section must be designed using the shear-friction method of Section 11.7, with shear-friction reinforcement provided as tie reinforcement for horizontal shear-transfer. For ties perpendicular to interfacing surfaces, required tie area is computed by Eq. (11-26). The friction coefficient μ is taken either as 1.0λ for interface intentionally roughened or, 0.6λ for interface not intentionally roughened. Note that the 0.6 factor is new with the '83 Code provisions for shear-friction, and increases the required tie area by 67 percent for unroughened surfaces. Note also that "intentionally roughened" by Section 11.7 means roughened to a full amplitude of 1/4 in., not just "roughened" as permitted by Section 17.5.2.1.

17.6 - Ties for Horizontal Shear

According to Section 17.6.3, ties are required to be "fully anchored" into interconnected elements "in accordance with Section 12.13." Application (and interpretation) of Section 12.13 for tie anchorage is difficult, if not impossible. In lieu of attempting to apply, much less satisfy, the provisions of Section 12.13, Fig. 19-1 shows some tie details that have been used successfully in testing and actual practice. Fig. 19-1(a) shows an extended stirrup detail used in tests of Reference 19.1. Use of an embedded "hairpin" tie, as illustrated in Fig. 19-1(b), is common practice in the precast-prestressed industry. Many precast products are manufactured in such a way that it is difficult to position tie reinforcement for horizontal shear before concrete is placed. Accordingly, the ties are embedded in the plastic concrete as permitted by Code Section 16.4.2.

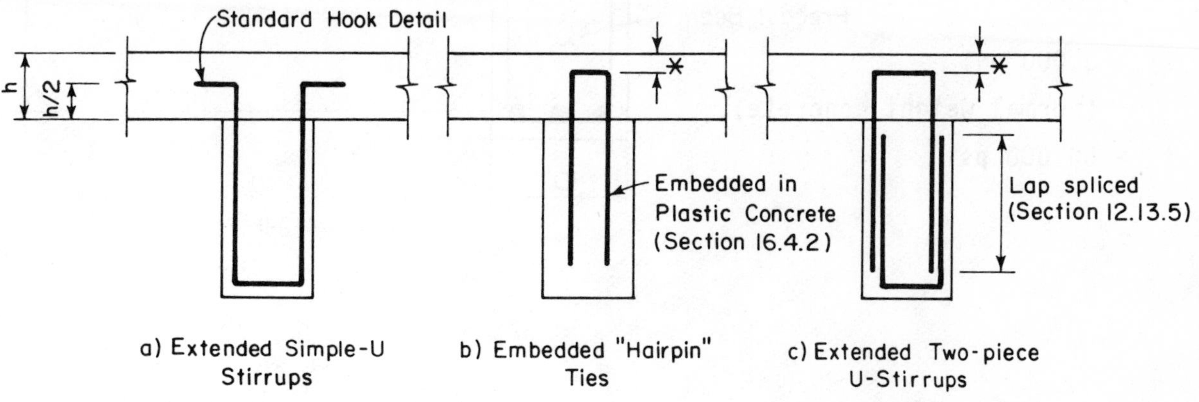

*Extended as close as cover and proximity of other reinforcement permit. 3 in. minimum projection into cast in place segment is common practice.

Fig. 19-1 Ties for Horizontal Shear

Selected References

19.1 Hanson, N.W., "Precast-Prestressed Concrete Bridges 2. Horizontal Shear Connections," Development Department Bulletin D35, Portland Cement Association, Skokie, 1960, 58 pp.

EXAMPLE 19.1 - Design for Horizontal Shear

For the composite slab and precast beam construction shown, design for transfer of horizontal shear at contact surface of beam and slab. Assume beam simply supported with a span of 30 feet.

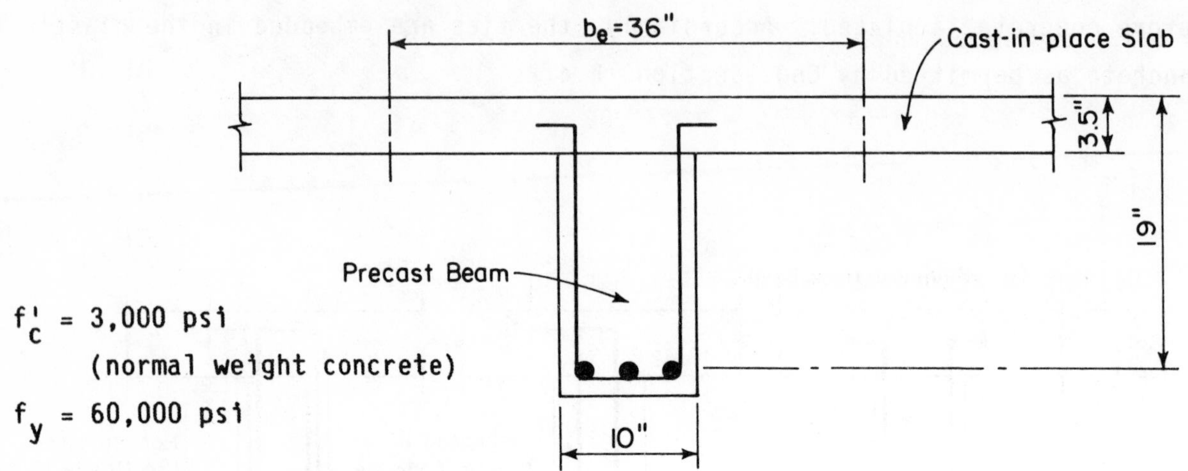

f'_c = 3,000 psi
(normal weight concrete)

f_y = 60,000 psi

Calculations and Discussion	Code Reference
<u>Case I</u>: Service dead load = 315 plf Service live load = 235 plf	
1. Determine factored shear force V_u at span end.	
$V_u = 1.4D + 1.7L$	Eq. (9-1)
$= 1.4(0.315)\ (30/2) + 1.7(0.235)\ (30/2)$	
$= 12.6^k$	

EXAMPLE 19.1 - Continued

Calculations and Discussion	Code Reference

2. Determine horizontal shear strength. 17.5.2

$$V_u \leq \varphi V_{nh}$$ Eq. (17-1)

$$\leq \varphi(80 b_v d)$$ 17.5.2.1 & 17.5.2.2

$$\leq 0.85(80 \times 10 \times 19)/1000$$

$$12.6^k < 12.9^k$$

Design in accordance with either Section 17.5.2.1 or 17.5.2.2:

If top surface of precast beam is intentionally roughened, no ties are required. 17.5.2.1

If top surface of precast beam is not intentionally roughened, minimum ties are required in accordance with Section 17.6. 17.5.2.2

Note: For either condition, top surface of precast beam must be cleaned and free of laitance prior to placing slab concrete.

Case II: Service dead load = 315 plf
Service live load = 1000 plf

1. Determine factored shear force V_u at span end.

$$V_u = 1.4(0.315)(15) + 1.7(1.0)(15)$$ Eq. (9-1)

$$= 32.1^k$$

2. Determine horizontal shear strength. 17.5.2

EXAMPLE 19.1 - Continued

Calculations and Discussion	Code Reference
$32.1^k > 12.9^k = (\varphi 80 b_v d)$	17.5.2.1 & 17.5.2.2
$V_u < \varphi(350 b_v d)$ $< 0.85(350 \times 10 \times 19)/1000$ $32.1^k < 56.5^k$	17.5.2.3
Design in accordance with Section 17.5.2.3:	
Contact surface must be intentionally roughened to "a full amplitude of approximately 1/4-in.," and minimum ties provided in accordance with Section 17.6.	17.5.2.3
3. Determine required tie area.	17.6
$A_v = \dfrac{50 b_w s}{f_y}$	Eq. (11-14)
where s = 4(3.5) = 14 in. < 24 in.	17.6.1
$A_v = \dfrac{50 \times 10 \times 14}{40000} = 0.175$ sq. in. @ 14 in. o.c. or 0.15 sq. in./ft	
4. Compare tie requirements with required vertical shear reinforcement at span end.	
$V_u = 32.1^k$	
$V_c = 2\sqrt{f'_c}\, b_w d = 2\sqrt{3000} \times 10 \times 1.9 = 20.8^k$	Eq. (11-3)
$V_u \le \varphi(V_c + V_s)$	Eq. (11-1)
$V_u \le \varphi V_c + \varphi A_v f_y \dfrac{d}{s}$	Eq. (11-17)
$\dfrac{A_v}{s} = \dfrac{(V_u - \varphi V_c)}{\varphi f_y d} = \dfrac{(32.1 - 0.85 \times 20.8)}{0.85 \times 40 \times 19}$	

EXAMPLE 19.1 - Continued

Calculations and Discussion	Code Reference

$= 0.0223\ in.^2/in.$

$s_{max} = 19/2 = 9.5\ in. < 24\ in.$ — 11.5.4.1

$A_v = 0.0223 \times 9.5 = 0.212\ in.^2$

#3 U-stirrups @ 9.5 in. o.c. (A_v = 0.28 sq. in./ft) which exceeds that required for horizontal shear. Provide #3 U-stirrups @ 9.5 in. o.c. The ties must be adequately anchored into the slab by embedment or hooks. See Fig. 17-1.

Case III: Service dead load = 315 plf
Service live load = 2000 plf

Determine factored shear force V_u at span end.

$$V_u = 1.4(0.315)\ (15) + 1.7(2.0)\ (15) = 57.6^k$$ Eq. (9-1)

$$57.6^k > 56.5^k = (\varphi 350 b_v d)$$ 17.5.2.4

Since V_u exceeds $\varphi(350 b_v d)$, design for horizontal shear must be in accordance with Section 11.7 - Shear-Friction. The relative displacement along the contact surface between beam and slab is resisted by shear-friction reinforcement across and perpendicular to the contact surface.

For the required area of tie reinforcement across the interface, the following three alternate methods are suggested:

EXAMPLE 19.1 - Continued

Calculations and Discussion	Code Reference

Alternative #1:

Using the shear-friction concept of Section 11.7, design for the total required shear-friction reinforcement from centerline to span end. Converting to a unit stress, the factored horizontal shear stress at span end is:

$$v_{uh} = \frac{V_u}{b_v d} = \frac{57.6}{10 \times 19} = 303 \text{ psi}$$

The shear "stress block" diagram may be shown as follows:

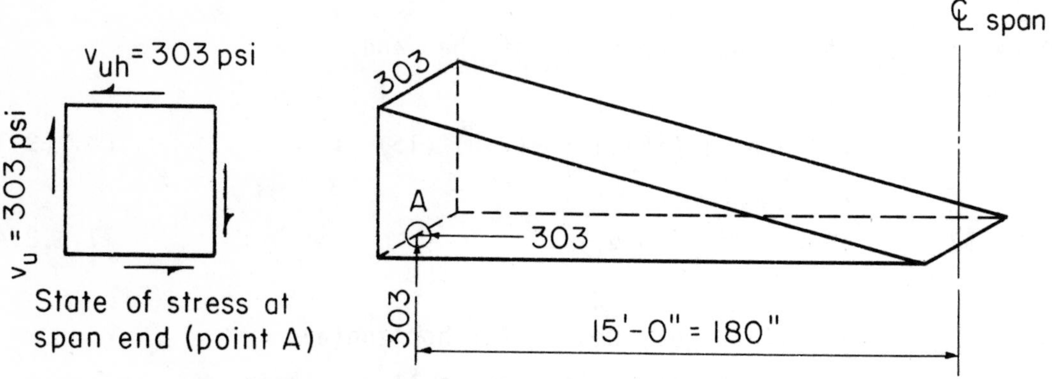

The total horizontal shear transfer between centerline and span end is:

$$V_{uh} = \frac{1}{2} \times 303 \times 180 \times 10 = 272.7^k$$

Required area of shear-friction reinforcement is computed by Eqs. (11-1) and (11-26):

EXAMPLE 19.1 - Continued

Calculations and Discussion	Code Reference

$V_{uh} \le \varphi V_n$ — Eq. (11-1)

$V_{uh} \le \varphi A_{vf} f_y \mu$ — Eq. (11-26)

$$A_{vf} = \frac{V_{uh}}{\varphi f_y \mu}$$

If top surface of precast beam is intentionally roughened to approximately 1/4 in., $\mu = 1.0$. — 11.7.4.3

$$A_{vf} = \frac{272.7}{0.85 \times 40 \times 1.0} = 8.02 \text{ in.}^2$$

$s_{max} = 4 \times 3.5 = 14 \text{ in.} < 24 \text{ in.}$ — 17.6.1

Space ties uniformly from centerline of span to end.
Use #5 double leg ties (U-stirrups).

$A_{vf} = 0.31 \times 2 = 0.62$ sq. in.

No. of ties required = 8.02/0.62 = 13
<u>Use 13 - #5 U-stirrups @ ± 14 in. o.c.</u>

If top surface of precast beam is not intentionally roughened, $\mu = 0.6$. — 11.7.4.3

$$A_{vf} = \frac{272.7}{0.85 \times 40 \times 0.6} = 13.37 \text{ in.}^2$$

No. of ties required = 13.37/0.62 = 22
<u>Use 22 - #5 U-stirrups @ ± 8 in. o.c.</u>

Note: Final selection of ties will depend on vertical shear requirements.

EXAMPLE 19.1 - Continued

Calculations and Discussion	Code Reference

Alternative #2:

If preferred, a varied tie spacing can be used, based on the actual shape of the horizontal shear distribution. The following method seems reasonable and has been used in the past:

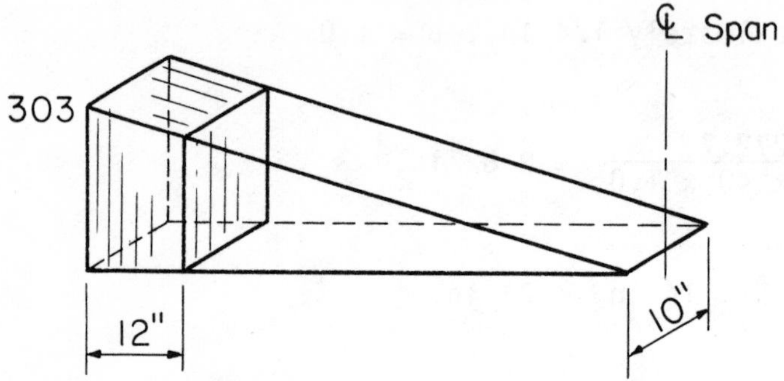

Referring to the horizontal shear stress block above, assume that the horizontal shear is uniform per foot of length, then the shear transfer force for the first foot is:

$$V_{uh} = 303 \times 10 \times 12 = 36.4^{k}$$

Using $\mu = 1.0$:

$$A_{vf} = \frac{36.4}{0.85 \times 40 \times 1.0} = 1.07 \text{ in.}^2/\text{ft}$$

With #5 double leg stirrups, $A_{vf} = 0.62 \text{ in.}^2$

$$s = \frac{0.62 \times 12}{1.07} = 6.95 \text{ in.} \simeq 7 \text{ in.}$$

EXAMPLE 19.1 - Continued

Calculations and Discussion	Code Reference
Use #5 U-stirrups @ 7 in. o.c. for the first 14 in. from span end.	
This method can be used to determine the tie spacing for each successive one-foot length. The shearing force will vary at each one-foot increment and the tie spacing can vary accordingly to a maximum of 14 in. toward the center of the span. This method will require more total ties than the 13 computed as based on a uniform spacing from span end to centerline.	
<u>Alternative #3</u>:	
Using the compressive force developed in the supported element:	17.5.3

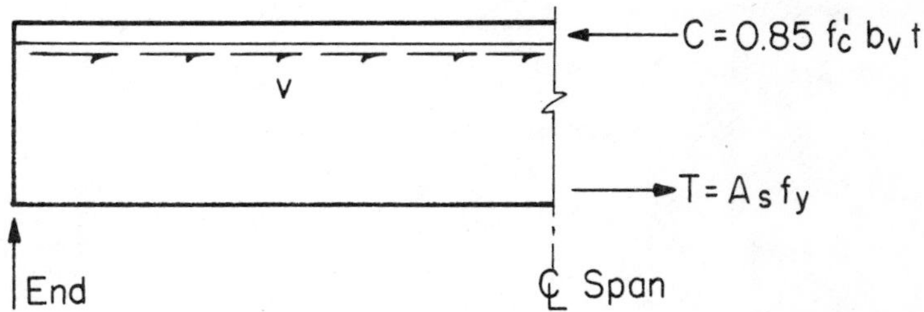

$$V_{uh} = C = 0.85 \times 3 \times 3.5 \times 36 = 321.3^k$$

Use the shear-friction method as in Alternate #1 to compute required ties. In actual design, use the smaller of either C or T. The 321.3^k is the maximum force that can be developed in the slab element; the actual depth of the stress block may be less than the 3.5-in. slab depth.

20

Two-way Slabs—Direct Design Method

Update for '83 Code

For ACI 318-83, three changes to Chapter 13 are significant:

(1) Use of the Direct Design Method for moment analysis of two-way slab systems is greatly simplified...by eliminating all stiffness calculations for determining design moments in an end span. A new table of moment coefficients for distribution of the total span moment in an end span (Section 13.6.3.3) replaces the expressions for distribution as a function of the stiffness ratio α. As a companion change, the approximate Eq. (13-4) for unbalanced moment transfer between the slab and an interior column is also simplified with elimination of the α term. With these changes, the Direct Design Method is now truly a Direct design procedure with all design moments determined directly from moment coefficients. Also, note especially new Section 13.6.3.6 addressing a special provision for moment transfer between a slab and edge column when the approximate moment coefficients of Section 13.6.3.6 are used. Application of this special requirement is illustrated in Design Example 20.1.

So that existing design aids and computer programs based on the original distribution as a function of the stiffness ratio α are still equally applicable for usage, Code Commentary Section 13.6.3.3 includes a "Modified Stiffness Method" reflecting the original distribution method.

(2) A performance statement on lateral load analysis of unbraced frames is added in a new Section 13.3.1.2; with a companion statement, Section 13.3.1.3, permitting moments from a Direct Design (or Equivalent Frame) analysis for gravity loading to be combined with moments from a lateral load analysis for proportioning the frame members.

(3) A new limitation "forming orthogonal frames" is added to Section 13.3.1.1 to further clarify that both design methods of Chapter 13 are limited in application to buildings with columns and/or walls laid out on a basically orthogonal grid...i.e., column lines taken longitudinally and transversely through the building are mutually perpendicular.

Introduction

The Direct Design Method is an approximate procedure for analyzing two-way slab systems of orthogonal frames subject to loads due to gravity only. Since it is an approximate procedure, the method is limited to slab systems with reasonable loading and continuity between panels, and meeting the limitations specified in Section 13.6.1. Two-way slab systems not meeting these limitations must be analyzed by more accurate procedures such as the Equivalent Frame Method, as specified in Section 13.7, unless it can be shown by analysis that the particular limitation does not apply to the structure. See Part 21 for discussion and design examples using the Equivalent Frame Method. For lateral load analysis, see discussion in Part 21, page 21-18.

Design Strip

For analysis, the slab system is divided into design strips consisting of a column strip and half middle strip(s) as defined in Sections 13.2.1 and 13.2.2 and illustrated in Fig. 20-1. Some judgment is required in applying the definitions given in Section 13.2.1 for column strips with varying span lengths along the design strip.

The reason for specifying that the column strip width be based on the shorter of ℓ_1 or ℓ_2 is to account for the tendency for moment to concentrate about the column line when the span length of the design strip is less than its width.

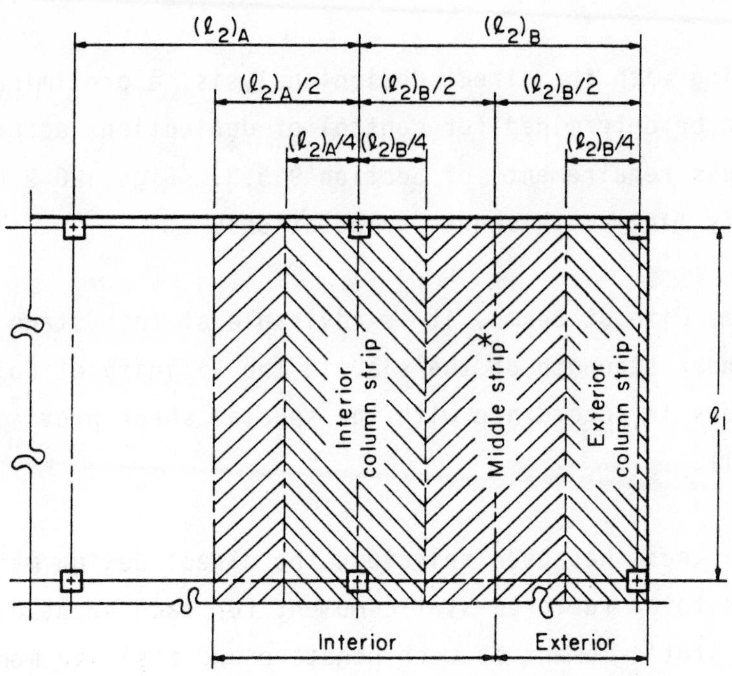

(a) Column strip for $\ell_2 \leq \ell_1$

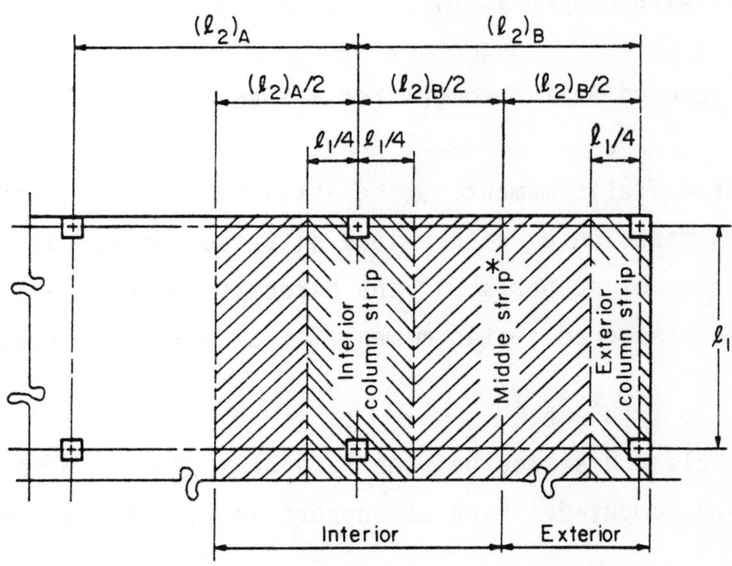

(b) Column strip for $\ell_2 > \ell_1$

Fig. 20-1 - Definition of Design Strip

---

*When edge of exterior design strip is supported by a wall, the factored moment resisted by this middle strip is defined in Section 13.6.6.3.

Preliminary Design

Before proceeding with the direct design analysis, a preliminary slab thickness h needs to be determined for control of deflections according to the minimum thickness requirements of Section 9.5.3. Figs. 20-2 and 20-3 can be used to simplify minimum thickness computations.

For slab systems without beams, it is advisable at this stage in the design to check the shear strength of the slab in the vicinity of columns or other support locations in accordance with the special shear provisions for slabs of Section 11.11.

Once a slab thickness has been selected, the direct design method involves determining the total factored static moment for each span, dividing the total factored static moment between negative and positive moments within each span, and distributing the negative and positive moment to the column and middle strip within each span.

13.6.2 Total Factored Static Moment for a Span

The total factored static moment is the absolute sum of positive and average negative moments based on a uniform load distribution across the design strip, between face of supports, in the direction moments are being determined. The total factored static moment for a span is expressed as:

$$M_o = w_u \ell_2 \ell_n^2/8 \qquad \text{Eq. (13-3)}$$

where ℓ_n is the clear distance between face of supports in the direction moments are being computed. Face of support is illustrated in Fig. 20-4.

13.6.3 Negative and Positive Factored Moments

The total static moment for a span is divided into negative and positive design moments as shown in Fig. 20-5.

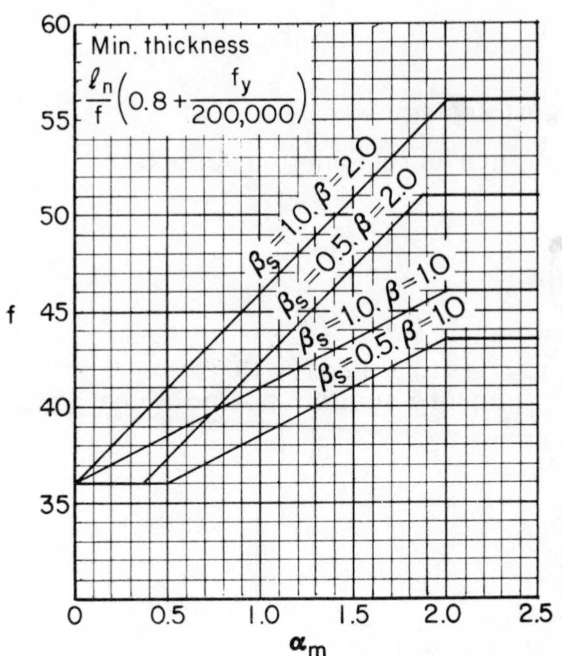

Fig. 20-2 - Minimum Thickness for Two-Way Slabs
(Code Eq. 9-11, 9-12, & 9-13)

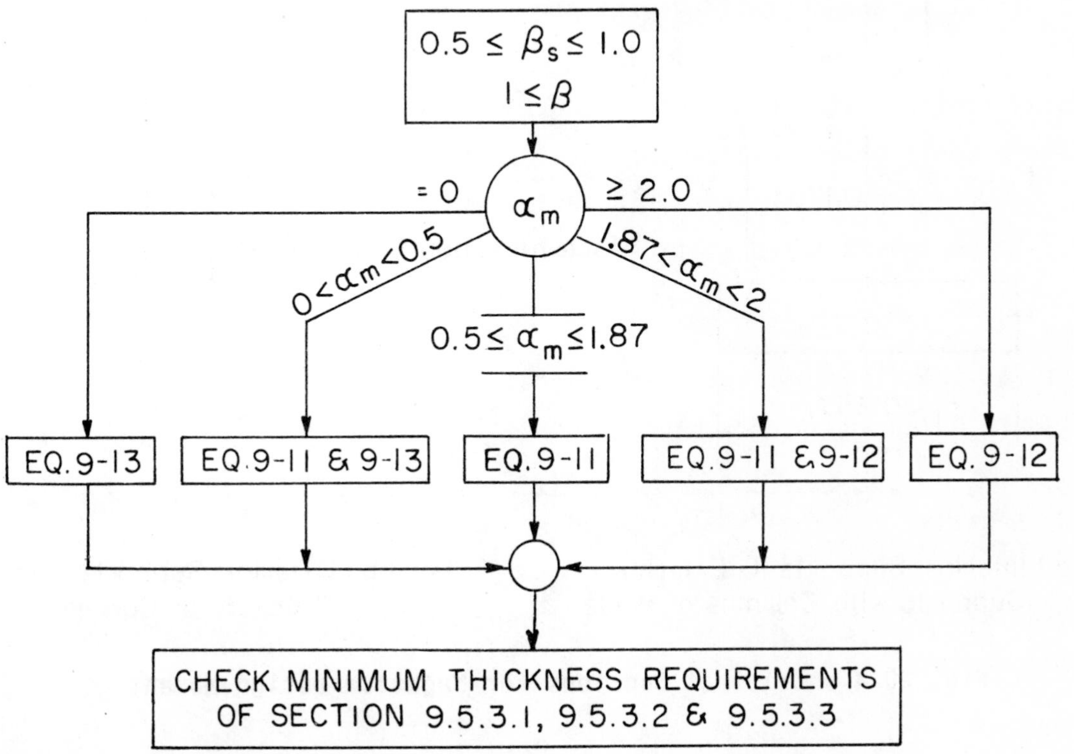

Fig. 20-3 - Flow Diagram for Two-Way Slab Thickness

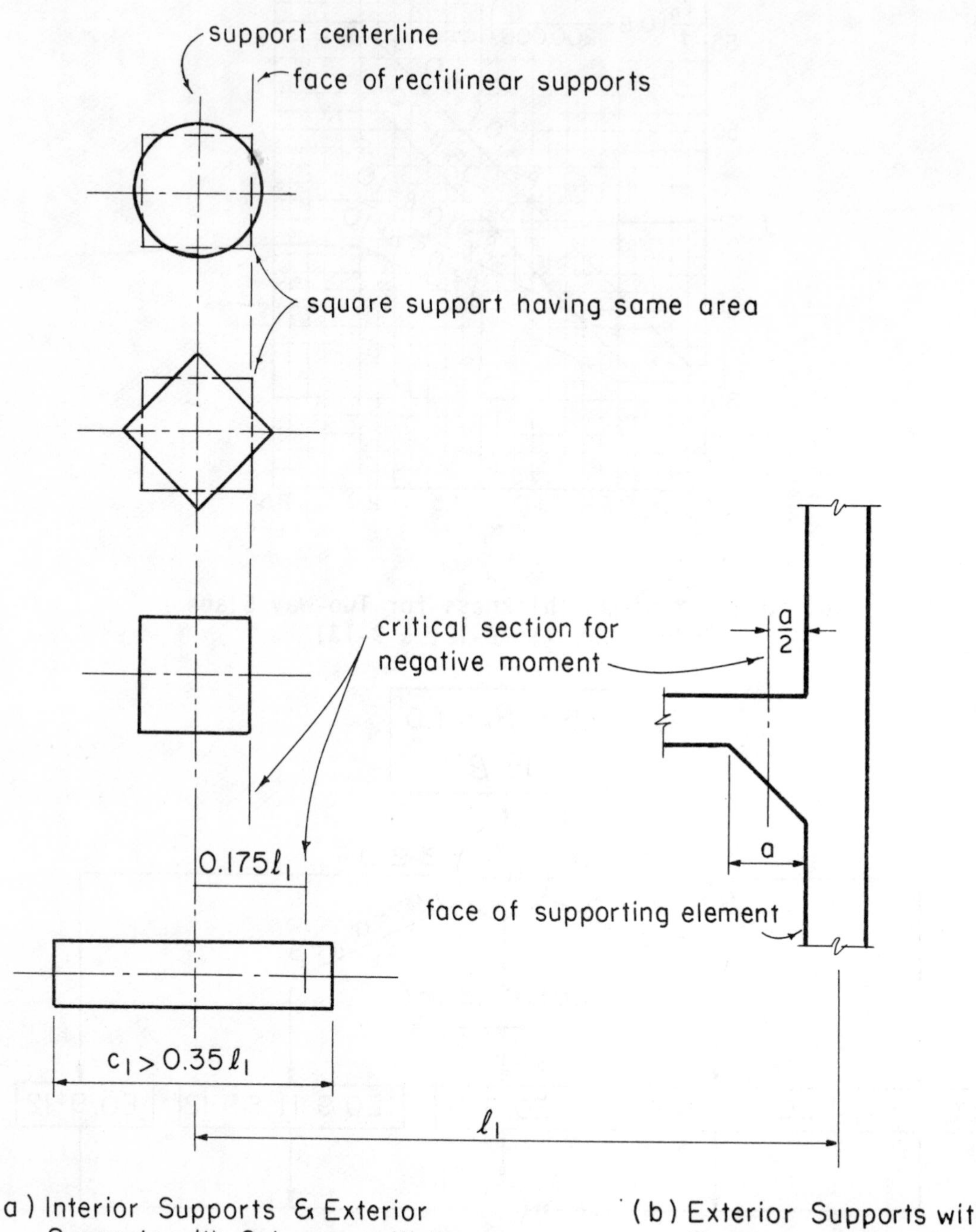

(a) Interior Supports & Exterior Supports with Columns or Walls

(b) Exterior Supports with Brackets or Corbels

Fig. 20-4 - Critical Sections for Negative Design Moment

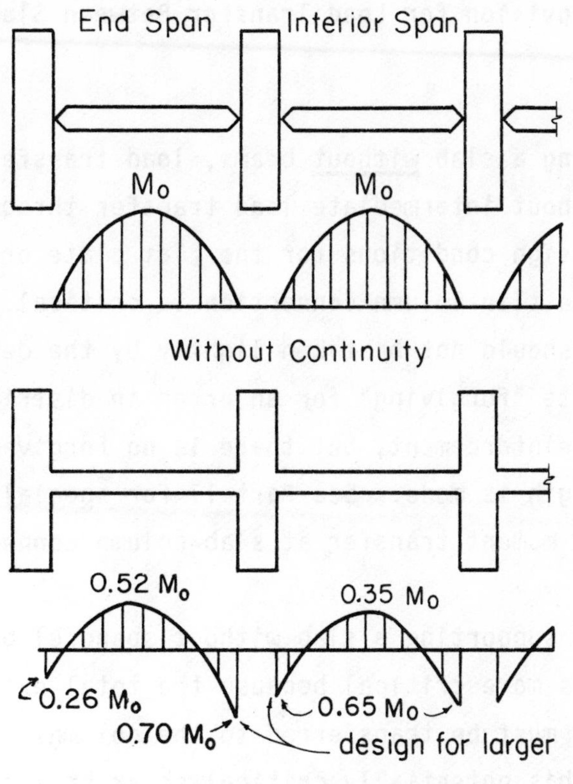

Fig. 20-5 - Design Strip Moments

End span moments are shown for a "flat plate or flat slab" without spandrels (slab system without beams between interior supports and without an edge beam). For other end span conditions, the total static moment M_o is distributed as follows. General characteristics of slab systems are shown that satisfy the intent of the end span conditions listed in Section 13.6.3.3.

	(1)	(2)	(3)	(4)	(5)
	Slab Simply Supported on Concrete or Masonry Wall	Two-Way Slabs	Flat Plates and Flat Slabs		Slab Monolithic With Concrete Wall
			Without Spandrel	With Spandrel	
Interior Negative	0.75	0.70	0.70	0.70	0.65
Positive	0.63	0.57	0.52	0.50	0.35
Exterior Negative	0	0.16	0.26	0.30	0.65

13.6.3.6 Special Provision for Load Transfer Between Slab and an Edge Column

For columns supporting a slab without beams, load transfer between a slab and column directly (without intermediate load transfer through a beam) is one of the more critical design conditions for the flat plate or flat slab system. Shear strength of the slab-column connection is critical. This aspect of two-way slab design should not be taken lightly by the designer. Two-way slab systems are quite "forgiving" for an error in distribution or even in amount of flexural reinforcement, but there is no forgiveness if a critical error in shear strength is made. See Part 17 for special shear provisions for direct shear and moment transfer at slab-column connections.

For exterior columns supporting a slab without spandrel beams, the load transfer condition is more critical because the total exterior negative moment from the slab must be transferred to the columns. A new Section 13.6.3.6 addresses this potentially critical shear strength condition between a slab and edge column. To ensure adequate shear strength when using the approximate end-span moment coefficient of Section 13.6.3.3, the full nominal moment strength M_n provided by the column strip must be used as the fraction of unbalanced moment transferred by eccentricity of shear ($\gamma_v M_n$) in accordance with Sections 11.12.2.3 and 11.12.2.4. (For end spans without spandrel beams, the column strip is proportioned to resist the total exterior negative factored moment.) The M_n requirement for evaluating slab shear strength resulting from moment transfer by eccentricity of shear is illustrated in Fig. 20-6. The total reinforcement provided in the column strip includes the additional reinforcement concentrated over the column to resist the fraction of unbalanced moment transferred by flexure $\gamma_f M_u = \gamma_f(0.26M_o)$, where the moment coefficient (0.26) is from Section 13.6.3.3, and γ_f is given by Eq. (13-1).

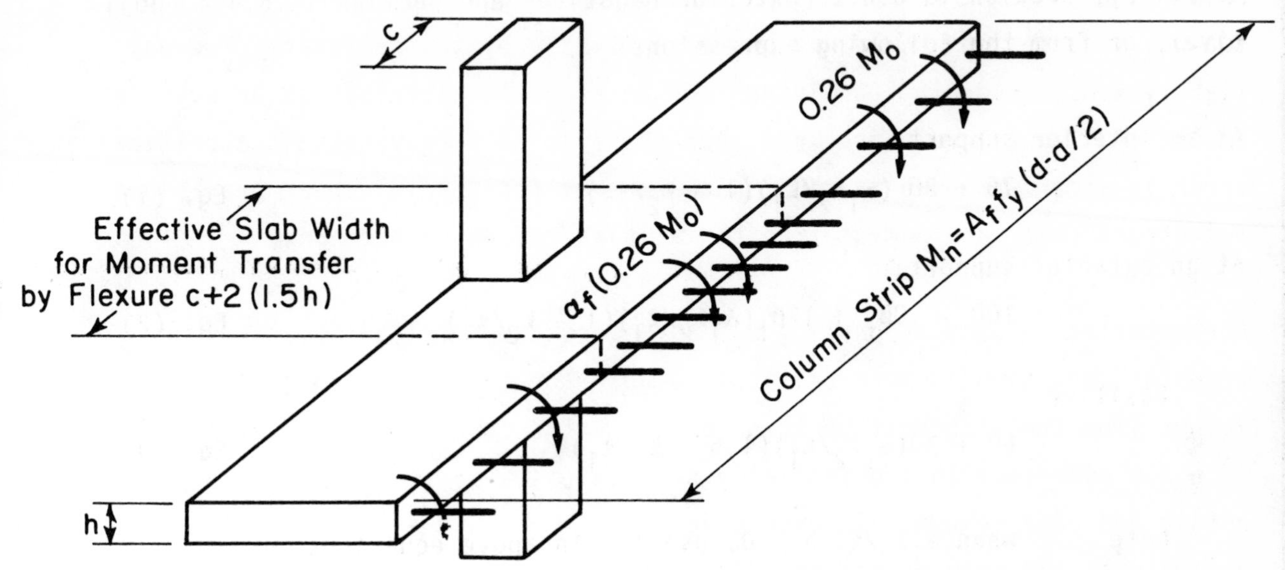

Fig. 20-6 Nominal Moment Strength of Column Strip to Evaluate $\gamma_v M_n$

13.6.4 Factored Moments in Column Strips

The amount of negative and positive factored moment to be resisted by a column strip, as defined in Fig. 20-1, depends on the relative stiffness of beam to slab and panel width-to-length in the direction of the span in which moments are being determined. An exception to this is when a support has a large transverse width.

The column strip at the exterior of an end span is required to resist the total factored negative moment in the design strip unless spandrel beams are provided.

When the transverse width of a support is equal to or greater than three fourths (3/4) of the design strip width, Section 13.6.4.3 requires that the negative factored moment be uniformly distributed across the design strip.

The percentage of total negative and positive moments to be resisted by a column strip may be determined from the tables in Section 13.6.4.1 (interior negative), Section 13.6.4.2 (exterior negative) and Section 13.6.4.4 (positive), or from the following expressions:

At an interior support:

$$75 + 30\,(\alpha_1 \ell_2/\ell_1)(1 - \ell_2/\ell_1) \qquad \text{Eq. (1)}$$

At an exterior support:

$$100 - 10\beta_t + 12\beta_t(\alpha_1 \ell_2/\ell_1)(1 - \ell_2/\ell_1) \qquad \text{Eq. (2)}$$

Positive:

$$60 + 30(\alpha_1 \ell_2/\ell_1)(1.5 - \ell_2/\ell_1) \qquad \text{Eq. (3)}$$

Note: When $\alpha_1 \ell_2/\ell_1 > 1.0$, use 1.0 in above equations.
When $\beta_t > 2.5$, use 2.5 in Eq. (2) above.

For slabs without beams between supports ($\alpha = 0$) and without edge beams ($\beta_t = 0$), distribution of total negative and positive moments to column strips is simply 75, 100, and 60 percent for interior, exterior, and positive moments, respectively. For slabs with beams between supports, distribution depends on stiffness ratio of beam to slab, and for edge beams, ratio of torsional stiffness of edge beam to flexural stiffness of slab. Figs. 20-7, 20-8, and 20-9 simplify evaluation of beam to slab ratio α. To evaluate β_t stiffness ratio for edge beams, Fig. 20-10 simplifies calculation for torsional constant C.

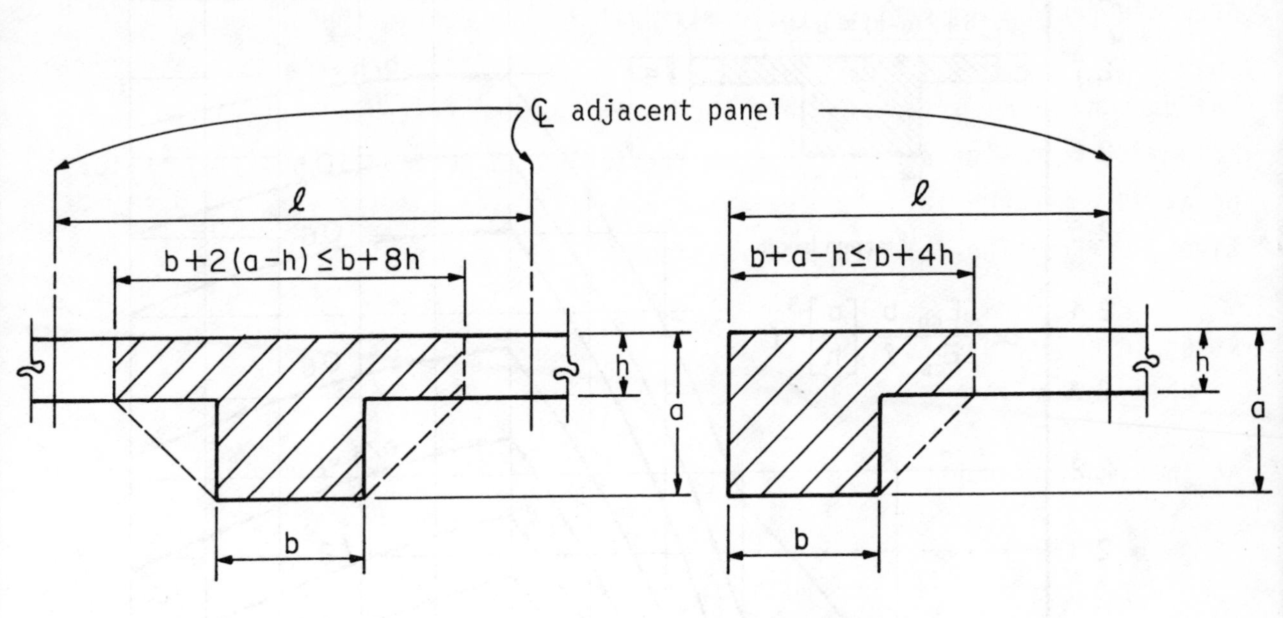

$I_b = ba^3/12)f$

$I_s = \ell h^3/12$

$\alpha = (E_{cb}I_b)/(E_{cs}I_s)$

$\quad = (E_{cb}/E_{cs})(b/\ell)(a/h)^3 f$

Fig. 20-8 is used to determine the factor "f" for interior beams.

Fig. 20-9 is used to determine the factor "f" for edge beams.

Fig. 20-7 - Design Aid for Computing Stiffness Ratio for Slab and Beam Sections

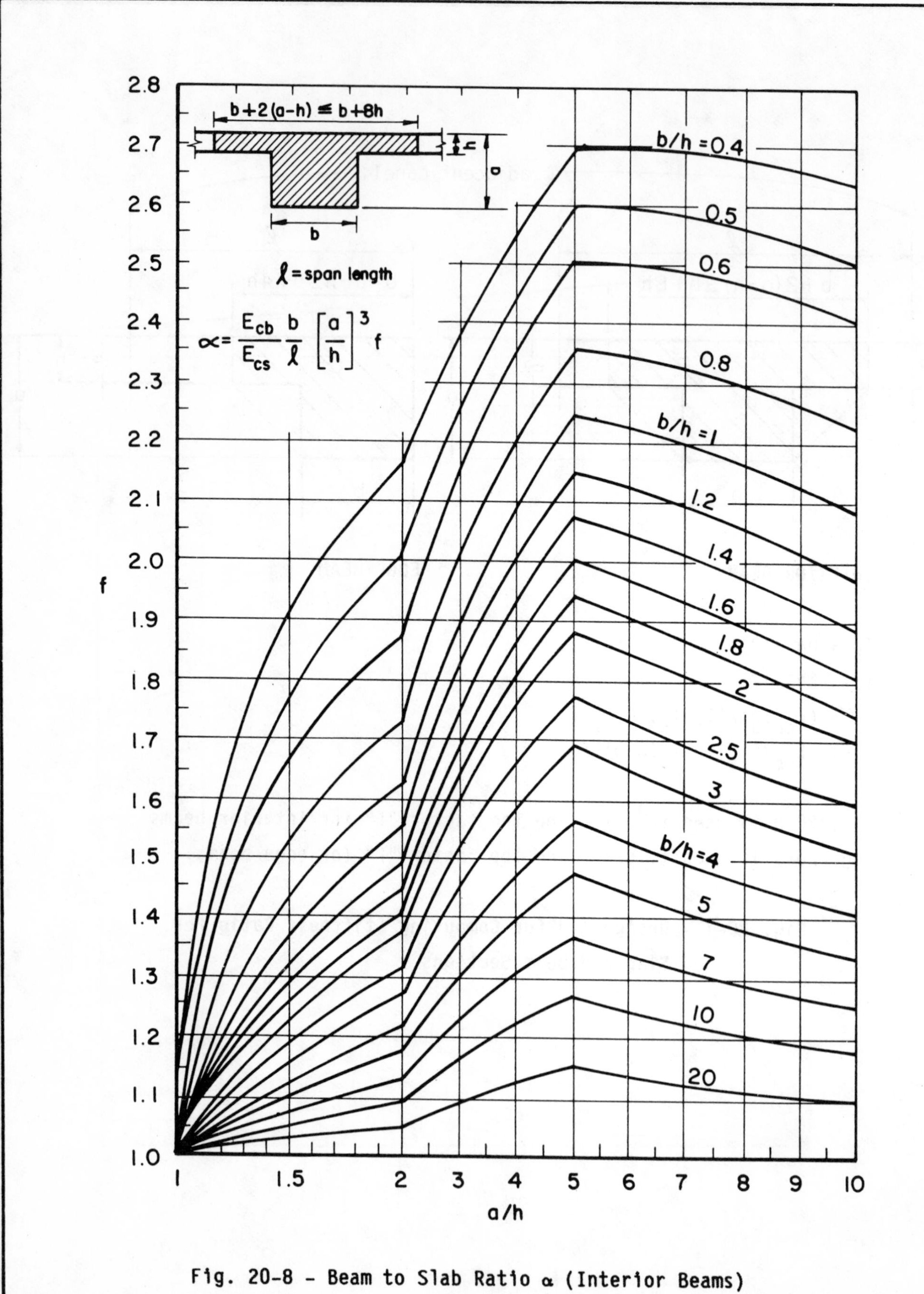

Fig. 20-8 - Beam to Slab Ratio α (Interior Beams)

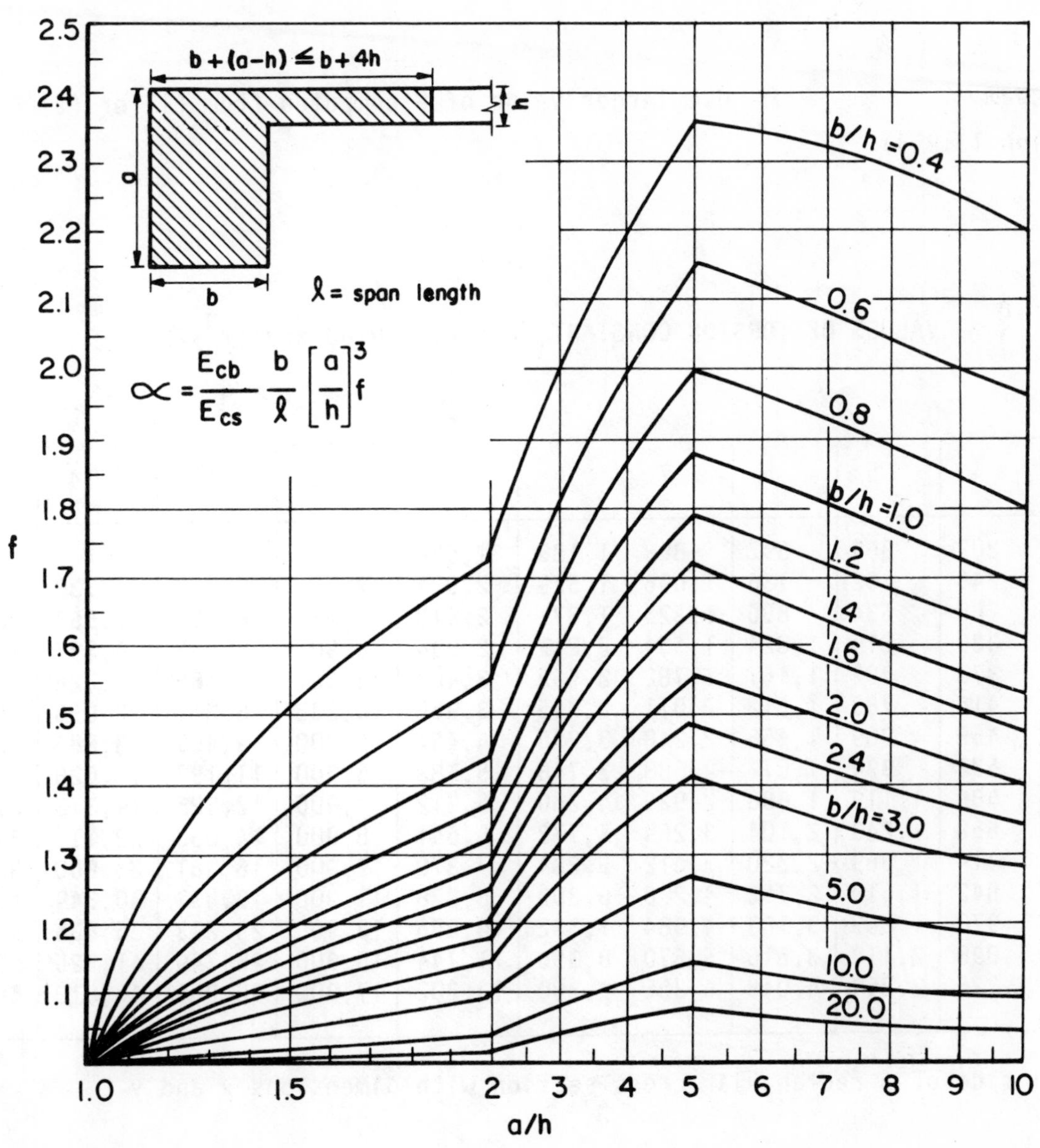

Fig. 20-9 - Beam to Slab Ratio α (Edge Beams)

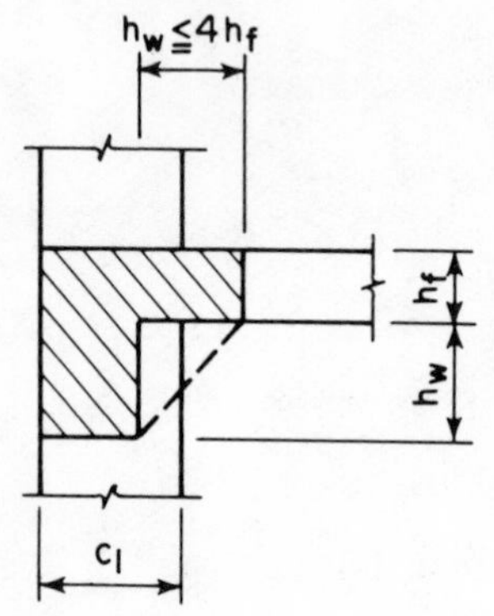

Edge Beam

(Section 13.2.4)

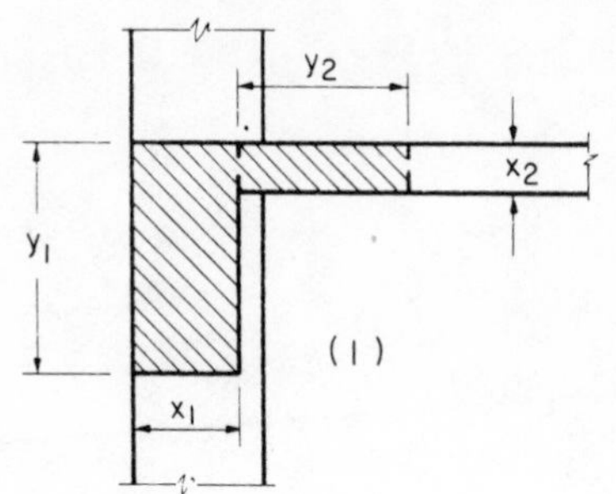

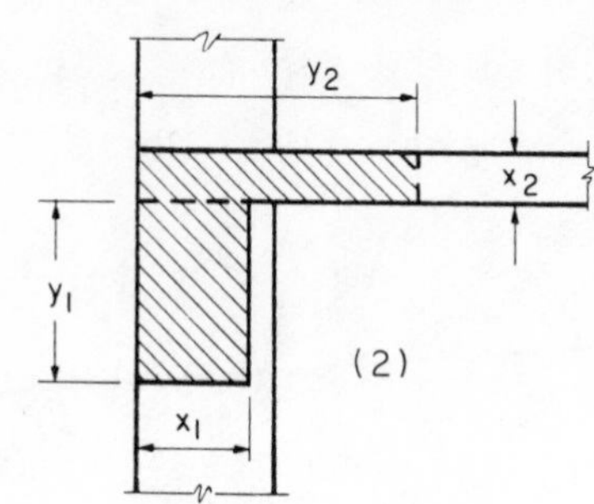

Use larger value of C computed from (1) or (2)

VALUES OF TORSION CONSTANT, $C = (1 - 0.63\ x/y)(x^3y/3)$

Y \ X*	4	5	6	7	8	9	10	12	14	16
12	202	369	592	868	1,188	1,538	1,900	2,557	--	--
14	245	452	736	1,096	1,529	2,024	2,566	3,709	4,738	--
16	288	534	880	1,325	1,871	2,510	3,233	4,861	6,567	8,083
18	330	619	1,024	1,554	2,212	2,996	3,900	6,013	8,397	10,813
20	373	702	1,167	1,782	2,553	3,482	4,567	7,165	10,226	13,544
22	416	785	1,312	2,011	2,895	3,968	5,233	8,317	12,055	16,275
24	458	869	1,456	2,240	3,236	4,454	5,900	9,469	13,885	19,005
27	522	994	1,672	2,583	3,748	5,183	6,900	11,197	16,628	23,101
30	586	1,119	1,888	2,926	4,260	5,912	7,900	12,925	19,373	27,197
33	650	1,243	2,104	3,269	4,772	6,641	8,900	14,653	22,117	31,293
36	714	1,369	2,320	3,612	5,284	7,370	9,900	16,381	24,860	35,389
42	842	1,619	2,752	4,298	6,308	8,828	11,900	19,837	30,349	43,581
48	970	1,869	3,183	4,984	7,332	10,286	13,900	23,293	35,836	51,773
54	1,098	2,119	3,616	5,670	8,356	11,744	15,900	26,749	41,325	59,965
60	1,226	2,369	4,048	6,356	9,380	13,202	17,900	30,205	46,813	68,157

*Small side of a rectangular cross section with dimensions x and y.

Fig. 20-10 - Design Aid for Computing Torsional Section Constant C

13.6.5 Factored Moments in Beams

When a design strip contains beams between columns, the factored moment assigned to the column strip must be distributed between the slab and beam section of the column strip. The amount of the column strip factored moment to be resisted by the beam varies linearly between zero and 85 percent as $\alpha_1 \ell_2/\ell_1$ varies between zero and 1.0. When $\alpha_1 \ell_2/\ell_1$ is equal to or greater than 1.0, 85 percent of the total column strip moment must be resisted by the beam. In addition, the beam section must resist loads applied directly to the beam, including weight of projecting beam stem above or below the slab. Note that Section 13.6.5.3, addressing loads applied directly to beams, has been revised for the '83 Code to further clarify design moments for beams.

13.6.6 Factored Moments in Middle Strips

Factored moments not assigned to column strips must be resisted by middle strips. An exception to this is a middle strip adjacent to and parallel with an edge supported by a wall where the moment to be resisted is twice the factored moment assigned to the half middle strip adjacent to the exterior design strip (See Fig. 20-1).

13.6.9 Factored Moments in Columns and Walls

Supporting columns and walls must resist any negative moments transferred from the slab system.

For interior columns (or walls), the approximate Eq. (13-4) may be used for unbalanced moment transfer by gravity loading, unless an analysis is made considering effects of pattern loading and unequal adjacent spans. For the '83 Code, Eq. (13-4) has been considerably simplified with elimination of the stiffness ratio α. The transfer moment is now computed directly as a function of span length and gravity loading. For the more usual case with equal transverse and adjacent spans, Eq. (13-4) reduces to

$$M_i = 0.07\ (0.5\ w_\ell \ell_2 \ell_n^2) \qquad \text{Eq. (4)}$$

Note that the load term w_ℓ is a factored value.

For exterior column or wall supports, the total exterior negative factored moment from the slab system (Section 13.6.3.3) is transferred directly to the supporting members. Due to the approximate nature of the moment coefficients, it seems unwarranted to consider the change in moment from face of support to centerline of support--use of the moment values from Section 13.6.3.3 directly is accurate enough.

Columns above and below the slab must resist a portion of the support moment based on the relative column stiffnesses--generally, in proportion to column lengths above and below the slab. Again, due to the approximate nature of the moment coefficients of the Direct Design Method, the refinement of considering the change in moment from centerline of slab-beam to top or bottom of column seems unwarranted.

13.6.10 Provisions for Effects of Pattern Loadings

When the ratio of unfactored dead-to-live load is less than 2 (high live-to-dead load ratio), the effect of increased moment due to pattern loading can be neglected in the analysis, if sufficiently stiff columns are provided. If the columns above and below the slab do not meet the minimum stiffness α_{min} of Table 13.6.10, and the load ratio is less than 2, the positive factored moments must be increased by the coefficient δ_s computed from Eq. (13-5).

EXAMPLE 20.1 - Two-Way Slab Without Beams Analyzed by Direct Design Method

Using the Direct Design Method, determine design moments for the slab system in the transverse direction for an intermediate floor.

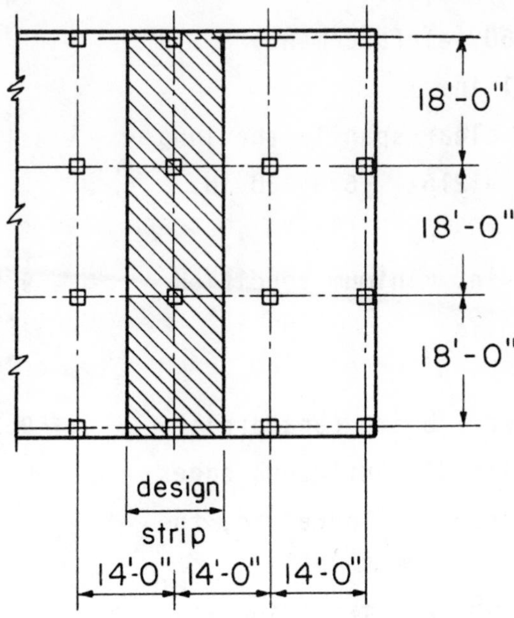

Story height = 9 ft

Columns = 16 in. x 16 in.

Lateral loads to be resisted by shear walls

No spandrel beams

Partition weight = 20 psf

Service live load = 40 psf

f'_c = 3,000 psi (for slab)

f'_c = 5,000 psi (for columns)

f_y = 60,000 psi

Calculations and Discussion	Code Reference
1. Preliminary design for slab thickness h	
a. Control of deflection:	
For slab systems without beams, the minimum overall thickness h is governed by	9.5.3

EXAMPLE 20.1 - Continued

Calculations and Discussion	Code Reference

$$h = \ell_n(800 + 0.005f_y)/36,000$$ Eq. (9-13)

$$= \ell_n/32.73 \text{ (for G60 reinforcement)}$$

$$= 200/32.73 = 6.11 \text{ in.}$$

where ℓ_n is the length of clear span in the long direction. Therefore, $\ell_n = 216 - 16 = 200$ in.

This is larger than the 5-in. minimum specified for slabs without drop panels. 9.5.3.1

For slab systems without edge beams (spandrels), the thickness of panels with discontinuous edges must be increased by 10 percent. Therefore, the minimum thickness is 9.5.3.3

$h = 6.11 \times 1.10 = 6.72$ in. <u>Try 7-in. slab</u>

b. Shear strength of slab:

Use an average effective depth, $d \simeq 5.75$ in. (3/4-in. cover and #4 bar)

Factored dead load, $w_d = (87.5 + 20)1.4 = 150.5$ 9.2.1

Factored live load, $w_\ell = 40 \times 1.7 = \underline{68.0}$

Total factored load, $w_u = 218.5$ psf

Investigation for wide beam action is made on a 12-in.-wide strip at "d" distance from face of support in the long direction (see Fig. 20-11). 11.11.1.1

$$V_u = 0.2185 \times 7.854 = 1.72 \text{ kips}$$

$$V_c = 2\sqrt{f'_c}\, b_w d$$ Eq. (11-3)

$$= 2\sqrt{3,000} \times 12 \times 5.75/1,000 = 7.56 \text{ kips}$$

$$\varphi V_c = 0.85 \times 7.56 = 6.43 \text{ kips}$$

$$V_u < \varphi V_c$$

EXAMPLE 20.1 - Continued

Calculations and Discussion	Code Reference

Since there are no shear forces at the centerline of adjacent panels (see Fig. 20-11), the shear strength in two-way action at "d/2" distance around a support is computed as follows:

$V_u = 0.2185(18 \times 14 - 1.81^2) = 54.3$ kips

$V_c = 4\sqrt{f'_c}\, b_o d$ (for square columns) — Eq. (11-36)

$= 4\sqrt{3{,}000} \times (4 \times 21.75) \times 5.75/1{,}000 = 109.6$ kips

$\varphi V_c = 0.85 \times 109.6 = 93.2$ kips

$V_u < \varphi V_c$

Therefore, preliminary design indicates that a 7 in. slab is adequate for control of deflection and shear strength.

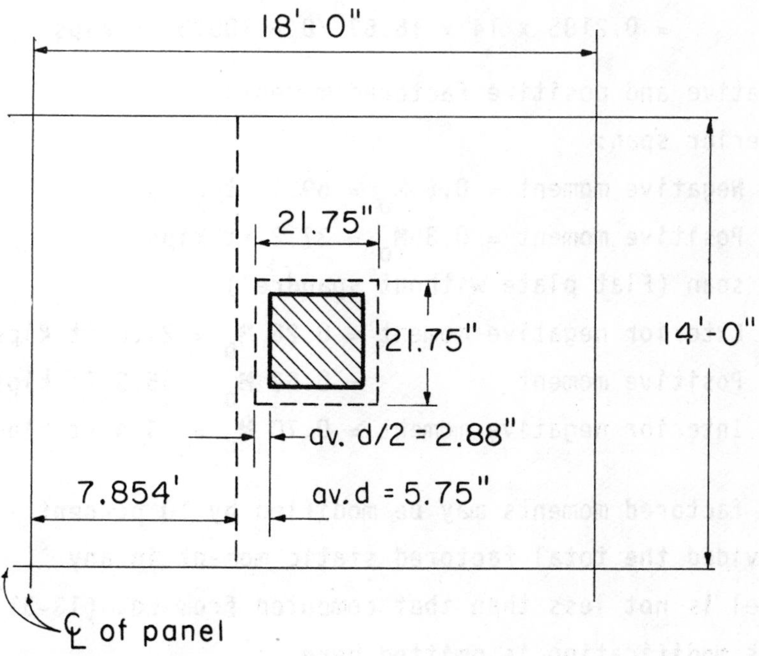

Fig. 20-11

EXAMPLE 20.1 - Continued

Calculations and Discussion	Code Reference

2. Check limitations for Direct Design Method — 13.6.1

- There are a minimum of three continuous spans in each direction, — 13.6.1.1
- Long span to short span ratio is 1.29 < 2.0, — 13.6.1.2
- Successive span lengths are equal, — 13.6.1.3
- Columns are not offset, — 13.6.1.4
- Loads are uniformly distributed with live to dead load ratio of 0.37 < 3.0, — 13.6.1.5
- Slab system is without beams — 13.6.1.6

3. Factored moments in slab

a. Total factored moment per span: — 13.6.2

$$M_o = w_u \ell_2 \ell_n^2/8 \qquad \text{Eq. (13-3)}$$

$$= 0.2185 \times 14 \times 16.67^2/8 = 106.3 \text{ ft kips}$$

b. Negative and positive factored moments — 13.6.3

Interior span: — 13.6.3.2

Negative moment = $0.65M_o$ = 69.1 ft kips

Positive moment = $0.35M_o$ = 37.2 ft kips

End span (Flat plate without spandrel): — 13.6.3.3

Exterior negative moment = $0.26\ M_o$ = 27.6 ft kips

Positive moment = $0.52\ M_o$ = 55.3 ft kips

Interior negative moment = $0.70\ M_o$ = 74.4 ft kips

Note: The factored moments may be modified by 10 percent provided the total factored static moment in any panel is not less than that computed from Eq. (13-3). This modification is omitted here. — 13.6.7

EXAMPLE 20.1 - Continued

Calculations and Discussion	Code Reference

4. Distribution of factored moments in column and middle strips 13.6.4 13.6.6

	Factored Moment		Column Strip		Two Half Middle Strips[3]
			Percent[2]	Moment	
End Span:					
Exterior negative	27.6	(42.1)[1]	100	27.6	0
Positive	55.3	(48.8)	60	33.2	22.1
Interior negative	74.4	(73.2)	75	55.8	18.6
Interior Span:					
Negative	69.1		75	51.8	17.3
Positive	37.2		60	22.3	14.9

[1]Values obtained using the "Modified Stiffness Method" of Com. Section 13.6.3.3.

[2]For slab systems without beams.

[3]That portion of the factored moment not resisted by the column strip is assigned to the two half middle strips.

5. Check for effects of pattern loading 13.6.10

Ratio of dead-to-live load:

$$\beta_a = 107.5/40 = 2.69$$

When $\beta_a \geq 2.0$, pattern loading effect may be neglected.

6. Factored moments in columns 13.6.9

a. Interior columns (with equal transverse and adjacent spans):

$$M_i = 0.07\ (0.5\ w_\ell \ell_2 \ell_n^{\ 2}) \qquad \text{Eq. (13-4)}$$

EXAMPLE 20.1 - Continued

Calculations and Discussion	Code Reference

$= 0.07\ (0.5 \times 1.7 \times 0.04 \times 14 \times 16.67^2)$

$= 9.3$ ft kips

With same column size and length above and below slab:

$M_c = 9.3/2 = 4.65$ ft kips

This moment is combined with the factored axial load (for each story) for design of the interior columns.

b. Exterior Columns:

Total exterior negative moment from slab must be transferred directly to the columns; $M_u = 27.6$ ft kips.

With same column size and length above and below the slab:

$M_c = 27.6/2 = 13.8$ ft kips

This moment is combined with the factored axial load (for each story) for design of the exterior columns.

7. Transfer of gravity load shear and moment at exterior column.

Check slab shear and flexural strength at edge column due to direct shear and unbalanced moment transfer. 11.12.2 13.3.3

a. Factored shear force transfer at exterior column:

$V_u = w_u \ell_1 \ell_2/2$

$= 0.2185 \times 14 \times 18/2 = 27.5$ kips

b. Unbalanced moment transfer at exterior column:

When the end span moments are determined using the approximate moment coefficients of Section 13.6.3.3, the special provision of Section 13.6.3.6 (moment transfer between slab and an edge column) requires that the fraction of unbalanced moment transferred by eccentricity of shear must be based on the full column strip nominal moment strength, M_n provided. The total reinforcement provided in the column strip includes the

Calculations and Discussion	Code Reference

additional reinforcement concentrated over the column to resist the fraction of unbalanced moment transferred by flexure $\gamma_f M_u$, where M_u is the exterior negative factored moment from the slab. For a slab without an edge beam, the total M_u = 27.6 ft kips is resisted by the column strip. (Minimum reinforcement per Section 13.4.1 is provided in the middle strip.)

For both middle strip and column strip:

$A_{s(min)}$ = 0.0018bh = 0.0018 x 84 x 7 = 1.06 $in.^2$ — 7.12.2.1(b)

where b is width of design strip = 14/2 = 7 ft = 84 in.

For #4 bars, total bars required = 1.06/0.20 = 5.3 bars

For s_{max} = 2h = 2 x 7 = 14 in., total bars required = 84/14 = 6 bars — 13.4.2

Check total reinforcement required for column strip negative moment M_u = 27.6 ft kips. Using Table 9-2, page 9-7:

$$\frac{M_u}{\varphi f'_c\ bd^2} = \frac{27.6 \times 12}{0.9 \times 3 \times 84 \times 5.75^2} = 0.0442$$

where d ≃ 7.0 - 1.25 = 5.75 in. (3/4-in. cover and #4 bar)

From Table 9-2; ω = 0.0454

$\rho = \omega f'_c/f_y$ = 0.0454 x 3/60 = 0.00227

$A_s = \rho bd$ = 0.00227 x 84 x 5.75 = 1.10 $in.^2$

For #4 bars, total bars required = 1.10/0.20 = 5.5 bars

6 bars for s_{max} = 14 in. governs.

EXAMPLE 20.1 - Continued

Calculations and Discussion	Code Reference
Use 6 #4 bars @ ± 14-in. spacing in middle strip and portion of column strip outside unbalanced moment transfer section c + 2(1.5h) = 16 + 2(1.5 x 7) = 37 in.	13.3.3.2
Additional reinforcement required over column within effective slab width of 37-in. to resist fraction of unbalanced moment transferred flexure is computed from Eq. (13-1). For square column, γ_f = 60%. (See Fig. 17-4.)	
$\gamma_f M_u$ = 0.60(27.6) = 16.6 ft kips must be transferred within the effective slab width of 37 in. Add 2 additional bars over column. Check moment strength for 4 #4 bars within 37-in. slab width. See sketch below.	

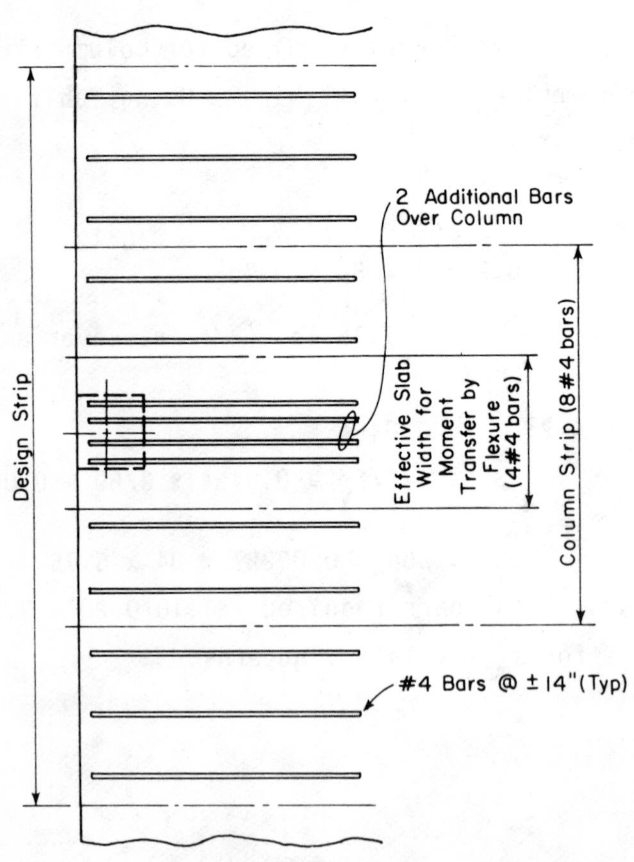

EXAMPLE 20.1 - Continued

Calculations and Discussion	Code Reference

For 4 #4 bars: $A_s = 4(0.20) = 0.80\ in.^2$

$\omega = A_s f_y / f'_c bd = 0.80 \times 60/3 \times 37 \times 5.75 = 0.0752$

From Table 9-2; $M_n / f'_c bd^2 = 0.0719$

$M_n = 0.0719 \times 3 \times 37 \times 5.75^2/12 = 22.0$ ft kips

$\varphi M_n = 0.9(22.0) = 19.8 > 16.6$ OK

Fraction of unbalanced moment transferred by eccentricity of shear must be based on full nominal moment strength M_n provided in column strip. 13.6.3.6

For 6 + 2 = 8 #4 bars: $A_s = 8(0.20) = 1.60\ in.^2$

$\omega = 1.60 \times 60/3 \times 84 \times 5.75 = 0.0663$

From Table 9-2; $M_n / f'_c bd^2 = 0.0637$

$M_n = 0.0637 \times 3 \times 84 \times 5.75^2/12 = 44.2$ ft kips

Assume transfer moment M_n at centroid of critical transfer section.

c. Combined shear stress at inside face of critical transfer section. For shear strength equations, see Part 17, page 17-6. 11.12.2.3

$v_u = V_u/A_c + \gamma_v M_n/(J/c)$

$= 27{,}500/342.2 + 0.4 \times 44.2 \times 12{,}000/2{,}358$

$= 80.4 + 90.0 = 170.4$ psi

where (referring to Fig. 17-6, edge column-bending perpendicular to edge)

EXAMPLE 20.1 - Continued

Calculations and Discussion	Code Reference

$a = c_1 + d/2 = 16 + 5.75/2 = 18.88$ in.

$b = c_2 + d = 16 + 5.75 = 21.75$ in.

$c = a^2/(2a + b) = 18.88^2/(2 \times 18.88 + 21.75) = 5.99$ in.

$A_c = (2a + b)d = 342.2$ in.^{2}

$J/c = [2ad(a + 2b) + d^3(2a + b)/a]/6$

$= 2{,}358$ in.^{3}

$\gamma_v = 0.40$ (See Fig. 17-4)

d. Combined shear stress at outside face of critical transfer section: — 11.12.2.3

$v_u = V_u A_c - \gamma_v M_n/(J/c')$

$= 27{,}500/342.2 - 0.4 \times 44.2 \times 12{,}000/1096$

$= 80.4 - 193.6 = 113.2$ psi

where $c' = a - c = 18.88 - 5.99 = 12.89$ in.

$J/c' = (J/c)(c/c') = 2{,}358 \times 5.99/12.89 = 1{,}096$ in.^{3}

e. Permissible shear stress — 11.12.2.4

$\varphi V_n = \varphi 4\sqrt{f'_c} = 0.85 \times 4\sqrt{3000}$

$= 186.2$ psi > 170.4 OK

EXAMPLE 20.2 - Two-Way Slab with Beams Analyzed by Direct Design Method

Using the Direct Design Method, determine design moments for the slab system in the transverse direction for an intermediate floor.

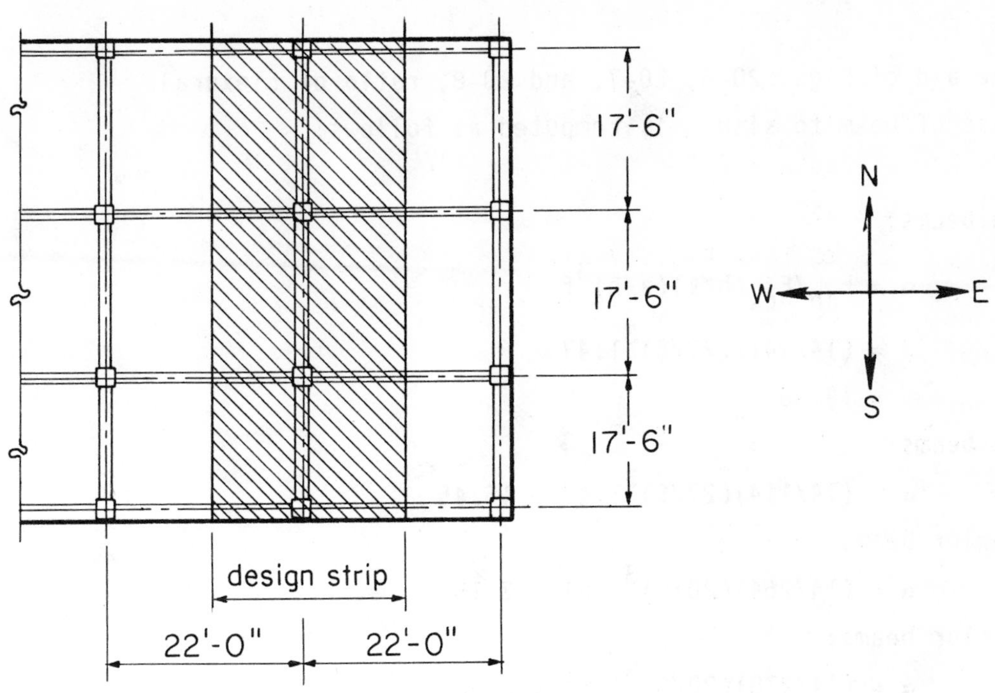

Story height = 12 ft

Spandrel beams = 14 in. x 27 in.

Interior beams = 14 in. x 20 in.

Columns = 18 in. x 18 in.

Slab = 6 in.

Service live load = 100 psf

f'_c = 4,000 psi (for all members) normal weight

f_y = 60,000 psi

EXAMPLE 20.2 - Continued

Calculations and Discussion	Code Reference

1. Preliminary design for slab thickness, h:
   (Control of deflection) — 9.5.3

With the aid of Figs. 20-6, 20-7, and 20-8, ratio of flexural stiffness of beam to slab α is computed as follows:

NS edge beams:

$$\alpha = E_{cb}/E_{cs}(b/\ell)(a/h)^3 f$$
$$= (14/141)(27/6)^3 1.47$$
$$= 13.30$$

EW edge beams:

$$\alpha = (14/114)(27/6)^3 1.47 = 16.45$$

NS interior beams:

$$\alpha = (14/264)(20/6)^3 1.61 = 3.16$$

EW interior beams:

$$\alpha = (14/210)(20/6)^3 1.61 = 3.98$$

Since all $\alpha > 2.0$, (see Fig. 20-3), Eq. (9-12) will control for minimum thickness. Also, $\beta_s = 0.5$ for a corner panel will be the smallest (and controlling) value of any panel (see Fig. 20-2).

Therefore,

$$h = \frac{\ell_n(800 + 0.005f_y)}{36{,}000 + 5{,}000\beta(1 + \beta_s)} = 5.93 \text{ in.}$$ Eq. (9-12)

where $\beta = 20.5/16 = 1.28$

$\beta_s = 0.5$

ℓ_n = clear span in long direction measured face to face of column = 20'-6" = 246 in.

<u>Use 6-in. slab thickness</u>

EXAMPLE 20.2 - Continued

Calculations and Discussion	Code Reference

2. Check limitations for Direct Design Method — 13.6.1

- There are a minimum of three continuous spans in each direction, — 13.6.1.1
- Long span to short span ratio is 1.26 < 2.0, — 13.6.1.2
- Successive span lengths are equal, — 13.7.1.3
- Columns are not offset, — 13.6.1.4
- Loads are uniformly distributed with live-to-dead ratio of 1.33 < 3.0 — 13.6.1.5
- Interior panel: $\alpha_1 \ell_2^2/\alpha_2 \ell_1^2 = 1.25$ — 13.6.1.6
- Exterior panel $\alpha_1 \ell_2^2/\alpha_2 \ell_1^2 = 0.30$

3. Factored moments in slab

a. Total factored moment per span: — 13.6.2

$$M_o = w_u \ell_2 \ell_n^2/8 \qquad \text{Eq. (13-3)}$$

$$= 0.288 \times 22 \times 16^2/8 = 202.8 \text{ ft kips}$$

where $w_u = w_d + w_\ell = 1.4(75 + 9.3) + 1.7(100) = 288$ psf
(9.3 psf is weight of beam stem per foot divided by ℓ_2)

b. Negative and positive factored moments — 13.6.3

Interior span: — 13.6.3.2

Negative moment = 0.65 M_o = 131.8 ft kips
Positive moment = 0.35 M_o = 71.0 ft kips

End span (Two-way slabs): — 13.6.3.3

Exterior negative moment = 0.16 M_o = 32.4 ft kips
Positive moment = 0.57 M_o = 115.6 ft kips
Interior negative moment = 0.70 M_o = 142.0 ft kips

EXAMPLE 20.2 - Continued

Calculations and Discussion	Code Reference

Note: The factored moments may be modified by 10 percent provided the total factored static moment in any panel is not less than that computed from Eq. (13-3). This modification is omitted here. 13.6.7

4. Distribution of factored moments in column and middle strips 13.6.4 13.6.6

a. Percentage of total negative and positive moments to column strip.

At interior support:

$75 + 30(\alpha_1 \ell_2/\ell_1)(1 - \ell_2/\ell_1)$ Eq. (1)

$75 + 30(1 - 1.26) = 67\%$

where α_1 (in direction of ℓ_1) is computed with the aid of Fig. 20-7:

$\alpha_1 = (b/\ell)(a/h)^3 f$

$= (14/264)(20/6)^3\ 1.61 = 3.16$

$\alpha_1 \ell_2/\ell_1 = 3.16 \times 22/17.5 = 3.98 > 1.0$ USE 1.0

at exterior support:

$100 - 10\beta_t + 12\beta_t\ (\alpha_1\ \ell_2/\ell_1)\ (1 - \ell_2/\ell_1)$ Eq. (2)

$100 - 10(1.88) + 12(1.88)(1 - 1.26) = 75\%$

where $\beta_t = C/(2I_s) = 17{,}868(2 \times 4{,}752) = 1.88$

$I_s = \ell_2 h^3/12 = 264 \times 6^3/12 = 4{,}752\ \text{in.}^4$

C is taken as the larger value computed (with the aid of Fig. 20-10) for the torsional member shown below.

x_1 =	14	x_2 =	6	x_1 =	14	x_2 =	6
y_1 =	21	y_2 =	35	y_1 =	27	y_2 =	21
C_1 =	11,141	C_2 =	2,248	C_1 =	16,628	C_2 =	1,240
$\Sigma C = 11{,}141 + 2{,}248 = 13{,}389\ \text{in.}^4$				$\Sigma C = 16{,}628 + 1{,}240 = 17{,}868\ \text{in.}^4$			

EXAMPLE 20.2 - Continued

Calculations and Discussion	Code Reference

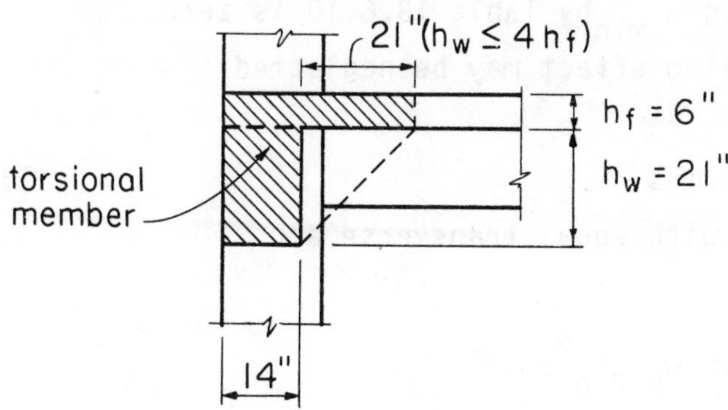

Positive moment:

$60 + 30(\alpha_1 \ell_2/\ell_1)(1.5 - \ell_2/\ell_1)$ Eq. (3)

$60 + 30(1.5 - 1.26) = 67\%$

Factored moments in column strips and middle strips are summarized as follows:

	Factored Moment	Column Strip		Two half Middle Strips[3]
		Percent	Moment[2]	
End span:				
Exterior negative	32.4 (64.6)[1]	75	24.3	8.1
Positive moment	115.6 (99.9)	67	77.5	38.1
Interior negative	142.0 (142.2)	67	95.1	46.9
Interior span:				
Negative	131.8	67	88.3	43.5
Positive	71.0	67	47.6	23.4

[1]Values obtained using the "Modified Stiffness Method" of Com. Section 13.6.3.3

[2]Since $\alpha_1 \ell_2/\ell_1 > 1.0$, beams must be proportioned to resist 85 percent of column strip moment as per Section 13.6.5.1.

[3]That portion of the factored moment not resisted by the column strip is assigned to the two half middle strips.

EXAMPLE 20.2 - Continued

Calculations and Discussion	Code Reference

5. Check for effects of pattern loading — 13.6.10

Ratio of dead-to-live load:

$\beta_a = 75/100 = 0.75 < 2.0$

Since $\alpha_1 > 2.0$, required α_{min} by Table 13.6.10 is zero.

Therefore, pattern loading effect may be neglected.

6. Factored moments in columns — 13.6.9

a. Interior columns (with equal transverse and adjacent spans):

$$M_i = 0.07(0.5\ w_\ell \ell_2 \ell_n^2)$$ Eq. (13-4)

$$= 0.07(0.5 \times 1.7 \times 0.1 \times 22 \times 16^2)$$

$$= 33.5 \text{ ft kips}$$

With same column size and length above and below slab:

$$M_c = 33.5/2 = 16.75 \text{ ft kips}$$

This moment is combined with the factored axial load (for each story) for design of the interior columns.

b. Exterior columns:

The total exterior negative moment from the slab/beam is transferred to the exterior columns; with same column size and length above and below the slab system:

$$M_c = 32.4/2 = 16.2 \text{ ft kips}$$

7. Shear strength

a. Beams. Since $\alpha_1\ell_2/\ell_1 = 3.98$, beams must resist total shear (b_w = 14 in., d = 17 in.) — 13.6.8.1

EXAMPLE 20.2 - Continued

Calculations and Discussion	Code Reference

NS Beams:

$$V_u = w_u\ell_1^2/4 = 0.288(17.5)^2/4 = 22.1 \text{ kips}$$

$$\varphi V_c = \varphi 2\sqrt{f'_c}\, b_w d$$

$$= 0.85 \times 2\sqrt{4{,}000} \times 14 \times 17 = 25.6 \text{ kips}$$

$$V_u < \varphi V_c$$

Only minimum shear reinforcement is required where $V_u > \varphi V_c/2$ in accordance with Section 11.5.5.

EW Beams:

$$V_u = w_u\ell_1(2\ell_2 - \ell_1)/4$$

$$= 0.288 \times 17.5(2 \times 22 - 17.5)/4 = 33.4 \text{ kips}$$

$$V_u > \varphi V_c$$

Required shear strength to be provided by shear reinforcement $V_s = (V_u - \varphi V_c)/\varphi = (33.4 - 25.6)/0.85$ $= 9.2$ kips.

b. Slab. (b_w = 12 in., d = 5.5 in.) 13.6.8.4

$$V_u = w_u\ell_1/2 = 0.275 \times 17.5/2 = 2.4 \text{ kips}$$

$$\varphi V_c = \varphi 2\sqrt{f'_c}\, b_w d$$

$$= 0.85 \times 2\sqrt{4{,}000} \times 12 \times 5.5 = 7.1 \text{ kips}$$

$$V_u < \varphi V_c$$

Shear strength of slab is adequate without shear reinforcement.

9. Spandrel beams must be designed to resist moment not transferred to exterior columns by parallel beams, in accordance with Section 11.6.

21

Two-way Slabs—Equivalent Frame Method

Update for '83 Code

For ACI 318-83, Section 13.7.4 has been rewritten eliminating the stiffness equation for the equivalent column in its entirety from the code. With the growing use of computers for two-way slab analysis by the equivalent frame stiffness procedure, the concept of combining stiffnesses of actual columns and torsional members into a single stiffness element is not practical with computerized frame analysis.

The "Equivalent Column" is, however, retained in the Commentary as an aid for analysis where slab-beams are analyzed separately for gravity loads, especially when using moment distribution or other hand calculation procedures for the analysis. See Commentary Section 13.7.4.

Preliminary Design

Before proceeding with the equivalent frame analysis, a preliminary slab thickness h needs to be determined for control of deflections according to the minimum thickness requirements of Section 9.5.3. Figs. 20-2 and 20-3 of Part 20 can be used to simplify minimum thickness computations.

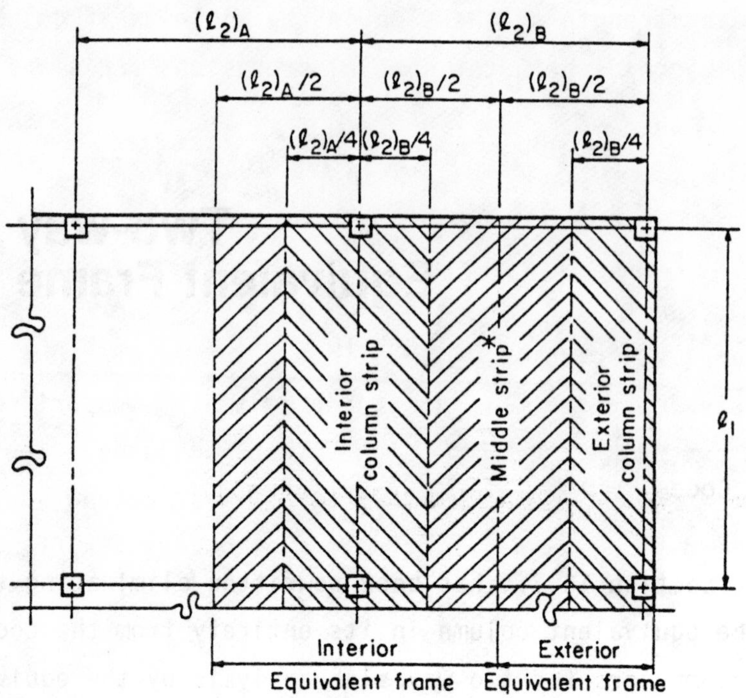

(a) Column strip for $\ell_2 \leq \ell_1$

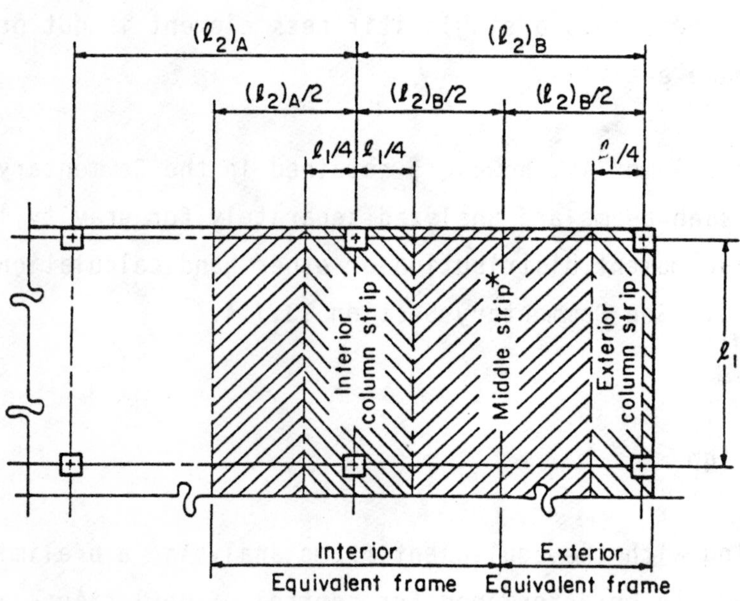

(b) Column strip for $\ell_2 > \ell_1$

Fig. 21-1 - Definition of Design Strip

*When edge of exterior design strip is supported by a wall the factored moment resisted by this middle strip is defined in Section 13.6.6.3.

For slab systems without beams, it is advisable at this stage in the design to check the shear strength of the slab in the vicinity of columns or other support locations according to the special provisions for slabs of Section 11.11.

13.7.2 Equivalent Frame

A three-dimensional building is divided into a series of two-dimensional frame bents (equivalent frames) centered on column or support centerlines, with each frame extending the full height of the building. The width of each equivalent frame is defined as the mid-point between column centerlines. The complete analysis of a slab system for a building consists of analyzing a series of equivalent (interior and exterior) frames spanning longitudinally and transversely through the building.

Application of the frame definitions given in Sections 13.2.1, 13.2.2, 13.7.2.2 and 13.7.2.3 is illustrated in Fig. 21-1. Some judgment is required in applying the definitions given in Section 13.2.1 for column strips with varying span lengths along the design strip.

The reason for specifying that the column strip width be based on the shorter of ℓ_1 or ℓ_2 is to account for the tendency for moment to concentrate about the column line when the span length of the design strip is less than its width.

Members of the equivalent frame are made up of slab-beams and torsional members (horizontal members) supported by columns (vertical members). The torsional members provide moment transfer between the slab-beams and columns. The equivalent frame members are illustrated in Fig. 21-2. The initial step in the frame analysis requires that the flexural stiffness of the equivalent frame members be determined.

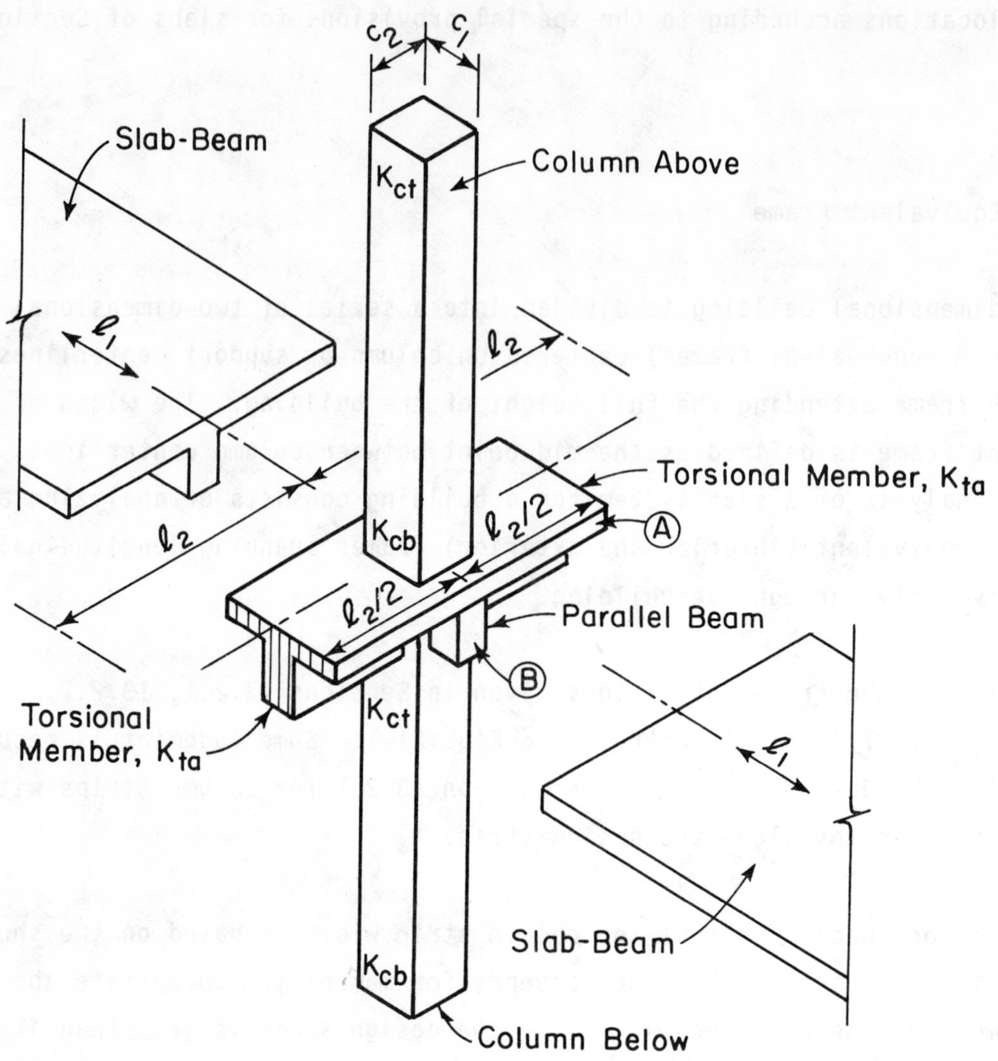

Fig. 21-2 - Equivalent Frame Members

13.7.3 - Slab Beams

Common types of slab systems with and without beams between supports are illustrated in Figs. 21-3 and 21-4. Cross sections for determining the stiffness of the slab-beam members K_{sb} between support centerlines are shown for each type. The equivalent slab-beam stiffness diagrams may be used to determine moment distribution constants and fixed-end moments for the equivalent frame analysis.

Make-up of the various areas for stiffness calculation is based on the following considerations:

(a) The moment of inertia of the slab-beam between face of supports is based on the gross cross-sectional area of the concrete. Variation in the moment of inertia along the axis of the slab-beam between supports is taken into account. (Section 13.7.3.2.)

(b) A support is defined as a column, capital, bracket or wall. Note that a beam is not considered a support member for the equivalent frame. (Section 13.7.3.3.)

(c) The moment of inertia of the slab-beam from the face of support to the centerline of support is assumed equal to the inertia of the slab-beam at the face of support divided by the quantity $(1 - c_2/\ell_2)^2$. (Section 13.7.3.3.)

The magnification factor $(1 - c_2/\ell_2)^2$ applied to the inertia between support face and centerline, in effect, makes each slab-beam at least a haunched member within its length. Consequently, stiffness and carryover factors and fixed-end moments based on the usual assumptions of uniform prismatic members cannot be applied to the slab-beam members.

Tables A1 through A6 in Appendix 21A give stiffness coefficients, carry-over factors, and fixed-end moment coefficients for different geometric and loading configurations. A wide range of column-to-span ratios in both longitudinal and transverse directions is covered in the tables. Table A1 can be used for flat plates and two-way slabs with beams. Tables A2 through A5 are intended to be used for flat slabs and waffle slabs with various drop (solid head) depths. Table A6 covers the unusual case of a flat plate combined with a flat slab. Fixed end moment coefficients are provided for both uniform and partially uniform loads. Partial load coefficients were developed for distribution over $0.2\ell_1$. However, loads acting over longer ranges can be handled by summing the effects of each $0.2\ell_1$ interval. For example, if the partial loading acts over $0.6\ell_1$, then the coefficients corresponding to three $0.2\ell_1$ intervals are to be added. This provides flexibility in the arrangement of loading. For concentrated loads, high intensity (w) of partial load can be taken at the appropriate location and assumed to be distributed over $0.2\ell_1$. For values in between, interpolation can be made. Stiffness diagrams are shown on each table and, with appropriate engineering judgment, different span conditions can be handled with the information given in these tables.

13.7.4 Columns

Common types of column and support conditions for slab systems are illustrated in Fig. 21-5. The column stiffness is based on a height of column ℓ_c measured from the mid-depth of the slab above to the mid-depth of the slab below. The column stiffness diagrams may be used to determine column flexural stiffness, K_c. Make up of the stiffness diagrams is based on the following considerations:

(a) The moment of inertia of the column outside the slab-beam joint is based on the gross cross-sectional area of the concrete. Variation in the moment of inertia along the axis of the column between slab-beam joints is taken into account. For columns with capitals, the moment of inertia is assumed to vary linearly from the base of the capital to the bottom of the slab-beam (Sec. 13.7.4.1 and 13.7.4.2).

(b) The moment of inertia is assumed infinite ($I = \infty$) from the top to the bottom of the slab-beam at the joint. As with the slab-beam members, the stiffness factor K_c for the columns cannot be based on the assumption of uniform prismatic members (Sec. 13.7.4.3).

Table A7 in Appendix 21A can be used to determine the actual column stiffness and carry-over factors.

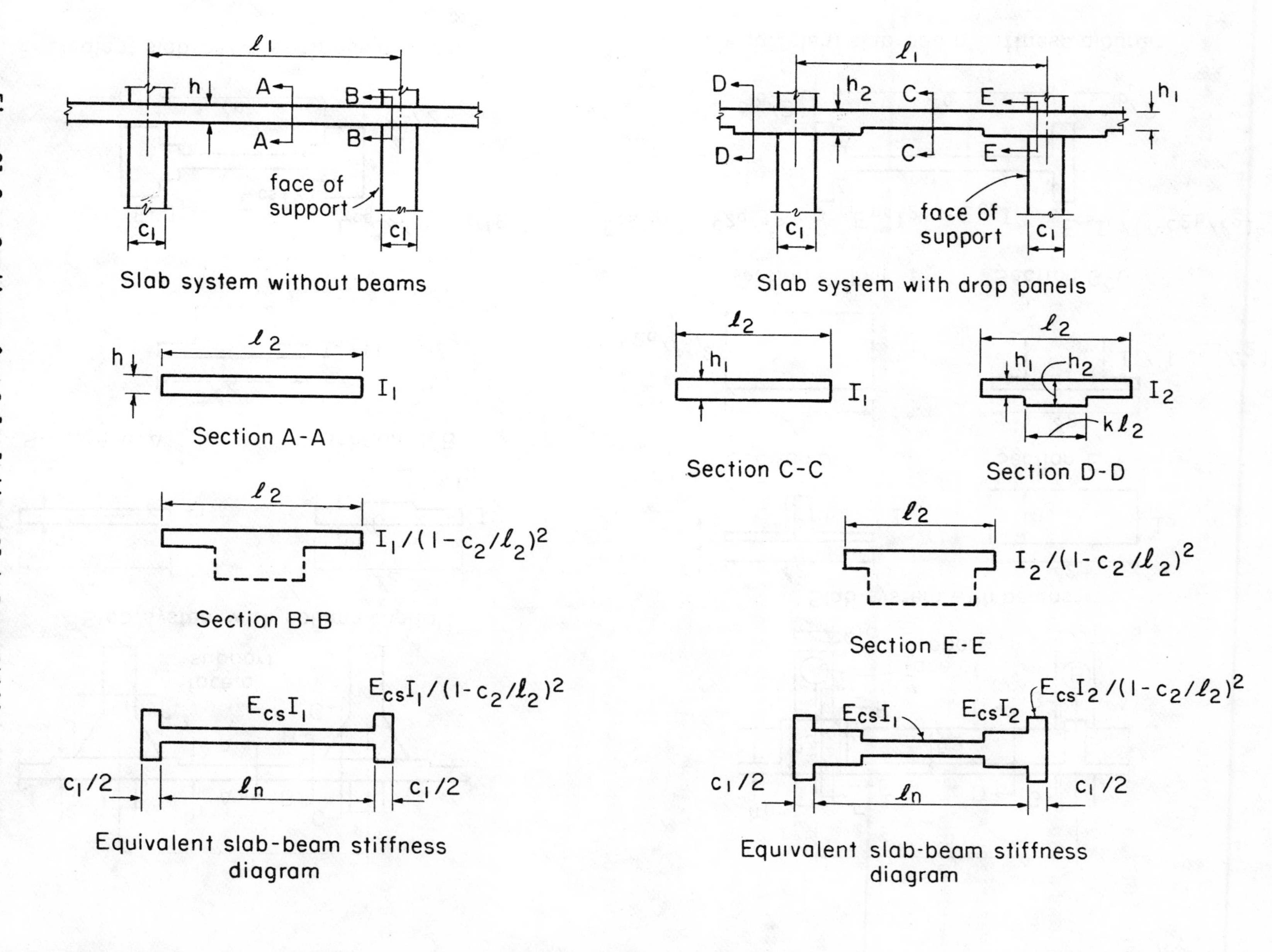

Fig. 21-3 - Sections for Calculating Slab-Beam Stiffness K_{sb}

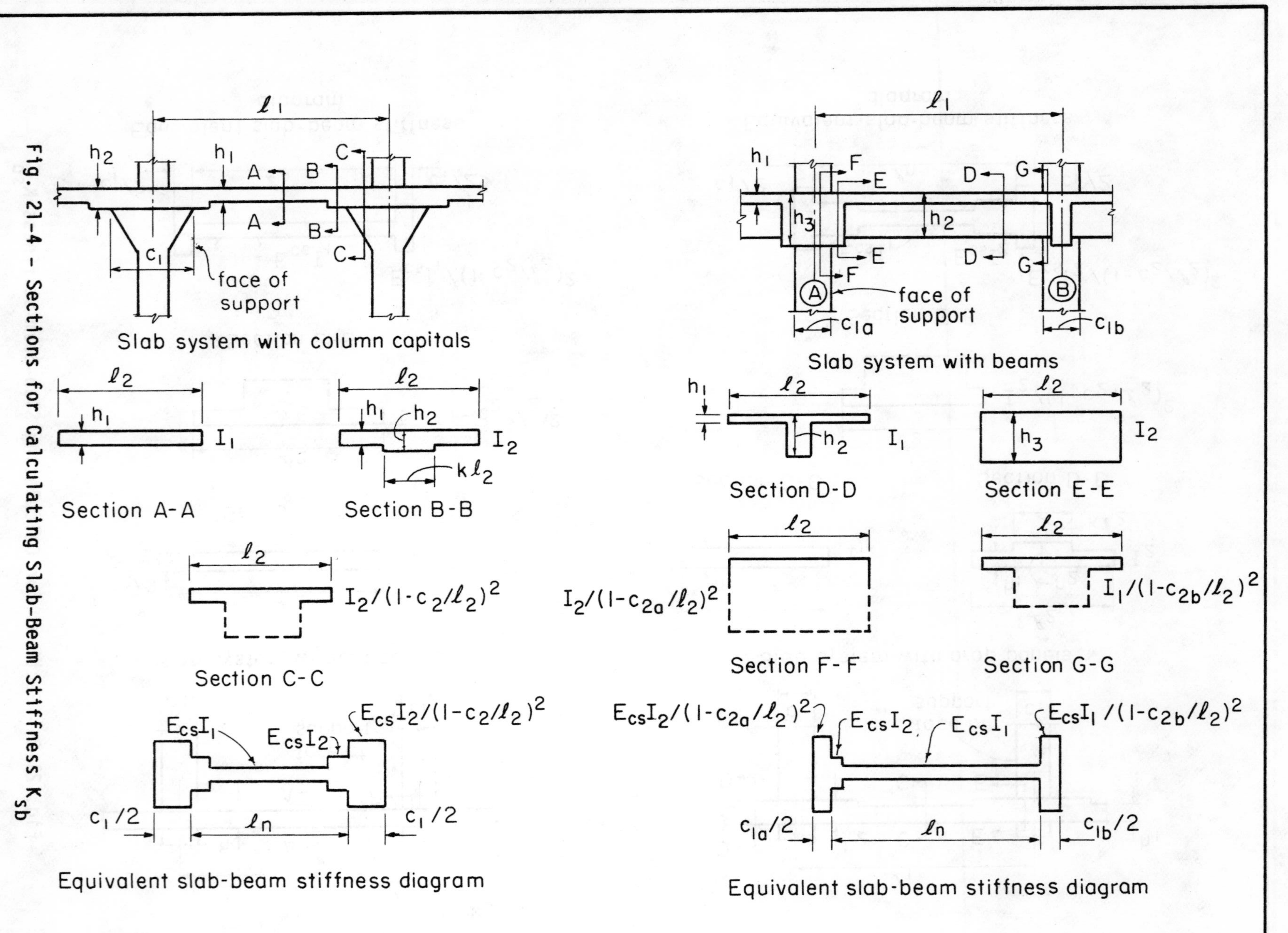

Fig. 21-4 - Sections for Calculating Slab-Beam Stiffness K_{sb}

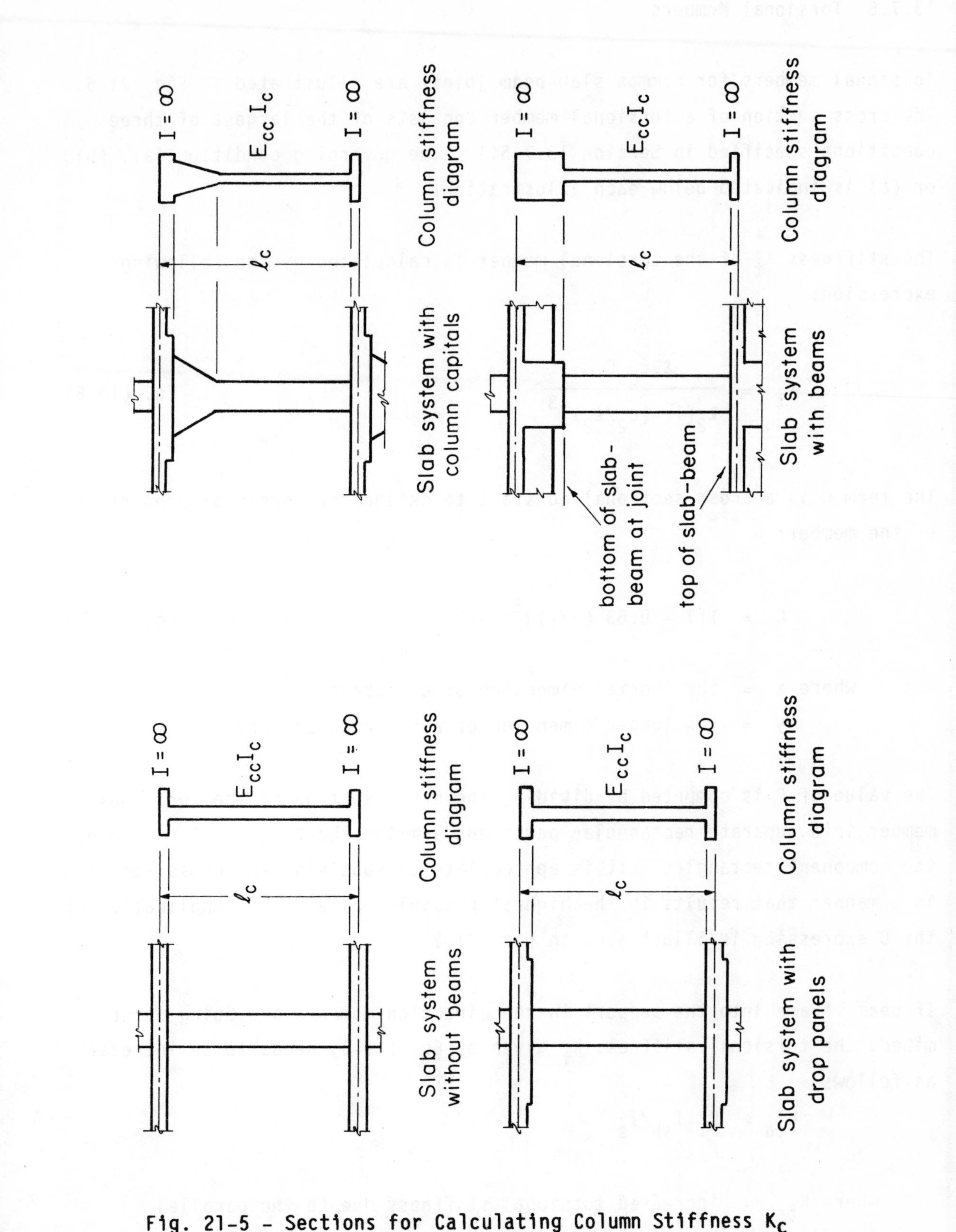

Fig. 21-5 - Sections for Calculating Column Stiffness K_c

13.7.5 Torsional Members

Torsional members for common slab-beam joints are illustrated in Fig. 21-6. The cross section of a torsional member consists of the largest of three conditions specified in Section 13.7.5.1. The governing condition (a), (b) or (c) is indicated below each illustration.

The stiffness K_t of the torsional member is calculated by the following expression:

$$K_t = \frac{\Sigma 9 E_{cs} C}{\ell_2 [1 - (c_2/\ell_2)]^3} \qquad \text{Eq. (13-6)}$$

The term C is a cross-sectional constant to define the torsional properties of the member:

$$C = \Sigma [1 - 0.63 (x/y)]^3 x\, y/3 \qquad \text{Eq. (13-7)}$$

where x = the shorter dimension of a rectangular part

y = the longer dimension of a rectangular part

The value of C is computed by dividing the cross section of the torsional member into separate rectangular parts and summing the C values for each of the component rectangles. It is appropriate to subdivide the cross section in a manner that results in the highest possible value of C. Application of the C expression is illustrated in Fig. 21-7.

If beams frame into the support in the direction moments are being determined, the torsional stiffness K_t given by Eq. (13-6) needs to be increased as follows:

$$K_{ta} = K_t\, I_{sb}/I_s$$

where K_{ta} = increased torsional stiffness due to the parallel beam (note parallel beam shown in Fig. 21-2)

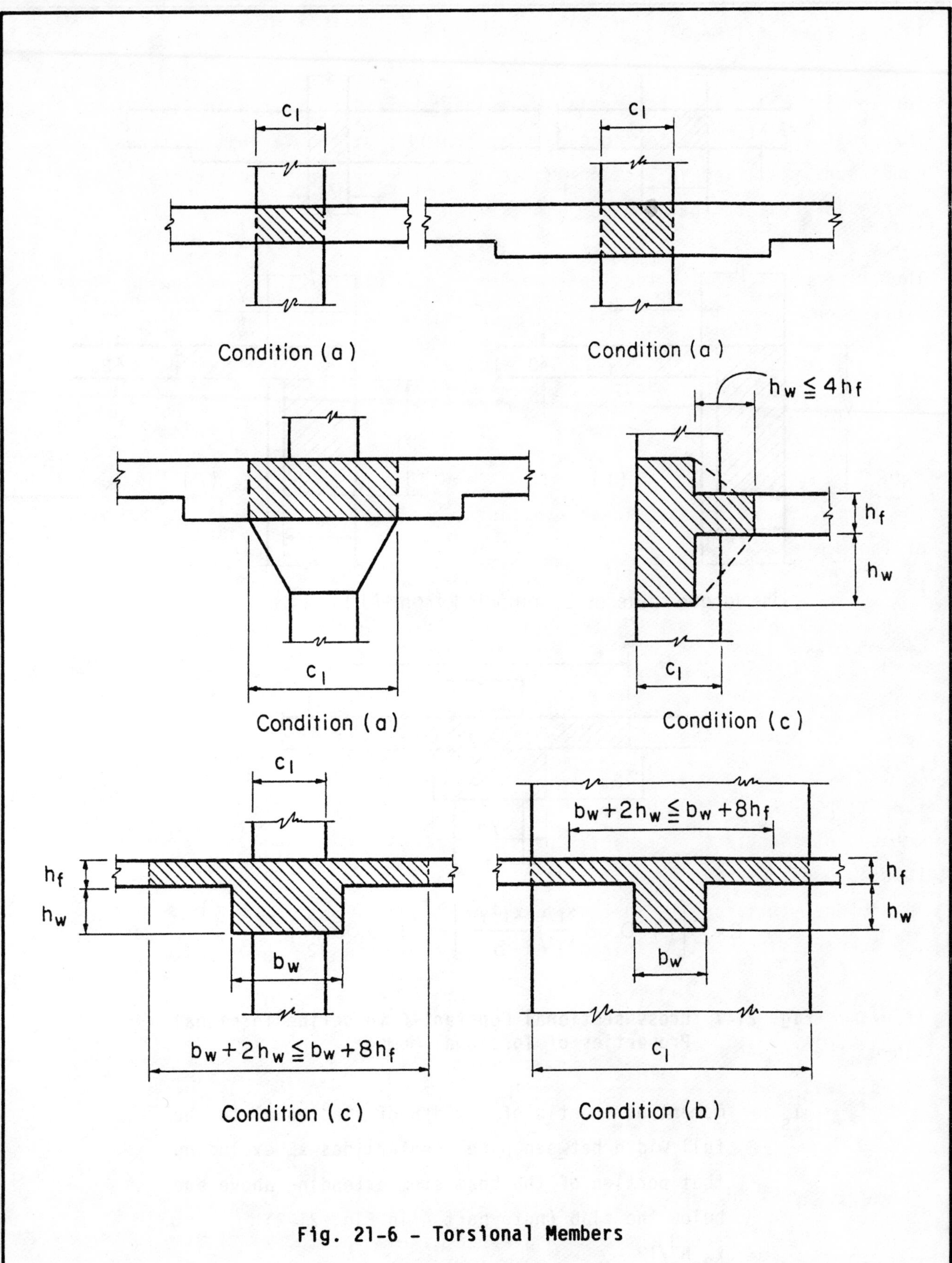

Fig. 21-6 - Torsional Members

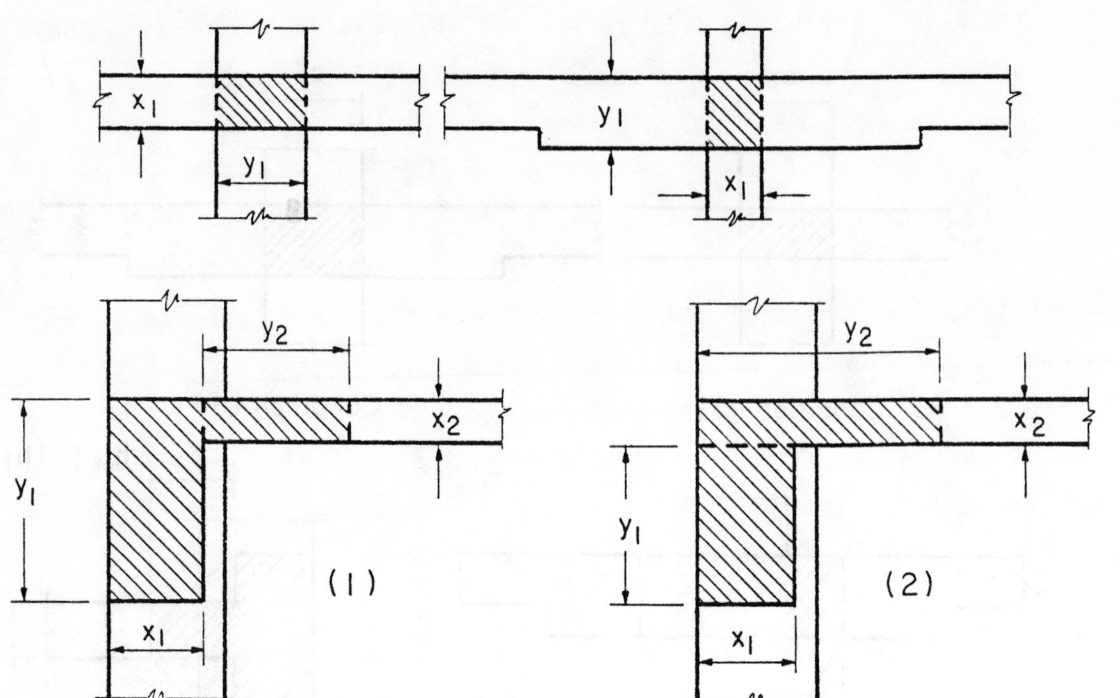

Use larger value of C computed from (1) or (2)

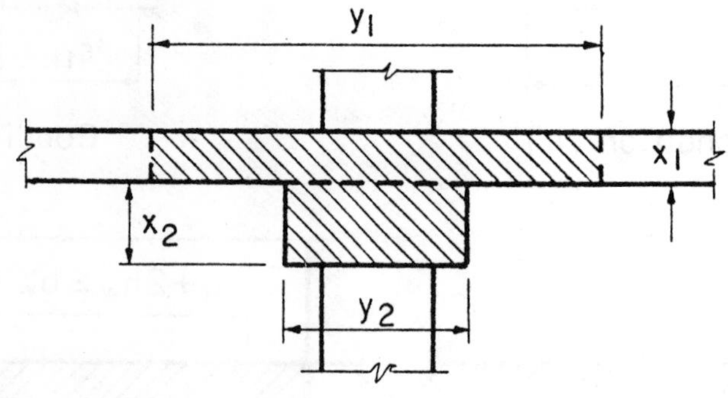

$$C = \sum \left[\left(1 - 0.63 \frac{x_1}{y_1}\right) \frac{x_1^3 y_1}{3} \right] + \left[\left(1 - 0.63 \frac{x_2}{y_2}\right) \frac{x_2^3 y_2}{3} \right]$$

Fig. 21-7 Cross sectional Constant C to define Torsional Properties of Torsional Member

I_s = moment of inertia of a width of slab equal to the full width between panel centerlines ℓ_2 excluding that portion of the beam stem extending above and below the slab (note part A in Fig. 21-2)

$= \ell_2 h^3/12$

I_{sb} = moment of inertia of the slab section specified for I_s including that portion of the beam stem extending above and below the slab (for the parallel beam illustrated in Fig. 21-2, I_{sb} is for the full tee section shown).

Equivalent Columns (Commentary Section 13.7.4)

When slab-beams are analyzed separately for gravity loads (especially when using moment distribution or other hand calculation procedures for the analysis), the concept of an equivalent column, combining the stiffness of the slab-beams and torsional members into a composite element is helpful. Both Design Examples 21.1 and 21.2 utilize the equivalent column concept with moment distribution for the gravity load analysis.

The equivalent column modifies the column stiffness to account for the torsional flexibility of the slab-to-column connection which reduces its efficiency for transmission of moments. An equivalent column is illustrated in Fig. 21-2. The equivalent column consists of the actual columns above and below the slab-beams plus "attached" torsional members on each side of the columns extending to the centerline of the adjacent panels. Note that for an edge frame the attached torsional member is located on one side only. The presence of parallel beams will also influence the stiffness of the equivalent column.

The flexural stiffness of the equivalent column K_{ec} is given in terms of its inverse or flexibility as follows:

$$1/K_{ec} = (1/\Sigma K_c) + (1/\Sigma K_t)$$

For purposes of computation, the designer may prefer that the above expression be given in terms of stiffness directly as follows:

$$K_{ec} = \Sigma K_c \times \Sigma K_t/(\Sigma K_c + \Sigma K_t)$$

Stiffness of the actual column K_c and torsional members K_t must comply with Sections 13.7.4 and 13.7.5.

After the values of K_c and K_t are determined, the equivalent column stiffness K_{ec} is computed. Using Fig. 21-2 for illustration,

$$K_{ec} = [(K_{ct} + K_{cb})(K_{ta} + K_{ta})] / [(K_{ct} + K_{cb}) + (K_{ta} + K_{ta})]$$

where K_{ct} = flexural stiffness at top of lower column framing into joint,
K_{cb} = flexural stiffness at bottom of upper column framing into joint,
K_{ta} = torsional stiffness of each torsional member, one on each side of the column, increased due to the parallel beam (if any).

13.7.6 Arrangement of Live Load

In the usual case where the exact loading pattern is not known, the maximum factored moments are developed with loading conditions illustrated by the three-span partial frame in Fig. 21-8 and described as follows:

(a) When the service live load does not exceed three-quarters of the service dead load, only loading pattern (1) with full factored live load on all spans need be analyzed for negative and positive factored moments.

(b) When the service live-to-dead load ratio exceeds three-quarters, the five loading patterns shown need to be analyzed to determine all factored moments in the slab-beam members. Loading patterns (2) through (5) consider a partially factored live load for determining factored moments. However, with partial live loading, the factored moments cannot be taken less than those occurring with full factored live load on all slab-beam spans; hence load pattern (1) is required to complete the analysis.

For slab systems with beams, loads supported directly by the beams (such as the weight of the beam stem or a wall supported directly by the beams) may be inconvenient to include in the frame analysis for the slab loads, $w_d + w_\ell$.

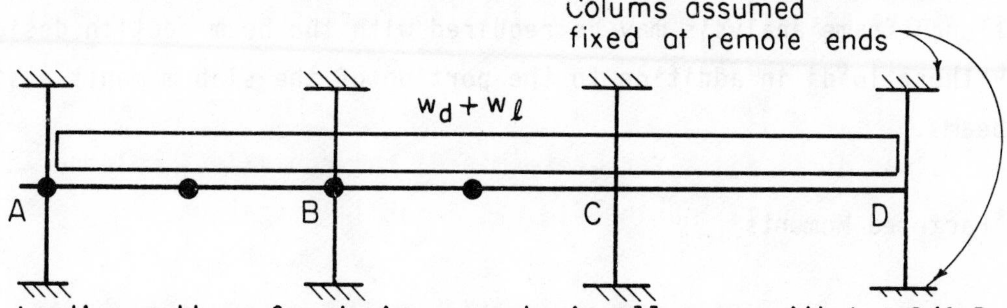

(1) Loading pattern for design moments in all spans with L ≤ 3/4 D

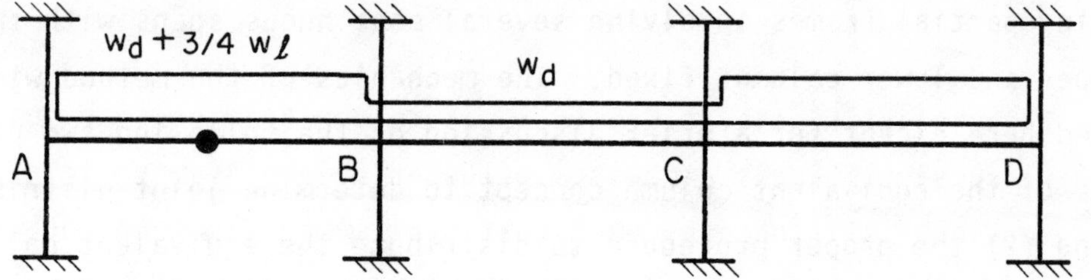

(2) Loading pattern for positive design moment in span AB*

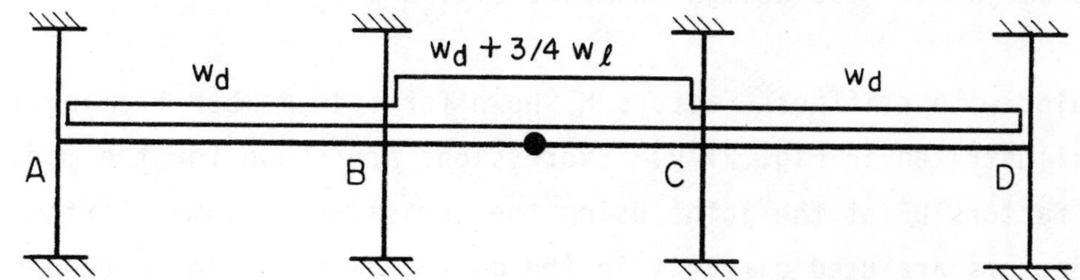

(3) Loading pattern for positive deisgn moment in span BC*

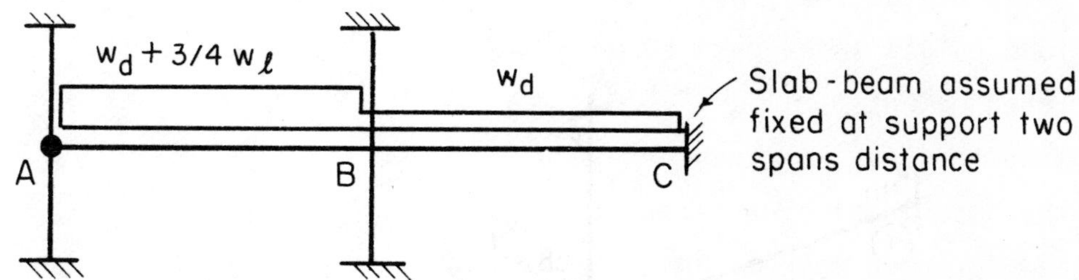

(4) loading pattern for negative design moment at support A*

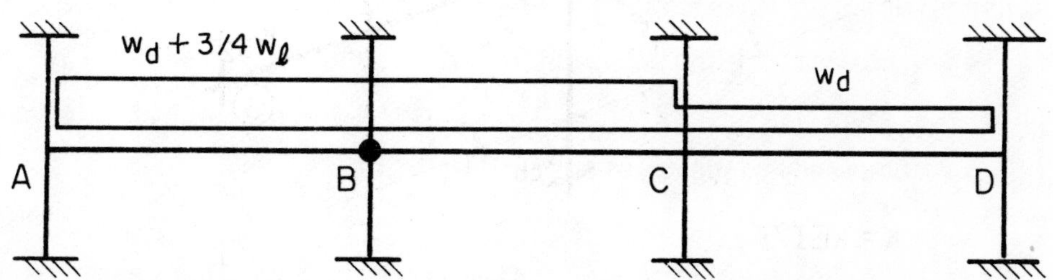

(5) Loading pattern for negative design moment at support B*

Fig. 21-8 - Partial Frame Analysis for Vertical Loading

An additional frame analysis may be required with the beam section designed to carry these loads in addition to the portion of the slab moments assigned to the beams.

13.7.7 Factored Moments

Moment distribution is probably the most convenient hand calculation method for analyzing partial frames involving several continuous spans with the far ends of upper and lower columns fixed. The mechanics of the method will not be described here except for a brief discussion of the following two points: (1) the use of the equivalent column concept to determine joint distribution factors; and (2) the proper procedure to distribute the equivalent column moment obtained in the frame analysis to the actual columns above and below the slab-beam joint. See Design Examples 21.1 and 21.2.

A frame joint with stiffness factors K shown for each member framing into the joint is illustrated in Fig. 21-9. Expressions are shown for the moment distribution factors DF at the joint using the equivalent column stiffness, K_{ec}. These DF factors are used directly in the moment distribution procedure.

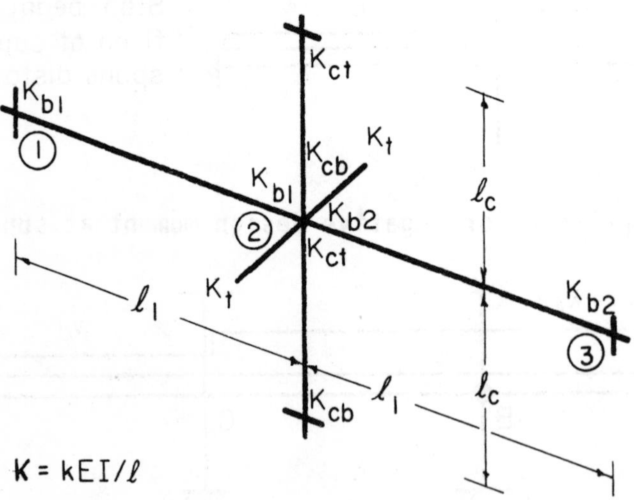

Fig. 21-9 - Moment Distribution Factors DF

Equivalent column stiffness,

$$K_{ec} = \Sigma K_c \times \Sigma K_t / (\Sigma K_c + \Sigma K_t)$$

$$= [(K_{cb} + K_{ct})(K_t + K_t)]/[(K_{cb} + K_{ct}) + (K_t + K_t)]$$

Slab-beam distribution factor,

$$\text{DF (span 2-1)} = K_{b1}/(K_{b1} + K_{b2} + K_{ec})$$

$$\text{DF (span 2-3)} = K_{b2}/(K_{b1} + K_{b2} + K_{ec})$$

Equivalent column distribution factor (unbalanced moment from slab-beam),

$$\text{DF} = K_{ec}/(K_{b1} + K_{b2} + K_{ec})$$

The unbalanced moment determined for the equivalent column in the moment distribution cycles is distributed to the actual columns above and below the slab-beam in proportion to the actual column stiffnesses at the joint. Referring to Fig. 21-9:

Portion of unbalanced moment to upper column,

$$= K_{cb}/(K_{cb} + K_{ct})$$

Portion of unbalanced moment to lower column,

$$= K_{ct}/(K_{cb} + K_{ct})$$

The "actual" columns are then designed for these moments.

Negative factored moments for design must be taken at faces of rectilinear support, but not at a distance greater than $0.175\ell_1$ from the center of support. This absolute value is a limit on long narrow supports in order to prevent undue reduction in design moment. The support member is defined as a column, capital, bracket or wall. Non-rectangular supports should be treated as square supports having the same area. Note that for slab systems with beams, the faces of beams are not considered a face-of-support location. Support conditions for locating the negative factored moment are illustrated in Fig. 21-10. Note the special requirements illustrated for exterior supports.

Should a designer choose to use the Equivalent Frame Method to analyze a slab system which meets the limitations of the Direct Design Method, then the factored moments may be reduced so that the total static factored moment (sum of average negative and positive moments) need not exceed M_o computed by Eq. (13-3). This permissible reduction is illustrated in Fig. 21-11.

Since the equivalent frame method of analysis is not an approximate method, the moment redistribution procedures allowed in Section 8.4 may be used. Excessive cracking can be anticipated if these provisions are imprudently applied. The burden of judgment is left to the designer to decide what, if any, redistribution is warranted.

The design moments may be distributed to the column strip and two half middle strips of the slab-beam in accordance with Sections 13.6.4, 13.6.5 and 13.6.6, provided the requirement of Section 13.6.1.6 is satisfied.

13.3.1.2 Lateral Load Analysis

While the equivalent frame defined in Section 13.7 is limited to gravity load analysis, it can be used for lateral load analysis, if modified to account for the loss of stiffness due to cracking in the slab-beams. Consideration of actual stiffness as affected by cracking is required for lateral load analysis because lateral displacement can significantly affect the moments in the columns, especially for tall unbraced frame buildings. Also, actual lateral displacement for a single story, or for total height of building is an important consideration for building stability and performance.

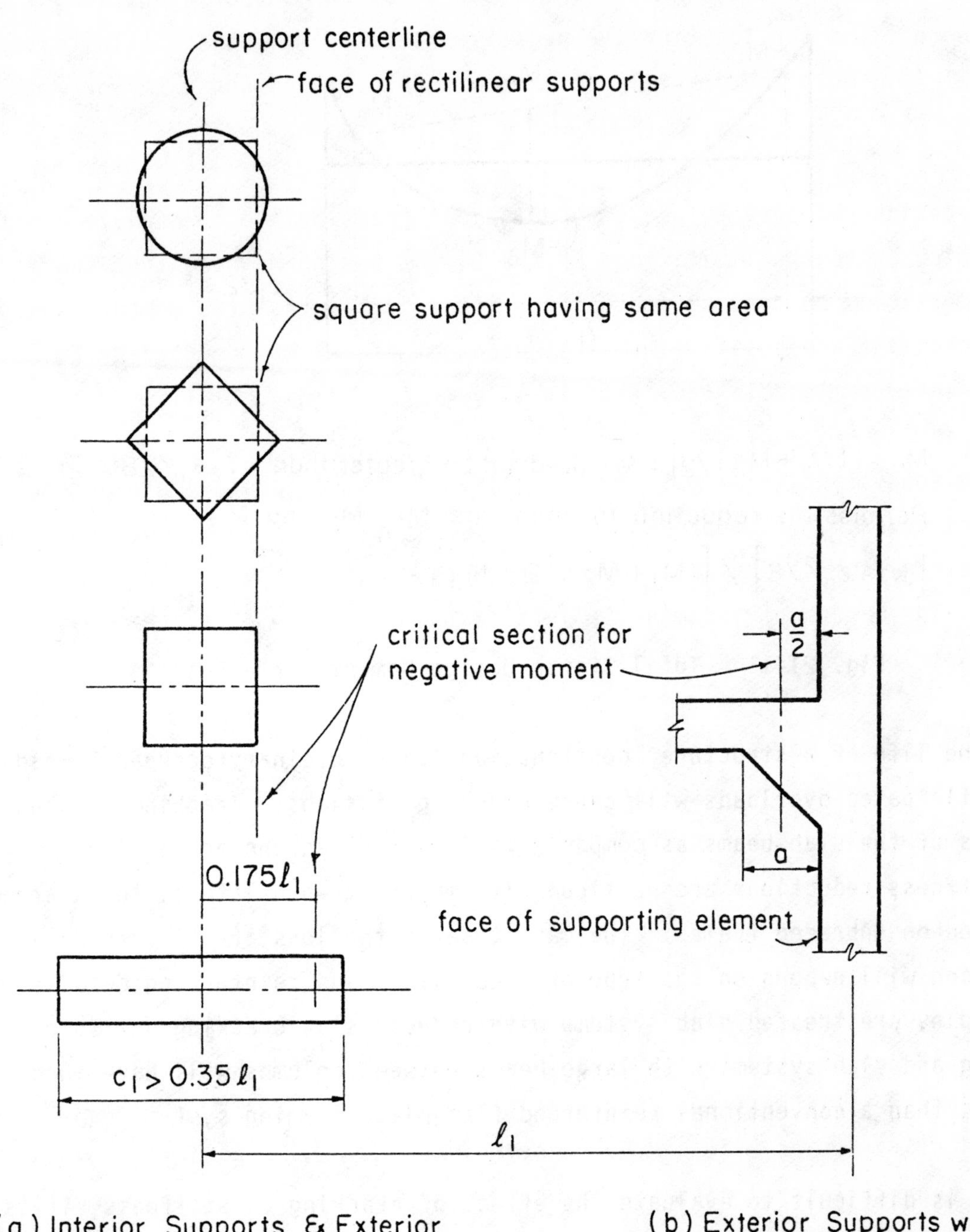

Fig. 21-10 - Critical Sections for Negative Design Moment

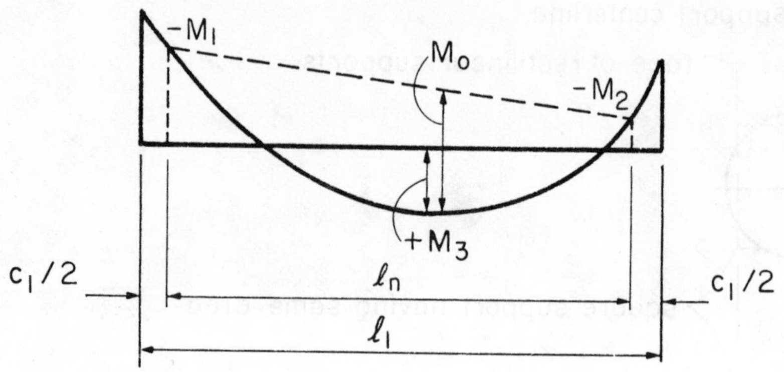

$M_o = \left[(M_1 + M_2)/2\right] + M_3$ need not be greater than $w\ell_2\ell_n^2/8$

Permissible reduction for moments M_1, M_2 and M_3 =

$\left[w\ell_2\ell_n^2/8\right] / \left[(M_1 + M_2)/2 + M_3\right]$

Fig. 21-11 - Total Static Design Moment for a Span

During the life of a structure, construction loads, ordinary occupancy loads, and unanticipated overloads will cause cracking of slabs. Cracking reduces stiffness of the slab-beams as compared with that of an uncracked floor. Such stiffness reductions are particularly important when lateral loads are considered on unbraced frames. The magnitude of the loss of stiffness due to cracking will depend on the type of slab system and reinforcing details. For example, prestressed slab systems with reduced slab cracking due to prestressing and slab systems with large beams between columns will have much less loss than a conventional reinforced flat-plate framing system.

Since it is difficult to evaluate the effect of cracking on stiffness, it is usually sufficient to use a lower bound for calculated stiffness. On the assumption of a fully cracked slab with minimum reinforcement at all locations, a stiffness for the slab-beam equal to one-fourth that based on the gross area of concrete ($K_{sb}/4$) should be reasonable. A detailed evaluation of the effect of cracking may also be made. Since slabs normally have more than minimum reinforcement and are not fully cracked, except under very unusual conditions, the one-fourth value should be expected to provide a safe lower bound for stiffness under lateral loads.

For both vertical and lateral load analyses, moments at critical sections of the slab-beams are distributed in accordance with Sections 13.6.4 (column strips) and 13.6.6 (middle strips).

Moments from an Equivalent Frame (or Direct Design) analysis for gravity loading may be combined with moments from a lateral load analysis (Section 13.3.1.3).

APPENDIX 21A

DESIGN AIDS FOR MOMENT DISTRIBUTION CONSTANTS

Table A1 - Moment Distribution Constants for Slab-Beam Members

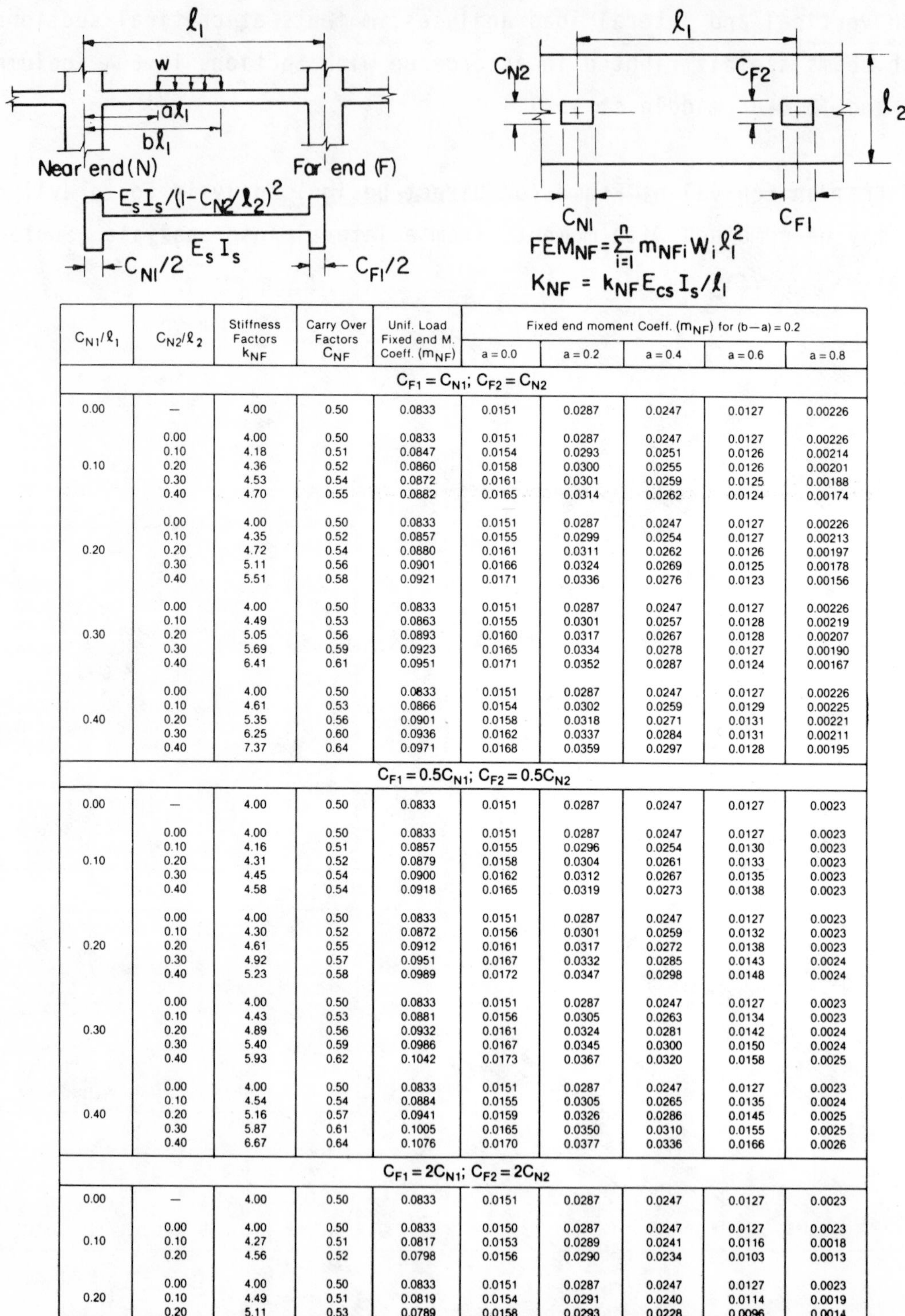

C_{N1}/ℓ_1	C_{N2}/ℓ_2	Stiffness Factors k_{NF}	Carry Over Factors C_{NF}	Unif. Load Fixed end M. Coeff. (m_{NF})	Fixed end moment Coeff. (m_{NF}) for (b−a) = 0.2				
					a = 0.0	a = 0.2	a = 0.4	a = 0.6	a = 0.8
$C_{F1} = C_{N1}$; $C_{F2} = C_{N2}$									
0.00	—	4.00	0.50	0.0833	0.0151	0.0287	0.0247	0.0127	0.00226
	0.00	4.00	0.50	0.0833	0.0151	0.0287	0.0247	0.0127	0.00226
	0.10	4.18	0.51	0.0847	0.0154	0.0293	0.0251	0.0126	0.00214
0.10	0.20	4.36	0.52	0.0860	0.0158	0.0300	0.0255	0.0126	0.00201
	0.30	4.53	0.54	0.0872	0.0161	0.0301	0.0259	0.0125	0.00188
	0.40	4.70	0.55	0.0882	0.0165	0.0314	0.0262	0.0124	0.00174
	0.00	4.00	0.50	0.0833	0.0151	0.0287	0.0247	0.0127	0.00226
	0.10	4.35	0.52	0.0857	0.0155	0.0299	0.0254	0.0127	0.00213
0.20	0.20	4.72	0.54	0.0880	0.0161	0.0311	0.0262	0.0126	0.00197
	0.30	5.11	0.56	0.0901	0.0166	0.0324	0.0269	0.0125	0.00178
	0.40	5.51	0.58	0.0921	0.0171	0.0336	0.0276	0.0123	0.00156
	0.00	4.00	0.50	0.0833	0.0151	0.0287	0.0247	0.0127	0.00226
	0.10	4.49	0.53	0.0863	0.0155	0.0301	0.0257	0.0128	0.00219
0.30	0.20	5.05	0.56	0.0893	0.0160	0.0317	0.0267	0.0128	0.00207
	0.30	5.69	0.59	0.0923	0.0165	0.0334	0.0278	0.0127	0.00190
	0.40	6.41	0.61	0.0951	0.0171	0.0352	0.0287	0.0124	0.00167
	0.00	4.00	0.50	0.0833	0.0151	0.0287	0.0247	0.0127	0.00226
	0.10	4.61	0.53	0.0866	0.0154	0.0302	0.0259	0.0129	0.00225
0.40	0.20	5.35	0.56	0.0901	0.0158	0.0318	0.0271	0.0131	0.00221
	0.30	6.25	0.60	0.0936	0.0162	0.0337	0.0284	0.0131	0.00211
	0.40	7.37	0.64	0.0971	0.0168	0.0359	0.0297	0.0128	0.00195
$C_{F1} = 0.5C_{N1}$; $C_{F2} = 0.5C_{N2}$									
0.00	—	4.00	0.50	0.0833	0.0151	0.0287	0.0247	0.0127	0.0023
	0.00	4.00	0.50	0.0833	0.0151	0.0287	0.0247	0.0127	0.0023
	0.10	4.16	0.51	0.0857	0.0155	0.0296	0.0254	0.0130	0.0023
0.10	0.20	4.31	0.52	0.0879	0.0158	0.0304	0.0261	0.0133	0.0023
	0.30	4.45	0.54	0.0900	0.0162	0.0312	0.0267	0.0135	0.0023
	0.40	4.58	0.54	0.0918	0.0165	0.0319	0.0273	0.0138	0.0023
	0.00	4.00	0.50	0.0833	0.0151	0.0287	0.0247	0.0127	0.0023
	0.10	4.30	0.52	0.0872	0.0156	0.0301	0.0259	0.0132	0.0023
0.20	0.20	4.61	0.55	0.0912	0.0161	0.0317	0.0272	0.0138	0.0023
	0.30	4.92	0.57	0.0951	0.0167	0.0332	0.0285	0.0143	0.0024
	0.40	5.23	0.58	0.0989	0.0172	0.0347	0.0298	0.0148	0.0024
	0.00	4.00	0.50	0.0833	0.0151	0.0287	0.0247	0.0127	0.0023
	0.10	4.43	0.53	0.0881	0.0156	0.0305	0.0263	0.0134	0.0023
0.30	0.20	4.89	0.56	0.0932	0.0161	0.0324	0.0281	0.0142	0.0024
	0.30	5.40	0.59	0.0986	0.0167	0.0345	0.0300	0.0150	0.0024
	0.40	5.93	0.62	0.1042	0.0173	0.0367	0.0320	0.0158	0.0025
	0.00	4.00	0.50	0.0833	0.0151	0.0287	0.0247	0.0127	0.0023
	0.10	4.54	0.54	0.0884	0.0155	0.0305	0.0265	0.0135	0.0024
0.40	0.20	5.16	0.57	0.0941	0.0159	0.0326	0.0286	0.0145	0.0025
	0.30	5.87	0.61	0.1005	0.0165	0.0350	0.0310	0.0155	0.0025
	0.40	6.67	0.64	0.1076	0.0170	0.0377	0.0336	0.0166	0.0026
$C_{F1} = 2C_{N1}$; $C_{F2} = 2C_{N2}$									
0.00	—	4.00	0.50	0.0833	0.0151	0.0287	0.0247	0.0127	0.0023
	0.00	4.00	0.50	0.0833	0.0150	0.0287	0.0247	0.0127	0.0023
0.10	0.10	4.27	0.51	0.0817	0.0153	0.0289	0.0241	0.0116	0.0018
	0.20	4.56	0.52	0.0798	0.0156	0.0290	0.0234	0.0103	0.0013
	0.00	4.00	0.50	0.0833	0.0151	0.0287	0.0247	0.0127	0.0023
0.20	0.10	4.49	0.51	0.0819	0.0154	0.0291	0.0240	0.0114	0.0019
	0.20	5.11	0.53	0.0789	0.0158	0.0293	0.0228	0.0096	0.0014

Table A2 - Moment Distribution Constants for Slab-Beam Members (Drop thickness = 0.25h)

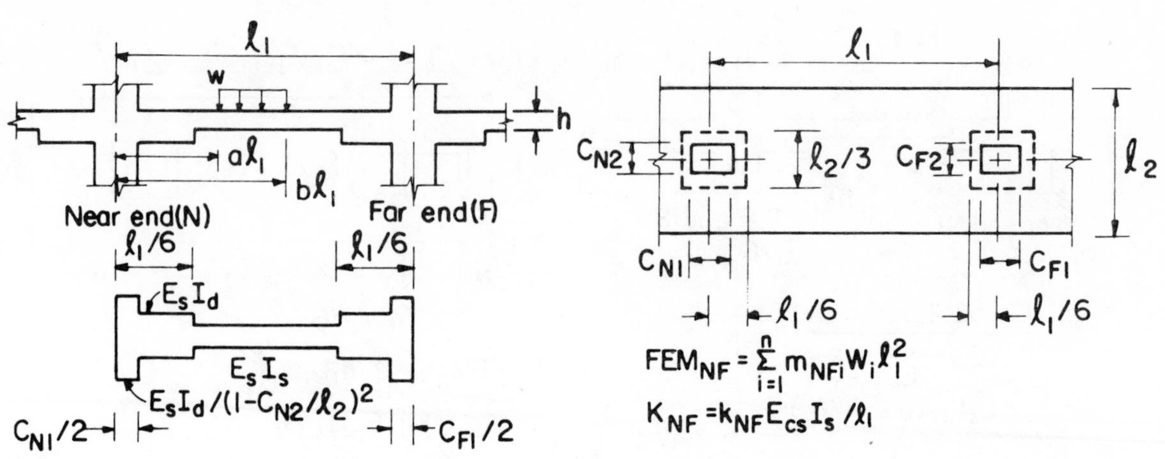

$$FEM_{NF} = \sum_{i=1}^{n} m_{NFi} w_i \ell_1^2$$

$$K_{NF} = k_{NF} E_{cs} I_s / \ell_1$$

C_{N1}/ℓ_1	C_{N2}/ℓ_2	Stiffness Factors k_{NF}	Carry Over Factors C_{NF}	Unif. Load Fixed end M. Coeff. (m_{NF})	Fixed end moment Coeff. (m_{NF}) for (b—a) = 0.2				
					a = 0.0	a = 0.2	a = 0.4	a = 0.6	a = 0.8
$C_{F1} = C_{N1}$; $C_{F2} = C_{N2}$									
0.00	—	4.79	0.54	0.0879	0.0157	0.0309	0.0263	0.0129	0.0022
0.10	0.00	4.79	0.54	0.0879	0.0157	0.0309	0.0263	0.0129	0.0022
	0.10	4.99	0.55	0.0890	0.0160	0.0316	0.0266	0.0128	0.0020
	0.20	5.18	0.56	0.0901	0.0163	0.0322	0.0270	0.0127	0.0019
	0.30	5.37	0.57	0.0911	0.0167	0.0328	0.0273	0.0126	0.0018
0.20	0.00	4.79	0.54	0.0879	0.0157	0.0309	0.0263	0.0129	0.0022
	0.10	5.17	0.56	0.0900	0.0161	0.0320	0.0269	0.0128	0.0020
	0.20	5.56	0.58	0.0918	0.0166	0.0332	0.0276	0.0126	0.0018
	0.30	5.96	0.60	0.0936	0.0171	0.0344	0.0282	0.0124	0.0016
0.30	0.00	4.79	0.54	0.0879	0.0157	0.0309	0.0263	0.0129	0.0022
	0.10	5.32	0.57	0.0905	0.0161	0.0323	0.0272	0.0128	0.0021
	0.20	5.90	0.59	0.0930	0.0166	0.0338	0.0281	0.0127	0.0019
	0.30	6.55	0.62	0.0955	0.0171	0.0354	0.0290	0.0124	0.0017
$C_{F1} = 0.5C_{N1}$; $C_{F2} = 0.5C_{N2}$									
0.00	—	4.79	0.54	0.0879	0.0157	0.0309	0.0263	0.0129	0.0022
0.10	0.00	4.79	0.54	0.0879	0.0157	0.0309	0.0263	0.0129	0.0022
	0.10	4.96	0.55	0.0900	0.0160	0.0317	0.0269	0.0131	0.0022
	0.20	5.12	0.56	0.0920	0.0164	0.0325	0.0276	0.0134	0.0022
0.20	0.00	4.79	0.54	0.0879	0.0157	0.0309	0.0263	0.0129	0.0022
	0.10	5.11	0.56	0.0914	0.0162	0.0323	0.0275	0.0133	0.0022
	0.20	5.43	0.58	0.0950	0.0167	0.0337	0.0286	0.0138	0.0022
$C_{F1} = 2C_{N1}$; $C_{F2} = 2C_{N2}$									
0.00	—	4.79	0.54	0.0879	0.0157	0.0309	0.0263	0.0129	0.0022
0.10	0.00	4.79	0.54	0.0879	0.0157	0.0309	0.0263	0.0129	0.0022
	0.10	5.10	0.55	0.0860	0.0159	0.0311	0.0256	0.0117	0.0017

Table A3 - Moment Distribution Constants for Slab-Beam Members (Drop thickness = 0.50h)

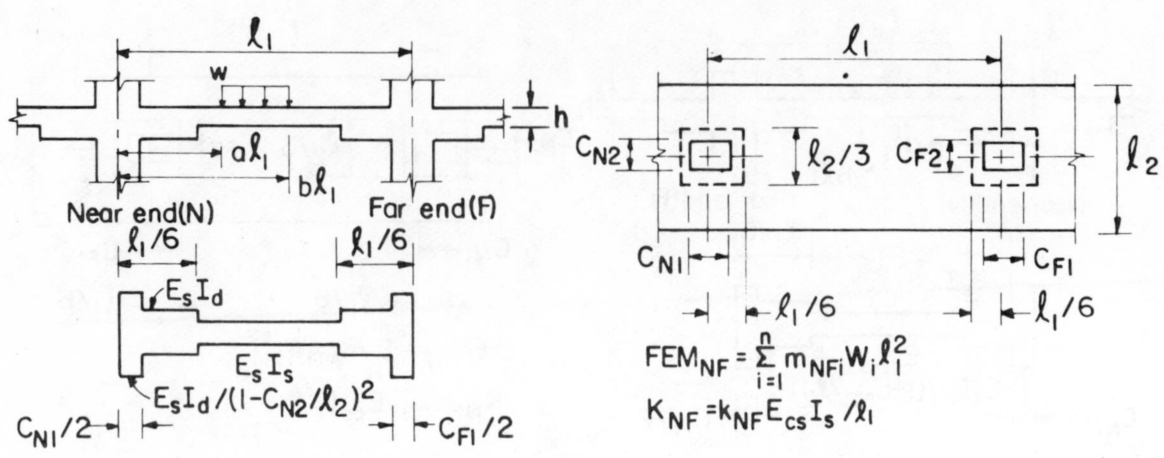

C_{N1}/ℓ_1	C_{N2}/ℓ_2	Stiffness Factors k_{NF}	Carry Over Factors C_{NF}	Unif. Load Fixed end M. Coeff. (m_{NF})	Fixed end moment Coeff. (m_{NF}) for (b—a) = 0.2				
					a = 0.0	a = 0.2	a = 0.4	a = 0.6	a = 0.8
$C_{F1} = C_{N1}$; $C_{F2} = C_{N2}$									
0.00	—	5.84	0.59	0.0926	0.0164	0.0335	0.0279	0.0128	0.0020
0.10	0.00	5.84	0.59	0.0926	0.0164	0.0335	0.0279	0.0128	0.0020
	0.10	6.04	0.60	0.0936	0.0167	0.0341	0.0282	0.0126	0.0018
	0.20	6.24	0.61	0.0940	0.0170	0.0347	0.0285	0.0125	0.0017
	0.30	6.43	0.61	0.0952	0.0173	0.0353	0.0287	0.0123	0.0016
0.20	0.00	5.84	0.59	0.0926	0.0164	0.0335	0.0279	0.0128	0.0020
	0.10	6.22	0.61	0.0942	0.0168	0.0346	0.0285	0.0126	0.0018
	0.20	6.62	0.62	0.0957	0.0172	0.0356	0.0290	0.0123	0.0016
	0.30	7.01	0.64	0.0971	0.0177	0.0366	0.0294	0.0120	0.0014
0.30	0.00	5.84	0.59	0.0926	0.0164	0.0335	0.0279	0.0128	0.0020
	0.10	6.37	0.61	0.0947	0.0168	0.0348	0.0287	0.0126	0.0018
	0.20	6.95	0.63	0.0967	0.0172	0.0362	0.0294	0.0123	0.0016
	0.30	7.57	0.65	0.0986	0.0177	0.0375	0.0300	0.0119	0.0014
$C_{F1} = 0.5C_{N1}$; $C_{F2} = 0.5C_{N2}$									
0.00	—	5.84	0.59	0.0926	0.0164	0.0335	0.0279	0.0128	0.0020
0.10	0.00	5.84	0.59	0.0926	0.0164	0.0335	0.0279	0.0128	0.0020
	0.10	6.00	0.60	0.0945	0.0167	0.0343	0.0285	0.0130	0.0020
	0.20	6.16	0.60	0.0962	0.0170	0.0350	0.0291	0.0132	0.0020
0.20	0.00	5.84	0.59	0.0926	0.0164	0.0335	0.0279	0.0128	0.0020
	0.10	6.15	0.60	0.0957	0.0169	0.0348	0.0290	0.0131	0.0020
	0.20	6.47	0.62	0.0987	0.0173	0.0360	0.0300	0.0134	0.0020
$C_{F1} = 2C_{N1}$; $C_{F2} = 2C_{N2}$									
0.00	—	5.84	0.59	0.0926	0.0164	0.0335	0.0279	0.0128	0.0020
0.10	0.00	5.84	0.59	0.0926	0.0164	0.0335	0.0279	0.0128	0.0020
	0.10	6.17	0.60	0.0907	0.0166	0.0337	0.0273	0.0116	0.0015

Table A4 - Moment Distribution Constants for Slab-Beam Members (Drop thickness = 0.75h)

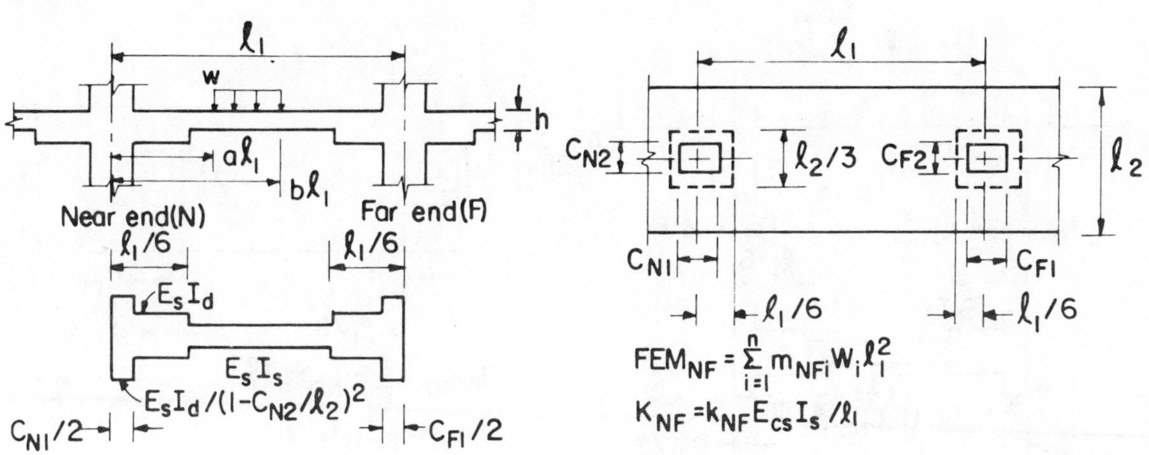

C_{N1}/ℓ_1	C_{N2}/ℓ_2	Stiffness Factors k_{NF}	Carry Over Factors C_{NF}	Unif. Load Fixed end M. Coeff. (m_{NF})	Fixed end moment Coeff. (m_{NF}) for (b—a) = 0.2				
					a = 0.0	a = 0.2	a = 0.4	a = 0.6	a = 0.8
$C_{F1} = C_{N1}$; $C_{F2} = C_{N2}$									
0.00	—	6.92	0.63	0.0965	0.0171	0.0360	0.0293	0.0124	0.0017
0.10	0.00	6.92	0.63	0.0965	0.0171	0.0360	0.0293	0.0124	0.0017
	0.10	7.12	0.64	0.0972	0.0174	0.0365	0.0295	0.0122	0.0016
	0.20	7.31	0.64	0.0978	0.0176	0.0370	0.0297	0.0120	0.0014
	0.30	7.48	0.65	0.0984	0.0179	0.0375	0.0299	0.0118	0.0013
0.20	0.00	6.92	0.63	0.0965	0.0171	0.0360	0.0293	0.0124	0.0017
	0.10	7.12	0.64	0.0977	0.0175	0.0369	0.0297	0.0121	0.0015
	0.20	7.31	0.65	0.0988	0.0178	0.0378	0.0301	0.0118	0.0013
	0.30	7.48	0.67	0.0999	0.0182	0.0386	0.0304	0.0115	0.0011
0.30	0.00	6.92	0.63	0.0965	0.0171	0.0360	0.0293	0.0124	0.0017
	0.10	7.29	0.65	0.0981	0.0175	0.0371	0.0299	0.0121	0.0015
	0.20	7.66	0.66	0.0996	0.0179	0.0383	0.0304	0.0117	0.0013
	0.30	8.02	0.68	0.1009	0.0182	0.0394	0.0309	0.0113	0.0011
$C_{F1} = 0.5C_{N1}$; $C_{F2} = 0.5C_{N2}$									
0.00	—	6.92	0.63	0.0965	0.0171	0.0360	0.0293	0.0124	0.0017
0.10	0.00	6.92	0.63	0.0965	0.0171	0.0360	0.0293	0.0124	0.0017
	0.10	7.08	0.64	0.0980	0.0174	0.0366	0.0298	0.0125	0.0017
	0.20	7.23	0.64	0.0993	0.0177	0.0372	0.0302	0.0126	0.0016
0.20	0.00	6.92	0.63	0.0965	0.0171	0.0360	0.0293	0.0124	0.0017
	0.10	7.21	0.64	0.0991	0.0175	0.0371	0.0302	0.0126	0.0017
	0.20	7.51	0.65	0.1014	0.0179	0.0381	0.0310	0.0128	0.0016
$C_{F1} = 2C_{N1}$; $C_{F2} = 2C_{N2}$									
0.00	—	6.92	0.63	0.0965	0.0171	0.0360	0.0293	0.0124	0.0017
0.10	0.00	6.92	0.63	0.0965	0.0171	0.0360	0.0293	0.0124	0.0017
	0.10	7.26	0.64	0.0946	0.0173	0.0361	0.0287	0.0112	0.0013

Table A5 - Moment Distribution Constants for Slab-Beam Members (Drop thickness = h)

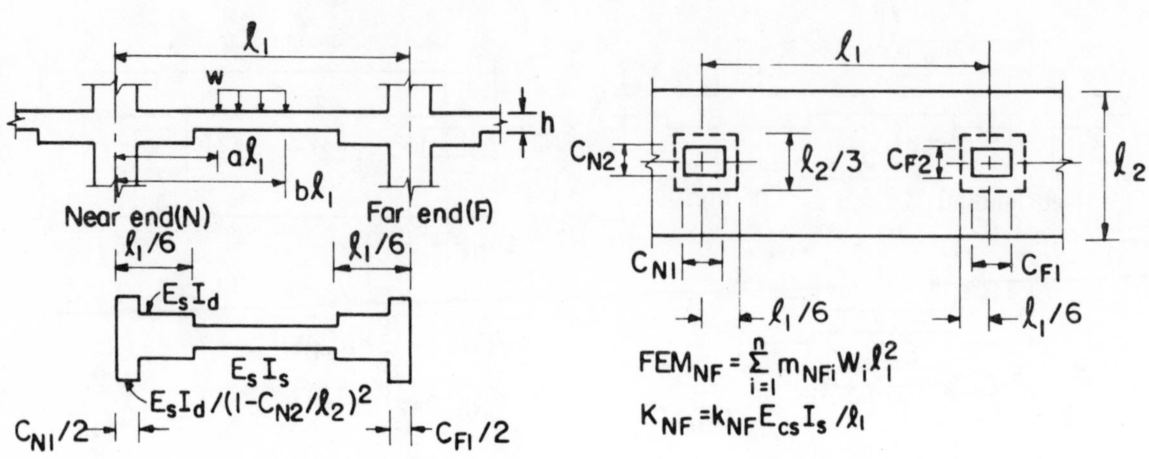

C_{N1}/ℓ_1	C_{N2}/ℓ_2	Stiffness Factors k_{NF}	Carry Over Factors C_{NF}	Unif. Load Fixed end M. Coeff. (m_{NF})	Fixed end moment Coeff. (m_{NF}) for (b—a) = 0.2				
					a = 0.0	a = 0.2	a = 0.4	a = 0.6	a = 0.8
$C_{F1} = C_{N1}$; $C_{F2} = C_{N2}$									
0.00	—	7.89	0.66	0.0993	0.0177	0.0380	0.0303	0.0118	0.0014
0.10	0.00	7.89	0.66	0.0993	0.0177	0.0380	0.0303	0.0118	0.0014
	0.10	8.07	0.66	0.0998	0.0180	0.0385	0.0305	0.0116	0.0013
	0.20	8.24	0.67	0.1003	0.0182	0.0389	0.0306	0.0115	0.0012
	0.30	8.40	0.67	0.1007	0.0183	0.0393	0.0307	0.0113	0.0011
0.20	0.00	7.89	0.66	0.0993	0.0177	0.0380	0.0303	0.0118	0.0014
	0.10	8.22	0.67	0.1002	0.0180	0.0388	0.0306	0.0115	0.0012
	0.20	8.55	0.68	0.1010	0.0183	0.0395	0.0309	0.0112	0.0011
	0.30	9.87	0.69	0.1018	0.0186	0.0402	0.0311	0.0109	0.0009
0.30	0.00	7.89	0.66	0.0993	0.0177	0.0380	0.0303	0.0118	0.0014
	0.10	8.35	0.67	0.1005	0.0181	0.0390	0.0307	0.0115	0.0012
	0.20	8.82	0.68	0.1016	0.0184	0.0399	0.0311	0.0111	0.0011
	0.30	9.28	0.70	0.1026	0.0187	0.0409	0.0314	0.0107	0.0009
$C_{F1} = 0.5C_{N1}$; $C_{F2} = 0.5C_{N2}$									
0.00	—	7.89	0.66	0.0993	0.0177	0.0380	0.0303	0.0118	0.0014
0.10	0.00	7.89	0.66	0.0993	0.0177	0.0380	0.0303	0.0118	0.0014
	0.10	8.03	0.66	0.1006	0.0180	0.0386	0.0307	0.0119	0.0014
	0.20	8.16	0.67	0.1016	0.0182	0.0390	0.0310	0.0120	0.0014
0.20	0.00	7.89	0.66	0.0993	0.0177	0.0380	0.0303	0.0118	0.0014
	0.10	8.15	0.67	0.1014	0.0181	0.0389	0.0310	0.0120	0.0014
	0.20	8.41	0.68	0.1032	0.0184	0.0398	0.0316	0.0121	0.0013
$C_{F1} = 2C_{N1}$; $C_{F2} = 0.5C_{N2}$									
0.00	—	7.89	0.66	0.0993	0.0177	0.0380	0.0303	0.0118	0.0014
0.10	0.00	7.79	0.66	0.0993	0.0177	0.0380	0.0303	0.0118	0.0014
	0.10	8.20	0.67	0.0981	0.0179	0.0382	0.0297	0.0113	0.0010

Table A6 - Moment Distribution Constants for Slab-Beam Members (Column dimensions assumed equal at rear end and far end - $c_{F1} = c_{N1}$, $c_{F2} = c_{N2}$)

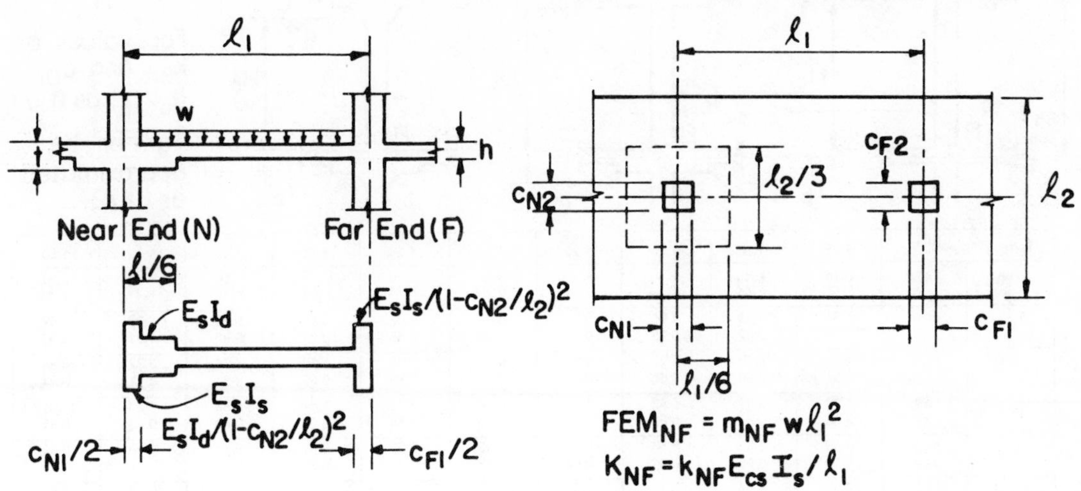

C_1/ℓ_1	C_2/ℓ_2	t = 1.5h						t = 2h					
		k_{NF}	C_{NF}	m_{NF}	k_{FN}	C_{FN}	m_{FN}	k_{NF}	C_{NF}	m_{NF}	k_{FN}	C_{FN}	m_{FN}
0.00	—	5.39	0.49	0.1023	4.26	0.60	0.0749	6.63	0.49	0.1190	4.49	0.65	0.0676
0.10	0.00	5.39	0.49	0.1023	4.26	0.60	0.0749	6.63	0.49	0.1190	4.49	0.65	0.0676
	0.10	5.65	0.52	0.1012	4.65	0.60	0.0794	7.03	0.54	0.1145	5.19	0.66	0.0757
	0.20	5.86	0.54	0.1012	4.91	0.61	0.0818	7.22	0.56	0.1140	5.43	0.67	0.0778
	0.30	6.05	0.55	0.1025	5.10	0.62	0.0838	7.36	0.56	0.1142	5.57	0.67	0.0786
0.20	0.00	5.39	0.49	0.1023	4.26	0.60	0.0749	6.63	0.49	0.1190	4.49	0.65	0.0676
	0.10	5.88	0.54	0.1006	5.04	0.61	0.0826	7.41	0.58	0.1111	5.96	0.66	0.0823
	0.20	6.33	0.58	0.1003	5.63	0.62	0.0874	7.85	0.61	0.1094	6.57	0.67	0.0872
	0.30	6.75	0.60	0.1008	6.10	0.64	0.0903	8.18	0.63	0.1093	6.94	0.68	0.0892
0.30	0.00	5.39	0.49	0.1023	4.26	0.60	0.075	6.63	0.49	0.1190	4.49	0.65	0.0676
	0.10	6.08	0.56	0.1003	5.40	0.61	0.085	7.76	0.62	0.1087	6.77	0.67	0.0873
	0.20	6.78	0.61	0.0996	6.38	0.63	0.092	8.49	0.66	0.1055	7.91	0.68	0.0952
	0.30	7.48	0.64	0.0997	7.25	0.65	0.096	9.06	0.68	0.1047	8.66	0.69	0.0991

Table A7 - Stiffness and Carry-Over Factors for Columns

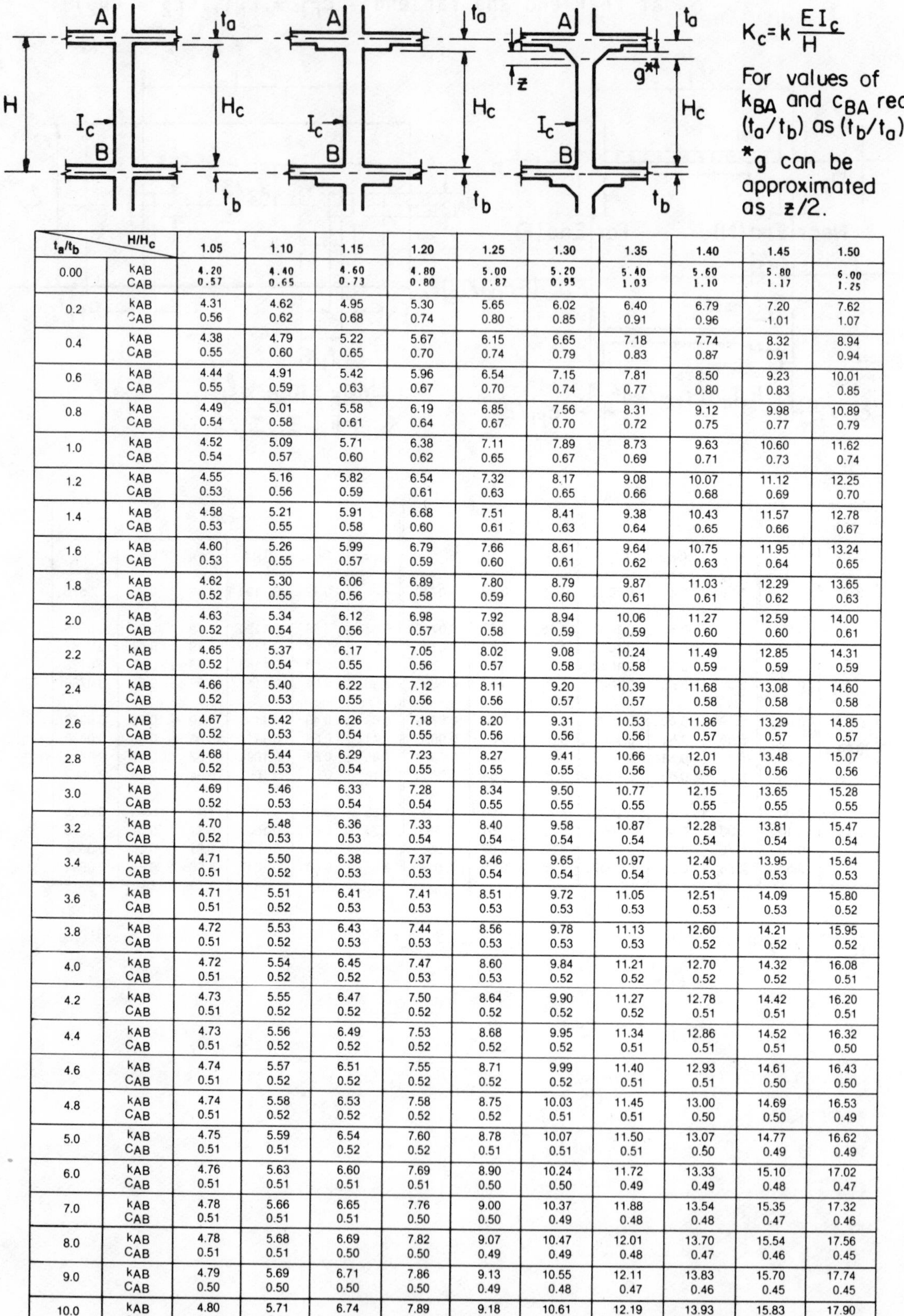

t_a/t_b	H/H_c	1.05	1.10	1.15	1.20	1.25	1.30	1.35	1.40	1.45	1.50
0.00	k_{AB}	4.20	4.40	4.60	4.80	5.00	5.20	5.40	5.60	5.80	6.00
	C_{AB}	0.57	0.65	0.73	0.80	0.87	0.95	1.03	1.10	1.17	1.25
0.2	k_{AB}	4.31	4.62	4.95	5.30	5.65	6.02	6.40	6.79	7.20	7.62
	C_{AB}	0.56	0.62	0.68	0.74	0.80	0.85	0.91	0.96	1.01	1.07
0.4	k_{AB}	4.38	4.79	5.22	5.67	6.15	6.65	7.18	7.74	8.32	8.94
	C_{AB}	0.55	0.60	0.65	0.70	0.74	0.79	0.83	0.87	0.91	0.94
0.6	k_{AB}	4.44	4.91	5.42	5.96	6.54	7.15	7.81	8.50	9.23	10.01
	C_{AB}	0.55	0.59	0.63	0.67	0.70	0.74	0.77	0.80	0.83	0.85
0.8	k_{AB}	4.49	5.01	5.58	6.19	6.85	7.56	8.31	9.12	9.98	10.89
	C_{AB}	0.54	0.58	0.61	0.64	0.67	0.70	0.72	0.75	0.77	0.79
1.0	k_{AB}	4.52	5.09	5.71	6.38	7.11	7.89	8.73	9.63	10.60	11.62
	C_{AB}	0.54	0.57	0.60	0.62	0.65	0.67	0.69	0.71	0.73	0.74
1.2	k_{AB}	4.55	5.16	5.82	6.54	7.32	8.17	9.08	10.07	11.12	12.25
	C_{AB}	0.53	0.56	0.59	0.61	0.63	0.65	0.66	0.68	0.69	0.70
1.4	k_{AB}	4.58	5.21	5.91	6.68	7.51	8.41	9.38	10.43	11.57	12.78
	C_{AB}	0.53	0.55	0.58	0.60	0.61	0.63	0.64	0.65	0.66	0.67
1.6	k_{AB}	4.60	5.26	5.99	6.79	7.66	8.61	9.64	10.75	11.95	13.24
	C_{AB}	0.53	0.55	0.57	0.59	0.60	0.61	0.62	0.63	0.64	0.65
1.8	k_{AB}	4.62	5.30	6.06	6.89	7.80	8.79	9.87	11.03	12.29	13.65
	C_{AB}	0.52	0.55	0.56	0.58	0.59	0.60	0.61	0.61	0.62	0.63
2.0	k_{AB}	4.63	5.34	6.12	6.98	7.92	8.94	10.06	11.27	12.59	14.00
	C_{AB}	0.52	0.54	0.56	0.57	0.58	0.59	0.59	0.60	0.60	0.61
2.2	k_{AB}	4.65	5.37	6.17	7.05	8.02	9.08	10.24	11.49	12.85	14.31
	C_{AB}	0.52	0.54	0.55	0.56	0.57	0.58	0.58	0.59	0.59	0.59
2.4	k_{AB}	4.66	5.40	6.22	7.12	8.11	9.20	10.39	11.68	13.08	14.60
	C_{AB}	0.52	0.53	0.55	0.56	0.56	0.57	0.57	0.58	0.58	0.58
2.6	k_{AB}	4.67	5.42	6.26	7.18	8.20	9.31	10.53	11.86	13.29	14.85
	C_{AB}	0.52	0.53	0.54	0.55	0.56	0.56	0.56	0.57	0.57	0.57
2.8	k_{AB}	4.68	5.44	6.29	7.23	8.27	9.41	10.66	12.01	13.48	15.07
	C_{AB}	0.52	0.53	0.54	0.55	0.55	0.55	0.56	0.56	0.56	0.56
3.0	k_{AB}	4.69	5.46	6.33	7.28	8.34	9.50	10.77	12.15	13.65	15.28
	C_{AB}	0.52	0.53	0.54	0.54	0.55	0.55	0.55	0.55	0.55	0.55
3.2	k_{AB}	4.70	5.48	6.36	7.33	8.40	9.58	10.87	12.28	13.81	15.47
	C_{AB}	0.52	0.53	0.53	0.54	0.54	0.54	0.54	0.54	0.54	0.54
3.4	k_{AB}	4.71	5.50	6.38	7.37	8.46	9.65	10.97	12.40	13.95	15.64
	C_{AB}	0.51	0.52	0.53	0.53	0.54	0.54	0.54	0.53	0.53	0.53
3.6	k_{AB}	4.71	5.51	6.41	7.41	8.51	9.72	11.05	12.51	14.09	15.80
	C_{AB}	0.51	0.52	0.53	0.53	0.53	0.53	0.53	0.53	0.53	0.52
3.8	k_{AB}	4.72	5.53	6.43	7.44	8.56	9.78	11.13	12.60	14.21	15.95
	C_{AB}	0.51	0.52	0.53	0.53	0.53	0.53	0.53	0.52	0.52	0.52
4.0	k_{AB}	4.72	5.54	6.45	7.47	8.60	9.84	11.21	12.70	14.32	16.08
	C_{AB}	0.51	0.52	0.52	0.53	0.53	0.52	0.52	0.52	0.52	0.51
4.2	k_{AB}	4.73	5.55	6.47	7.50	8.64	9.90	11.27	12.78	14.42	16.20
	C_{AB}	0.51	0.52	0.52	0.52	0.52	0.52	0.52	0.51	0.51	0.51
4.4	k_{AB}	4.73	5.56	6.49	7.53	8.68	9.95	11.34	12.86	14.52	16.32
	C_{AB}	0.51	0.52	0.52	0.52	0.52	0.52	0.51	0.51	0.51	0.50
4.6	k_{AB}	4.74	5.57	6.51	7.55	8.71	9.99	11.40	12.93	14.61	16.43
	C_{AB}	0.51	0.52	0.52	0.52	0.52	0.52	0.51	0.51	0.50	0.50
4.8	k_{AB}	4.74	5.58	6.53	7.58	8.75	10.03	11.45	13.00	14.69	16.53
	C_{AB}	0.51	0.52	0.52	0.52	0.52	0.51	0.51	0.50	0.50	0.49
5.0	k_{AB}	4.75	5.59	6.54	7.60	8.78	10.07	11.50	13.07	14.77	16.62
	C_{AB}	0.51	0.51	0.52	0.52	0.51	0.51	0.51	0.50	0.49	0.49
6.0	k_{AB}	4.76	5.63	6.60	7.69	8.90	10.24	11.72	13.33	15.10	17.02
	C_{AB}	0.51	0.51	0.51	0.51	0.50	0.50	0.49	0.49	0.48	0.47
7.0	k_{AB}	4.78	5.66	6.65	7.76	9.00	10.37	11.88	13.54	15.35	17.32
	C_{AB}	0.51	0.51	0.51	0.50	0.50	0.49	0.48	0.48	0.47	0.46
8.0	k_{AB}	4.78	5.68	6.69	7.82	9.07	10.47	12.01	13.70	15.54	17.56
	C_{AB}	0.51	0.51	0.50	0.50	0.49	0.49	0.48	0.47	0.46	0.45
9.0	k_{AB}	4.79	5.69	6.71	7.86	9.13	10.55	12.11	13.83	15.70	17.74
	C_{AB}	0.50	0.50	0.50	0.50	0.49	0.48	0.47	0.46	0.45	0.45
10.0	k_{AB}	4.80	5.71	6.74	7.89	9.18	10.61	12.19	13.93	15.83	17.90
	C_{AB}	0.50	0.50	0.50	0.49	0.48	0.48	0.47	0.46	0.45	0.44

EXAMPLE 21.1 - Two-Way Slab Without Beams Analyzed By Equivalent Frame Method

Using the equivalent frame method, determine design moments for the slab system in the transverse direction for an intermediate floor.

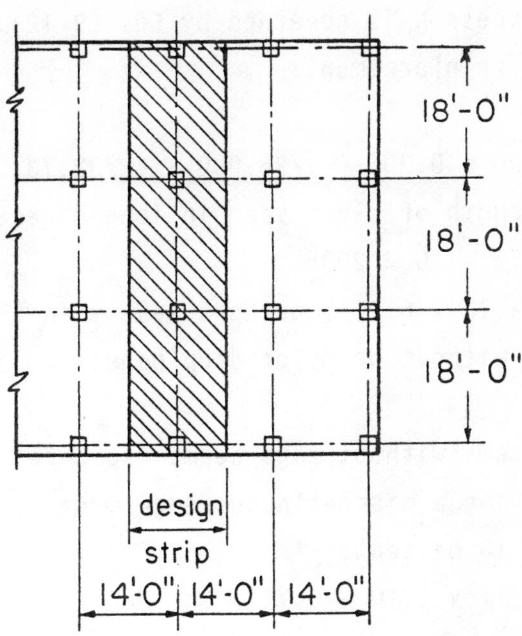

Story height = 9 ft

Columns = 16 in. x 16 in.

Lateral loads to be resisted by shear walls

No spandrel beams

Partition weight = 20 psf

Service live load = 40 psf

f'_c = 3,000 psi (for slab)

f'_c = 5,000 psi (for columns)

f_y = 60,000 psi

EXAMPLE 21.1 - Continued

Calculations and Discussion	Code Reference

1. Preliminary design for slab thickness h.

(a) Control of deflections:

For slab systems without beams, the minimum overall thickness h is governed by Eq. (9-12). For Grade 60 reinforcement: — 9.5.3

$h = \ell_n (800 + 0.005 f_y)/36{,}000 = \ell_n/32.73$ — Eq. (9-13)

ℓ_n = length of clear span in long direction

= 216 - 16 = 200"

h = 200/32.73 = 6.11", but not less than 5" for slabs without beams or drop panels

For slab systems without edge beams (spandrels), the panel having a discontinuous edge must be increased by 10 percent: — 9.5.3.3

h = 6.11 x 1.10 = 6.72"

<u>Try 7" slab for all panels</u> (weight = 87.5 psf)

(b) Shear strength of slab:

Use average effective depth d ≃ 5.75"
(3/4" Cover + #4 bar)

Factored dead load, w_d = (87.5 + 20) 1.4 = 150.5 — 9.2.1
Factored live load, w_ℓ = 40 x 1.7 = 68.0
Total factored load = 218.5 psf

Wide beam action: Wide beam action is investigated for a 12 in. wide strip taken at d distance from the face of support in the long direction. (Note Section 11.11.1.1 and Fig. 21-12.) — 11.11.1.1

EXAMPLE 21.1 - Continued

Calculations and Discussion	Code Reference
$V_u = 0.2185 \times 7.854 = 1.72$ kips	
$V_c = 2\sqrt{f'_c}\, b_w d$	Eq. (11-3)
$\varphi V_c = 0.85 \times 2\sqrt{3{,}000} \times 12 \times 5.75/1000 = 6.43$ kips	
$V_u < \varphi V_c$ OK	
Two-way action: Since there are no shear forces at the centerline of adjacent panels (see Fig. 21-12), the shear strength at d/2 distance around the support reaction is computed as follows:	11.11.1.2
$V_u = 0.2185\,(18 \times 14 - 1.81^2) = 54.3$ kips	
$V_c = 4\sqrt{f'_c}\, b_o d$ (for sq. column)	Eq. (11-36)
$= 4\sqrt{3000} \times 4 \times 21.75 \times 5.75/1000 = 109.6$ kips	
$\varphi V_c = 0.85 \times 109.6 = 93.2$ kips	
$V_u < \varphi V_c$ OK	
Preliminary design indicates that a 7 in. overall slab thickness is adequate for control of deflections and shear strength.	

Calculations and Discussion	Code Reference

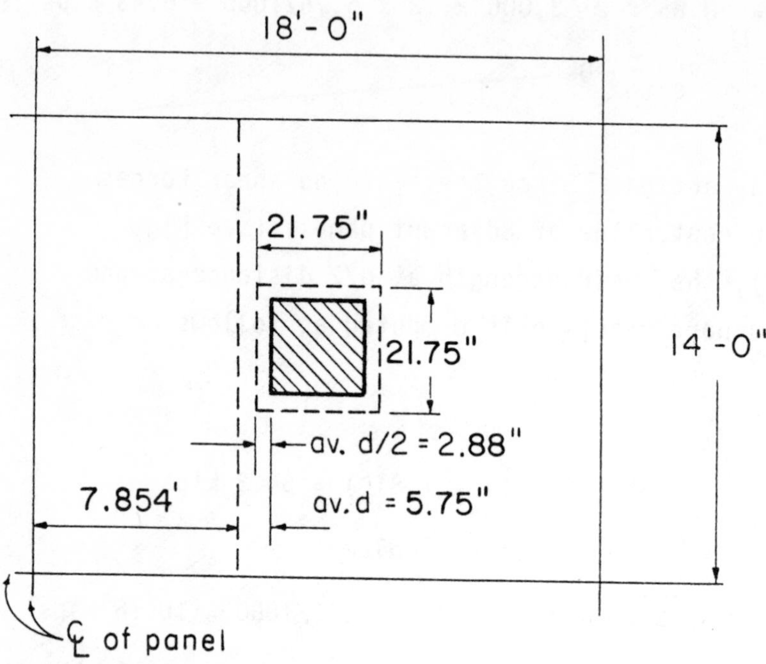

Fig. 21-12

2. Frame members of equivalent frame.

Determine moment distribution constants and fixed-end moments for the equivalent frame members. The moment distribution procedure will be used to analyze the partial frame. Stiffness factors k, carry over factors COF, and fixed-end moment factors FEM, for the slab-beams and column members are determined by conjugate beam procedures. These calculations are not shown here.

EXAMPLE 21.1 - Continued

Calculations and Discussion	Code Reference

(a) Slab-beam, flexural stiffness at both ends K_{sb}:

K_{sb} = $4.13\ E_{cs}\ I_s/\ell_1$
= $4.13 \times 3.12 \times 10^6 \times 4802/216 = 286 \times 10^6$ in. lbs

where I_s = $\ell_2 h^3/12 = 168(7)^3/12 = 4802$ in.^{4}
E_{cs} = $57{,}000\sqrt{3{,}000} = 3.12 \times 10^6$ psi — 8.5.1
ℓ_1 = 18'-0" = 216 in.

Carry-over factor COF = 0.509

Fixed-end moment FEM = $0.0843\ w\ell_2\ell_1^2$

(b) Column members, flexural stiffness at both ends K_c:

K_c = $4.74\ E_{cc} I_c/\ell_c$
= $4.74 \times 4.03 \times 10^6 \times 5{,}461/108 = 966 \times 10^6$ in. lbs

where I_c = $c^4/12 = (16)^4/12 = 5461$ in.^{4}
E_{cc} = $57{,}000\sqrt{5{,}000} = 4.03 \times 10^6$ psi — 8.5.1
ℓ_c = 9'-0" = 108 in.

(c) Torsional members, torsional stiffness K_t:

K_t = $9\ E_{cs}\ C/[\ell_2\ (1 - c_2/\ell_2)^3]$ — Eq. (13-6)
= $9 \times 3.12 \times 10^6 \times 1{,}325/[168\ (0.905)^3]$
= 299×10^6 in. lbs.

where C = $\Sigma(1 - 0.63\ x/y)(x^3y/3)$ — Eq. (13-7)
= $(1 - 0.63 \times 7/16)\ (7^3 \times 16/3) = 1{,}325$ in.^{4}
c_2 = 16", and ℓ_2 = 14'-0" = 168 in.

EXAMPLE 21.1 - Continued

Calculations and Discussion	Code Reference

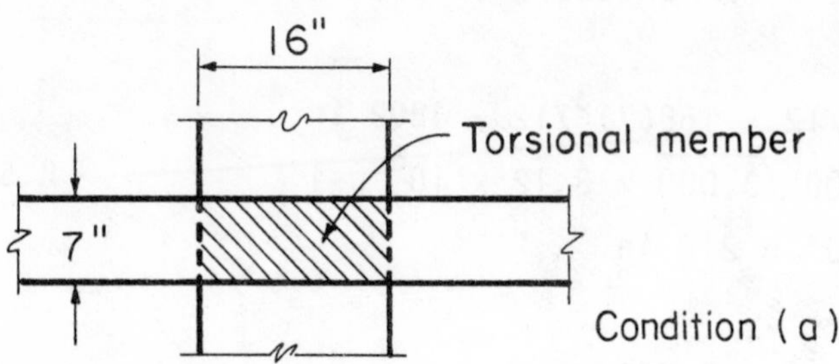

(d) Equivalent column stiffness K_{ec}:

$$K_{ec} = \Sigma K_c \times \Sigma K_t/(\Sigma K_c + \Sigma K_t)$$

$$= (2 \times 966)(2 \times 299)/\ (2 \times 966) + (2 \times 299)$$

$$= 457 \times 10^6 \text{ in. lb}$$

where ΣK_t is for two torsional members, one each side of column, and ΣK_c is for the upper and lower columns at the slab-beam joint of an intermediate floor.

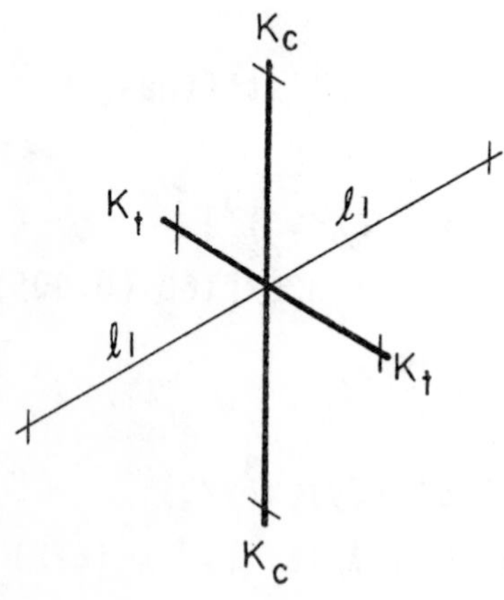

EXAMPLE 21.1 - Continued

Calculations and Discussion	Code Reference

(e) Slab-beam joint distribution factors DF:

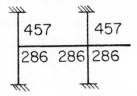

At exterior joint:

DF = 286/(286 + 457) = 0.385

At interior joint:

DF = 286/(286 + 286 + 457) = 0.278

COF for slab-beam = 0.509

3. Partial frame analysis of equivalent frame:

Determine maximum negative and positive moments for the slab-beams using the moment distribution method. Since the service live load does not exceed three-quarters of the service dead load, design moments are assumed to occur at all critical sections with full factored live load on all spans. 13.7.6.2

L/D = 40/(87.5 + 20) = 0.37 < 3/4

(a) Factored load and fixed-end moments:

Factored dead load per unit area, w_d = 1.4 (87.5 + 20)
= 150.5 psf

Factored live load per unit area, w_ℓ = 1.7 (40) = 68 psf

Factored load per unit area, $w_d + w_\ell$ = 218.5 psf

FEM's for slab-beams = $0.0843\, w\ell_2\ell_1^2$

FEM due to $w_d + w_\ell$ = $0.0843\,(0.2185 \times 14)18^2$ = 83.6 kips

Calculations and Discussion	Code Reference

(b) Moment distribution is shown in Table 21-1. Counter clockwise rotational moments acting on the member ends are taken as positive. Positive span moments are determined from the following equation:

$$M(\text{midspan}) = M_s - \frac{1}{2}(M_L + M_R)$$

where M_S is the moment at midspan for a simple beam.

When the end moments are not equal, the maximum moment in the span does not occur at midspan, but its value is close to that at midspan.

Positive moment in span 1-2:

$$+M = (0.2185 \times 14)\, 18^2/8 - \frac{1}{2}(53.0 + 95.1)$$
$$= 49.8 \text{ ft kips}$$

Positive moment in span 2-3:

$$+M = (0.2185 \times 14)\, 18^2/8 - \frac{1}{2}(86.4 + 86.4)$$
$$= 37.5 \text{ ft kips}$$

4. Design moments.

Positive and negative factored moments for the slab system in the transverse direction are shown in Fig. 21-13. The negative design moments are taken at the face of rectilinear supports but not greater than a distance of $0.175\ \ell_1$ from the center of supports. — 13.7.7.1

0.67 ft < 0.175 x 18 (Use face of support location)

EXAMPLE 21.1 - Continued

Calculations and Discussion	Code Reference

TABLE 21-1 - MOMENT DISTRIBUTION FOR PARTIAL FRAME

(transverse direction)

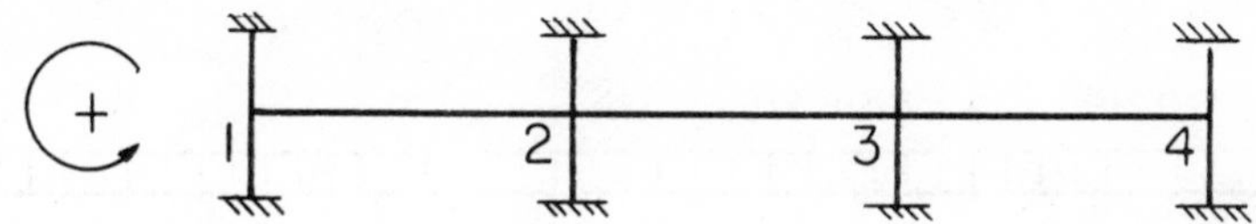

Joint	1	2		3		4
Member	1-2	2-1	2-3	3-2	3-4	4-3
DF	0.385	0.278	0.278	0.278	0.278	0.385
COF	0.509	0.509	0.509	0.509	0.509	0.509
FEM	+83.6	-83.6	+83.6	-83.6	+83.6	-83.6
COM*	0	-16.4	0	0	+16.4	0
"	+ 2.3	0	- 2.3	+ 2.3	0	- 2.3
"	+ 0.3	- 0.5	- 0.3	+ 0.3	+ 0.5	- 0.3
Σ	+86.2	-100.5	+81.0	-81.0	+100.5	-86.2
DM**	-33.2	+ 5.4	+ 5.4	- 5.4	- 5.4	+33.2
Neg. M	+53.0	- 95.1	+86.4	-86.4	+ 95.1	-53.0
***	+53.6	- 95.8	+87.0	-87.0	+ 95.8	-53.6
M @ ℄ of span	49.8 50.1***		37.5 37.8***		49.8 50.1***	

* Carry-over moment, COM, is the negative product of the distribution factor, carry-over factor and unbalanced joint moment carried to the opposite end of span.

** Distributed moment, DM, is the negative product of the distribution factor and the unbalanced joint moment.

*** Moments from computer program for the analysis and design of slab systems by the Portland Cement Association.

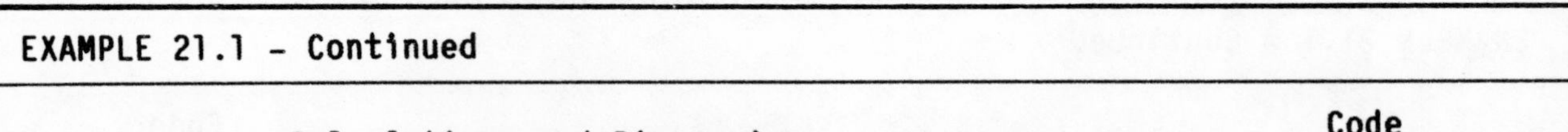

EXAMPLE 21.1 - Continued

Calculations and Discussion	Code Reference

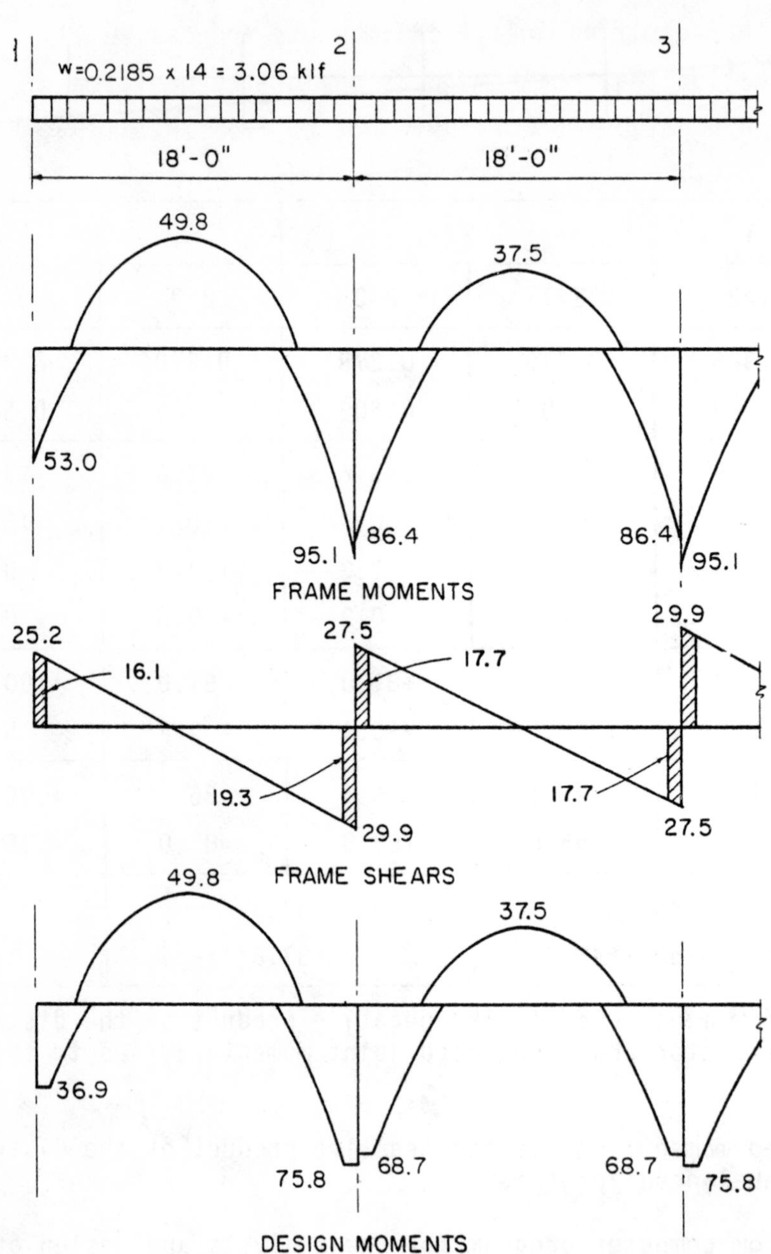

Fig. 21-13 - Positive and Negative Design Moments for Slab-Beam (all spans loaded with full factored live load)

EXAMPLE 21.1 - Continued

Calculations and Discussion	Code Reference

5. Total factored moment per span.

Slab systems within the limitations of Section 13.6.1 may have the resulting analytical moments reduced in such proportion that the numerical sum of the positive and average negative moments need not be greater than, (13.7.7.4)

$$M_o = w\ell_2\ell_n^2/8 = 0.2185 \times 14 \times (16.67)^2/8$$ Eq. (13-3)

$$= 106.3 \text{ ft kips}$$

End span: 49.8 + (36.9 + 75.8)/2 = 106.2 ft kips ≃ 106.3

Interior span: 37.5 + (68.7 + 68.7)/2 = 106.2 ft kips ≃ 106.3

It may be seen that the total design moments from the equivalent frame analysis are very close indeed to the static moment expression used with the Direct Design Method.

6. Distribution of design moments across slab-beam strip. (13.7.7.5)

The negative and positive factored moments at critical sections may be distributed to the column strip and two half middle strips of the slab-beam according to the proportions specified in Sections 13.6.4 and 13.6.6. The requirement of Section 13.6.1.6 does not apply for slab systems without beams, $\alpha = 0$. Distribution of factored moments at critical sections is summarized in Table 21-2.

EXAMPLE 21.1 - Continued

Calculations and Discussion	Code Reference

7. Column moments.

The unbalanced moment from the slab-beams at the supports of the equivalent frame are distributed to the actual columns above and below the slab-beam in proportion to the relative stiffness of the actual columns. Referring to Table 21-1, the unbalanced moments at joints 1 and 2 are:

Joint 1 = +53.0 ft kips
Joint 2 = -95.1 + 86.4 = -8.7 kips

The stiffness and carry-over factors of the actual columns and the distribution of the unbalanced moments to the exterior and interior columns are shown in Fig. 21-14. The design moments for the columns may be taken at the juncture of column and slab. Summarizing:

Design moment in exterior column = 25.2 ft
Design moment in interior column = 4.13 ft kips

TABLE 21-2 - DISTRIBUTION OF FACTORED MOMENTS

	Factored moments	Column strip		Two half middle strips**
		Percent*	Moment	
End span:				
Exterior negative	36.9	100	36.9	0.0
Positive	49.8	60	29.9	19.9
Interior negative	75.8	75	56.9	18.9
Interior span:				
Negative	68.7	75	51.5	17.2
Positive	37.5	60	22.5	15.0

* For slab systems without beams

** That portion of the factored moment not resisted by the column strip is assigned to the two half middle strips

EXAMPLE 21.1 - Continued

Calculations and Discussion	Code Reference

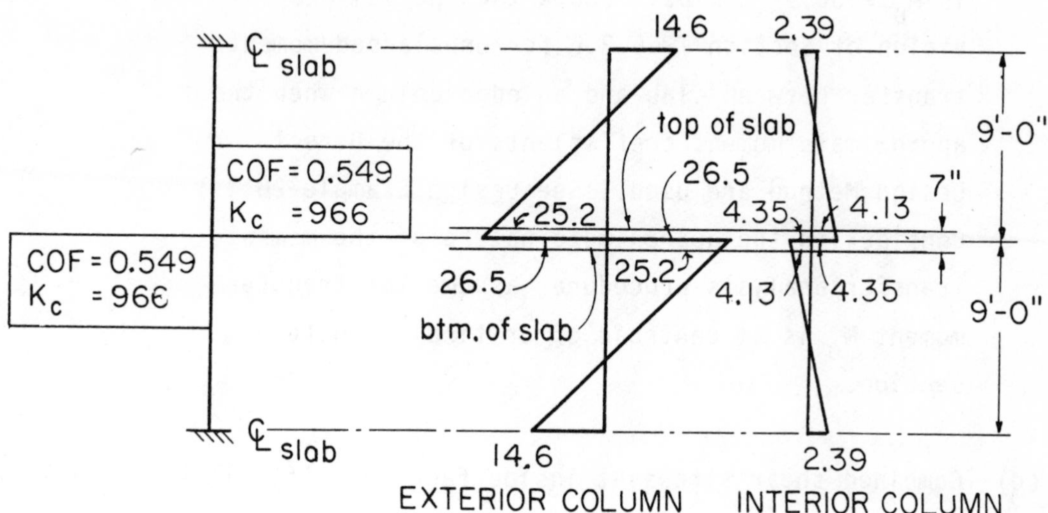

Fig. 21-14 - Column Moments (unbalanced moments from slab-beam)

8. Transfer of gravity load shear and moment at exterior column.

Check slab shear and flexural strength at edge column due to direct shear and unbalanced moment transfer — 11.12.2, 13.3.3

(a) Factored shear force transfer at exterior column:

$V_u = w_u \ell_1 \ell_2/2$

$= 0.2185 \times 14 \times 18/2 = 27.5$ kips

(b) Unbalanced moment transfer at exterior column:

When factored moments are determined by a more accurate method of frame analysis, considering actual member stiffnesses, such as the Equivalent Frame Analysis procedure, transfer moment is taken

EXAMPLE 21.1 - Continued

Calculations and Discussion	Code Reference

directly from the results of the frame analysis. Unbalanced moment at exterior column (Table 21-2) is M_u = 36.9 ft kips. (Note the special provision of Section 13.6.3.6 for unbalanced moment transfer between slab and an edge column when the approximate moment coefficients of the Direct Design Method are used. See Design Example 20.1.) Considering the approximate nature of the moment transfer analysis procedure, assume the transfer moment M_u is at centroid of critical transfer section.

(c) Combined shear stress at inside face of critical transfer section. For shear strength equations, see Part 17, page 17-6. — 11.12.2.3

$$v_u = V_u/A_c + \gamma_v M_u/(J/c)$$

$$= 27{,}500/342.2 + 0.4 \times 36.9 \times 12{,}000/2358$$

$$= 80.4 + 75.1 = 155.5 \text{ psi}$$

where fraction of unbalanced moment transfered by eccentricity of shear at slab-column connections γ_v = 0.40 (see Fig. 17-4). For section properties A_c and (J/c), see Design Example 20.1.

(d) Combined shear stress at outside face of critical transfer section. — 11.12.2.3

$$v_u = 27{,}500/342.2 - 0.4 \times 36.9 \times 12{,}000/1096$$

$$= 80.4 - 161.6 = 81.2 \text{ psi}$$

(e) Permissible shear stress — 11.12.2.4

EXAMPLE 21.1 - Continued

Calculations and Discussion	Code Reference

$\varphi V_n = \varphi 4\sqrt{f'_c} = 0.85 \times 4\sqrt{3000}$

$= 186.2$ psi > 155.5 OK

(f) Design for unbalanced moment transfer by flexure for both middle strip and column strip: 13.3.3

$A_{s(min)} = 0.0018bh = 0.0018 \times 84 \times 7 = 1.06$ in.^{2} 7.12.2.1(b)

where b is width of design strip = 14/2 = 7 ft = 84 in.

For #4 bars, total bars required = 1.06/0.20 = 5.3 bars

For $s_{max} = 2h = 2 \times 7 = 14$ in., total bars required = 84/14 = 6 bars 13.4.2

Check total reinforcement required for column strip negative moment M_u = 36.9 ft kips. Using Table 9-2, page 9-7:

$$\frac{M_u}{\varphi f'_c bd^2} = \frac{36.9 \times 12}{0.9 \times 3 \times 84 \times 5.75^2} = 0.0591$$

where $d \simeq 7.0 - 1.25 = 5.75$ in.

from Table 9-2; $\omega = 0.0613$

$A_s = \omega f'_c bd/f_y = 0.0613 \times 3 \times 84 \times 5.75/60 = 1.48$ in.^{2}

For #4 bars, total bars required = 1.48/0.20 = 7.4 bars

EXAMPLE 21.1 - Continued

Calculations and Discussion	Code Reference

Use 6 #4 bars @ ±14 in. spacing in middle strip and portion of column strip outside unbalanced moment transfer section c+2(1.5h)=16+2(1.5x7)=37 in. 13.3.3.2

Additional reinforcement required over column within effective slab width of 37 in. to resist fraction of unbalanced moment transfered by flexure is computed from Eq. (13-1). For square column, γ_f = 60% (See also Fig. 17-4).

$\gamma_f M_u$ = 0.60 (36.9) = 22.1 ft kips must be transferred within the effective slab width of 37 in. Try 2 additional bars over column. Check moment strength for 4 #4 bars within 37 in. width.

For 4 #4 bars: $A_s = 4\ (0.20) = 0.80\ \text{in.}^2$

$\omega = A_s f_y / f'_c bd = 0.80 \times 60/3 \times 37 \times 5.75 = 0.0752$

From Table 9-2; $M_n/f'_c bd^2 = 0.0719$

$M_n = 0.0719 \times 3 \times 37 \times 5.75^2/12 = 22.0$ ft kips

$\varphi M_n = 0.9(22.0) = 19.8 < 22.1$ NG

Try 3 additional bars. Moment strength for 5 #4 bars within 37 in. width:

$A_s = 5(0.20) = 1.00\ \text{in.}^2$

$\omega = 1.00 \times 60/3 \times 37 \times 5.75 = 0.0928$

From Table 9-2; $M_n/f'_c bd^2 = 0.0877$

$\varphi M_n = 0.90(0.0877 \times 3 \times 37 \times 5.75^2/12)$

$= 24.1$ ft kips > 22.1 OK

EXAMPLE 21.1 - Continued

Detail bars as shown below. Total bars in column strip = 6+3 = 9 bars > 7.4 required for total column strip negative moment. OK

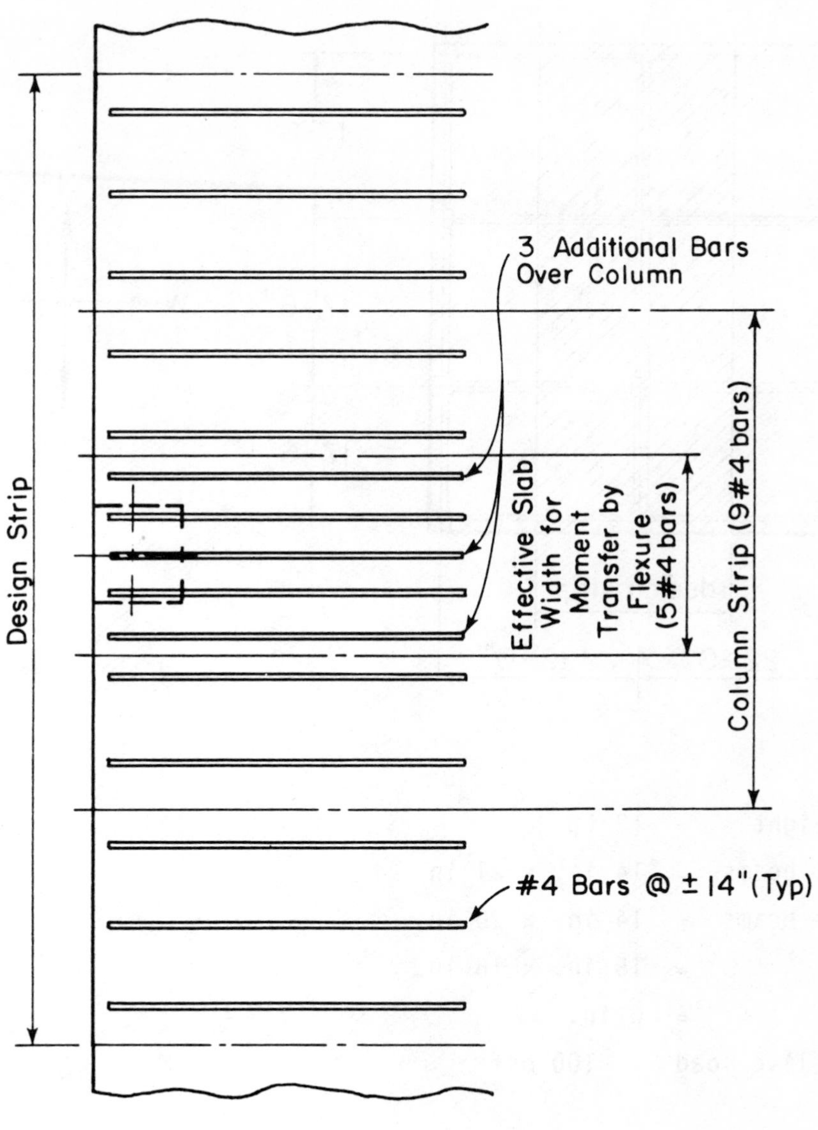

EXAMPLE 21.2 - Two-way Slab with Beams Analyzed by Equivalent Frame Method

Using the equivalent frame method, determine design moments for the slab system in the transverse direction for an intermediate floor.

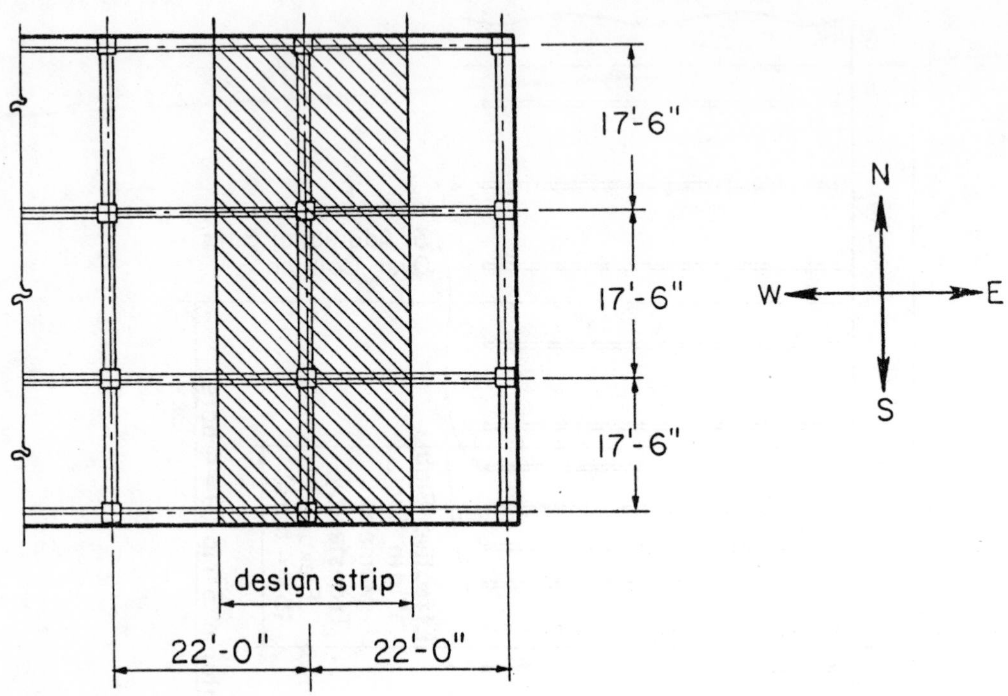

Story height = 12 ft

Spandrel beams = 14 in. x 27 in.

Interior beams = 14 in. x 20 in.

Columns = 18 in. x 18 in.

Slab = 6 in.

Service live load = 100 psf

f'_c = 4,000 psi (for all members)

f_y = 60,000 psi

Calculations and Discussion	Code Reference

1. Preliminary design for slab thickness h:

Control of deflection: 9.5.3

The ratio of flexural stiffness of beam to slab α:

α = 13.30 (NS edge beam)*
= 16.45 (EW edge beam)*
= 3.16 (NS interior beam)*
= 3.98 (EW interior beam)*

Since all $\alpha > 2.0$ (see Fig. 20-3, Part 20), Eq. (9-12) will control. Also, $\beta_s = 0.5$ at the corner panel will be the smaller value of any panel. This means that the corner panel will control (see Fig. 20-2, Part 20). Therefore,

$$h = \frac{\ell_n (800 + 0.005 f_y)}{36{,}000 + 5{,}000\beta \ (1 + \beta_s)} = 5.93"$$ Eq. (9-12)

where β = 20.5/16.0 = 1.28
β_s = 0.5
ℓ_n = clear span in long direction measured face to face of columns = 20'-6" = 246"

<u>Use 6" slab thickness</u>.

2. Frame members of equivalent frame.

Determine moment distribution constants and fixed-end moment coefficients for the equivalent frame members. The moment distribution procedure will be used to analyze the partial frame for vertical loading. Stiffness factors k, carry over factors COF, and fixed-end moment factors FEM,

*Calculations for α terms are given in Design Example 20.2, Part 20.

EXAMPLE 21.2 - Continued

Calculations and Discussion	Code Reference

for the slab-beams and column members are determined by conjugate beam procedures. These calculations are not shown here.

(a) Slab-beams, flexural stiffness at both ends K_{sb}:

$K_{sb} = 4.11\ EI_{sb}/\ell_1 = 4.11 \times 25{,}387E/210 = 497E$

where I_{sb} = moment of inertia of slab-beam section shown in Fig. 21-15 and computed with the aid of Fig. 21-20

$= 2.72\ (14 \times 20^3/12) = 25{,}387\ \text{in.}^4$

$\ell_1 = 17'\text{-}6" = 210"$

Carry-over factor COF = 0.507

Fixed-end moment, FEM = $0.0842\ w\ell_2\ell_1^{\ 2}$

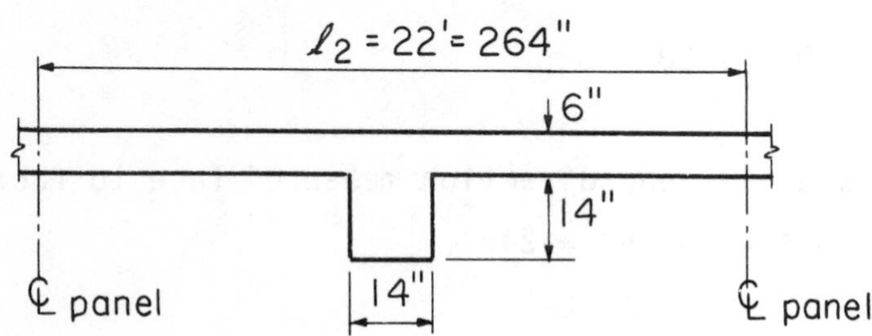

Fig. 21-15

(b) Column members, flexural stiffness K_c:

For interior columns:

$K_{ct} = 6.82\ EI_c/\ell_c = 6.82 \times 8748E/144 = 414E$

$K_{cb} = 4.99\ EI_c/\ell_c = 4.99 \times 8748E/144 = 303E$

EXAMPLE 21.2 - Continued

Calculations and Discussion	Code Reference

For exterior columns:

$K_{ct} = 8.57\ EI_c/\ell_c = 8.57 \times 8748E/144 = 521E$

$K_{cb} = 5.31\ EI_c/\ell_c = 5.31 \times 8748E/144 = 323E$

where $I_c = (c)^4/12 = (18)^4/12 = 8748\ \text{in.}^4$

$\ell_c = 12'\text{-}0'' = 144''$

(c) Torsional members, torsional stiffness K_t:

$K_t = 9EC/\ell_2\ (1 - c_2/\ell_2)^3$ — Eq. (13-6)

where $C = \Sigma(1 - 0.63\ x/y)(x^3y/3)$ — Eq. (13-7)

For interior columns:

$K_t = 9E \times 11{,}698/[264\ (0.932)^3] = 493E$

where C is taken as the larger value computed (with the aid of Table 21-3) for the torsional member shown in Fig. 21-16.

$x_1 = 14$	$x_2 = 6$	$x_1 = 14$	$x_2 = 6$
$y_1 = 14$	$y_2 = 42$	$y_1 = 20$	$y_2 = 14$
$C_1 = 4738$	$C_2 = 2752$	$C_1 = 10{,}226$	$C_2 = 736$
$\Sigma C = 4738 + 2752 = 7{,}490\ \text{in.}^4$		$\Sigma C = 10{,}226 + 736 \times 2 = 11{,}698\ \text{in.}^4$	

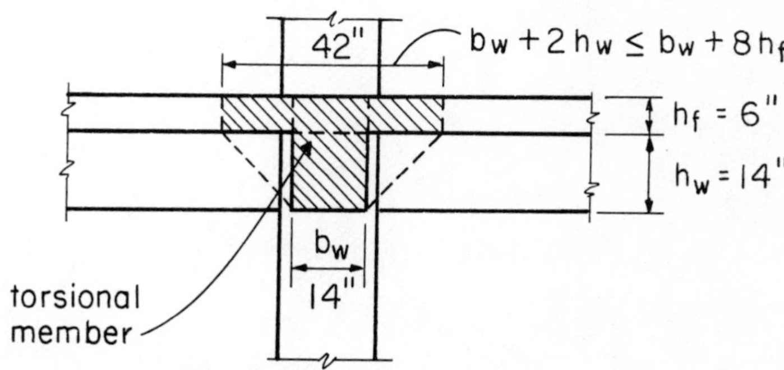

Fig. 21-16

EXAMPLE 21.2 - Continued

Calculations and Discussion	Code Reference

For exterior columns:

$K_t = 9E \times 17,868/[264\ (0.932)^3] = 752E$

where C is taken as the larger value computed (with the aid of Table 21-3) for the torsional member shown in Fig. 21-17.

x_1	=	14	x_2	=	6	x_1	=	14	x_2	=	6
y_1	=	21	y_2	=	35	y_1	=	27	y_2	=	21
C_1	=	11,141	C_2	=	2,248	C_1	=	16,628	C_2	=	1,240
ΣC	=	11,141 + 2,248 = 13,389 in.^{4}				ΣC	=	16,628 + 1,240 = 17,868 in.^{4}			

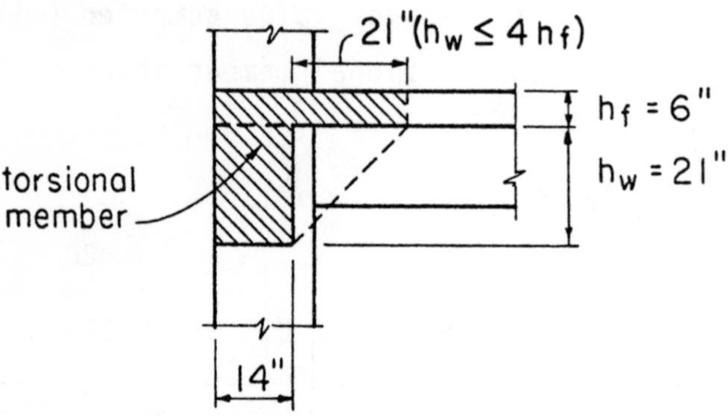

Fig. 21-17

TABLE 21-3 - VALUES OF TORSION CONSTANT, C*

y \ x**	4	5	6	7	8	9	10	12	14	16
12	202	369	592	868	1,188	1,538	1,900	2,557	-	-
14	245	452	736	1,096	1,529	2,024	2,566	3,709	4,738	-
16	288	534	880	1,325	1,871	2,510	3,233	4,861	6,567	8,083
18	330	619	1,024	1,554	2,212	2,996	3,900	6,013	8,397	10,813
20	373	702	1,167	1,782	2,553	3,482	4,567	7,165	10,226	13,544
22	416	785	1,312	2,011	2,895	3,968	5,233	8,317	12,055	16,275
24	458	869	1,456	2,240	3,236	4,454	5,900	9,469	13,885	19,005
27	522	994	1,672	2,583	3,748	5,183	6,900	11,197	16,628	23,101
30	586	1,119	1,888	2,926	4,260	5,912	7,900	12,925	19,373	27,197
33	650	1,243	2,104	3,269	4,772	6,641	8,900	14,653	22,117	31,293
36	714	1,369	2,320	3,612	5,284	7,370	9,900	16,381	24,860	35,389
42	842	1,619	2,752	4,298	6,308	8,828	11,900	19,837	30,349	43,581
48	970	1,869	3,183	4,984	7,332	10,286	13,900	23,293	35,836	51,773
54	1,098	2,119	3,616	5,670	8,356	11,744	15,900	26,749	41,325	59,965
60	1,226	2,369	4,048	6,356	9,380	13,202	17,900	30,205	46,813	68,157

* $\Sigma C = (1 - 0.63\ x/y)(x^3 y/3)$

** Small side of a rectangular cross section with dimensions x and y.

(d) Parallel beams, increased torsional stiffness K_{ta}:

For interior columns:

$$K_{ta} = K_t\ I_{sb}/I_s = 493E \times 25{,}387/4{,}752 = 2{,}634E$$

For exterior columns:

$$K_{ta} = 752E \times 25{,}387/4{,}752 = 4{,}017E$$

where I_s = moment of inertia of slab-section shown in Fig. 21-18.

= $264\ (6)^3/12 = 4{,}752$ in.^{4}

I_{sb} = moment of inertia of full "T" section shown in Fig. 21-18 and computed with the aid of Fig. 21-20.

= $2.72\ (14 \times 20^3/12) = 25{,}387$ in.^{4}

EXAMPLE 21.2 - Continued

Calculations and Discussion	Code Reference

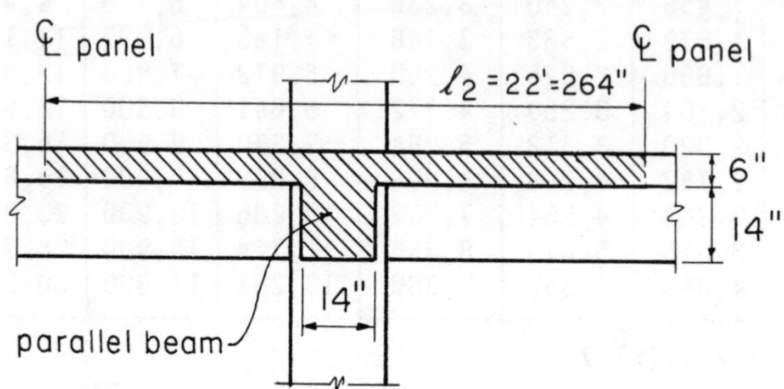

Fig. 21-18

(e) Equivalent column stiffness, K_{ec}:

$$K_{ec} = \Sigma K_c \times \Sigma K_{ta} / \Sigma K_c + \Sigma K_{ta}$$

where ΣK_{ta} is for two torsional members, one each side of column, and ΣK_c is for the upper and lower columns at the slab-beam joint of an intermediate floor.

For interior columns:

$$K_{ec} = \frac{(303E + 414E)\ (2 \times 2634E)}{(303E + 414E) + (2 \times 2634E)}$$

$$= 631E$$

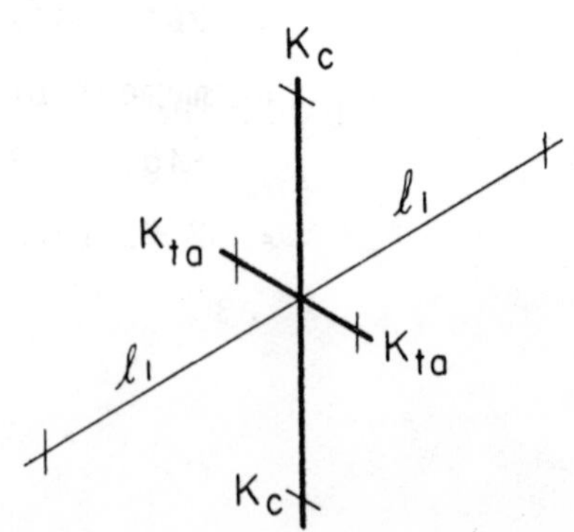

Calculations and Discussion	Code Reference

For exterior columns:

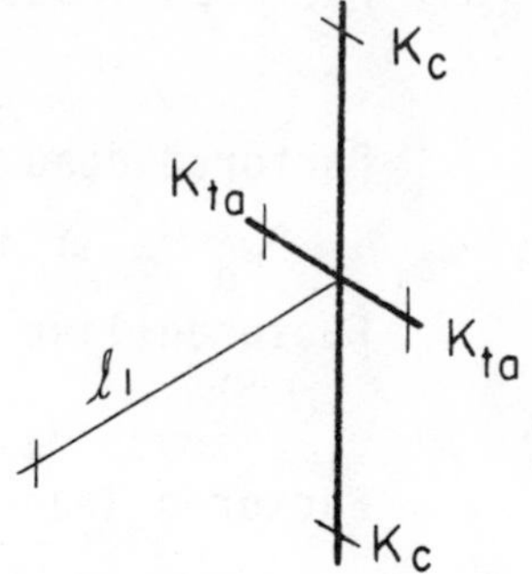

$$K_{ec} = \frac{(323E + 521E)\ (2 \times 4{,}017E)}{(323E + 521E) + (2 \times 4{,}017E)}$$

$$= 764E$$

(f) Slab-beam joint distribution factors DF:

At exterior joint:

DF = 497E/(497E + 764E)

= 0.394

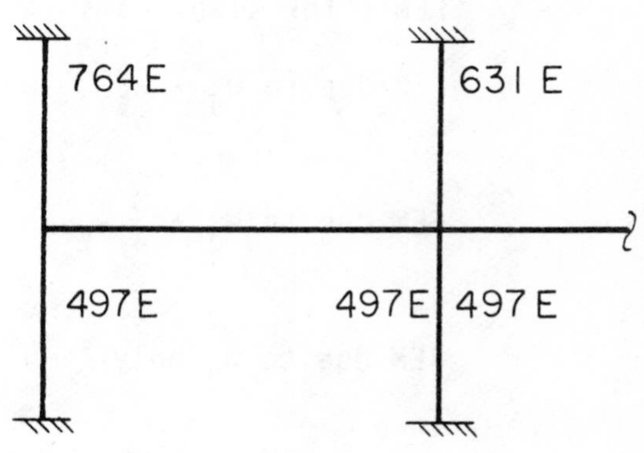

At interior joint:

DF = 497E/(497E + 497E + 631E)

= 0.306

COF for slab-beam = 0.507

3. Partial frame analysis of equivalent frame.

Determine maximum negative and positive moments for the slab-beams using the moment distribution method.

With a service live-to-dead load ratio:

$L/D = 100/75 = 1.33 > \frac{3}{4}$,

the frame will be analyzed for five loading conditions with pattern loading and partial live load as allowed by Section 13.7.6.3. (See Fig. 21-8 for an illustration of the five load patterns considered.) 13.7.6.3

EXAMPLE 21.2 - Continued

Calculations and Discussion	Code Reference

(a) Factored loads and fixed-end moments:

Factored dead load per unit area,

$$w_d = 1.4\ (75 + 9.3)^* = 118 \text{ psf}$$

Factored live load per unit area,

$$w_\ell = 1.7(100) = \underline{170 \text{ psf}}$$

Factored load per unit area,

$$w_d + w_\ell = 288 \text{ psf}$$

$$\text{FEM's for slab-beams} = 0.0842\ w\ell_2\ell_1^2$$

$$\text{FEM due to } w_d + w_\ell = 0.0842\ (0.288 \times 22)\ 17.5^2 = 163.4 \text{ ft kips}$$

$$\text{FEM due to } w_d + \frac{3}{4} w_\ell = 0.0842\ (0.2455 \times 22)\ 17.5^2 = 139.3 \text{ ft kips}$$

$$\text{FEM due to } w_d \text{ only} = 0.0842\ (0.118 \times 22)\ 17.5^2 = 66.9 \text{ ft kips}$$

(b) Moment distribution for the five loading conditions is shown in Table 21-4. Counter clockwise rotational moments acting on the member ends are taken as positive. Positive span moments are determined from the equation:

$$M_{(midspan)} = M_s - \frac{1}{2}(M_L + M_R)$$

where M_s is the moment at midspan for a simple beam.

When the end moments are not equal, the maximum moment in the span does not occur at midspan, but its value is close to that at midspan.

* 9.3 psf is weight of beam stem per foot divided by ℓ_2

Calculations and Discussion	Code Reference

Positive moment in span 1-2 for loading (1):

$+M = (0.288 \times 22)\ 17.5^2/8 - \frac{1}{2}(102.8 + 185.0) = 98.7$ ft kips

The following moment values for the slab-beams may be obtained by summarizing the results in Table 21-4. Note that according to Section 13.7.6.3, the design moments shall be taken not less than those occurring with full factored live load on all spans.

Maximum positive moment in end span
= the larger of 98.7 or 92.2 = 98.7 ft kips

Maximum positive moment in interior span*
= the larger of 73.6 or 78.4 = 78.4 ft kips

Maximum negative moment at end support
= the larger of 102.8 or 94.0 = 102.8 ft kips

Maximum negative moment at interior support for exterior span
= the larger of 185.0 or 161.5 = 185.0 ft kips

Maximum negative moment at interior support for interior span
= the larger of 169.4 or 154.1 = 169.4 ft kips

* This is the only moment governed by the pattern loading with partial live load. All other maximum moments occur with full factored live load on all spans.

Calculations and Discussion	Code Reference

TABLE 21-4 - MOMENT DISTRIBUTION FOR PARTIAL FRAME
(transverse direction)

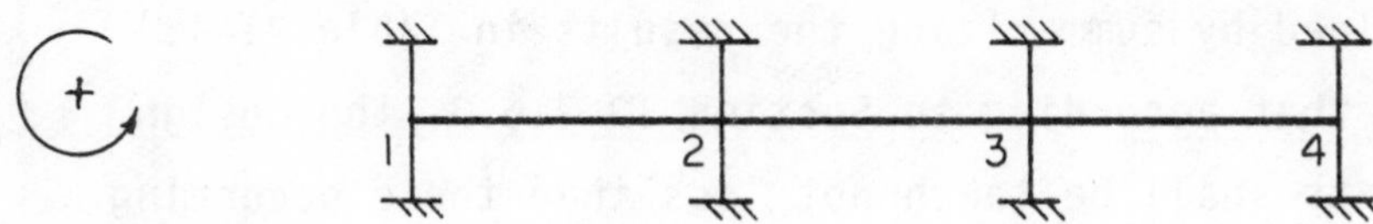

Joint	1	2		3		4
Member	1 - 2	2 - 1	2 - 3	3 - 2	3 - 4	4 - 3
DF	0.394	0.306	0.306	0.306	0.306	0.394
COF	0.507	0.507	0.507	0.507	0.507	0.507

(1) All spans loaded with full factored live load

	1 - 2	2 - 1	2 - 3	3 - 2	3 - 4	4 - 3
FEM	+163.4	-163.4	+163.4	-163.4	+163.4	-163.4
COM*		- 32.6			+ 32.6	
"	+ 5.1		- 5.1	+ 5.1		- 5.1
"	+ 0.8	- 1.0	- 0.8	+ 0.8	+ 1.0	- 0.8
"	+ 0.3	- 0.2	- 0.3	+ 0.3	+ 0.2	- 0.3
Σ	+169.6	-197.2	+157.2	-157.2	+197.2	-169.6
DM**	- 66.8	+ 12.2	+ 12.2	- 12.2	- 12.2	- 66.8
Total	+102.8	-185.0	+169.4	-169.4	+185.0	-102.8
℄ Moment	98.7		73.6		98.7	

(2) First and third spans loaded with 3/4 factored live load

	1 - 2	2 - 1	2 - 3	3 - 2	3 - 4	4 - 3
FEM	+139.3	-139.3	+ 66.9	- 66.9	+139.3	-139.3
COM	+ 11.2	- 27.8	- 11.2	+ 11.2	+ 27.8	- 11.2
"	+ 6.1	- 2.2	- 6.1	+ 6.1	+ 2.2	- 6.1
"	+ 1.3	- 1.2	- 1.3	+ 1.3	+ 1.2	- 1.3
"	+ 0.4	- 0.3	- 0.4	+ 0.4	+ 0.3	- 0.4
Σ	+158.3	-170.8	+ 47.9	- 47.9	+170.8	-158.3
DM	- 62.4	+ 37.6	+ 37.6	- 37.6	- 37.6	+ 62.4
Total	+ 95.9	-133.2	+ 85.5	- 85.5	+133.2	- 95.9
℄ Moment	92.2				92.2	

Calculations and Discussion	Code Reference

TABLE 21-4 - Continued

(3) Center span loaded with 3/4 factored live load

FEM	+ 66.9	- 66.9	+139.3	-139.3	+ 66.9	- 66.9
COM	- 11.2	- 13.4	+ 11.2	- 11.2	+ 13.4	+ 11.2
"	+ 0.3	+ 2.2	- 0.3	+ 0.3	- 2.2	- 0.3
"	- 0.3	- 0.1	+ 0.3	- 0.3	+ 0.1	+ 0.3
Σ	+ 55.7	- 78.2	+150.5	-150.5	+ 78.2	- 55.7
DM	- 21.9	- 22.1	- 22.1	+ 22.1	+ 22.1	+ 21.9
Total	+ 33.8	-100.3	+128.4	-128.4	+100.3	- 33.8
℄ Moment			78.4			

(4) First span loaded with 3/4 factored live load
and beam-slab assumed fixed at support two span distance

FEM	+139.3	-139.3	+ 66.9	- 66.9		
COM	+ 11.2	- 27.8		+ 11.2		
"	+ 4.3	- 2.2		+ 4.3		
"	+ 0.3	- 0.9		+ 0.3		
Σ	+155.1	-170.2	+ 66.9			
DM	- 61.1	+ 31.6	+ 31.6			
Total	+ 94.0	-138.6	+ 98.5	- 51.1		

(5) First and second span loaded with 3/4 factored live load

FEM	+139.3	-139.3	+139.3	-139.3	+ 66.9	- 66.9
COM		- 27.8	+ 11.2		+ 13.4	+ 11.2
"	+ 2.6		- 2.1	+ 2.6	- 2.2	- 2.1
"	+ 0.3	- 0.2	- 0.1	+ 0.3	+ 0.4	- 0.1
Σ	+142.2	-167.3	+148.3	-136.4	+ 78.5	- 57.9
DM	- 56.0	+ 5.8	+ 5.8	+ 17.7	+ 17.7	+ 22.8
Total	+ 86.2	-161.5	+154.1	-118.7	+ 96.2	- 35.1

* Carry-over moment, COM, is the negative product of the distribution factor, carry-over factor and unbalanced joint moment carried to the opposite end of span.

** Distributed moment, DM, is the negative product of the distribution factor and the unbalanced joint moment.

EXAMPLE 21.2 - Continued

Calculations and Discussion	Code Reference
4. Design moments.	
Positive and negative factored moments for the slab-system in the transverse direction are shown in Fig. 21-19. The negative factored moments are taken at the face of rectilinear supports but not greater than a distance of $0.175\ \ell_1$ from the center of supports. $0.75 < 0.175 \times 17.5$ (Use face of support location)	13.7.7.1
5. Total factored moment per span.	13.7.7.4
Slab systems within the limitations of Section 13.6.1 may have the resulting analytical moments reduced in such proportion that the numerical sum of the positive and average negative moments are not greater than the total static moment M_o given by Eq. (13-3). Check limitations of Section 13.6.1.6 for relative stiffness of beams in two perpendicular directions.	
For interior panel:	
$\alpha_1 \ell_2^2 / \alpha_2 \ell_1^2 = 3.16\ (22)^2 / 3.98(17.5)^2 = 1.25$	
$0.2 < 1.25 < 5.0$ OK	13.6.1.6
For exterior panel:	
$3.16(22)^2 / 16.45(17.5)^2 = 0.30$	
$0.2 < 0.30 < 5.0$ OK	
All limitations of Section 13.6.1 are satisfied and the provisions of Sec. 13.7.7.4 may be applied.	
$M_o = w \ell_2 \ell_n^2 / 8 = 0.288 \times 22 \times 16^2 / 8 = 202.8$ ft kips	Eq. (13-3)

EXAMPLE 21.2 - Continued

Calculations and Discussion	Code Reference

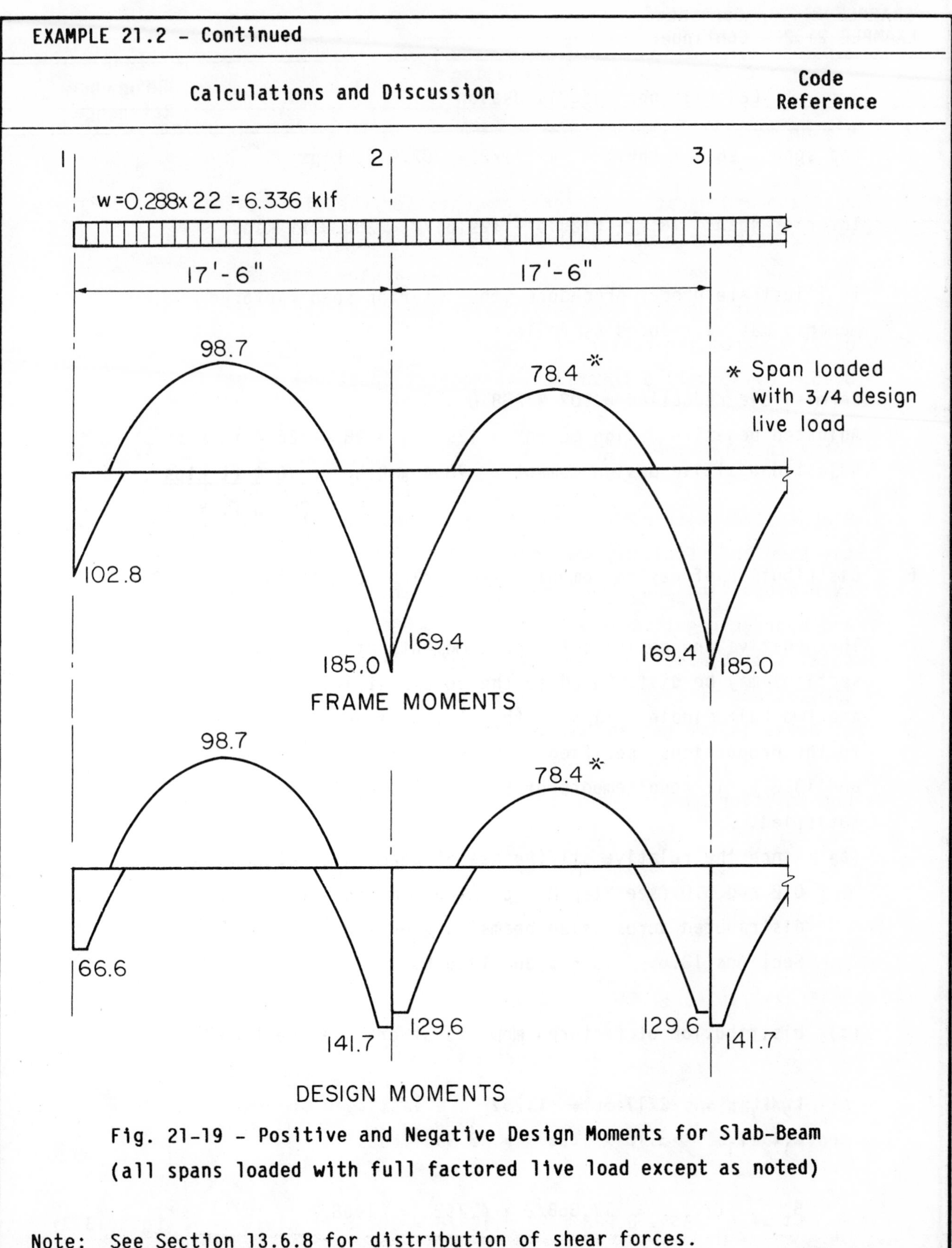

Fig. 21-19 - Positive and Negative Design Moments for Slab-Beam (all spans loaded with full factored live load except as noted)

Note: See Section 13.6.8 for distribution of shear forces.

Calculations and Discussion	Code Reference

End span: 98.7 + (66.6 + 141.7)/2 = 202.9 ft kips

Interior span: 78.4 + (129.6 + 129.6)/2 = 208.0 ft kips

To illustrate proper procedure, the interior span factored moments may be reduced as follows:

Permissible reduction = 202.9/208.0 = 0.98
Adjusted negative design moment = 129.6 x 0.98 = 126.4 ft kips
Adjusted positive design moment = 78.4 x 0.98 = 76.4 ft kips
M_o = 202.8 ft kips

6. Distribution of design moments across slab-beam strip. 13.7.7.5

The negative and positive factored moments at critical sections may be distributed to the column strip, beam and two half middle strips of the slab-beam according to the proportions specified in Sections 13.6.4, 13.6.5 and 13.6.6, if requirements of Section 13.6.1.6 are satisfied.

(a) Since the relative stiffnesses of beams are between 0.2 and 5.0 (see Step No. 5), the moments can be distributed across slab-beams as specified in Sections 13.6.4, 13.6.5 and 13.6.6.

(b) Distribution of factored moments at critical section:

$\ell_2/\ell_1 = 22/17.5 = 1.257$
$\alpha_1\ell_2/\ell_1 = 3.16 \times 1.257 = 3.97$

$\beta_t = C/2I_s = 17{,}868/2 \times 4{,}752 = 1.88$

Calculations and Discussion	Code Reference

where I_s = $22 \times 12 \times 6^3/12$ = 4,752 $in.^4$

C = 17,868 $in.^4$ (see Fig. 21-17)

Factored moments at critical sections are summarized in Table 21-5.

TABLE 21-5 - DISTRIBUTION OF DESIGN MOMENTS

	Factored moments	Column strip		Two half middle strips**
		Percent	Moment*	
End span:				
Exterior negative	66.6	75	50.0	16.6
Positive	98.7	67	66.1	32.6
Interior negative	141.7	67	94.9	46.8
Interior span:				
Negative	129.6	67	86.8	42.8
Positive	78.4	67	52.5	25.9

* Since $\alpha_1 \ell_2/\ell_1 > 1.0$, beams must be proportioned to resist 85 percent of column strip moment as per Section 13.6.5.1.

** That portion of the factored moment not resisted by the column strip is assigned to the two half middle strips.

7. Calculations for shear in beams and slab are performed in Example 20.2, Part 20.

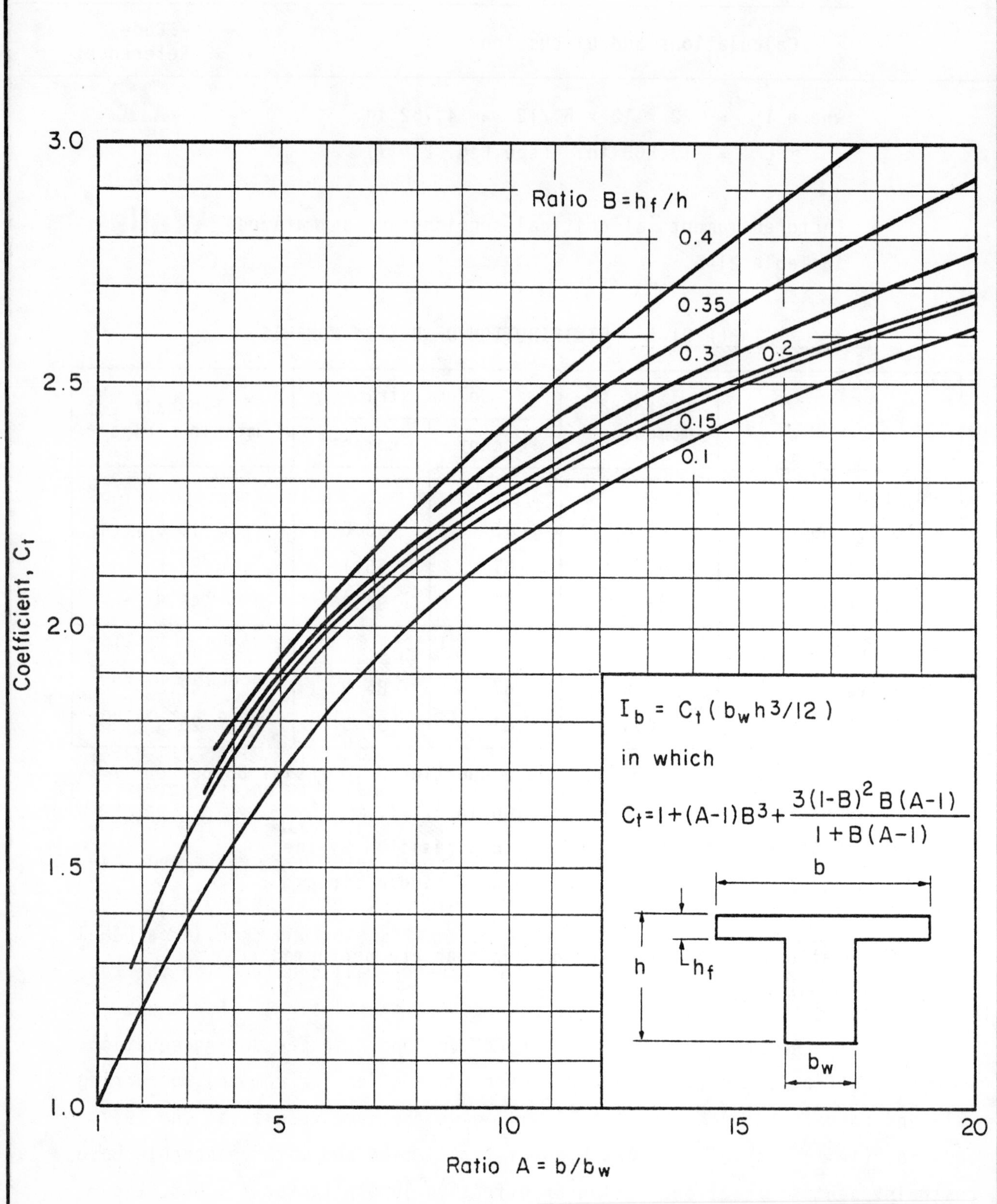

Fig. 21-20 - Coefficient C_t for Gross Moment of Inertia of Flanged Sections (flange on one or two sides)

22

Walls

General Considerations

With publication of ACI 318-83, two changes were made to the Code provisions that affect design of walls. One first appeared in the 1980 Code Supplement and one is new in the 1983 Code edition.

With the 1980 Supplemental Revisions, the empirical wall design Eq. (14-1) was modified to reflect the general range of end conditions encountered in wall design. The wall strength equation in the 1977 Code edition was based on the assumption of a wall with top and bottom fixed against lateral movement and with moment restraint at one end, corresponding to an effective length factor between 0.8 and 0.9. Axial load strength values determined from the original equation can be unsafe unless at least one of the ends is restrained against rotation. This pinned-pinned end condition can exist in certain walls, particularly precast tilt-up construction and some large panel wall systems. In addition, in some other types of wall construction the top of the wall is free standing (not braced against translation). For these cases it is necessary to reflect the effective length in the design equation. Values of effective vertical length factor k are given for commonly occurring wall end conditions. New Eq. (14-1) will give the same results as the 1977 Code Eq. (14-1) for walls braced against translation and with reasonable base restraint against rotation. Using an effective length factor $k = 0.8$, the slenderness function becomes,

$$\left[1 - \left(\frac{0.8\ell_c}{32h}\right)^2\right] = \left[1 - \left(\frac{\ell_c}{40h}\right)^2\right]$$ ... same as ACI 318-77

Reasonable base restraint against rotation implies attachment to a member having a flexural stiffness EI/ℓ at least equal to that of the wall. Selection of the proper k for a particular set of support end conditions is left to the judgment of the engineer.

It is now clearly stated that the empirical design method applies to load bearing walls, and only to walls of solid rectangular cross section. Retaining walls are designed by the flexural design provisions of Chapter 10. Load bearing walls of nonrectangular cross section, such as ribbed wall panels, must be designed by Section 14.4.

Also with the 1980 Supplemental Revisions, a new Section 14.2.3 was added to ensure that the special shear provisions for walls of Chapter 11 are not overlooked. For some cases, shear forces must be considered in the design of walls; required shear reinforcement may exceed the minimum wall reinforcement of Section 14.3. Design Example 22.4 illustrates shear design for a wall.

New for the <u>'83 Code Edition</u> is a complete reformat of Chapter 14 to locate all code design requirements for walls within Chapter 14. In the 1971 and 1977 Code editions, the wall provisions were divided between Chapter 10 (Section 10.15) and Chapter 14. Section 10.15 addressed walls designed as compression members and Chapter 14 essentially addressed the empirical design of walls. Both design methods are now integrated into new Chapter 14. The provisions have also been reorganized to improve clarity.

One additional item concerning walls for '83: the code provisions for force transfer at footings (Section 15.8) now specifically address force transfer between a "wall" and footing, with Section 15.8.2.2 requiring a minimum amount of reinforcement not less than the minimum wall reinforcement.

14.1 Scope

In accordance with the code, the designer has two options for design of walls: walls designed as compression members (Section 14.4) using the strength design provisions for flexure and axial loads of Chapter 10; or the empirical design method of Section 14.5. It should be noted that the provisions of Section 14.2 and 14.3 apply for walls designed by either method. Also, no minimum wall thicknesses are stated for bearing walls designed by Section 14.4.

14.4 Walls Designed as Compression Members

Where wall geometry and loading conditions do not satisfy the limitations of Section 14.5, and especially where lateral loads are present, walls must be designed as compression members by the strength design provisions for flexure and axial loads of Chapter 10. Minimum reinforcement by Section 14.3 applies for walls designed by strength design provisions of Chapter 10. Especially note that the vertical wall reinforcement need not be enclosed by lateral ties (as for columns) if the conditions of Section 14.3.6 are satisfied. All other code provisions for compression members apply to walls designed by Chapter 10.

As with columns, design of walls is difficult without the use of design aids. Wall design is further complicated by the fact that slenderness is a consideration in the design of practically all walls. Two methods for slenderness consideration are specified in the Code. The so-called rigorous analysis, which takes into account variable wall stiffness, is specified in Section 10.10.1. In lieu of that procedure, the approximate evaluation of slenderness effects described in Section 10.11 may be used.

Considering the approximate method, note that Eqs. (10-10) and (10-11) for EI originally were not derived for members with a single layer of reinforcement. An equation for EI for walls (with a single layer of reinforcement) has been suggested by MacGregor:[22.1]

$$EI = \left(\frac{E_c I_g}{\beta}\right)\left(0.5 - \frac{e}{h}\right) \geq 0.1\left(\frac{E_c I_g}{\beta}\right) \tag{1}$$

$$\leq 0.4 \left(\frac{E_c I_g}{\beta}\right)$$

where $\beta = 0.9 + (0.5\,\beta_d^2) - 12\,\rho \geq 1.0$

and β_d = ratio of dead load moment to total load moment

Comparison of EI by Eq. (1) and Code Eq. (10-11) is shown in Fig. 22-1. Values of EI in terms of $E_c I_g$ are plotted as a function of eccentricity e/h for several values of β_d with reinforcement ratio ρ constant at 0.0015. Values obtained using Eq. (10-11) are constant for each value of β_d since e/h is not a variable. For walls with higher load eccentricity, Code Eq. (10-11) appears to overestimate wall stiffness. For walls designed by Chapter 10 with slenderness evaluation by Section 10.11, Eq. (1) is recommended in lieu of Code Eq. (10-11) for wall stiffness. Design Example 22.2 illustrates application of Code Section 10.11 using Eq. (1) for wall slenderness evaluation.

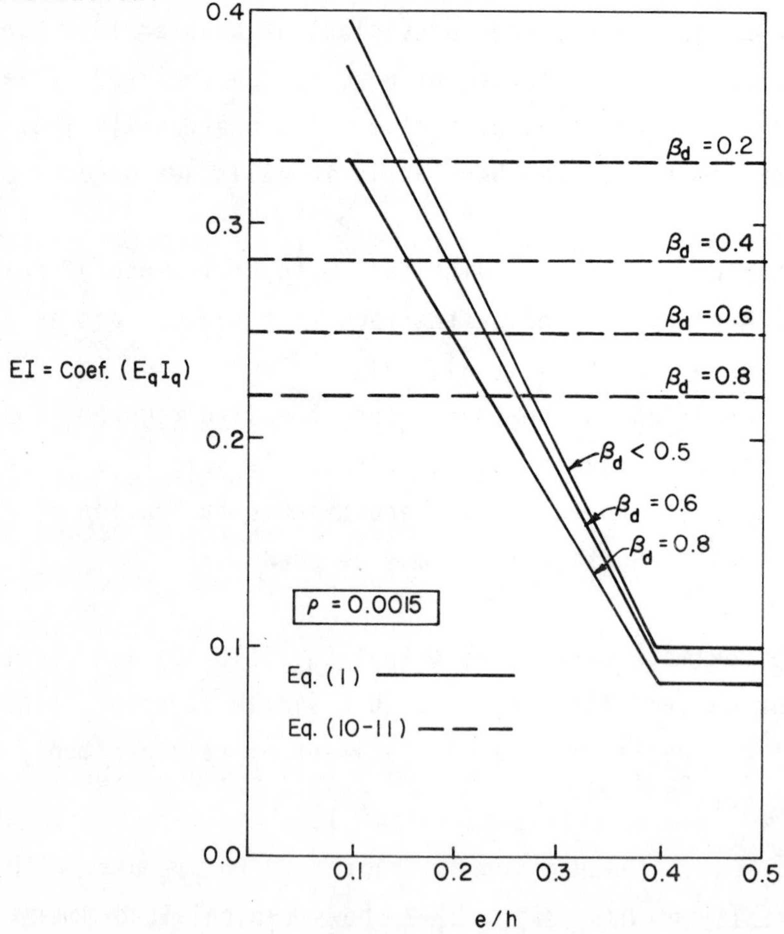

Fig. 22-1 - Stiffness EI for Walls

When wall slenderness exceeds the limit for application of the approximate slenderness evaluation method of Section 10.11 ($k\ell_u/h \geq 30$), a more detailed evaluation of wall slenderness effects is required, as defined in Section 10.10.1. The slender load-bearing concrete wall panels currently used in some building systems, especially tilt-up wall construction, are in this high slenderness category requiring the more detailed analysis of Section 10.10.1. Such an analysis should account for the influence of variable wall stiffness, effects of deflections on the moments and forces, and effects of the duration of the loads. Such an analysis is presented in a PCA design aid EB074D "Tilt-up Load-Bearing Walls."[22.2] The design aid presents load capacities of slender panels ($20 < k\ell_u/h < 50$) with wall thickness varying from 5-1/2 to 9-1/2 in. and having single or double layers of reinforcement. Design assistance is given in the form of load capacity coefficient tables. A description of how the design tables were developed is included in the publication. The design aid is simple to use, requiring only a minimum amount of design calculations. Design Example 22.3 illustrates application of Design Aid EB074D.

14.5 Empirical Design Method

The empirical method specified in Section 14.5 may be used for design of load bearing walls if the resultant of the vertical loads is located within the middle one-third of the wall thickness, and the thickness is at least 1/25 the unsupported height or length (whichever is less). Note that the effect of any lateral loads on the wall must be included to determine the "effective" eccentricity of the resultant vertical load. Also the method applies only to walls of solid rectangular cross-section. The empirical method is a straightforward design procedure for these limited cases, requiring only a single strength calculation to determine the design axial load strength for a wall.

The strength equation for φP_{nw} considers both load eccentricity and slenderness effects. The eccentricity factor 0.55 was selected to give strengths comparable with those given by Chapter 10 for members with an axial load at an eccentricity of h/6. Fig. 22-2 shows typical load-moment strength curves for 8-, 10-, and 12-in. walls.[22.3] Strength design by Chapter 10 yields eccentricity factors (ratio of strength under eccentric loading to that under concentric loading) of 0.562, 0.568, and 0.563 for the 8-, 10-, and 12-in. walls with ρ = 0.0015.

Note that the minimum wall reinforcement required by Section 14.3.2 (ρ = 0.0015) does not substantially increase the strength of a wall above that for a plain concrete wall. The minimum wall reinforcement as required by Section 14.3 is provided primarily for control of cracking due to shrinkage and temperature stresses.

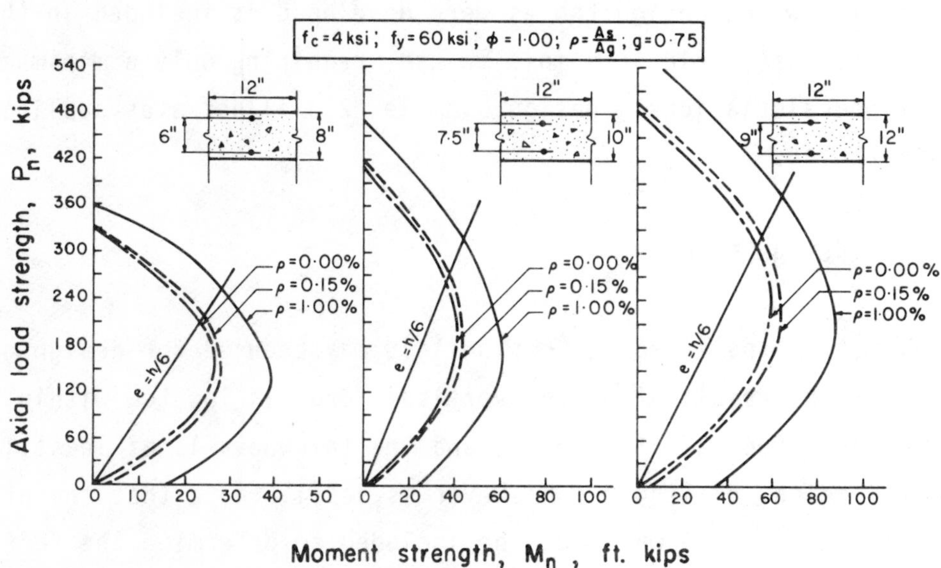

Fig. 22-2 - Typical Load-Moment Strength Curves for 8-, 10-, and 12-in. Walls

For wall slenderness effects, Eq. (14-1) results in strengths comparable with those given by the slenderness evaluation procedure of Section 10.11 for members with different braced and restrained end conditions. This is illustrated in Fig. 22-3. Note that the slenderness part of the wall strength equation was revised with the 1980 Code Supplement to allow for a wider range of application as well as for conformance with current experimental evidence.

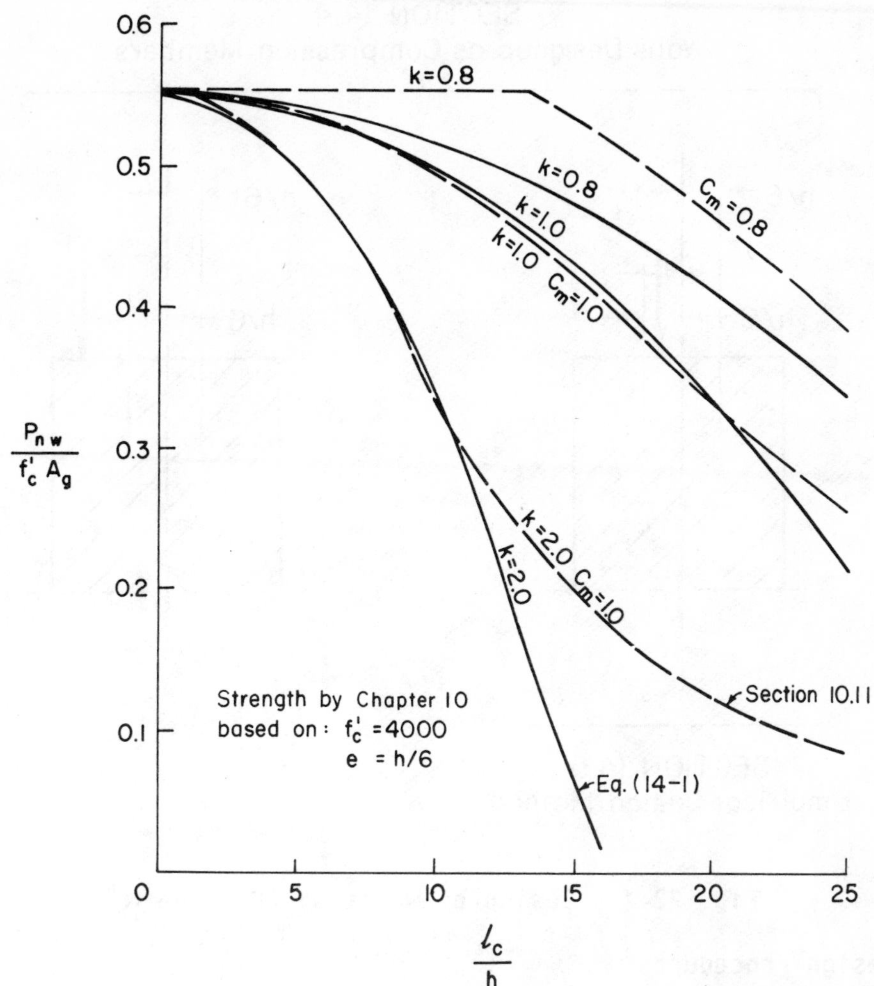

Fig. 22-3 - Wall Strength - Eq. (14-1) vs. Section 10.11

Principal application of the empirical design method is for relatively short walls spanning vertically, and subject to vertical loads only, such as those resulting from the reaction of floor or roof systems supported on a wall. Application becomes extremely limited when lateral loads must be considered, because of the "effective" load eccentricity limitation of h/6. Other than short walls carrying "reasonably concentric" loads, walls should be designed as compression members for axial load and flexure by provisions of Chapter 10 (Section 14.4). Fig. 22-4 illustrates design by Eq. (14-1) versus Chapter 10.

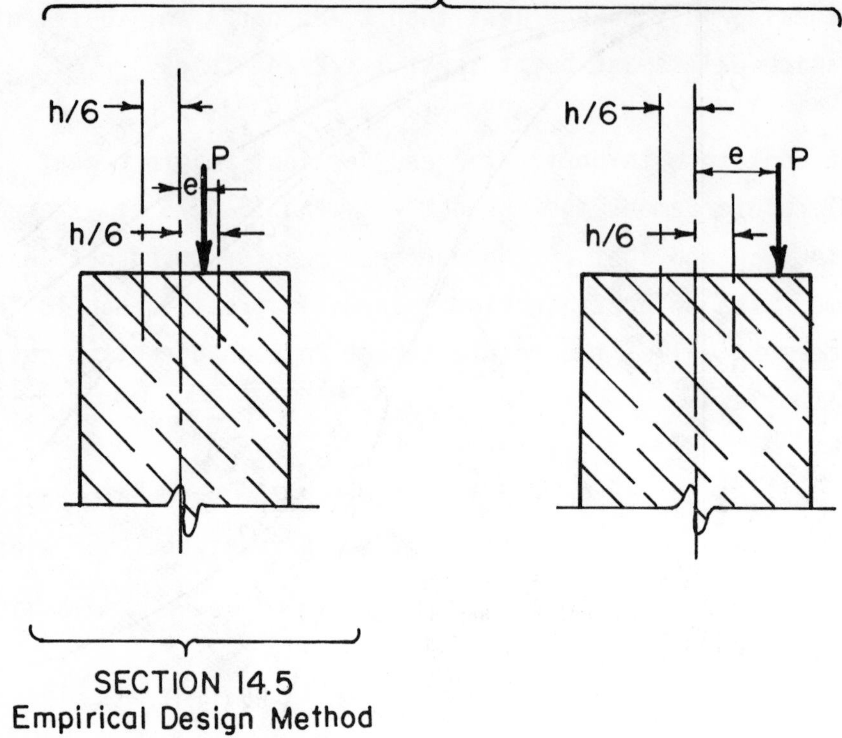

Fig. 22-4 - Design of Walls by ACI Code

Emperical Design Procedure

The empirical design method may be used for design of bearing walls carrying a "reasonably concentric" vertical load. This condition is considered to exist when the resultant of the vertical loads falls within the middle third of the wall thickness h. Consequently, when the load eccentricity e does not exceed h/6 the design is performed considering P_u as a concentric load. The factored axial load P_u must be less than the design axial load strength φP_{nw} computed by Eq. (14-1), or

$$P_u \leq \varphi P_{nw}$$

$$\leq 0.55 \ \varphi f'_c A_g \left[1 - \left(\frac{k\ell_c}{32h}\right)^2\right] \qquad \text{Eq. (14-1)}$$

Use of Eq. (14-1) is further limited to the following design conditions:

(1) Wall thickness h must not be less than $\ell_c/25$ nor 4 in. Basement walls and foundation walls must be at least 7-1/2 in. thick.

(2) Walls must contain both horizontal and vertical reinforcement. Area of horizontal reinforcement must be not less than 0.0025 times area of vertical section and that of the vertical reinforcement not less than 0.0015 times area of cross section. In walls greater than 10 in. thick (except basement walls) the reinforcement in each direction must be placed in two layers.

(3) Length of wall to be considered as effective for each beam reaction must not exceed center-to-center distance between reactions, nor width of bearing plus 4h.

(4) The wall must be anchored to the floors or to columns and other structural elements of the building.

Design Example 22.1 illustrates application of the empirical design method to a bearing wall supporting precast floor beams.

It should be noted that the reinforcement and minimum thickness requirements of Sections 14.3 and 14.5.3 may be waived where structural analysis shows adequate strength and wall stability (Section 14.2.7). Structural analysis may be satisfied by a design using ACI Standard 318.1 "Building Code Requirements for Structural Plain Concrete." See Reference 22.4.

Design Summary

A trial procedure for wall design is suggested: first assume a wall thickness h and reinforcement ratio ρ, then check the trial wall for the applied loading conditions.

It is not within the scope of Part 22 to include design aids for a broad range of wall and loading conditions. The intent is to present examples of design options and aids. Using programmable calculators, the designer can, with reasonable effort, produce design aids to fit the range of conditions usually encountered in practice and in a form of his choice. For example, strength interaction diagrams such as those plotted in Fig. 22-5 (ρ = 0.0015) and Fig. 22-6 (ρ = 0.0025) can be helpful design aids in evaluation of wall strength. "Blow-ups" of the lower portions of the strength interaction diagrams are shown for specific walls (h = 6.5 in.). Load charts, such as Fig. 22-7 can also be developed for specific walls. Design aids such as the one shown in Fig. 22-8 may facilitate selection of wall reinforcement. The designer can tailor his design aids to the "normal construction practice" in his area.

Prestressed walls are not covered specifically in Part 22. Prestressing of walls is advantageous for handling (precast panels) and for increased buckling resistance. For design of prestressed walls the designer is referred to Reference 22.5.

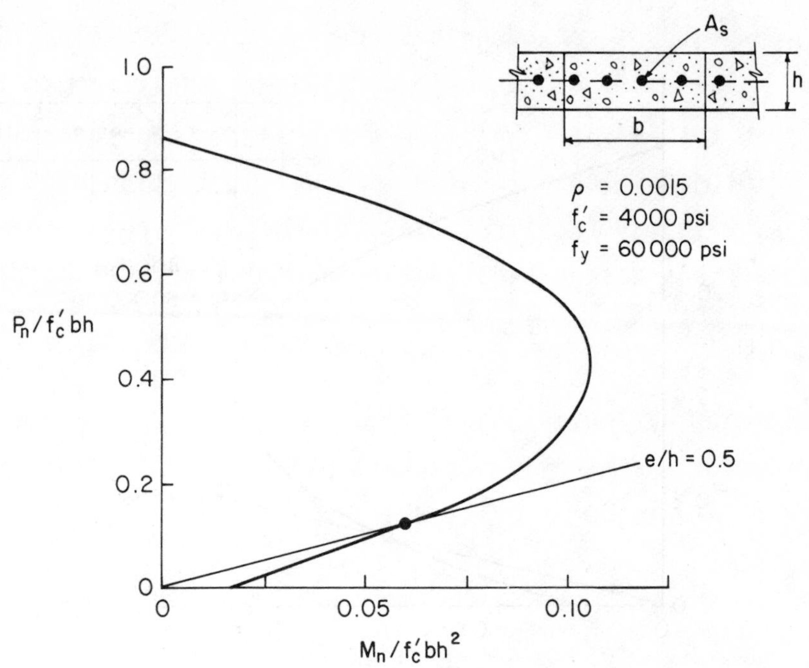

(a) Nondimensional Interaction Strength

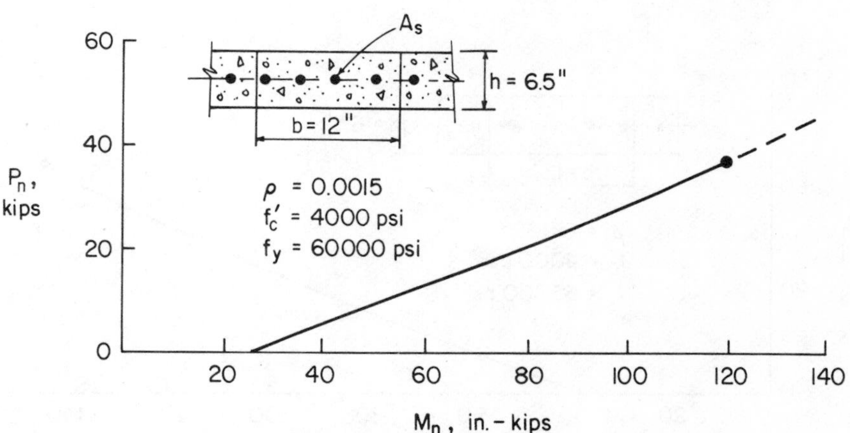

(b) Load-Moment Strength (e/h > 0.5)

Fig. 22-5 - Load-Moment Interaction Strength (Single Layer ρ = 0.0015)

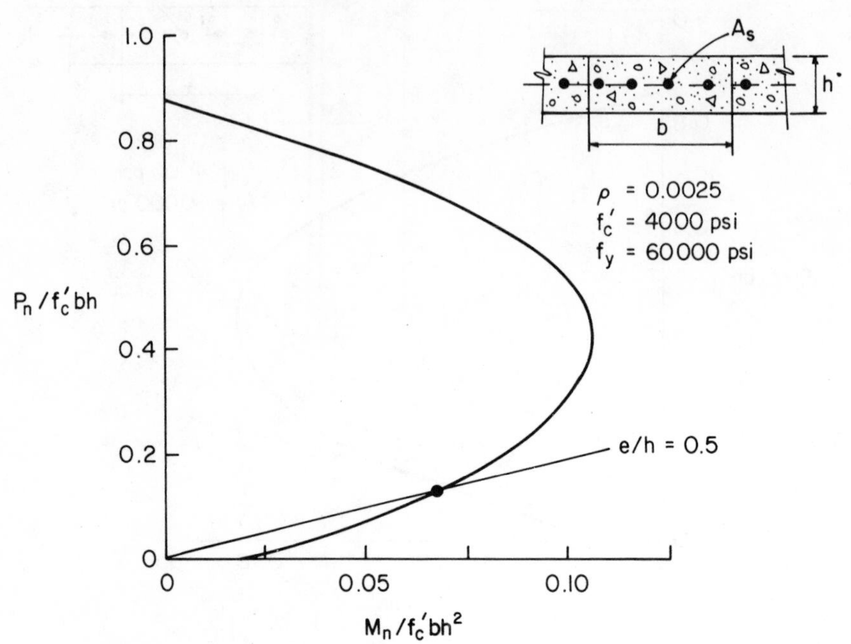

(a) Nondimensional Interaction Strength

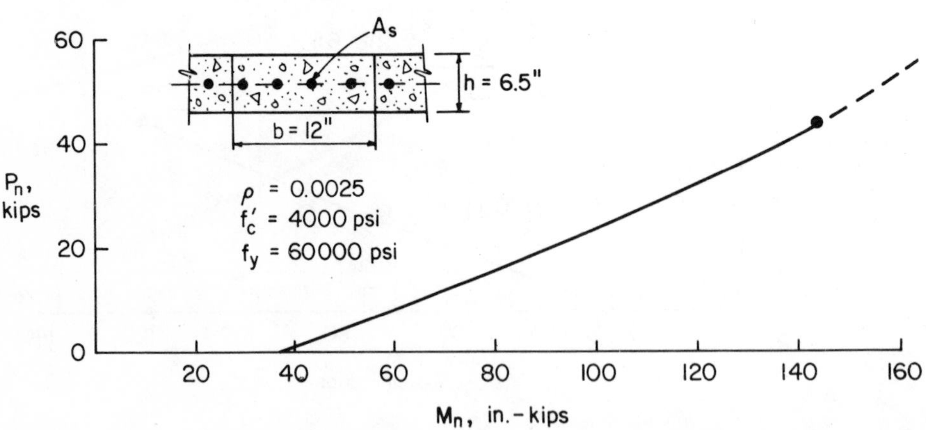

(b) Load-Moment Strength (e/h > 0.5)

Fig. 22-6 - Load-Moment Interaction Strength (Single Layer ρ = 0.0025)

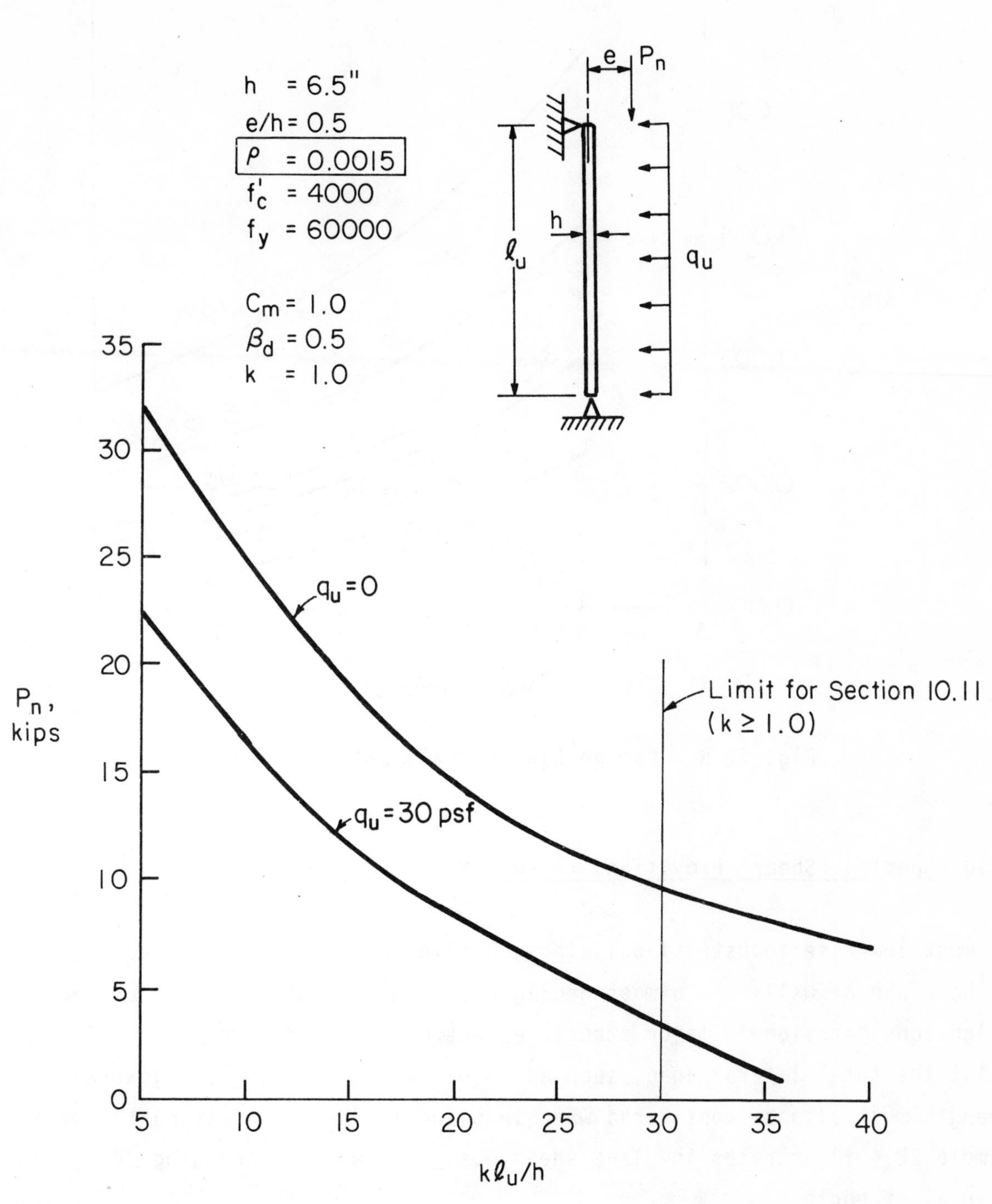

Fig. 22-7 - Design Chart for 6.5-in. Wall

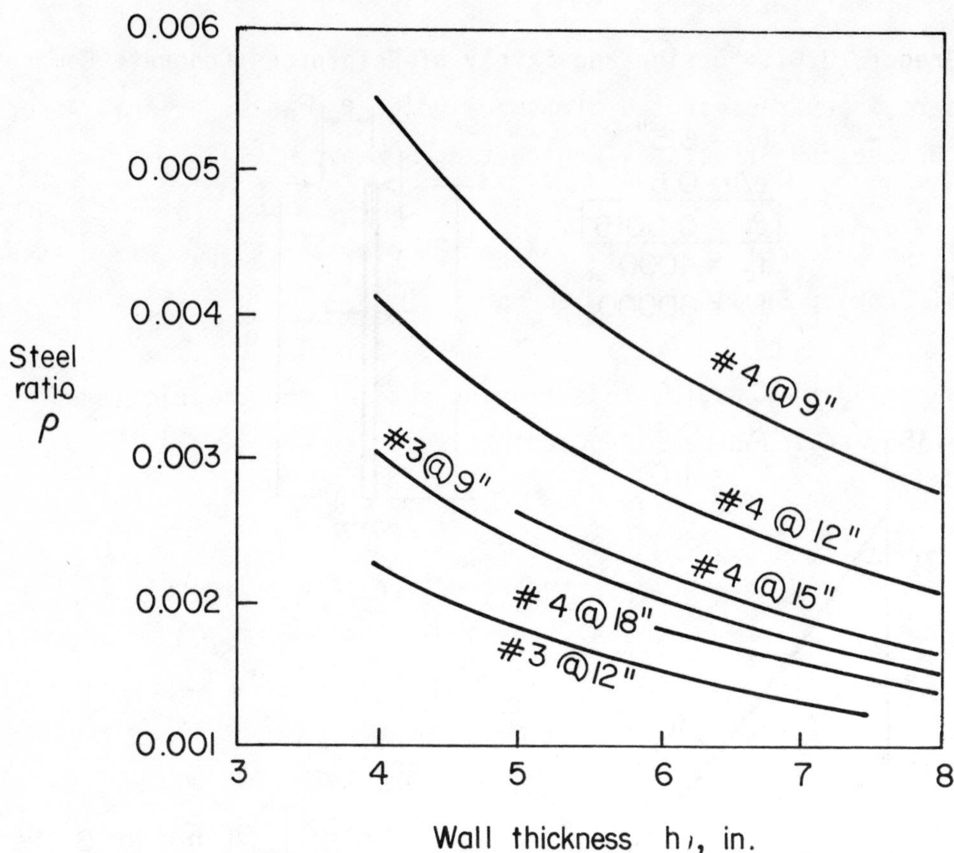

Fig. 22-8 - Design Aid for Wall Reinforcement

11.10 Special "Shear" Provisions for Walls

For most low-rise industrial buildings, horizontal "shear" forces acting in the plane of walls are almost negligible. Such in-plane forces become a design consideration in major structures where a limited number of walls resist the total lateral load, such as in high-rise buildings. Flexural strength must also be considered when in-plane loads are significant. Design Example 22.4 illustrates in-plane shear design of walls, including design for flexural strength.

Selected References

22.1 MacGregor, J.G., "Design and Safety of Reinforced Concrete Compression Members," paper presented at Symposium of the International Association for Bridge and Structural Engineering, Quebec, 1974.

22.2 "Tilt-Up Load Bearing Walls - A Design Aid," Portland Cement Association, Skokie, EB074D, 1980, 28 pp.

22.3 Kripanaryanan, K.M., "Interesting Aspects of the Empirical Wall Design Equation," ACI Journal, Proceedings Vol. 74, No. 5, May 1977, pp. 204-207.

22.4 "Building Code Requirements for Structural Plain Concrete (ACI 318.1-83)," American Concrete Institute, Detroit, 1983.

22.5 "PCI Design Handbook - Precast and Prestressed Concrete," Prestressed Concrete Institute, Chicago, 2nd Edition, 1978, 380 pp.

EXAMPLE 22.1 - Design of Bearing Wall by Empirical Design Method (Section 14.5)

A concrete bearing wall supports a floor system of precast single tees spaced 8 ft on centers. Stem of each tee section is 8 in. wide. Tees have full bearing on wall. Height of wall is 15'-0". Wall considered laterally restrained at top.

Design Data:

Floor beam reactions:

dead load = 28 kips

live load = 14 kips

f'_c = 4,000 psi

f_y = 60,000 psi

Neglect self weight of wall

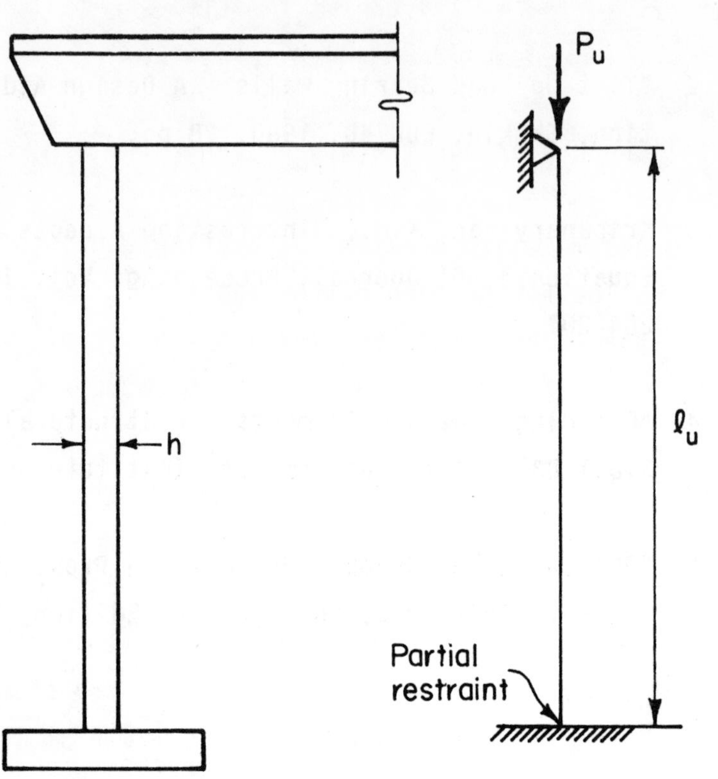

Calculations and Discussion	Code Reference
The general design procedure is to select a trial wall thickness h, then check the trial wall for the applied loading conditions.	
1. Select trial wall thickness h.	
$h > \ell_u/25$ but not less than 4 in.	14.5.3.1
$> 15 \times 12/25 = 7.2$ in.	14.2.6
Try h = 7.5 in.	

EXAMPLE 22.1 - Continued

Calculations and Discussion	Code Reference

2. Calculate factored loading.

$P_u = 1.4D + 1.7L$ — Eq. (9-1)

$= 1.4\ (28) + 1.7\ (14)$

$= 39.2 + 23.8 = 63$ kips

3. Check bearing strength on concrete. — 10.15

Assume width of stem for bearing equal to 7 in. to allow for beveled bottom edges.

Loaded area $= A_1 = 7 \times 7.5 = 52.5$ in.^{2}

$\varphi(0.85\ f'_c\ A_1) = 0.70\ (0.85 \times 4 \times 52.5) = 125$ kips

$63 < 125$ (bearing strength OK)

4. Calculate design strength of wall.

Effective horizontal length of wall per tee reaction is controlled by bearing width of tee stem plus 4 times wall thickness. $7 + 4 \times 7.5 = 37$ in. — 14.2.4

c/c distance between tee stems is greater than 37 in., and does not govern in this case.

$$\varphi P_{nw} = 0.55\ \varphi f'_c\ A_g \left[1 - \left(\frac{k\ell_c}{32h}\right)^2\right]$$ — Eq. (14-1)

$$= 0.55 \times 0.70 \times 4\ (37 \times 7.5) \left[1 - \left(\frac{0.8 \times 15 \times 12}{32 \times 7.5}\right)^2\right]$$

$= 273.5$ kips

$P_u \leq \varphi P_{nw}$

$63 < 273.5$ OK. 7.5-in.-thick wall is adequate with sufficient margin for possible effect of load eccentricity.

EXAMPLE 22.1 - Continued

Calculations and Discussion	Code Reference
5. Select reinforcement. Provide single layer of reinforcement.	
Based on 1-ft width of wall, and Grade 60 reinforcement:	
horizontal A_s = 0.0020 x 12 x 7.5 = 0.180 $in.^2$/ft	14.3.3
vertical A_s = 0.0012 x 12 x 7.5 = 0.108 $in.^2$/ft	14.3.2
Spacing = 3h, but not greater than 18 in.	7.6.5
= 3 x 7.5 = 22.5 in. (18 in. governs)	
Horizontal A_s - Use #4 @ 12-in. centers (A_s = 0.20 $in.^2$)	
Vertical A_s - Use #4 @ 18-in. centers (A_s = 0.13 $in.^2$)	
Design aids such as Fig. 22-8 may be used to select reinforcement directly.	

EXAMPLE 22.2 - Design of Tilt-up Wall Panel by Chapter 10 (Section 14.4)

Roof dead load (8'-0" double tees) = 50 psf

Roof live load = 20 psf

Wind load = 35 psf

ℓ_u = 16'-0"

k = 1.0

f'_c = 4,000 psi

w_c = 150 pcf

f_y = 60,000 psi

Wall considered laterally restrained at top.

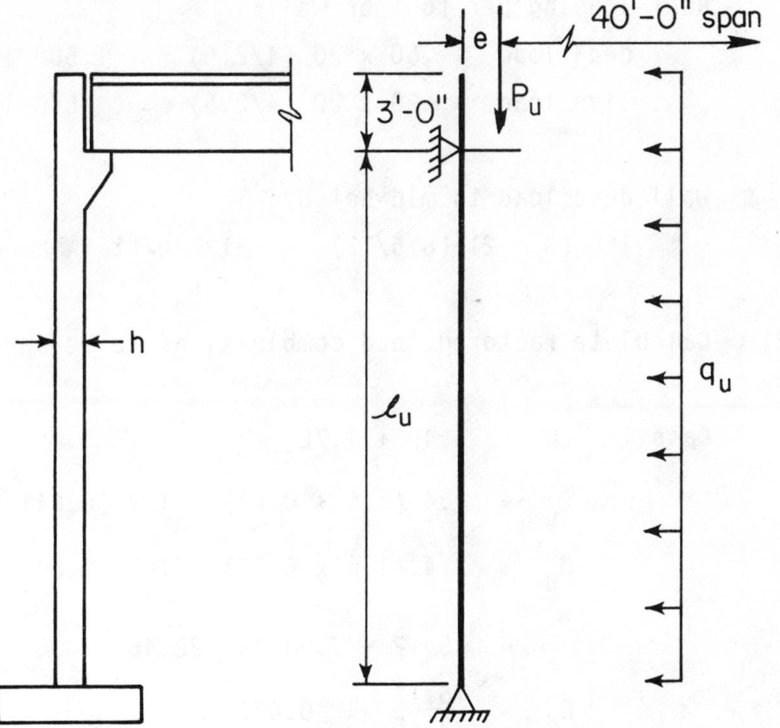

Calculations and Discussion	Code Reference

1. Select trial wall section and reinforcement.

Try h = 6.5 in. (e = 6.75 in.)

A_s = #4 @ 12 in. = 0.20 in.^{2}/ft

ρ = A_s/bh = 0.20/12 x 6.5 = 0.00256

[Design for 1-ft (b = 12 in.) wall section]

Effective length of wall for roof reaction:

(4-in. tee stems @ 4'-0" o.c.)

4 in. + 4 (6.5) = 30 in. (governs) — 14.2.4

distance between stems = 48 in.

EXAMPLE 22.2 - Continued

Calculations and Discussion	Code Reference

Roof loading per foot of wall:

dead load = 50 x 20 (4/2.5) = 1,600 lb/ft

live load = 20 x 20 (4/2.5) = 640 lb/ft

Wall dead load to mid-height:

150 (8 + 2) (6.5/12) = 813 lb/ft

2. Calculate factored load combinations to be investigated.

Case 1: $U = 1.4D + 1.7L$ Eq. (9-1)

$P_u = 1.4\ (1.6 + 0.81) + 1.7\ (0.64) = 4.47^k$

$M_u = 1.4\ (1.6 \times 6.75) + 1.7\ (0.64 \times 6.75)$

$= 15.12 + 7.34 = 22.46^{"k}$

$\beta_d = \frac{15.12}{22.46} = 0.673$

Case 2: $U = 0.75\ (1.4D + 1.7L + 1.7W)$ Eq. (9-2)

$P_u = 0.75\ [1.4\ (1.6 + 0.81) + 1.7\ (0.64)] = 3.35^k$

$M_u = 0.75\ [15.12 + 7.34$

$+ 1.7\ (0.035 \times 16^2 \times 12/8)]$

$= 11.34 + 5.51 + 17.13 = 33.98^{"k}$

$\beta_d = \frac{11.34}{33.98} = 0.334$

Case 3: $U = 0.9D + 1.3W$ Eq. (9-3)

$P_u = 0.9\ (1.6 + 0.81) = 2.17^k$

$M_u = 0.9\ (1.6 \times 6.75) + 1.3\ (0.035 \times 16^2 \times 12/8)$

EXAMPLE 22.2 - Continued

Calculations and Discussion	Code Reference

$$= 9.72 + 17.47 = 27.19^{"k}$$

$$\beta_d = \frac{9.72}{27.19} = 0.357$$

3. Check wall slenderness. — 10.11.4

$k\ell_u/r = 1.0\ (16 \times 12)/0.3 \times 6.5 = 98.5$ — 10.11.3

$98.5 < 100$ Approximate evaluation of slenderness effects by Section 10.11 may be used. — 10.11.4.3

4. Calculate moment magnification by Section 10.11.5 using Eq. (1) for EI.

$$EI = \left(\frac{E_c\ I_g}{\beta}\right)\left(0.5 - \frac{e}{h}\right) \geq 0.1\left(\frac{E_c\ I_g}{\beta}\right)$$ Eq. (1)

$$\leq 0.4\left(\frac{E_c\ I_g}{\beta}\right)$$

$$E_c = 57{,}000\sqrt{4{,}000} = 3.605 \times 10^6 \text{ psi}$$ 8.5.1

$$I_g = \frac{12\ (6.5)^3}{12} = 274.6 \text{ in.}^4$$

$$\frac{e}{h} = \frac{6.75}{6.5} = 1.04$$

$$EI = 0.1\left(\frac{3.605 \times 10^6 \times 274.6}{\beta}\right) = \frac{99 \times 10^6}{\beta}$$

$$P_c = \frac{\pi^2\ EI}{(k\ell_u)^2} = \frac{\pi^2}{(16 \times 12)^2}\left(\frac{99 \times 10^6}{\beta}\right) = \frac{26.5^k}{\beta}$$ Eq. (10-9)

$$\beta = 0.9 + 0.5\ \beta_d^{\ 2} - 12\rho \geq 1.0$$

$$\beta = 0.9 + 0.5\ \beta_d^{\ 2} - 12\ (0.00256)$$

$$\beta = 0.869 + 0.5\ \beta_d^{\ 2} \geq 1.0$$

EXAMPLE 22.2 - Continued

Calculations and Discussion	Code Reference

$$\delta_b = \frac{1.0}{1 - (P_u/\varphi P_c)} \geq 1.0$$ Eq. (10-7)

Calculate increased φ factor for largest $P_u = 4.47^k$ 9.3.2.2(b)

$\varphi = 0.9 - 0.2\ P_u/(0.1\ f'_c A_g) \geq 0.70$

$= 0.9 - 0.2 \times 4.47/(0.1 \times 4 \times 12 \times 6.5) = 0.87$

$M_c = \delta_b M_u$ Eq. (10-6)

Moment Magnification

Load Case	P_u kips	M_u in.-kips	β_d	$\beta \geq 1.0$	EI	P_c kips	δ_b	M_c in.-kips
1	4.47	22.46	0.673	1.10	90×10^6	24.1	1.27	28.52
2	3.35	33.98	0.334	1.0	99×10^6	26.5	1.17	39.76
3	2.17	27.19	0.357	1.0	99×10^6	26.5	1.10	29.91

5. Compare strength required vs. strength provided.

Referring to load-moment interaction strength, Fig. 22-6(b):

Case 1: Required nominal strength $P_u/\varphi = 4.47/0.87 = 5.14^k$

$M_c/\varphi = 28.52/0.87 = 32.78^{"k}$

From Fig. 22-6(b) at $P_n = 5.14^k$, read $M_n = 51^{"k} > 32.78$ OK

EXAMPLE 22.2 - Continued

Calculations and Discussion	Code Reference
Case 2: Required nominal strength P_u/φ = 3.35/0.87 = 3.85^k	
M_c/φ = 39.76/0.87 = $45.7^{"k}$	
From Fig. 22-6(b) at P_n = 3.85^k, read M_n = $46^{"k}$ > 45.7 OK	
Case 3: Required nominal strength P_u/φ = 2.17/0.87 = 2.49^k	
M_c/φ = 29.91/0.87 = $34.4^{"k}$	
From Fig. 22-6(b) at P_n = 2.49^k, read M_n = $42^{"k}$ > 34.4 OK	

EXAMPLE 22.3 - Design of Tilt-up Wall Panel Using Design Aid EB074D

Roof dead load (8'-0" double tees) = 50 psf

Roof live load = 20 psf

Wind load = 20 psf

Unsupported length ℓ_u = 18'-0"

k = 1.0

f'_c = 4,000 psi

f_y = 60,000 psi

Wall considered laterally restrained at top.

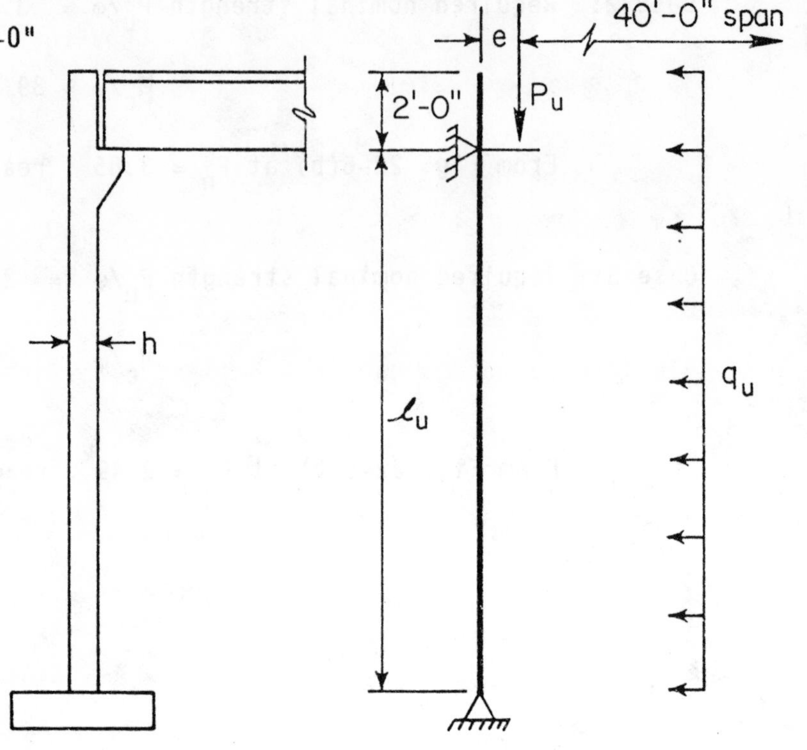

Calculations and Discussion	Code Reference

1. Select trial wall section and reinforcement.

Try h = 6.5 in. (e = 6.75 in.)

A_s = #4 @ 12 in. = 0.20 in.^{2}/ft

ρ = A_s/bh = 0.20/12 x 6.5 = 0.00256

(Design for 1-ft (b = 12 in.) wall section)

Effective length of wall for roof reaction:

(4-in. tee stems @ 4'-0" o.c.)

EXAMPLE 22.3 - Continued

Calculations and Discussion	Code Reference
$4 \text{ in.} + 4\,(6.5) = 30 \text{ in.}$ (governs)	14.2.4
distance between stems = 48 in.	
Roof loading per foot of wall:	
dead load = $50 \times 20\,(4/2.5) = 1{,}600$ lb/ft	
live load = $20 \times 20\,(4/2.5) = 640$ lb/ft	
2. Calculate factored load combinations to be investigated.*	
Case 1: $U = 1.4D + 1.7L$	Eq. (9-1)
$P_u = 1.4\,(1{,}600) + 1.7\,(640) = 3{,}328$ lb/ft	
Case 2: $U = 0.75\,(1.4D + 1.7L + 1.7W)$	Eq. (9-2)
$P_u = 0.75\,(1.4 \times 1{,}600 + 1.7 \times 640) = 2496$ lb/ft	
$q_u = 0.75\,(1.7 \times 20) = 25.5$ psf	
Case 3: $U = 0.9D + 1.3W$	Eq. (9-3)
$P_u = 0.9\,(1{,}600) = 1{,}440$ lb/ft)	
$q_u = 1.3\,(20) = 26$ psf) Case 2 governs	

* Strength tables of Design Aid EB074D include weight of wall.

EXAMPLE 22.3 - Continued

Calculations and Discussion	Code Reference

Calculate increased φ permitted for lightly loaded members: 9.3.2.2(b)

$$P_u = 1.4\ (1{,}600 + 975) + 1.7\ (640) = 4{,}693\ \text{lb/ft}$$

$$\varphi = 0.9 - 0.2\ P_u/(0.1 f'_c\ A_g) \geq 0.70$$

$$= 0.9 - 0.2 \times 4.69/(0.1 \times 4 \times 12 \times 6.5) = 0.87$$

Although slight variations from this value will occur for different load combinations, this single value is considered adequate.

3. <u>Using Design Aid EB074D</u>:

The tabulated strength values in EB074D are nominal strengths P_n. When designing with this aid, the factored loading P_u must be divided by the φ factor to enter the strength tables directly.

Case 1: Required $P_u/\varphi = 3{,}328/0.87 = 3{,}825$ lb/ft

$q_u/\varphi = 0$

From Strength Table A5 (see Table 22-1) for h = 6-1/2 in., $\rho \simeq 0.0025$, e = 6.75 in., $q_u/\varphi = 0$, and

$$k\ell_u/h = 1.0 \times 18 \times 12/6.5 = 33$$

By interpolation, coeff. = 0.0184

$$P_n = 0.0184 \times 4{,}000 \times 12 \times 6.5 = 5{,}740\ \text{lb/ft}$$

$$P_n \geq P_u/\varphi$$

$$5740 > 3825 \quad \text{OK}$$

EXAMPLE 22.3 - Continued

Calculations and Discussion	Code Reference

Case 2: Required P_u/φ = 2,496/0.87 = 2,870 lb/ft

q_u/φ = 25.5/0.87 = 29.3 ≃ 30 psf

From Strength Table A6 (see Table 22-2) for h = 6-1/2 in., $\rho \simeq 0.0025$, e = 6.75 in., q_u/φ = 30 psf, and $k\ell_u/h = 33$

By interpolation, coeff. = 0.011

P_n = 0.011 x 4,000 x 12 x 6.5 = 3,432 lb/ft

$P_n \geq P_u/\varphi$

3432 > 2870 OK

Therefore, h = 6-1/2 in. and A_s = #4 @ 12 in. is adequate for load combinations investigated.

Calculations and Discussion	Code Reference

Table A5 **Load Capacity Coefficients of Tilt-up Concrete Walls*** (h = 6½" and q_u/φ = 0 or 15 psf)

h = 6½"

P_u/φ = (coeff.) $b_1 h f'_c$

w = 150 pcf, $f'_c \le$ 4,000 psi, f_y = 60 ksi

$\rho = \frac{A_s \times 100}{b_1 \times h}$	End eccentricity, e, in.	q_u/φ = 0 psf, Slenderness ratio, $k\ell_u/h$ = 20	30	40	50	q_u/φ = 15 psf, Slenderness ratio, $k\ell_u/h$ = 20	30	40	50
0.15	1.00	0.498	0.347	0.227	0.155	0.468	0.331	0.191	0.085
	3.25	0.094	0.042	0.018	0.013	0.087	0.021	0.005	**
	6.75	0.018	0.014	0.005	0.003	0.017	0.009	0.003	**
0.25	1.00	0.498	0.347	0.227	0.155	0.468	0.331	0.191	0.090
	3.25	0.110	0.050	0.026	0.018	0.105	0.037	0.011	0.003
	6.75	0.029	0.022	0.010	0.006	0.025	0.015	0.006	0.002
0.50	1.00	0.498	0.347	0.227	0.155	0.483	0.331	0.191	0.100
	3.25	0.128	0.066	0.034	0.022	0.124	0.055	0.023	0.011
	6.75	0.049	0.034	0.020	0.012	0.045	0.029	0.016	0.009
0.75	1.00	0.498	0.347	0.227	0.155	0.498	0.331	0.191	0.110
	3.25	0.146	0.082	0.042	0.026	0.142	0.073	0.035	0.019
	6.75	0.069	0.046	0.030	0.018	0.065	0.044	0.026	0.016

Table 22-1 - Reproduced from page 15 of EB074D

Table A6. **Load Capacity Coefficients of Tilt-up Concrete Walls*** (h = 6½" and q_u/φ = 30 or 45 psf)

h = 6½"

P_u/φ = (coeff.) $b_1 h f'_c$

w = 150 pcf, $f'_c \le$ 4,000 psi, f_y = 60 ksi

$\rho = \frac{A_s \times 100}{b_1 \times h}$	End eccentricity, e, in.	q_u/φ = 30 psf, Slenderness ratio, $k\ell_u/h$ = 20	30	40	50	q_u/φ = 45 psf, Slenderness ratio, $k\ell_u/h$ = 20	30	40	50
0.15	1.00	0.468	0.316	0.035	–	0.438	0.110	–	–
	3.25	0.079	0.011	**	–	0.067	**	–	–
	6.75	0.016	0.005	**	–	0.014	**	–	–
0.25	1.00	0.468	0.316	0.151	0.030	0.438	0.301	0.065	–
	3.25	0.101	0.026	0.006	**	0.092	0.016	**	–
	6.75	0.024	0.014	0.004	**	0.023	0.009	**	–
0.50	1.00	0.483	0.316	0.151	0.040	0.453	0.301	0.070	0.010
	3.25	0.121	0.046	0.016	0.004	0.114	0.036	0.010	0.003
	6.75	0.042	0.028	0.013	0.003	0.040	0.024	0.009	0.002
0.75	1.00	0.498	0.316	0.151	0.050	0.468	0.301	0.070	0.020
	3.25	0.141	0.066	0.026	0.009	0.137	0.056	0.021	0.006
	6.75	0.061	0.042	0.023	0.007	0.059	0.039	0.020	0.005

Table 22-2 - Reproduced from page 16 of EB074D

EXAMPLE 22.4 - Shear Design of Walls

Investigate the shear and flexural strength for the wall shown.

$h = 8$ in.

$f'_c = 3{,}000$ psi

$f_y = 60{,}000$ psi

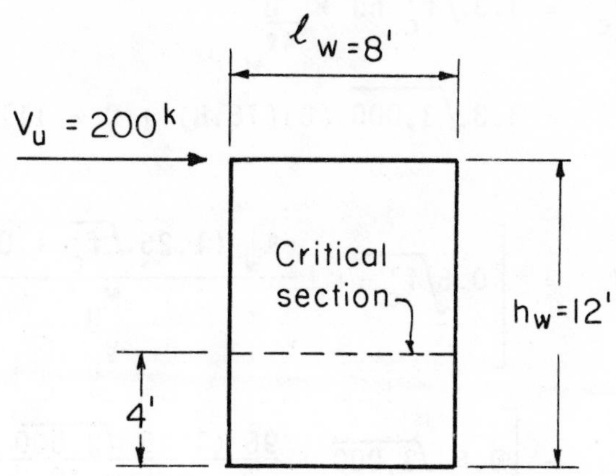

Calculations and Discussion	Code Reference

1. Calculate maximum shear strength permitted.

$V_u \leq \varphi V_n$ — Eq. (11-1)

$\leq \varphi 10\sqrt{f'_c}\ hd$ — 11.10.3

$\leq 0.85 \times 10\sqrt{3{,}000} \times 8\ (0.8 \times 96) = 286^k$

$200^k \leq 286^k$

8-in. wall section is adequate

2. Calculate shear strength provided by concrete V_c. — 11.10.6

Critical section for shear: — 11.10.7

$\ell_w/2 = 8/2 = 4$ ft (governs)

$h_w/2 = 12/2 = 6$ ft

EXAMPLE 22.4 - Continued

Calculations and Discussion	Code Reference

$$V_c = 3.3\sqrt{f'_c}\,hd + \frac{N_u d}{4\ell_w}$$ Eq. (11-32)

$$= 3.3\sqrt{3{,}000}\,(8)(76.8) + 0 = 111^k$$

or

$$V_c = \left[0.6\sqrt{f'_c} + \frac{\ell_w\,(1.25\sqrt{f'_c} + 0.2\,N_u/\ell_w h)}{\frac{M_u}{V_u} - \frac{\ell_w}{2}}\right] hd$$ Eq. (11-33)

$$= \left[0.6\,\sqrt{3{,}000} + \frac{96\,(1.25\,\sqrt{3{,}000} + 0)}{96 - 48}\right] 8 \times 76.8 = 104^k \text{ (governs)}$$

where $d = 0.8\ell_w = 0.8(96) = 76.8$ in. 11.10.4

$N_u = 0$

$M_u = (12-4)\,V_u = 8V_u\,'^k = 96\,V_u\,''^k$

$V_u = 200^k > \varphi V_c/2 = 0.85\,(104)/2 = 44.2^k$ 11.10.8

Shear reinforcement must be provided in accordance with Section 11.10.9.

3. Calculate required horizontal shear reinforcement. 11.10.9.1

$V_u \leq \varphi V_n$ Eq. (11-1)

$\leq \varphi(V_c + V_s)$ Eq. (11-2)

$\leq \varphi V_c + \varphi\dfrac{A_v f_y d}{s_2}$ Eq. (11-34)

Solving for $\dfrac{A_v}{s_2} = \dfrac{(V_u - \varphi V_c)}{\varphi f_y d}$

$$= \frac{(200 - 0.85 \times 104)}{0.85 \times 60 \times 76.8} = 0.0285$$

EXAMPLE 22.4 - Continued

Calculations and Discussion	Code Reference
For 2 #3: $s_2 = \dfrac{2 \times 0.11}{0.0285} = 7.72$ in.	
2 #4: $s_2 = \dfrac{2 \times 0.20}{0.0285} = 14.04$ in.	
2 #5: $s_2 = \dfrac{2 \times 0.31}{0.0285} = 21.76$ in.	
Maximum spacing not greater than the smaller of:	11.10.9.3
$\ell_w/5 = 8 \times 12/5 = 19.2$ in.	
$3h = 3 \times 8 = 24.0$ in.	
or 18 in.	
<u>Use 2 #4 @ 14 in.</u>	
Check $\rho_h \geq 0.0025$	11.10.9.2
$\rho_h = \dfrac{A_v}{A_g} = \dfrac{2 \times 0.20}{8 \times 14} = 0.0036 > 0.0025$ (OK)	
4. Calculate vertical shear reinforcement	11.10.9.4
$\rho_n = 0.0025 + 0.5\left(2.5 - \dfrac{h_w}{\ell_w}\right)(\rho_h - 0.0025)$	Eq. (11-35)
$= 0.0025 + 0.5\ (2.5 - 1.5)\ (0.0036 - 0.0025)$	
$= 0.0031$	
<u>Use 2 #4 @ 14 in.</u>	
5. Design for flexure.	
$M_u = V_u h_w = 200 \times 12 = 2400^{'k}$	

Calculations and Discussion	Code Reference

Using Table 9-2, page 9-7:

Compute, $\dfrac{M_u}{\varphi f'_c bd^2} = \dfrac{2400 \times 12}{0.9 \times 3 \times 8 \times 76.8^2} = 0.2261$

where $d = 0.8\ \ell_w = 0.8 \times 96 = 76.8$ in. 11.10.4

A larger value of d could be used if determined by a strain compatibility analysis.

From Table 9-2, read $\omega \simeq 0.269$

$$A_s = \rho hd = \frac{\omega f'_c hd}{f_y} = \frac{0.269 \times 3 \times 8 \times 76.8}{60} = 8.26 \text{ in.}^2$$

Use 11 #8 bars each side

(A_s = 8.69 in.^{2})

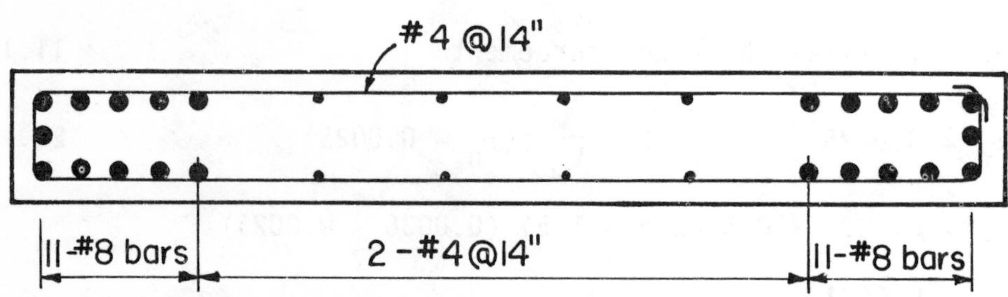

Reinforcement Details

23

Footings

Update for '83 Code

For ACI 318-83, Section 15.8 addressing transfer of force between a footing and supported member (column, wall, or pedestal) is extensively revised to address both cast-in-place and precast construction. New Section 15.8 consists of three parts: Section 15.8.1 gives general requirements applicable to both cast-in-place and precast construction; Section 15.8.2 gives additional rules for cast-in-place construction; and Section 15.8.3 gives additional rules for precast construction.

For force transfer between a footing and a precast column or wall, anchor bolts or mechanical connectors are now specifically addressed in Section 15.8.3. (Prior to '83, connections between a precast member and footing were required to have either longitudinal bars or dowels crossing the interface . . . obviously, contrary to common practice.) See Design Example 23.9.

Also note that walls are now specifically addressed in Section 15.8 for force transfer to footings.

There is one significant deletion from Chapter 15: with publication of "Building Code Requirements for Structural Plain Concrete" ACI 318.1-83, '77 Code Section 15.11 addressing plain concrete pedestals and footings is deleted. Design of plain concrete footings is now in accordance with ACI 318.1. See Design Example 23.8.

15.2 Loads and Reactions

The first step in design is to determine required footing base area on the basis of permissible soil pressures or pile bearing loads and actual unfactored service loads in whatever combination governs the design. See Design Example 23.1.

When the plan dimensions of the footing have been established, the depth and required reinforcement follow. For this purpose, the contact pressures and all loads are increased by the appropriate load factors specified in Section 9.2 - Required Strength. The factored loads or related internal moments and shears are then used to proportion the footing for required shear and moment strength.

15.5 Shear in Footings

Shear strength of a footing in the vicinity of the supported member (column or wall) must be determined for the more severe of the two conditions stated in Section 11.11.

Both beam action (Section 11.11.1.1) and two-way action (Section 11.11.1.2) for the footing must be checked to determine required footing depth. See Design Example 23.2. Beam action assumes the footing acts as a wide beam with a critical section across its entire width. If this condition is the more severe, design for shear proceeds in accordance with Sections 11.1 through 11.5. Two-way action for the footing checks "punching" shear strength. The critical section for punching shear is a perimeter b_o around the supported member with the shear strength computed in accordance with Eq. (11-36).

The shear strength for two-way action is a function of support size β_c, with a reduction in shear strength from $4\sqrt{f'_c}\, b_o d$ to $2\sqrt{f'_c}\, b_o d$, depending upon the β_c ratio. (β_c is the ratio of long-to-short side of the column or support area). Fig. 23-1 illustrates the shear strength reduction as a function of β_c.

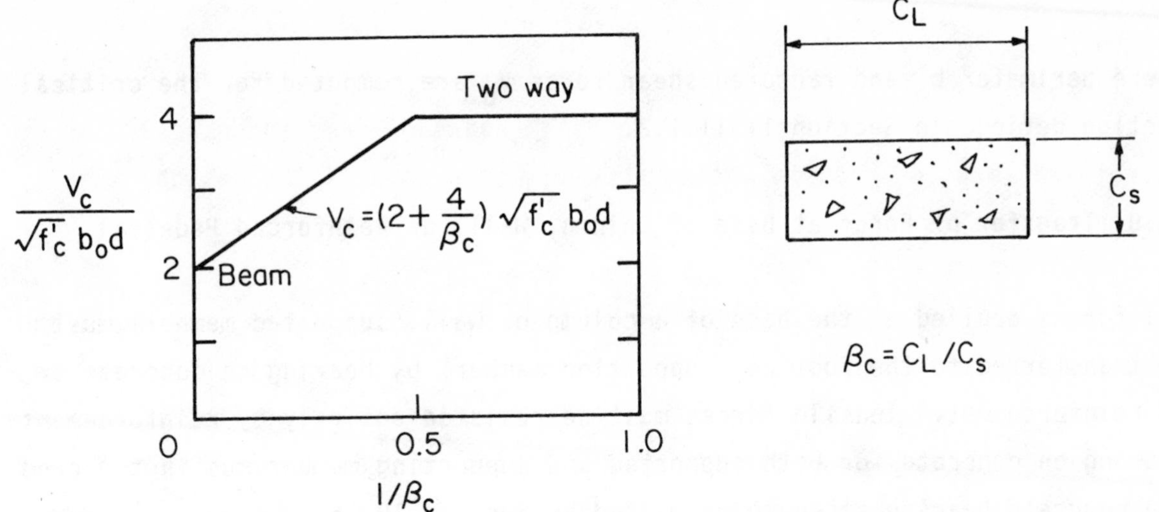

Fig. 23-1 - Shear Strength of Footings

If the factored shear force V_u at the critical section exceeds the shear strength φV_c given by Eq. (11-36), shear reinforcement must be provided. If shear reinforcement consisting of bars or wires is used, the shear strength may be increased to a maximum value of $6\sqrt{f'_c}\, b_o d$. However, shear reinforcement must be designed to carry shear in excess of $2\sqrt{f'_c}\, b_o d$. This limit is one-half that permitted by Eq. (11-36) with a β_c ratio of 2 or less.

For footing design (without shear reinforcement), the shear strength equations may be summarized as follows:

Beam Action

$$V_u \leq \varphi V_n \qquad \text{Eq. (11-1)}$$

$$V_u \leq \varphi(2\sqrt{f'_c}\, b_w d) \qquad \text{Eq. (11-3)}$$

where footing width b_w and factored shear force V_u are computed for the critical section defined in Section 11.11.1.1.

Two-Way Action

$$V_u \leq \varphi V_n \qquad \text{Eq. (11-1)}$$

$$V_u \leq \varphi\, (2 + \frac{4}{\beta_c})\sqrt{f'_c}\, b_o d \qquad \text{Eq. (11-36)}$$

but not greater than $4 \varphi \sqrt{f'_c}\, b_o d$

where perimeter b_o and factored shear force V_u are computed for the critical section defined in Section 11.11.1.2.

15.8 Transfer of Force at Base of Column, Wall, or Reinforced Pedestal

All forces applied at the base of a column or wall (supported member) must be transferred to the footing (supporting member) by bearing on concrete or by reinforcement. Tensile forces must be resisted entirely by reinforcement. Bearing on concrete for both supported and supporting member must not exceed the concrete bearing strength permitted by Section 10.15.

For a supported column

$$\varphi P_{nb} = \varphi(0.85 f'_c A_1) \qquad 10.15.1$$

where f'_c is strength of column concrete.

For the usual case of a supporting footing with a total area considerably greater than the supported column, $\sqrt{A_2/A_1} > 2$

$$\varphi P_{nb} = 2\,[\varphi(0.85 f'_c A_1)] \qquad 10.15.2$$

where f'_c is strength of footing concrete.

Where bearing strength is exceeded, reinforcement must be provided to transfer the excess. A minimum area of reinforcement must be provided across the interface of column or wall and footing, even where concrete bearing strength is not exceeded. With the new force transfer provisions addressing both cast-in-place and precast construction, including force transfer between a wall and footing, the minimum reinforcement requirements are based on type of supported member.

CIP Columns	$A_s = 0.005\, A_g$
CIP Walls	A_s = minimum vertical wall reinforcement (Section 14.3.2)
P/C Columns	$A_s = 200\, A_g/f_y$
P/C Walls	$A_s = 50\, A_g/f_y$

For cast-in-place construction, reinforcement may consist of extended reinforcing bars or dowels; for precast construction, reinforcement may consist of anchor bolts or mechanical connectors. Unfortunately the Code does not give any specific data for design of anchor bolts or mechanical connectors. See Design Example 23.9.

The shear-friction method of Section 11.7 should be used to design for horizontal force transfer between column and footing. See Design Examples 23.4 and 23.5. Consideration of some of the lateral force being transferred by shear through a formed shear key is questionable. Considerable slip is required to develop a shear key. Shear keys, if provided, should be considered as an added mechanical factor of safety only, with no design shear force assigned to the shear key. See Design Example 23.6.

With publication of ACI 318-83, design of plain concrete pedestals and footings is now in accordance with ACI Standard 318.1-83. See Design Example 23.8.

EXAMPLE 23.1 - Design for Base Area of Footing

Determine base area A_f for a square spread footing with the following design conditions:

Service dead load = 350^k

Service live load = 275^k

Service surcharge = 100 psf

Assume average weight of soil and concrete above footing base = 130 pcf

Permissible soil pressure = 4.5 ksf

Column = 30 in. x 12 in.

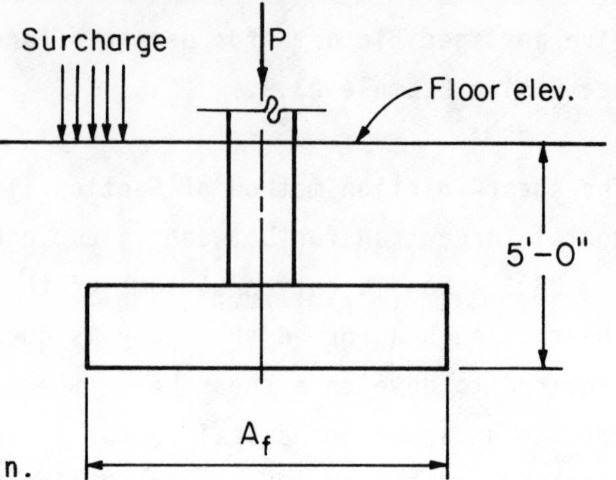

Calculations and Discussion	Code Reference
1. Total weight of surcharge:	
$0.130 \times 5 + 0.100 = 0.750$ ksf	
2. Net permissible soil pressure:	
$4.5 - 0.75 = 3.75$ ksf	
3. Footing base area:	
$A_f = \dfrac{350 + 275}{3.75} = 167\ ft^2$	15.2.2
Use 13 ft x 13 ft square footing ($A_f = 169\ ft^2$)	
Note that base area of footing is determined using service loads (unfactored loads) with the permissible soil pressure.	

EXAMPLE 23.1 - Continued

Calculations and Discussion	Code Reference

4. Factored loads and soil reaction due to factored loading:

$$U = 1.4\,(350) + 1.7\,(275) = 957.5^{k}$$ Eq. (9-1)

$$q_s = \frac{U}{A_f} = \frac{957.5}{169} = 5.70 \text{ ksf}$$

To proportion the footing for strength (depth and required reinforcement) factored loads must be used. 15.2.1

EXAMPLE 23.2 - Design for Depth of Footing

For the design conditions of Example 23.1, determine overall thickness of footing.

f'_c = 3000 psi

P_u = 957.5^k

q_s = 5.70 ksf

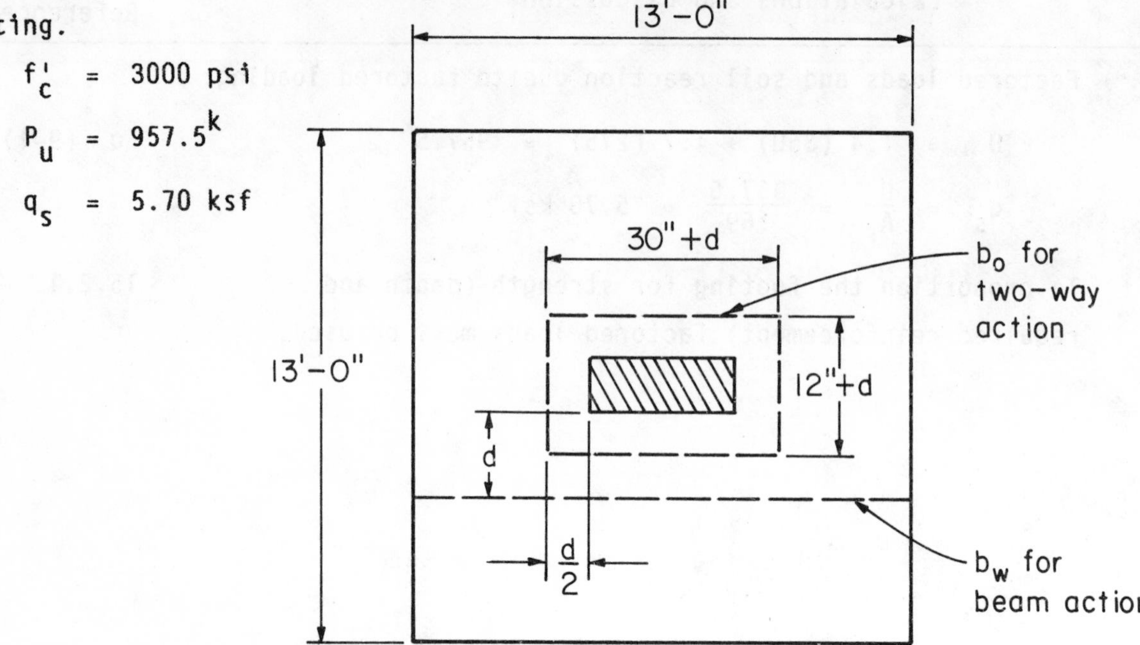

Calculations and Discussion	Code Reference
Determine depth based on shear strength without shear reinforcement. Depth required for shear usually controls footing thickness. Both beam action and two-way action for the footing must be investigated. Assume overall thickness of 33 in. Average d ≃ 28 in.	11.11
1. Beam action for footing:	11.11.1.1
$V_u \leq \varphi V_n$	Eq. (11-1)
$V_u < \varphi(2\sqrt{f'_c}\ b_w d)$	Eq. (11-3)
$V_u = 5.70\ (6.0 - 2.33)(13) = 272^k$	
b_w = 13 ft = 156 in.	

EXAMPLE 23.2 - Continued

Calculations and Discussion	Code Reference
$V_u \leq 0.85\ (2\sqrt{3000} \times 156 \times 28)/1000$	
$272^k < 407^k$ OK	
2. Two-way action for footing:	11.11.1.2
$V_u \leq \varphi V_n$	Eq. (11-1)
$V_u \leq \varphi\ (2 + \frac{4}{\beta_c})\sqrt{f'_c}\ b_o d$	Eq. (11-36)
but not greater than $\varphi\ 4\sqrt{f'_c}\ b_o d$	
$V_u = 5.70\ (169 - 4.83 \times 3.33) = 872^k$	
$b_o = 2\ (30 + 28) + 2\ (12 + 28) = 196$ in.	
$\beta_c = 30/12 = 2.5$	
$V_u \leq 0.85\ (2 + \frac{4}{2.5})\sqrt{3000} \times 196 \times 28/1000$	
$872^k < 920^k$ OK	
therefore, the 28 in. effective depth is adequate for shear.	

EXAMPLE 23.3 - Design for Footing Reinforcement

For the design conditions of Example 23.1, determine required footing reinforcement.

f'_c = 3,000 psi

f_y = 60,000 psi

P_u = 957.5^k

q_s = 5.70 ksf

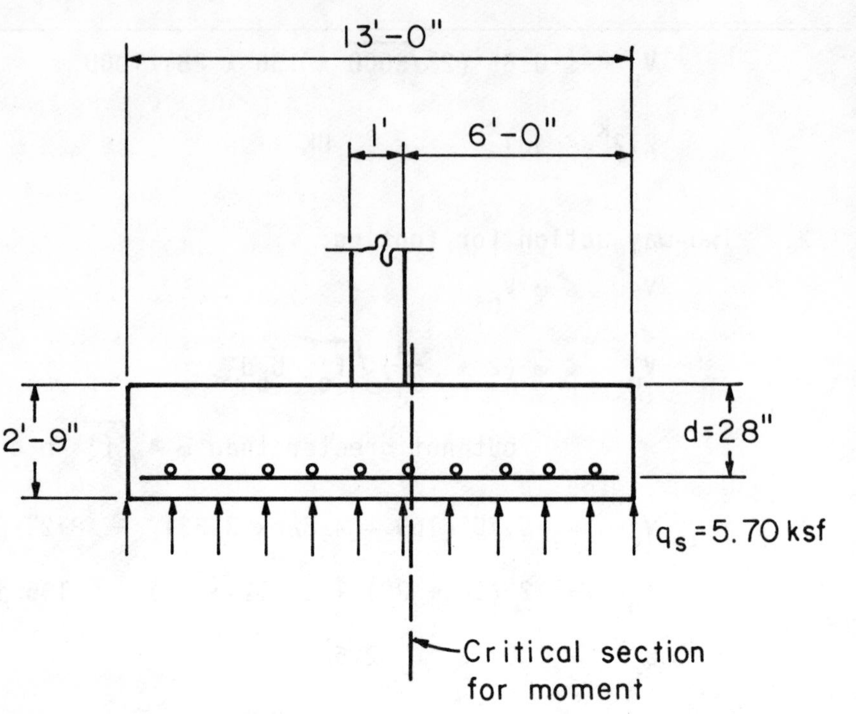

Calculations and Discussion	Code Reference
1. Critical section for moment is at face of column:	15.4.2
$M_u = 5.70 \times 13 \times 6^2/2 = 1334^{'k}$	
2. Compute A_s required using formula (4) of Part 9:	
Required $R_u = \frac{M_u}{\varphi bd^2} = \frac{1334 \times 12 \times 1000}{0.9 \times 156 \times 28^2} = 145$ psi	
$\rho = \frac{0.85 f'_c}{f_y}\left(1 - \sqrt{1 - \frac{2 R_u}{0.85 f'_c}}\right)$	
$= \frac{0.85 \times 3}{60}\left(1 - \sqrt{1 - \frac{2 \times 145}{0.85 \times 3000}}\right) = 0.0025$	
Check minimum required for structural slabs of uniform thickness; for Grade 60:	10.5.3
$\rho_{min} = 0.0018 < 0.0025$ OK	7.12.2

EXAMPLE 23.3 - Continued

Calculations and Discussion	Code Reference

Earlier editions of the Code were subject to interpretation regarding whether the minimum reinforcement requirements of Section 10.5 should apply to footings. A minimum reinforcement is required to guard against a mode of failure that can occur in very lightly reinforced members; if the moment strength of a cracked section is less than the moment strength of the uncracked section, the member will fail immediately upon formation of a crack. It seems reasonable that this mode of failure applies to footings as well as beams. The lesser value for slabs (Section 10.5.3) is considered adequate because of the lateral distribution of any overload possible with slab-like members. Commentary Section 10.5.3 is more explicit, stating that slabs (footings) which help support the structure vertically should meet the requirements of Section 10.5.3.

Required $A_s = \rho bd$

$A_s = 0.0025 \times 156 \times 28 = 10.92$ sq in.

Use 14 #8 bars (A_s = 11.06 sq. in.) each way*

*A lesser amount of reinforcement is required in the perpendicular direction. For ease of placement use same reinforcement each way.

3. Check development of reinforcement. 15.6

Critical section for development is the same as that for moment (at face of column). 15.6.3

EXAMPLE 23.3 - Continued

Calculations and Discussion	Code Reference
For #8 bars: $\ell_d = (0.04\ A_b f_y / \sqrt{f'_c})\ 0.8$	12.2.2 12.2.4.1
$\ell_d = (0.04 \times 0.79 \times 60{,}000/\sqrt{3000})\ 0.8$	
$\ell_d = 27.7$ in. < 60 in. (short projection) OK	

EXAMPLE 23.4 - Design for Transfer of Force at Base of Column

For the design conditions of Example 23.1, check force transfer at interface of column and footing.

f'_c (column) = 5,000 psi

f'_c (footing) = 3,000 psi

f_y = 60,000 psi

P_u = 957.5^k

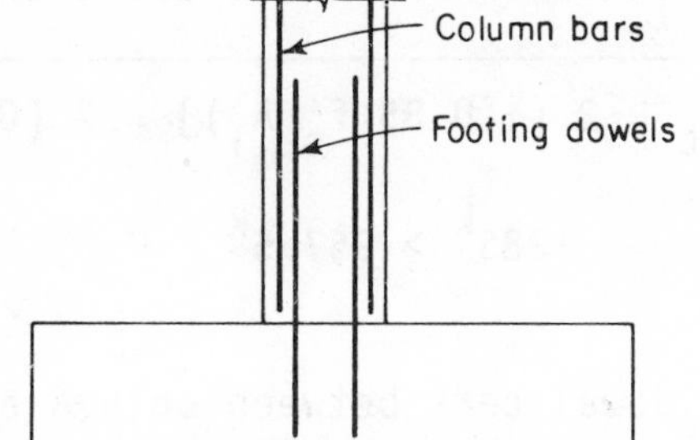

Calculations and Discussion	Code Reference
1. Bearing strength on column concrete, f'_c = 5,000 psi:	15.8.1.1
$\varphi P_{nb} = \varphi(0.85\ f'_c\ A_1)$ = 0.70 (0.85 x 5 x 12 x 30)	10.15.1
= $1071^k > 957.5^k$ OK	9.3.2.4
2. Bearing strength on footing concrete, f'_c = 3,000 psi:	15.8.1.1
For bearing on footing concrete, the bearing strength is increased due to the large footing area permitting a greater distribution of the column load. The increase permitted varies between 1 and 2 in accordance with the expression $\sqrt{A_2/A_1} \leq 2$, where A_1 is the column area (loaded area) and A_2 is the maximum area of the portion of the footing area that is geometrically similar to and concentric with the column area. For the 30 in. x 12 in. column supported on the 13 ft x 13 ft square footing.	10.15.2

$$\sqrt{A_2/A_1} = \sqrt{138 \times 156/30 \times 12} = 7.7 > 2$$

Note: When the loaded area A_1 is half or less of the supported area A_2, as for footings, the bearing strength will be increased by a factor of 2. Also, bearing on column concrete will always govern until the strength of the column concrete exceeds twice that of the footing.

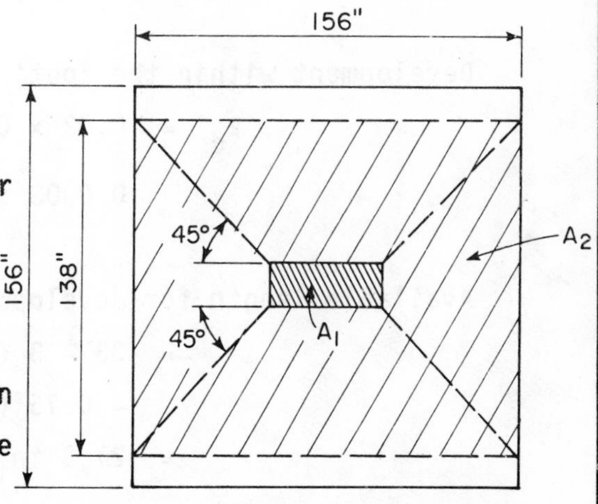

EXAMPLE 23.4 - Continued

Calculations and Discussion	Code Reference

$\varphi P_{nb} = 2\ [\varphi(0.85\ f'_c\ A_1)] = 2\ [0.70\ (0.85 \times 3 \times 12 \times 30)]$

$= 1285^k > 957.5^k$ OK

3. Required dowel bars between column and footing: 15.8.2

Even though bearing strength on column and footing concrete is adequate to transfer the factored loading, a minimum area of reinforcement is required across the interface. 15.8.2.1

$A_s\ (\text{min}) = 0.005\ (30 \times 12) = 1.80$ sq in.

Provide 4 #6 bars as dowels (A_s = 1.76 sq in.)

4. Development of dowel reinforcement:

For #6 bars: $\ell_d = 0.02\ d_b f_y/\sqrt{f'_c}$ 12.3.2
but not less than $0.0003\ d_b f_y$

Development within the column,

$\ell_d = 0.02 \times 0.75 \times 60{,}000/\sqrt{5{,}000} = 12.7$ in.

$0.0003 \times 0.75 \times 60{,}000 = 13.5$ in. (Controls)

Development within the footing,

$\ell_d = 0.02 \times 0.75 \times 60{,}000/\sqrt{3{,}000} = 16.4$ in. (Controls)

$0.0003 \times 0.75 \times 60{,}000 = 13.5$ in.

Available length for development above footing reinforcement,

= 33 - 3 (cover) - 2 x 1.0 (footing bars)
- 0.75 (dowels)

= 27.3 in. > 16.4 in. OK

EXAMPLE 23.5 - Design for Transfer of Force by Reinforcement

For the design conditions given below, provide for transfer of force between column and footing.

12 x 12 tied reinforced column
with 4 #14 longitudinal bars

f'_c = 4,000 psi (column & footing)

f_y = 60,000 psi

P_D = 200^k

P_L = 100^k

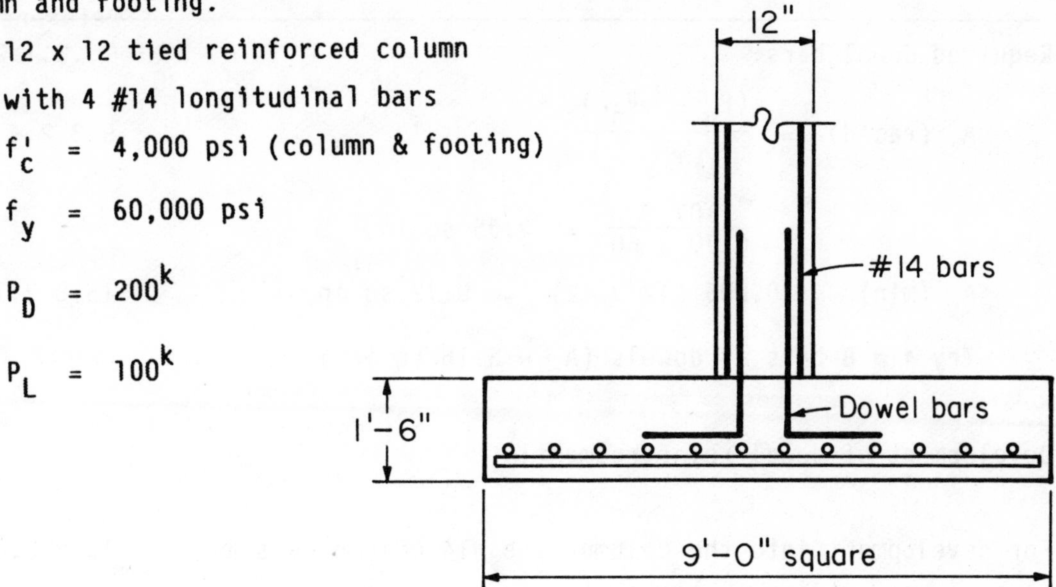

Calculations and Discussion	Code Reference
1. Bearing strength on column concrete:	15.8.1.1
$\varphi P_{nb} = \varphi(0.85\ f'_c\ A_1) = 0.70\ (0.85 \times 4 \times 12 \times 12)$	10.15.1
$= 342.7^k$	
$P_u = 1.4 \times 200 + 1.7 \times 100 = 450^k$	Eq. (9-1)
$450^k > 342.7^k$ N.G.	
The column load cannot be transferred by bearing on concrete alone. The excess load = 450 - 342.7 = 107.3^k must be transferred by reinforcement.	15.8.1.2
2. Bearing strength on footing concrete:	15.8.1.1
$\sqrt{A_2/A_1} > 2$	
$\varphi P_{nb} = 2\ (342.7) = 685.4^k > 450^k$ OK	

EXAMPLE 23.5 - Continued

Calculations and Discussion	Code Reference

3. Required dowel bars: 15.8.1.2

$$A_s\ (\text{req'd}) = \frac{(P_u - \varphi P_{nb})}{\varphi f_y}$$ 9.3.2.4

$$= \frac{107.3}{0.70 \times 60} = 2.55 \text{ sq in.}$$

$$A_s\ (\text{min}) = 0.005\ (12 \times 12) = 0.72 \text{ sq in.}$$ 15.8.2.1

Try 4 # 8 bars as dowels (A_s = 3.16 sq in.)

4. Development of dowel reinforcement:

For development into the column, the #14 column bars may be lap spliced with the #8 footing dowels. The dowels must extend into the column a distance not less than the development length of the #14 column bars or the lap splice length of the #8 footing dowels, whichever is greater. 15.8.2.4

For #14 bars: $\ell_d = 0.02\ d_b f_y / \sqrt{f'_c}$ 12.3.2

but not less than $0.0003\ d_b f_y$

$= 0.02 \times 1.693 \times 60{,}000/\sqrt{4{,}000} = 32.1$ in. (Controls)

$0.0003 \times 1.693 \times 60{,}000 = 30.5$ in.

For #8 bars: lap $= 0.02\ d_b f_y / \sqrt{f'_c}$ 12.16.1

but not less than $0.0005\ d_b f_y$

$= 0.02 \times 1.0 \times 60{,}000/\sqrt{4{,}000} = 19.0$ in.

$0.0005 \times 1.0 \times 60{,}000 = 30.0$ in.

The #8 dowel bars must extend not less than 32 in. into the column.

EXAMPLE 23.5 - Continued

Calculations and Discussion	Code Reference
For development into the footing, the #8 dowels must extend a full development length.	15.8.2.4
$\ell_d = 0.02 \times 1.0 \times 60{,}000/\sqrt{4{,}000} = 19.0$ in.	12.3.2
$0.0003 \times 1.0 \times 60{,}000 = 18.0$ in.	
ℓ_d may be reduced to account for excess area.	12.3.3.1
A_s req'd./A_s prov'd $= 2.55/3.16 = 0.81$	
$\ell_d = 19 \times 0.81 = 15.4$ in.	
If the footing dowels are bent for placement on top of the footing reinforcement (as shown in the sketch), the bent portion cannot be considered effective for developing the bars in compression. Available length for development above footing reinforcement $\simeq 18 - 6 = 12$ in. < 15.4 required. Either the footing depth must be increased or a larger number of smaller-sized dowels used.	12.1
Try 6 #7 bars as dowels ($A_s = 3.60$ sq in.)	
$\ell_d = 0.02 \times 0.875 \times 60{,}000/\sqrt{4{,}000} = 16.6$ in.	
$0.0003 \times 0.875 \times 60{,}000 = 15.8$ in.	
ℓ_d required $= 16.6 \times \frac{2.55}{3.60}$	
$= 11.8$ in. < 12 is available OK	
Provide 6 #7 bars as dowels, extended 32 in. into the column and bent 90 deg. for placement on top of the footing reinforcement. Total vertical length = 32 + 12 = 44 in., say 3 ft-9in.	

EXAMPLE 23.6 - Design for Transfer of Horizontal Force at Base of Column

For the design conditions of Example 23.5, provide for transfer of a horizontal factored force of 95^k acting at the base of column. f'_c = 4,000 psi and f_y = 60,000 psi.

Calculations and Discussion	Code Reference

1. The shear-friction design method of Section 11.7.4 will be used to design for horizontal force transfer. — 15.8.1.4

$V_u \leq \varphi V_n$ — Eq. (11-1)

$V_u \leq \varphi (A_{vf} f_y \mu)$ — Eq. (11-26)

Use μ = 0.6 (concrete not intentionally roughened) — 11.7.4.3

and φ = 0.85 (shear) — 9.3.2.3

$$A_{vf} = \frac{95}{0.85 \times 60 \times 0.6} = 3.10 \text{ sq in.}$$

2. The 6 #7 footing dowels (A_s = 3.60 sq in.) provided for vertical force transfer may also act as shear-friction reinforcement. Therefore, no additional reinforcement is required for the horizontal force transfer. If the 6 #7 dowels were not adequate, a 40% reduction in required A_{vf} is possible if the footing concrete in contact with the column concrete is roughened "to an amplitude of approximately 1/4 in." With the roughened surface, μ = 1.0, and required A_{vf} = 95/0.85 x 60 x 1.0 = 1.86 sq in.

Tensile development length of the dowels needs to be checked.

For #7 bars: $\ell_d = 0.04 A_b f_y / \sqrt{f'_c}$ — 12.2.2

but not less than $0.0004 d_b f_y$

$\ell_d = 0.04 \times 0.60 \times 60{,}000/\sqrt{4{,}000} = 22.8$ in.

$0.0004 \times 0.875 \times 60{,}000 = 21.0$ in.

EXAMPLE 23.6 - Continued

Calculations and Discussion	Code Reference
To account for excess reinforcement:	12.2.4.2
ℓ_d = 22.8 x 1.86/3.60 = 11.8 in.	
Bar lengths as provided for vertical force transfer, OK.	
3. Check maximum shear transfer strength permitted.	11.7.5
$V_u \leq \varphi\ (0.2\ f'_c\ A_c)$ but not greater than $\varphi\ (800\ A_c)$	
$\leq 0.85\ (0.2 \times 4 \times 12 \times 12) = 97.9^k$	
$0.85\ (800 \times 12 \times 12)/1000 = 97.9^k$	
$95^k < 97.9^k$ OK	
The top of the footing at the interface between column and footing must be clean and free of laitance before placement of the column concrete.	11.7.9

EXAMPLE 23.7 - Design for Depth of Footing with Piles

For the footing supported on piles shown, determine thickness of footing (pile cap).

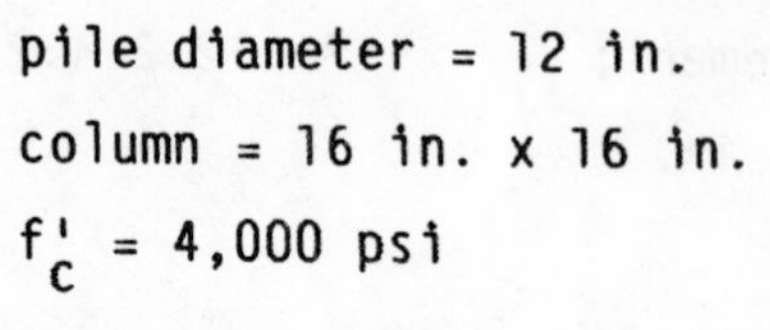

pile diameter = 12 in.

column = 16 in. x 16 in.

f'_c = 4,000 psi

Load per pile:

$P_D = 20^k$

$P_L = 10^k$

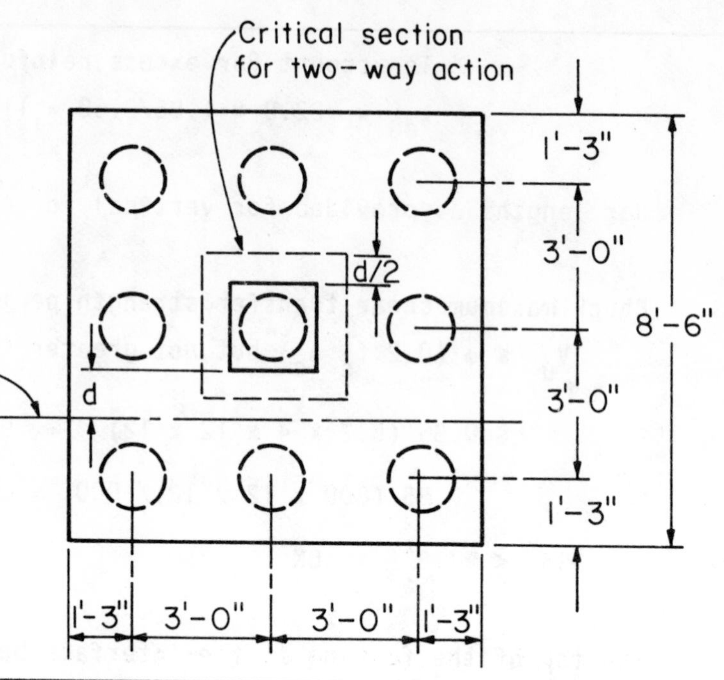

Calculations and Discussion	Code Reference
A. Depth required for shear usually controls footing thickness. Both beam action and two-way action for the footing must be investigated. Assume an overall thickness of 1 ft-9 in. Average d ≃ 14 in.	11.11 15.7
Factored pile loading:	
$P_u = 1.4\ (20) + 1.7\ (10) = 45^k$	Eq. (9-1)
1. Beam action for footing:	11.11.1.1
$V_u \leq \varphi V_n$	Eq. (11-1)
$V_u \leq \varphi\ (2\sqrt{f'_c}\ b_w d)$	Eq. (11-3)
V_u (neglecting footing wt.) = 3 x 45 = 135^k	

EXAMPLE 23.7 - Continued

Calculations and Discussion	Code Reference

$b_w = 8\text{ ft-6 in.} = 102\text{ in.}$

$V_u \leq 0.85\ (2\sqrt{4{,}000} \times 102 \times 14)/1000$

$135^k < 153.5^k$ OK

2. Two-way action for footing: 11.11.1.2

$V_u \leq \varphi V_n$ Eq. (11-1)

$V_u \leq \varphi\ (2 + 4/\beta_c)\sqrt{f'_c}\ b_o d$ Eq. (11-36)

but not greater than $\varphi\ 4\sqrt{f'_c}\ b_o d$

$V_u = 8 \times 45^k = 360^k$

$b_o = 4\ (16 + 14) = 120\text{ in.}$

$\beta_c = 1.0, \quad \varphi V_n = \varphi\ 4\sqrt{f'_c}\ b_o d$

$V_u \leq 0.85 \times 4\sqrt{4{,}000} \times 120 \times 14/1000$

$360^k < 361.3^k$ OK

Therefore, the 14 in. effective depth is adequate for shear.

B. Check "punching" shear strength at piles.

With piles spaced at 3 ft-0 in. cts., critical perimeters do not overlap. $V_u = 45^k$ per pile. $b_o = \pi\ (12 + 14) = 81.7\text{ in.}$

$V_u \leq \varphi\ 4\sqrt{f'_c}\ b_o d$ 11.11.2

$\leq 0.85 \times 4\sqrt{4{,}000} \times 81.7 \times 14/1000$

$45^k < 246^k$ OK

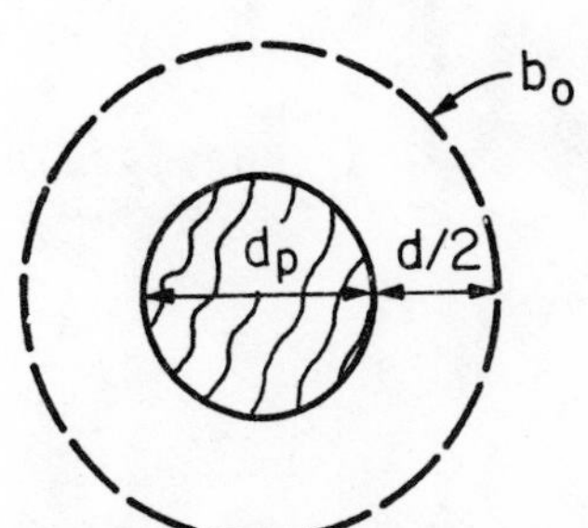

EXAMPLE 23.8 - Design of Plain Concrete Footing

Proportion a plain concrete square footing for the following design conditions:

Service dead load = 40^k

Service live load = 60^k

Service surcharge = 0

Supported member (pedestal) = 12 in. x 12 in.

Permissible soil pressure = 4.0 ksf

f'_c = 3,000 psi (footing & pedestal)

Design to be in accordance with ACI 318.1-83

Calculations and Discussion	318.1 Reference
1. Footing base area:	
$A_f = \dfrac{40 + 60}{4.0} = 25 \text{ ft}^2$	7.2.2
Use 5 ft x 5 ft square footing (A_f = 25 ft^2)	
Note that the base area is determined using service loads (unfactored) with the permissible soil pressure. To proportion footing for strength, factored loads must be used.	
$U = 1.4\,(40) + 1.7\,(60) = 158^k$	7.2.1
$q_s = \dfrac{U}{A_f} = \dfrac{158}{25} = 6.32 \text{ ksf}$	
2. Determine footing depth. For plain concrete, flexural strength will usually control thickness. Referring to sketch, the critical section for moment is at face of pedestal.	7.2.5

EXAMPLE 23.8 - Continued

Calculations and Discussion	Code Reference

$$M_u = q_s \frac{b}{2} \left(\frac{b-c}{2}\right)^2$$

$$= 6.32 \times 2.5 (2)^2 = 63.2^{'k}$$

$$f_t \geq \frac{Mc}{I} = \frac{6 M_u}{bh^2}$$

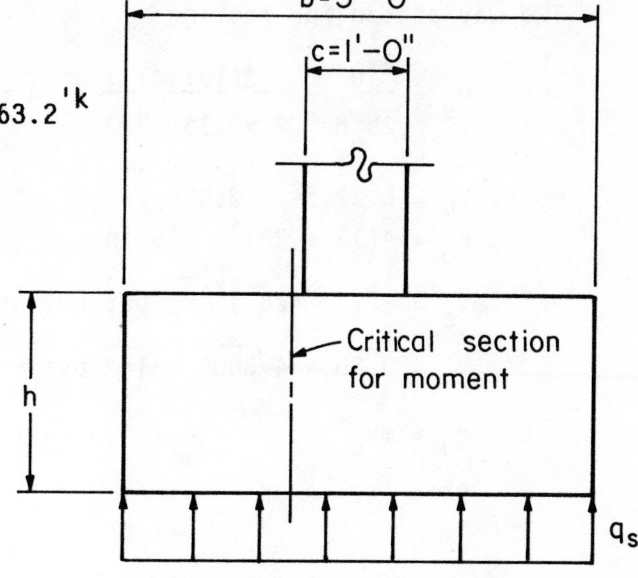

Permissible flexural stress $f_t = 5 \varphi \sqrt{f'_c}$ 6.2.1(a) 6.2.2

$= 5 \times 0.65 \sqrt{3,000} = 178$ psi.

$$178 \geq \frac{6(63.2 \times 12 \times 1000)}{(5 \times 12)h^2}$$

Solving for h = 20.6 in.

For concrete cast against soil, bottom 2 in. of concrete can not be considered for strength computations. (The reduced overall thickness is to allow for uneveness of excavation and for some contamination of the concrete adjacent to the soil.) Use overall footing depth of 24 in. 6.3.5

3. Check shear strength for 24 in. footing depth. Use effective depth for shear $h_{eff} = 24 - 2 = 22$ in.

The critical section for beam action (effective depth distance from face of pedestal) is located only 2.0 - 1.83 = 0.17 ft from edge of footing; therefore, not critical. 7.2.6.2(a)

EXAMPLE 23.8 - Continued

Calculations and Discussion	Code Reference
Two-way action for footing:	7.2.6.2(b)
$v_u = \dfrac{3V_u}{2b_oh} = \dfrac{3(107.4)}{2 \times 136 \times 22} = 54$ psi	Eq. (7-2)
where $V_u = 6.32(5^2 - 2.83^2) = 107.4^k$	
$b_o = 4(12 + 22) = 136$ in.	7.0
$\varphi v_c = \varphi(2 + 4/\beta_c)\sqrt{f'_c}$ but not greater than $\varphi 4\sqrt{f'_c}$	6.2.1(c)
$= 0.65 \times 4\sqrt{3000} = 142$ psi	6.2.2
$v_u \leq \varphi v_c$	
54 < 142 OK	
Therefore, the 22 in. effective depth is adequate for shear. Shear stress rarely will control for plain concrete members.	
4. Bearing stress on pedestal	7.3.3
$f_b = 0.85\varphi f'_c = 0.85 \times 0.65 \times 3000 = 1658$ psi	6.2.1(d) 6.2.2
$158^k/12 \times 12 = 1097$ psi < 1658 OK	

EXAMPLE 23.9 - Design for Transfer of Force at Base of Precast Column

For the 18 in. x 18 in. precast column and base plate detail shown, check force transfer between column and pedestal for a factored load $P_u = 1050^k$.

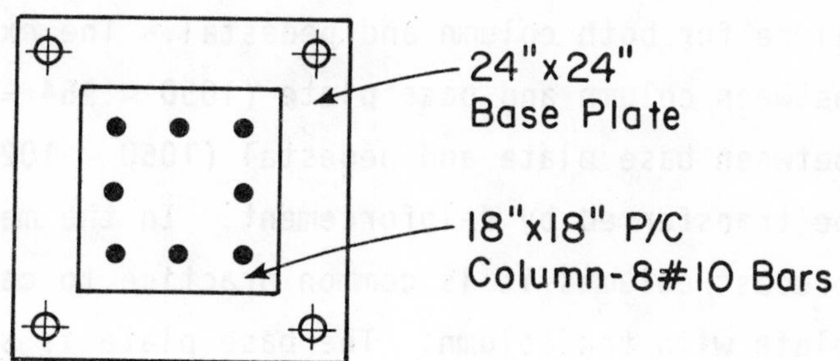

$f'_c = 5000$ psi (column)
$f'_c = 3000$ psi (pedestal)
$f_y = 60{,}000$ psi

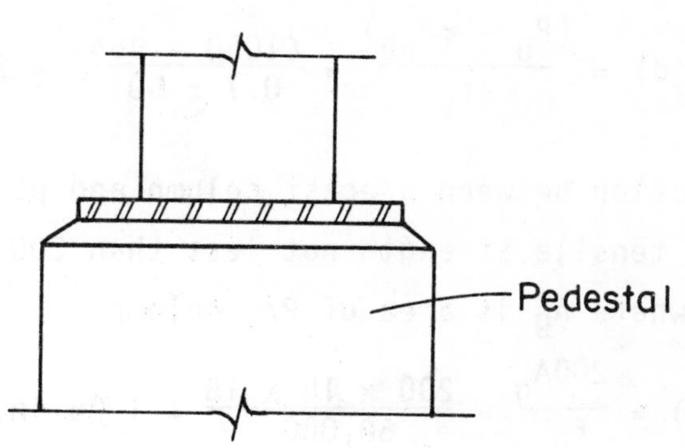

Calculations and Discussion	Code Reference
1. Bearing strength on column concrete (between P/C column and base plate), $f'_c = 5000$ psi	
$\varphi P_{nb} = \varphi(0.85\ f'_c\ A_1)$	10.15.1
$= 0.70(0.85 \times 5 \times 18 \times 18) = 964^k < 1050^k$	9.3.2.4
2. Bearing strength on pedestal concrete (between base plate and pedestal), $f'_c = 3000$ psi	
$\varphi P_{nb} = 0.70(0.85 \times 3 \times 24 \times 24) = 1028^k < 1050^k$	10.15.1

EXAMPLE 23.9 - Continued

Calculations and Discussion	Code Reference

3. Column load cannot be transferred by bearing on concrete alone for both column and pedestal. The excess load between column and base plate ($1050 - 964 = 86^k$), and between base plate and pedestal ($1050 - 1028 = 22^k$) must be transferred by reinforcement. In the manufacture of precast columns it is common practice to cast the base plate with the column. The base plate is secured to the column either by deformed bar anchors (dowels) or column bars welded to the base plate. Required area of dowel bars

$$A_s(\text{req'd}) = \frac{(P_u - \varphi P_{nb})}{\varphi f_y} = \frac{(1050 - 964)}{0.7 \times 60} = 2.04 \text{ in.}^2$$ 9.3.2.4

Also, connection between precast column and pedestal must have a tensile strength not less than 200 A_g in pounds, where A_g is area of P/C column 15.8.3.1

$$A_s(\min) = \frac{200A_g}{f_y} = \frac{200 \times 18 \times 18}{60{,}000} = 1.08 \text{ in.}^2 < 2.04$$

Required #5 deformed bar anchors $= \dfrac{2.04}{0.31} = 6.6$

Development of bar anchors:

$$\ell_d = 0.02 d_b f_y / \sqrt{f'_c} = 0.02 \times 0.625 \times 60{,}000/\sqrt{5000}$$ 12.3.2

$$= 10.6 \text{ in.}$$

but not less than $0.0003 d_b f_y = 0.003 \times 0.625 \times 60{,}000$

$$= 11.25 \text{ in.}$$

for excess area provided (Say 8 #5 bar anchors) 12.3.3

$$\ell_d = 11.25 \times 2.04/(8 \times 0.31) = 9.25 \text{ in.}$$

Use 8 #5 x 10 in. deformed bar anchors. Anchors are automatically welded (similar to headed studs) to base plate. The base plate and bar anchor assembly is then cast with the column.

EXAMPLE 23.9 - Continued

Calculations and Discussion	Code Reference

4. Excess load between base plate and pedestal (1050 - 1028 = 22^k) must also be transferred by reinforcement, with an area not less than $200A_g/f_y$ (Section 15.8.3.1). Check 4 anchor bolts, ASTM A36 steel.

$$A_s(\text{req'd}) = \frac{(1050 - 1028)}{0.7 \times 36} = 0.873 \text{ in.}^2$$

but not less than $A_s(\text{min})^* = 1.08$ in.^{2}

*Note: The code minimum of 200 A_g/f_y applies also to the connection between base plate and pedestal.

Required 3/4 in. anchor bolts = 1.08/0.44 = 2.5

The anchor bolts must be embedded into the pedestal to develop their design strength in bond. Determine embedment length of smooth anchor bolt as 2 times the embedment length for a deformed bar.

$\ell_d = 2(0.02 \times 0.75 \times 60{,}000/\sqrt{3000}) = 32.8$ in. — 12.3.2

but not less than $2(0.0003 \times 0.75 \times 60{,}000) = 27.0$ in.

for excess area provided (Say 4-3/4 in. anchor bolts) — 12.3.3

$\ell_d = 32.8 \times 1.08/(4 \times 0.44) = 20.1$ in.

Use 4-3/4 in. x 1 ft-9 in. anchor bolts. Enclose anchor bolts with 4 #3 ties at 3 in. centers. See Connection Detail.

EXAMPLE 23.9 - Continued

Calculations and Discussion	Code Reference

Note: The reader should refer to the PCI Design Handbook for an indepth treatise on design and construction details for precast column connections. Design for the base plate thickness is also addressed in the PCI Handbook.

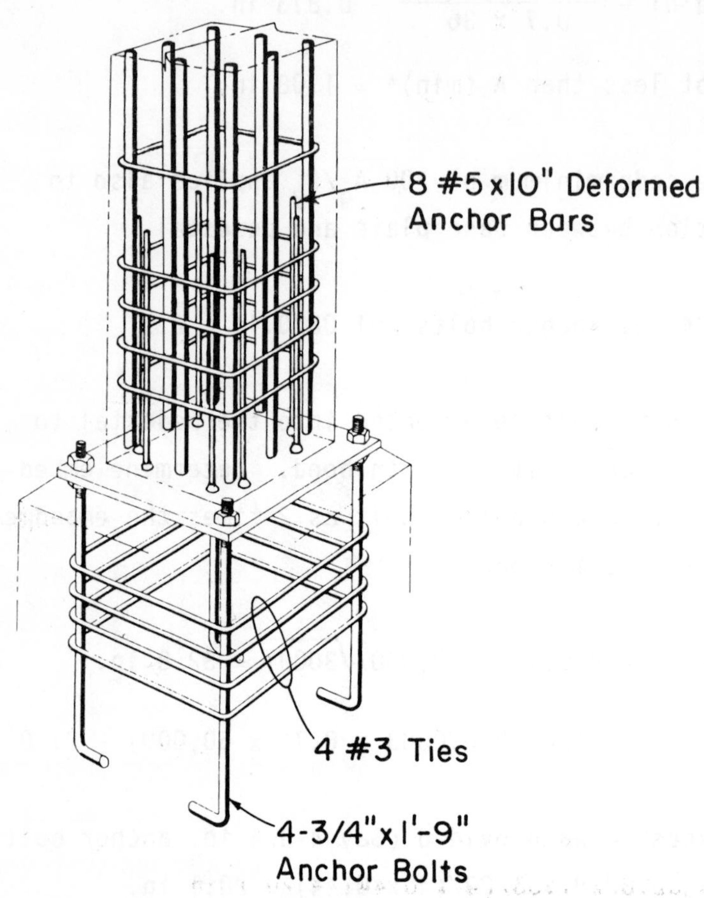

Connection Detail

24

Precast Concrete

General Considerations

A concrete element is considered to be precast when the element is cast elsewhere than in its final position in the structure. With the 1983 Code Edition, the Scope for Chapter 16 was revised to require that all precast concrete, whether produced in a plant or on site, meet the requirements of the Chapter. As the scope originally read, the provisions of Chapter 16 applied only to precast concrete members manufactured under controlled plant conditions. There are no liberalizations in Chapter 16 that would require meeting "plant-controlled conditions." This revision is actually a clarification of intent.

Design and detailing of joints and connections for precast concrete structures can be a very specialized task. Sections 16.2.2 and 16.2.4 provide typical design considerations related to joints and connections. In many cases, connections in precast concrete structures represent a discontinuity in the elastic properties of the structure. Typically, the precast members themselves are much stiffer than the joints connecting them, resulting in an abrupt change in behavior at the joints. This effect can be significant in determining the forces transferred through the joints from one member to another, as well as in computing deflections in the structure. Moment connections for rigid frame buildings often fall into this category. Fig. 24-1 shows a simple frame with moment connections, plus a schematic diagram which models the above effect.

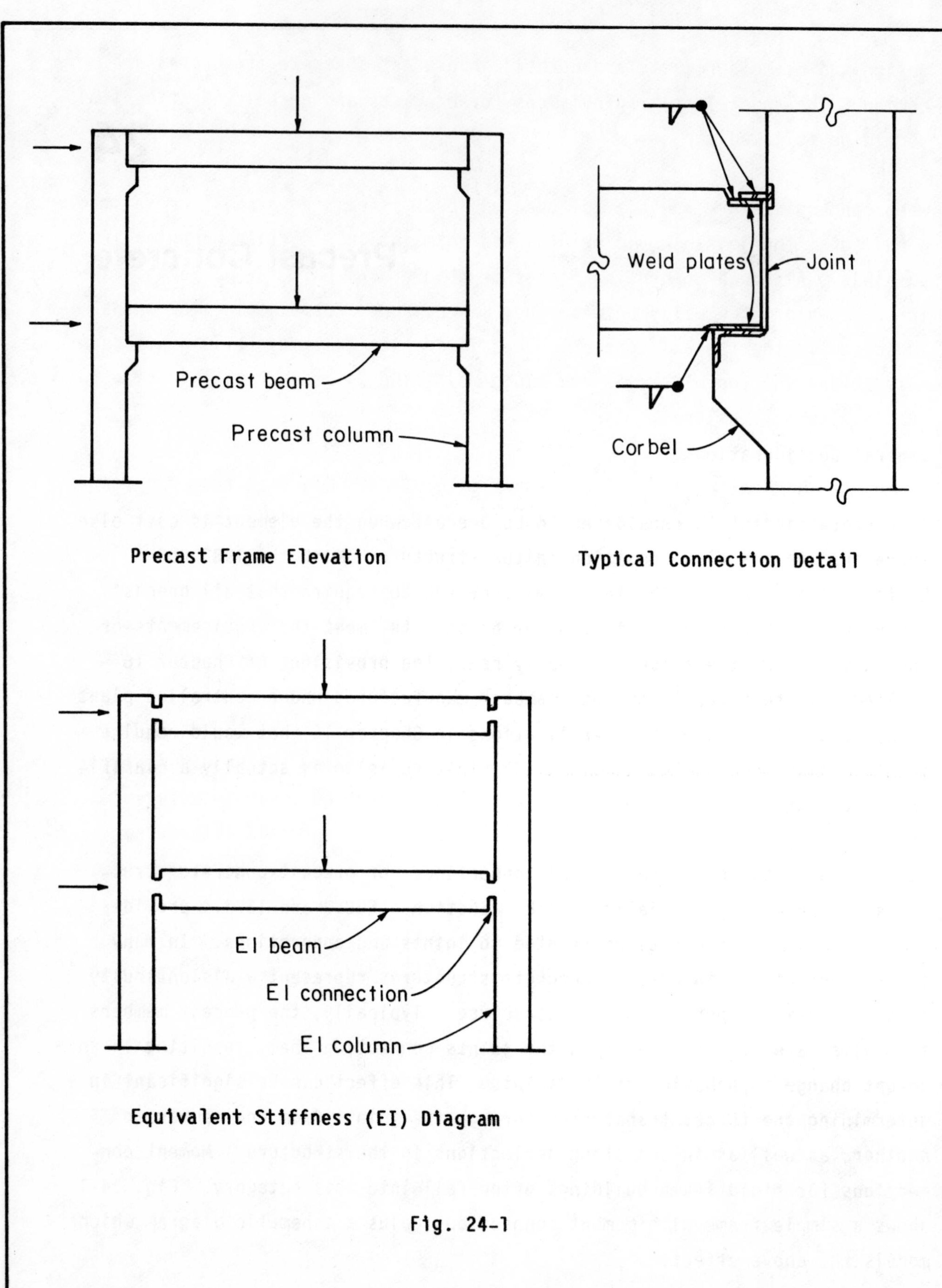
Precast beam
Precast column
Precast Frame Elevation
Weld plates
Joint
Corbel
Typical Connection Detail
EI beam
EI connection
EI column
Equivalent Stiffness (EI) Diagram

Fig. 24-1

Design and use of brackets and corbels in precast concrete structures are common. The special shear provisions for brackets and corbels, Section 11.9, should be followed.

When connections are subject to repeated loads, stress reversals, or seismic conditions, the joint should be designed to maintain a ductile behavior if possible. Although connecting plates or reinforcing bars may be ductile in their behavior, the welds between them may not be. For example, the welds indicated in the typical connection detail in Fig. 24-1 may fail in a sudden mode before the top plate yields. Such welds should be increased in size to preclude such a failure.

Since standard connection methods for typical joints may vary from one manufacturer to another, the designer may consider indicating the general scheme of connecting the elements, specifying the required forces and stiffness to be transferred through the joint. If this is done, the designer should request details and calculations from the manufacturer to verify project requirements. Industry standards, such as References 24.1, 24.2, and 24.3, should be followed.

The design of precast concrete wall panels should follow the requirements of Chapters 10 and 14 of the Code, and the recommendations given in References 24.1 to 24.3. Minimum reinforcing as specified in Chapter 14 is necessary only when the empirical design of reinforced walls is used. Otherwise, reinforcing or prestressing shall be determined by a more exact analysis. Walls acting as deep flexural members should be analyzed as such, and should conform to all applicable provisions of the Code, especially Sections 10.7 and 11.8.

Shop drawings should include information related to the fabrication, storage, shipping, and erection of the precast elements. In addition to those items mentioned in Sections 16.4.1 and 16.6.1, the designer may require that the following items be included on the shop drawings:

(1) Concrete strengths at removal from forms, at the time of prestressing, and at 28 days;

(2) Handling methods and methods of support during storage, shipping, and erection; and

(3) Initial and final camber for prestressed concrete members.

Selected References

24.1 PCI Design Handbook -- Precast and Prestressed Concrete, 2nd Edition, Prestressed Concrete Institute, Chicago, 1978, 380 pp. (Contains load tables, graphical design aids, and examples. Enables selection of sections and systems, design of components, and verification of design.)

24.2 Connections for Precast-Prestressed Concrete Buildings - Including Earthquake Resistance, Prestressed Concrete Institute, Chicago, 1982, 297 pp. (Updates available information on design connections, evaluates over 100 connections typically used in the industry, and includes design aids and extensive references.)

24.3 PCI Manual for Structural Design of Architectural Precast Concrete, Prestressed Concrete Institute, Chicago, 1977, 448 pp. (Provides comprehensive structural design information for precast concrete wall systems--precast elements, support systems, and connections. Design examples are included.)

25

Prestressed Concrete—Flexure

Update for '83 Code

For ACI 318-83, five changes to Chapter 18 are noteworthy:

(1) Permissible Stresses in Prestressing Tendons--Permissible tendon stresses (Section 18.5) are revised to recognize the higher yield strength and reduced relaxation losses for low-relaxation tendons. Also, with post-tensioning tendons, initial stresses along the tendon can differ from tendon stresses at anchorages due to friction. This condition, not clearly indicated in the 1977 Code, is now clarified in Section 18.5.1(c) by specifying "at anchorages and couplers," and by requiring that Section 18.5.1(b) apply to both pretensioned and post-tensioned tendons.

Note especially that when specifying low-relaxation tendons, the higher permitted jacking forces in the prestressing beds dictate special quality control and safety procedures to ensure safety of workmen during tendon jacking operations.

(2) Loss of Prestress -- The lump sum values for prestress losses recommended in Commentary Section 18.6.1 (35,000 psi for pretensioning and 25,000 psi for post-tensioning) are now considered obsolete and have been deleted. Also, the lump sum values may not be adequate for some design conditions. Reasonably accurate estimates of prestress losses can be easily calculated from Reference 25.2. See Appendix 25A.

(3) Flexural Strength -- For members with bonded prestressing tendons, design stress Eq. (18-3) has been revised for application to members reinforced with a combination of prestressed and nonprestressed reinforcement (partially prestressed members), and to take into account effects of any compression reinforcement, high strength concrete, and low-relaxation tendons.

For members with unbonded prestressing tendons, a new Eq. (18-5) has been added for application to members with high span-to-depth ratios (post-tensioned one-way slabs, flat plates, and flat slabs).

(4) Limits for Reinforcement -- With the '83 Code, an exception to the minimum reinforcement requirement of Section 18.8.3 is added to provide for those cases when reinforcement required to develop 1.2 times the cracking strength would be excessive. The exception waives the 1.2 strength requirement when the strength provided is at least twice the flexural and shear strength required for factored load conditions. See further discussion under Section 18.8.3.

(5) Prestressed Slab Systems -- Section 18.12 has been extensively revised to specifically permit the Equivalent Frame Method of Chapter 13 for analysis of two-way prestressed slab systems. Also, new guidelines are given for tendon distribution and bonded reinforcement requirements that will permit the use of banded tendon distribution.

Introduction

In prestressed members, compressive stresses are introduced into the concrete to reduce tensile stresses resulting from applied loads. Prestressing tendons such as wire, strands, or bars impart compressive stresses to the concrete. Pretensioning is a method of prestressing in which the tendons are tensioned before concrete is placed. Post-tensioning is a method of prestressing in which the tendons are tensioned after the concrete has hardened.

The act of prestressing a member introduces "prestressing loads" to the member. The induced prestressing loads, acting in conjunction with the externally applied loads, must provide serviceability and strength to the member beginning at the moment of prestressing transfer and continuing throughout the life of the member.

Prestressed structures must be analyzed taking into account prestressed loads, service loads, temperature, creep, shrinkage and the structural properties of all materials involved. However, the Code indicates that empirical or simplified analytical methods should be avoided since such approximate methods may not account adequately for prestressing forces. Thus, various sections of the Code are excluded although the "entire Code applies to prestressed concrete structures" (Sections 18.1.2 and 18.1.3).

18.2 General

The Code requires fulfillment of strength and serviceability requirements that are basic to any structure, prestressed or nonprestressed. It also calls attention to certain structural aspects more common in prestressed concrete structures, such as stress concentrations at anchorages (Section 18.2.3), compatibility of deformation with the adjoining structure (Section 18.2.4), and the possibility of buckling of thin flanges as well as of any part of the member between points where concrete and prestressing tendons are in contact (Section 18.2.5). Regarding the effect of prestressing on adjoining parts of the structure, it is frequently necessary to calculate column moments due to axial shortening of prestressed floors. For the analysis of buckling, the engineer should refer to recognized texts on the subject. At the present time minimum "width to thickness ratios" are not given by the Code. In computing section properties, Section 18.2.6 requires that effect of loss of area due to open ducts for post-tensioning must be considered in design.

18.3 Design Assumptions

In applying fundamental structural principles (equilibrium, stress-strain relations, and geometric compatibility) to prestressed structures, certain simplifying assumptions can be made. For computation of strength (Section 18.3.1), the basic assumptions are the same as for nonprestressed members. An exception is that Section 10.2.4 refers only to nonprestressed reinforcement. For behavior at service conditions, the "straight line theory" (referring to the straight line variation of stress with strain) may be used. For analysis at service conditions, the modulus of elasticity for concrete and reinforcement (nonprestressed and prestressed) are given in Section 8.5.

18.4 and 18.5 Permissible Stresses

Both concrete and prestressing tendon stresses are limited to ensure proper behavior at service loads and immediately after prestress transfer. The Code allows a concrete tension of $6\sqrt{f'_c}$ at the ends of simply supported members. Concrete stress limitations at service loads are based on the definition of a "precompressed" concrete zone. Fig. 25-1 illustrates the definition of "precompressed" area. Under Section 18.4.2, for a tensile stress of $12\sqrt{f'_c}$, a 50% increase in concrete cover is required by Section 7.7.3.2. For this stress, deflection requirements must comply with Section 9.5.4.

Deflections of prestressed members calculated according to Section 9.5.4 should not exceed the values in Table 9.5(b) of the Code. According to Section 9.5.1, prestressed concrete members, like any other concrete member, should be designed to have adequate stiffness to prevent deformations which may adversely affect the strength or serviceability of the structure.

With the 1983 Code edition, permissible tendon stresses have been revised to recognize the higher yield strength of low-relaxation tendons. For ordinary tendons, the tendon stresses are the same as permitted in the 1977 Code. Permissible stresses in terms of the specified minimum tensile strength f_{pu} are as follows:

(a) Due to tendon jacking force:

low-relaxation wire and strands ($f_{py} = 0.90\ f_{pu}$)................0.85 f_{pu}

ordinary wire, strands, and bars ($f_{py} = 0.85\ f_{pu}$)...............0.80 f_{pu}

bar tendons ($f_{py} = 0.80\ f_{pu}$).................................0.75 f_{pu}

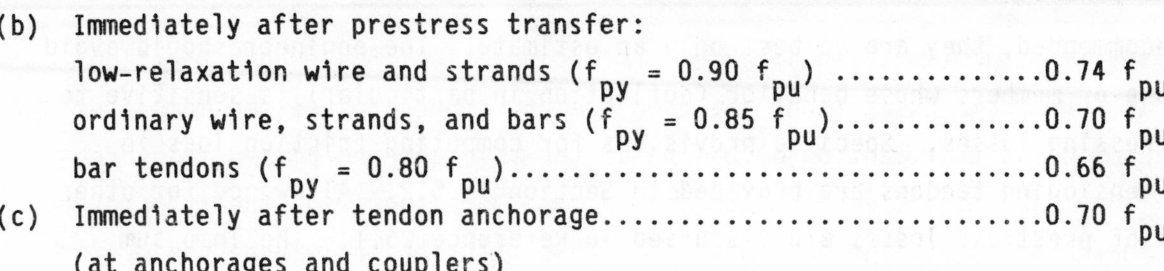

(b) Immediately after prestress transfer:

low-relaxation wire and strands ($f_{py} = 0.90 f_{pu}$) ....0.74 f_{pu}

ordinary wire, strands, and bars ($f_{py} = 0.85 f_{pu}$)..............0.70 f_{pu}

bar tendons ($f_{py} = 0.80 f_{pu}$)......................................0.66 f_{pu}

(c) Immediately after tendon anchorage.................................0.70 f_{pu}

(at anchorages and couplers)

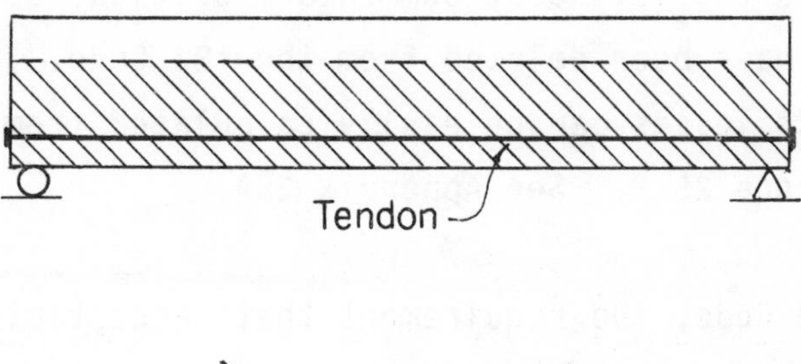

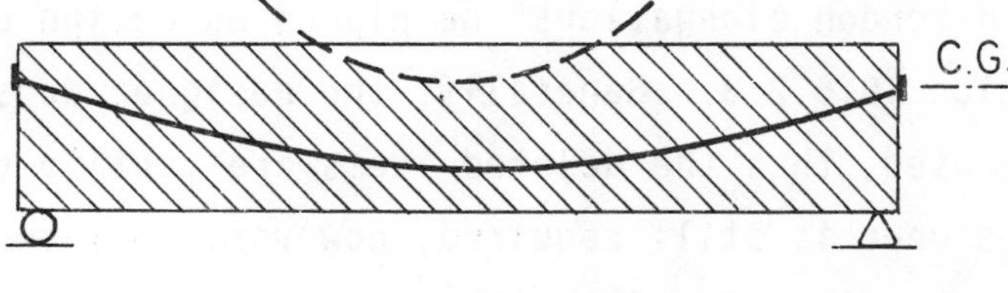

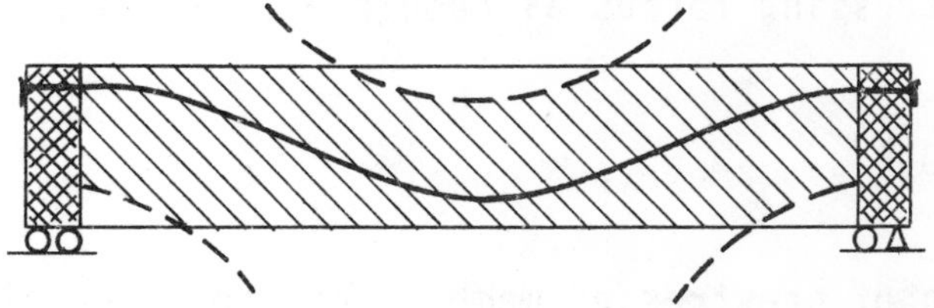

☐ = Nonprecompressed area

▨ = Precompressed area

Fig. 25-1 - Precompressed Areas in Prestressed Members

18.6 Loss of Prestress

Another factor which must be considered in design of prestressed members is the loss of prestress due to various causes. These losses can drastically affect the behavior of a member at service loads. Although calculation procedures and certain values of creep strain, friction factors, etc., may be recommended, they are at best only an estimate. The engineer should avoid the use of members whose behavior (deflection in particular) is sensitive to prestressing losses. Specific provisions for computing friction loss in post-tensioning tendons are provided in Section 18.6.2. Allowance for other types of prestress losses are discussed in Reference 25.1. The lump sum values of prestress losses (for both pretensioned and post-tensioned members) that were indicated in the 1977 Code Commentary Section 18.6.1 are now considered obsolete and have been deleted from the '83 Code. Reasonably accurate estimates of prestress losses can be easily calculated from the procedures recommended in Reference 25.2. See Appendix 25A.

Note that for the '83 Code, the requirement that "acceptable ranges of tendon jacking forces and tendon elongations" be placed on design drawings has been deleted from Section 18.6.2.3. Generally, the designer does not know which materials will be used, thus the deleted items are usually given on shop drawings. The designer is still required, however, to show on the design drawings his assumed values for friction coefficients along with magnitude and location of prestressing forces as required by Section 1.2.1(g).

18.7 Flexural Strength

The flexural strength of prestressed members can be calculated using the same assumptions as for nonprestressed members. Prestressing tendons, however, do not have a well defined yield point as does mild reinforcement. As a prestressed cross-section reaches its flexural strength, (defined by a maximum compressive concrete strain of 0.003), stress in the prestressed reinforcement at nominal strength f_{ps} will vary depending on the amount of prestressing. For the '83 Code, design stress Eq. (18-3), for members with bonded prestressing tendons, is expanded for application to members reinforced with a combination of prestressed and nonprestressed reinforcement (partially prestressed members), and to take into account effects of any compression reinforcement, high strength concrete, and low-relaxation tendons. For a fully prestressed member, Eq. (18-3) reduces to:

$$f_{ps} = f_{pu}\left(1 - \frac{\gamma_p}{\beta_1}\rho_p\frac{f_{pu}}{f'_c}\right)$$

where γ_p = 0.40 for ordinary wire and strands

= 0.28 for low-relaxation wire and strands

and β_1, as defined in Section 10.2.7.3,

β_1 = 0.85 for $f'_c \leq 4000$ psi

= 0.80 for $f'_c = 5000$ psi

= 0.75 for $f'_c = 6000$ psi

For members with unbonded prestressing tendons, a new Eq. (18-5) has been added for application to members with high span-to-depth ratios (post-tensioned one-way slabs, flat plates, and flat slabs.) With the value of f_{ps} known, the nominal moment strength can be calculated as follows:

$$M_n = A_{ps} f_{ps}\left(d_p - 0.59\frac{A_{ps} f_{ps}}{bf'_c}\right) \quad (1)$$

or in nondimensional terms:

$$R_n = \omega_p(1 - 0.59\,\omega_p) \quad (2)$$

where,

$$R_n = \frac{M_n}{bd_p^2 f'_c} \text{ and } \omega_p = \frac{A_{ps} f_{ps}}{bd_p f'_c} \quad (3)$$

The value of f_{ps} can be obtained using the conditions of equilibrium, stress-strain relations, and strain compatibility. However, the analysis is quite cumbersome, especially in the case of unbonded prestressing. For bonded prestressing, the compatibility of strains can be considered at an individual section, while for unbonded prestressing, compatibility relations can be written only at the anchorage points and will depend on the entire cable profile and member loading. To avoid such lengthy calculations, the Code allows f_{ps} to be obtained by the approximate Eqs. (18-3), (18-4), and (18-5). Eq. (18-3) can be written in nondimensional form as follows:

$$\omega_p = \omega_{pu}\left(1 - \frac{\gamma_p}{\beta_1}\omega_{pu}\right) \quad (4)$$

where

$$\omega_{pu} = \frac{A_{ps} f_{pu}}{bd_p f'_c} \quad (5)$$

Section 18.7.2 also allows a more accurate determination of f_{ps} based on strain compatibility analysis. Design Example 25.3 illustrates the procedure.

18.8 Limits for Reinforcement of Flexural Members

The requirements for percentage of reinforcement are illustrated in Fig. 25-2. Note that reinforcement can be added to provide a reinforcement index higher than $0.36\beta_1$; however, this added reinforcement cannot be assumed to contribute to the moment strength.

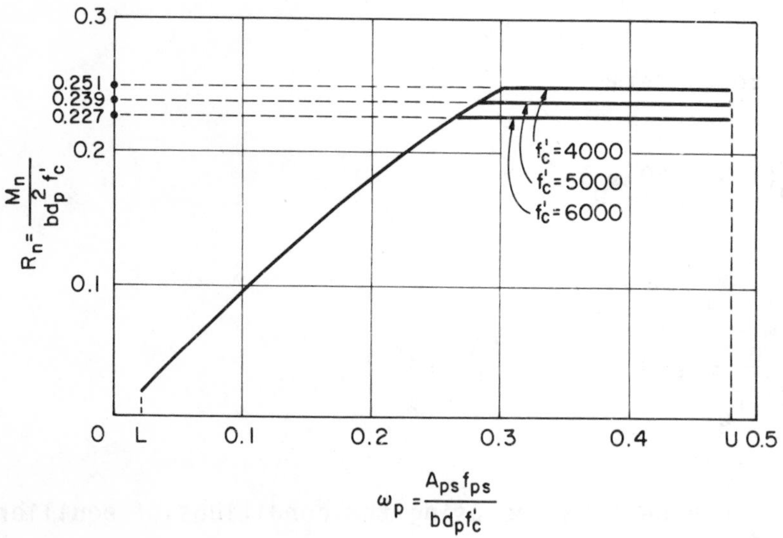

Note: The lower limit (L) is determined by the cracking moment (Section 18.8.3) and the upper limit (U) is determined by the permissible serviceability stresses (Section 18.4). Actual values for these limits differ for each problem.

Fig. 25-2 - Permissible Limits of Prestressed Reinforcement and Influence on Moment Strength

Section 18.8.3 requires the total amount of prestressed and nonprestressed reinforcement of flexural members to be adequate to develop a design moment strength at least equal to 1.2 times the cracking moment strength ($\varphi M_n \geq 1.2M_{cr}$), where M_{cr} is computed by elastic theory using a modulus of rupture equal to $7.5\sqrt{f'_c}$. The provisions of Section 18.8.3 are analogous to Section 10.5 for nonprestressed members; a precaution against abrupt flexure failure resulting from rupture of the prestressing tendons when failure occurs immediately after cracking. The provision ensures that cracking will occur before flexural strength is reached, and by a large enough margin so that significant deflection will occur to warn that the ultimate capacity is being approached. The typical prestressed member will have a fairly large margin between cracking strength and flexural strength, but the designer must be certain by checking it.

Cracking moment for a prestressed member is determined by summing all the moments that will cause a stress in the bottom fiber equal to the modulus of rupture f_r. Referring to Fig. 25-3, for a prestressed composite member

$$-\left(\frac{P_{se}}{A_c}\right) - \left(\frac{P_{se}.e}{S_b}\right) + \left(\frac{M_d}{S_b}\right) + \left(\frac{M_a}{S_c}\right) = +f_r$$

Solving for $M_a = \left(f_r + \frac{P_{se}}{A_c} + \frac{P_{se}.e}{S_b}\right) S_c - (M_d)\frac{S_c}{S_b}$

Since $\quad M_{cr} = M_d + M_a$

$$M_{cr} = \left(f_r + \frac{P_{se}}{A_c} + \frac{P_{se}.e}{S_b}\right) S_c - M_d\left(\frac{S_c}{S_b} - 1\right)$$

For a prestressed member alone (without composite slab), $S_c = S_b$, and M_{cr} reduces to

$$M_{cr} = \left(f_r + \frac{P_{se}}{A_c}\right)S_b + P_{se}.e$$

Design Examples 25.5 and 25.6 illustrate cracking moment strength of prestressed members.

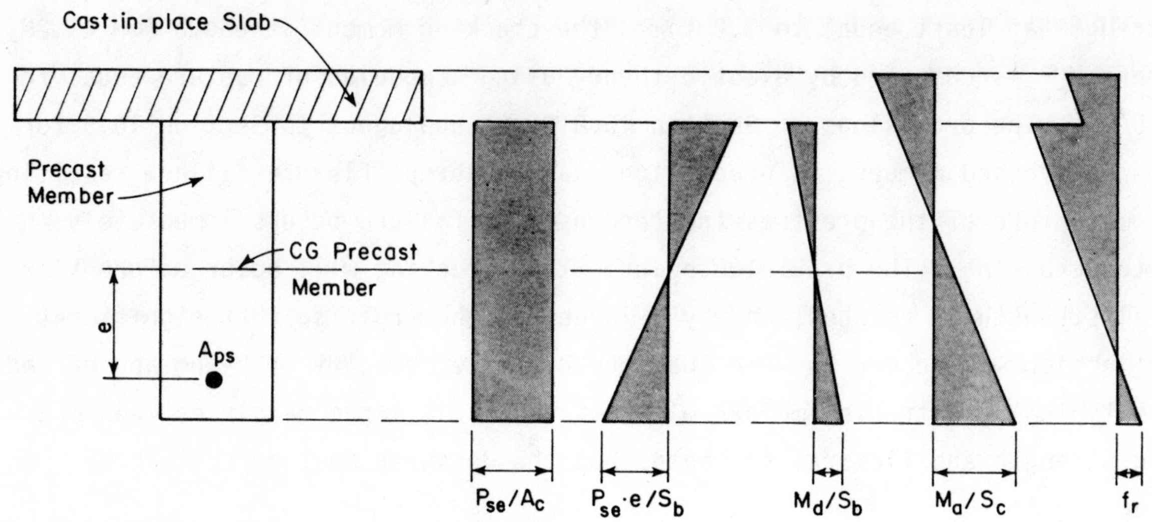

A_{ps} = area of prestressed reinforcement
A_c = area of precast member
S_b = section modulus for bottom of precast member
S_c = section modulus for bottom of composite member
P_{se} = effective prestress force
e = eccentricity of prestress force
M_d = dead load moment of composite member
M_a = additional moment to cause a stress in bottom fiber equal to modulus of rupture f_r

Fig. 25-3 Stress Conditions for Evaluating Cracking Moment Strength

Note that for ACI 318-83, an exception waives the 1.2 strength requirement for those cases where the strength provided is at least twice the flexural and shear strength required by Section 9.2.

For flexural strength: $\varphi M_n \geq 2M_u \geq 2(1.4M_d + 1.7\ M_\ell)$

For shear strength: $\varphi V_n \geq 2V_u \geq 2(1.4V_d + 1.7\ V_\ell)$

The $1.2M_{cr}$ provision often requires excessive reinforcement for certain prestressed flexural members. The exception is intended to limit the amount of additional reinforcement required to amounts comparable to the similar requirements for nonprestressed members in Section 10.5.2

18.9 Minimum Bonded Reinforcement

A minimum amount of bonded reinforcement is desirable in members with unbonded tendons. Reference to Code Commentary discussion for Section 18.9 is suggested.

For all flexural members with unbonded prestressing tendons, except two-way solid slabs of uniform thickness, a minimum area of bonded reinforcement computed by Eq. (18-6) must be uniformly distributed over the precompressed tensile zone as close as practicable to the extreme tension fiber. Fig. 25-4 illustrates application of Eq. (18-6).

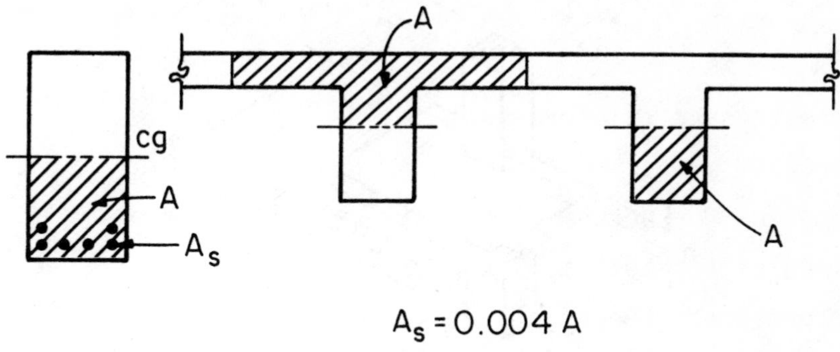

Fig. 25-4 - Bonded Reinforcement for Flexural Members

For solid slabs of uniform thickness, the special provisions of Section 18.9.3 apply. Depending on the tensile stress in the concrete at service loads, the requirements for positive moment areas of solid slabs are illustrated in Fig. 25-5(a).

Fig. 25-5(b) illustrates the minimum bonded reinforcement requirements for the negative moment areas at column supports. The bonded reinforcement must be located within the width $C_2 + 2\ (1.5h)$ as shown, with 4 bars minimum spaced not greater than 12 inches.

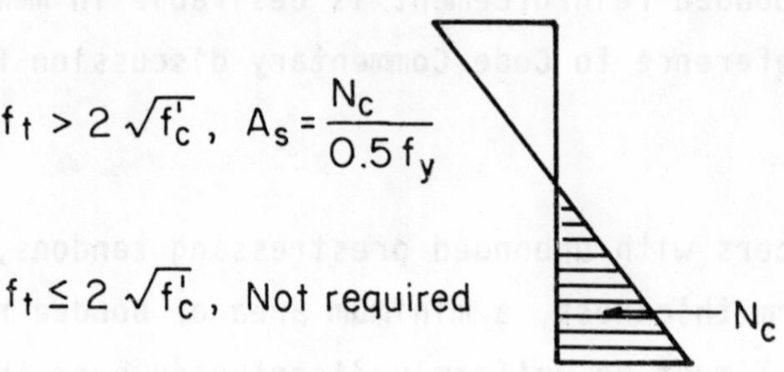

$f_t > 2\sqrt{f'_c}$, $A_s = \frac{N_c}{0.5 f_y}$
$f_t \le 2\sqrt{f'_c}$, Not required
N_c

(a) Positive Moment Areas

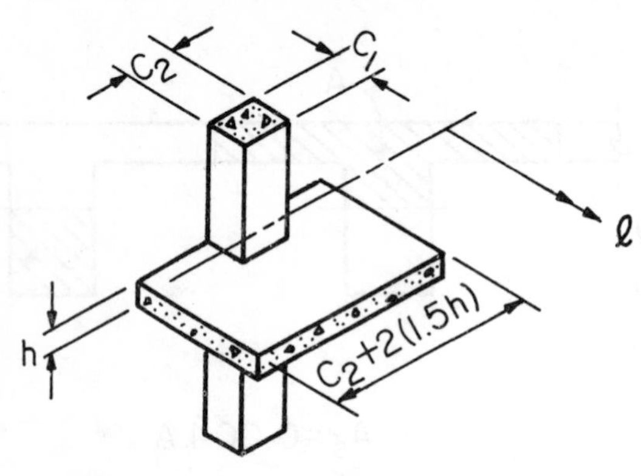

$A_s = 0.00075h\ell$

(b) Negative Moment Areas

Fig. 25-5 - Bonded Reinforcement for Flat Plates

18.11 Compression Members - Combined Flexure and Axial Loads

Provisions of the Code for calculating the strength of prestressed compression members are the same as for members without prestressing. The only two additional considerations are (a) accounting for prestressing strains, and (b) using an appropriate stress-strain relation for the prestressing tendons. Design Example 25.6 illustrates the calculation procedure.

For compression members with an average concrete stress due to prestressing of less than 225 psi, minimum nonprestressed reinforcement must be provided.

Selected Reference

25.1 PCI Committee on Prestress Losses, "Recommendations for Estimating Prestress Losses," PCI Journal, Vol. 20, No. 4, July-August 1975, pp. 43-75.

25.2 Fia, Paul, Preston, H. Kent, Scott, Norman L., and Workman, Edwin B., "Estimating Prestress Losses," Concrete International: Design and Construction, Vol. 1, No. 6, June 1979, pp. 32-38.

Appendix 25A - Estimating Prestress Losses

Reference 25.2 presents a reasonably accurate procedure for estimating prestress losses due to various causes for pretensioned and post-tensioned members with bonded and unbonded tendons. The procedure is intended for practical design applications under normal design conditions. The various sources of loss of prestress and equations for computing each effect (taken from Ref. 25.2) are summarized below. The simple equations enable the designer to estimate the various types of prestress loss rather than a lump sum value. The reader is referred to Reference 25.2 for in-depth discussion of the procedure, including sample computations for typical prestressed concrete beams.

COMPUTATION OF LOSSES

Elastic Shortening of Concrete (ES)

For members with bonded tendons:

$$ES = K_{es} E_s \frac{f_{cir}}{E_{ci}} \qquad (1)$$

in which

K_{es} = 1.0 for pretensioned members

K_{es} = 0.5 for post-tensioned members when tendons are tensioned in sequential order to the same tension. With other post-tensioning procedures, the value for K_{es} may vary from 0 to 0.5.

$$f_{cir} = K_{cir} f_{cpi} - f_g \qquad (2)$$

in which K_{cir} = 1.0 for post-tensioned members

K_{cir} = 0.9 for pretensioned members.

For members with unbonded tendons:

$$ES = K_{es} E_s \frac{f_{cpa}}{E_{ci}} \qquad (1a)$$

in which f_{cpa} = average compressive stress in the concrete along the member length at the center of gravity of the tendons immediately after the prestress has been applied to the concrete.

Creep of Concrete (CR)

For members with bonded tendons:

$$CR = K_{cr} \frac{E_s}{E_c} (f_{cir} - f_{cds}) \quad (3)$$

in which

K_{cr} = 2.0 for pretensioned members

K_{cr} = 1.6 for post-tensioned members

For members made of sand lightweight concrete the foregoing values of K_{cr} should be reduced by 20 percent.

For members with unbonded tendons:

$$CR = K_{cr} \frac{E_s}{E_c} f_{cpa} \quad (3a)$$

Shrinkage of Concrete (SH)

$$SH = 8.2 \times 10^{-6} K_{sh} E_s (1 - 0.06 \frac{V}{S}) (100 - RH) \quad (4)$$

in which

K_{sh} = 1.0 for pretensioned members

or

K_{sh} is taken from Table 1 for post-tensioned members.

Appendix 25A - Continued

TABLE 1 - Values of K_{sh} for post-tensioned members

Time, days*	1	3	5	7	10	20	30	60
K_{sh}	0.92	0.85	0.80	0.77	0.73	0.64	0.58	0.45

*Time after end of moist curing to application of prestress

Relaxation of Tendon Stress (RE)

$$RE = [K_{re} - J(SH + CR + ES)]C \qquad (5)$$

in which the values of K_{re}, J and C are taken from Tables 2 and 3.

Appendix 25A - Continued

TABLE 2 - Values of K_{re} and J

Type of tendon	K_{re}	J
270 Grade stress-relieved strand or wire	20,000	0.15
250 Grade stress-relieved strand or wire	18,500	0.14
240 or 235 Grade stress-relieved wire	17,600	0.13
270 Grade low-relaxation strand	5,000	0.040
250 Grade low-relaxation wire	4,630	0.037
240 or 235 Grade low-relaxation wire	4,400	0.035
145 or 160 Grade stress-relieved bar	6,000	0.05

Appendix 25A - Continued

TABLE 3 - Values of C

f_{pi}/f_{pu}	Stress relieved strand or wire	Stress-relieved bar or low relaxation strand or wire
0.80		1.28
0.79		1.22
0.78		1.16
0.77		1.11
0.76		1.05
0.75	1.45	1.00
0.74	1.36	0.95
0.73	1.27	0.90
0.72	1.18	0.85
0.71	1.09	0.80
0.70	1.00	0.75
0.69	0.94	0.70
0.68	0.89	0.66
0.67	0.83	0.61
0.66	0.78	0.57
0.65	0.73	0.53
0.64	0.68	0.49
0.63	0.63	0.45
0.62	0.58	0.41
0.61	0.53	0.37
0.60	0.49	0.33

Friction

Computation of friction losses is covered in ACI Section 18.6.2. When the tendon is tensioned, the friction losses computed can be checked with reasonable accuracy by comparing the measured elongation and the prestressing force applied by the tensioning jack.

MAXIMUM LOSS

The total amount of prestress loss due to elastic shortening, creep, shrinkage, and relaxation need not be more than the values given below if the tendon stress immediately after anchoring does not exceed 0.83 f_{py}:

Type of strand	Maximum Loss psi	
	Normal Concrete	Lightweight Concrete
Stress relieved strand	50,000	55,000
Low-relaxation strand	40,000	45,000

SUMMARY OF NOTATION

A_c = area of gross concrete section at the cross section considered

A_{ps} = total area of prestressing tendons

CR = stress loss due to creep of concrete

e = eccentricity of center of gravity of tendons with respect to center of gravity of concrete at the cross section considered

E_{ci} = modulus of elasticity of concrete at time prestress is applied

E_c = modulus of elasticity of concrete at 28 days

E_s = modulus of elasticity of prestressing tendons. Usually 28,000,000 psi

ES = stress loss due to elastic shortening of concrete

f_{cds} = stress in concrete at center of gravity of tendons due to all superimposed permanent dead loads that are applied to the member after it has been prestressed

f_{cir} = net compressive stress in concrete at center of gravity of tendons immediately after the prestress has been applied to the concrete.

f_{cpa} = average compressive stress in the concrete along the member length at the center of gravity of the tendons immediately after the prestress has been applied to the concrete

f_{cpi} = stress in concrete at center of gravity of tendons due to P_{pi}

f_g = stress in concrete at center of gravity of tendons due to weight of structure at time prestress is applied

f_{pi} = stress in tendon due to P_{pi}, $f_{pi} = P_{pi}/A_{ps}$

f_{pu} = specified tensile strength of prestressing tendon, psi

I_c = moment of inertia of gross concrete section at the cross section considered

M_d = bending moment due to dead weight of member being prestressed and to any other permanent loads in place at time of prestressing

P_{pi} = prestressing force in tendons at critical location on span after reduction for losses due to friction and seating loss at anchorages but before reduction for ES, CR, SH, and RE.

RE = stress loss due to relaxation of tendons

RH = average relative humidity surrounding the concrete member. See Fig. 1.

SH = stress loss due to shrinkage of concrete

V/S = volume to surface ratio. Usually taken as gross cross-sectional area of concrete member divided by its perimeter.

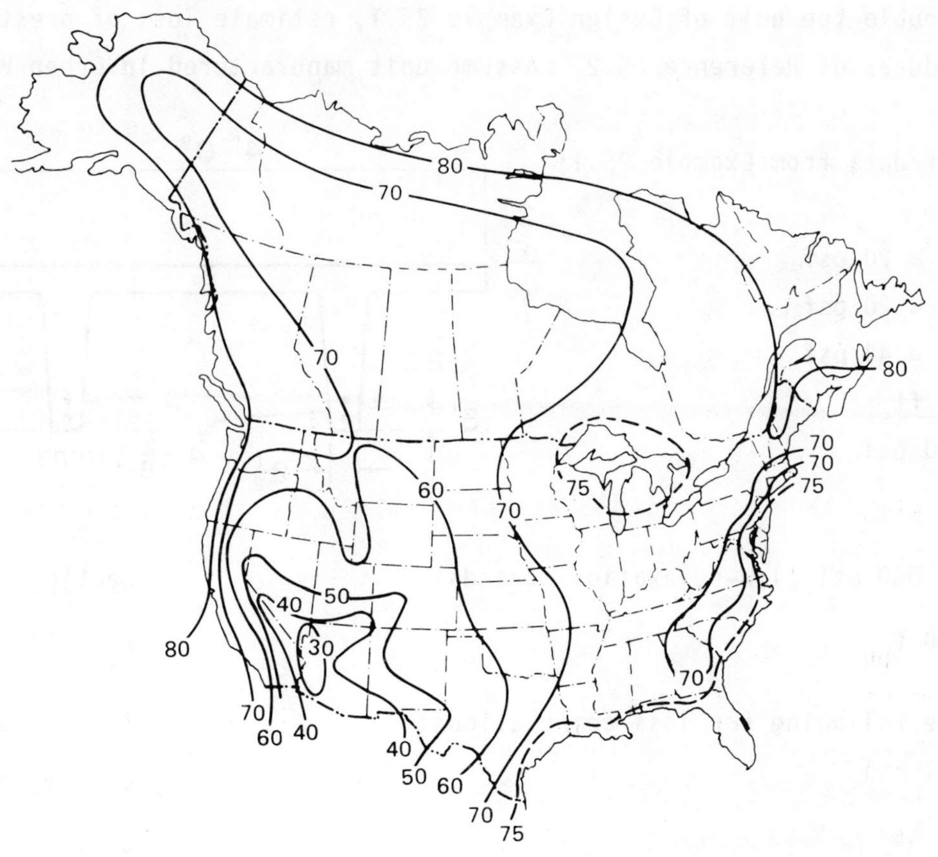

Fig. 1 - Annual Average Ambient Relative Humidity

Appendix 25A - Continued

Sample Computation - Estimating Prestress Losses

For the double-tee unit of Design Example 25.1, estimate loss of prestress using the procedures of Reference 25.2. Assume unit manufactured in Green Bay, Wisc.

Summary of data from Example 25.1:

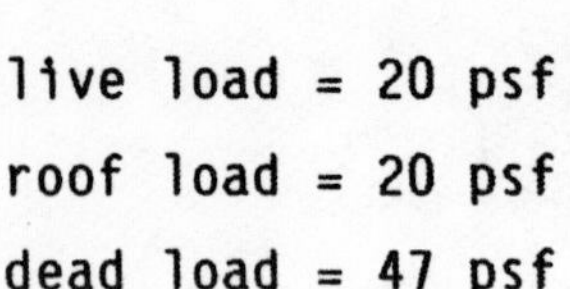

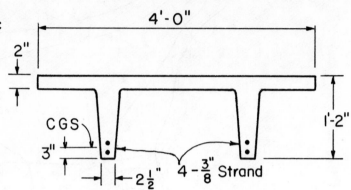

live load = 20 psf

roof load = 20 psf

dead load = 47 psf

span = 32 ft

f'_{ci} = 3500 psi

f'_c = 5000 psi

f_{pu} = 270,000 psi (low-relaxation strands)

$f_{py} = 0.90\, f_{pu}$

Assume the following for loss computations:

E_{ci} = 3587 ksi

E_c = 4287 ksi

E_s = 28,000 ksi

Section Properties

$A_c = 180 \text{ in.}^2$

$I_c = 2{,}864 \text{ in.}^4$

y_b = 10.0 in.

y_t = 4.0 in.

$A_{ps} = 0.32 \text{ in.}^2$

e = 7.0 in.

P_i = 63.9 kips

Calculations and Discussion	Code Reference

1. Elastic Shortening of Concrete (ES)

$ES = K_{es}\ E_s f_{cir}/E_{ci}$ — Eq. (1)

$= 1.0(28,000)\ 0.6/3587 = \underline{4.68\ ksi}$

where $f_{cir} = K_{cir} f_{cpi} - f_g$

$= K_{cir}(P_i/A_c + P_i e^2/I_c) - M_d\ e/I_c$

$= 0.9\ (\frac{63.9}{180} + \frac{63.9 \times 7^2}{2864}) - \frac{288.8 \times 7}{2864} = 0.60\ ksi$

$M_d = 0.047 \times 4 \times 32^2 \times 12/8 = 288.8$ in. kips (dead load of unit)

and for pretensioned members, $K_{es} = 1.0$

$K_{cir} = 0.9$

2. Creep of Concrete (CR)

$CR = K_{cr} \frac{E_s}{E_c}\ (f_{cir} - f_{cds})$ — Eq. (3)

$= 2.0 \times \frac{28,000}{4287}\ (0.60 - 0.30) = \underline{3.92\ ksi}$

where $f_{cds} = M_{ds} \cdot e/I = 122.9 \times 7/2864 = 0.30\ ksi$

$M_{ds} = 0.02 \times 4 \times 32^2 \times 12/8 = 122.9$ in. kips (roof load only)

and $K_{cr} = 2.0$ for pretensioned members.

3. Shrinkage of Concrete (SH)

$SH = 8.2 \times 10^{-6}\ K_{sh}\ E_s\ (1 - 0.06\ \frac{V}{S})(100 - RH)$ — Eq. (4)

$= 8.2 \times 10^{-6} \times 1.0 \times 28,000\ (1 - 0.06 \times 1.22)(100 - 75) = \underline{5.32\ ksi}$

Sample Computation - Continued

where $V/S = 180/(4 \times 12 \times 2 + 2 \times 2 + 12 \times 4) = 1.22$

RH = average relative humidity surrounding the concrete member from Fig. 1. For Green Bay, Wisc., RH = 75%

and $K_{sh} = 1.0$ for pretensioned members.

4. Relaxation of Tendon Stress (RE)

$$RE = [K_{re} - J(SH + CR + ES)]\ C \qquad \text{Eq. (5)}$$

$$= [5 - 0.04\ (5.32 + 3.92 + 4.68)]\ 0.95 = \underline{4.22\ \text{ksi}}$$

where, for 270 Grade low-relaxation strand:

$K_{re} = 5$ ksi (Table 2)

$J = 0.040$ (Table 2)

$C = 0.95$ (Table 3 for $f_{pi}/f_{pu} = 0.74$)

5. Total allowance for loss of prestress — 18.6.1

$$ES + CR + SH + RE = 4.68 + 3.92 + 5.32 + 4.22$$

$$= \underline{18.14\ \text{ksi}}$$

6. Effective prestress stress f_{se} and effective prestress force P_e

$f_{se} = 0.74\ f_{pu}$ - allowance for all prestress losses

$$= 0.74\ (270) - 18.14 = 181.7\ \text{ksi}$$

$$P_e = f_{se}\ A_{ps} = 181.7 \times 0.32 = 58.1\ \text{kips}$$

EXAMPLE 25.1 - Investigation of Stresses at Prestress Transfer and at Service Load

For the simply supported double-tee shown below, check all permissible concrete stresses immediately after prestress transfer and at service load assuming the unit is used for roof framing.

Live load (20) + roofing (20) = 40 psf

Dead load = 47 psf

Span = 32 ft

f'_{ci} = 3,500 psi

f'_c = 5,000 psi

f_{pu} = 270,000 psi

(low-relaxation strands

$f_{py} = 0.9\ f_{pu}$)

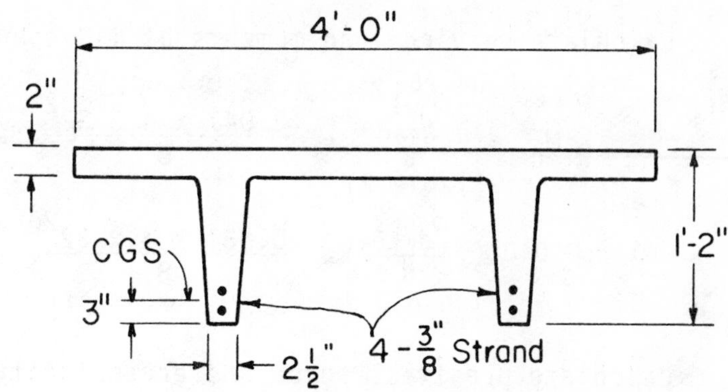

Section Properties

A = 180 in.^{2}

I = 2,864 in.^{4}

y_b = 10.0 in.

y_t = 4.00 in.

A_{ps} = 0.32 in.^{2}

Calculations and Discussion	Code Reference
1. Calculate permissible stresses in concrete.	18.4
At prestress transfer	18.4.1
Tension: $6\sqrt{f'_{ci}}$ = 354 psi*	
Compression: $0.60\ f'_{ci}$ = 2,100 psi	

*At ends of simply supported members; otherwise $3\sqrt{f'_{ci}}$

EXAMPLE 25.1 - Continued

Calculations and Discussion	Code Reference

<u>At service load</u> 18.4.2

Tension: $6\sqrt{f'_c}$ = 424 psi

Compression: $0.45\ f'_c$ = 2,250 psi

2. Calculate service load moments at mid span:

$$M_d = \frac{w_d \ell^2}{8} = \frac{0.047 \times 4 \times 32^2}{8} = 24.0 \text{ ft kips}$$

$$M_{d+\ell} = \frac{w_{d+\ell} \ell^2}{8} = \frac{0.087 \times 4 \times 32^2}{8} = 44.6 \text{ ft kips}$$

3. Calculate prestress force and eccentricity:

At transfer: $P_i = 0.74\ f_{pu}\ A_{ps}$ 18.5.1

$= 0.74 \times 270 \times 0.32$

$= 63.9$ kips

At service load:

Allowance for loss of prestress = 18.1 ksi 18.6

(estimate of loss of prestress calculated by method of Reference 25.2. See Appendix 25A for calculations of prestress loss for this Example)

Effective prestress stress: $f_{se} = 0.74\ f_{pu}$ - prestress loss

$= 0.74 \times 270 - 18.1 = 181.7$ ksi

$P_e = f_{se}\ A_{ps} = 181.7 \times 0.32 = 58.1$ kips

Eccentricity: $e = y_b - 3 = 10 - 3 = 7$ in.

EXAMPLE 25.1 - Continued

Calculations and Discussion	Code Reference

4. Calculate extreme fiber stresses by "straight line theory" which leads to the following well known formulas:

$$f_t = P/A + My_t/I - Pey_t/I$$

$$f_b = P/A - My_b/I + Pey_b/I$$

Stresses at Prestress Transfer (psi)

	Support		Mid Span	
	Top	Bottom	Top	Bottom
P_i/A	+ 355	+ 355	+ 355	+ 355
P_iey/I	- 623	+1562	- 623	+1562
My/I	--	--	+ 402	-1006
Total	- 268 (OK)	+1917 (OK)	+ 134 (OK)	+ 911 (OK)
Permissible	- 354	+2100	+2100	+2100

Compression (+)
Tension (-)

Stresses at Service Load (psi)

	Support		Mid Span	
	Top	Bottom	Top	Bottom
P_e/A	+ 323	+ 323	+ 323	+ 323
P_eey/I	- 568	+1420	- 568	+1420
My/I	--	--	+ 747	-1869
Total	- 245	+1743 (OK)	+ 502 (OK)	- 126 (OK)
Permissible	No limit	+2250	+2250	- 424

Compression (+)
Tension (-)

EXAMPLE 25.1 - Continued

Calculations and Discussion	Code Reference

Notes:

1. The violation of permissible stresses is allowed under Section 18.4.3 in certain cases.

2. The tension stresses computed at the support will always be less than the calculated values since the assumption of plane sections remaining plane cannot hold at this point. As a rough guide, a distance d (section depth) from the end of the member is required for this assumption to apply with a good degree of accuracy.

3. In computing stresses at the support, reduction of prestress to account for the bond transfer length of the wire can be used.

EXAMPLE 25.2 - Flexural Strength of Prestress Member Using Approximate Value for f_{ps}

Calculate the nominal moment strength of the prestressed member shown below.

f'_c = 5,000 psi

f_{pu} = 270,000 psi (ordinary strands $f_{py} = 0.85\ f_{pu}$)

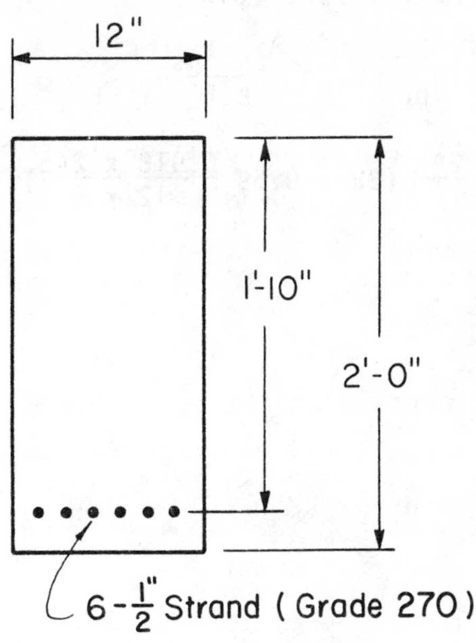

Calculations and Discussion	Code Reference
1. Calculate stress in prestressed reinforcement at nominal strength using approximate value for f_{ps}. For a fully prestressed member, Eq. (18-3) reduces to:	
$f_{ps} = f_{pu}\ (1 - \frac{\gamma_p}{\beta_1}\ \rho_p\ \frac{f_{pu}}{f'_c})$	Eq. (18-3)
$= 270\ (1 - \frac{0.40}{0.80}\ \times 0.00348\ \times\ \frac{270}{5}) = 245$ ksi	
where $\gamma_p = 0.40$ for $f_{py}/f_{pu} = 0.85$	18.0
$\beta_1 = 0.80$ for $f'_c = 5000$ psi	10.2.7.3
$\rho_p = A_{ps}/bd_p = 0.918/12 \times 22 = 0.00348$	

EXAMPLE 25.2 - Continued

Calculations and Discussion	Code Reference

2. Calculate reinforcement index ω_p. 18.8.1

$$\omega_p = \frac{A_{ps}\ f_{ps}}{bd_p\ f'_c} = \frac{0.918 \times 245}{12 \times 22 \times 5} = 0.170 < 0.36\beta_1 = 0.36\ (0.80) = 0.288 \quad \text{OK}$$

3. Calculate nominal moment strength.

$$M_n = A_{ps}\ f_{ps}\ (d_p - 0.59\ \frac{A_{ps}\ f_{ps}}{b\ f'_c})$$

$$M_n = \frac{0.918 \times 245}{12}\ (22 - 0.59\ \frac{0.918 \times 245}{12 \times 5}) = 371 \text{ ft kips}$$

EXAMPLE 25.3 - Flexural Strength of Prestressed Member Based on Strain Compatibility

The rectangular beam section shown below reinforced with a combination of prestressed and nonprestressed strands, calculate the nominal moment strength.

f'_c = 5,000 psi

f_{pu} = 270,000 psi (low-relaxation strand; $f_{py} = 0.9\ f_{pu}$)

E_{ps} = 28,000 ksi

losses = 31.7% ksi (calculated by method of Reference 25.2, excluding allowance for shortening of concrete. See Appendix 25A for procedure)

Calculations and Discussion	Code Reference
1. Calculate effective strain in prestressing steel. $\varepsilon = (0.74\ f_{pu} - f_{se})/E_{ps} = (0.74 \times 270 - 31.7)/28{,}000 = 0.0060$. Note: This value of ε should not include losses due to shortening of concrete.	
2. Draw strain diagram at nominal moment strength (defined by a maximum concrete compressive strain of 0.003). For f'_c = 5,000 psi, β_1 = 0.80.	18.3.1

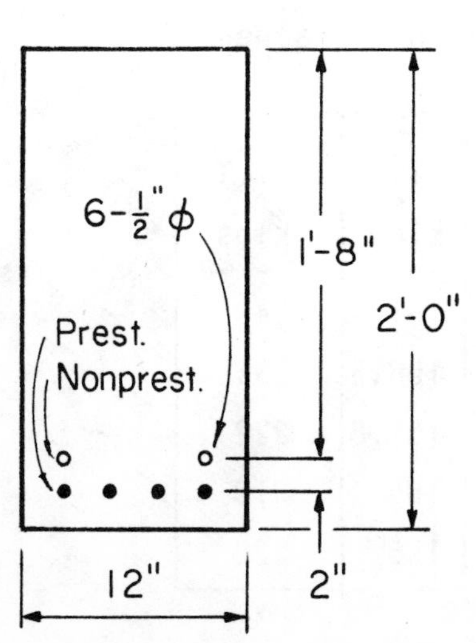

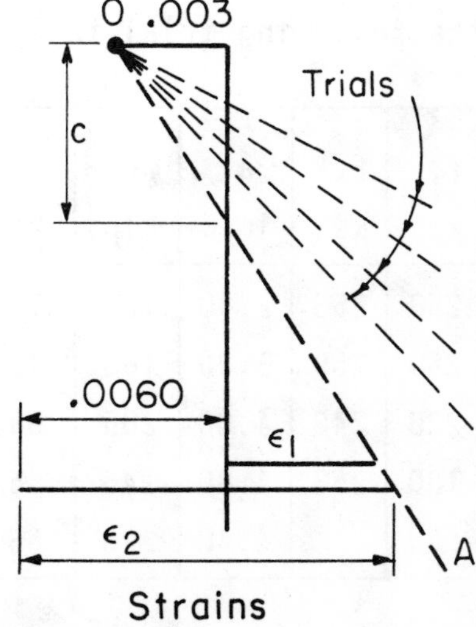

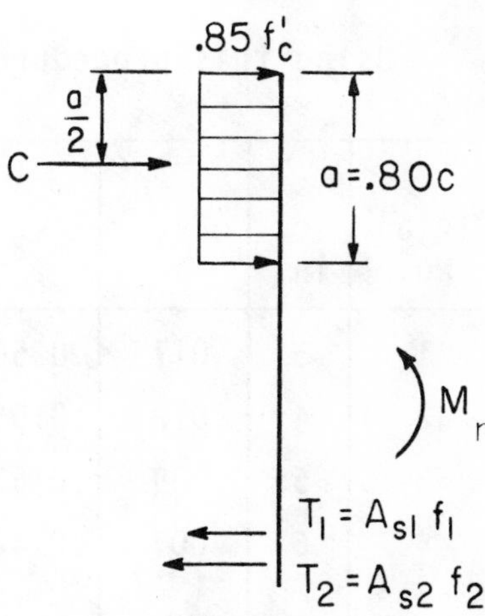

EXAMPLE 25.3 - Continued

Calculations and Discussion	Code Reference

3. Obtain equilibrium of horizontal forces.

The "strain line" drawn above from point O, must be such that one has equilibrium of horizontal forces.

$$C = T_1 + T_2$$

In computing T_1 and T_2, the strain ε_1, ε_2 and stresses f_1, f_2 must satisfy the stress-strain relation for the strand (see Fig. 25-6). Equilibrium can be obtained by a trial-and-error procedure as follows:

1. assume c (location of neutral axis)
2. compute ε_1 and ε_2 (a graphical procedure is very convenient)
3. obtain f_1 and f_2 from stress-strain curve (see Fig. 5).
4. compute $a = \beta_1 c$
5. compute $C = 0.85\ f'_c ab$
6. compute T_1 and T_2
7. check equation $C = T_1 + T_2$
8. if (7) is not verified go to (1).

Using this procedure the following trial table may be obtained:

Trial No.	c in.	ε_1	ε_2	f_1 ksi	f_2 ksi	a in.	C kips	T_1 kips	T_2 kips	$T_1 + T_2$ kips
1	3	.017	.0250	258	263	2.40	122	78.9	161	240
2	4	.012	.0195	250	259	3.20	163	76.5	158.5	235
3	5	.009	.0162	228	248	4.00	205	69.8	151.8	222
4	6	.007	.0140	180	253	4.80	245	55.1	155	210
*						4.40	220	65.0	155	220

*By interpolation

EXAMPLE 25.3 - Continued

Calculations and Discussion	Code Reference

4. Calculate nominal moment strength.

Using C = 220 kips, T_1 = 65 kips and T_2 = 155 kips, the nominal moment strength can be calculated as follows:

Taking moments about T_2:

$$M_n = (19.8 \times 220 - 2 \times 65)/12 = 352 \text{ ft kips}$$

EXAMPLE 25.3 - Continued

Calculations and Discussion	Code Reference

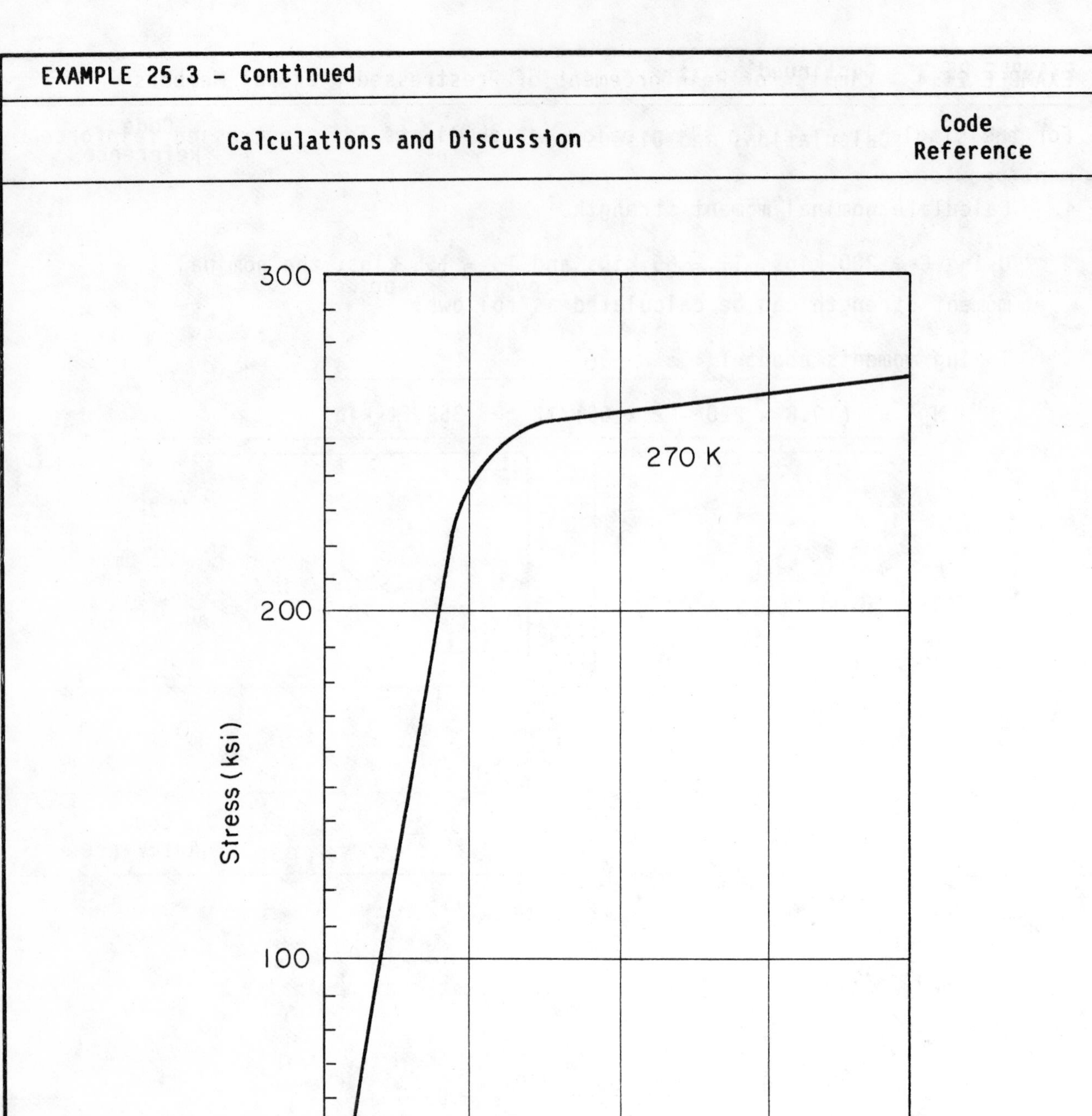

Fig. 25-6 Stress-Strain Relation for 270k Low-Relaxation Strand

EXAMPLE 25.4 - Limits for Reinforcement of Prestressed Flexural Members

For the single tee section shown below, check limits for prestressed reinforcement provided.

f'_c = 5,000 psi

f_{pu} = 270,000 psi (ordinary strands; $f_{py} = 0.85 f_{pu}$)

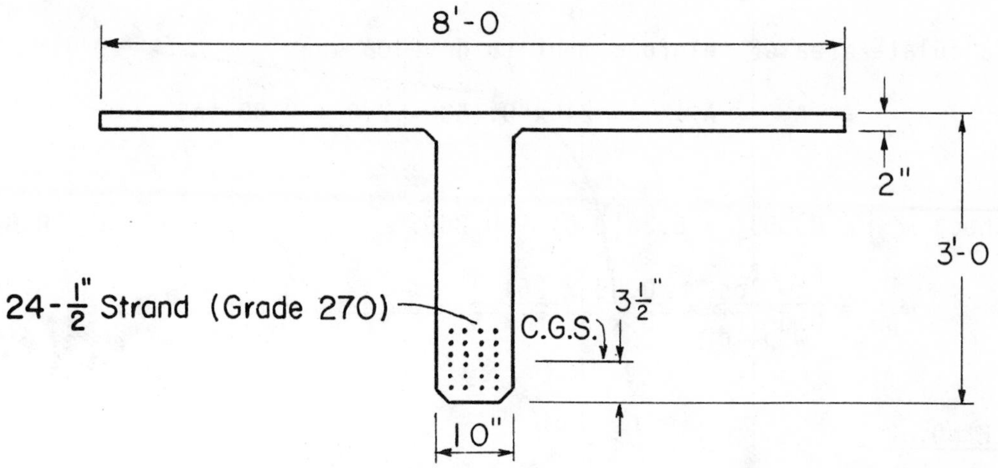

Calculations and Discussion	Code Reference

EXAMPLE No. 1

1. Calculate stress in prestressed reinforcement at nominal strength.

$$\omega_{pu} = \frac{A_{ps}\ f_{pu}}{bd_p\ f'_c} = \frac{24 \times 0.153 \times 270}{96 \times 32.5 \times 5} = 0.0636$$

$$f_{ps} = f_{pu}\left(1 - \frac{\gamma_p}{\beta_1}\omega_{pu}\right) = 270\left(1 - \frac{0.4}{0.8} \times 0.0636\right)$$ Eq. (18-3)

$$= 261 \text{ ksi}$$

2. Calculate required depth of concrete stress block.

$$a = \frac{24 \times .153 \times 261}{96 \times .85 \times 5} = 2.35 \text{ in.} > h_f = 2 \text{ in.}$$

EXAMPLE 25.4 - Continued

Calculations and Discussion	Code Reference

3. Calculate area of reinforcement to develop flange.

$$A_{pf} = \frac{0.85\ h_f\ (b-b_w)\ f'_c}{f_{ps}} = \frac{0.85 \times 2(96 - 10)\ 5}{261} = 2.8\ \text{in.}^2$$

4. Calculate area of reinforcement to develop web.

$$A_{pw} = A_{ps} - A_{pf} = 24 \times 0.153 - 2.8 = 0.88\ \text{in.}^2$$

5. Check $\omega_{pw} \leq 0.36\beta_1 = 0.36(0.8) = 0.288$: 18.8.1

$$\omega_{pw} = \frac{A_{pw}\ f_{ps}}{b_w d_p\ f'_c} = \frac{0.88 \times 261}{10 \times 32.5 \times 5} = 0.142 < 0.288 \quad \text{(OK)}$$

EXAMPLE NO. 2

In the previous problem check the limits of reinforcement assuming a 3 in. thick flange.

1. $f_{ps} = 261$ ksi

2. $a = 2.35 < h_f = 3$ in.

Since the stress block is entirely within the flange, the section acts effectively as a rectangular section.

3. Check limit $\omega_p \leq 0.36\beta_1$: 18.8.1

$$\omega_p = \frac{A_{ps}\ f_{ps}}{bd\ f'_c} = \frac{24 \times 0.153 \times 261}{96 \times 32.5 \times 5} = 0.0615 < 0.288\ \text{(OK)}$$

EXAMPLE 25.4 - Continued

Calculations and Discussion	Code Reference

Note: The limitation on ω_{pw} used in Example No. 1 needs to be checked only when the depth of the equivalent stress block is larger than the depth of the flange (see Fig. 25-7).

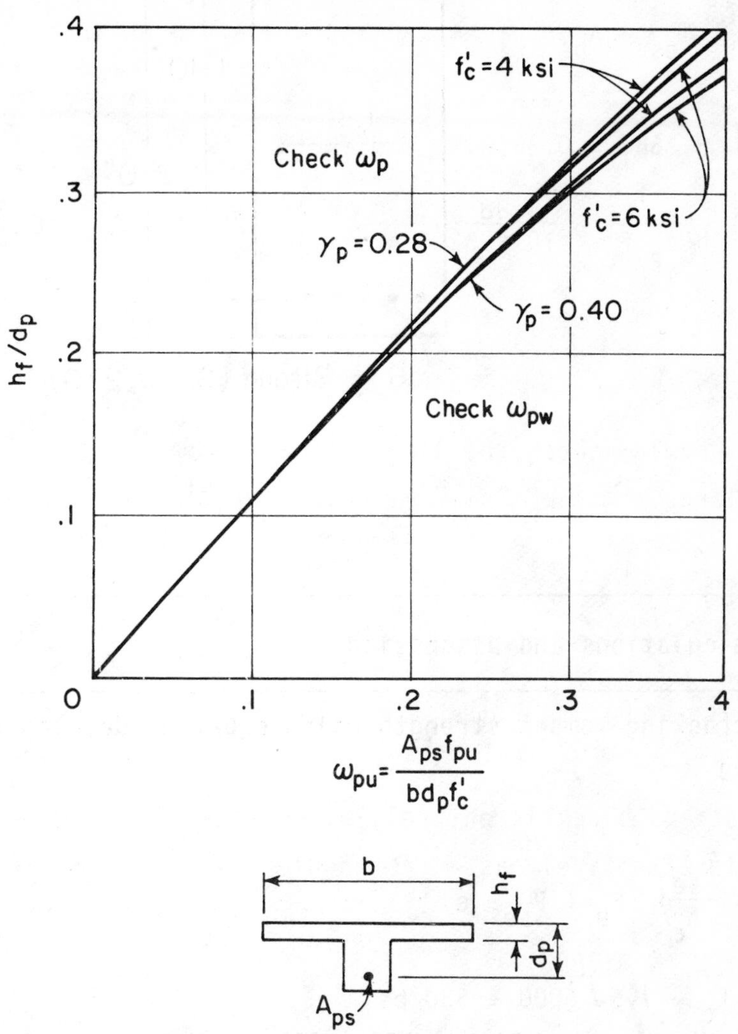

Fig. 25-7 - Relation Between Flange Thickness and Steel Percentage for Controlling ω_{pw}. Based on Depth of Rectangular Stress Block

EXAMPLE 25.5 - Cracking Moment Strength of Prestressed Member

For the prestressed member of Example 25.2, calculate the cracking moment strength and compare with the design moment strength.

f'_c = 5,000 psi

f_{pu} = 270,000 psi

losses = assume 20%

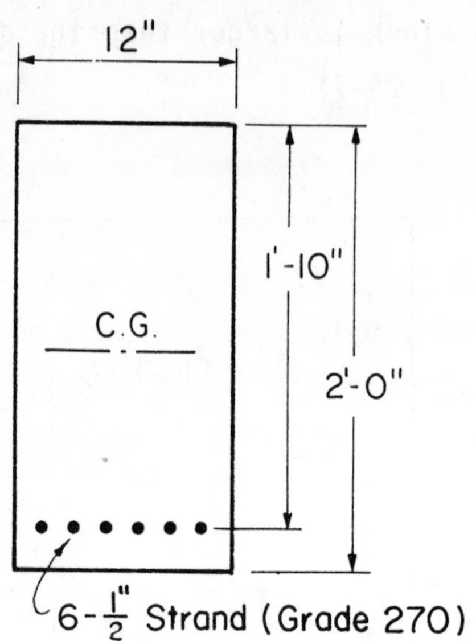

Calculations and Discussion	Code Reference

1. Calculate cracking moment strength using equation developed on page 25-9.

$$M_{cr} = (f_r + \frac{P_{se}}{A_c})\ S_b + P_{se} \cdot e$$

$f_r = 7.5\sqrt{f'_c} = 7.5\sqrt{5000} = 530$ psi — Eq. (9-9)

$P_{se} = 0.8 \times 6 \times 0.153 \times 0.7 \times 270 = 138$ kips (assuming 20% losses)

$S_b = bh^2/6 = 12 \times 24^2/6 = 1152$ in.^{3}

$A_c = bh = 12 \times 24 = 288$ in.^{2}

Calculations and Discussion	Code Reference

$e = 12 - 2 = 10$ in.

$M_{cr} = (0.530 + \frac{138}{288})1152 + 138 \times 10 = 2690$ in. kips

$= 224.2$ ft kips

Note: Cracking moment strength needs to be determined for checking maximum reinforcement per Section 18.8.3.

2. Section 18.8.3 requires that the total reinforcement (prestressed and nonprestressed) must be adequate to develop a design moment strength at least equal to 1.2 times the cracking moment strength. From Example 25.2, $M_n = 371$ ft kips.

$$\varphi M_n \geq 1.2M_{cr}$$

$$0.9(371) > 1.2\,(224.2)$$

$$334 > 269 \qquad \text{OK}$$

EXAMPLE 25.6 - Cracking Moment Strength of Prestressed Composite Member

For the 6 in. solid flat slab with 2 in. composite topping, calculate the cracking moment strength. Slab supported on bearing walls with 15 ft-0 span.

Section properties
per foot of width

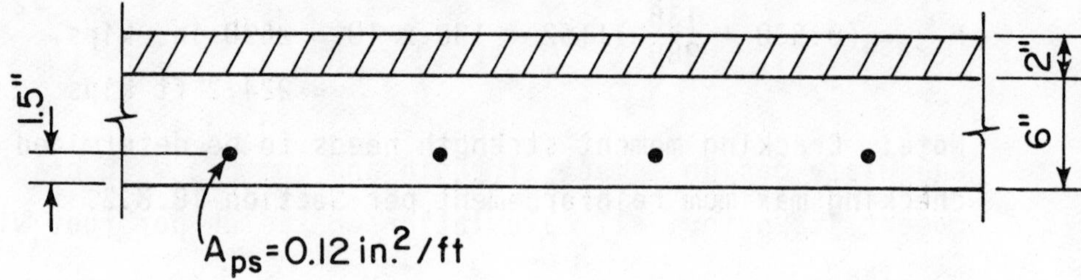

A_c = 72 in.^{2}

S_b = 72 in.^{3}

S_c = 132.7 in.^{3}

w_c = 83 psf

f'_c = 5000 (lightweight concrete)

f_{pu} = 250,000 psi

assume 25% losses

Calculations and Discussion	Code Reference

1. Calculate cracking moment strength using equation developed on page 25-9. All calculations based on per foot width of slab.

$$M_{cr} = (f_r + \frac{P_{se}}{A_c} + \frac{P_{se}\cdot e}{S_b})\ S_c - (M_{db} + M_{ds})\ (\frac{S_c}{S_b} - 1)$$

f_r = 0.75 $(7.5\sqrt{5000})$ = 398 psi

P_{se} = 0.75(0.12 x 0.7 x 250) = 15.75^k

e = 3 - 1.5 = 1.5 in.

M_{db+ds} = $w\ell^2/8$ = 0.083 x $15^2/8$ = $2.33^{'k}$ = $28.0^{''k}$

EXAMPLE 25.6 - Continued

Calculations and Discussion	Code Reference

$$M_{cr} = (0.398 + \frac{15.75}{72} + \frac{15.75 \times 1.5}{72})\ 132.7 - 28.0\ (\frac{132.7}{72} - 1)$$

$$M_{cr} = 125.4 - 23.6 = 101.8^{"k}$$

2. Calculate design moment strength and compare with cracking moment strength. All calculations based on per foot width of slab.

$A_{ps} = 0.12\ in.^2 \quad d_p = 8.0 - 1.5 = 6.5\ in.$ 10.0

$$\rho_p = \frac{A_{ps}}{bd_p} = \frac{0.12}{12 \times 6.5} = 0.00154$$

For $f_{pu} = 250$ ksi and $f'_c = 5000$ psi, Eq.(18-3) reduces to:

$$f_{ps} = f_{pu}\ (1 - 0.5\rho_p \frac{f_{pu}}{f'_c}) = 250\ (1 - 0.5 \times 0.00154 \times \frac{250}{5}) = 240.4\ ksi$$

$$a = \frac{A_{ps}\ f_{ps}}{0.85f'_c b} = \frac{0.12 \times 240.4}{0.85 \times 5 \times 12} = 0.57\ in.$$

$$M_n = A_{ps}\ f_{ps}\ (d_p - a/2) = 0.12 \times 240.4\ (6.5 - 0.57/2) = 179.3^{"k}$$

$$\varphi M_n = 0.9\ (179.3) = 161.4^{"k}$$

$\varphi M_n \geq 1.2\ (M_{cr})$ 18.8.3

$161.4 > 1.2\ (101.8)$

$161.4 > 122.2$ OK

EXAMPLE 25.7 - Prestressed Compression Members

For the short column shown below, calculate the nominal moment strength M_n for a nominal axial load P_n = 30 kips.

f'_c = 5,000 psi

f_{pu} = 270,000 psi (ordinary strand)

losses = assume 10%

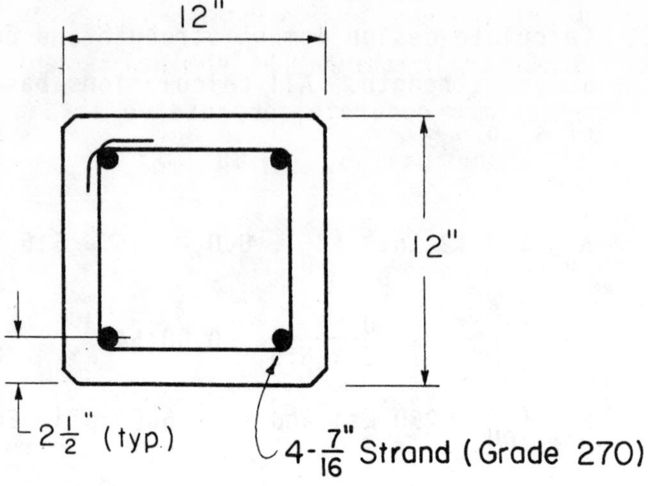

Calculations and Discussion	Code Reference

The same "strain compatibility" procedure used for flexure must be used here. The only difference is that for columns the load P_n must be included in the equilibrium of axial forces.

1. Calculate effective prestress.

$$f_{se} = 0.9 \times 0.7 f_{pu} = 0.9 \times 0.7 \times 270 = 170 \text{ ksi}$$

$$P_e = A_{ps} f_{se} = 4 \times 0.115 \times 170 = 78.2 \text{ kips}$$

2. Calculate average prestress on the column. — 18.11.2.1

$$f_{pc} = \frac{P_e}{A_g} = \frac{78.2}{12^2} = 0.542 > 0.225 \text{ ksi}$$

$\therefore$ no minimum reinforcement as per Section 10.9.1.

EXAMPLE 25.7 - Continued

Calculations and Discussion	Code Reference

3. Calculate effective strain in prestressing steel.

$$\varepsilon = \frac{f_{se}}{E_{ps}} = \frac{170}{28,000} = 0.0061$$

4. Draw strain diagram at nominal moment strength (defined by a maximum concrete compressive strain of 0.003). For f'_c = 5000 psi, β_1 = 0.80.

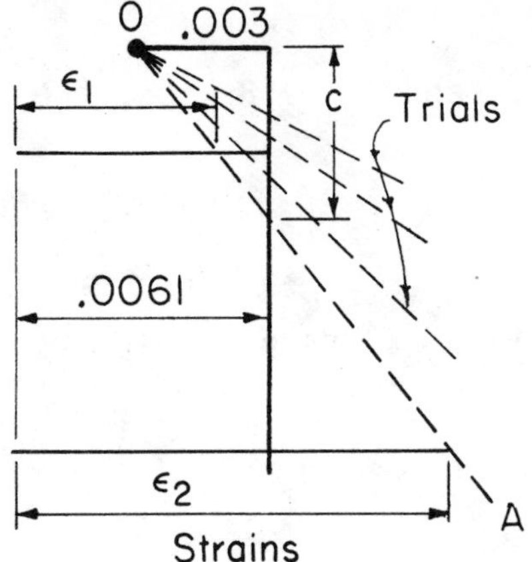

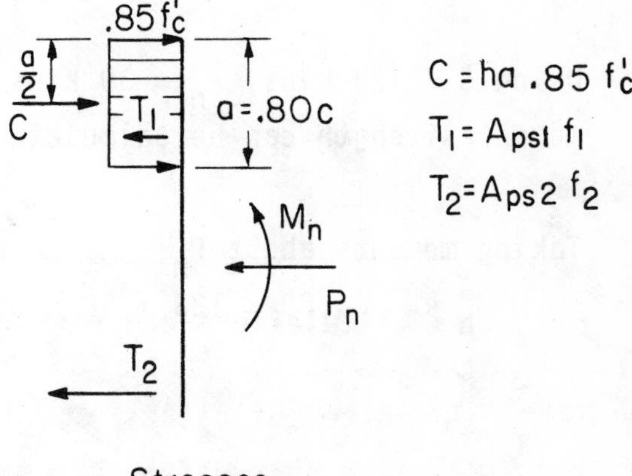

5. Obtain equilibrium of axial forces. The strain line OA drawn above, must be such that equilibrium of axial forces exists.

$$C = T_1 + T_2 + P_n$$

This can be done by trial-and-error as outlined in Design Example 25.3. Assuming different values of c, the following trial table is obtained:

EXAMPLE 25.7 - Continued

Calculations and Discussion	Code Reference

Trial No.	c in.	ε_1	ε_2	f_1* ksi	f_2* ksi	a in.	C kips	T_1 kips	T_2 kips	$T_1 + T_2 + P_n$ kips
1	1.5	0.0084	0.0217	215	260	1.20	61.2	49.4	59.6	139
2	2	0.0071	0.0175	187	256	1.60	81.7	43.0	58.8	132
3 OK	3	0.0057	0.0127	152	250	2.40	122	35.0	57.6	123

*From Fig. 25-6 (Design Example 25.3).

6. Calculate nominal moment strength.

Using C = 123 kips, P_n = 30 kips, T_1 = 35 kips, and T_2 = 58 kips, the moment strength can be calculated as follows:

Taking moments about P_n,

$$M_n = (4.80 \times 123 + 58 \times 3.5 - 35 \times 3.5)/12 = 55.9 \text{ ft kips}$$

26

Prestressed Concrete— Shear

General Considerations

The structural analysis of a prestressed member involves the determination of member stresses and strengths (both flexural and shear) and deformations at all critical sections of the member. Such an analysis is to be done at all load stages that may be significant during the life of the structure. The three specific loading stages indicated in the Code are as follows: (1) initial load stage, (2) design load stage, and (3) service load stage. Section 11.4 gives shear strength provisions for prestressed members and refers only to the "design load stage" of the member. Specific procedures of Section 11.4 are illustrated in Design Examples 26.1 and 26.2.

11.1 Shear Strength for Prestressed Members

Shear is expressed in terms of the factored shear force V_u directly, using the basic shear strength equality:

Required Shear Strength ≤ Design Shear Strength

$$V_u \leq \varphi V_n \quad \text{Eq. (11-1)}$$

$$\leq \varphi V_c + \varphi V_s \quad \text{Eq. (11-2)}$$

where the design shear strength φV_n is simply the sum of the shear strength provided by concrete (Section 11.4) plus the shear strength provided by shear reinforcement (Section 11.5). (Beginning with the 1977 Code, shear design provisions are presented in terms of shear forces ($V_n = V_c + V_s$) to better clarify application of the material strength reduction factor φ for shear design. In force format, the φ factor is correctly applied to the material strengths φV_c and φV_s directly, with only the load factor U applied to the required shear strength V_u.)

11.4 Shear Strength Provided by Concrete for Prestressed Members

Like the whole of Chapter 11, shear strength provided by concrete for prestressed members is presented in terms of nominal shear strength V_c, with the material understrength factor φ included with the basic shear strength Eq. (11-1). Substituting Eq. (11-2), the design shear strength is equal to φV_c where V_c is taken from the appropriate expressions of Section 11.4.

Section 11.4 is arranged with the simplified (Section 11.4.1) followed by the more complex option (Section 11.4.2):

Section 11.4.1 - Simplified

$$\varphi V_c = \varphi(2\sqrt{f'_c}\, b_w d)$$

$$\text{or } \varphi V_c = \varphi\left(0.6\sqrt{f'_c} + 700\,\frac{V_u d}{M_u}\right) b_w d \qquad \text{Eq. (11-10)}$$

Section 11.4.2 - More complex option

$$\text{The lesser of } \varphi V_{ci} = \varphi\left[0.6\sqrt{f'_c}\, b_w d + V_d + \frac{V_i M_{cr}}{M_{max}}\right] \qquad \text{Eq. (11-11)}$$

$$\text{or } \varphi V_{cw} = \varphi\left[(3.5\sqrt{f'_c} + 0.3\, f_{pc})\, b_w d + V_p\right] \qquad \text{Eq. (11-13)}$$

11.4.1 - The simplified V_c expressions of Section 11.4.1 are limited to members with an effective total prestress force at least equal to 40% of the tensile strength of the flexural reinforcement. Note that the limitation on "d" for Eq. (11-10) applies only to the $(V_u d/M_u)$ term of Eq. (11-10).

The value of "d" in the term ($b_w d$) is as defined in Section 11.0. Note also the additional limits on Eq. (11-10) in the end regions of pretensioned members as provided in Sections 11.4.3 and 11.4.4. Actually, Section 11.4.4 is new with the '83 Code, to ensure that the effect on shear strength of reduced prestress is properly taken into account when bonding of some of the tendons is prevented (debonding) near the ends of a pretensioned member, as permitted by Section 12.9.3.

11.4.2 - The optional V_c expressions of Section 11.4.2 are difficult to apply without design aids, and should be used only when V_c by Section 11.4.1 is not adequate. Shear strength by Section 11.4.2 is governed by the lesser value resulting from either flexural-shear cracking (V_{ci}) or web-shear cracking (V_{cw}).

11.4.2.1 - V_{ci} usually governs for members subject to uniform loading. The total shear strength V_{ci} is the sum of three parts: (1) the shear force required to transform a flexural crack into an inclined crack---$0.6 \sqrt{f'_c}\, b_w d$; (2) the <u>unfactored</u> dead load shear force--V_d; and (3) the portion of the remaining <u>factored</u> shear force that will cause a flexural crack to initially occur--$V_i\ M_{cr}/M_{max}$.

For non-composite members, V_d is the shear force caused by the <u>unfactored</u> selfweight of the member only. For composite members, V_d is computed using the <u>unfactored</u> selfweight plus <u>unfactored</u> superimposed dead load. The value V_i is the <u>factored</u> shear force resulting from the externally applied loads. V_i is determined by subtracting V_d from the shear force resulting from the total factored loads, V_u. Similarly, $M_{max} = M_u - M_d$. The load combination used to determine V_i and M_{max} is the one that causes maximum moment at the section under consideration. When calculating the cracking moment M_{cr}, the load used to determine f_d is the same unfactored load used to compute V_d.

11.4.2.2 - The expression for web shear strength V_{cw} usually governs for heavily prestressed beams with thin webs, especially when the beam is subject to large concentrated loads near simple supports. Eq. (11-13) predicts the shear strength at first web-shear cracking. An alternate value of V_{cw} can be computed as the shear force corresponding to dead load plus live load that results in a principal tensile stress of $4\sqrt{f'_c}$ at the centroid of the axis of the member, or at the interface of web and flange when the centroidal axis is located in the flange. This alternate method may be advantageous when designing members where shear is critical. Note the limitation on V_{cw} in the end regions of pretensioned members as provided in Sections 11.4.3 and 11.4.4.

11.5 Shear Strength Provided by Shear Reinforcement for Prestressed Members

Where the factored shear force V_u exceeds the shear strength φV_c, shear reinforcement must be provided to satisfy the basic shear strength Eq. (11-1) and Eq. (11-2). For the usual case with shear reinforcement perpendicular to the axis of the member, the design shear strength provided by shear reinforcement is equal to

$$\varphi V_s = \varphi(A_v f_y d/s) \qquad \text{Eq. (11-17)}$$

where A_v is the area of shear reinforcement within a distance s.

For design of shear reinforcement, required area A_v is computed directly from the basic shear strength Eq. (11-1) and Eq. (11-2),

$$V_u \leq \varphi V_n \qquad \text{Eq. (11-1)}$$

$$\leq \varphi V_c + \varphi V_s \qquad \text{Eq. (11-2)}$$

$$V_u \leq \varphi V_c + \varphi A_v f_y \frac{d}{s} \qquad \text{Eq. (11-17)}$$

$$\text{Solving for } A_v = \frac{(V_u - \varphi V_c)\,s}{\varphi f_y d}$$

Note that the φ factor is correctly applied to the strength provided by concrete and shear reinforcement.

Putting it all together, the shear strength equality for the load combination of dead load and live load can be summarized as follows:

Required Shear Strength ≤ Design Shear Strength

$$V_u \le \varphi V_n$$

$$\le \varphi (V_c + V_s)$$

$$1.4 V_d + 1.7 V_\ell \le \varphi \left[(0.6\sqrt{f'_c} + 700 \frac{V_u d}{M_u}) b_w d + A_v f_y \frac{d}{s} \right]$$

11.5.5.4 - For prestressed members, minimum shear reinforcement may be computed by either Eq. (11-14) or (11-15); however, Eq. (11-14) will generally give a higher minimum than Eq. (11-15). Note that Eq. (11-15) may not be used for members with an effective prestress force less than 40 percent of the tensile strength of the flexural reinforcement.

As permitted by Section 11.5.5.2, shear reinforcement may be omitted in any member if shown by physical tests that the required strength can be developed without shear reinforcement. For ACI 318-83, Section 11.5.5.2 is revised to clarify conditions for appropriate tests. Also, Commentary discussion has been added to give further guidance on appropriate tests to meet the intent of Section 11.5.5.2. The Commentary also calls attention to the need for sufficient stirrups in all thin-web, post-tensioned members to support the tendons in the design profile, and to provide reinforcement for tensile stresses in the webs resulting from local deviations of the tendons from the design tendon profile. (There have been some instances of horizontal cracking in webs of post-tensioned beams, joists, and waffle slabs when all stirrups were omitted.)

EXAMPLE 26.1 - Design for Shear (Section 11.4.1)

For the prestressed single tee shown, determine shear requirements using V_c by Eq. (11-10):

Precast concrete: f'_c = 5,000 psi (sand lightweight)
Topping concrete: f'_c = 4,000 psi (normal weight)
Prestress: Thirteen 1/2-in. dia. 270^k strand
Single depression at midspan
Span = 60 ft (simple)
Dead load, WDL = 723 lb/ft (includes topping)
Live load, WLL = 600 lb/ft
f_{se} (after all losses) = 125 ksi

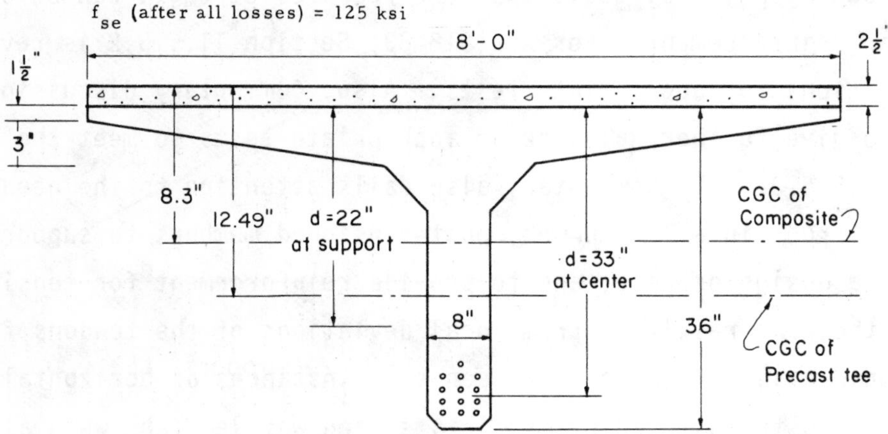

Calculations and Discussion	Code Reference

1. Determine factored shear force V_u at various locations along the span. The results are shown in Fig. 26-1.

Calculations and Discussion	Code Reference

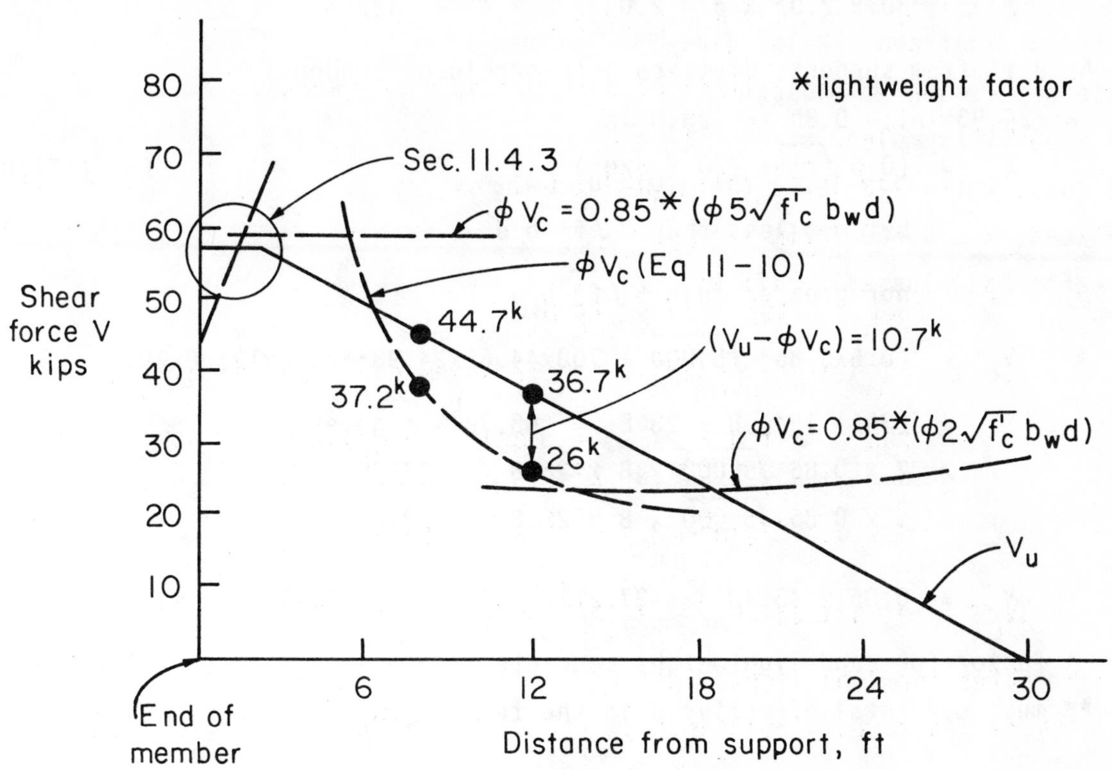

Fig. 26-1 Shear Force Variation Along Member

2. Determine shear strength provided by concrete V_c using Eq. (11-10). The effective prestress f_{se} is greater than 40 percent of f_{pu} (125 psi > 0.40 x 270 = 108 psi). Note that the value of "d" need not be taken less than 0.8h for shear strength computations. Typical computations using (Eq. 11-10) for a section 8 feet from support are as follows, assuming the shear is entirely resisted by the web of the precast section: — 11.4.1; 11.0

EXAMPLE 26.1 - Continued

Calculations and Discussion	Code Reference
$w_u = 1.4\ (0.723) + 1.7\ (0.600) = 2.03k/ft$	Eq. (9-1)
$V_u = [(60/2) - 8]\ 2.03 = 44.66^k$	
$M_u = 30 \times 2.03 \times 8 - 2.03 \times 8 \times 4 = 422^{'k}$	
At 8 ft from support, distance d (centroid of tendons) = 24.93 in.; 0.8h = 28.8 in.	
$V_c = (0.6\sqrt{f'_c} + 700\ V_u d/M_u)\ b_w d$ but not less than $2\sqrt{f'_c}\ b_w d$ nor greater than $5\sqrt{f'_c}\ b_w d$	Eq. (11-10)
$V_c = (0.6 \times 0.85^* \sqrt{5,000} + 700 \times 44.66 \times 24.93^{**}/422 \times 12)\ 8 \times 28.8$ $= (36 + 154)\ 8 \times 28.8 = 43.78^k$ (governs) $\geq 2 \times 0.85\sqrt{5,000} \times 8 \times 28.8 = 27.7^k$ $\leq 5 \times 0.85\sqrt{5,000} \times 8 \times 28.8 = 69.2^k$	
$\varphi V_c = 0.85 \times 43.78 = 37.21^k$ (see Fig. 26-1)	
* factor for sand-lightweight concrete.	11.2.1.2
** must use total effective d in the term $V_u d/M_u$.	11.4.1
Note: For members simply supported and subject to uniform loading, $V_u d/M_u$ in. Eq. (11-10) becomes a simple function of d/ℓ, where ℓ is the span length,	
$V_c = [0.6\sqrt{f'_c} + 700\ (\ell - 2x)/x\ (\ell - x)]\ b_w d$	Eq. (11-10)
where x is the distance from section being investigated to support. At 8 feet from support,	
$V_c = [0.6 \times 0.85\sqrt{5,000} + 700 \times 24.93(60-16)/8(60-8)12] 8 \times 28.8$ $= 43.78^k$	

EXAMPLE 26.1 - Continued

Calculations and Discussion	Code Reference

3. In the end regions of pretensioned members, the shear strength provided by concrete V_c may be limited by the provisions of Section 11.4.3. For this design, Section 11.4.3 does not apply because the section at h/2 is farther out into the span than the bond transfer length (see Fig. 26-2). The following will, however, illustrate typical calculations to satisfy Section 11.4.3. Compute V_c at 10 in. from end of member.

Bond transfer length for 1/2-in. diameter strand
= 50 (0.5) = 25 in. Prestress force at 10-in. location
= (10/25) 125 x 0.153 x 13 = 99.4^k

Vertical component of prestress force at 10-in. location,
V_p = 3.0 kips.

Distance d = 22.3 in, use 0.8h = 28.8 in. — 11.4.2.3

M_d (unfactored weight of precast unit + topping) = $215.8^{"k}$

Distance of composite section centroid from cgc of precast unit, c = 4.19 in.

Tendon eccentricity, e = 22.3 - 12.49 + 2.5
= 12.31 in. below cg.

f_{pc} (see notation definition) = $P_h/A_g - (P_h e)C/I_g + M_d c/I_g$
= 112.1 psi.

where A_g and I_g are for the precast section alone.

EXAMPLE 26.1 - Continued

Calculations and Discussion	Code Reference

V_{cw} = $(3.5\sqrt{f'_c} + 0.3\ f_{pc})\ b_w d + V_p$ Eq. (11-13)

= $(3.5 \times 0.85\sqrt{5{,}000} + 0.3 \times 112.1)\ 8 \times 28.8 + 3{,}000$

= 59.1^k

φV_{cw} = 0.85×59.1 = 50.2^k

The results of such an analysis are shown graphically in Fig. 26-2.

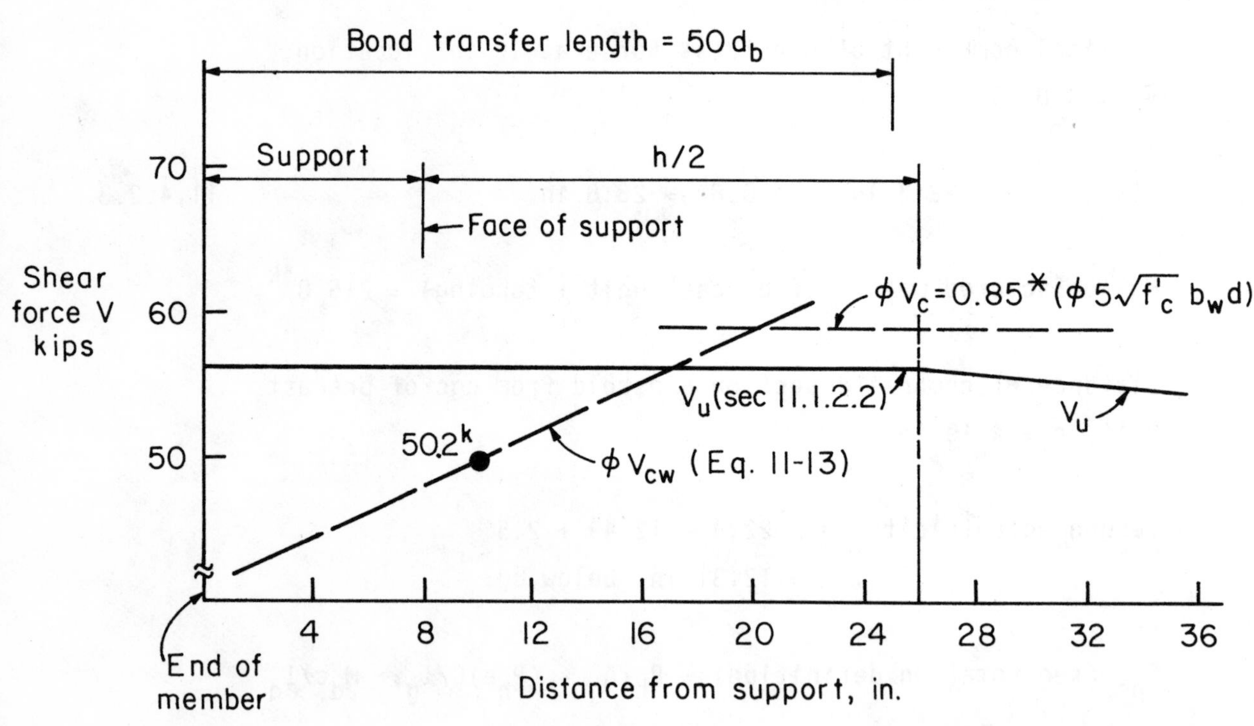

Fig. 26-2 Shear Force Variation at End of Member

EXAMPLE 26.1 - Continued

Calculations and Discussion	Code Reference

4. Compare factored shear V_u with shear strength provided by concrete φV_c. Where $V_u > \varphi V_c$, shear reinforcement must be provided to carry the excess, otherwise provide minimum shear reinforcement.

Shear reinforcement required at 11.9 ft from support is calculated as follows:

$$V_u = [(60/2) - 11.9]\ 2.03 = 36.74^k$$

$$\varphi V_c = 26.0^k \quad \text{(see Fig. 26-1)}$$

$$A_v = \frac{(V_u - \varphi V_c)s}{\varphi f_y d} = \frac{(36.74 - 26.0)\ 12}{0.85 \times 60 \times 28.8} = 0.087 \text{ in.}^2\text{/ft}$$

Check minimum required by Section 11.5.5.4. Use Eq. (11-15) since it generally requires a smaller amount for typical building members.

$$A_v\ (\text{min}) = \frac{A_{ps}}{80}\ \frac{f_{pu}}{f_y}\ \frac{s}{d}\sqrt{\frac{d}{b_w}}$$ Eq. (11-15)

$$= \frac{1.99}{80} \times \frac{270}{40} \times \frac{12}{28.8} \times \sqrt{\frac{28.8}{8}} = 0.133 \text{ in.}^2\text{/ft}$$
(governs)

Use #3 bar stirrups @ 18 in. for entire member length.

$$(A_v = 0.147 \text{ in.}^2\text{/ft})$$

EXAMPLE 26.2 - Design for Shear (Section 11.4.2)

For the pretensioned double tee shown, determine shear requirements using V_c by Eqs. (11-11) and (11-13).

Prestress:

3-3/8-in. dia. 250^k strands per stem

Force/strand after all losses = 11.4^k

Single point depression at midspan

Span = 30 ft 0 in. (simple)

d = 8.5 in. at end (11.5-in. at midspan)

Loading:

Ceiling load = 10 psf

Live load = 60 psf

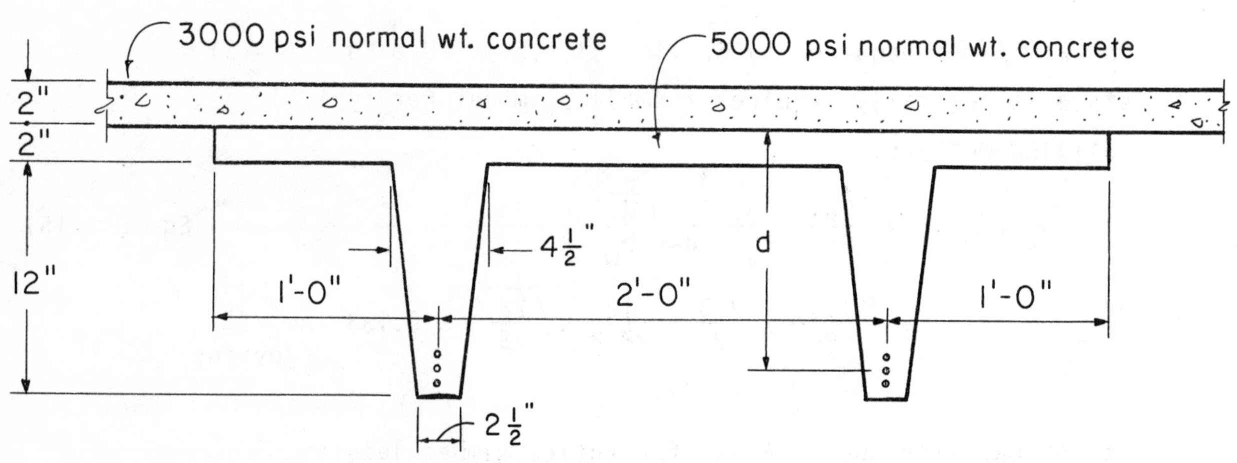

Properties	Area, in.^{2}	Weight psf	I, in.^{4}	Y_b, in.
Precast	180	47	2,864	10.0
Composite		25	4,203*	11.45*

* Corrected for difference in concrete strengths.

EXAMPLE 26.2 - Continued

Calculations and Discussion	Code Reference
1. Determine factored shear V_u at various locations along the span. The results are shown in Fig. 26-3.	

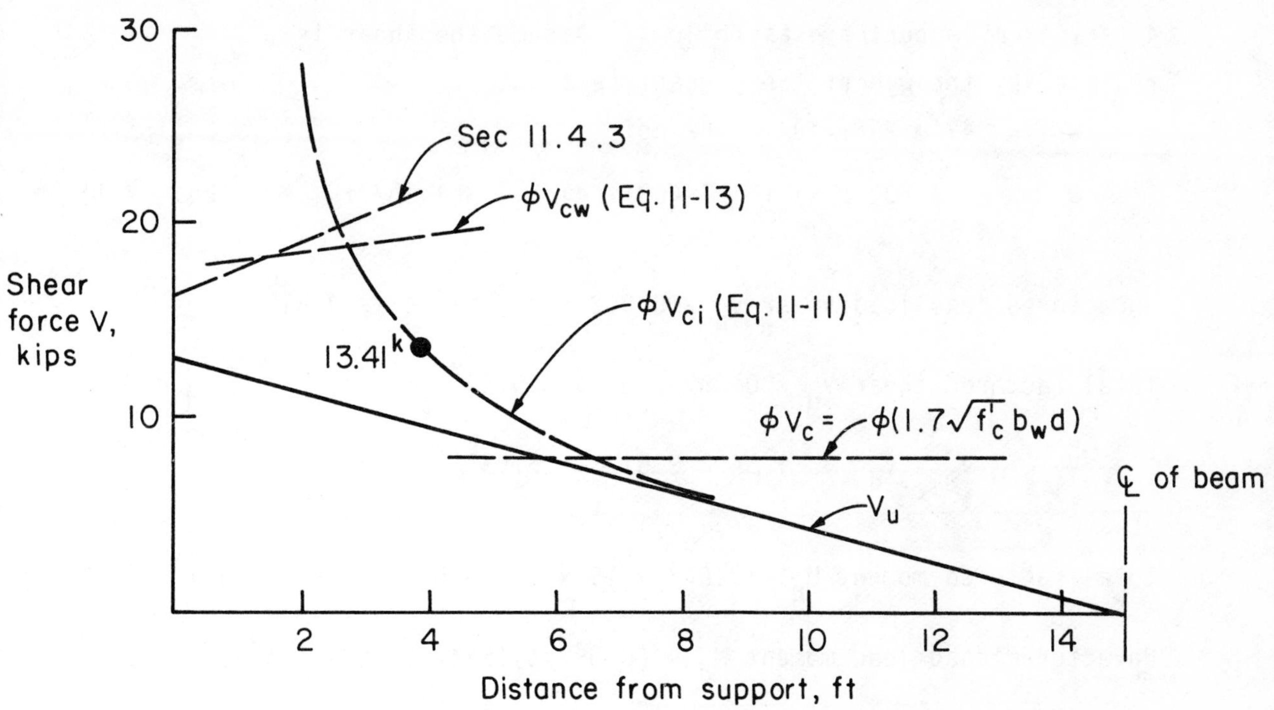

Fig. 26-3 Shear Force Variation Along Member

2. Determine shear strength provided by concrete V_{ci} using Eq. (11-11). Note that the value of "d" need not be taken less than 0.8h for shear strength computations. Values of V_d and M_d [f_d in Eq. (11-12)] are for <u>unfactored</u> dead load (member dead load plus superimposed dead load for composite members). The value of V_i is equal to the shear caused by total factored load minus V_d. Likewise,	11.4.2.3

Calculations and Discussion	Code Reference

$M_{max} = M_u - M_d$. For f_d in Eq. (11-12), the designer needs to know the construction sequence. For an unshored composite member, the beam and slab dead loads are resisted by the precast unit while the ceiling loads are resisted by the composite section. Typical computations for a section 4 feet from support are as follows. Assume the shear is resisted by the web of the precast section.

$w_d = 47 + 25 + 10 = 82$ psf

$w_u = 1.4\ (82 \times 4) + 1.7\ (60 \times 4) = 867$ lb/ft — Eq. (9-1)

Unfactored dead load shear $V_d = 0.082 \times 4\ (15 - 4) = 3.61^k$

Total factored shear $V_u = 0.867\ (15\text{-}4) = 9.54^k$

$V_i = V_u - V_d = 9.54 - 3.61 = 5.93^k$

Total factored moment $M_u = 0.867 \times 15 \times 4 - 0.867 \times 4^2/2 = 45.1^{'k}$

Unfactored dead load moment $M_d = (0.082\text{x}4)15\text{x}4\text{-}(0.082\text{x}4)\text{x}4^2/2 = 17.1^{'k}$

$M_{max} = M_u - M_d = 45.1 - 17.1 = 28.0^{'k}$

Note that both V_i and M_{max} result from the total factored loads less the unfactored dead loads (member plus superimposed dead load).

Dead load moment carried by precast section:

$M_{d1} = (47 + 25)\ 4 \times 15 \times 4 - (47 + 25)\ 4 \times 4^2/2 = 14.9^{'k}$

EXAMPLE 26.2 - Continued

Calculations and Discussion	Code Reference

Dead load stress:

$$f_{d1} = M_{d1}Y_b/I_g = 14.9 \times 12 \times 10/2864 = 0.627 \text{ ksi}$$

For composite sections, dead load moments carried by precast unit at any section consists of dead load of slab and beam only. Dead load moment carried by composite section consists of ceiling dead load:

$$M_{d2} = (10 \times 4)\ 15 \times 4 - (10 \times 4)\ 4^2/2 = 2.08^{'k}$$

Dead load stress:

$$f_{d2} = M_{d2}Y_{bc}/I_{gc} = 2.08 \times 12 \times 11.45/4203 = 0.068 \text{ ksi}$$

Therefore, dead load stress $f_d = f_{d1} + f_{d2} = 0.695$ ksi.

Eccentricity of prestressing tendons:

$$e = 10 - [14 - (4 \times 3/15 + 8.5)] = 5.3 \text{ in.}$$

Horizontal component of prestressing force $P_h = 68^k$

Concrete stress due to prestress:

$$f_{pe} = P_h/A_g + P_h e\ Y_b/I_g$$

$$= 68/180 + 68 \times 5.3 \times 10/2864 = 1.64 \text{ ksi}$$

Cracking moment:

$$M_{cr} = (I/Y_t)\ (6\sqrt{f'_c} + f_{pe} - f_d)$$ Eq. (11-12)

$$= (4203/11.45)\ (6\sqrt{5,000} + 1640 - 695)/12$$

$$= 41.75^{'k}$$

Note I and Y_t are properties of composite section.

Average width of two stems b_w = 7 in.

Effective depth d = 4x3/15+8.5 = 9.3 in.

Distance 0.8h = 11.2 in. (governs) 11.4.2.3

EXAMPLE 26.2 - Continued

Calculations and Discussion	Code Reference

$$V_{ci} = 0.6\sqrt{f'_c}\, b_w d + V_d + \frac{V_i M_{cr}}{M_{max}}$$ Eq. (11-11)

but not less than $1.7\sqrt{f'_c}\, b_w d$

$$V_{ci} = 0.6\sqrt{5{,}000} \times 7 \times 11.2 + 3.61 + \frac{5.93 \times 41.75}{28.0}$$

$$= 3.33 + 3.61 + 8.84 = 15.78^k \quad \text{(governs)}$$

$$\geq 1.7\sqrt{5{,}000} \times 7 \times 11.2 = 9.42^k$$

$$\varphi V_{ci} = 0.85 \times 15.78 = 13.41^k$$ (see Fig. 26-3)

3. Determine shear strength provided by concrete V_{cw} using Eq. (11-13). Note the use of a reduced prestress force in the end regions (Section 11.4.3). Computations using Eq. (11-13) are similar to those presented for Design Example 26.1. The results of such an analysis are shown graphically in Fig. 26-3.

4. Compare factored shear force V_u with shear strength provided by concrete φV_c. Where $V_u > \varphi V_c$, shear reinforcement must be provided to carry the excess, otherwise provide minimum shear reinforcement. Referring to Fig. 26-3, only minimum shear reinforcement by Section 11.5.5.4 is required for this design. Using Eq. (11-15):

$$A_v = \frac{A_{ps}}{80}\,\frac{f_{pu}}{f_y}\,\frac{s}{d}\sqrt{\frac{d}{b_w}}$$ Eq. (11-15)

$$= \frac{6 \times 0.08}{80} \times \frac{250}{60} \times \frac{12}{11.2}\sqrt{\frac{11.2}{7}} = 0.034 \text{ in.}^2\text{/ft}$$

$$= 0.017 \text{ in.}^2\text{/ft/stem}$$

Use 6 x 6 - W1.4 x W1.4 W.W.F. (A_v = 0.028 in.^{2}/ft)

EXAMPLE 26.2 - Continued

Calculations and Discussion	Code Reference
5. Check horizontal shear strength between precast unit and topping slab.	
$V_u = 0.867 \times 15 = 13.0^k$	
$\varphi V_{nh} = \varphi(80\ b_w d)$	17.5.2.1
$= 0.85\ (80 \times 48 \times 12.8) = 41.8^k$	
$V_u \leq \varphi V_{nh}$	Eq. (17-1)
$13.0^k < 41.8^k$ (OK)	
Contact surface must be clean, free of laitance, and intentionally roughened (broom finish).	17.5.2.1

27

Prestressed Slab Systems

Introduction

ACI 318-83 includes significant revisions to four Code sections with respect to analysis and design of prestressed slab systems:

Section 11.11.2 - Shear strength of prestressed slabs.

Section 11.12.2.4 - Shear strength of prestressed slabs with moment transfer.

Section 18.7.2 - f_{ps} for calculation of flexural strength.

Section 18.12 - Slab systems.

Discussion of each of these Code revisions are presented below, followed by Design Example 27.1 for a post-tensioned flat plate. The design example illustrates the Code revisions as well as general applicability of the Code to analysis and design of post-tensioned flat plates.

11.11.2 Shear Strength

For the first time, the ACI Code includes specific provisions for calculation of shear strength in two-way prestressed concrete systems. At columns of two-way prestressed slabs (and footings) utilizing unbonded tendons and meeting the bonded reinforcement requirements of Section 18.9.3, the shear strength V_n must not be taken greater than the shear strength V_c computed in accordance with Sections 11.11.2.1 or 11.11.2.2 unless shear reinforcement is provided in accordance with Sections 11.11.3 or 11.11.4. Section 11.11.2.2 is new for '83, and gives the following value of the shear strength V_c at columns of two-way prestressed slabs:

$$V_c = (3.5\sqrt{f'_c} + 0.3\ f_{pc})\ b_o d + V_p \qquad \text{Eq. (11-37)}$$

where b_o is perimeter of critical section d/2 from the face of the column, f_{pc} is average value of f_{pc} for the two directions, and V_p is vertical component of all effective prestress forces crossing the critical section. If shear strength is computed by Eq. (11-37), the following must be satisfied; otherwise, Eq. (11-36) applies:

(a) no portion of the column cross section can be closer to a discontinuous edge than 4 times the slab thickness, and

(b) f'_c must not be taken greater than 5000 psi, and

(c) f_{pc} in each direction must not be less than 125 psi, nor be taken greater than 500 psi.

In accordance with the above limitations, Eq. (11-36) is applicable to columns closer to the discontinuous edge than 4 times the slab thickness:

$$V_c = (2 + 4/\beta_c)\sqrt{f'_c}\ b_o d \qquad \text{Eq. (11-36)}$$

but not greater than $4\sqrt{f'_c}\ b_o d$. β_c is the ratio of long side to short side of column cross section and b_o is perimeter of critical section as defined above.

Results of tests at the University of Illinois[27.1] on edge column - prestressed flat plate joints, published after the 1983 Code provisions were adopted, indicate that Eq. (11-37) is applicable for edge columns as well as interior columns. However, unless a waiver is obtained from local building code officials for use of Eq. (11-37) on the basis of the University of Illinois tests, Eq. (11-36) should be applied in evaluation of shear strength of exterior column-slab connections.

11.12.2.4 Shear Strength With Moment Transfer

For moment transfer calculations, the controlling shear stress at columns of two-way prestressed slabs with bonded reinforcement in accordance with Section 18.9.3 is governed by Eq. (11-42) as follows:

$$v_c = \varphi(3.5\sqrt{f'_c} + 0.3\, f_{pc} + V_p/b_o d) \qquad \text{Eq. (11-42)}$$

where b_o is perimeter of critical section d/2 from the perimeter of the column and V_p is vertical component of all effective prestress forces crossing the critical section. If permissible shear stress is computed by Eq. (11-42), the following must be satisfied; otherwise, Eq. (11-41) applies:

(a) no portion of column cross section can be closer to a discontinuous edge than 4 times the slab thickness, and

(b) f'_c must not be taken greater than 5000 psi, and

(c) f_{pc} in each direction can not be less than 125 psi, nor be taken greater than 500 psi.

Eq. (11-41) limits the shear stress for exterior columns under moment transfer conditions to the same value as for nonprestressed slabs as follows:

$$v_c = \varphi(2 + 4/\beta_c)\sqrt{f'_c} \qquad \text{Eq. (11-41)}$$

but not greater than $\varphi 4\sqrt{f'_c}$.

The University of Illinois test results[27.1] also indicate that Eq. (11-41) is much too conservative for exterior column connections of prestressed slabs. However, Eq. (11-41) applies unless a waiver is obtained on the basis of the University of Illinois tests.

Application of Eqs. (11-41) and (11-42) are illustrated in Design Example 27.1.

18.7.2 f_{ps} for Unbonded Tendons

A new Eq. (18-5) is introduced in the '83 Code pertaining to prestressed elements with unbonded tendons having a span/depth ratio greater than 35:

$$f_{ps} = f_{se} + 10{,}000 + \frac{f'_c}{300\rho_p} \qquad \text{Eq. (18-5)}$$

but not greater than f_{py}, nor $(f_{se} + 30{,}000)$.

Nearly all prestressed one-way slabs and flat plates will have span/depth ratios greater than 35. Eq. (18-5) provides values of f_{ps} which are generally 15,000 to 20,000 psi lower than the values of f_{ps} given by Eq. (18-4) which was devised primarily from results of beam tests. These lower values of f_{ps} are more compatible with values of f_{ps} obtained in more recent tests of prestressed one-way slabs and flat plates. Application of Eq. (18-5) is illustrated in Design Example 27.1.

18.12 Slab Systems

The revision of Section 18.12 incorporates major modifications of analysis and design procedures for two-way prestressed slab systems as follows:

(1) Use of the Equivalent Frame Method of Section 13.7, or more detailed analysis procedures, is required for determination of factored moments and shears in prestressed slab systems. The 1977 Code permitted use of "any procedure satisfying conditions of equilibrium and geometric compatibility," although use of the equivalent frame model was recommended in the 1977 Code Commentary.

(2) Limits on tendon spacing are included in the Code for the first time. Spacing of tendons <u>in one direction</u> must not exceed 8 times the slab thickness nor 5 ft. Spacing of tendons must also provide a minimum average prestress, after allowance for all prestress losses, of 125 psi

on the slab section tributary to the tendon or tendon group. Special consideration of tendon spacing must be provided for slabs with concentrated loads.

(3) A minimum of two tendons must be provided in each direction through the critical shear section over columns. This provision, in conjunction with the limits on tendon spacing outlined in item 2 provides specific guidance for distributing tendons in prestressed flat plates in accordance with the "banded" pattern illustrated in Fig. 27-1. This method of tendon installation greatly simplifies detailing and installation procedures and it has been used successfully on a large volume of construction since 1969.

Calculation of equivalent frame properties is illustrated in Design Example 27.1. Tendon distribution is also discussed in Example 27-1.

Reference 27.2 also illustrates application of ACI 318 requirements for design of one-way and two-way post-tensioned slabs. Detailed design examples are also presented in Reference 27.2

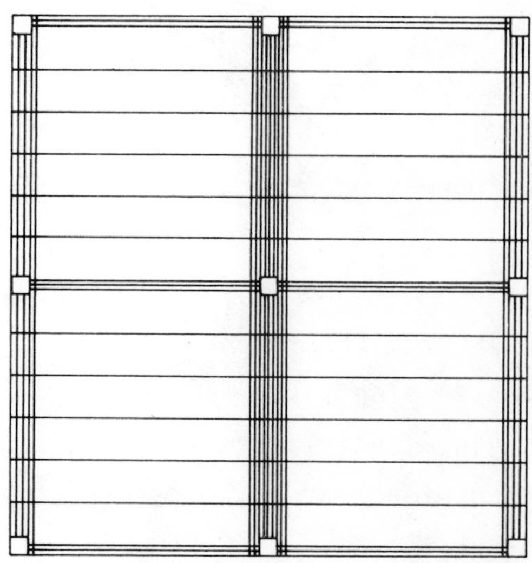

Fig. 27-1 Banded Tendon Distribution

Selected References

27.1 Sunidja, Harianto, Foutch, Douglas A., and Gamble, William A., "Responses of Prestressed Concrete Plate-Edge Column Connection," Structural Research Series No. 498, Report No. UILU-ENG-82-2006, University of Illinois at Urbana-Champaign, Urbana, Illinois, March 1982.

27.2 "Design of Post-Tensioned Slabs," Post-Tensioning Institute, Phoenix, Ariz., 1977, 52 pp.

EXAMPLE 27.1 - Two-Way Prestressed Slab-System

Design a typical transverse strip for the prestressed flat plate partial plan and section shown below.

f'_c = 4000 psi (slabs and columns)

f_y = 60,000 psi

f_{pu} = 270,000

live load = 40 psf

partitions = 15 psf

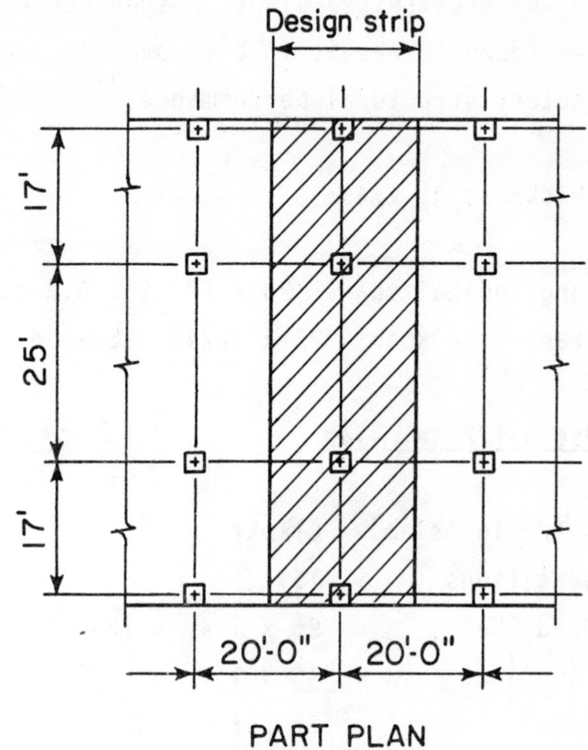

PART PLAN

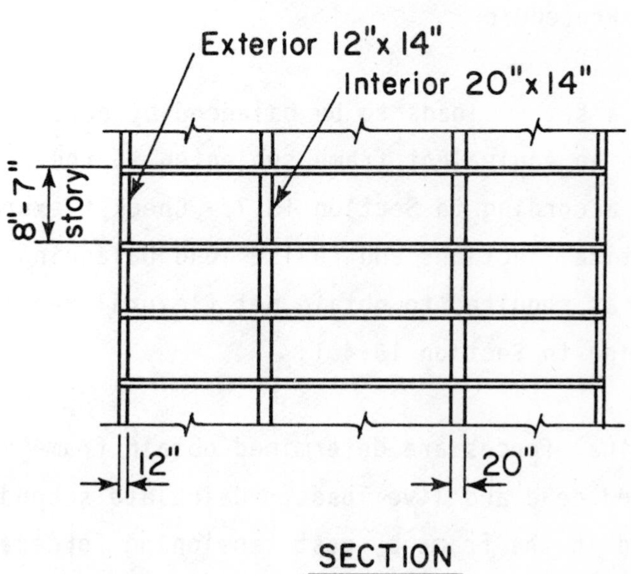

SECTION

EXAMPLE 27.1 - Continued

Calculations and Discussion	Code Reference

1. Slab thickness

For two-way prestressed slabs, a span/depth ratio of ℓ/45 has been found to result in best overall economy and provide satisfactory structural performance.[27.2]

Slab thickness at ℓ/45:

longitudinal span = 20 x 12/45 = 5.3 in.

transverse span = 25 x 12/45 = 6.7 in.

<u>Use 6-1/2 in. slab</u>

6-1/2 in. slab	=	81 psf		
partitions	=	15		
Dead load	=	96 x 1.4	=	134
Live load	=	40 x 1.7	=	68
Total	=	136 psf	=	202 psf

2. Design Procedure

Assume a set of loads to be balanced by parabolic tendons. Analyze an equivalent frame subjected to the net downward loads, according to Section 13.7. Check flexural stresses at critical sections and revise load balancing tendon forces as required to obtain net flexural tension stresses according to Section 18.4.1.

When final forces are determined obtain frame moments for factored dead and live loads. Calculate secondary moments induced in the frame by post-tensioning forces, and combine with factored load moments to obtain design factored moments. Provide minimum reinforcement in accordance with Section 18.9.

Calculations and Discussion	Code Reference

Check design flexural strength and increase nonprestressed reinforcement if required by strength criteria. Investigate shear strength, including shear due to vertical load and due to moment transfer, compare total to permissible values calculated in accordance with Section 11.11.2.

3. Load Balancing

Arbitrarily, a force corresponding to an average compressive stress of 175 psi, with maximum parabolic tendon profile, will be used for the initial estimate of balanced load.

Then F_e = 0.175 x 6.5 x 12
= 13.65 kips/ft

Assuming 1/2 in. diameter, 270 ksi strand tendons and 30 ksi long term losses, effective force per tendon is 0.153 x (0.7 x 270 - 30) = 24.33^k.

For a 20 foot bay, 20 x 13.65/24.33 = 11.2 tendons.

Use 11-1/2 in. (270 ksi) tendons/bay

Then F_e = 11 x 24.33/20 = 13.38^k/ft
F_e/A = 13.38/78 = 0.172 ksi

EXAMPLE 27.1 - Continued

Calculations and Discussion	Code Reference

4. Tendon Profile

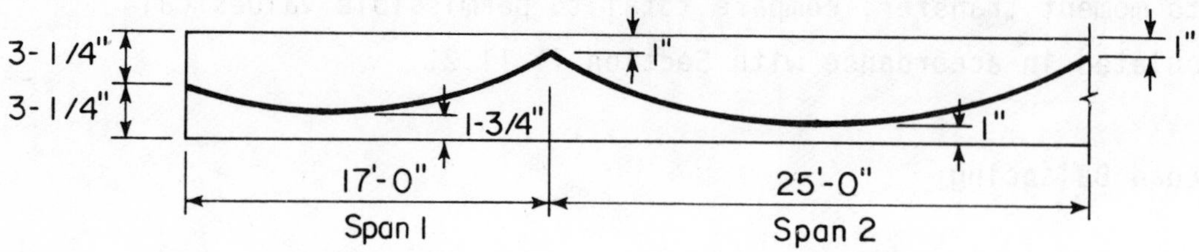

For spans 1 and 3:

$a = (3.25 + 5.5)/2 - 1.75 = 2.625$ in.

$w_{bal} = 8F_e a/12\ell^2 = 8 \times 13.38 \times 2.625/12 \times 17^2 = 0.081$ ksf

net load causing bending $= w_{net} = 0.136 - 0.081 = 0.055$ ksf

For span 2:

$a = 6.5 - 1 - 1 = 4.5$ in.

$w_{bal} = 0.064$ ksf

$w_{net} = 0.072$ ksf

5. Equivalent Frame Properties — 13.7

(a) Column Stiffness — 13.7.4

Column stiffness, including effects of "infinite" stiffness within the slab-column joint, may be calculated by classical methods or by simplified methods which are in close agreement. The following approximate stiffness K_c will give results within five percent of "exact" values.[27.2]

$K_c = 4EI/(\ell - 2h)$

where ℓ = center to center column height

h = slab thickness.

EXAMPLE 27.1 - Continued

Calculations and Discussion	Code Reference

For exterior columns (14 x 12):

$I = 14 \times 12^3/12 = 2016$ in.^{4}

$E_{col}/E_{slab} = 1.0$

$K_c = (4 \times 1.0 \times 2016)/(103 - 2 \times 6.5) = 90 \times 2 = 180$ (joint total)

Stiffness of torsional members is calculated as follows: — 13.7.5

$C = (1 - 0.63x/y)x^3y/3$ — Eq. (13-7)

$= (1 - 0.63 \times 6.5/12)6.5^3 \times 12/3 = 724$

$K_t = \dfrac{\Sigma 9CE}{\ell_2(1 - c_2/\ell_2)^3}$ — Eq. (13-6)

$= \dfrac{9 \times 724 \times 1}{20 \times 12(1 - 1.17/20)^3} = 32.5 \times 2 = 65$ (joint total)

Equivalent column stiffness, $1/K_{ec} = 1/\Sigma K_t + 1/\Sigma K_c$ — Com. 13.7.4

$K_{ec} = (1/65 + 1/180)^{-1} = 48$

For interior columns (14 x 20):

$I = 14 \times 20^3/12 = 9333$ in.^{4}

$K_c = (4 \times 1.0 \times 9333)/(103 - 2 \times 6.5) = 415 \times 2 = 830$ (joint total)

$C = (1 - 0.63 \times 6.5/20)6.5^3 \times 20/3 = 1456$

$K_t = \dfrac{9 \times 1456}{240(1 - 1.17/20)^3} = 65 \times 2 = 130$ (joint total)

$K_{ec} = (1/830 + 1/130)^{-1} = 112$

(b) Slab-beam stiffness — 13.7.3

Slab stiffness, including effects of infinite stiffness within slab-column joint, can be calculated by the following approximate expression[27.2]

$K_s = 4EI/(\ell_1 - c_1/2)$

where ℓ_1 = span center-to-center of supports

c_1 = column size in direction of ℓ_1

EXAMPLE 27.1 - Continued

Calculations and Discussion	Code Reference

At exterior column:

$K_s = (4 \times 1 \times 20 \times 6.5^3)/(12 \times 17 - 12/2) = 111$

At interior column (spans 1 & 3):

$K_s = (4 \times 1 \times 20 \times 6.5^3)/(12 \times 17 - 20/2) = 110$

For simplicity, use single value of 111 for both ends of span 1 and 3

At interior span 2:

$K_s = (4 \times 1 \times 20 \times 6.5^3)/(12 \times 25 - 20/2) = 76$

(c) Distribution factors for analysis by moment distribution.

Slab distribution factor at exterior joint = 111/(111 + 48) = 0.70

At interior joints for spans 1 and 3 = 111/(111 + 76 + 111) = 0.37

Span 2 = 76/298 = 0.25

6. Moment Distribution - Net loads

Since the nonprismatic section causes only very small effects on fixed-end moments and carryover factors, fixed-end moments will be calculated from $FEM = w\ell^2/12$ and carryover factors taken as COF = 1/2.

For span 1 and 3, net load $FEM = 0.055 \times 17^2/12 = 1.32$ ft-kips

For span 2, $FEM = 0.072 \times 25^2/12 = 3.75$ ft-kips

EXAMPLE 27.1 - Continued

Calculations and Discussion	Code Reference

Moment Distribution - Net Loads

DF	0.70	0.37	0.25
FEM	-1.32	-1.32	-3.75
dist.	+0.92	-0.90	+0.61
CO	+0.45	-0.46	-0.30
dist.	-0.32	+0.06	-0.05
	-0.27	-2.62	-3.49

7. Check net tensile stresses

(a) At face of column:

Moment at column face is centerline moment + Vc/3

$$-M_{max} = -3.49 + \frac{20}{3 \times 12}(12.5 \times 0.072) = -2.99 \text{ ft-kips}$$

$$s = bh^2/6 = 12 \times 6.5^2/6 = 84.5 \text{ in.}^3 \times \text{ft}/12 \text{ in.} = 7.04 \text{ in.}^2 \text{ ft}$$

$$f_t = 2.99/7.04 - 0.172 = 0.425 - 0.172 = +.253 \text{ ksi}$$

$6\sqrt{f'_c} = 0.380$ ksi > 0.253 OK — 18.4.2

(b) At midspan:

$$+M_{max} = 5.62 - 3.49 = +2.13 \text{ ft-kips}$$

$$f_t = 2.13/7.04 - 0.172 = +0.130 \text{ ksi}$$

$$2\sqrt{f'_c} = 2\sqrt{4000} = 0.126 \text{ ksi} < 0.130$$

When tensile stress of $2\sqrt{f'_c}$ is exceeded, Section 18.9.3 requires that the total tensile force be replaced by bonded reinforcement at a stress of $f_y/2$. — 18.9.3.2

EXAMPLE 27.1 - Continued

Calculations and Discussion	Code Reference
$f_c = -0.302 - 0.172 = -0.474 < 0.45\ f'_c$	18.4.2
$y = 6.5 \times 0.130/(0.130 + 0.474) = 1.40$ in.	
$T = 0.130 \times 1.40 \times 12/2 = 1.09^{k/ft}$	
$A_s = 1.09/(60/2) = 0.036$ in.^{2}/ft	Eq. (18-7)

<u>Use 4 #4 bars at 60 in. o.c. bottom of midspan 2</u>

(A_s = 0.04 in.^{2}/ft)

This completes the service load portion of the design. The design strength in flexure and shear must still be verified to complete the design.

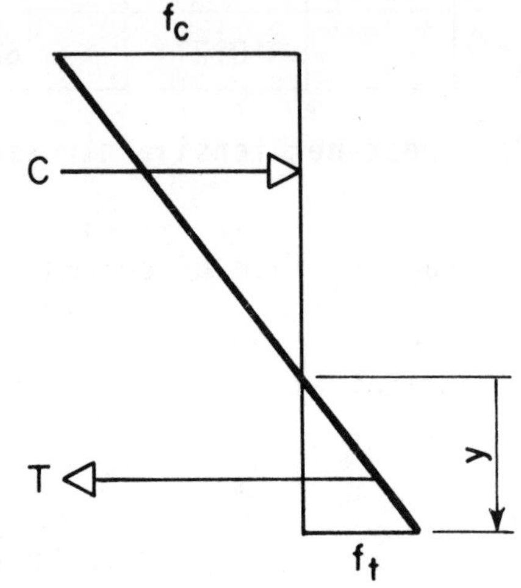

8. Flexural Strength

(a) Calculation of design moments

Design moments for statically indeterminate post-tensioned members are determined by combining frame moments due to factored dead and live loads with secondary moments induced into the frame by the tendons. The load balancing approach directly includes both primary and secondary effects, so that for service conditions only "net loads" need be considered.

At design flexural strength, the balanced load moments are used to determine secondary moments by subtracting the primary moment, which is simply F x e, at each support. For multistory buildings where typical vertical load design is combined with varying moments due to lateral loading, an efficient design approach would be

Calculations and Discussion	Code Reference

to analyze the equivalent frame under each case of dead, live, balanced, and lateral loads, and combine these cases for each design condition with appropriate load factors. For this example the balanced load moments are determined by moment distribution as follows:

For span 1 and 3, balanced load FEM = $0.081 \times 17^2/12 = 1.95$ ft-kips

For span 2, FEM = $0.064 \times 25^2/12 = 3.33$ ft-kips

Moment Distribution - Balanced Loads

DF	0.70	0.37	0.25
FEM	+1.95	+1.95	+3.33
dist.	-1.37	+0.51	-0.35
CO	-0.25	+0.68	+0.17
dist.	+0.18	-0.19	+0.13
	+0.51	+2.95	+3.28

Since load balancing accounts for both primary and secondary moment directly, secondary moments can be found from the following relationship:

$$M_{bal} = M_1 + M_2, \text{ then } M_2 = M_{bal} - M_1$$

The primary moment M_1 equals F x e at each support.
Thus, at exterior columns, the secondary moment M_2 equals:

$$M_2 = 0.51 - 13.38 \times 0 \text{ in.}/(12 \text{ in./ft}) = 0.51 \text{ ft-kips}$$

At interior column:

Span 1 and 3, $M_2 = 2.95 - 13.38(3.25 - 1.0)/12 = 0.44$ ft-kips

EXAMPLE 27.1 - Continued

Calculations and Discussion	Code Reference

Span 2, $M_2 = 3.28 - 13.38 \times 2.25/12 = 0.77$ ft-kips

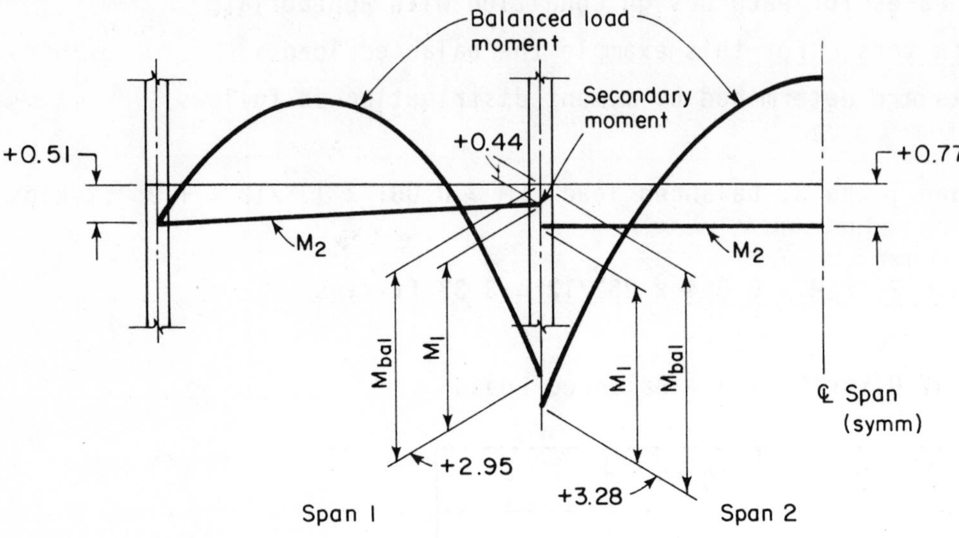

Factored load moments $w_u = 202$ psf:

For span 1 and 3, FEM = $0.202 \times 17^2/12 = 4.86$ ft-kips

For span 2, FEM = $0.202 \times 25^2/12 = 10.52$ ft-kips

Moment Distribution - Factored Loads

DF	0.70	0.37	0.25
FEM	-4.86	-4.86	-10.52
dist.	+3.40	-2.09	+ 1.42
CO	+1.05	-1.70	- 0.71
dist.	-0.74	+0.37	- 0.25
	-1.15	-8.28	-10.06

Calculations and Discussion	Code Reference

Combine factored load and secondary moments to obtain design moments:

	Span 1		Span 2
Factored load moments Secondary moments	-1.15 +0.51	-8.28 +0.44	-10.06 + 0.77
Moments at Col ℄ Moment reduction to face of column. Vc/3	-0.64 +0.42	-7.84 +1.19	- 9.29 + 1.40
Design moments at face of col.	-0.22	-6.65	- 7.89

Calculate design moments at midspan:

For span 1, V_{ext} = 0.202 x 17/2 - (8.28 - 1.15)/17
= 1.72 - 0.42 = 1.30 kips/ft

V_{center} = 1.72 + 0.42 = 2.14 kips/ft

Point of zero shear and maximum moment,

x = 1.30/0.202 = 6.45 ft from exterior column.

End span positive moment = 1.30 x 6.45 = +8.40

$-0.202(6.45)^2$ = -4.20

end moment = -1.15

+3.05

M_2 = 0.43

$+M_{max}$ = +3.48 ft-kips/ft

EXAMPLE 27.1 - Continued

Calculations and Discussion	Code Reference

For span 2, $V = 0.202 \times 25/2 = 2.52$ kips/ft

$$+\text{Moment} = 0.202 \times 25^2/8 = +15.78$$

$$\text{end moment} = \underline{-10.06}$$

$$+\ 5.72$$

$$M_2 = \underline{+\ 0.77}$$

$$+M_{max} = +\ 6.49 \text{ ft-kips/ft}$$

(b) Calculation of flexural strength

Check interior support section. Section 18.9.3 requires a minimum amount of bonded reinforcement in negative moment areas at column supports regardless of service load stress conditions or strength, unless more than the minimum is required for flexural strength. This minimum amount is to help ensure the integrity of the punching zone so that full shear strength can be developed.

$$A_s = 0.00075h\ell \qquad \text{Eq. (18-7)}$$

$$= 0.00075 \times 6.5 \times (17 + 25)/2 \times 12 = 1.22 \text{ in.}^2$$

Say 6 #4 bars x 9 ft. Space at maximum 6 in. centers so that bars are placed within column width plus 1.5 slab thickness each side of column.

For average one foot strip:

$$A_s = 6 \times 0.20/20 = 0.06 \text{ in.}^2/\text{ft}$$

Initial check of flexural strength will be made considering this reinforcement.

EXAMPLE 27.1 - Continued

Calculations and Discussion	Code Reference

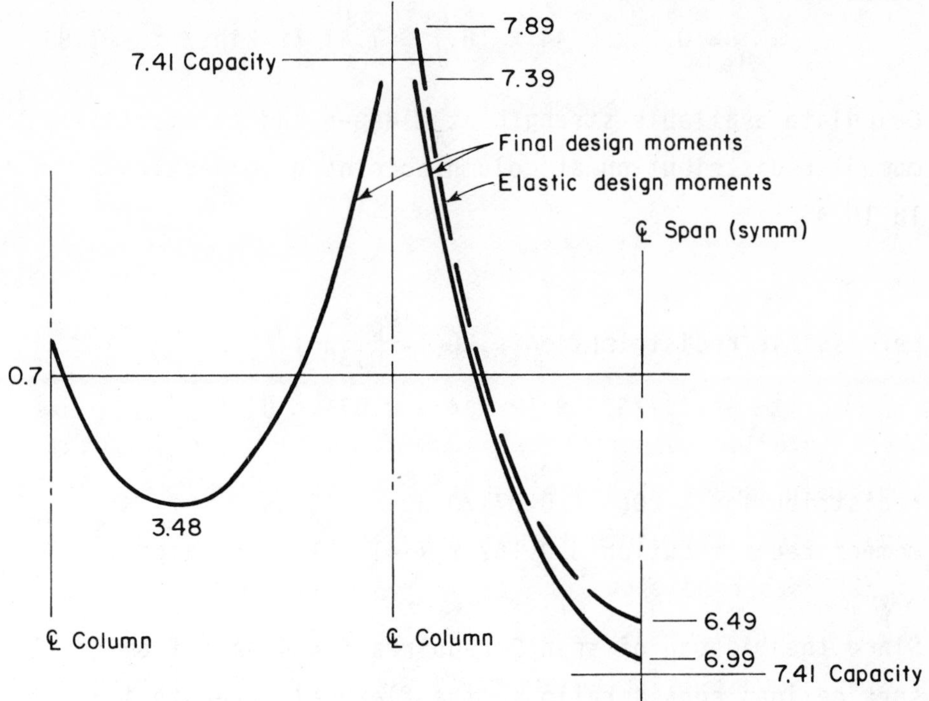

Calculate stress in tendons at nominal strength:

$$f_{ps} = f_{se} + 10{,}000 + \frac{f'_c}{300\rho_p}$$ Eq. (18-5)

With 11 tendons in 20 ft bay:

$$\rho_p = A_{ps}/bd = 11 \times 0.153/20 \times 12 \times 5.5 = 0.00127$$

$$f_{se} = 0.7 \times 270 - 30 \text{ ksi (losses)} = 159 \text{ ksi}$$

$$f_{ps} = 159 + 10 + 4/300(0.00127) = 169 + 10.5 = 179.5 \text{ ksi}$$

$$A_{ps}f_{ps} = 179.5 \times 0.153 \times 11/20 = 15.10 \text{ kip/ft}$$

$$A_sf_y = 0.06 \times 60 = \underline{3.60}$$

$$18.70 \text{ kip/ft}$$

$$a = \frac{A_{ps}f_{ps} + A_sf_y}{0.85 f'_c b} = \frac{18.70}{0.85 \times 4 \times 12} = 0.46 \text{ in.}$$

Since bars and tendons are in same layer:

$$(d-a/2) = (5.5 - \frac{0.46}{2})/12 = 0.44 \text{ ft}$$

EXAMPLE 27.1 - Continued

Calculations and Discussion	Code Reference

At column centerline:

$\varphi M_n = 0.9 \times 0.44 \times 18.7 = 7.41$ ft-kips/ft < 7.89 — 9.3.2.1

Calculate available strength at midspan and permissible moment redistribution at column according to Section 18.10.4.

$$\text{permissible redistribution} = 20(1 - \frac{\omega_p + \omega}{0.30})$$

$\Sigma\omega = 18.7/5.5 \times 12 \times 4 = 0.071 < 0.20$ — 18.10.4.3

redistribution = 20(1 - 0.071/0.30) = 15.2%

moment redistribution = 0.152 x 7.41 = 1.13 ft-kips

Since the midspan of span 2 requires 4 # 4 bars from service load considerations, the flexural strength is:

$$A_{ps}f_{ps} = 15.10$$

$$A_s f_y = 0.04 \times 60 = \underline{2.40}$$

$$17.50 \text{ kips/ft}$$

$a = 17.50/0.85 \times 4 \times 12 = 0.43$ in.

$$(d-a/2) = (5.5 - \frac{0.43}{2})/12 = 0.44 \text{ ft}$$

at center of span, $\varphi M_n = 0.9 \times 0.44 \times 17.50 = 6.93$ ft-kips/ft

The required moment strength = 6.49 ft-kips, which leaves 0.44 ft-kips available to accommodate moment redistributed from the support section. If 0.44 is redistributed:

-M = -7.89 + 0.44 = -7.45 > 7.41 NG

+M = +6.49 + 0.44 = +6.93 = 6.93 OK

EXAMPLE 27.1 - Continued

Calculations and Discussion	Code Reference

Thus, minimum rebars plus tendons are not adequate for flexural strength at column support. Addition of 2 #4 bars at midspan will make midspan strength identical to strength at column (which also has 6 #4 bars) = 7.41 ft-kips/ft. For this case, 7.41 - 6.49 or 0.92 ft-kips/ft are available for redistribution. Redistributing 0.50 ft-kips,

$$-M = -7.89 + 0.50 = -7.39 < 7.41 \quad \text{OK}$$

$$+M = +6.49 + 0.50 = +6.99 < 7.41 \quad \text{OK}$$

With 2 #4 bars added at midspan and redistribution of 0.50 ft-kips from the negative moment section to midspan, both negative and positive moment sections are adequate. Midspan sections of span 1 and 3 have more than adequate strength by comparison with span 2.

The flexural strength at exterior columns is governed by moment transfer requirements. Since moment transfer also involves shear, the two aspects will be treated under shear strength considerations.

9. Shear and moment transfer strength at exterior column. 11.12.2
13.3.3

(a) Shear and moment transferred at exterior column

$$V_u = 0.202 \times 17/2 - (8.28 - 1.15)/17 = 1.30 \text{ kips/ft}$$

Assume building enclosure is masonry and glass = 0.40 kips/ft

Total slab shear at exterior column:

$$V_u = (1.4 \times 0.40 + 1.30)20 = 37.2 \text{ kips}$$

Transfer moment = 20(0.64) = 12.8 ft-kips

EXAMPLE 27.1 - Continued

Calculations and Discussion	Code Reference

(b) Combined shear stress at inside face of critical transfer section.

For shear strength equations:
see Part 17, page 17-6 and Fig. 17-6.

$$v_u = V_u/A_c + \gamma_v M_u/(J/c)$$
$$= 37{,}200/252 + 0.39 \times 12.8 \times 12{,}000/1419$$
$$= 148 + 42 = 190 \text{ psi}$$

where (referring to Fig. 17-6: edge column-bending perpendicular to edge)

assume $d = 0.8 \times 6.5 = 5.2$ in.

$c_1 = 12$ in.

$c_2 = 14$ in.

$a = c_1 + d/2 = 14.6$ in.

$b = c_2 + d = 19.2$ in.

$c = a^2/(2a + b) = 4.40$ in.

$A_c = (2a + b)d = 252 \text{ in.}^2$

$J/c = [2ad(a + 2b) + d^3(2a + b)/a]/6$
$= 1419 \text{ in.}^3$

$$\gamma_v = 1 - \frac{1}{1 + 2/3\sqrt{\dfrac{c_1 + d}{c_2 + d}}} = 0.39$$ Eq. (11-40)

(c) Permissible shear stress — 11.12.2.4

For edge columns, Eq. (11-41) applies:

$$v_c = \varphi 4\sqrt{f'_c} = 0.85 \times 4\sqrt{4000} = 215 \text{ psi} > 190 \quad \text{OK}$$

(d) Check moment transfer strength — 13.3.3

Although the transfer moment is small, for illustrative purposes, calculate the moment strength of the effective slab width for moment transfer (width of column plus 1.5 — 13.3.3.2

Calculations and Discussion	Code Reference

slab thickness each side). Assume that of the 11 tendons required for the 20 ft bay width, 3 tendons are anchored within the column and are bundled together across the building. This amount should be noted on the design drawings. Besides providing flexural strength, this prestress will act directly on the critical section for shear and improve shear strength. As previously shown, a minimum amount of bonded reinforcement is required at all columns. For the exterior column, the required area is

$$A_s = 0.00075\ h\ell = 0.00075 \times 6.5 \times 17 \times 12$$ Eq. (18-7)

$$= 1.0\ \text{in.}^2$$

<u>Say 5 #4 bars x 5 ft</u>(including standard end hook).

Calculate stress in tendons:

Effective slab width = 14 + 2(1.5 x 6.5) = 33.5 in.

Effective prestress $f_{pe} = \dfrac{33.5 \times 3.25 \times 4}{300 \times 0.153 \times 3} + 169 = 172$ ksi

Effective prestress force = (172/159)24.33 x 3 = 79.0 kips

$A_s f_y = 5 \times 0.20 \times 60 = \underline{60.0}$

139.0 kips

$$a = \frac{139.0}{0.85 \times 4 \times 33.5} = 1.22\ \text{in.}$$

tendon $j_u d = (3.25 - 1.22/2)/12 = 0.22$ ft

rebar $j_u d = (5.5 - 1.22/2)/12 = 0.41$ ft

$\varphi M_n = 0.9(79 \times 0.22 + 60 \times 0.41) = 37.4$ ft-kips

$$\gamma_f = \frac{1}{1 + 2/3\sqrt{\dfrac{c_1 + d}{c_2 + d}}} = 0.61$$ Eq. (13-1)

$\gamma_f M_u = 0.61(12.8) = 7.8$ ft-kips « 37.4 OK

EXAMPLE 27.1 - Continued

Calculations and Discussion	Code Reference

10. Shear and moment transfer strength at interior column. 11.12.2 13.3.3

(a) Shear and moment transferred at interior column.

Direct shear left and right of interior columns is calculated in Step 8 above.

$$V_u = (2.14 + 2.52)20 = 93.2 \text{ kips}$$

Transfer moment = 20(9.35 - 7.84) = 30.2 ft-kips

(b) Combined shear stress at face of critical transfer section.

For shear strength equations: See Part 17, page 17-6 and Fig. 17-5.

$$v_u = V_u/A_c + \gamma_v M_u/(J/c)$$
$$= 93{,}200/462 + 0.43 \times 30.2 \times 12{,}000/3664$$
$$= 202 + 43 = 245 \text{ psi}$$

where (referring to Fig. 17-5: interior column)

$d \simeq 0.8 \times 6.5 = 5.2$ in.

$c_1 = 20$ in.

$c_2 = 14$ in.

$a = c_1 + d = 25.2$ in.

$b = c_2 + d = 19.2$ in.

$A_c = 2(a + b)d = 462 \text{ in.}^2$

$J/c = [ad(a + 3b) + d^3]/3 = 3664 \text{ in.}^3$

$$\gamma_v = 1 - \frac{1}{1 + 2/3\sqrt{\frac{c_1+d}{c_2+d}}} = 0.43$$ Eq. (13-40)

(c) Permissible shear stress

For interior columns, Eq. (11-42) applies:

$$v_c = \varphi(3.5\sqrt{f'_c} + 0.3 f_{pc} + V_p/b_o d)$$ Eq. (11-42)

EXAMPLE 27.1 - Continued

Calculations and Discussion	Code Reference

V_p is the shear carried through critical transfer section by tendons. For thin slabs, the V_p term must be carefully evaluated, as field placing practices can have a great effect on the profile of the tendons through the critical section. Conservatively, this term may be taken as zero.

$$v_c = 0.85(3.5\sqrt{4000} + 0.3 \times 172) = 232 \text{ psi} < 245^*$$

(d) Check moment transfer strength — 13.3.3

$$\gamma_f = 1 - \gamma_v = 1 - 0.43 = 0.57$$ — Eq. (13-1)

Moment transferred by flexure within width of column plus 1.5 slab thickness on each side = 0.57(30.2) = 17.2 ft-kips — 13.3.3.2

Effective slab width = 14 + 2(1.5 x 6.5) = 33.5 in.
Say effective prestress force = 79 kips (same as exterior column)

$$A_s = 0.00075h\ell = 0.00075 \times 6.5 \times 21 \times 12 = 1.22 \text{ in.}^2$$

<u>Use 6 #4 bars</u>(A_s = 1.20 in.^{2}), $A_sf_y = 1.20 \times 60 = 72$ kips

$A_{ps}f_{ps} + A_sf_y = 79 + 72 = 151$ kips

$$a = \frac{151}{0.85 \times 4 \times 33.5} = 1.33 \text{ in.}$$

$$(d - a/2) = (5.5 - 1.33/2)/12 = 0.40$$

---

*The V_p component of tendon force crossing the critical transfer section will make up the slight deficiency.

EXAMPLE 27.1 - Continued

Calculations and Discussion	Code Reference

$\varphi M_n = 0.9(151 \times 0.40) = 54.4$ ft-kips

Moment transfer strength in flexure of 54.4 is much greater than the required transfer moment of 17.2.

11. Distribution of tendons

In accordance with Section 18.12, the 11 tendons per 20 ft bay will be distributed in a group of 3 tendons directly through the column with the remaining 8 tendons spaced at 2 ft 3 in. centers (about 4 x slab thickness). Tendons in the perpendicular direction will be placed in a narrow band through and immediately adjacent to the columns.

28

Shells and Folded Plate Members

New for '83

Chapter 19, addressing shell and plate members, is a completely updated version for ACI 318-83. It reflects the current state-of-knowledge and improvements in shell design practices that have occurred since specific code provisions for shell and folded plate members were first introduced into the 1971 code edition. New Chapter 19 clarifies provisions, includes more positive guidance on proper analysis methods appropriate for different types of shell structures, and provides more specific guidance on design and proper placement of shell reinforcement. The code commentary for Chapter 19 should be more helpful to designers. Its expanded discussion reflects current information--including an extended reference listing.

General Considerations

Even with considerable improvement in clarity of provisions for new Chapter 19, the code requirements for shells and folded plates must, by necessity, be somewhat general in nature as compared to the provisions for other types of structures where the practice of design has been firmly established. Chapter 19 is specific in only a few critical areas peculiar to shell design; otherwise, it refers to standard provisions of the code. It should be noted that strength design is permitted for shell structures, even though most of the shells in this country have been built using working stress design procedures.

The code, the commentary, and the list of references are an excellent source of information and guidance on shell design. The list of references does not exhaust the possible sources of design assistance.

Following are some factors that should be considered in studying the provisions of Chapter 19 and in approaching the design of a shell structure:

(1) Chapter 19 covers the design of a large class of concrete structures that are quite different from the ordinary slab, and beam and column construction. Structural action varies from shells with considerable bending in the shell portions (folded plates and barrel shells) to those with very little bending except at the junction of shell and support (hyperbolic paraboloids and domes of revolution). The problems of shell design, therefore, cannot be lumped together as each type has its own peculiar attributes which must be thoroughly understood by the designer. Even shells classified under one type such as the hyperbolic paraboloid, vary greatly in their structural action. Recent studies have shown that gabled hyperbolic paraboloids, for example, are much more complex than the simple membrane theory would indicate. This is one explanation for the lack of a rigid set of rules in the code for the design of shells and folded plate structures.

(2) For the reasons given above, design of a shell requires considerable lead time to gain an understanding of the design problems for the particular type of shell. An attempt to design a shell without proper study may invite poor performance. Design of shell structures requires the ability to think in terms of three dimensional space; this is only gained by study and experience. The conceptual stage is the most critical period in shell design, since this is when vital decisions on form and dimensions must be made.

(3) Strength of shell structures is inherent in their shape and is not created by increasing the performance of materials to their limit as in the case of other types of concrete structures such as conventional and prestressed concrete beams. Therefore, the design stresses in the concrete should not be raised to their fullest except where required for very large structures. Deflections are normally not a problem if the stresses are low.

(4) Shell size is a very important determinant in the analytical precision required for its design. Short spans (up to 60 ft) can be designed using approximate methods such as the beam method for barrel shells, provided the exterior shell elements are properly supported by beams and columns. However, the limits and approximations of any method must be thoroughly understood. Large spans may require much more elaborate analyses. For example, a large hyperbolic paraboloid (150-ft span or more) may require a finite element analysis. Again, a long lead time is necessary to obtain the program and learn how to run it.

Application of the following Code provisions warrants further explanation.

19.2 Analysis and Design

19.2.6 - The components of force produced by prestressing tendons draped in a thin shell must be taken into account in the design. In the case of a barrel shell, it should be noted that the tendon does not lie in one plane, as shown in Fig. 28-1.

19.2.7 - The strength design method is permitted for the design of shells but it should be noted that for slab elements intersecting at an angle, and having high tensile stresses at inside corners, the ultimate strength is greatly reduced from that at the center of a concrete slab. Therefore, special attention should be given to the reinforcement used in these details, and thicknesses should be greater than the minimum allowed by the strength method.

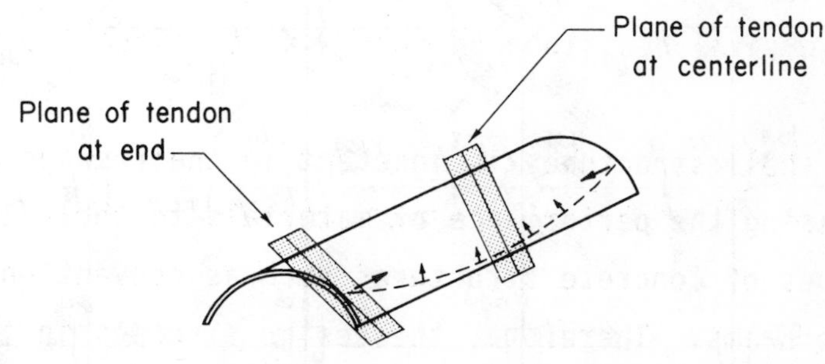

Fig. 28-1 Draped Prestressing Tendon in Barrel Shell

19.4 Shell Reinforcement

19.4.6 - For shells with essentially membrane forces, such as hyperbolic paraboloids and domes of revolution, it is usually convenient to place the reinforcing in the direction of the principal forces. Even though folded plates and barrel shells act essentially as longitudinal beams (traditionally having vertical stirrups as shear reinforcement), an orthogonal pattern of reinforcing (diagonal bars) is much easier to place and also assures end anchorage in the barrel or folded plate. With diagonal bars, five layers of reinforcement may be required at some points.

The direction of principal stresses near the supports is usually about 45 degrees so equal areas of reinforcing will be required in each direction to satisfy the requirements of Section 19.4.4. For illustration, Fig. 28-2 shows a plot of the principal membrane forces in a barrel shell with a span of 60 ft, a rise of 6.3 ft, a thickness of 3.5 in. and a snow load of 25 psf, and a roof load of 10 psf. Forces are shown in kips per linear foot.

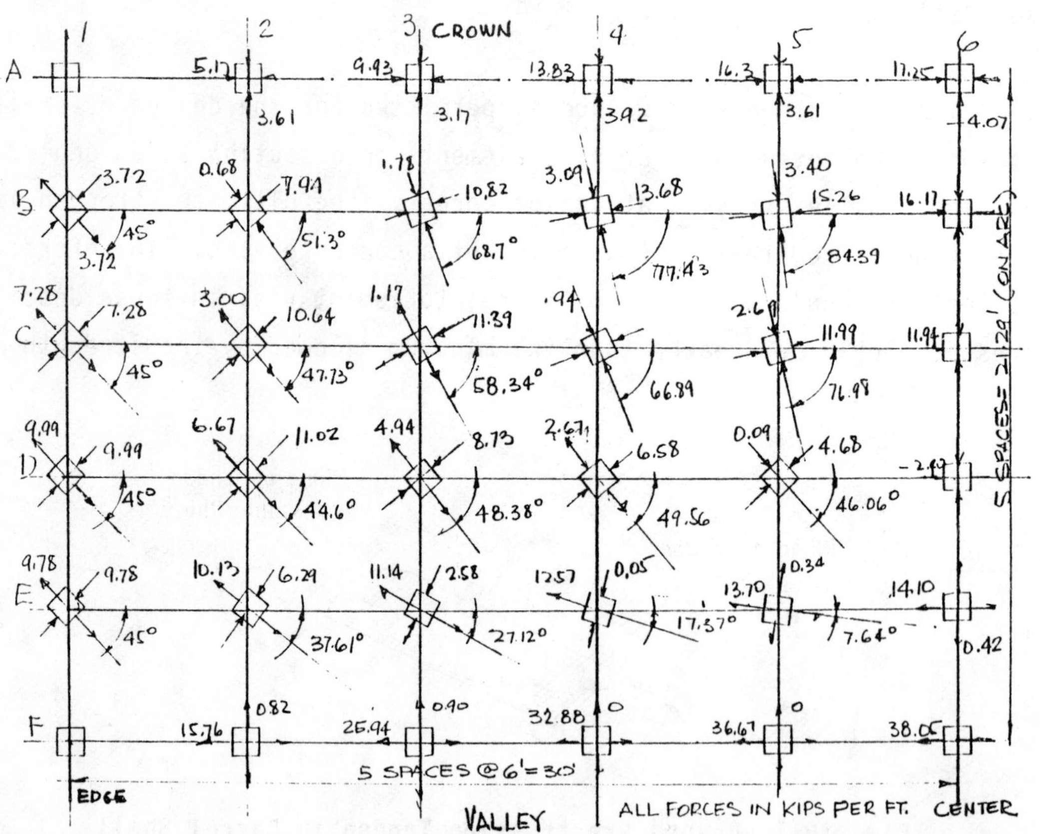

Fig. 28-2 Membrane Forces and Direction for 60-ft-span Barrel Shell

19.4.8 - In the case of long barrel shells (or domes) it is often desirable to concentrate tensile reinforcement near the edges rather than distribute the reinforcement over the entire tensile zone. When this is done, a minimum amount of reinforcement equal to 0.0035 bh must be distributed over the remaining portion of the tensile zone, as shown in Fig. 28-3. This amount in practical terms is twice the minimum steel requirement for shrinkage and temperature stresses.

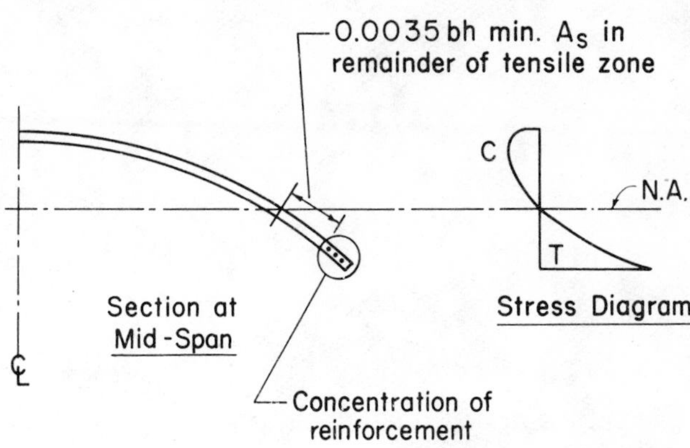

Fig. 28-3 Concentration of Shell Reinforcement

19.4.10 - Spacing of reinforcement up to 5 times the shell thickness or 18 in. is the maximum permissible. Therefore, for shells less than 3.6 in. thick, 5 times the thickness controls. For thicker shells, the spacing of bars shall not exceed 18 in.

29

Strength Evaluation of Existing Structures

20.1 Strength Evaluation - General

Strength evaluation of existing structures brings into play one of the most important attributes in the design field today, that is, experience and sound judgment. Chapter 20 sets specific criteria for testing and evaluation of flexural members only. Other members may be evaluated by analysis or load tests, or a combination of both.

Load definitions:

D = Service dead load supported by the member being tested as defined by the general building code (without load factors).

L = Service live load supported by the member being tested as specified by the general building code (without load factors). L may include live load reductions if permitted by the general building code.

20.4 Load Tests of Flexural Members

Criteria

(1) Portion of structure subject to load test must be at least 56 days old . . . (20.3.2).

(2) Forty-eight hours prior to load test, apply full service dead load, D . . . (20.3.4).

(3) Immediately prior to application of test load, take data readings for measurements of deflections . . . (20.4.2).

(4) In addition to full service dead load, D, apply test load equal to (0.2D + 1.45L) in four equal increments. Total load acting must equal 0.85(1.4D + 1.7L) . . . (20.4.3 and 20.4.4).

(5) Take deflection readings after 24 hours with test load in place . . . (20.4.5).

(6) Remove test load equal to (0.2D + 1.45L) . . . (20.4.6).

(7) Take final deflection readings 24 hours after removal of test load . . . (20.4.6).

Acceptance

If visible evidence of failure has occurred, the tested portion has failed, and no re-testing is permitted. If the structure shows no visible evidence of failure, the following conditions must be satisfied:

(1) When maximum deflection "a" exceeds $\ell_t^2/20{,}000h$, percentage recovery must be 75% after 24 hours (80% for prestressed members) . . . (20.4.8).

(2) When maximum deflection "a" is less than $\ell_t^2/20{,}000h$, recovery requirement is waived . . . (20.4.8).

Figs. 29-1 and 29-2 illustrate application of the limiting deflection criteria.

(3) Members failing to meet the 75% recovery factor may be re-tested . . . (20.4.10).

(4) Before re-testing, 72 hours must have elapsed after load removal and on re-test the percentage recovery must be 80% . . . (20.4.10).

(5) Prestressed members shall not be re-tested . . . (20.4.11).

20.5 Members Other Than Flexural Members

An analytical method to investigate these members is preferred. In testing compression members, generally a wall or column, the test would have to be carried out to destruction.

20.6 Provision for Lower Load Rating

A lower load rating may be approved by the building official when a structure does not satisfy the analytical or percentage recovery tests. It should be recognized that the lower load rating will generally prohibit the use of the structure for the purpose originally intended.

Summary

Chapter 20 does not cover load testing for the approval of new or novel design or construction methods, nor the demonstration of quality of prefabricated units, unless these units already form a part of the structure under test, nor was the chapter written to settle private disputes or litigation over construction quality.

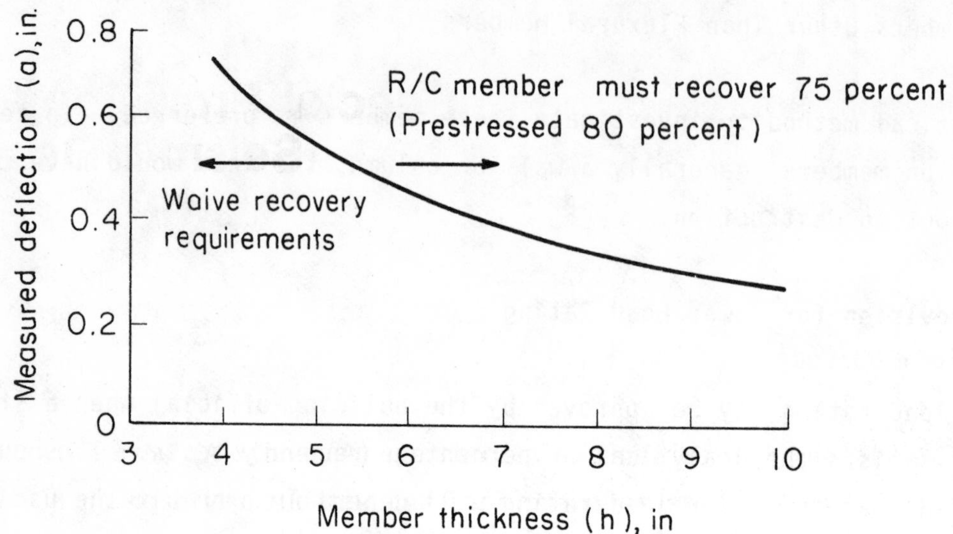

Fig. 29-1. Load Testing Acceptance Criteria for Members With Span Length of ℓ_t = 20 ft

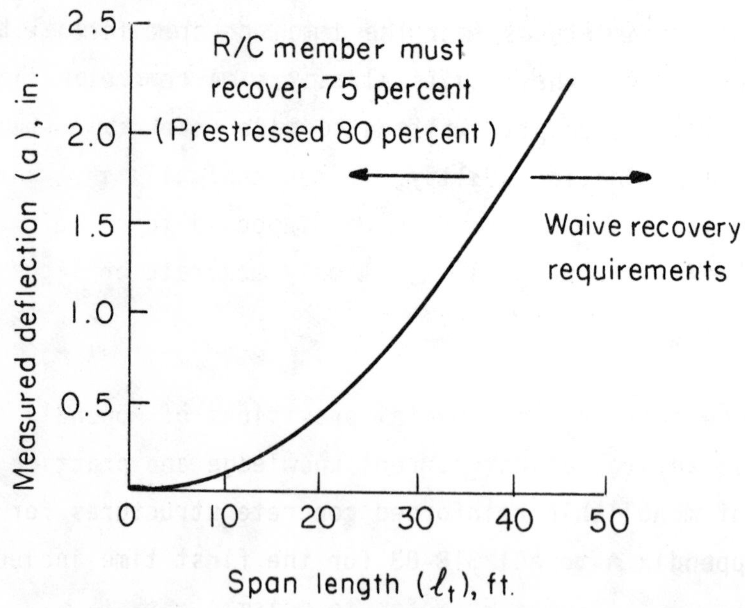

Fig. 29-2. Load Testing Criteria for Members With Overall Thickness of h = 6-1/2 in.

30

Special Provisions for Seismic Design

Update for '83 Code

Special provisions for earthquake resistance (Appendix A) were first introduced in the 1971 ACI Code and were included without revision in ACI 318-77. The original provisions of Appendix A were intended to apply only to reinforced concrete structures located in regions of high seismicity, and designed with a substantial reduction in total lateral seismic forces (as compared with the elastic response forces), in anticipation of inelastic structural behavior. Also, with publication of the 1971 Code edition, several changes were incorporated into the main body of the Code specifically to improve toughness, in order to increase the resistance of concrete structures to earthquakes or other catastrophic loads. While Appendix A was meant for application to lateral load resisting frames and walls in regions of high seismicity, the main body of the Code was supposed to be sufficient for regions where there is a probability of only moderate or light earthquake damage.

For the 1983 Code edition, the special provisions of Appendix A have been extensively revised, to reflect current knowledge and practice of the design and detailing of monolithic reinforced concrete structures for earthquake resistance. Appendix A to ACI 318-83 for the first time includes special detailing for frames in zones of moderate seismic risk.

For buildings located in regions of low seismic risk, no special design or detailing is required; the general requirements of the main body of the Code apply. Concrete structures proportioned by the main body of the Code are expected to provide a level of toughness adequate for low earthquake intensities.

For buildings located in regions of moderate seismic risk, reinforced concrete moment frames proportioned to resist earthquake effects require some special reinforcing details, as specified in Section A.9 of Appendix A. The special details apply only to frames (beams, columns, and slabs) to which earthquake-induced forces have been assigned in design. The special reinforcing details will serve to accommodate a suitable level of inelastic behavior if the frame is subjected to an earthquake of such intensity as to require it to perform inelastically. There are no special requirements for structural walls provided to resist lateral effects of wind and earthquakes, or for nonstructural components of buildings located in regions of moderate seismic risk. Structural walls proportioned by the main body of the Code are considered to have sufficient toughness at drift levels anticipated in regions of moderate seismicity.

For buildings located in regions of high seismic risk, where damage to construction has a high probability of occurrence, all building components, structural and nonstructural, must satisfy requirements of Sections A.2 through A.8 of Appendix A. The special proportioning and detailing provisions of Appendix A are intended to provide a monolithic reinforced concrete structure with adequate toughness to respond inelastically under severe earthquake motions.

Seismic risk level is usually designated by zones or areas of equal risk or probability of damage, such as Zone 0--no damage; Zone 1--minor damage; Zone 2--moderate damage; and Zone 3--major damage. Areas within Zone 3 that are close to major fault systems are assigned to Seismic Zone 4. Seismic risk levels (Seismic Zone Maps) are under the jurisdiction of general building codes rather than ACI 318. In the absence of a general building code that addresses earthquake loads and seismic zoning, it is intended that local authorities (engineers, geologists, and building code officials) should decide on the need and proper application of the special provisions for seismic design.

Introduction

Economical earthquake-resistant design should aim at providing appropriate dynamic characteristics in structures so that acceptable response levels would result under the design earthquake(s). The structural properties over which the designer exercises some degree of control, which he can modify to achieve the desired results, are the magnitude and distribution of stiffness and mass, relative strengths of members and their deformabilities.

In some structures, such as slender free-standing towers or smoke stacks which depend for their stability on the stiffness of the single element making up the structure, or in nuclear containment buildings where a more-than-usual conservatism in design is required, yielding of the principal elements in the structure cannot be tolerated. In such cases, the design needs to be based on an essentially elastic response to moderate-to-strong earthquakes, with the critical stresses limited to the range below yield.

In most buildings, particularly those consisting of frames and other multiply-redundant systems, however, economy is achieved by allowing yielding to take place in some members under moderate-to-strong earthquake motion.

The performance criteria implicit in most earthquake code provisions require that a structure be able to:[30.1]

a. resist earthquakes of minor intensity without damage; a structure would be expected to resist such frequent but minor shocks within its elastic range of stresses;

b. resist moderate earthquakes with minor structural and some non-structural damage; with proper design and construction, it is expected that structural damage due to the majority of earthquakes will be repairable; and

c. resist major catastrophic earthquakes without collapse.

The above performance criteria allow only for the effects of a typical ground shaking. The effects of slides, subsidence or active faulting in the immediate vicinity of the structure, which may accompany an earthquake, are not considered.

While no clear quantitative definition of the above earthquake intensity ranges has been given, their use implies the consideration not only of the actual intensity level but also of their associated probability of occurrence with reference to the expected life of a structure.

The principal concern in earthquake-resistant design is the provision of adequate strength and ductility to assure life safety, i.e., prevention of collapse under the most intense earthquake that may reasonably be expected at a site during the life of a structure. Observations of building behavior in recent earthquakes, however, have made engineers increasingly aware of the need to ensure that buildings housing facilities essential to post-earthquake operations, such as hospitals, power plants, fire stations and communication centers, not only survive without collapse but remain operational after an earthquake. This means that such buildings should suffer a minimum of damage. Thus, damage control has been added to life safety as a second design criterion.

Often, damage control becomes desirable from a purely economic point of view. The extra cost of preventing severe damage to the nonstructural components of a building, such as partitions, glazing, ceiling, elevators and other mechanical systems, may be justified by the savings realized in replacement costs and from continued use of a building after a strong earthquake.

The principal steps involved in the aseismic design of a typical concrete structure according to building code provisions are as follows:

(1) Determination of design earthquake forces:
   (a) calculation of base shear corresponding to computed or estimated fundamental period of vibration of the structure (a preliminary design of the structure is assumed here);

(b) distribution of the base shear over the height of the building.

(2) Analysis of the structure under the (static) lateral forces calculated in step 1, as well as under gravity and wind loads, to obtain member design forces.

(3) Designing members and joints for the most unfavorable combination of gravity and lateral loads, and detailing them for ductile behavior.

The design base shear represents the total horizontal seismic force (service load level) that may be assumed acting parallel to the axis of the structure considered. The force in the other horizontal direction is assumed to act non-concurrently. Vertical seismic forces are not considered because the building is already designed for the vertical acceleration of gravity; the additional vertical acceleration due to an earthquake would normally be only a fraction of g. It can be accommodated by the factor of safety on gravity loads.

The code-specified design lateral forces have the same general distribution as the typical envelope of maximum horizontal shears indicated by an elastic dynamic analysis. However, the code forces, which are assumed to be resisted by a structure within its elastic (working) range of stresses, are substantially smaller than those which would be developed in a structure subjected to an earthquake of intensity equal to that of the 1940 El Centro, if the structure were to respond elastically to such ground excitation. Thus, buildings designed under the present codes would be expected to undergo fairly large deformations (four to six times the lateral displacements resulting from the code-specified, statically applied shears) when subjected to an earthquake with the intensity of the 1940 El Centro. These large deformations will be accompanied by yielding in many members of the structure, and in fact, such is the intent of the codes. The acceptance of the fact that it is economically unwarranted to design buildings to resist major earthquakes elastically and the recognition of the capacity of structures possessing adequate strength and ductility to withstand major earthquakes by responding inelastically to these, lies behind the relatively low forces specified by the codes. These reduced force levels must be and are coupled with additional requirements for the design and detailing of members and their connections in order to ensure sufficient deformation capacity in the inelastic range.

The capacity of a structure to deform in a ductile manner, that is to deform beyond the yield limit without significant loss of strength, allows such a structure to absorb a major portion of the energy from an earthquake without serious damage. Laboratory tests have demonstrated that reinforced concrete members and their connections, designed and detailed by the present codes, do possess the necessary ductility to allow a structure to respond inelastically to earthquakes of major intensity without significant loss of strength.

Because of the relatively large inelastic deformations which a building designed by the present codes may undergo during a strong earthquake, proper provisions must be made to ensure that the structure does not become unstable under the vertical loads. The codes thus prescribe the so-called strong column-weak beam design with the intent of confining yielding to the beams while the columns remain elastic throughout their seismic response. It is required that the sum of the moment strengths of the columns meeting at a joint, under the design axial loads, be greater than the sum of the moment strengths of the beams framing into the joint in the same plane.

The design provisions contained in the main body of the ACI Building Code as well as the regular provisions in most other codes do in fact provide some ductility which should be sufficient for structures subjected only to minor earthquakes that may occur frequently, such as those associated with UBC[30.2] or ANSI[30.3] Zone 1 areas. For structures which may be subjected to earthquakes of moderate intensity (Zone 2) some additional confinement, anchorage and shear reinforcement details may be required. Some such details for lateral load resisting frames, for the first time, are given in Appendix A to the 1983 edition of the ACI Code (Section A.9). For structures that may be subjected to strong intensity earthquakes (Zones 3 and 4), appreciable inelastic deformations can be expected, so that substantial ductility is required. The design provisions contained in Appendix A to ACI 318-83 (Sections A.2 - A.8) are primarily intended to provide this additional ductility.

A.2 General Requirements

A.2.2 Analysis and Proportioning of Structural Members

The interaction of all structural and nonstructural components affecting linear and nonlinear structural response are to be considered in analysis (A.2.2.1). Consequences of failure of structural and nonstructural components not forming part of the lateral force resisting system should also be considered (A.2.2.2). The intent of A.2.2.1 and A.2.2.2 is to draw attention to the influence of nonstructural components on structural response and to hazards from falling objects.

Section A.2.2.3 alerts the designer to the fact that the base of the structure as defined in analysis may not necessarily correspond to the foundation or ground level. It requires that structural members below base, which transmit forces resulting from earthquake effects to the foundation, shall also comply with the requirements of Appendix A.

Even though some element(s) of a structure may not be considered part of the lateral force resisting system, the effect on all elements of displacements several times those caused by Code forces should be considered (A.2.2.4). The only exception should be when complete failure of the element would not result in loss of the vertical load carrying capacity of the structure.

A.2.3 Strength Reduction Factors

The strength reduction factors of Chapter 9 are not based on the observed behavior of reinforced concrete members under load or displacement cycles simulating earthquake effects. Some of those factors have been modified in Appendix A in view of the effects on strength of large reversing displacements into the inelastic range of response.

Section A.2.3.1 refers to brittle members such as low-rise walls or portions of walls between openings which are proportioned such as to make it impractical to raise their nominal shear strength above the shear corresponding to nominal flexural strength for the pertinent loading conditions. The provision does not apply to beam-column joints.

Section A.2.3.2 is intended to discourage the use of tied columns for resisting earthquake induced forces.

A.2.4, A.2.5 Limitations on Material Strength

A minimum specified concrete strength, f'_c, of 3000 psi and a maximum specified yield strength of reinforcement, f_y, of 60,000 psi are required. These limits are imposed as reasonable bounds on the variation of material properties, particularly with respect to their unfavorable effects on the sectional ductility of members in which they are used. A decrease in the concrete strength and an increase in the yield strength of the tensile reinforcement tend to decrease the ultimate curvature and hence the sectional ductility of a member subjected to flexure. Also, an increase in the yield strength of reinforcement is generally accompanied by a decrease in the ductility--as measured by the maximum deformation--of the material itself.

Appendix A requires that reinforcement for resisting flexure and axial forces in frame members and wall boundary elements be ASTM 706 Grade 60 low alloy steel intended for application where welding or bending, or both, are important. However, ASTM 615 billet steel bars of Grade 40 or 60 may be used in these members if the following two conditions are satisfied:

$$\text{actual } f_y \leq \text{specified } f_y + 18{,}000 \text{ psi}$$

$$\frac{\text{actual ultimate tensile stress}}{\text{actual } f_y} \geq 1.25$$

The first requirement helps to limit the magnitude of the actual shears that can develop in a flexural member above that computed on the basis of the specified yield value when plastic hinges form at the ends of a beam. The second requirement is intended to ensure steel with a sufficiently long yield plateau.

In the "strong column-weak beam" frame intended by the Code, the relationship between the moment capacities of columns and beams may be upset if the beams turn out to have much greater moment capacity than intended by the designer. Thus, the substitution of 60-ksi steel of the same area for specified 40-ksi steel in beams can be detrimental. The shear strength of beams and columns, which is generally based on the condition of plastic hinges forming at the ends of the members, may become inadequate if the moment capacity of member ends should be greater than intended as a result of the steel having a substantially greater yield strength than specified.

A.3 Flexural Members of Frames

These include members, having a clear span greater than 4 times the effective depth, that are subjected to a factored axial compressive force not exceeding ($A_g f'_c/10$), where A_g is the gross cross-sectional area (A.3.1.1, A.3.1.2). The significant provisions relating to flexural members are

a. Limitations on section dimensions (A.3.1.3, A.3.1.4)

width-to-depth ratio $\geq$ 0.3
width $\geq$ 10 in.
$\leq$ [width of supporting column + 1.5 (depth of beam)]

These limitations have been guided by experience with test specimens subjected to cyclic inelastic loading.

b. Limitations on flexural reinforcement ratio (A.3.2.1 and Fig. 1):

$\rho_{min} = 200/f_y$, with at least two continuous bars at both top and bottom of member

$\rho_{max} = 0.025$

Because the ductility of a flexural member decreases with increasing values of the reinforcement ratio, Appendix A limits the maximum reinforcement ratio to 0.025. The use of a limiting value

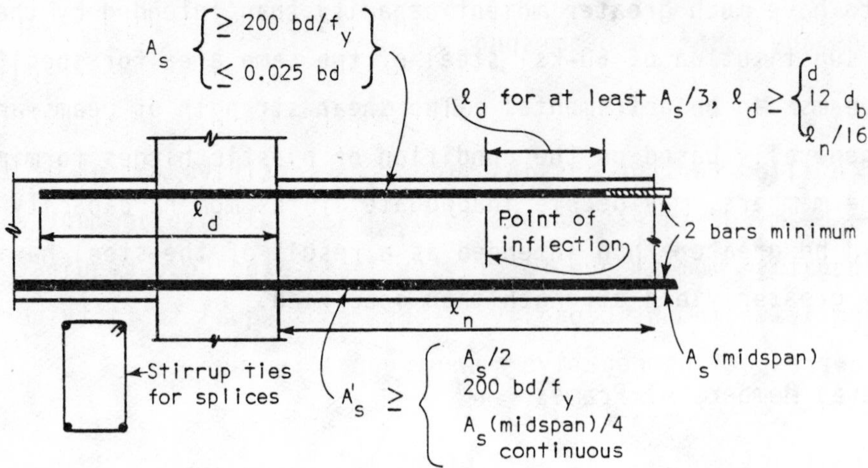

Fig. 30-1 Reinforcement requirements for flexural members

based on the "balanced reinforcement ratio," as given in the main body of the Code, while applicable to members subjected to monotonically increasing loads, fails to describe conditions in a flexural member subjected to reversals of inelastic deformation. The limiting ratio of 0.025 is based mainly on considerations of steel congestion and also on limiting shear stresses in beams of typical proportions. From a practical standpoint, lower steel ratios should be used whenever possible. The requirement of at least two bars, top and bottom, is dictated by construction rather than behavioral requirements.

The selection of the size, number, and arrangement of flexural reinforcement should be made with full consideration of construction requirements. This is particularly important in relation to beam-column connections, where construction difficulties can arise as a result of reinforcement congestion. The preparation of large-scale drawings of the connections, showing all beam, column, and joint reinforcement, will help eliminate unanticipated problems in the field. Such large-scale drawings will pay dividends in terms of lower bid prices and a smooth-running construction job.

c. Moment capacity requirements (Section A.3.2.2):

At beam ends, $M_y^+ \geq 0.50\ M_y^-$

At any point in beam span,

M_y^+ or $M_y^- \geq 0.25\ M_y^{max}$ at beam ends

To allow for the possibility of the positive moment at the end of a beam due to earthquake-induced lateral displacement exceeding the negative moment due to gravity loads, the Code requires a minimum positive moment capacity at beam ends equal to fifty percent of the corresponding negative moment capacity.

d. Restrictions on lap splices (A.3.2.3):

Lap splices shall not be used
- within joints
- within 2d from face of support, where d is depth of beam
- at locations of potential plastic hinging

Lap splices are to be confined by hoops or spiral reinforcement with maximum spacing or pitch of d/4 or 4 in.

e. Restrictions on welding of longitudinal reinforcement (A.3.2.4):

Welded splices and mechanical connectors may be used provided
- they are used only on alternate bars in each layer at any section
- the distance between splices of adjacent bars is at least 24 in.

Welded splices and mechanical connectors shall conform to the requirements given in the main body of the Code. A major requirement is that the splice should develop at least 125 percent of the specified yield strength of the bar.

f. Transverse reinforcement requirements for confinement and shear (A.3.3):

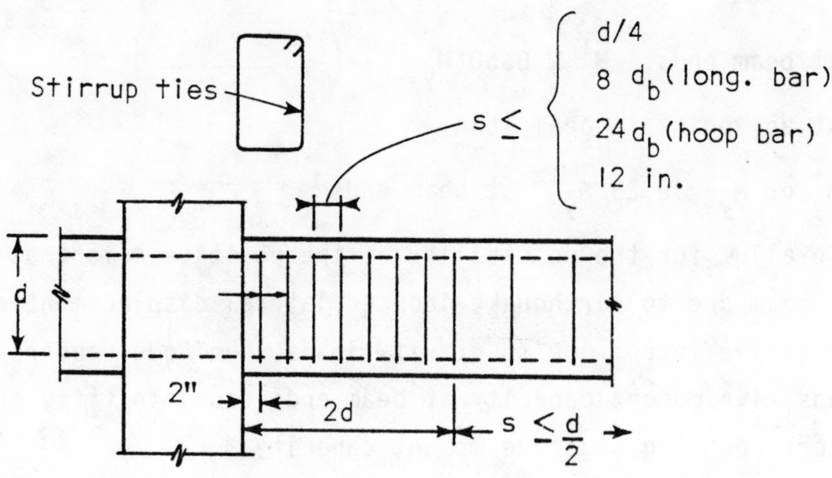

Fig. 30-2 Limitations on transverse reinforcement in beams. Min. bar size = #3

Transverse reinforcement in beams must satisfy requirements associated with their dual function as confinement reinforcement and shear reinforcement (Fig. 2).

Confinement reinforcement in the form of hoops is required

- over a distance 2d from faces of support (where d is the effective depth of the member)
- over distances 2d on both sides of sections where flexural yielding may occur due to earthquake loading.

Hoop spacing must satisfy the following requirements

- first hoop at 2 in. from face of support
- maximum spacing ≤ d/4
  8 x (diameter of smallest longitudinal bar)
  24 x (diameter of hoop bar)
  12 in.

Where hoops are not required, tie spacing ≤ d/2.

Shear reinforcement is to be provided so as to preclude shear failure prior to the development of plastic hinges at beam ends. Design shears for determining shear reinforcement are to be based on a condition where plastic

$$V_a = \frac{M_{p1} + M_{p2}}{\ell} + 0.75 \left(\frac{1.4D + 1.7L}{2} \right)$$

$$V_b = \frac{M_{p1} + M_{p2}}{\ell} - 0.75 \left(\frac{1.4D + 1.7L}{2} \right)$$

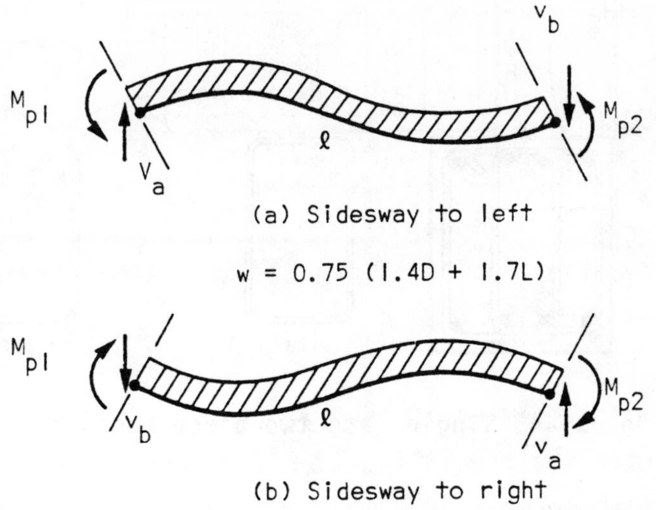

M_{p1} V_b M_{p2} ℓ V_a

(a) Sidesway to left

$w = 0.75\ (1.4D + 1.7L)$

M_{p1} M_{p2} ℓ V_b V_a

(b) Sidesway to right

Fig. 30-3 Loading cases for design of shear reinforcement in beams - uniform gravity loads

hinges occur at the beam ends due to the combined effects of lateral displacements and factored gravity loads (Fig. 3). The "probable flexural strength", M_p, associated with plastic hinging is to be computed using a strength reduction factor, $\varphi = 1.0$, and assuming a stress in the tensile reinforcement, $f_s = 1.25\ f_y$.

In determining the required shear reinforcement, the contribution of the concrete, V_c, is to be neglected if the earthquake-induced shear is greater than one-half of the total design shear and the factored axial compressive force including earthquake effects is less than $(A_g f'_c/20)$.

Shear reinforcement shall be in the form of hoops in regions where confinement is also required, as indicated above for confinement reinforcement (see Fig. 4). Otherwise, stirrups or ties may be used.

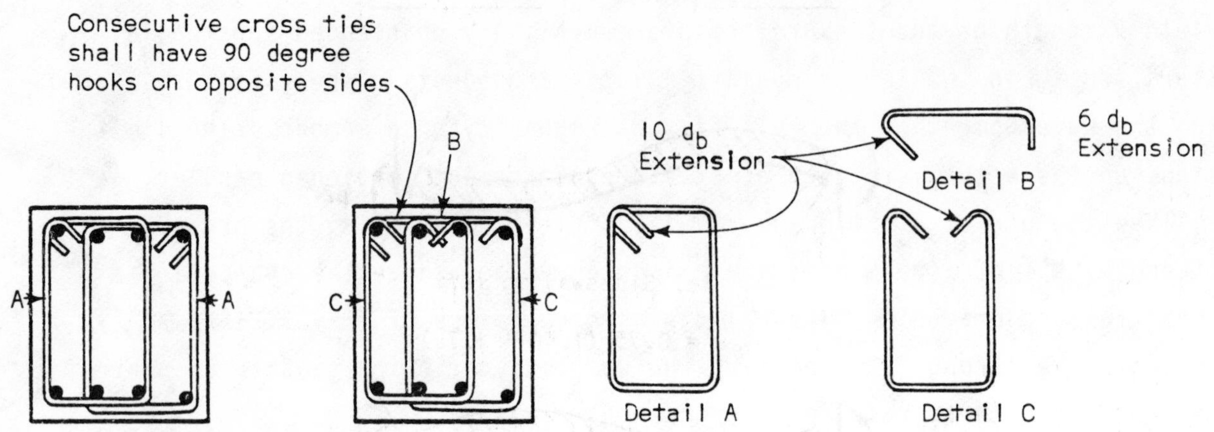

Fig. 30-4 Single- and two-piece hoops

The transverse reinforcement provided shall satisfy the requirement for confinement or shear, whichever is larger.

Because the ductile behavior of earthquake-resistant frames designed to current codes is premised on the ability of the beams to develop plastic hinges with adequate rotational capacity, it is essential to ensure that shear failure does not occur before the flexural capacity of the beams has developed. Transverse reinforcement is required for two related functions: (a) to provide sufficient shear strength so that the full flexural capacity of a member can be developed, and (b) to help ensure adequate rotation capacity at plastic hinging regions by confining the concrete in the compression zone and providing lateral support to the compression steel. To be equally effective with respect to both functions under load reversals, the transverse reinforcement should be placed perpendicular to the longitudinal reinforcement.

Shear reinforcement in the form of stirrups or stirrup-ties is designed for the shear due to the factored gravity loads and the shear corresponding to plastic hinges forming at both ends. Plastic end moments associated with lateral displacements in either direction should be considered (Fig. 3). It is important to note that the required shear strength in beams (as in columns) is determined by the flexural strength of the frame member (as well as the factored loads acting on the member) rather than by the factored shear force calculated from a lateral load analysis.

Because of the direct dependence of the required web reinforcement on the yield strength of the flexural reinforcement, any unintended substantial overstrength in the latter could result in a nonductile shear failure preceding the development of the full flexural capacity of a member. The limitations on the actual strength of steel reinforcement mentioned earlier, as well as the use of $\varphi = 1.0$ and $f_s = 1.25\ f_y$ in calculating the probable strength, M_p, of a beam end, are all designed to reduce the chances of a shear failure preceding flexural yielding. The use of $f_s = 1.25\ f_y$ reflects the strong likelihood of the deformation in the tensile reinforcement entering the strain-hardening range.

To allow for load combinations unaccounted for in design, a minimum amount of web reinforcement is required throughout the length of all flexural members. Within regions of potential hinging, stirrup-ties or hoops are required. Hoops may be made of two pieces of reinforcement: a stirrup having 135-degree hooks with ten-bar-diameter extensions anchored in the confined core and a crosstie to make a closed hoop (Fig. 4). Consecutive crossties should have their 90-degree hooks on opposite sides of the flexural member.

A.4 Frame Members Subjected to Bending and Axial Load

Appendix A makes the distinction between columns or beam-columns and flexural members on the basis of the magnitude of the factored axial load imposed on a member. Thus, when the factored axial load does not exceed ($A_g f'_c/10$), the member falls under the category of flexural members, as discussed in the preceding section. When the factored axial force on a member resisting earthquake-induced forces exceeds ($A_g f'_c/10$), the member is considered a beam-column (A.4.1). The design of such members is governed by the requirements given below.

a. Limitations on section dimensions (A.4.1.1, A.4.1.2):

Shortest cross-sectional dimension ≥ 12 in. (measured on line passing through geometric centroid)

$$\frac{\text{Shortest dimension}}{\text{Perpendicular dimension}} \geq 0.4$$

b. Limitations on longitudinal reinforcement (A.4.3.1):

$\rho_{min} = 0.01, \quad \rho_{max} = 0.06$

Appendix A specifies a reduced upper limit for the reinforcement ratio in columns from the 8% of Chapter 10 of the Code to 6%. However, construction considerations will in most cases place the practical upper limit on the reinforcement ratio, ρ, near 4%. Convenience in detailing and placing reinforcement in beam-column connections makes it desirable to keep the column reinforcement low.

The minimum reinforcement ratio is intended to provide for the effects of time-dependent deformations in concrete under axial loads as well as maintain a sizable difference between cracking and yield moments.

c. Flexural strength of columns relative to beams framing into a joint ("strong column-weak beam" provision) (A.4.2.2):

$$\Sigma M_e \geq \frac{6}{5} \Sigma M_g \qquad \text{(A-1)}$$

where

ΣM_e = sum of design flexural strengths of columns framing into joint. Column flexural strength shall be calculated for the factored axial force, consistent with the direction of lateral loading considered, which results in the lowest flexural strength.

ΣM_g = sum of design flexural strengths of beams framing into joint.

To ensure the stability of a frame and maintain its vertical load carrying capacity while it undergoes large lateral displacements, the Code requires that inelastic deformations be generally restricted to the beams. This is accomplished by requiring that the sum of the flexural strengths of the columns meeting at a joint, under the design axial loads, be equal to or greater than 6/5 times the sum of the moment strengths of the beams framing into the joint in the same plane. As indicated in Fig. 5, the signs of the bending moments in columns and beams are to be such that the column moments oppose the beam moments. Also, the "strong column-weak beam" relationship has to be satisfied for beam moments acting in both directions.

If Eq. (A-1) is not satisfied at a joint, columns supporting reactions from that joint are to be provided with transverse reinforcement over their full height. Columns not satisfying Eq. (A-1) are to be ignored in calculating the strength and stiffness of the structure. However, since such columns contribute to the stiffness of the structure before they suffer severe loss of strength due to plastic hinging, they should not be ignored if neglecting them results in unconservative estimates of design forces. This may occur in determining the design base shear or in calculating the effects of torsion in a structure. Columns not satisfying Eq. (A-1) should satisfy the minimum requirements for "members not proportioned to resist earthquake-induced forces", discussed under Section A.8.

d. Restriction on lap splices (A.4.3.2):

Lap splices are to be used only within the center half of the column length and shall be designed as tension splices.

$$(M^P_{ct} + M^P_{cb}) \geq \frac{6}{5}(M^P_{bl} + M^P_{br})$$

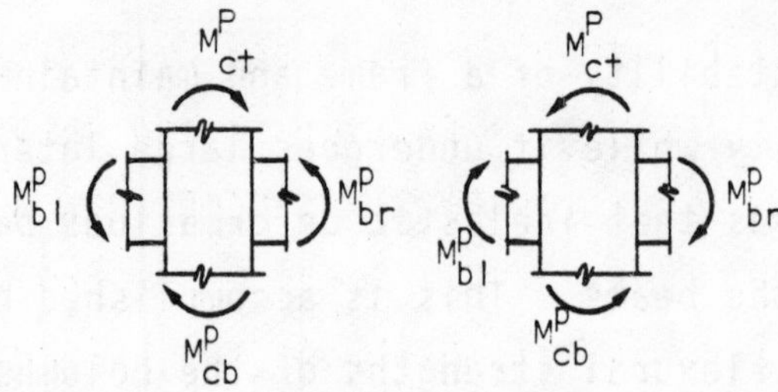

Fig. 30-5 "Strong column-weak beam" frame requirements

e. Welded splices or mechanical connectors in longitudinal reinforcement (A.4.3.2):

Welded splices or mechanical connectors may be used at any section of the column, provided

- they are used only on alternate longitudinal bars at a section
- the distance between splices along the longitudinal axis of reinforcement ≥ 24 in.

f. Transverse reinforcement for confinement and shear (Section A.4.4):

As in beams, transverse reinforcement in columns must provide confinement of the concrete core as well as shear resistance. In columns, however, the transverse reinforcement must all be in the form of closed hoops or continuous spiral reinforcement. Sufficient reinforcement should be provided to satisfy the requirement for confinement or shear, whichever is larger.

Confinement requirements (Fig. 6) are

Volumetric ratio or spiral or circular hoop reinforcement, $\rho_s \geq \begin{cases} 0.12 \dfrac{f'_c}{f_{yh}} & \text{(A-2)} \\ 0.45\left(\dfrac{A_g}{A_{ch}} - 1\right) \dfrac{f'_c}{f_{yh}} & \text{(10-5)} \end{cases}$

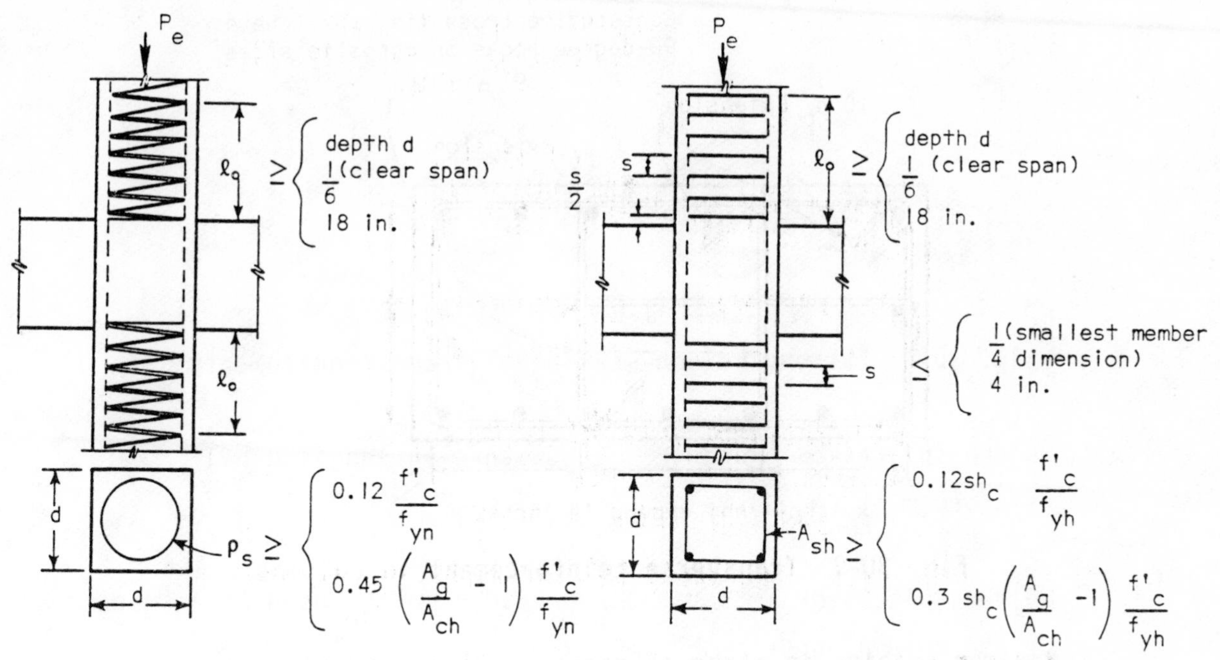

Fig. 30-6 Confinement requirements at column ends

where f_{yh} = specified yield strength of transverse reinforcement, in psi

A_{ch} = core area of column section, measured to the outside of transverse reinforcement, in sq. in.

For rectangular hoop reinforcement, total cross-sectional area within spacing s,

$$A_{sh} \geq \begin{cases} 0.12\ s h_c \dfrac{f'_c}{f_{yh}} & \text{(A-4)} \\ 0.3\ s h_c \left(\dfrac{A_g}{A_{ch}} - 1\right) \dfrac{f'_c}{f_{yh}} & \text{(A-3)} \end{cases}$$

where h_c = cross-sectional dimension of column core, measured center-to-center of confining reinforcement, in in.

and s = spacing of transverse reinforcement measured along longitudinal axis of member, in in.

$s_{max} = \frac{1}{4}$ (smallest cross-sectional dimension of member), or 4 in., whichever is smaller

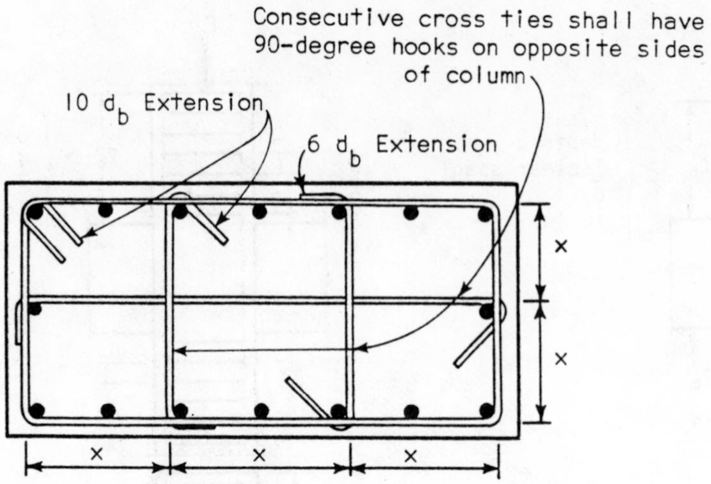

Consecutive cross ties shall have 90-degree hooks on opposite sides of column

$10\ d_b$ Extension

$6\ d_b$ Extension

x shall not exceed 14 inches

Fig. 30-7 Transverse reinforcement in columns

Maximum spacing in plane of cross-section between legs of overlapping hoops or cross-ties (Fig. 7) must not exceed 14 in.

Confinement reinforcement is to be provided over a length ℓ_o from each joint face or over distances ℓ_o on both sides of any section where flexural yielding may occur in connection with inelastic lateral displacements of the frame, where

$$\ell_o \geq \begin{cases} \text{depth, d, of member} \\ \frac{1}{6}\text{ (clear span of member)} \\ \text{18 in.} \end{cases}$$

Transverse reinforcement for shear in columns is to be determined for the shear associated with the largest nominal moment strengths at the column ends (using $\varphi = 1.0$) calculated for the factored axial compressive force resulting in the largest moment strengths.

Circular ties represent the most efficient form of confinement reinforcement. The extension of such spirals into the beam-column joint, however, may cause some construction difficulties.

Rectangular hoops, when used in place of spirals, are considered less effective with respect to confinement of the concrete core. Their effectiveness is increased, however, by the use of supplementary crossties, each end of which has to engage a peripheral longitudinal bar. Crossties of the same bar size and spacing as the hoops may be used. Consecutive crossties are to be alternated end for end along the longitudinal reinforcement (Fig. 7). Crossties or legs of overlapping hoops are to be spaced no farther than 14 in. apart in a direction perpendicular to the longitudinal axis of the member.

In addition to satisfying confinement requirements, the transverse reinforcement in columns must resist the maximum shear associated with the formation of plastic hinges in the frame. Although the "strong column-weak beam" provision governing the relative moment strengths of beams and columns is intended to have most of the inelastic deformation occur in the beams of a frame, the Code recognizes that hinging can occur in the columns. Thus, the shear reinforcement in columns is to be based on the condition that the nominal moment strength (i.e., plastic or yield moment, with strength reduction factor, $\varphi = 1.0$) is developed at the ends of a column. The value of these plastic moments - obtained from the P-M interaction diagram for the column section - is to be the maximum consistent with the possible factored axial compressive forces on the column. Moments associated with lateral displacements of the structure in both directions are to be considered, as indicated in Fig. 8. The axial load corresponding to the maximum moment capacity should then be used in computing the permissible shear stress in concrete, v_c.

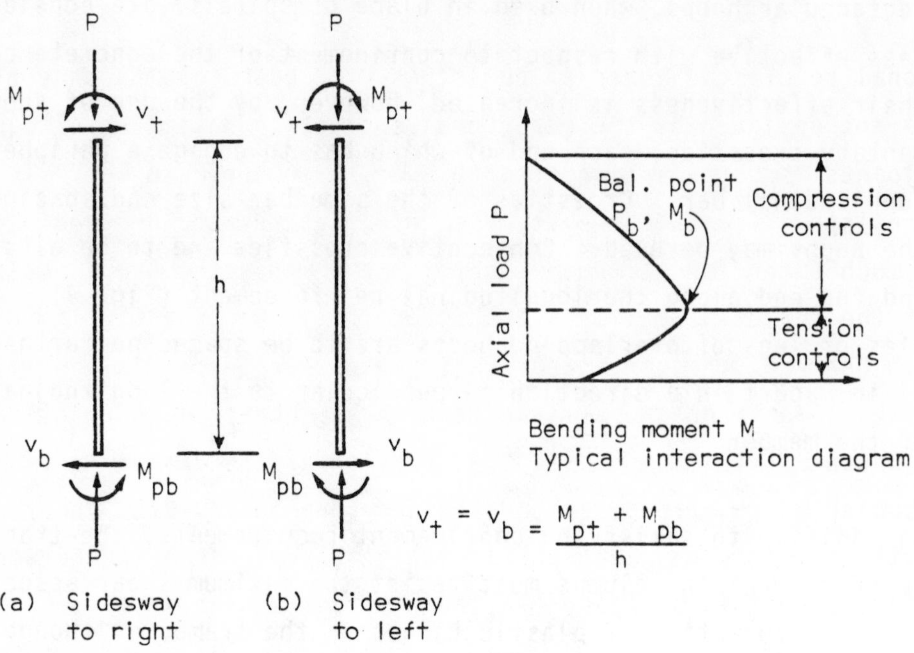

Fig. 30-8 Loading cases for design of shear reinforcement in columns

g. Columns supporting discontinued walls (A.4.4.5):

Columns supporting discontinued shearwalls or stiff partitions tend to be subjected to large shear and compressive forces and can be expected to suffer significant inelastic deformations during strong earthquakes. In recognition of this, the Code requires confinement reinforcement throughout the height of such columns (Fig. 9) whenever the axial compressive force due to earthquake effects exceeds $(A_g f'_c/10)$.

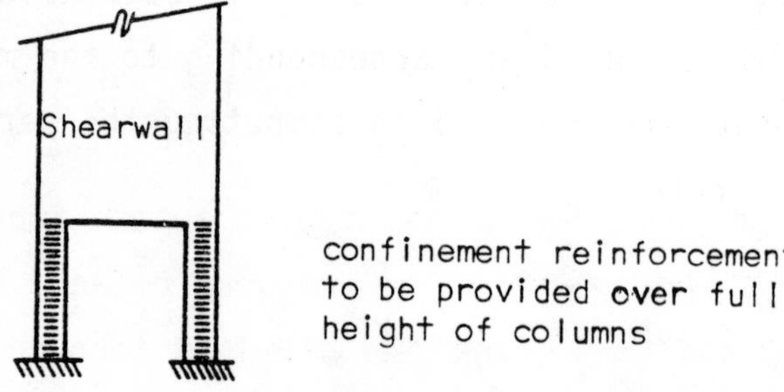

Fig. 30-9 Columns supporting discontinued shearwall

A.6 Joints of Frames

In conventional reinforced concrete buildings, the beam-column connections normally are not designed by the structural engineer. Detailing of bars within the joints is usually relegated to a draftsman or detailer. In earthquake-resistant frames, however, the design of beam-column connections requires as much attention as the design of the members themselves, since the integrity of the structure may well depend on the proper functioning of such connections. Beam-column joints represent regions of geometric and stiffness discontinuities in a frame and as such tend to be subjected to relatively high force concentrations. A substantial portion of the damage in frame structures subjected to strong earthquakes has been observed to occur at these connections. This has been particularly evident where inadequate attention had apparently been given to their proper design.

Because of the congestion of reinforcement that may occur as a result of too many bars converging within the limited space of the joint, the proportioning of the frame columns and beams should be undertaken with due regard to the design of the beam-column connection. Little difficulty is usually encountered if the amount of longitudinal reinforcement used in the frame members is kept low. Also, the preparation of large-scale detailed drawings showing bar arrangements within the joints will help much in avoiding unexpected difficulties in the field.

The provisions of Appendix A relating to beam-column connections have to do mainly with:

a. Transverse reinforcement for confinement (Section A.6.2):

   Minimum confinement reinforcement of the same amount required for potential hinging regions in columns, as defined by Eqs. (10-5), (A-2), (A-3), and (A-4), must be provided through beam-column joints around the column reinforcement.

For "confined" joints, a 50% reduction in the amount of confinement reinforcement is allowed. A confined joint is one with beams framing into all four sides and where each beam has a width equal to at least three-fourths of the width of the column face into which it frames.

The transverse reinforcement in a beam-column connection is intended to provide adequate confinement of the concrete to ensure its ductile behavior and allow it to maintain its vertical load-carrying capacity even after spalling of the outer shell. It also helps resist the shear forces transmitted by the framing members and improves the bond between steel and concrete within the connection.

The minimum amount of confinement reinforcement, as given by Eqs. (10-5), (A-2), (A-3), and (A-4), must be provided through the joint regardless of the magnitude of the calculated shear force in the joint. The 50% reduction in the amount of confinement reinforcement allowed for joints having horizontal members framing into all four sides recognizes the beneficial effect provided by these members in resisting the bursting pressures generated within the joint.

b. Design for shear (A.6.1.1, A.6.1.2, A.6.3):

Shear force in a joint is to be calculated by assuming stress in tensile reinforcement of framing beams equal to 1.25 f_y. Shear strength of the connection is to be computed as

φV_c = $\varphi 20 \sqrt{f'_c}\, A_j$ for confined joints

$\varphi 15 \sqrt{f'_c}\, A_j$ for unconfined joints

(for normal-weight concrete)

where φ = 0.85 for shear in joints

A_j = minimum cross-sectional area of joint in a plane parallel to the axis of the reinforcement generating the shear force.

For lightweight concrete, V_c, shall be taken as three-fourths the value given above for normal weight concrete.

As indicated in Fig. 10, the design shear is based on the most critical combination of horizontal shears transmitted by the framing beams and columns. Tests have indicated that plastic hinging at the ends of beams, for deformations associated with response to strong earthquakes, impose strains in the flexural reinforcement well in excess of the yield strain. Because of the likelihood of strains in the tensile reinforcement going into the strain-hardening range and to allow for the actual yield strength of the steel exceeding the specified value, the Code requires that the horizontal shear in the joint be determined by assuming the stress in the flexural tensile steel to be equal to 1.25 f_y.

Test results indicate that the shear strength of joints is not too sensitive to the amount of transverse (shear) reinforcement. Based on these results, the 1983 edition of ACI Appendix A makes the shear strength of beam-column connections a function only of the cross-sectional area of the joint, A_j, and f'_c.

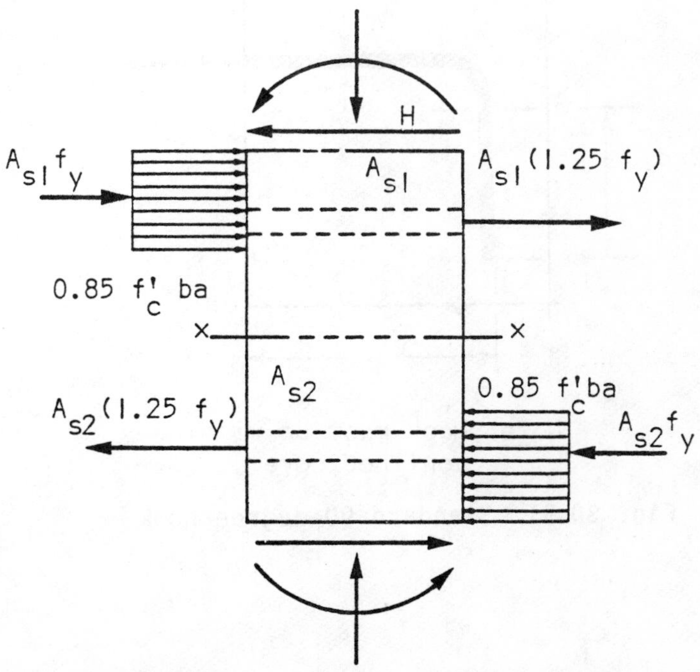

Fig. 30-10 Horizontal shear in beam-column connection

When the factored shear in the joint exceeds the shear strength of the concrete, the designer may either increase the column size or increase the depth of the beams. The former will increase the shear capacity of the joint section while the latter will tend to reduce the required amount of flexural reinforcement in the beams, with accompanying decrease in the shear transmitted to the joint.

c. Anchorage of longitudinal beam reinforcement within confined column core (A.6.1.3):

Beam longitudinal reinforcement terminated in a column is to be extended to the far face of the confined column core and anchored in tension according to Section A.6.4 (discussed below) and in compression according to Chapter 12 of the Code.

d. Development length for reinforcement in tension (A.6.4):

For bar sizes #3 through #11 with standard 90-degree hooks (Fig. 11) in normal weight concrete,

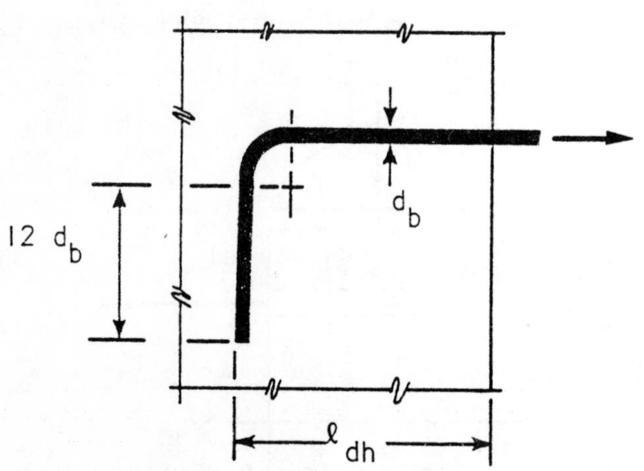

Note: Hook must be within confined core

Fig. 30-11 Standard 90-degree hook

development length, $\ell_{dh} \geq \begin{cases} f_y\, d_b/65\,\sqrt{f'_c} \\ 8\, d_b \quad (d_b \text{ is bar diameter}) \\ 6 \text{ in.} \end{cases}$

When bars of the same size are embedded in lightweight-aggregate concrete, ℓ_{dh} is to be at least 1.25 times the above value.

The 90-degree hook shall be located within the confined core of a column or other boundary element.

For straight bars of sizes #3 through #11,

ℓ_{dh} = 2.5 (ℓ_{dh} specified for bars with 90-degree hooks) when the depth of concrete cast in one lift beneath the bar $\leq$ 12 in.

= 3.5 (ℓ_{dh} specified for bars with 90-degree hooks) when the depth of concrete cast in one lift beneath the bar > 12 in.

If a bar is not anchored by means of a 90-degree hook within the confined column core but is extended into a boundary element, the portion of the required straight development length not located within the confined core shall be increased by a factor of 1.6.

The expression for ℓ_{dh} given above already includes the coefficients 0.7 (for concrete cover) and 0.80 (for ties) that are normally applied to the required basic development length, ℓ_{hb}. This is because Appendix A requires that hooks be embedded in the confined core of the column or other boundary element. The expression for ℓ_{dh} also includes a factor of about 1.4, representing an increase over the development length required for conventional structures, to provide for the effect of load reversals.

Except in very large columns, it is usually not possible to develop the yield strength of a reinforcing bar from a framing beam within the width of a column. Where beam reinforcement can extend through a column, its capacity is developed by embedment in the column and within the compression zone of the beam on the far side of the connection (Fig. 1). Where no beam is present on the opposite side of a column, such as in exterior columns, the flexural reinforcement in a framing beam has to be developed within the confined region of the column. This is usually done by means of a standard 90-degree hook plus whatever extension is necessary to develop the bar, the development length being measured from the near face of the column (Fig. 11).

Appendix A makes no provision for the use of #14 and #18 bars because of a lack of information on the behavior of anchorages of such bars when subjected to load reversals simulating earthquake effects.

A.5 Structural Walls, Diaphragms, and Trusses
A.7 Shear-Strength Requirements

When properly proportioned so that they possess adequate lateral stiffness to reduce interstory distortions due to earthquake-induced motions, shear walls or structural walls reduce the likelihood of damage to the nonstructural elements of a building. When used with rigid frames, walls form a system that combines the gravity-load-carrying efficiency of the rigid frame with the lateral-load-resisting efficiency of the structural wall.

Observations of the comparative performance of rigid-frame buildings and buildings stiffened by structural walls during recent earthquakes have pointed to the consistently better performance of the latter. The performance of buildings stiffened by properly designed structural walls has been better with respect to both safety and damage control. The need to ensure that critical facilities remain operational after a major tremor and the need to reduce economic losses from structural and nonstructural damage, in addition to the primary requirement of life safety, i.e., no collapse, has focused attention on the desirability of introducing greater lateral stiffness into earthquake-resistant multistory structures. Structural walls, which have long been used in designing for wind resistance, offer a logical and efficient solution to the problem of lateral stiffening of multistory buildings.

Shear walls are normally much stiffer than regular frame elements and are therefore subjected to correspondingly greater lateral forces during response to earthquake motions. Because of their relatively greater depth, the lateral deformation capacities of walls are limited, so that, for a given amount of lateral displacement, shear walls tend to exhibit greater apparent distress than frame members. However, over a broad period range, a shear wall structure, which is substantially stiffer and hence has a shorter period than a frame structure, will suffer less lateral displacement than the frame, when subjected to the same ground motion intensity. Shear walls with a height-to-depth ratio of about 3 behave essentially as vertical cantilever beams and should therefore be designed as flexural members, with their strength governed by flexure rather than by shear.

Primarily because of the greater stiffness of shear wall structures, but also becasue of earlier concerns about the deformation capacities of shear walls, codes have specified larger design lateral forces for such structures.

Isolated shear walls or individual walls connected to frames will tend to yield first at the base where the moment is the greatest. Coupled walls, i.e., two or more walls linked by short, rigidly-connected beams at the floor levels, on the other hand, have the desirable feature that significant energy dissipation through inelastic action in the coupling beams can be made to precede hinging at the bases of the walls.

The principal provisions of ACI Appendix A relating to structural walls (and diaphrams) are as follows (Fig. 12):

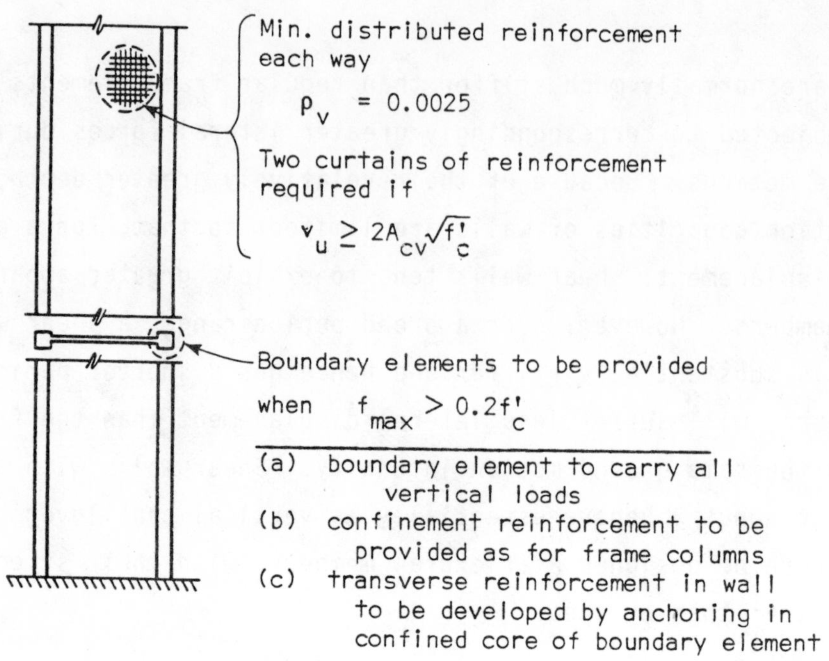

Fig. 30-12 Structural wall requirements

a. Reinforcement (A.5.2.1, A.5.2.2):

Walls and (diaphragms) are to be provided with shear reinforcement in two orthogonal directions in the plane of the wall. Minimum reinforcement ratios for both longitudinal and transverse directions,

$$\rho_v = \frac{A_{sv}}{A_{cv}} = \rho_n \geq 0.0025$$

with reinforcement continuous and distributed uniformly across the shear area

where

A_{cv} = net area of concrete section, i.e., product of thickness and length of section in direction of shear considered

A_{sv} = projection on A_{cv} of area of shear reinforcement crossing the plane of A_{cv}

and ρ_n = ratio of distributed shear reinforcement on a plane perpendicular to plane of A_{cv}

The spacing of reinforcement must not exceed 18 in.

At least two curtains of reinforcement - each having bars running in the longitudinal and transverse directions - are to be provided if the in-plane factored shear force assigned to the wall exceeds $2A_{cv}\sqrt{f'_c}$.

The use of two curtains of reinforcement in walls subjected to significant shears ($> 2 A_{cv}\sqrt{f'_c}$) serves to reduce fragmentation and premature deterioration of the concrete under load reversals into the inelastic range. Distributing the reinforcement uniformly across the height and horizontal length of the wall helps control the width of inclined cracks.

It should be noted that the vertical reinforcement in the boundary elements (or reinforcement concentrated near the edges of the wall when no boundary elements are used) for resisting flexure in the wall is not to be included in determining satisfaction of the requirements for ρ_v or ρ_n.

b. Shear strength of walls (and diaphragms) (A.7.3):

For walls with height-to-horizontal length ratio, $h_w/\ell_w \geq 2.0$, the shear strength is to be determined from the expression

$$\varphi V_n = \varphi A_{cv} \left(2\sqrt{f'_c} + \rho_n f_y\right)$$

where

- φ = 0.60, unless nominal shear strength provided exceeds the shear corresponding to development of nominal flexural capacity of wall
- A_{cv} = as defined earlier
- ρ_n = as defined earlier
- h_w = height of entire wall or of segment of wall considered
- ℓ_w = length of entire wall or of segment of wall considered in direction of shear force

For walls with $h_w/\ell_w < 2.0$, the shear strength may be determined from

$$\varphi V_n = \varphi A_{cv}(\alpha_c\sqrt{f'_c} + \rho_n f_y)$$

where the coefficient α_c varies linearly from a value of 3.0 for $h_w/\ell_w = 1.5$ to 2.0 for $h_w/\ell_w = 2.0$.

Where a wall is divided into several segments by openings, the value of the ratio h_w/ℓ_w to be used in calculating V_n for any segment shall not be less than the corresponding ratio for the entire wall.

The nominal shear strength, V_n, of all wall segments or piers resisting a common lateral force shall not exceed $8\ A_{cv}\sqrt{f'_c}$, where A_{cv} is the total cross-sectional area of the walls. The nominal shear strength of any individual vertical or horizontal segment of wall shall not exceed $10\ A_{cp}\sqrt{f'_c}$, where A_{cp} is the cross-sectional area of the wall segment.

Appendix A allows calculation of the shear strength of any structural wall using a coefficient $\alpha_c = 2.0$. However, advantage can be taken of the greater observed shear strength of walls with low height-to-horizontal length (h_w/ℓ_w) ratios by using an α_c-value of up to 3.0 for $h_w/\ell_w = 1.5$ or less.

Appendix A limits the average nominal unit shear strength of structural walls to $8\sqrt{f'_c}$, with allowance for exceeding this average in any individual wall in a group of walls or wall segments, provided that the unit shear in the individual wall does not exceed $10\sqrt{f'_c}$. This upper bound on strength that may be developed in any individual segment is intended to limit the degree of shear redistribution among several connected wall segments. A wall segment refers to a part of wall bounded by openings or by an opening and an edge.

It is important to note that Section A.2.3.1 of ACI 318-83 requires the use of a strength reduction factor (φ) for shear of 0.6 for all members (except joints) where the nominal shear strength is less than the shear corresponding to the development of the nominal flexural strength of the member. In the case of beams, the design shears are obtained by assuming plastic end moments corresponding to a tensile steel stress of 1.25 f_y (A.7.1.1). Similarly, for a column, the design shears are determined not by applying load factors to shears obtained from a lateral load analysis, but from the consideration of maximum developable moments, consistent with the axial force on the column, occurring at the column ends. This approach to shear design is intended to ensure that even when flexural hinging occurs at member ends due to earthquake-induced deformations, no shear failure would develop. Under the above conditions, the Code allows the use of the normal strength reduction factor for shear of 0.85. When design shears are not based on the condition of flexural strength being developed at member ends, the Code requires the use of a lower shear strength reduction factor to achieve the same result, i.e., prevention of premature shear failure.

In the case of structural walls, a condition similar to that used for the shear design of beams and columns is not as readily established. This is primarily because the magnitude of the shear at the base of a wall (or at any level above) is influenced significantly by the forces and deformations beyond the particular level considered. Appendix A thus prescribes that the design shear force, V_u, for a structural wall shall be obtained from lateral load analysis of the structure containing the wall in accordance with the factored loads and combinations specified in Section 9.2 (A.7.1.3). Unlike the flexural behavior of beams and columns in a frame, which can be considered as close-coupled systems, i.e., with the forces and deformations in the members determined primarily by the displacements in the end joints, the state of flexural deformation at any section of a structural wall (a far-coupled system) is influenced significantly by the displacements of points far removed from the section considered. Results of dynamic inelastic analyses of isolated structural walls under earthquake loads also indicate that the base shear in such walls is strongly influenced by the higher modes of response.

A distribution of static lateral forces along the height of a wall, corresponding to the fundamental mode, such as is assumed by most current seismic codes, may produce flexural yielding at the base if the section at the base of the wall is designed for such yielding. However, other distributions of lateral forces, having a resultant closer to the base, can produce yielding at the base only if the magnitude of the resultant horizontal force, and hence the base shear, is increased. Results of research on isolated walls, which would also apply to frame-shear wall systems in which the frame is flexible relative to the wall, in fact indicate that for a wide range of wall properties and input motions, the resultant of the dynamic horizontal forces producing yielding at the base of the wall generally occurs well below the two-thirds-of-total height level associated with fundamental mode response. This would imply significantly larger base shears than those due to lateral forces distributed according to the fundamental mode response. Research on isolated walls indicates ratios of maximum dynamic shears to "fundamental mode shears" (i.e., shears associated with horizontal forces distributed according to the seismic codes) ranging from 1.3 to 4.0, the value of the ratio increasing with fundamental period.

c. Development length and splices (Sections A.5.2.4, A.5.3.5):

All continuous reinforcement is to be anchored or spliced in accordance with the provisions for reinforcement in tension (Section A.6.4).

Where boundary elements are present, the transverse reinforcement in walls is to be anchored within the confined core of the boundary element to develop the yield stress in tension of the transverse reinforcement.

Actual forces in longitudinal bars of stiff members may exceed calculated forces. Because of this likelihood, and the importance of maintaining the flexural capacity of a wall, the Code requires that all continuous reinforcement be developed fully.

Similarly, the horizontal reinforcement in walls requiring boundary elements are called upon to function as web reinforcement. Because of this, the Code requires that such bars be fully anchored in the boundary elements (which act as flanges of vertical cantilever beams). Standard 90-degree hooks should be used whenever possible. Such hooks minimize the loss of bond that may otherwise result due to the occurrence of large transverse cracks in the boundary elements when subjected to large inelastic deformations.

d. Boundary elements (Section A.5.3):

Boundary elements are to be provided, both along vertical boundaries of a wall and around the edges of openings, if any, when the maximum extreme-fiber stress in the wall due to factored forces including earthquake forces exceeds 0.2 f'_c. The boundary members may be discontinued when the calculated compressive stress is less than 0.15 f'_c.

Boundary elements need not be provided if the <u>entire</u> wall is reinforced in accordance with the provisions governing transverse reinforcement for members subjected to axial load and bending, as given by Eqs. (10-5), (A-2), (A-3), (A-4) and the related spacing requirements.

Boundary members of structural walls are to be designed to carry all the factored vertical loads on the wall, including self-weight and gravity loads tributary to the wall, as well as the vertical force required to resist the overturning moment due to factored earthquake loads. Such boundary elements are to be provided with confinement reinforcement in accordance with Eqs. (10-5), (A-2), (A-3), (A-4) and the related spacing requirements.

The Code uses a concrete stress of $0.2\ f'_c$, calculated using a linearly elastic model based on gross sections of structural members and factored forces, as indicative of significant compression. Structural walls subjected to compressive stresses exceeding this value are generally required to have boundary elements.

The condition assumed in requiring that boundary members be designed for all gravity loads as well as the vertical forces associated with overturning of the wall due to earthquake forces is illustrated in Fig. 13. This requirement assumes that the boundary member alone may have to carry all the vertical (compressive) forces at the critical wall section when the maximum horizontal earthquake force acts on the wall. Under load reversals, this condition imposes severe demands on the concrete in the boundary element. Hence the requirement for confinement reinforcement similar to those for members subjected to axial load and bending.

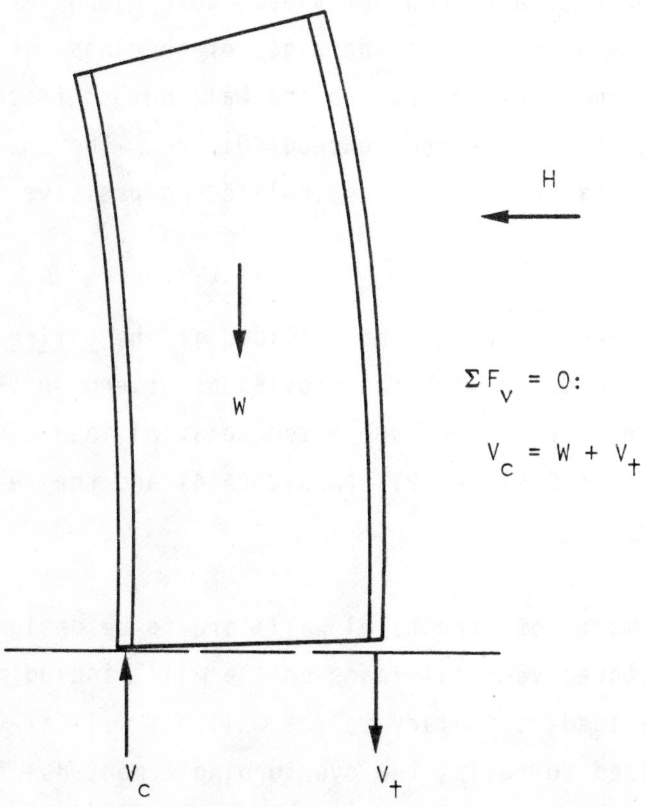

Fig. 30-13 Loading condition assumed for design of structural wall boundary element

The design of boundary element is carried out by considering it as an axially loaded short column subjected to the factored compressive axial force at the critical section.

Diaphrams of reinforced concrete, such as floor slabs, that are designed to transmit horizontal forces through bending and shear in their plane, are treated in much the same manner as structural walls.

Truss elements of reinforced concrete are also covered, although very briefly, in the 1983 edition of Appendix A. A major requirement for truss elements relates to the provision of special transverse reinforcement when the compressive stress exceeds 0.2 f'_c.

A.8 Frame Members not Proportioned to Resist Forces Induced by Earthquake Motions

Frame members that are not relied on to provide lateral resistance to earthquake-induced forces need not satisfy the stringent requirements governing lateral-load-resisting elements. This refers mainly to the requirements for transverse reinforcement for confinement and shear.

Except for the requirements noted below, non-lateral-load-resisting elements, whose primary function is the transmission of vertical loads to the foundation, need comply only with the minimum reinforcement requirements of Appendix A, and applicable provisions in the main body of the Code.

A special requirement for non-lateral-load-resisting elements is that they be checked for adequacy with respect to a lateral displacement twice that calculated for the structure under the factored lateral forces. This requirement should enable the gravity-load system to maintain its vertical-load-carrying capacity, without reduction, under the specified lateral forces. Although elements of the gravity-load system need not be designed for moments related to the lateral forces, they may have to be provided with adequate confinement reinforcement in regions where plastic hinging can occur.

For gravity-load frame members subjected to factored axial compressive forces exceeding ($A_c f'_c/10$), the following requirements relating to transverse reinforcement have to be satisfied:

Maximum tie spacing, s_o (over length ℓ_o from face of joint) $\leq$
- 8 (diameter of smallest longitudinal bar)
- 24 tie diameter
- 1/2 least cross-sectional dimension of column

where $\ell_o \geq$
- 1/6 clear height of column
- maximum cross-sectional dimension of column
- 18 in.

The first tie is to be located within a distance of $s_o/2$ from face of joint. Maximum tie spacing in any part of the column = 2 s_o.

A.9 Requirements for Frames in Regions of Moderate Seismic Risk

Although ACI Appendix A does not define "moderate seismic risk" in terms of a commonly accepted quantitative measure, it assumes that the probable ground motion intensity in such regions would be a fraction of that expected in a high seismic risk zone, to which the bulk of Appendix A is addressed. By the above description, an area of moderate seismic risk would correspond to Zone 2 as defined in UBC-82[30.2] or ANSI-82.[30.3]

For regions of moderate seismic risk, the provisions for the design of structural walls given in the main body of the ACI Code are considered sufficient to provide the necessary toughness. The requirements of Appendix A for structures in moderate-risk areas relate mainly to frames.

The distinction between flexural members and columns based on an axial compressive force of $A_g f'_c/10$, as used in high seismic risk zones, also applies in regions of moderate seismicity (A.9.2).

For shear design of beams, columns or two-way slabs resisting earthquake effects, the magnitude of the design shear (A.9.3) should not be less than either

a) the sum of the shear associated with the development of nominal moment strength at each restrained end and that due to factored gravity loads. This is similar to the corresponding requirements for high-risk zones illustrated in Fig. 3, except that the stress in the flexural reinforcement is taken as f_y rather than 1.25 f_y.

or b) the maximum factored shear corresponding to the application of design gravity and earthquake forces, but with the earthquake effect taken at twice the calculated value. Thus, if the critical load combination consists of dead load (D) + live load (L) + earthquake effects (E), then the design shear is to be computed from

$$U = 0.75[1.4D + 1.7L + 2(1.87E)]$$

Detailing requirements for beams (A.9.4) - The positive moment strength at the face of a joint shall not be less than 1/3 the negative moment capacity at the same section. (This compares with 1/2 for beams in areas of high seismic risk, Fig. 1.) The moment strength - positive or negative - at any section is not to be less than 1/5 the maximum moment strength at either end of the beam.

Stirrup spacing requirements are identical to those for beams in regions of high seismic risk (Fig. 2).

Detailing requirements for columns (A.9.5) - The tie spacing requirements for columns are identical to those for gravity load carrying members as given in Section A.8 (see above).

Detailing requirements for two-way slabs without beams (A.9.6) - It is worth noting that requirements for two-way slabs without beams are included in Appendix A for frames in regions of moderate seismic risk only. This suggests that the Code considers the use of properly designed two-way slabs without beams as acceptable components of the lateral-load-resisting system in regions of moderate seismic risk only.

The requirements for slabs without beams are illustrated in Figs. 14 and 15. The moment M_s in Fig. 14 is the portion of the factored slab moment balanced by the support moment. The factor γ_f represents the fraction of the unbalanced moment at a joint transferred by flexure, as defined in Chapter 13 of the Code, i.e.

$$\gamma_f = \frac{1}{1 + \frac{2}{3}\sqrt{\frac{c_1 + d}{c_2 + d}}}$$

where

c_1 = dimension of column cross-section in the direction of the span for which moments are determined

c_2 = dimension of column cross-section measured transverse to direction of span

d = effective depth of slab

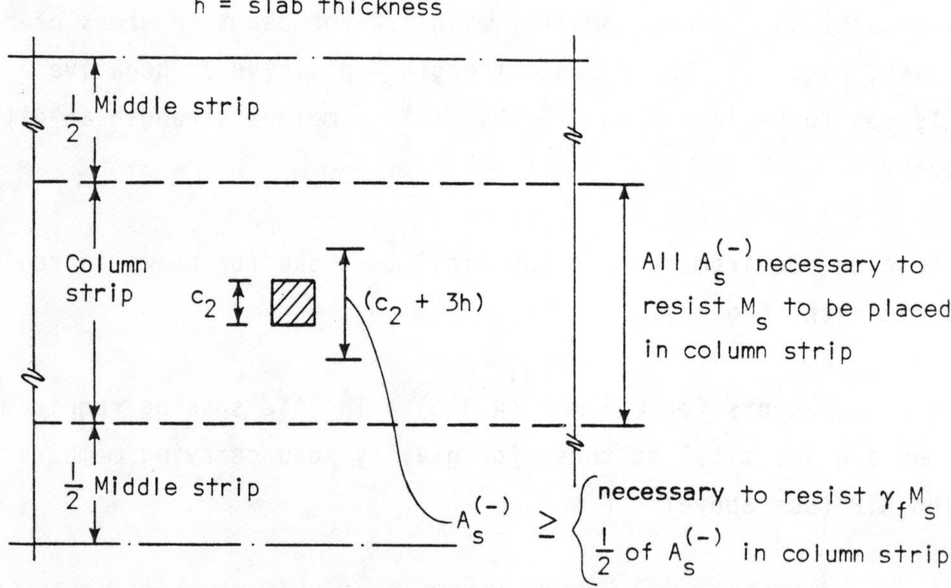

Fig. 30-14 Requirements relating to location of reinforcement in slabs without beams in regions of moderate seismic risk

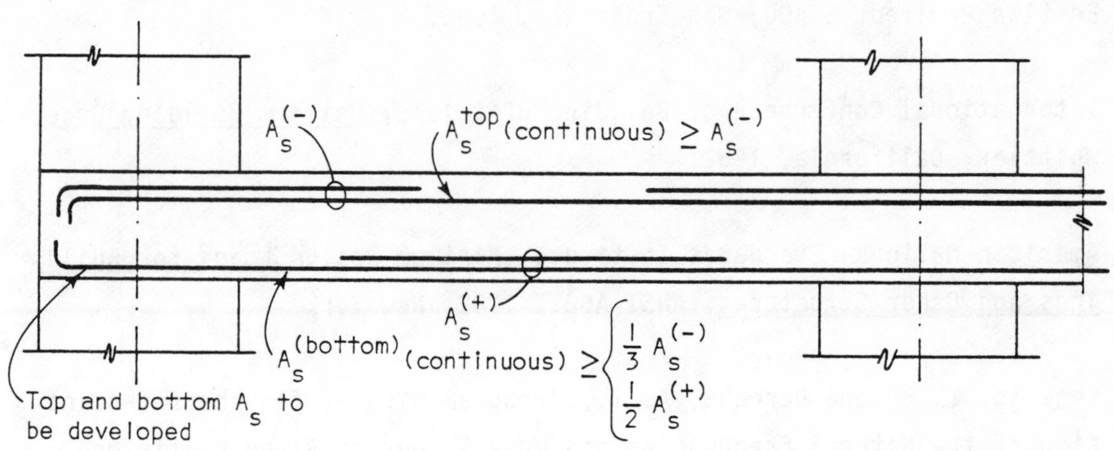

(a) Column strip

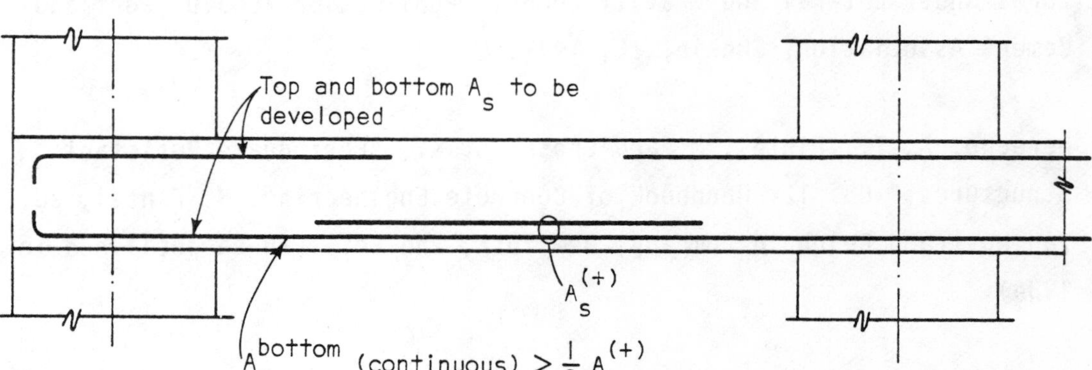

(b) Middle strip

Fig. 30-15 Requirements relating to arrangement of reinforcement in slabs without beams in regions of moderate seismic risk

References

30.1 Seismology Committee of the Structural Engineers Association of California, Recommended Lateral Force Requirements and Commentary, 4th Edition Revised, SEAOC, San Francisco, 1980.

30.2 International Conference of Building Officials, Uniform Building Code, Whittier, California, 1982.

30.3 American National Standards Institute, Minimum Design Loads for Buildings and Other Structures, ANSI A58.1-1982, New York.

30.4 Fugelso, L. E. and Derecho, A. T., "Program DYFRQ - for the Determination of the Natural Frequencies and Mode Shapes of Plane Multistory Structures," Portland Cement Association, Skokie, IL, 1975 (unpublished).

30.5 Derecho, A. T., "Analysis of Plane Multistory Frame-Shear Wall Structures Under Lateral and Gravity Loads," Publication XL097D, Portland Cement Association, Skokie, IL, 1971.

30.6 Derecho, A. T., Fintel, M. and Ghosh, S. K., "Earthquake-Resistant Structures," Ch. 12, Handbook of Concrete Engineering, M. Fintel, ed., Van Nostrand Reinhold Company, New York, 2nd Ed., (to be published in 1984).

30.7 Design Handbook in Accordance with the Strength Design Method of ACI 318-77: Vol.2 - Columns, Publication SP-17A(78), American Concrete Institute, Detroit, 1978.

EXAMPLE 30.1 - Sample Designs for Elements of a 12-Story Frame-Shear Wall Building

Application of the earthquake-resistant design provisions of Appendix A of ACI 318-83 to the design of typical members of a 12-story frame-shear wall building located in Seismic Zone 3 is illustrated below. The design loads are taken from ANSI A58.1-1982.[30.3] Except for minor variations in local areas, the division of the United States into different seismic zones (i.e., 0, 1, 2, 3 and 4), as adopted by ANSI-82,[30.3] is very similar to that used in UBC-82.[30.2] In UBC Seismic Zone 3, major damage associated with a Modified Mercalli intensity of VII or greater is expected. Included in Seismic Zone 4 are those areas within Zone 3 that are close to certain major fault systems.

Typical plan and elevation of the structure considered are shown in Figs. 16a and b. The columns and structural walls have constant cross-sections throughout the height of the building[1], the bases of the lowest story segments being assumed fixed. The beams and slabs also have the same dimensions at all floor levels. Although the element dimensions in this example are within the practical range, the structure itself is a hypothetical one and has been chosen mainly for illustrative purposes. Other pertinent design data are as follows:

Service loads - vertical:

Live load:	50 psf additional average value to allow for heavier load on corridors = 25 psf thus, total average live load = 75 psf
Superimposed dead load:	average for partitions = 20 psf ceiling and mechanical = 10 psf thus, total average superimposed dead load = 30 psf

Material properties:

Concrete: f'_c = 4000 psi, w_c = 145 pcf

Reinforcement: f_y = 60 ksi

[1]The uniformity in member dimensions used in this example has been adopted mainly for simplicity. Such uniform stiffness along height has a beneficial effect on the dynamic response of a structure.

EXAMPLE 30.1 - Continued

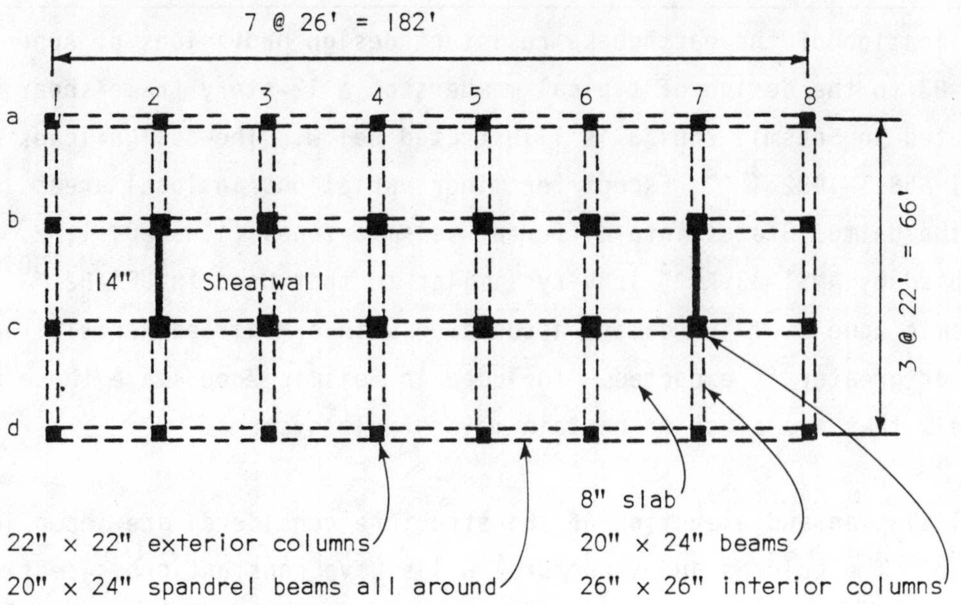

Fig. 30-16a Typical plan of building considered

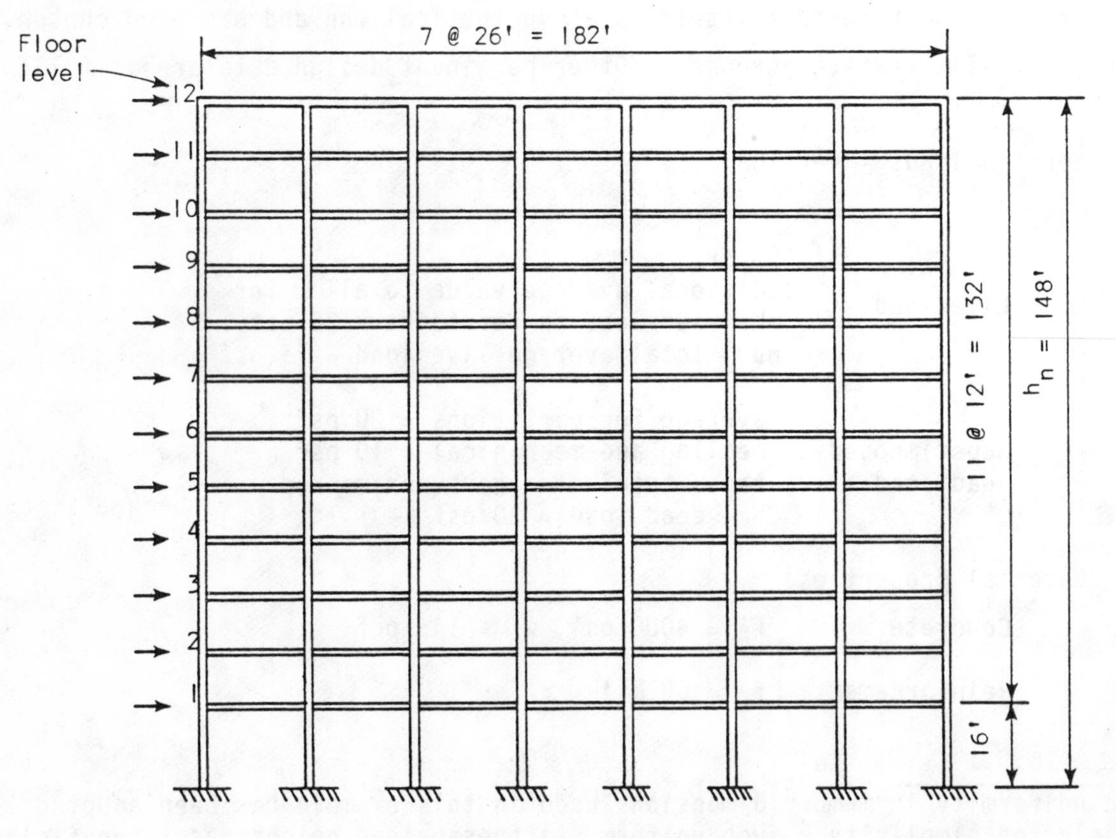

Fig. 30-16b Longitudinal section

EXAMPLE 30.1.1 - Determination of Design Lateral Forces

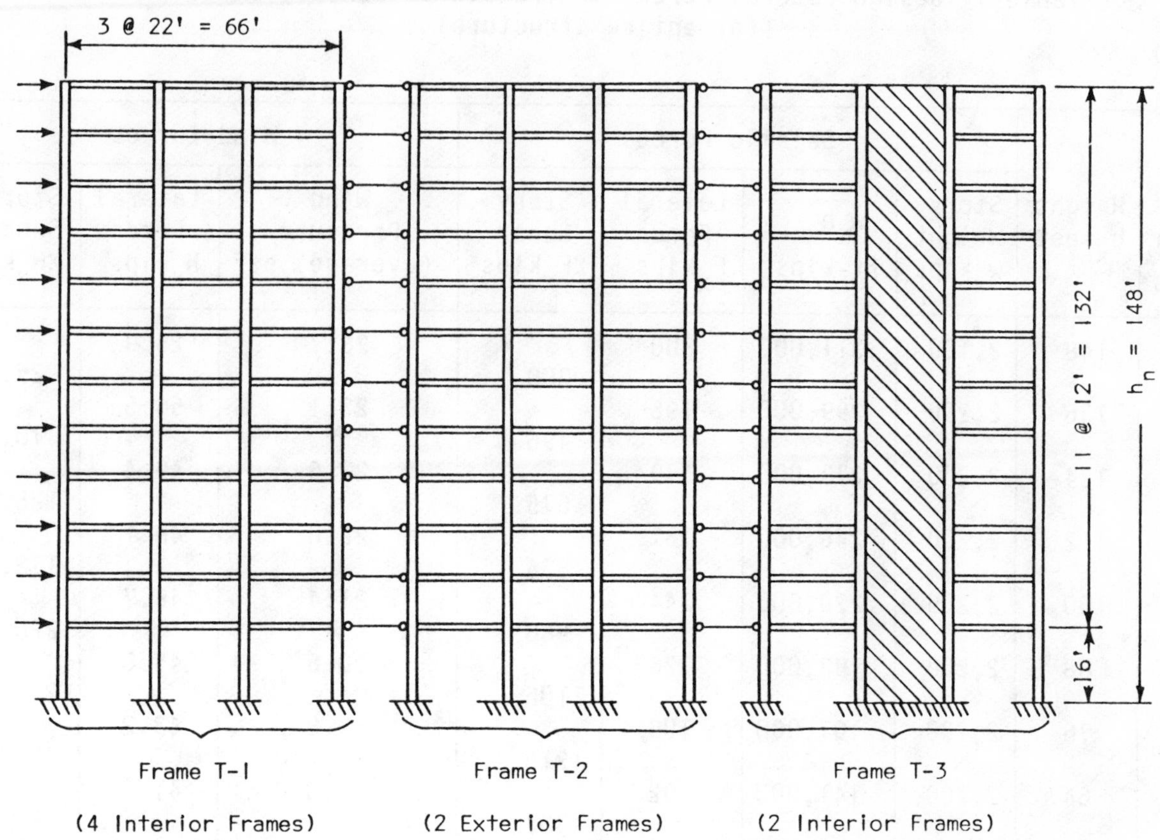

Fig. 30-16c Analytical model of building for lateral load analysis in transverse direction

On the basis of the given data and the dimensions shown in Fig. 16, the weights of a typical floor[1] and the roof were estimated and are listed in Tables 1 and 2. The calculation of the base shear, V, for the transverse and longitudinal directions are shown at the bottom of Tables 1 and 2, respectively. For this example, the importance factor, I, and the soil factor, S, have been assigned values of unity. The period of the structure in the transverse direction is computed using Eq. (9A) of ANSI A58.1-1982,[30.3] while that in the longitudinal direction is obtained using Eq. (9C). Note that a value of K = 0.8 has been

[1]The weight of a typical floor includes that of all elements located between two imaginary parallel planes passing through the midheight of columns above and below the floor considered.

EXAMPLE 30.1.1 - Continued

Table 1 Design Lateral Forces in Transverse (Short) Direction (for entire structure)

Floor Level (from base)	Height h_x feet	Seismic Forces: Story Weight w_x kips	$w_x h_x$ ft-kips	Lateral Force F_x kips	Story Shear ΣF_x kips	Wind Forces: Wind Pressure (average) psf	Lateral Force H_x kips	Story Shear ΣH_x kips
12 (Roof)	148	2,100	311,000	300*	300	23.7	25.9	25.9
11	136	2,200	299,000	196	496	23.1	50.5	76.4
10	124	2,200	273,000	179	675	22.6	49.4	125.8
9	112	2,200	246,000	161	836	22.0	48.1	173.9
8	100	2,200	220,000	144	980	21.4	46.7	220.6
7	88	2,200	193,000	126	1106	20.6	45.0	265.6
6	76	2,200	167,000	109	1215	19.8	43.2	308.8
5	64	2,200	141,000	92	1307	19.0	41.5	350.3
4	52	2,200	114,000	75	1382	18.2	39.8	390.1
3	40	2,200	88,000	58	1440	17.1	37.3	427.4
2	28	2,200	61,500	40	1480	15.9	34.7	462.1
1	16	2,200	30,800	20	1500	14.4	36.7	498.8
Σ =		26,300	2,144,300	1500			498.8	

*representing the sum $(F_t + F_{12})$

Base shear, V = ZIKCSW, where $C = \dfrac{1}{15\sqrt{T}}$ and $T = \dfrac{0.05\,h_n}{\sqrt{D}}$

In transverse direction, h_n = 148' and D = 66'→ T = 0.91 sec and C = 0.07

Thus, V = (1.0)(1)(0.8)(0.07)(1)W = 0.056W = 0.056(26,300) = 1473, say <u>1500 kips</u>

F_t = 0.07 TV = (.07)(.91)(1500) = 96 kips

EXAMPLE 30.1.1 - Continued

Table 2 Design Lateral Forces in Longitudinal Direction (for entire structure)

Floor Level (from base)	Height h_x feet	Seismic Forces: Story Weight w_x kips	$w_x h_x$ ft-kips	Lateral Force F_x kips	Story Shear ΣF_x kips	Wind Forces: Wind Pressure (average) psf	Lateral Force H_x kips	Story Shear ΣH_x kips
12 (Roof)	148	2,100	311,000	239*	239	19.7	7.8	7.8
11	136	2,200	299,000	149	388	19.3	15.3	23.1
10	124	2,200	273,000	136	524	18.8	14.9	38.0
9	112	2,200	246,000	122	646	18.0	14.3	52.3
8	100	2,200	220,000	109	755	17.2	13.6	65.9
7	88	2,200	193,000	96	851	16.6	13.2	79.1
6	76	2,200	167,000	83	934	16.0	12.7	91.8
5	64	2,200	141,000	70	1004	15.0	11.9	103.7
4	52	2,200	114,000	57	1061	14.0	11.1	114.8
3	40	2,200	88,000	44	1105	13.0	10.3	125.1
2	28	2,200	61,500	30	1135	11.8	9.4	134.5
1	16	2,200	30,800	15	1150	10.1	9.3	143.8
Σ =		26,300	2,144,300	1150			143.8	

*$(F_t + F_{12})$

In longitudinal direction, $T = C_T h_n^{3/4}$, where C_T (concrete frames) = 0.025

$$= 0.025(148)^{3/4} = 1.06 \text{ sec}$$

$$C = \frac{1}{15\sqrt{T}} = \frac{1}{15\sqrt{1.06}} = 0.65$$

Base shear, $V = ZIKCSW = (1.0)(1)(0.67)(0.065)(1)W = 0.0436W$

$= 0.0436(26{,}300) = 1146$, say <u>1150 kips</u>

$F_t = 0.07\ TV = (.07)(1.06)(1150) = 85$ kips.

EXAMPLE 30.1.1 - Continued

used in the transverse direction where one has a frame-shear wall structural system,[1] while a value of K = 0.67 has been used in the longitudinal direction, in which the structure consists of moment-resisting frames.

Calculation[2] of the undamped natural periods of vibration of the structure in the transverse direction, using the story weights listed in Table 1 and member stiffnesses based on gross concrete sections, gave a value for the fundamental period of 1.34 sec compared to the T value of 0.91 sec given by the approximate formula in the Code. The calculated mode shapes as well as the corresponding periods of the first five modes of vibration of the structure in the transverse direction are shown in Fig. 17. Note that the mode shapes indicate only the relative displacements of the story masses (assumed concentrated at the floor levels). The maximum displacement for each mode has been set equal to unity.

The lateral seismic design forces resulting from the distribution of the base shear in accordance with Eqs. (10), (11) and (12) of ANSI A58.1-1982[30.3] are listed in Tables 1 and 2. As an example, the seismic lateral force F_x at the 10th floor level in the transverse direction is given by

$$F_{10} = \frac{(V - F_t)\, w_x h_x}{\sum_{i=1}^{n} w_i h_i} = \frac{(1500 - 96)(124)(2200)}{2{,}144{,}300} = 179 \text{ kips}$$

Also shown in the tables are the story shears corresponding to the distributed seismic forces.

For comparison, the wind forces and story shears corresponding to a basic wind speed of 75 mph and Exposure B (urban and suburban areas), computed as prescribed in ANSI-82[30.3], are shown for each direction in Tables 1 and 2.

---

[1]The requirements relating to the use of K = 0.8 for dual systems are that the moment-resisting space frame be capable of resisting at least 25% of the prescribed seismic forces and that the individual elements satisfy the requirements of ACI Appendix A for special ductile moment-resisting frames and structural walls.

[2]Using the computer program described in Ref. 30.4.

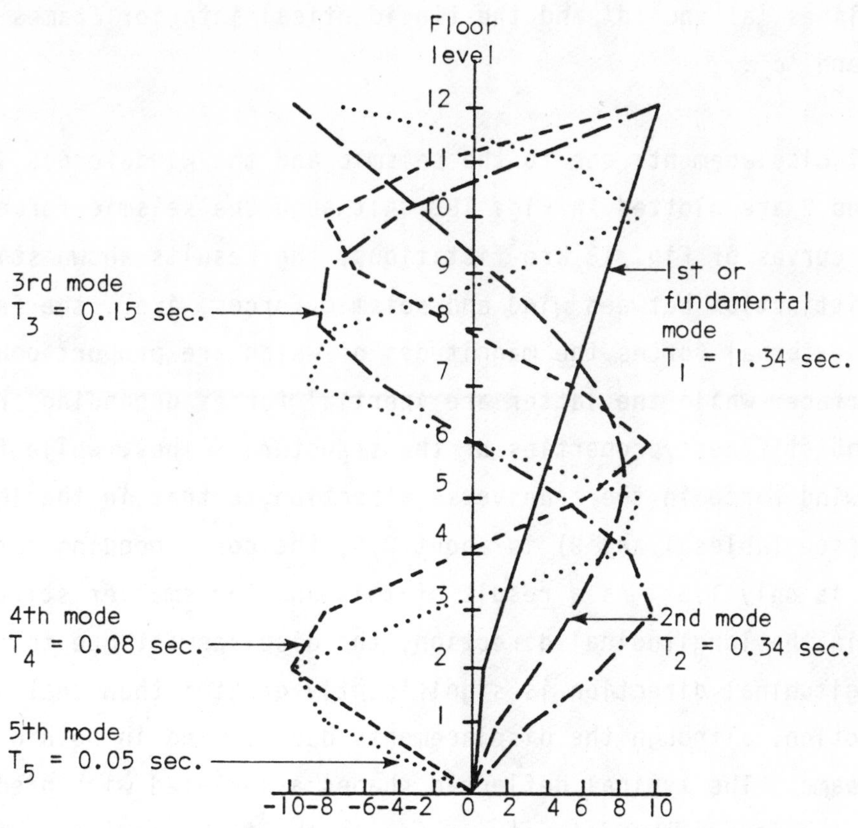

Fig. 30-17 Undamped natural modes and periods of vibration of structure in transverse direction

Analyses of the structure in both directions under the respective seismic and wind loads, and assuming no torsional effects, were carried out using a plane frame computer program.[30.5] For the purpose of analyzing the structure in the transverse direction, the model shown in Fig. 16c was used. This model consists of three different frames linked by hinged rigid bars at the floor levels to impose equal horizontal displacements at these levels. (This device is used to model the effect of floor slabs which may be assumed as rigid in their own planes.) Frame T-1 represents the four identical interior frames along lines 3, 4, 5 and 6 which have been lumped together in this single frame, while Frame T-2 represents the two exterior frames along lines 1 an 8. The third frame, T-3, represents the two identical frame-shear wall systems along lines 2 and 7.

EXAMPLE 30.1.1 - Continued

In the longitudinal direction, two linked frames, each similar to the frame shown in Fig. 16b, were used to represent the two identical exterior frames L-1 along lines 'a' and 'd' and the two identical interior frames L-2 along lines 'b' and 'c'.

The lateral displacements due to the seismic and the wind forces listed in Tables 1 and 2 are plotted in Fig. 18. Although the seismic forces used to obtain the curves of Fig. 18 are fictitious, the results shown still serve to draw the distinction between wind and seismic forces, i.e., the fact that the former are external forces the magnitudes of which are proportional to the exposed surface, while the latter are inertial forces depending primarily on the mass and stiffness properties of the structure. Thus, while the ratio of the total wind force in the transverse direction to that in the longitudinal direction (see Tables 1 and 2) is about 3.5, the corresponding ratio for seismic forces is only 1.3. As a result of this and the smaller stiffness of the structure in the longitudinal direction, the displacement due to seismic forces in the longitudinal direction is significantly greater than that in the transverse direction, although the displacements due to wind in both directions are about the same. The typical deflected shapes associated with predominantly cantilever or flexure type structures (as in the transverse direction) and shear (open frame) type buildings (as in the longitudinal direction) are evident in Fig. 18. The average deflection indices, i.e., the ratio of the lateral displacement at the top to the total height of the structure, are 1/4790 for wind and 1/1170 for seismic loads in the transverse direction. The corresponding values in the longitudinal direction are 1/7510 for wind and 1/760 for seismic loads.

An idea of the distribution of lateral loads among the different frames making up the structure in the transverse direction may be obtained from Table 3, which lists the portion of the total story shear at each level resisted by each of the three lumped frames. Note that at the top (12th story level), the lumped frame T-1 takes 121% of the total story shear. This reflects the fact that in frame-shear wall systems of average proportions, interaction between frame and wall under lateral loads results in the frame "supporting" the wall at the top while

EXAMPLE 30.1.1 - Continued

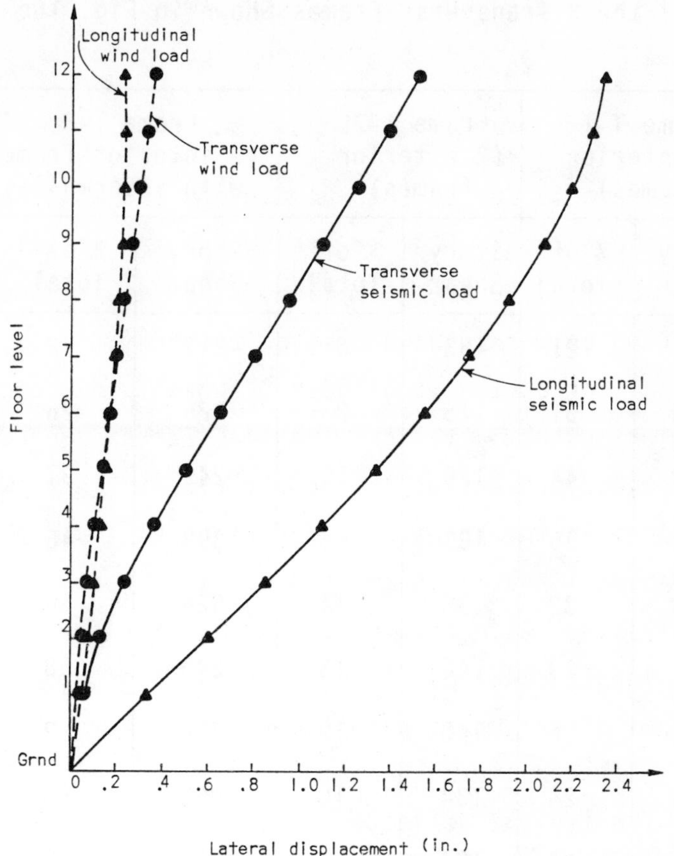

Fig. 30-18 Lateral displacements under seismic and wind loads

at the base most of the horizontal shear is taken by the wall. Table 3 indicates that for the structure considered, the two frames with walls take 96% of the shear at the base in the transverse direction.

To illustrate the design of two typical beams on the sixth floor of an interior frame, the results of analysis of the structure in the transverse direction under seismic loads have been combined for the beams, using Eqs. (9-1), (9-2) and (9-3) of ACI 318-83, with results obtained from a gravity load analysis of a single-story bent including those beams (using the computer program described in Ref. 30.5). The results are listed in Table 4. Similar values for typical exterior and interior columns on the second floor of the same interior frame are shown in Table 5. Corresponding design forces for the structural wall section at the first floor level of Frame T-3 (Fig. 16c) are listed in Table 6. The

EXAMPLE 30.1.1 - Continued

Table 3 Distribution of Horizontal Seismic Story Shears Among the 3 Transverse Frames Shown in Fig. 16c

Story Level	Frame T-1 (4 interior frames)		Frame T-2 (2 exterior frames)		Frame T-3 (2 interior frames with shearwalls)		Total Story Shear (kips)
	Story Shear	% of Total	Story Shear	% of Total	Story Shear	% of Total	
12	364	121	153	51	-217	-72	300
11	255	51	115	23	126	26	496
10	297	44	129	19	249	37	675
9	304	36	133	16	399	48	836
8	316	32	138	14	526	54	980
7	321	29	140	13	645	58	1106
6	319	26	140	12	756	62	1215
5	307	24	134	10	866	66	1307
4	285	21	124	9	973	70	1382
3	242	17	105	7	1093	76	1440
2	205	14	88	6	1187	80	1480
1	48	3	20	1	1432	96	1500

last column in Table 6 lists the axial load on the boundary elements (the 26 in. x 26 in. columns forming the flanges of the structural walls) calculated in accordance with the ACI requirement that these be designed to carry all factored vertical loads on the walls, including self-weight, gravity loads and vertical forces due to earthquake-induced overturning moments. The loading condition associated with this requirement is illustrated in Fig. 13. In both Tables 5 and 6, the additional forces due to the effects of horizontal torsional moments

EXAMPLE 30.1.1 - Continued

Table 4 Summary of Design Moments for Typical Beams on 6th Floor of Interior Transverse Frames along Lines 3 through 6 (Fig. 16a)

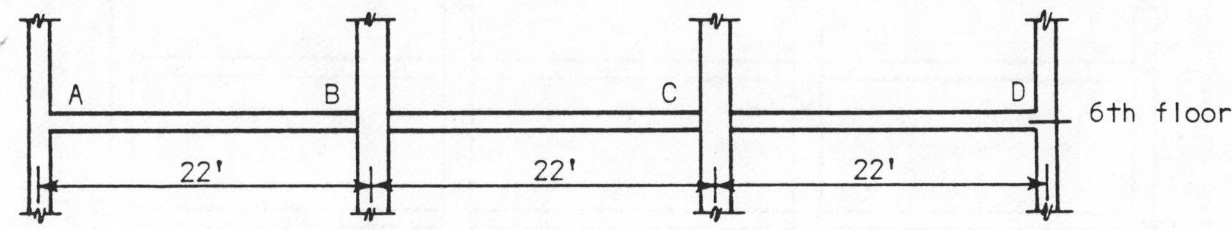

+M

$$U = \begin{cases} 1.4D + 1.7L \\ 0.75\ (1.4D + 1.7L \pm 1.87E) \\ 0.9D \pm 1.43E \end{cases}$$

Eq. (9-1) of ACI 318-83
Eq. (9-2)
Eq. (9-3)

Beam AB		Design Moment in ft-kips		
		A	near midspan	B
Eq. (9-1)		- 86	+130	-179
Eq. (9-2)	sidesway to right	+127	+126	-324
	sidesway to left	-257	+ 70	+ 56
Eq. (9-3)	sidesway to right	+162	+ 79	-265
	sidesway to left	-230	+ 22	+125

Beam BC		B	near midspan	C
Eq. (9-1)		-179	+ 93	-179
Eq. (9-2)	sidesway to right	+ 89	+ 80	-357
	sidesway to left	-357	+ 60	+ 89
Eq. (9-3)	sidesway to right	+157	+ 47	-297
	sidesway to left	-297	+ 26	+157

EXAMPLE 30.1.1 - Continued

Table 5 Summary of Design Moments and Axial Loads for Typical Columns on 2nd Floor of Interior Transverse Frames along Lines 3 through 6 (Fig. 16a)

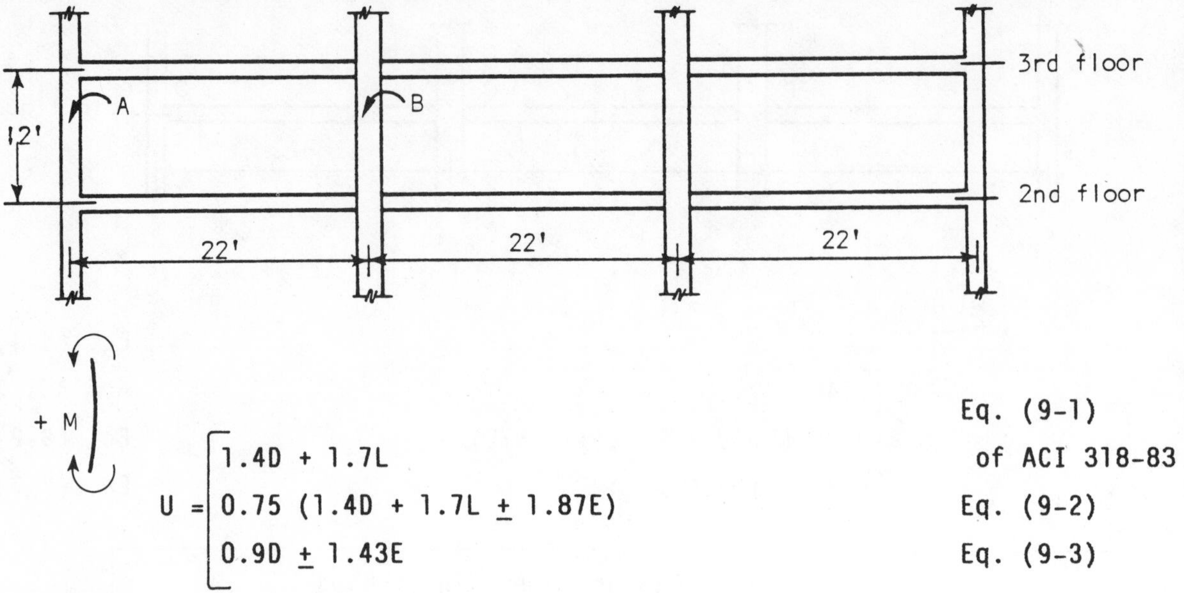

+ M

$$U = \begin{cases} 1.4D + 1.7L & \text{Eq. (9-1) of ACI 318-83} \\ 0.75\,(1.4D + 1.7L \pm 1.87E) & \text{Eq. (9-2)} \\ 0.9D \pm 1.43E & \text{Eq. (9-3)} \end{cases}$$

		Exterior Column A			Interior Column B		
		Axial Load kips	Moment, ft-k*		Axial Load kips	Moment, ft-k*	
			Top	Bottom		Top	Bottom
Eq. (9-1)		-930	- 95	+ 95	-1678	+ 53	-53
Eq. (9-2)	sidesway to right	-526	- 15	- 9	-1225	+173	-207
	sidesway to left	-870	-133	+151	-1293	- 93	+127
Eq. (9-3)	sidesway to right	-328	+ 25	- 44	- 857	+146	-181
	sidesway to left	-680	-101	+120	- 925	-126	+161

*Including moments corresponding to horizontal torsional shears due to minimum story shear eccentricity.

EXAMPLE 30.1.1 - Continued

Table 6 Summary of Design Loads on Structural Wall Section at First Floor Level of Transverse Frame Along Line 2 (or 7) (Fig. 16a)

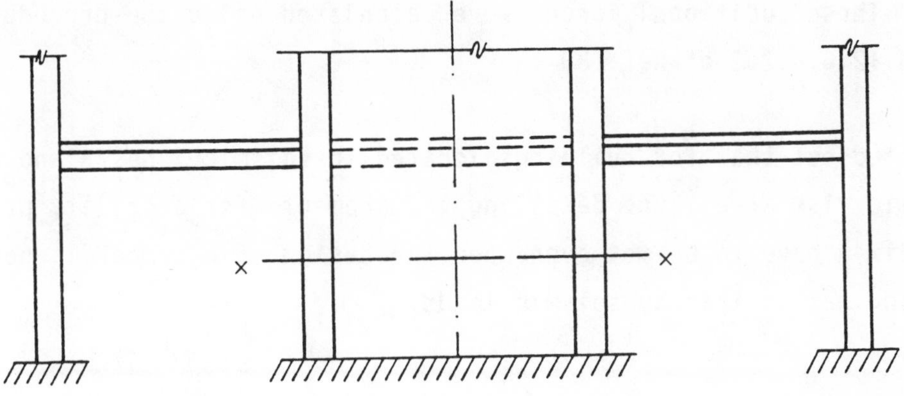

$$U = \begin{cases} 1.4D + 1.7L & \text{Eq. (9-1) of ACI 318-83} \\ 0.75\,(1.4D + 1.7L \pm 1.87E) & \text{Eq. (9-2)} \\ 0.9D \pm 1.43E & \text{Eq. (9-3)} \end{cases}$$

Loading Condition	Design Forces Acting on Entire Structural Wall			Axial Load** on Boundary Element (kips)
	Axial Load (kips)	Bending* (Overturning) Moment (ft-kips)	Horizontal* Shear (kips)	
Eq. (9-1)	4598	nominal	nominal	2326
Eq. (9-2)	3449	53,280	1129	4166*
Eq. (9-3)	2506	54,464	1153	3744*

*Including effect of horizontal torsion due to accidental eccentricity of story shear (ecc. = 0.05 x 182 = 9.1 ft)
**Based on loading condition illustrated in Fig. 13.

EXAMPLE 30.1.1 - Continued

corresponding to the minimum ANSI-prescribed eccentricity of 5% of the building dimension perpendicular to the direction of the applied forces have been included. These additional forces were calculated using the procedure described in Section 12.6.8.2.2 of Ref. 30-6.

It is pointed out that for buildings located in seismic Zones 3 and 4, (i.e., high seismic risk areas) the detailing requirements for ductility provided in ACI Appendix A have to be met even when the design of a member is governed by wind loading rather than by seismic loads.

EXAMPLE 30.1.2 - Design of Flexural Members

The aim is to determine the flexural and shear reinforcement for the beam AB on the sixth floor of a typical interior transverse frame. The beam carries a dead load of 3.7 kips/ft of span and a live load of 1.95 kips/ft. The design (factored) moments are as indicated in the figure below (see Table 4). The beam has dimensions b = 20 in. and d = 21.5 in. The slab is 8 in. thick. Use f'_c = 4,000 psi and f_y = 60,000 psi.

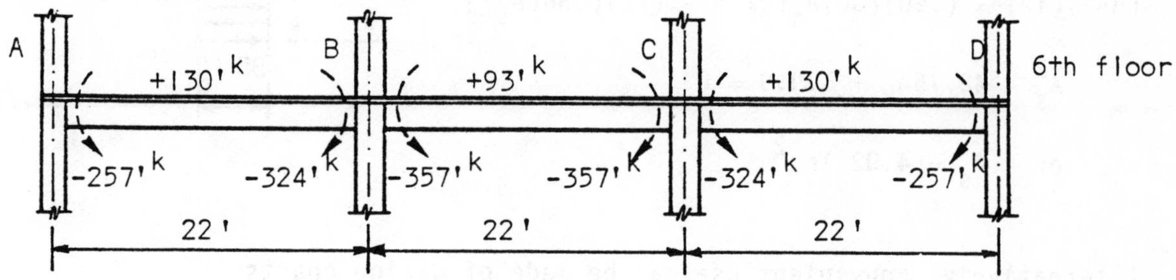

Calculations and Discussion	Code Reference

a. Check satisfaction of limitations on section dimensions. — A.3.1.3, A.3.1.4

$$\frac{\text{width}}{\text{depth}} = \frac{20}{21.5} = 0.93 > 0.3 \quad \text{OK}$$

width = 20 in. ≥ 10 in. OK
≤ [width of supporting column + 1.5 x depth of beam]
= 26 + 1.5(21.5) = 58.25 in. OK

b. Determine required flexural reinforcement.

(1) Negative moment reinforcement at support B:
Since the negative flexural reinforcement for both beams AB and BC at joint B will be provided by the same continuous bars, the larger negative moment at joint B will be used. Thus, M_u = 357 ft-kips from Table 4. In the following calculations, the effect of any compressive reinforcement will be neglected.

EXAMPLE 30.1.2 - Continued

Calculations and Discussion	Code Reference

from

$$C = 0.85f'_c ba = T = A_s f_y$$

$$a = \frac{A_s f_y}{.85f'_c b} = \frac{60A_s}{(.85)(4)(20)} = 0.882\ A_s$$

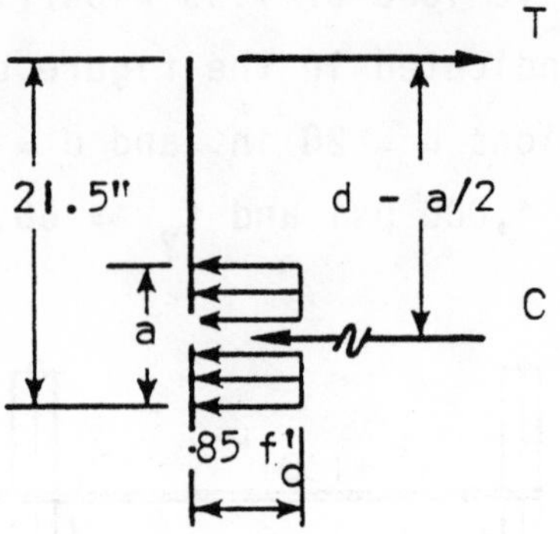

$$M_u \leq \varphi M_n = \varphi A_s f_y[d - a/2]$$

$$(357)(12) = (.90)(60)A_s[21.5 - (.5)(.882A_s)]$$

$$A_s^2 - 48.75A_s + 179.9 = 0$$

or $A_s = 4.02\ \text{in.}^2$

d = 21.5"

d - a/2

T

C

a

.85 f'c

Alternatively, convenient use may be made of design charts for singly-reinforced flexural members with rectangular cross-section, given in standard references. These charts generally present curves relating the coefficient of resistance, $R_u = M_u/bd^2$ and the reinforcement ratio, $\rho = A_s/bd$.

<u>Use 4 No. 9 bars, $A_s = 4.00\ \text{in.}^2$</u> This gives a negative moment capacity at support B of $\varphi\ M_n = 355$ ft-kips.

Check satisfaction of limitations on reinforcement ratio -

$$\rho = \frac{A_s}{bd} = \frac{4.0}{(20)(21.5)} = 0.0093 > \rho_{min} = \frac{200}{f_y} = 0.0033$$ A.3.2.1

and $< \rho_{max} = 0.025$ OK

(2) Negative moment reinforcement at support A:

$M_u = 257$ ft-kips

As at support B, $a = 0.882A_s$. Substitution into

$M_u = \varphi A_s f_y[d - a/2]$

yields $A_s = 2.82\ \text{in.}^2$

EXAMPLE 30.1.2 - Continued

Calculations and Discussion	Code Reference

Use 3 No. 9 bars, A_s = 3.0 in.^{2} This gives a negative moment capacity at support A of φM_n = 272 ft-kips.

(3) Positive moment reinforcement at supports:

A positive moment capacity at the supports equal to at least 50% of the corresponding negative moment capacity is required, i.e., A.3.2.2

min M_u^+ (at support A) = $\frac{272}{2}$ = 136 ft-kips, which is less than M_{max}^+ at A (see Table 4) = 162 ft-kips.

min M_u^+ (at support B for both spans AB and BC) = $\frac{355}{2}$ = 178 ft-kips.

Note that the above required capacities are greater than the design positive moments near the midspans of both beams AB and BC.

(4) Positive moment reinforcement at midspan - to be made continuous to supports:

$$a = \frac{A_s f_y}{.85 f'_c b} = \frac{60 A_s}{(.80)(4)(60)} = 0.294\ A_s$$

M_u^+ = (162)(12) = $\varphi A_s f_y [d - a/2]$
yields A_s^+ required at A = 1.7 in.^{2}

Similarly, corresponding to M_u^+ (required capacity) at support B = 178 ft-kips, A_s^+ required = 1.85 in.^{2}

Use 3 No. 7 bars continuous through both spans. A_s = 1.80 in.^{2} This provides a positive moment capacity of 172 ft-kips - about 3% less than that required at support B.

Check

$$\rho = \frac{1.8}{(20)(21.5)} = 0.0042 > \rho_{min} = \frac{200}{f_y} = 0.0033 \text{ OK}$$ 10.5.1

EXAMPLE 30.1.2 - Continued

Calculations and Discussion	Code Reference

c. Calculate required length of anchorage of flexural reinforcement in exterior column.

Development length, ℓ_{dh} ≥ (plus standard 90° hook located in confined region of column)
$$\begin{cases} f_y d_b / 65 \sqrt{f'_c} \\ 8\ d_b \\ 6\ \text{in.} \end{cases}$$
A.6.4.1

For the No. 9 (top) bars (bend radius ≥ 5 d_b)

$$\ell_{dh} \geq \begin{cases} (60{,}000)(1.128)/65\sqrt{4000} = \underline{17\ \text{in.}} \\ (8)(1.128) = 9\ \text{in.} \\ 6\ \text{in.} \end{cases}$$

For the No. 7 (bottom) bars (bend radius ≥ 4 d_b)

$$\ell_{dh} \geq \begin{cases} (60{,}000)(.875)/65\sqrt{4000} = \underline{13\ \text{in.}} \\ (8)(.875) = 7\ \text{in.} \\ 6\ \text{in.} \end{cases}$$

See figure below for detail of flexural reinforcement anchorage in exterior column. Note that the development length ℓ_{dh} is measured from the near face of the column to the far edge of the vertical 12-bar-diameter-extension (see Fig. 2).

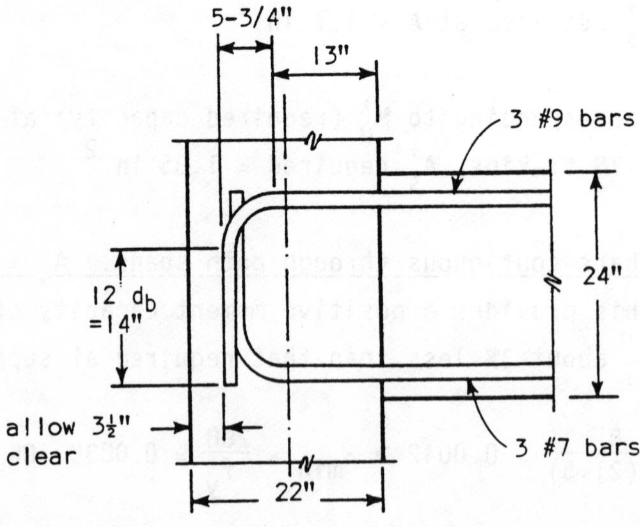

EXAMPLE 30.1.2 - Continued

Calculations and Discussion	Code Reference

d. Determine shear reinforcement requirements.

(1) Design for shears corresponding to end moments obtained by assuming the stress in the tensile flexural reinforcement equal to 1.25 f_y and a strength reduction factor, φ = 1.0, plus factored gravity loads (see Fig. 3). A.7.1.1

Table 7 shows values of design end shears corresponding to the two loading cases to be considered. In the table

$$w_u = 0.75(1.4w_D+1.7w_L) = 0.75[1.4(3.7)+1.7(1.95)]$$
$$= 6.37 \text{ kips/ft}$$

Appendix A requires that the contribution of concrete to shear resistance, V_c, be neglected if the earthquake-induced shear force (corresponding to the "probable flexural strengths" at beam ends calculated using f_s = 1.25 f_y and φ = 1.0) is greater than one-half of the total design shear and if the axial compressive force including earthquake effects is less than $A_g f'_c/20$. A.7.2.1

For sidesway to right, the shear at end B due to the plastic end moments in the beam (see Table 7),

$$V_B = \frac{238+482}{20} = 36 \text{ kips}$$

which is less than 50% of the total design shear, V_u = 99.7 kips. Therefore, the contribution of concrete to shear resistance can be considered in determining shear reinforcement requirements.

Table 7 Determination of Design Shear Forces for Beam Spans

Loading	$V_u = \frac{M_{AB}^{\pm} + M_{BA}^{\mp}}{\ell} + 0.75 \frac{w_u \ell}{2}$	
	A	B
A, B; w_u = 6.37 k/'; 238'^k; 482'^k Sidesway to right	27.7 kips	99.7 kips
A, B; w_u; 238'^k; 372'^k Sidesway to left	94.2 kips	33.2 kips

*when compression reinforcement is considered, a value about 0.5 percent greater than the above value is obtained.

Shear Diagram

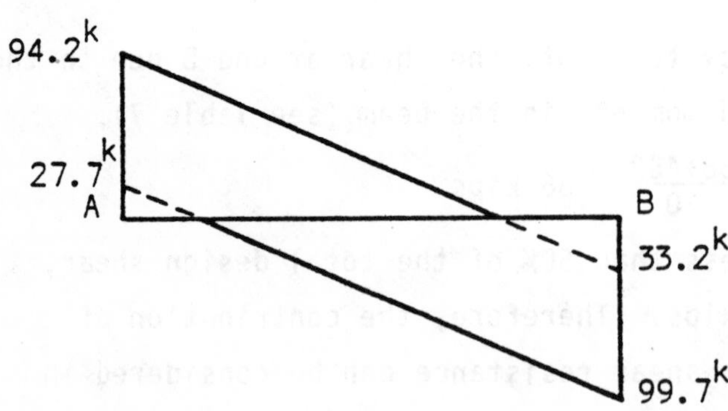

EXAMPLE 30.1.2 - Continued

Calculations and Discussion	Code Reference

At right end B, $V_u = 99.7$ kips.

Using $V_c = 2\sqrt{f'_c}\, b_w d = 2\sqrt{4000}\,(20)(21.5)/1000 =$ 54.4 kips 11.3.1.1

$\varphi V_s = V_u - \varphi V_c = 99.7 - 0.85 \times 54.4 = 53.5$ kips

$V_s = 62.9$ kips

Required spacing of #3 closed stirrups (hoops) -

A_v (2 legs) = 0.22.in.^{2}

$$s = \frac{A_v f_y d}{V_s} = \frac{(.22)(60)(21.5)}{62.9} = 4.5 \text{ in.}$$ 11.5.6.2

Maximum allowable hoop spacing within distance 2d = 2(21.5) = 43 in. from faces of supports,

s_{max} =
- d/4 = 21.5/4 = 5.4 in. A.3.3.2
- 8 x (dia. of smallest long. bar) = 8(.875) = 7 in.
- 24 x (dia. of hoop bars) = 24(.375) = 9 in.
- 12 in.

Beyond distance 2d from the supports, maximum spacing of stirrups

$s_{max} = d/2 = 10.5$ in. A.3.3.4

<u>Use #3 stirrups spaced as shown below</u>

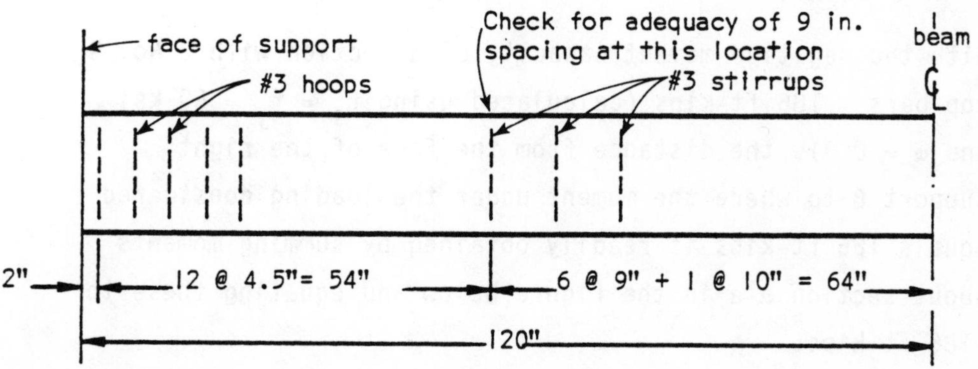

EXAMPLE 30.1.2 - Continued

Calculations and Discussion	Code Reference

Where the loading is such that inelastic deformation may occur at intermediate points within the span, e.g., due to concentrated loads near midspan, the spacing of hoops will have to be determined in a manner similar to that used above for regions near supports. In the present example, the maximum positive moment near midspan (see Table 4) is well below the positive moment capacity provided by the 3 No. 7 continuous bars. A.3.3.1

Note that lap splices in longitudinal reinforcement should not be used within joints or within a distance of 2d from the faces of supports. Where used outside of these regions, such splices are to be confined over the length of the lap by hoops or spirals with a maximum spacing or pitch of d/4 or 4 in. A.3.2.3

e. Negative reinforcement cutoff points.

For the purpose of determining cutoff points for the negative reinforcement, a moment diagram corresponding to plastic end moments and 0.9 times the dead load will be used. The cutoff point for two of the four No. 9 bars at the top, near support B of beam AB, will be determined.

With the negative moment capacity of a section with 2 No. 9 top bars = 186 ft-kips (calculated using $f_s = f_y = 60$ ksi and $\varphi = 0.9$), the distance from the face of the right support B to where the moment under the loading considered equals 186 ft-kips is readily obtained by summing moments about section a-a in the figure below and equating these to -186 ft-kips.

EXAMPLE 30.1.2 - Continued

Calculations and Discussion	Code Reference

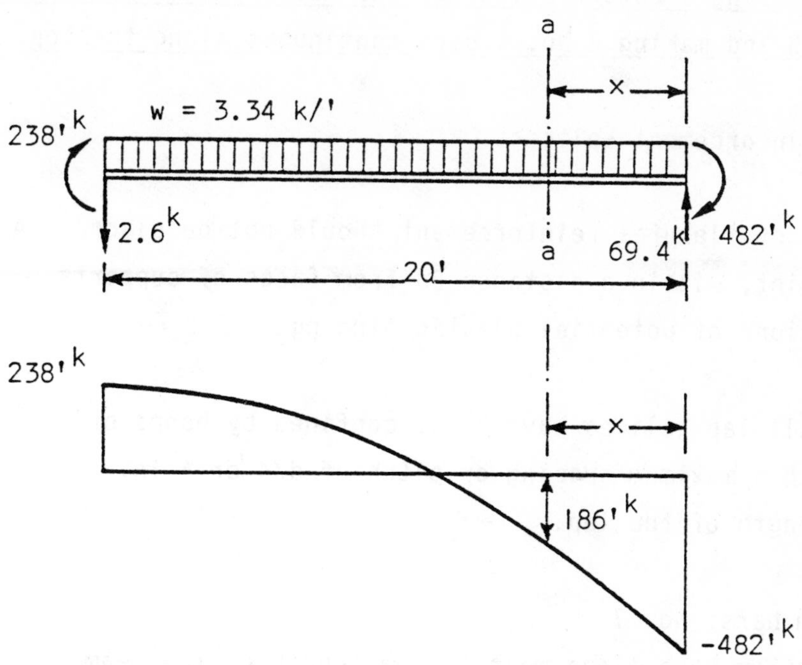

Thus, $69.4x - 482 - 3.34\frac{x^2}{2} = -186$

Solution of the above equation gives x = 4.82 ft. Hence, two of the four No. 9 top bars near support B may be cut off at (noting that for a No. 9 bar, ℓ_{dh} = 17 in.)

$x + 3.5\ \ell_{dh} = 4.8 + (3.5)(17)/12 = 9.8$ ft from face of right support B — A.6.4.2

Since this indicated cutoff point is practically at midspan, examine the alternative of cutting only one of the four No. 9 top bars near support B and making the 3 No. 9 bars continuous along the top of the beam.

Calculations and Discussion	Code Reference

By a procedure similar to that given above, the cutoff point for one of the 4 No. 9 top bars is obtained as 8.25 ft. This scheme will be adopted, using a cutoff point for one No. 9 top bar 8.5 ft from the face of support B and making 3 No. 9 bars continuous along the top.

f. Flexural reinforcement splices.

Lap splices of flexural reinforcement should not be placed within a joint, within a distance 2d from faces of supports or at locations of potential plastic hinging. A.3.2.3

Note that all lap splices have to be confined by hoops or spirals with a maximum spacing or pitch of d/4 or 4 in. over the length of the lap. A.3.2.3

(1) Bottom bars, No. 7
The bottom bars along most of the length of the beam may be subjected to maximum stress. Use Class C splice.

Required length of splice = $1.7\ \ell_d \geq 12$ in. 12.16.1

where

$$\ell_d = 0.04A_b f_y / \sqrt{f'_c} \geq 0.0004\ d_d f_y$$ 12.2.2

$$= (.04)(.60)(60{,}000)/\sqrt{4000} \geq (.0004)(.875)(60{,}000)$$

$$= 22.8 \text{ in. (governs)} > 21 \text{ in.}$$

Class C splice length = (1.7)(22.8) = 39 in.

(2) Top bars, No. 9
Since the midspan portion of the span is always subject to a positive bending moment (see Table 4), splices in the top bars should be located at or near midspan.

EXAMPLE 30.1.2 - Continued

Calculations and Discussion	Code Reference
Required length of Class A splice = 1.4 ℓ_d ≥ 12 in.	12.16.1
where, by using the same expression given above,	12.2.2
one obtains ℓ_d = 38 in.	12.2.3
Required splice length = 1.4(38) = 53 in.	

g. Detail of beam.

See Fig. 19.

EXAMPLE 30.1.2 - Continued

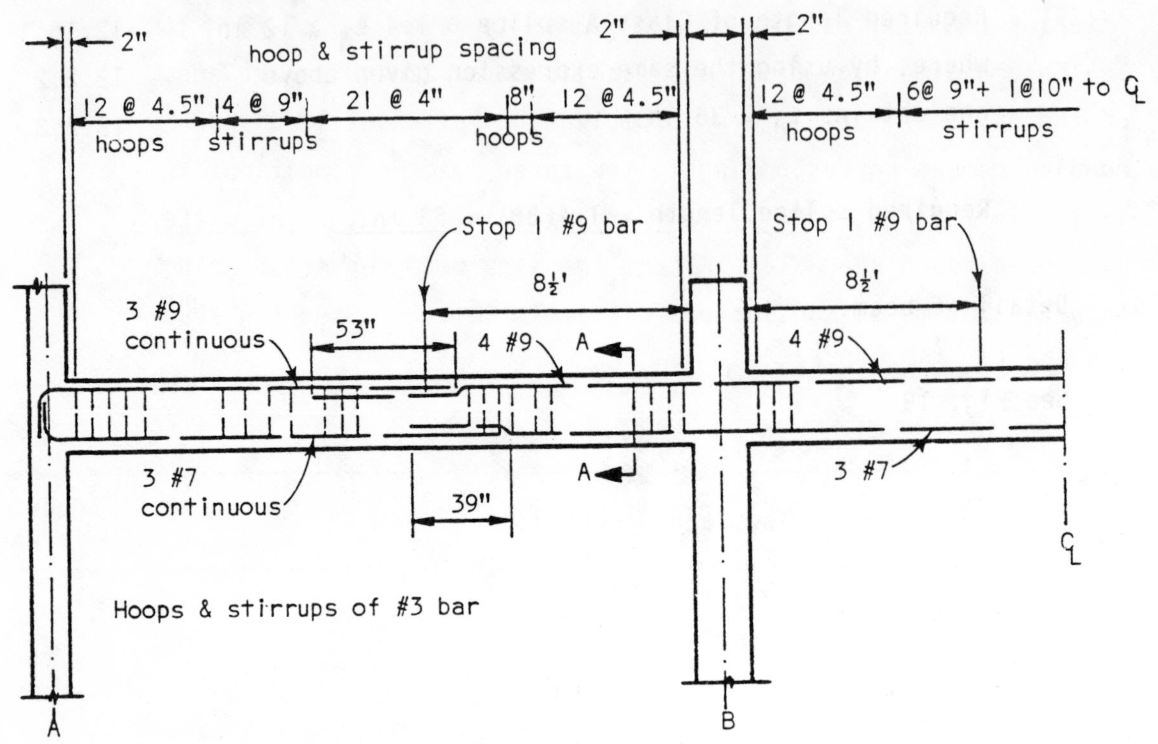

Fig. 30-19a Flexural member reinforcement

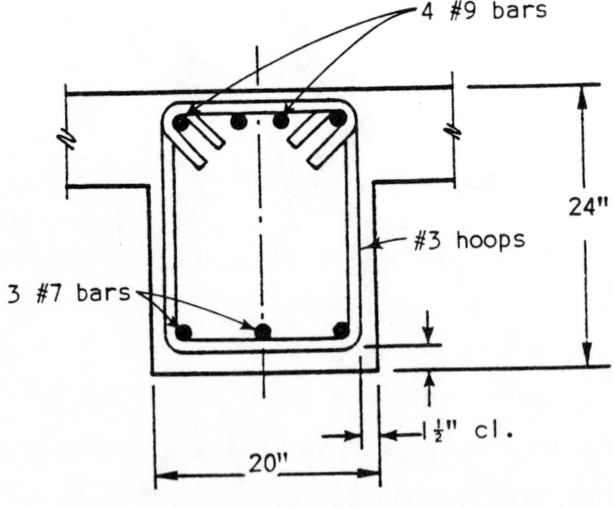

Fig. 30-19b Beam section A-A

EXAMPLE 30.1.3 - Design of Frame Columns

The aim is to design the transverse reinforcement for the exterior tied column on the second floor of a typical transverse interior frame, i.e., one of the frames in Frame T-1 of Fig. 16c. The column dimension has been established as 22 in. square and, on the basis of the different combinations of axial load and bending moment corresponding to the three loading conditions listed in Table 5, 8 No. 7 bars arranged in a symmetrical pattern have been found adequate. Assume the same beam section framing into the column as considered in Example 30.1.2. Use $f'_c = 4000$ psi and $f_y = 60,000$ psi.

Calculations and Discussion	Code Reference

From Table 5, P_u(max) = 930 kips

$P_u = 930 \text{ kips} > A_g f'_c/10 = (22)^2(4)/10 = 194 \text{ kips}$ — A.4.1

Thus, ACI Appendix A provisions governing members subjected to bending and axial load apply.

a. Check satisfaction of vertical reinforcement limitations and moment capacity requirements.

(1) Reinforcement ratio -

$$0.01 \leq \rho \leq 0.06$$

$$\rho = \frac{A_{st}}{A_g} = \frac{8(.60)}{(22)(22)} = 0.01 \quad \text{OK}$$ — A.4.3.1

(2) Moment strength of columns relative to that of framing beams in transverse direction -

$$\Sigma M_e \text{ (columns)} \geq \frac{6}{5} \Sigma M_g \text{ (beams)}$$ — A.4.2.2

From Example 30.1.2, φM_n^- of beam at A = 272 ft-kips, corresponding to sidesway to left.

EXAMPLE 30.1.3 - Continued

Calculations and Discussion	Code Reference

From Table 5, maximum axial load on Column A at the 2nd floor level for sidesway to left, P_u = 870 kips. Using the P-M interaction charts given in ACI SP-17A, the moment capacity of the column section, correspond to $P_u = \varphi P_n$ = 870 kips, f'_c = 4 ksi, f_y = 60 ksi, γ^* = 0.75 and ρ = 0.01, is obtained as $\varphi M_n = M_e$ = 284 ft-kips. 30.7

With the same size column above and below the beam, total moment capacity of columns = 2(284) = 568 ft-kips. Thus,

$$\Sigma M_e = 568 > \frac{6}{5} M_g = (6)(272)/5 = 326 \text{ ft-kips} \quad \text{OK}$$

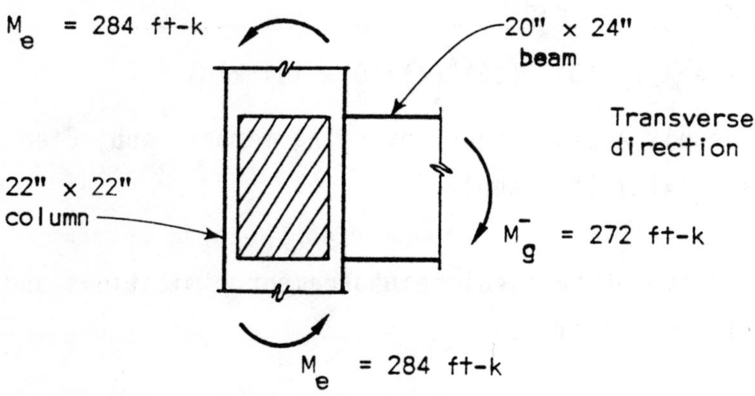

(3) Moment strength of columns relative to that of framing beams in longitudinal direction -

Since the columns considered here are located in the center portion of the exterior longitudinal frames, the axial forces due to seismic loads in the longitudinal direction are negligible. (Analysis of the longitudinal frames under seismic loads indicated practically zero axial forces in the exterior columns of

*γ = ratio of distance between centroids of outer rows of bars and dimension of cross-section, in the direction of bending.

Calculations and Discussion	Code Reference

the four transverse frames represented by Frame T-1 in Fig. 16c). Under an axial load of 0.75[1.4D + 1.7L + 1.87(E=0)] = (.75)(930) = 698 kips (see Table 5), the moment capacity of the column section with 8 No. 7 bars is obtained as $\varphi M_n = M_e = 319$ ft-kips.

If we assume a ratio for the negative moment reinforcement of approximately 0.0075 in the beams of the exterior longitudinal frames (b_w = 20 in., d = 21.5 in.),

$A_s = \rho b_w d$ = approx.(.0075)(20)(21.5) = 3.23 in.^{2}

Assume 4 No. 8 bars, A_s = 3.16 in.^{2}

Negative moment capacity of beam:

$$a = \frac{A_s f_y}{0.85 f'_c b_w} = \frac{(3.16)(60)}{(.85)(4)(20)} = 2.79 \text{ in.}$$

$$\varphi M_n^- = M_g = \varphi A_s f_y (d-a/2) = (.90)(3.16)(60)[(21.5-1.39)/12] = 286 \text{ ft-kips}$$

Assume a positive moment capacity of the beam on the opposite side of the column equal to one-half the negative moment capacity calculated above, or 143 ft-kips.

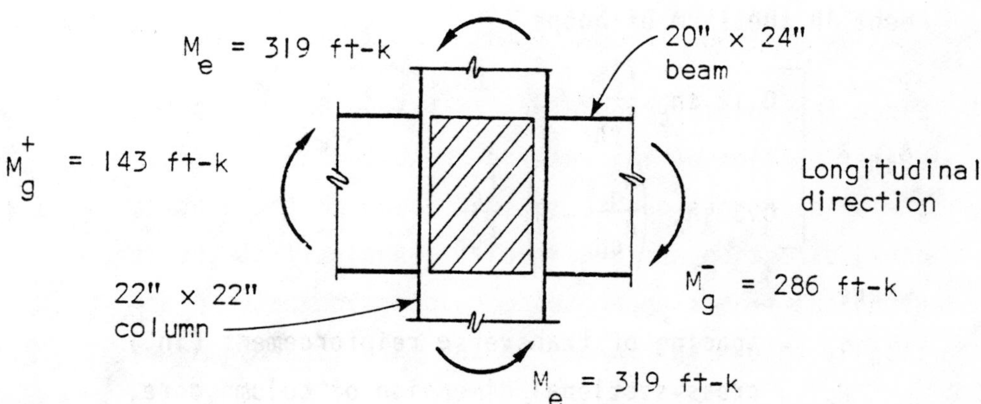

EXAMPLE 30.1.3 - Continued

Calculations and Discussion	Code Reference

Total moment capacity of beams framing into joint in longitudinal direction, for sidesway in either direction,

$\Sigma M_g = 284 + 143 = 429$ ft-kips.

$\Sigma M_e = 2(319) = 638 \text{ ft-kips} > \frac{6}{5}\Sigma M_g = \frac{6}{5}(429) =$ 515 ft-kips, OK — A.4.2.2

b. Determine transverse reinforcement requirements.

(1) Confinement reinforcement (see Fig. 6) - Transverse reinforcement for confinement is required over distance ℓ_o from column ends, where

$$\ell_o \geq \begin{cases} \text{depth of member} = \underline{22\text{ in.}}\ \text{(governs)} \\ \frac{1}{6}\text{ clear height} = (10 \times 12)/6 = 20\text{ in.} \\ 18\text{ in.} \end{cases}$$ A.4.4.4

Maximum allowable spacing of rectangular hoops

$$s_{max} = \begin{cases} \frac{1}{4}\text{ smallest dimension of column} = 22/4 = 5.5\text{ in.} \\ \underline{4\text{ in.}}\ \text{(governs)} \end{cases}$$ A.4.4.2

Required cross-sectional area of confinement reinforcement in the form of hoops

$$A_{sh} \geq \begin{cases} 0.12\, sh_c \dfrac{f'_c}{f_{yh}} \\ 0.3\, sh_c \left(\dfrac{A_g}{A_{ch}} - 1\right) \dfrac{f'_c}{f_{yh}} \end{cases}$$ A.4.4.1

where

s = spacing of transverse reinforcement (in.)

h_c = cross-sectional dimension of column core, measured c-c of confining reinforcement (in.)

EXAMPLE 30.1.3 - Continued

Calculations and Discussion	Code Reference

A_{ch} = core area of column section, measured outside to outside of transverse reinforcement (in.^{2})

f_{yh} = specified yield strength of transverse reinforcement (psi)

For a hoop spacing of 4 in., f_{yh} = 60,000 psi, and tentatively assuming No. 4 bar hoops (for the purpose of estimating h_c and A_{ch}), required cross-sectional area,

$$A_{sh} \geq \begin{cases} (.12)(4)(18.4)(4000)/60000 = \underline{0.59 \text{ in.}^2} \text{ (governs)} \\ (.3)(4)(18.4)\left(\frac{484}{357} - 1\right)\frac{4000}{60000} = 0.52.\text{in.}^2 \end{cases}$$

No. 4 hoops with one crosstie, as shown in the figure below, provide A_{sh} = 3(.20) = 0.60 in.^{2} A.4.4.3

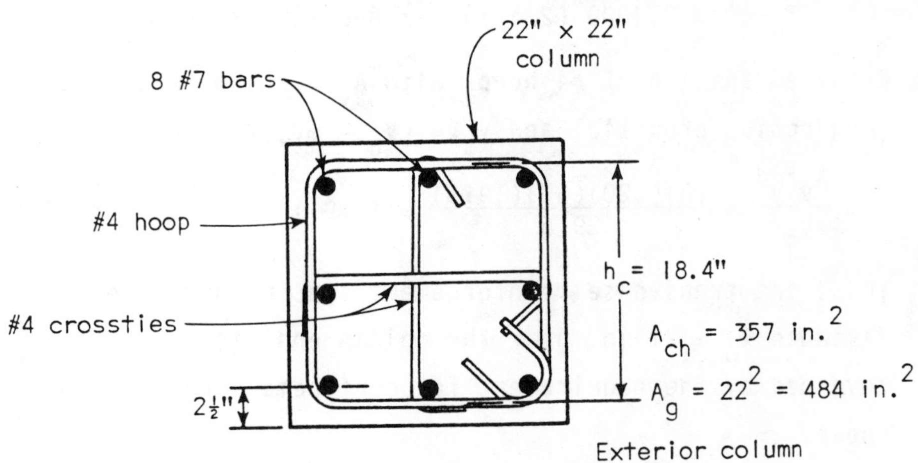

(2) Transverse reinforcement for shear -
As in the design of shear reinforcement for beams, the design shear in columns is based not on the factored shear forces obtained from a lateral load analysis but rather on the nominal flexural strength provided in the columns. ACI Appendix A requires that the shear A.7.1.2

Calculations and Discussion	Code Reference

be determined from the largest nominal moment strength consistent with the estimated axial forces on the column.

Assume that an axial force close to φP_b = 484 kips (corresponding to the "balanced point" on the interaction diagram for the column section considered - which would yield close to if not the largest moment strength) can occur (see Table 5). On this basis,

M_u (at column ends) = φM_b = 355 ft-kips,

from which (see Fig. 9)

$$V_u = \frac{2M_u}{\ell} = \frac{2(355)}{10} = 71 \text{ kips.}$$

Assuming $V_c = 2\sqrt{f'_c}\, bd$*

$= 2\sqrt{4000}\,(22)(19.5)/1000 = 54$ kips

Required spacing of #4 hoops with $A_v = 2(.20) = 0.40$ in.^{2} (neglecting crosstie) and $V_s = (V_u - \varphi V_c)/\varphi = 29.5$ kips,

$$s = \frac{A_v f_y d}{V_s} = \frac{(2)(.20)(60)(19.5)}{29.5} = 15.9 \text{ in.}$$ 11.5.6.2

Thus, the transverse reinforcement spacing over the distance ℓ_o = 22 in. near the column ends is governed by the requirement for confinement rather than shear.

Maximum allowable spacing of shear reinforcement = d/2 or 9.7 in. 11.5.4.1

*This lower bound value for V_c is assumed here primarily for convenience. Advantage can be taken of the increased shear resistance of concrete due to axial compression, when needed, by using the appropriate expression given in ACI 318-83.

EXAMPLE 30.1.3 - Continued

Calculations and Discussion	Code Reference

Use No. 4 hoops and crossties spaced at 4 in. within a distance of 24 in. from the column ends and No. 4 hoops spaced at 9 in. or less over the remainder of the column.

c. Minimum length of lap splices of column vertical bars.

ACI Appendix A limits the location of lap splices of column bars within the center half of the member length, the splices to be designed as tension splices. A.4.3.2

Since generally all of the column bars will be spliced at the same location, a Class C splice will be required. 12.16.2

Required length of splice = 1.7 ℓ_d 12.16.1

where

$\ell_d = 0.04A_b f_y/\sqrt{f'_c} \geq 0.0004\ d_b f_y$ 12.2.2

$= (.04)(.60)(60{,}000)/\sqrt{4000} \geq (.0004)(.875)(60{,}000)$

$= 23$ in. (governs) > 21 in.

Thus, required splice length = 1.7(23) = <u>39 in.</u>

Use 40 in. lap splices.

d. Detail of column.

See Figure 20.

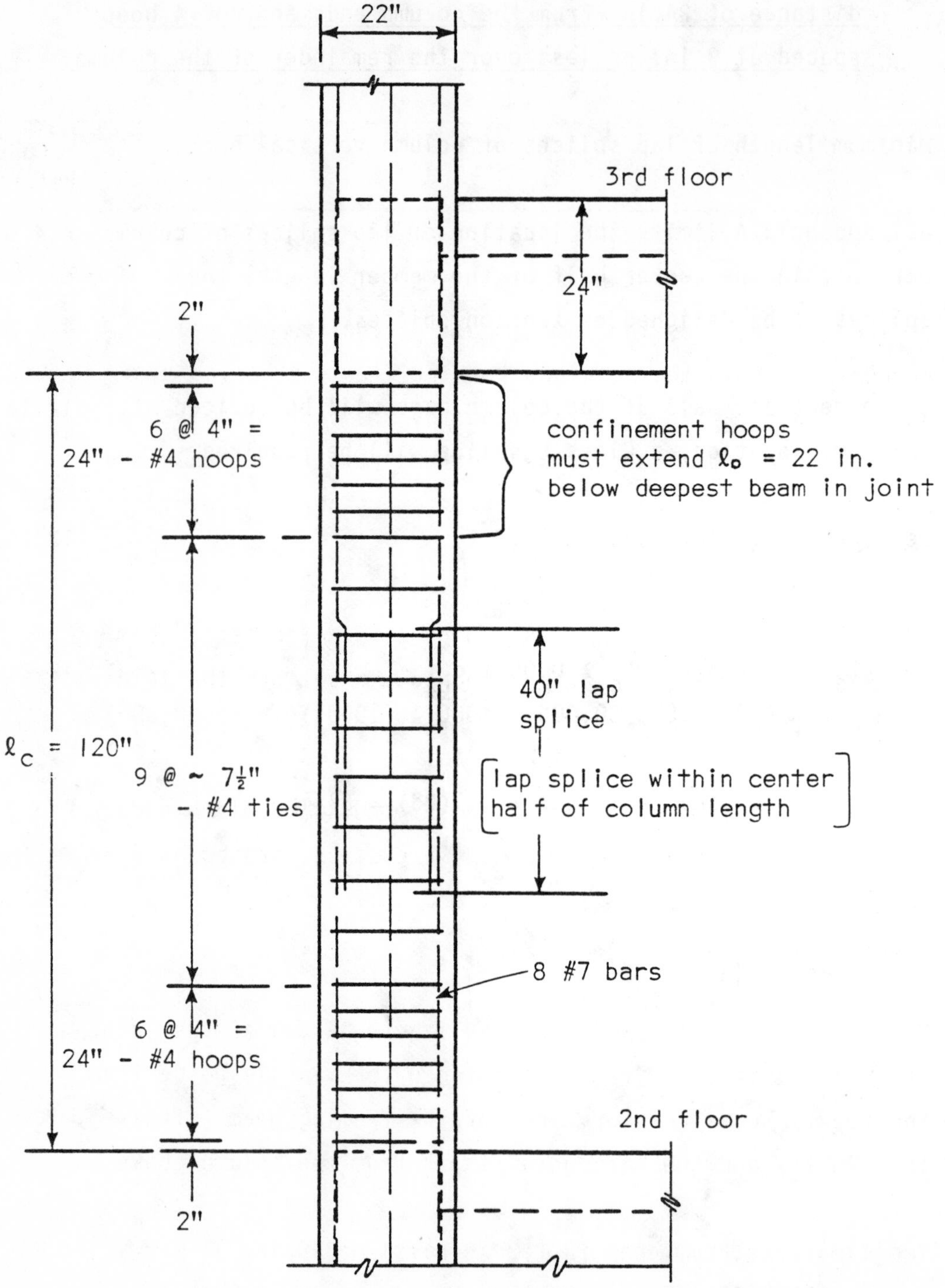

Fig. 20 Column reinforcing details

EXAMPLE 30.1.4 - Design of Exterior Beam-Column Connection

The aim is to determine the transverse reinforcement and shear strength requirements for the exterior beam-column connection between the beam considered in Example 30.1.2 and the column of Example 30.1.3. Assume the joint to be located at the sixth floor level.

Calculations and Discussion	Code Reference
a. Transverse reinforcement for confinement.	
Appendix A requires the same amount of confinement reinforcement within the joint as for the length ℓ_o at column ends, unless the joint is confined by beams framing into all vertical faces of the column. In the latter case, only one-half of the reinforcement required for unconfined joints need be provided.	A.6.2.1 A.6.2.2
In the case of the beam-column joint considered here, beams frame into only three sides of the column so that the joint is considered unconfined.	
In Example 30.1.3, confinement requirements at column ends were satisfied by No. 4 hoops with crossties, spaced at 4 in.	
b. Check shear strength of joint.	
The shear across section x-x (see figure below) of the joint is obtained as the difference between the tensile force from the top flexural reinforcement of the framing beam (stressed to 1.25 f_y) and the horizontal shear from the column above.	
Tensile force from beam [A_s (3 #9 bars) = 3.0 in.^{2}] = (3)(1.25)(60) = 225 kips	

EXAMPLE 30.1.4 - Continued

Calculations and Discussion	Code Reference

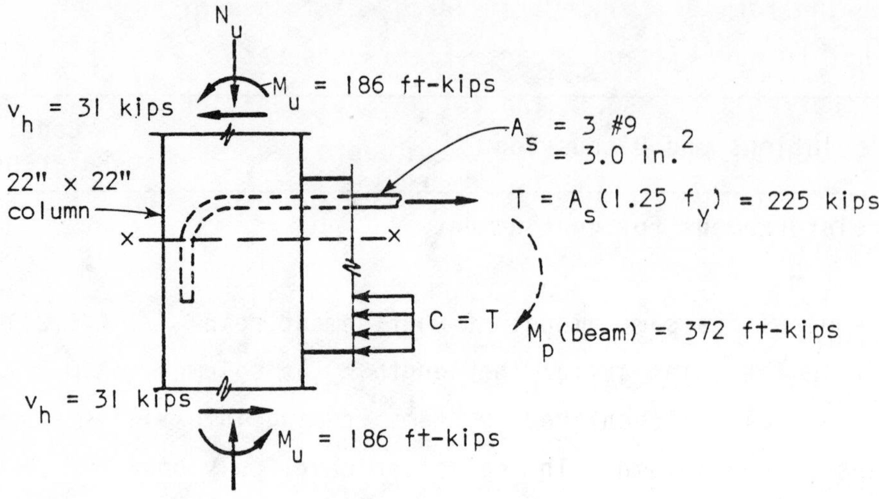

An estimate of the horizontal shear from the column, V_h, can be obtained by assuming that the beams in the adjoining floors are also deformed so that plastic hinges form at their junctions with the column, with M_p (beam) = 372 ft-kips (see Table 7, for sidesway to left). By further assuming that the plastic moments in the beams are resisted equally by the columns above and below the joint, one obtains for the horizontal shear at the column ends

$$V_h = \frac{M_p \text{ (beam)}}{\text{story height}} = \frac{372}{12} = 31 \text{ kips}$$

Thus, net shear at section x-x of joint = 225-31 = 194 kips

ACI Appendix A makes the nominal shear strength of a joint a function only of the area of the joint cross-section, A_j, and the degree of confinement by framing beams. For the unconfined joint considered here

$$\varphi V_c = \varphi 15\sqrt{f'_c}\ A_j$$

$= (.85)(15)(\sqrt{4000})(22)^2/1000$ A.2.3.1

$= 390 \text{ kips} > V_u = 194 \text{ kips}$, OK A.6.3.1

EXAMPLE 30.1.4 - Continued

Calculations and Discussion	Code Reference

Note that if the shear strength of the concrete in the joint as calculated above were inadequate, any adjustment would have to take the form of (since transverse reinforcement is considered not to have a significant effect on shear strength) either an increase in the column cross-section (and hence A_j) or an increase in the beam depth (to reduce the amount of flexural reinforcement required and hence the tensile force T).

c. Detail of joint.

See Fig. 21. (The detail should be checked for adequacy in the longitudinal direction).

Note: The use of crossties within the joint may cause some placement difficulties. To relieve the congestion, No. 6 hoops spaced at 4 in. but without crossties may be considered as an alternative. Although the cross-sectional area of confinement reinforcement provided by No. 6 hoops at 4 in. (A_{sh} = 0.88 $in.^2$) exceeds the required amount (0.59 $in.^2$), the requirement of Section A.4.4.3 relating to a maximum spacing of 14 in. between crossties or legs of overlapping hoops (see Fig. 8) will not be satisfied. However, it is believed that this should not be a serious shortcoming in this case since the joint is restrained by beams on three sides.

EXAMPLE 30.1.4 - Continued

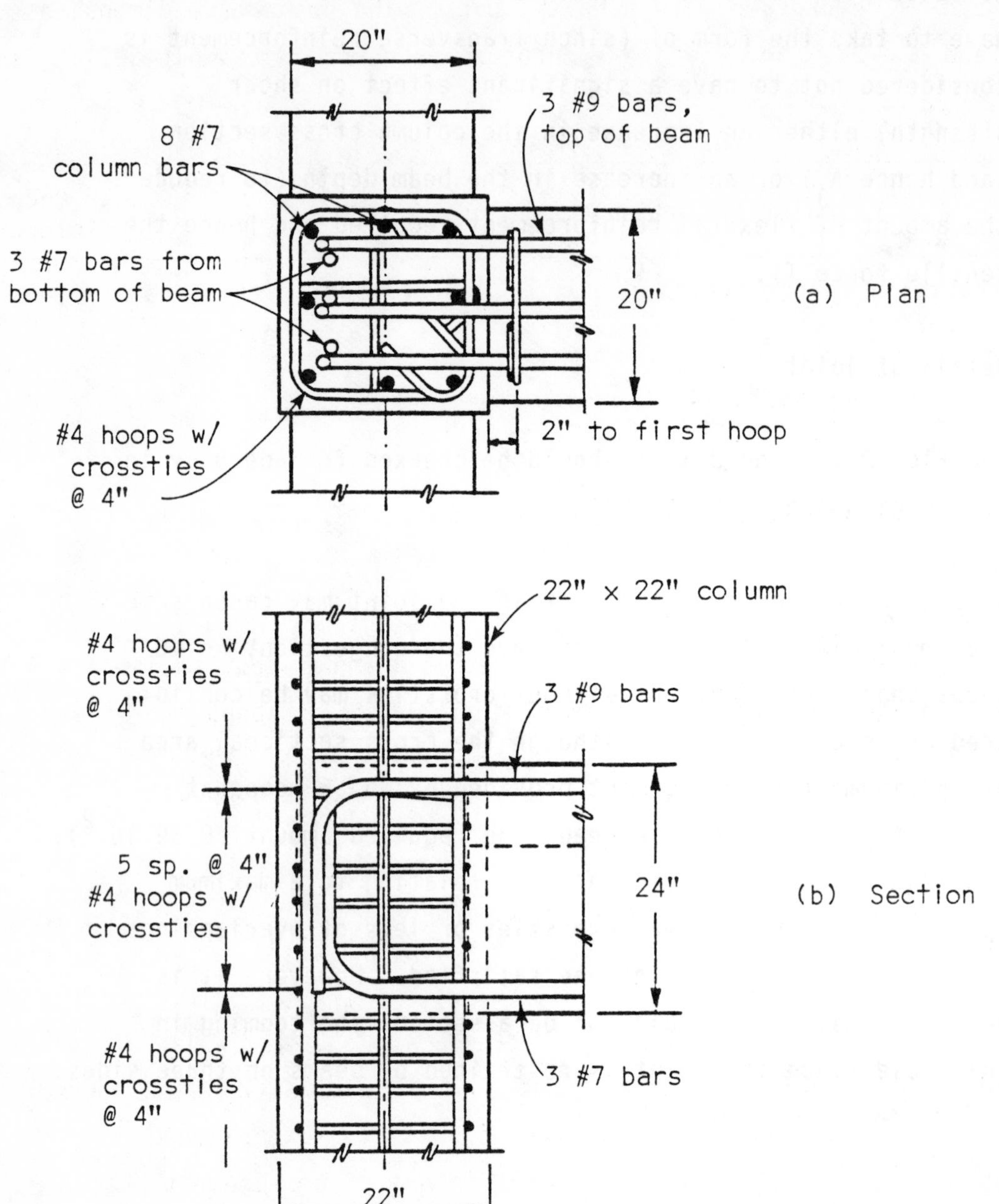

Fig. 21 Detail of exterior beam-column connection

EXAMPLE 30.1.5 - Design of Interior Beam-Column Connection

The objective is to determine the transverse reinforcement and shear strength requirements for the interior beam-column connection at the sixth floor of the interior transverse frame considered in previous examples. The column is 26 in. square and is reinforced with 8 No. 11 bars. The beams have dimensions b = 20 in. and d = 21.5 in. and are reinforced as noted in Example 30.1.2 (see Fig. 19).

Calculations and Discussion	Code Reference

a. Transverse reinforcement requirements (for confinement).

$$s_{max} = \begin{cases} \frac{1}{4} \text{ smallest dimension of column} = 26/4 = 6.5 \text{ in.} \\ \underline{4 \text{ in.}} \text{ (governs)} \end{cases}$$ A.4.4.2

For the column cross-section considered and assuming No. 4 hoops, h_c = 21.9 in., $A_{ch} = (22.4)^2 = 502$ in.^{2} and $A_g = (26)^2 = 676$ in.^{2}

With a hoop spacing of 4 in., the required cross-sectional area of confinement reinforcement in the form of hoops A.4.4.1

$$A_{sh} \geq \begin{cases} 0.12\, sh_c \dfrac{f'_c}{f_{yh}} = (.12)(4)(21.9)(4000)/(60000) \\ \qquad \underline{0.70 \text{ in.}^2} \text{ (governs)} \\ 0.3\, sh_c \left(\dfrac{A_g}{A_{ch}} - 1\right) \dfrac{f'_c}{f_{yh}} = (.3)(4)(21.9) \\ \qquad \left(\dfrac{676}{502} - 1\right) \dfrac{4000}{60000} = 0.61 \text{ in.}^2 \end{cases}$$ A.4.4.1

Since the joint is framed by beams [having widths (20 in.) ≥ 3/4 width of column] on all four sides, it is considered confined and a 50% reduction in the amount of confinement reinforcement indicated above is allowed. Thus, A_{sh}(required) ≥ 0.35 in.^{2} A.6.6.2

<u>No. 4 hoops with crossties spaced at 4 in. o.c.</u> provide $A_{sh} = 0.60$ in.^{2} (see Note at the end of Example 30.1.4).

EXAMPLE 30.1.5 - Continued

Calculations and Discussion	Code Reference

b. Check shear strength of joint.

Following the same procedure used in Example 30.1.4, the forces affecting the horizontal shear across a section near mid-depth of the joint shown in the figure below are obtained.

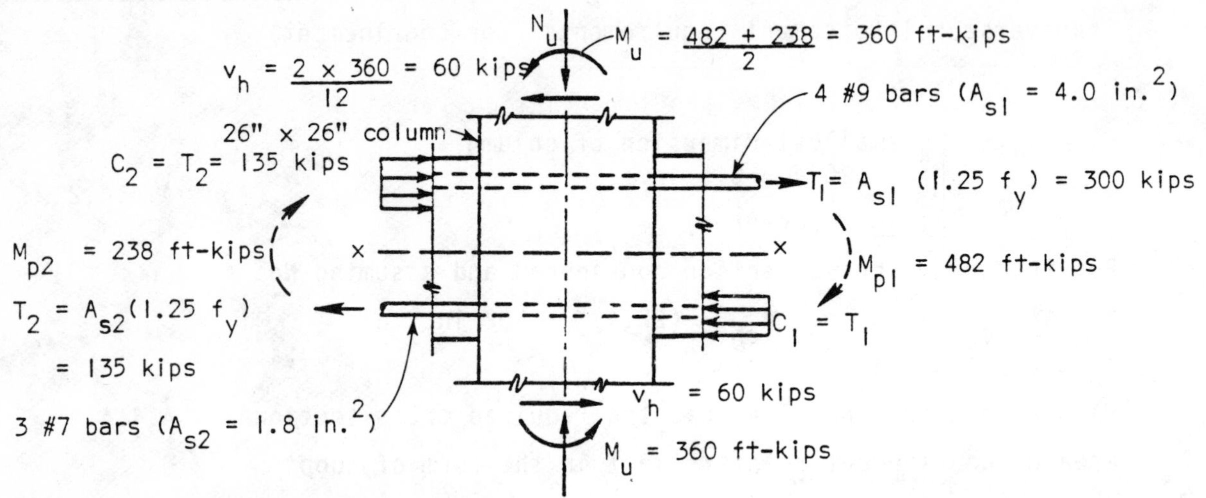

Net shear across section x-x $= T_1 + C_2 - V_h$

$= 300+135 - 60 = 375$ kips $= V_u$

Shear strength of joint, noting that joint is confined,

$\varphi V_c = \varphi 20 \sqrt{f'_c}\, A_j$ A.6.3.1

$= (.85)(20)(\sqrt{4000})(26)^2/1000 = 726$ kips $> V_u = 375$ kips OK

EXAMPLE 30.1.6 - Design of Structural Wall

The aim is to design the structural wall (shear wall) section at the first floor of one of the identical frame-shear wall systems in Frame T-3 (Fig. 16c). The preliminary design, as shown in Fig. 16, is based on a 14-in. thick wall with 26 in. square vertical boundary elements, each of the latter being reinforced with 8 No. 11 bars.

Calculations and Discussion	Code Reference

Preliminary calculations indicated that the cross-section of the structural wall at the lower floor levels needed to be increased. In the following, a 20-in. thick wall section with 32 in. x 50 in. boundary elements reinforced with 24 No. 11 bars is investigated, and other reinforcement requirements determined.

The design forces on the structural wall at the first floor level are listed in Table 6. Note that because the axis of the shear wall coincides with the centerline of the transverse frame of which it is a part, lateral loads do not induce any vertical (axial) forces on the wall.

The calculation of the maximum axial force on the boundary element corresponding to Eq. (9-2) of ACI 318-83, P_u = 4166 kips, shown in Table 6, involved the following steps:

At base of wall,

dead load, D = 2823 kips

live load, L = 412 kips

moment at base of wall due to seismic load (from lateral load analysis of transverse frames), including a moment of 1806 ft-kips due to accidental eccentricity, M_{base} = 38,056 ft-kips

EXAMPLE 30.1.6 - Continued

Calculations and Discussion	Code Reference

Referring to Fig. 13, and noting the load factors used in Eq. (9-2) of ACI 318-83,

$$W = 0.75\ (1.4D + 1.7L)$$
$$= (.75)[(1.4)(2823) + (1.7)(412)] = 3489 \text{ kips}$$
$$Ha = 1.4\ M_{base} = (1.4)(38056) = 53{,}278 \text{ ft-kips}$$
$$C_v = W/2 + Ha/d$$
$$= 3489/2 + 53278/22 = 4166 \text{ kips}$$

a. Check if boundary elements are required.

Appendix A requires boundary elements to be provided if the maximum compressive extreme-fiber stress under factored forces exceeds 0.2 f_c', unless the entire wall is reinforced to satisfy Sections A.4.4.1 through A.4.4.3 (relating to confinement reinforcement). A.5.3.1

It will be assumed that the wall will not be provided with confinement reinforcement over its entire section. For a homogeneous rectangular wall 26.17 ft long (horizontally) and 20 in. (1.67 ft) thick,

$$I_{n.a.} = \frac{(1.67)(26.17)^3}{12} = 2494 \text{ ft}^4$$

$$A_g = (1.67)(26.17) = 43.7 \text{ ft}^2$$

Extreme-fiber compressive stress under M_u = 53,280 ft-kips and P_u = 3449 kips (see Table 6),

$$f_c = \frac{P_u}{A_g} + \frac{M_u \ell_w/2}{I_{n.a.}} = \frac{3449}{43.7} + \frac{(53280(26.17)/2}{2494}$$
$$= 358.4 \text{ ksf or } 2.49 \text{ ksi} > 0.2\ f_c' = (.2)(4) = 0.8 \text{ ksi}$$

EXAMPLE 30.1.6 - Continued

Calculations and Discussion	Code Reference

Therefore, boundary elements are required, subject to the confinement and special loading requirements specified in Appendix A.

b. Determine minimum longitudinal and transverse reinforcement requirements in wall.

(1) Check if two curtains of reinforcement are required. Appendix A requires that two curtains of reinforcement be provided in a wall if the in-plane factored shear force assigned to the wall exceeds $2A_{cv}\sqrt{f'_c}$, where A_{cv} is the cross-sectional area bounded by the web thickness and the length of section in the direction of the shear force considered. From Table 6, the maximum factored shear force on the wall at the first floor level is V_u = 1153 kips. — A.5.2.2

$2A_{cv}\sqrt{f'_c}$ = (2)(20)(26.17 x 12)($\sqrt{4000}$)/1000 = 839 kips < V_u = 1153 kips

Therefore, two curtains of reinforcement are required.

(2) Required longitudinal and transverse reinforcement in wall. Minimum required reinforcement ratio

$$\rho_v = \frac{A_{sv}}{A_{cv}} = \rho_n \geq 0.0025 \quad \text{(max. spacing = 18 in.)}$$ — A.5.2.1

With A_{cv} (per foot of wall) = (20)(12) = 240 in.^{2}, required area of reinforcement in each direction per foot of wall = (0.0025)(240) = 0.60 in.^{2}/ft.

Required spacing of No. 5 bars [in two curtains, A_s = 2(.31) = 0.62 in.^{2}],

$$s(\text{required}) = \frac{2(.31)}{.60}(12) = 12.4 \text{ in.} < 18 \text{ in.}$$

EXAMPLE 30.1.6 - Continued

Calculations and Discussion	Code Reference

c. Determine reinforcement requirements for shear. [Refer to Section A.7 of ACI 318-83, shear strength design for structural walls].

To allow for increased shear reinforcement requirements above the minimum indicated above, assume two curtains of No. 6 bars spaced at 14 in. o.c. both ways.

Shear strength of wall (h_w/ℓ_w = 148/26.17 = 5.66 > 2),

$$\varphi V_n = \varphi A_{cv}(2\sqrt{f'_c} + \rho_n f_y)$$ A.7.3.2

where

φ = 0.60

A_{cv} = (20)(26.17 x 12) = 6271 in.^{2}

$\rho_n = \dfrac{2(.44)}{20(12)} \times \dfrac{12}{14} = 0.00314$

Thus,

φV_n = (.60)(6281)[$2\sqrt{4000}$ + (.00314)(60000)]/1000

= 3768.6[126.4 + 188.4]/1000 = 1186 kips > V_u = 1153 kips OK

Therefore, USE two curtains of No. 6 bars spaced at 14 in. o.c. in both horizontal and vertical directions. A.7.3.5

d. Check adequacy of boundary element acting as a short column under factored vertical forces due to gravity and lateral loads (see Fig. 13). A.5.3.3

From Table 6, maximum compressive axial load on boundary element, P_u = 4166 kips.

With boundary elements having dimensions 32 in. x 50 in. and reinforced with 24 No. 11 bars,

EXAMPLE 30.1.6 - Continued

Calculations and Discussion	Code Reference

$A_g = (32)(50) = 1600$ in.^{2}

$A_{st} = (24)(1.56) = 37.4$ in.^{2}

$\rho_{st} = 37.4/1600 = 0.0234$

Axial load capacity of boundary element acting as a short column

$\varphi P_{n(max)} = 0.80\ \varphi[0.85f'_c(A_g - A_{st}) + f_y A_{st}]$ 10.3.5.2

$= (.80)(.70)[(.85)(4)(1600-37.4) + (60)(37.4)]$

$= (.56)[5313 + 2246] = 4233$ kips $> P_u = 4166$ kips OK

e. Check adequacy of structural wall section at base under combined axial load and bending in the plane of the wall.

From Table 6, the following combinations of factored axial load and bending moment at the base of the wall are listed, corresponding to Eqs. (9-1), (9-2) and (9-3) of ACI 318-83:

a. Eq. (9-1): $P_u = 4598$ kips, M_u = small

b. Eq. (9-2): $P_u = 3449$ kips, $M_u = 53{,}280$ ft-kips

c. Eq. (9-3): $P_u = 2506$ kips, $M_u = 54{,}464$ ft-kips

Fig. 22 shows the φP_n-φM_n interaction diagram (obtained using a computer program for generating P-M diagrams) for a structural wall section having a 20-in. thick web reinforced with two curtains of reinforcement each having No. 6 horizontal and vertical bars spaced at 14 in. o.c. and 32 in. x 50 in. boundary elements reinforced with 24 No. 11 vertical bars, with $f'_c = 4000$ psi, f_y = 60,000 psi (see Fig. 23). The design load combinations listed above are shown plotted in the figure. The point

Calculations and Discussion	Code Reference

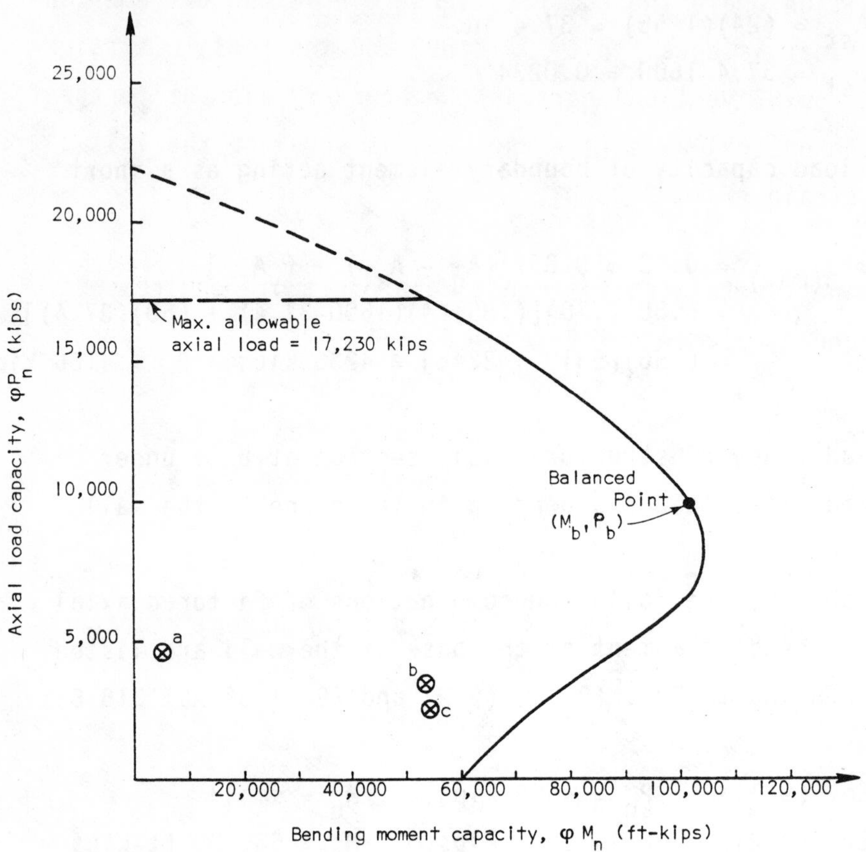

Fig. 30-22 Axial load-moment interaction diagram for structural wall section

marked "a" represents the P-M combination corresponding to Eq. (9-1), with similar notation for the other two combinations.

It is seen in Fig. 22 that the three design load combinations represent points inside the interaction diagram for the structural wall section considered. <u>Therefore, the section is adequate with respect to combined bending and axial load</u>.

EXAMPLE 30.1.6 - Continued

Calculations and Discussion	Code Reference

Incidentally, the "balanced point" in Fig. 22 corresponds to a condition where the compressive strain in the extreme concrete fiber is equal to ε_{cu} = 0.003 and the tensile strain in the row of vertical bars in the boundary element farthest from the neutral axis (see Fig. 23) is equal to the initial yield strain, ε_y = 0.00207.

f. Determine lateral (confinement) reinforcement requirements for boundary elements (see Fig. 23).

max, spacing, s_{max} = { 1/4 smallest dimension of boundary element = 32/4 = 8 in. — A.5.3.2; 4 in. (governs) — A.4.4.2 }

(1) Required cross-sectional area of confinement reinforcement in short direction

$$A_{sh} \geq \begin{cases} 0.12\, sh_c \dfrac{f'_c}{f_{yh}} \\ 0.3\, sh_c \left(\dfrac{A_g}{A_{ch}} - 1\right) \dfrac{f'_c}{f_{yh}} \end{cases}$$ A.4.4.1

Assuming No. 5 hoops and crossties spaced at 4 in. o.c. and a distance from centerline of No. 11 vertical bars to face of column of 3 in., we have

h_c (for short direction) = 44 + 1.41 + 0.625 = 46.04 in.

A_{ch} = (46.04 + 0.625)(26 + 1.41 + 1.25) = 1337 in.^{2}

$$A_{sh}\text{(required in short direction)} \geq \begin{cases} (.12)(4)(46.04)(4/60) = \underline{1.47 \text{ in.}^2} \text{ (governs)} \\ (.3)(4)(46.04)\left(\dfrac{(32)(50)}{1337} - 1\right)\dfrac{4}{60} = 0.72 \text{ in.}^2 \end{cases}$$

EXAMPLE 30.1.6 - Continued

Calculations and Discussion	Code Reference

With 3 crossties (i.e., 5 legs, including outside hoop),

A_{sh}(provided) = 5(.31) = 1.55 in.^{2} OK

(2) Required cross-sectional area of confinement reinforcement in long direction

h_c (for long direction) = 26 + 1.41 + 0.625 = 28.04 in.

A_{ch} = 1337 in.^{2}

$$A_{sh}(\text{required in long direction}) \geq \begin{cases} (.12)(4)(28.04)(4/60) = \underline{0.90 \text{ in.}^2} \text{ (governs)} \\ (.3)(4)(28.04)(1.196 - 1)(4/60) = 0.44 \text{ in.}^2 \end{cases}$$

With one crosstie (i.e., 3 legs, including outside hoop),

A_{sh}(provided) = 0.93 in.^{2} OK

g. Determine required development and splice lengths.

ACI Appendix A requires that all continuous reinforcement in structural walls be anchored or spliced in accordance with the provisions for reinforcement in tension as given in the Appendix. — A.5.2.4

(1) Lap splice for No. 11 vertical bars in boundary elements. (The use of mechanical connectors may be considered as an alternative to lap splices for these large bars).

Assuming that 50% or less of the vertical bars are spliced at any one location, a Class B splice may be used. — 12.16.2

Required length of splice = 1.3 ℓ_d

where

EXAMPLE 30.1.6 - Continued

Calculations and Discussion	Code Reference

$$\ell_d \geq 2.5 \times \begin{cases} f_y d_b / 65\sqrt{f'_c} = (60000)(1.41)/(65)(\sqrt{4000}) \\ \quad = \underline{21\text{ in.}}\text{ (governs)} \\ 8\,d_b = (8)(1.41) = 12\text{ in.} \\ 6\text{ in.} \end{cases}$$

A.6.4.2

Thus required splice length = (1.3)(2.5)(21) = <u>68 in.</u>

(2) Lap splice for No. 6 vertical bars in wall "web". Again assuming no more than 50% of bars spliced at any one level so that a Class B splice may be used, and using the same expression for ℓ_d as above, ℓ_d = 11 in.

Hence, required length of splice = (1.3)(2.5)(11) = <u>36 in.</u>

(3) Development length for No. 6 horizontal bars in wall, assuming no hooks are used within boundary element.

Since it is reasonable to assume that the depth of concrete cast in one lift beneath a horizontal bar will be greater than 12 in., the required factor of 3.5 to be applied to the development length, ℓ_{dh}, required for a 90°-hooked bar will be used. — A.6.4.2

$$\ell_d = 3.5\,\ell_{dh} \geq 3.5 \times \begin{cases} f_y d_b / 65\sqrt{f'_c} = (60000)(.75)/(65)(\sqrt{4000}) \\ \quad = \underline{11\text{ in.}}\text{ (governs)} \\ 8\,d_b = (8)(.75) = 6\text{ in.} \\ 6\text{ in.} \end{cases}$$

Thus, required development length ℓ_d = 3.5(11) = <u>39 in.</u>

This length can be accommodated within the confined core of the boundary element so that no hooks are needed, as assumed.

EXAMPLE 30.1.6 - Continued

Calculations and Discussion	Code Reference

Required lap splice length for these No. 6 horizontal bars, assuming 50% or less of bars spliced at any one location, = 1.3 ℓ_d = (1.3)(39) = 51 in.

h. Detail of structural wall.

See Fig. 23.

It will be noted in Fig. 23 that the No. 6 vertical wall "web" reinforcement, required for shear resistance, has been carried into the boundary elements. The Commentary to ACI Appendix A specifically states that the concentrated reinforcement provided at wall edges for bending shall not be included in determining shear reinforcement requirements. The area of vertical shear reinforcement located within the boundary elements could, if desired, be considered as contributing to axial load and bending capacity.

EXAMPLE 30.1.6 - Continued

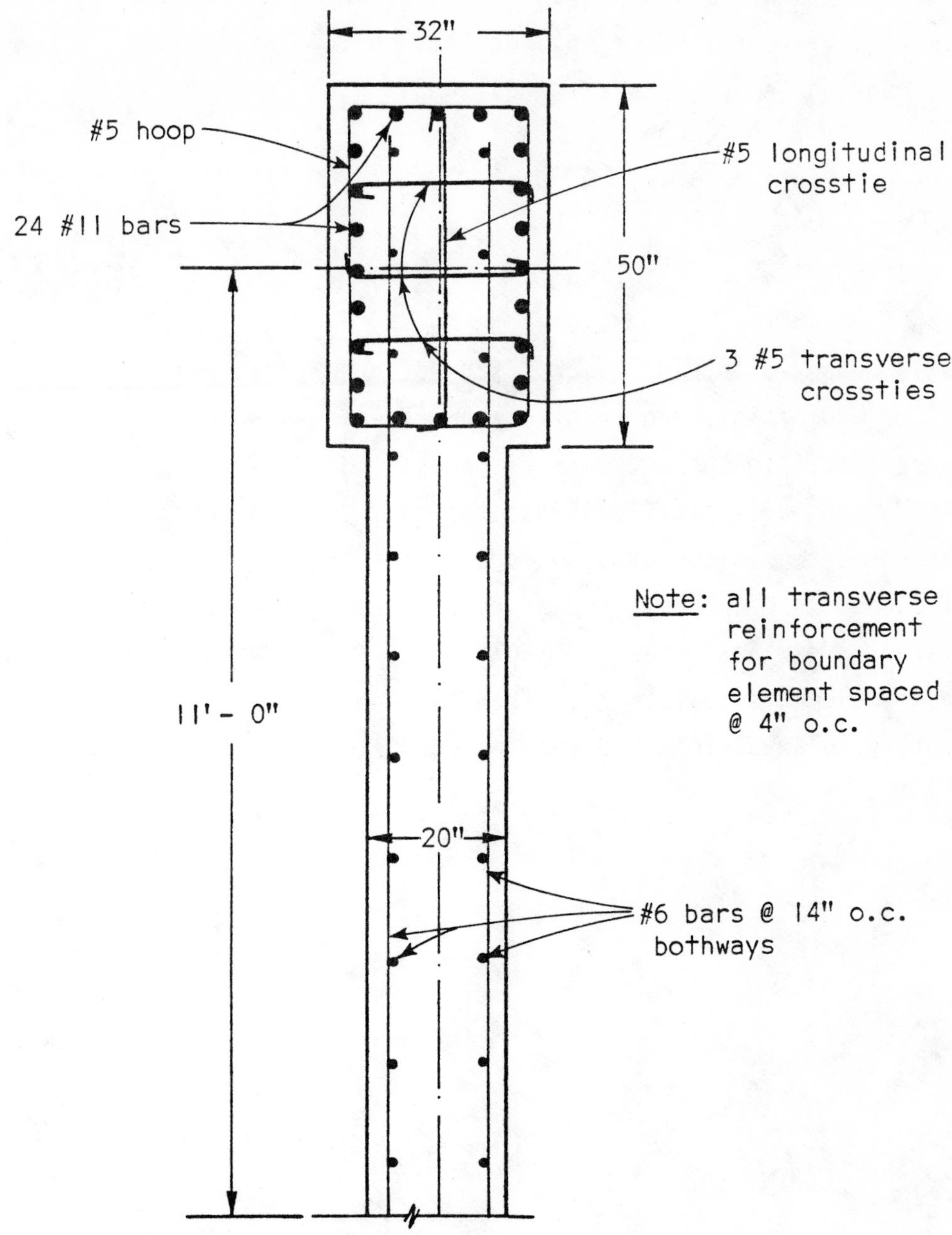

Fig. 30-23 Half-section of structural wall at base